PROGRESS IN PESTICIDES RESEARCH

PROGRESS IN PESTICIDES RESEARCH

CHRISTOPHER M. KANZANTZAKIS
EDITOR

Nova Science Publishers, Inc.
New York

For permission to use material from this book please contact us:
Telephone 631-231-7269; Fax 631-231-8175
Web Site: http://www.novapublishers.com

LIBRARY OF CONGRESS CATALOGING-IN-PUBLICATION DATA

Available upon request.

ISBN: 978-1-60456-201-9

Published by Nova Science Publishers, Inc. ✝ New York

CONTENTS

Preface vii

Short Communications

A Comparative Toxicity of Three Pesticides Affecting Non-Target Terrestrial Isopod Porcellio scaber (Organophosphorus Pesticide Diazinon, Neonicotinoids Imidacloprid and Acetamiprid) 1
Polonca Trebše, Damjana Drobne, Katja Stanek, Mateja Blažič, Anita Jemec and Vladka Lešer

B Determination of Pesticide Residues in Food Using Large Volume Injection with Programmable Temperature Vaporiser and Gas Chromatography – Mass Spectrometry 15
Aniko Kende, Kornel Torkos, Izabella Csernyak, Krisztina Vind and Tamas Rikker

Research and Review Studies

Chapter 1 The Evaluation Multi-residues Method for Determination of 858 Pesticides and Related Chemical in Grains and Cereals by GC-MS and LC-MS-MS 35
Guo-Fang Pang, Yong-ming Liu,Yan-Zhong Cao, Chun-Lin Fan, Yu-Jing Lian, Feng Zheng, Jin-Jie Zhang, Xue-Min Li and Zeng-Yin Li

Chapter 2 Primary and Secondary Targets of Action-Response to Anticholinesterase Pesticide Exposure in Fish: Trends in Underlying Molecular Mechanisms 233
Ana Ferrari, Ana María Pechen de D'Angelo and Andrés Venturino

Chapter 3 New Trends Towards Environmentally Friendly Microextraction Techniques of Pesticide Residues 269
R. Romero-González, A. Garrido Frenich and J. L. Martínez Vidal

Chapter 4	Environmental Friendly Methods for the Extraction and Determination of Organophosphorus and Organochlorine Pesticides *J.J. Santana Rodríguez, Z. Sosa Ferrera and D. Vega Moreno*	**307**
Chapter 5	Residue Level of Organochlorine Pesticides in Fish from Italian Fishery and Aquaculture *Orban Elena and Masci Maurizio*	**349**
Chapter 6	Validation of Neurobehavioral Studies for Evaluating the Perinatal Effects of Single and Mixture Exposure to Pesticides *Vera Lucia S. S. de Castro and Paulo Afonso Silva*	**385**
Chapter 7	Tracking the Dispersal of DDT Residue-Bearing Sediments onto the Floor of Santa Monica Basin *C. K. Paull', W. Ussler III, W. Normark and R. Keaten*	**411**
Chapter 8	Recent Progress in Methods for Determination and AOM Removal of Organophosphorus and Neonicotinoid Insecticides *Polonca Trebše, Mladen Franko, Mojca Bavcon Kralj, Urh Černigoj and Aleš Boškin*	**429**
Chapter 9	Solid-phase Microextraction (SPME) for the Analysis of Pesticide Residues in Food: An Overview *Gianni Sagratini, Dario Giardinà and Sauro Vittori*	**453**
Chapter 10	Pesticide Degradation by Laccases *Susana Rodríguez Couto and José L. Toca Herrera*	**475**
Chapter 11	Cytotoxic Effects of Pesticides on Microalgae Determined by Flow Cytometry *Carmen Rioboo, Raquel Prado, Concepción Herrero and Angeles Cid*	**491**
Index		**507**

PREFACE

This new book focuses on pesticide research. A pesticide is a substance or mixture of substances used for preventing, controlling, or lessening the damage caused by a pest. A pesticide may be a chemical substance, biological agent (such as a virus or bacteria), antimicrobial, disinfectant or device used against any pest. Pests include insects, plant pathogens, weeds, mollusks, birds, mammals, fish, nematodes (roundworms) and microbes that compete with humans for food, destroy property, spread or are a vector for disease or cause a nuisance. Many pesticides are poisonous to humans.

Short Communication A – The authors present a comparative toxicity of three pesticides affecting cholinergic synapses: organophosphorus pesticide diazinon, and two neonicotinoids imidacloprid and acetamiprid. Diazinon is registered to control foliage and soil pests of many fruit, nut, vegetable and ornamental crops. Imidacloprid is a possible alternative for the commonly used organophosphorus pesticides, which has gained registration in about 120 countries for use in over 140 agricultural crops. Acetamiprid is a new insecticide used to kill the insects resistant to organophosphorus insecticides. Only few data are available on the toxicity of these insecticides for soil invertebrates. Especially scarce are toxicity data for neonicotinoids. In our study we compared the effects of three pesticides on a terrestrial isopod *Porcellio scaber* (Isopoda, Crustacea). We assessed the effects on biomarkers at different levels of biological complexity. Imidacloprid and acetamiprid were more toxic to *P. scaber* than diazinon, tested at the same experimental conditions and using a similar set of toxicity endpoints. Literature data on toxicity of these three pesticides on honeybees were compared; imidacloprid appeared more toxic than diazinon and than acetamiprid, respectively. At actual environmental concentrations, diazinon poses higher risk to both, *P. scaber* and bees, compared to the two neonicotinoids. However, due to imidacloprid's increasing use in crop protection and higher persistence in soil, it might potentially be more dangerous after long-term application. Our results indicate that toxicity testing with *P. scaber* provides relevant, repeatable, reproducible and comparable toxicity data that are useful for the risk assessment of pesticides in the terrestrial environment.

Short Communication B - The use of large volume injection is becoming widespread in the gas chromatographic analysis of pesticides in food since this technique allows the user to improve two major issues of the field: increase method sensitivity and simplify sample preparation. By increasing the amount of sample injected, the detection limit of the method can be decreased, which is necessary to comply with the new maximum residue levels of legislation. Large volume injection with solvent venting involves on-line sample concentration, which may replace the sample concentration step often used at the end of sample preparation procedures.

This chapter reports on large volume injection optimisation with temperature programmable vaporiser (PTV) discussing continuous slow injection. After method optimisation fortified food samples were analysed using multi residue sample preparation methods DFG S19 and QuEChERs. Along with the evaluation of a number of false negatives and reproducibility, the impact of each method on instrument maintenance was examined to determine suitability for routine analytical screening.

Chapter 1 - In this chapter, a systematic study was conducted regarding the mass spectrometric characteristics of GC-MS and LC-MS-MS and GPC behaviors characteristics for 858 pesticides, and a GC-MS analytical database has been established for 541 pesticides, a LC-MS-MS analytical database for 629 pesticides and a GPC behavioral characteristic database for 751 pesticides.

Under the GC-MS and LC-MS-MS optimized analytical conditions based on these datebase, a series of the comparison testes were completed. 486 pesticides and related chemicals in 5 kinds of grain samples such as wheat, barley, rice, maize and oats were pretreated by ASE extraction and SPE cleanup as well as 465 pesticides were pretreated by homogeneous and shaking extraction and GPC cleanup. A GC-MS analytical method has been established for determination of 475 pesticides in grains and cereals, and same as LC-MS-MS analytical method for determination of 372 pesticides and related chemicals.

Of these, 263 pesticides can be analyzed by both GC-MS and LC-MS-MS methods, with 212 pesticides only suitable for determination by GC-MS and 109 pesticides only by LC-MS-MS. The total quantifying determination reached 584 pesticides, accounting for 68% of the pesticides studied, while the rest made up 32 % considered unfit for determination by the method described. With 475 pesticides analyled by GC-MS, grain samples were mixed with Celit 545, which was then extracted on accelerated solvent extractor with acetonitrile; the extracts were cleaned up via Envi-18 cartridge and then cleaned up via Envi-Carb and Sep-Pak NH_2 cartridge in series, with the pesticides eluted with acetonitrile +toluene (3+1) , and the eluates were concentrated and exchanged with hexane before being used for analysis. For 372 pesticides analyzed by LC-MS-MS, they were extracted with cyclohexane+ ethyl acetate (1+1) mixed solvent on homogenizer and shaker, cleaned up on GPC equipped with Bio-beads S-X3 column, collecting 22-44min fractions, concentrated and volumed with acetonitrile +water (3+2) before being used for analysis.

For GC-MS method, 5 kinds of grain samples were chosen for fortified recovery and precision tests at low, medium and high levels, with average recoveries of 475 pesticides and related chemicals falling between 60%-130% for 453 pesticides, accounting for 95%; 30%-60% for 22 pesticides, making up the total number of 5%; the relative standard deviation (RSD) falling within 30% involved 451 pesticides, accounting for 95%; those within 30%-40% involved 24 pesticides, making up the total number of 5%.

For LC-MS-MS, 5 kinds of grain samples were also selected for fortified recoveries and precision tests at two levels, with the average recoveries of 372 pesticides and related chemicals falling within 60% -130% for 322 pesticides, accounting for 86 %; Recoveries for 31 pesticides at low levels were unstable, with RSD% larger than 25% but high-levels recoveries were larger than 60% with RSD% less than 25%, accounting for 8%; recoveries at both low levels and high levels for 9 pesticides ranged between 40%~60%, with RSD% less than 30%, making up 3%; only low-level fortified tests were carried out for 10 pesticides because of their limited quantity, with satisfactory results obtained for both recoveries and RSD%, accounting for 3%. There were 50 pesticides of such kind in total, accounting for 14%. Recoveries and precision tests proved that such methods are capable of quantifying 584 pesticides in the above-mentioned 5 grain samples. Limit of detection of GC-MS method is between 0.0025mg/kg-0.8000mg/kg and that of LC-MS-MS method is 0.04 μg/kg – 8.00mg/kg, which is subject to the nature of the pesticides.

Chapter 2 - Organophosphorus (OP) and carbamate (CB) pesticides are known to inhibit the enzyme acetylcholinesterase (AChE) in vertebrates and invertebrates, leading to the accumulation of the neurotransmitter acetylcholine and the subsequent over-stimulation of cholinergic receptors. In fish, the effects of these compounds on AChE and different esterases have been extensively studied. In most cases, OP are more effective and persistent AChE inhibitors than CB, due to both differences in reactivation rates and the aging process. Brain AChE inhibition has been proposed as the major determinant of OP and CB acute toxicity in mammals and other organisms. However, many fish species are able to survive high percentages of brain AChE inhibition indicating that there are other determinant targets of anticholinesterase pesticides acute toxicity such as muscular ChE. The protective action of carboxylesterases seems to be also variable among fish in relation to the one observed in other organisms.

In addition to esterases, secondary targets are currently studied in order to elucidate molecular mechanisms modulating the action of OP and CB and the cellular response to exposure. In fish, different OP and CB have been reported to induce oxidative stress and affect glutathione, glutathione-related enzymes, and other antioxidant enzymes such as catalase and superoxide dismutase. It is known that reactive oxygen species modulate cell signaling and transcription factor activity, affecting gene expression. These effects related to pesticide exposure have been mainly reported in mammalian cell lines, whereas research on fish is scarce. Some OP and CB induce glutathione-S transferase enzymes in fish, and this induction may be related to the nuclear factor Nrf-2 binding to the antioxidant response element (ARE). In addition, some CB are weak agonists of the aryl hydrocarbon receptor (AhR), activating the xenobiotic response element (XRE) and inducing cytochrome P450 1A (CYP1A). The differential modulation of gene expression in OP and CB exposures may be crucial to cell fate towards apoptosis or proliferation and sublethal/ chronic toxicity. Thus, the study of these molecular targets in fish is highly relevant for the elucidation of molecular mechanisms of toxicity and the searching of earlier and more sensitive biomarkers of effect and response.

Chapter 3 - Analytical process is comprised of several steps such as sampling, sample preparation, separation and quantification. Despite of much attention has been focused on developing advanced determination techniques such as liquid or gas chromatography coupled to mass spectrometric detectors, which provides a rapid and powerful determination, it is well known that sample preparation is still one of the most critical steps in the determination of trace pollutants such as pesticide residue analysis in different matrices. Actually, sample preparation step still influences the reliable and accurate analysis of pesticides in food and environmental samples. Bearing in mind that traditional extraction techniques based on liquid-liquid or solid-phase extraction are time consuming, labor intensive and require the use of large amounts of solvents, which are expensive and generate considerable waste, a lot of progress has been made toward the development of faster, safer and more environmentally friendly techniques for sample extraction and extraction clean up prior to instrumental analysis. Although different alternatives such as QuEChERS (Quick, Easy, Cheap, Effective, Rugged and Safe) method based on acetonitrile extraction and dispersive solid phase extraction have been proposed, recent trends in sample preparation are based on miniaturization, and simplification of sample preparation methods, automation, high throughput performance, on-line coupling with analytical instruments such as liquid, gas chromatographs and capillary electrophoresis, and reduction in solvent volume and time, providing simple, inexpensive, and virtually solvent-free extraction techniques.

Thus, techniques such as solid-phase and liquid-phase microextraction (single-drop microextraction and hollow-fiber supported liquid membrane), and stir bar sorptive extraction have been developed in the last few years. These new microextraction techniques, which usually integrates sampling, extraction and concentration into a single step, are defined as non-exhaustive sample preparation techniques, and they only use microliters of extracting solvent or are solvent-free techniques.

This chapter includes a discussion of the different microextraction techniques, presenting their advantages and shortcomings, extraction devices available and commercial equipment. The main parameters for method optimization and practical considerations such as differences between operating in the kinetic extraction regime or under equilibrium conditions will be discussed as well as quantification procedures used for a reliable determination. Finally the potential of these techniques to be used in monitoring purposes and their application in pesticide residue analysis in environmental and food samples will be given, highlighting some common characteristics such as low detection limits and high selectivity.

Chapter 4 - Organochlorine pesticides have been used during decades as insecticides and for the control of vector-borne diseases. In 1980s these pesticides were banned in developed countries and were replaced by other kind of compounds like organophosphorus pesticides; however, they are usually more toxic than organochlorine pesticides. Nowadays these compounds are present in the environment; thus the long-term persistence and their metabolites are presented in water, sediments, soils and vegetables.

The methods usually used for the extraction of these compounds from environmental samples are often too time consuming, involve multi-step procedures that are prone to loss of analytes and use high amounts of toxic organic solvents.

Modern strategies in analytical chemistry have been developed towards the simplification of sample's preparation, as well as the minimization of organic solvent used on the extraction process. An alternative to organic solvents is the use of micellar systems as extractants of organic pollutants in liquid and solid environmental samples. Micellar media facilitate the solubilization of different compounds present in different kind of samples. Benefits obtained when micellar media are used, are lesser toxicity and cost than organic solvents.

Cloud-point extraction (CPE) process in liquid samples and microwave assisted micellar extraction (MAME) in solid samples, have been developed for the extraction of pesticides prior to their determination by high performance liquid chromatography. Moreover, during extraction process, many interfering components are usually co-extracted from solid samples together with target analytes. For increasing sensitivity and selectivity, MAME procedure can be coupled with other extraction and preconcentration techniques, introducing a clean-up step for removing substances that could interfere with the final determination of target analytes in HPLC system. This coupling can be done with solid phase microextraction (SPME), called MAME-SPME, and with solid phase extraction (SPE), MAME-SPE, producing also an intensification of the analytes signals by means of extract clean-up and preconcentration.

In this study an overview of the use of micellar systems well alone, or well combined with the aforementioned extraction/preconcentration techniques, for the determination of organochlorine and organophosphorus pesticides in liquid and solid samples is presented. Fundaments, experimental, variables optimization, quantitative procedures and applications are described and discussed, making emphasis on the advantages of these green methodologies with regard to traditional extraction techniques.

Chapter 5 - The risk analysis represents the basis of any food safety policy. The three components of the risk analysis are the following: risk assessment, risk management (through regulations and food inspections) and risk communication. As regards the first point, in the Fish Quality Unit of the Italian National Institute for Food and Nutrition Research (INRAN) works on fishery safety are carried out. Since 2001 our Laboratory has been involved in checking for the presence of 34 selected Organochlorine Pesticides (OC Pesticides) in fish samples collected either from the wild environment (Mediterranean Sea, Lakes, Rivers) or from Italian Sea Farms.

This type of trace analysis requires high-level equipment, well-planned sample preparation techniques and regular Quality Controls to ensure that accurate results are produced. A powerful instrumental procedure developed in our Laboratory and routinely used is described.

Pollutants selected are included in the current Italian regulatory status and are also cited in the Stockholm Conference as "POP's" (Persistent Organic Pollutants). These Chlorine compounds, largely used in the past, are still present in the Food Chain.

The aquatic environment is almost always the final destination of all waste products; therefore the fishery is particularly exposed to this kind of contamination. Most of the Pesticides we analyse have been banned for use since the 1980s; nevertheless they are not degradable compounds and now represent a big problem. Some of them are well-known substances such as DDT, Dieldrin, Endrin, Chlordane, Aldrin, Lindane, and Heptachlor. Their toxicity was confirmed by a great number of publications and it acts with "short term effects" (acute effects derived from exposition at high concentrations) and "long term effects" (chronicle action, included carcinogenesis).

In the authors' studies they observed different fish contamination levels (from low to medium) depending on the sample origin, size, species, and age.

Fish from Italian Sea Farms seem to be of poor contamination, mainly due to the small age of these products. We found that a relationship between the contamination of farmed fish and the contamination of fish feedstuffs probably exists.

Wild fish collected in three lakes of Central Italy were of good safety, but sometimes a medium contamination was observed especially when the fat content was high. This was the case, for example, of Eels (*Anguilla anguilla*)

Finally an evaluation of OC Pesticide content in Tunas deriving from the new practice of "Bluefin Tuna Sea Ranching" is reported.

Chapter 6 - Laboratory studies have clearly revealed neurotoxic damage after exposure to specific pesticides and in some studies at concentrations equivalent to the environmental exposure. There are numerous opportunities during gestation where pesticides can alter the purpose of a cell, tissue, organ, or system function in the CNS. It is not uncommon for many classes of pesticides, such as insecticides, herbicides, and fungicides, to be used concomitantly. Additional work with environmentally-based mixtures is needed to test the hypothesis of dose-additivity or synergistic effects. Also, the maternal exposure to pesticides during the reproductive life may lead to animal developmental damage. Consequently, the evaluation of the possible effects due to this exposure is important to be evaluated during different developmental periods, especially those considered critical by the influence on nervous system. However, the quality and quantity of the data about the risk posed to humans by individual pesticides vary considerably and few are know about mixture pesticides effects. Consequently, sometimes it is difficult to quantify and compare neurodevelopmental impairment. The potential of pesticides mixtures to affect human toxicity or to pose a reproductive hazard to female rats exposed during critical periods of development needs more research attention. Facing this, the present study provides an overview of the theoretical discussion on potential neurodevelopmental effects of perinatal exposure to single or various chemical substances, considering different experimental situations and their limitations. Knowledge of data quality aids in the correct interpretation of the potential effects of a chemical pollutant exposure and opens perspectives of its use in public health. In this way, some criterions selected to the data evaluation and to guarantee the quality control are discussed, especially the dose-response effect and the variability obtained in a laboratorial test. Based on the findings from the data, a methodology was written with an overall decision rule for the acceptable behavior presented by the experimental units. It is also important to observe the possible influence of others factors interfering on results may cause experimental bias and establish a laboratorial historical control for test procedures. The criticism on the evaluation of them needs to be assessed in light of the end use of neurobehavioral results for improving the process in risk analysis to simple and mixture exposure.

Chapter 7 - The concentrations of the pesticide DDT and its residues (DDTr) were measured in sediment samples collected from the seafloor using a remotely operated vehicle in 286 to 875 m water depths offshore Los Angeles, California. The northwestward decrease in DDTr concentrations from the Whites Point outfall off Palos Verdes, California suggests dispersal pathways are linked to a concentrated point source of DDTr-contaminated particles. This outfall is known to have carried DDTr released from a DDT production facility. The concentration of DDTr in these sediment samples show that DDTr-bearing sediment extends farther down slope into the adjacent 900 m deep Santa Monica Basin and northwestward along the continental slope than previously known. DDTr concentrations in near-surface sediment are higher in samples from the northeastern flank of the Santa Monica Basin (200 to 900 m water depths) than on the adjacent continental shelf (water depths <200 m) at equivalent distances from the point source. Thus, previous surveys to track the dispersion of DDTr from the Whites Point outfall have not encompassed a broad enough area to properly outline and define the dispersion pathways and depocenters. Even with the new observations from deep water, the area that has been surveyed to date is still inadequate to fully understand the dispersion of DDTr released from the Whites Point outfall.

Chapter 8 - This chapter summarizes the most recent investigations in the field of insecticides and methods for their degradation as well as determination in waters and some foodstuff. Due to their mode and extent of application insecticides pose a serious threat to the environment and humans. This is why the insecticides are extensively survailed groups of chemicals and are subject to several residue monitoring programs. Organophosphorus compounds (OP) were more than fifty years applied in crop, fruit and vegetable production as well as in veterinary more than fifty years. Because of this, they appear so frequently (within first three positions) on the top lists of pesticide residues monitoring in European and

American programs. However, during last years there are strong attempts to replace them with neonicotinoids.

The application of several different AOMs in the removal of insecticides is described in this contribution. The photodegradation processes were evaluated not only in terms of kinetics, but also product identification (GC-MS, HPLC-DAD, HPLC-MS) and the degree of mineralisation is compared.

The research in the field of insecticide degradation was, in addition, combined with the development of analytical methods and tools for the determination of trace amounts of pesticide residues in different environmental samples. Very low concentration of OPs (malathion, diazinon, chlorpyrifos, azinphos-methyl) and neonicotinoides (thiamethoxam, imidacloprid, acetamiprid and thiachloprid) were determined by coupling chromatographic techniques and flow injection bioanalytical techniques to a very sensitive detector system based on thermal lens spectrometry (TLS). The achived LODs for both OPs and neonicotinoids are well below 50 ppb, which is approx. 10-times lower than maximal residue limits (MRLs) set in EU and USA legislation. Beside already mentioned classical analytical techniques, the newly developed analytical techniques were also applied in the degradation studies of selected pesticides.

Chapter 9 - The analysis of pesticide residues in food has received great attention in the last years, as testified by the high consideration of the international food safety authorities on the problem and by the large number of papers published in this field. Incorrect use of pesticides can threat the consumer's health, especially if the food is directed to baby population, and for this reason robust and sensible analytical methods are required to detect pesticide traces in food.

Depending on the physical-chemical characteristics of the analytes, instrumental analysis has been normally carried out using gas-chromatography (GC), liquid chromatography (LC) or capillary electrophoresis (CE), each one preceeded by a preliminary extraction process, such as liquid-liquid extraction (LLE), supercritical fluid extraction (SFE), solid-phase extraction (SPE) and matrix solid-phase dispersion (MSPD). A faster extraction technique with respect to those listed above, that doesn't require a long and expensive treatment of sample and that can integrate into a single step sampling, extraction, pre-concentration, and sample introduction, is the solid-phase microextraction (SPME), introduced by Pawliszyn and coworkers in 1990. SPME uses a fused silica fiber externally coated with an appropriate stationary phase such as poly(dimethylsiloxane) (PDMS), poly(acrylate) (PA), PDMS/divinylbenzene (DVB), Carboxen/PDMS. Two basic types of extractions can be performed using SPME: direct immersion (DI) and headspace (HS); the choice depends on sample matrix composition and on volatility of the analytes. Different kind of pesticides, as organophosporus, organochlorine, carbamate and phenylurea, have been monitored in fruit and fruit juices matrices using SPME coupled with GC/MS system or, using a specific interface, with LC/MS and LC/MS/MS systems. In addition, phenylurea and aniline herbicides were analyzed by SPME/GC/MS in solid vegetable matrices as potatoes, carrots and onion, whereas carbamate and phenylurea were monitored in liquid food samples as wine or drinking water by SPME/LC/MS. The aim of this review is to describe the current trends of SPME applied to analysis of pesticide residues in food, underlining all positive and negative aspects that this technique, coupled with chromatographic systems, posses with respect to other more classic extraction procedures.

Chapter 10 - Pesticides are the only chemicals intentionally made to be toxic and introduced directly into the environment. Only 5% of pesticides reach the target organism. The rest runs off into water or dissipates in the soil or air. Such compounds are extremely resistant to biodegradation by native micro-organisms. In addition, the implementation of more and more stringent environmental regulations on hazardous wastes has impelled the search for novel and eco-friendly treatment technologies to complement or substitute the conventional ones. Biodegradation appears a promising technology, particularly the use of oxidative enzymes as biocatalysts, included in a micro-organism or used as a free enzyme.

White-rot fungi produce various types of oxidative enzymes, which are useful for the remediation of environmental contaminants. Laccase [EC: 1.10.3.2] is one of such enzymes. Compared to other oxidative enzymes that need various additives (e.g., hydrogen peroxide for peroxidases), dissolved oxygen is the only requirement for laccase-catalysed reactions. Enzyme-mediated degradation of pesticides has two advantages: (i) the degradation reaction proceeds under relatively mild reaction conditions and (ii) the reaction rate is faster than that by micro-organisms. In addition, they also present some other interesting properties such as low energy requirements, easy process control and operation over a wide range of pH, temperature and ionic strength. Furthermore, enzymes are active in the presence of high concentrations of organic solvents in which hydrophobic molecules are soluble. The purpose of this paper is to review the use of laccases in pesticides degradation.

Chapter 11- As much as *ca.* 99.7% of the applied load of pesticides is dispersed in the environment, not reaching the target pests. In this way, pesticides enter into aquatic ecosystems from agricultural runoff or leaching and, as a consequence, have become some of the most frequently ocurring organic pollutants in aquatic ecosystems.

Most phytotoxicological research with herbicides has been conducted on target plants. The sensitivity of algae to many herbicides is very high, and a better understanding of their environmental effects is probably acquired by using test species representing non-target groups. Because of their short generation times, microalgae respond rapidly to environmental changes. Furthermore, microalgal tests are generally sensitive, rapid and low-cost effective. For these reasons, the use of microalgal toxicity tests is increasing, and today these tests are frequently required by authorities for notifications of chemicals and are also increasingly being used to manage chemical discharges.

Cytotoxic effects of aquatic pollutants on microalgae are very heterogeneous, and they are influenced by the environmental conditions and the test species. Growth, photosynthesis, chlorophyll fluorescence and others parameters reflect the toxic effects of pollutants on microalgae; however, other relevant endpoints are less known because experimental difficulties, especially under *in vivo* conditions.

Flow cytometry offers multiparametric analysis of cells on a cell-by-cell basis in near *in vivo* conditions. Applied in the study of the aquatic environments since the 80's, this technique has achieved extensive use for the study of microalgae and has been introduced as an alternative to more traditional techniques of analyzing cells in culture and from natural populations.

Several parameters determined by flow cytometry, such as cell volume and granularity, chlorophyll *a* fluorescence, cell viability, cell proliferation, oxidative stress, membrane potential or intracellular calcium level, were employed to assess changes in the physiological status of different microalgae as a consequence of the toxic action of herbicides. The variety of results obtained in the present study reveals that flow cytometry is a useful tool in the toxicity tests with microalgae.

In: Progress in Pesticides Research
Editor: Christopher M. Kanzantzakis
ISBN 978-1-60456-201-9

Short Communication A

COMPARATIVE TOXICITY OF THREE PESTICIDES AFFECTING NON-TARGET TERRESTRIAL ISOPOD *PORCELLIO SCABER* (ORGANOPHOSPHORUS PESTICIDE DIAZINON, NEONICOTINOIDS IMIDACLOPRID AND ACETAMIPRID)

Polonca Trebše[1*], Damjana Drobne[2], Katja Stanek[3], Mateja Blažič[3], Anita Jemec[4], Vladka Lešer[2]

[1]University of Nova Gorica, Laboratory for Environmental Research, P.O.B. 301, SI-5001 Nova Gorica, Slovenia

[2]University of Ljubljana, Biotechnical Faculty, Department of Biology, Večna pot 111, SI-1000 Ljubljana, Slovenia

[3]University of Nova Gorica, School of Environmental Sciences, P.O.B. 301, SI-5001 Nova Gorica, Slovenia

[4]National Institute of Chemistry, Hajdrihova 19, SI-1000 Ljubljana, Slovenia

ABSTRACT

We present a comparative toxicity of three pesticides affecting cholinergic synapses: organophosphorus pesticide diazinon, and two neonicotinoids imidacloprid and acetamiprid. Diazinon is registered to control foliage and soil pests of many fruit, nut, vegetable and ornamental crops. Imidacloprid is a possible alternative for the commonly used organophosphorus pesticides, which has gained registration in about 120 countries for use in over 140 agricultural crops. Acetamiprid is a new insecticide used to kill the insects resistant to organophosphorus insecticides. Only few data are available on the toxicity of these insecticides for soil invertebrates. Especially scarce are toxicity data for

[*] Corresponding author: polonca.trebse@p-ng.si

neonicotinoids. In our study we compared the effects of three pesticides on a terrestrial isopod *Porcellio scaber* (Isopoda, Crustacea). We assessed the effects on biomarkers at different levels of biological complexity. Imidacloprid and acetamiprid were more toxic to *P. scaber* than diazinon, tested at the same experimental conditions and using a similar set of toxicity endpoints. Literature data on toxicity of these three pesticides on honeybees were compared; imidacloprid appeared more toxic than diazinon and than acetamiprid, respectively. At actual environmental concentrations, diazinon poses higher risk to both, *P. scaber* and bees, compared to the two neonicotinoids. However, due to imidacloprid's increasing use in crop protection and higher persistence in soil, it might potentially be more dangerous after long-term application. Our results indicate that toxicity testing with *P. scaber* provides relevant, repeatable, reproducible and comparable toxicity data that are useful for the risk assessment of pesticides in the terrestrial environment.

Keywords: neonicotinoids, organophosphates, terrestrial environment, pesticides, risk assessment, toxicity test, *Porcellio scaber*

INTRODUCTION

1.1. General Introduction

World pesticide amount used exceeded 5.0 billion pounds in 2000 (see Table 1). Herbicides accounted for the largest portion of total use. Total world pesticide amount used decreased in 2001 for all pesticide types.

Council Directive 91/414/EEC of 15 July 1991 regulates the field of phytopharmaceutical compounds in European Union and all necessary documents for each of the active compounds. Due to the registration procedure by the end of July 2003, almost 300 active substances were cut off from the list of all registered compounds. These compounds should be replaced in the future by new ones. In Slovenia, 30 compounds were withdrawn from the list of insecticides, among them 8 organophosphates. These compounds are supposed to be replaced by new ones, such as neonicotinoids.

Table 1. World use of pesticide active ingredient at user level by pesticide type, 2000

	Year 2000	World Market
Type	Million kg (lbs) of a.i.	%
Herbicides	0.88 (1.944)	36%
Insecticides	0.615 (1.355)	25%
Fungicides	0.234 (0.516)	10%
Other	0.70 (1.536)	29%
Total	2.427 (5.351)	100%

Note: Table does not cover wood preservatives, especially biocides, and chlorine/hypochlorites. The abbreviation "a.i." stands for active ingredient (US EPA, 2001).

Since there are only few data available on the toxicity of these new insecticides for non-target soil invertebrates, the question appears whether a replacement of the existing pesticides is always justified.

1.2. Organophosphates

Organophosphates (OPs) is the term that includes all insecticides containing phosphorus. Other names used, but no longer in vogue, are *organic phosphates, phosphorus insecticides, nerve gas relatives,* and *phosphoric acid esters.* All organophosphates are derived from one of the phosphorus acids, and as a class are generally the most toxic of all pesticides to vertebrates. Because of the similarity of OP chemical structures to the "nerve gases", their modes of action are also similar. Their insecticidal qualities were first observed in Germany during World War II in the study of the extremely toxic OP nerve gases *sarin*, *soman,* and *tabun.* Initially, the discovery was made in search of substitutes for nicotine, which was heavily used as an insecticide but in short supply in Germany.

The OPs are chemically unstable or nonpersistent, which brought them into agricultural use as substitutes for the persistent *organochorines*. Because of the relatively high toxicity of the OPs, U.S. EPA, under provisions of the Food Quality Protection Act (1996), undertook an extensive reappraisal of the entire class beginning in the late 1990's. Many OPs were voluntarily cancelled (U.S. EPA, 2004).

All OPs are esters of phosphorus having varying combinations of oxygen, carbon, sulfur and nitrogen attached, resulting in six different subclasses: phosphates, phosphonates, phosphorothioates, phosphorodithioates, phosphorothiolates and phosphoramidates. These subclasses are easily identified by their chemical names.

The OPs are generally, with regard to the so called »leaving group«, divided into three subgroups - *aliphatic, phenyl,* and *heterocyclic* derivatives.

Heterocyclic derivatives - The term *heterocyclic* means that the ring structures are composed of different or unlike atoms, e.g. oxygen, nitrogen or sulfur. The first of this group was diazinon, introduced in 1952. Other examples in this group are azinphos-methyl (Guthion®), azinphos-ethyl (Acifon®, Gusathion®), chlorpyrifos (Dursban®, Lorsban®, Lock-On®), methidathion (Supracide®), phosmet (Imidan®), isazophos (Brace®, Triumph®), and chlorpyrifos-methyl (Reldan®).

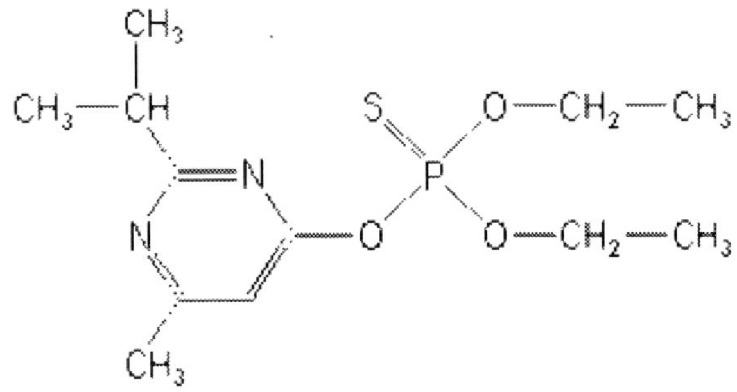

Diazinon: *O,O*-diethyl-*O*-(2-isopropyl-6-methyl-4-pyrimidyl) phosphorothioate.

Table 2. Summary of diazinon, imidacloprid and acetamiprid toxicity to adult *Porcellio scaber* in two-week dietary exposure tests, Lowest-observed Effect (Adverse) Concentrations (LO(A)EC) on the basis of concentrations in food (µg/g leaf)

	Diazinon (adults)	**Imidacloprid (adults)**	**Acetamiprid (adults)**
Weight change	LOAEC ≥ 100 µg/g	LOAEC > 25 µg/g	LOAEC > 25 µg/g
Food assimilation efficiency	LOAEC > 100 µg/g	LOAEC = 10 µg/g	LOAEC = 10 µg/g
Biochemical responses	LOEC = 10 µg/g AChE inhibition	LOEC = 25 µg/g GST inhibition	LOEC = 25 µg/g GST inhibition

The LOAEC indicates adverse effects on higher-level endpoints (survival, weight gain, feeding rate); while LOEC indicates effects on lower-level biomarkers (decrease of acetylcholinesterase or glutathione S-transferase activity).

The US EPA created a candidate list of ten alternatives to diazinon and chlorpyrifos (TDC Environmental, 2003). Among them, a neonicotinoid imidacloprid is suggested to be a possible alternative for diazinon.

1.3. Nicotinoids

The nicotinoids are a newer class of insecticides. They have been previously referred to as *nitro-quanidines, neonicotinyls, neonicotinoids, chloronicotines,* and more recently as the *chloronicotinyls*. Just as the synthetic pyrethroids are similar to and modelled after the natural pyrethrins, so too, are the nicotinoids similar to and modelled after the natural nicotine. Imidacloprid was introduced in Europe and Japan in 1990 and first registered in the U.S. in 1992. It is currently marketed as several proprietary products worldwide, e.g., Admire®, Confidor®, Gaucho®, Merit®, Premier®, Premise® and Provado®. Very possibly it is used in the greatest amount globally of all insecticides.

Imidacloprid is used as a soil, seed or foliar treatment in cotton, rice cereals, peanuts, potatoes, vegetables, pome fruits, pecans and turf, for the control of sucking insects, soil insects, whiteflies, termites, turf insects and the Colorado potato beetle, with long residual control. Imidacloprid has the appearance of a reduced risk pesticide with its comparatively low hazards and low exposure potential for non-target organisms (Agrichemical and Environmental News, 2001).

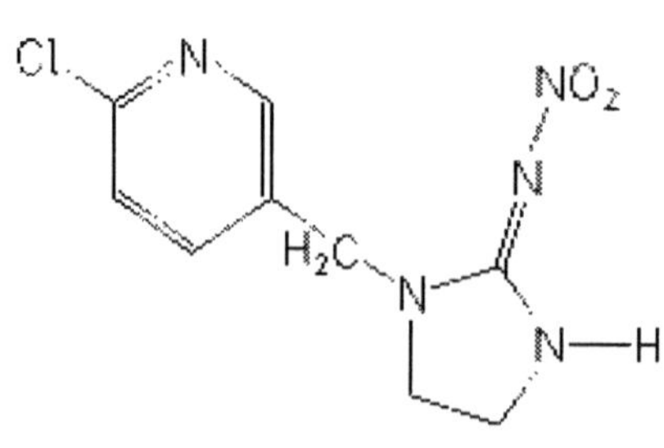

Imidacloprid: 1-(6-chloro-3-pyridylmethyl)-*N*-nitroimidazolidin-2-ylideneamine.

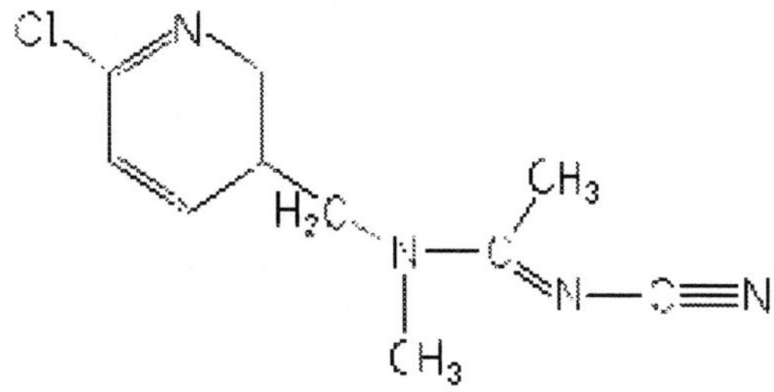

Acetamiprid: (*E*)-*N*[1]-[(6-chloro-3-pyridyl)methyl]-*N*[2]-cyano-*N*[1]-methyl acetamidine.

Other nicotinoids include acetamiprid (Assail®), thiamethoxam (Actara®, Platinum®), nitenpyram (Bestguard®), clothianidin (Poncho®), dinotefuran (Starke®) and thiacloprid. U.S. registrations for acetamiprid, thiamethoxam and thiacloprid were granted in 2002 and clothianidin in 2003 (Ware and Whitacre, 2004).

1.4. Mode of Action of Organophosphates and Nicotinoids

The target site of both, organophosphates and neonicotinoids are acetylcholine synapses.

Diazinon's primary toxicity arises from the inhibition of the enzyme acetylcholinesterase (AChE). This inhibition results in the reduction of the cholinesterase activity level by phosphorilation of AChE and the consequent accumulation of acetylcholine in nerve tissues and effector organs. All organophosphorous pesticide compounds potentially have this mechanism of toxicity in common (Pope, 1999).

Imidacloprid and acetamiprid are nicotinic acetylcholine receptor (nAChR) agonists (Buckingham et al., 1997). The mechanism of action of nicotinoids has been extensively studied, and is relatively well known. In essence, they activate nAChR through binding at or near the sites where nicotine and AChE binds, resulting in dysfunction of the nervous system, immobilization or death. The selectivity of vertebrates and invertebrates nAChR is attributed to structural heterogeneity of neuronal nAChRs (Tomizawa and Casida, 2005).

Due to the high insecticidal activity and low mammalian toxicity, neonicotinoids are considered safer as traditional OP, carbamate and pyrethroid insecticides. In addition, they can be applied at very low rates. However, very little toxicity data exist for neonicotinoids on non-target organisms, especially those inhabiting terrestrial systems. A few comparative toxicity studies indicate the high species-specific response to imidacloprid, which suggests that toxicity data on neonicotinoids may not be generalized (Jemec et al., 2007; Drobne et al., submitted).

1.5. Toxicity Tests for the Terrestrial Environment

The soil litter subsystem is probably the most complicated one in terms of taxonomic diversity and trophic relations. Here, in order to conduct toxicity studies, it is crucial to accept some level of simplification and concentrate on those taxonomic groups and ecological relations that are most important for the functioning of that system.

Terrestrial isopods are widely accepted as organisms of choice in terrestrial ecotoxicology and ecophysiology (Drobne, 1997; Løkke and van Gestel, 1998). They are used also for assessing sustainability of the terrestrial environment (Paoletti and Hassall, 1999). Terrestrial isopods represent an important level in the detritus food chain. Over the last decade they have been used in numerous toxicity studies. In most of them, the effects of elevated concentrations of chemicals were studied at physiological and behavioural level. Fewer provided data on the effects at cellular and subcellular parameters or the population level. Only in some rare studies the linkage of a target enzyme inhibition to integrated organism response variables was demonstrated (Ribeiro et al., 1999; Engenheiro et al., 2005). Beside laboratory studies, terrestrial isopods were used also in microcosm experiments and field studies (Vink and Van Straalen, 1999; Lapanje et al., 2007). It appears from a bulk of published data that a thorough toxicity study along the levels of biological complexity in both juveniles and adults, in the laboratory or field study would be a promising approach to provide comprehensive toxicity information. An integrated biomarker study conducted in parallel with different life stages of terrestrial isopods has not been conducted yet.

During the past decade a number of research efforts have been made to develop terrestrial toxicity tests with invertebrates. At present, several tests are at various stages of formulation. Acute toxicity tests on earthworms and Collembola have reached the broadest international applications. »Dietary exposure tests« are a step behind the »substratum exposure test« in standardisation.

The most extensive toxicity studies of dietary exposure in terrestrial invertebrates have been done with terrestrial isopods. One of the first testing protocols was developed for a terrestrial isopod, *Porcellio scaber* (Drobne, 1997). The test was later improved and assessed also as a toxicity test for pesticides in the terrestrial environment (Stanek et al., 2006).

1.6. Our Work

The aim of our work summarised here was to test the toxicity of three pesticides, all affecting acetylcholine synapses (organophosphorus pesticide diazinon, neonicotinoids imidacloprid and acetamiprid). We compared the toxicity endpoints at biochemical and organismal levels. We compared our toxicity data on *P. scaber* with those on honeybees published in the literature. The investigated pesticides are ranged according to their effect. We discuss the need to conduct comparative toxicity analyses with different toxicity endpoints and compare them with the expected exposure levels before the pesticides are classified according to their toxicity.

2. Material and Methods

2.1. Chemicals

Diazinon (*O,O*-diethyl-*O*-(2-isopropyl-6-methyl-4-pyrimidyl) phosphorotioate) (98.9%) was provided by Pestanal, Riedel-de Haën. Imidacloprid (1-(6-chloro-3-pyridylmethyl)-*N*-nitroimidazolidin-2-ylideneamine) (99.8%) was provided by Bayer Cropscience Slovenija and acetamiprid ((*E*)-N^1-[(6-chloro-3-pyridyl)metyl]-N^2-cyano-N^1-methylacetamidine) (99.9%) by Riedel-de Haën.

Ethyl acetate, methanol, and chloroform were purchased from Riedel-de Haën, acetylthiocholine iodide, dithiobisnitrobenzoic acid and bovine serum albumin (albumin fraction V) from Merck. Bradford reagent, 1-chloro-2,4-dinitrobenzene (CDNB), and L-glutathione (reduced form) (GSH) were purchased from Sigma Aldrich and concentrated sulphuric acid from Carlo Erba. Sodium hydrogen phosphate (Na_2HPO_4) and potassium dihydrogen phosphate (KH_2PO_4) were purchased from Fluka. Absolute ethanol was obtained from J.T.Baker.

2.2. Test Organisms

Terrestrial isopods (*Porcellio scaber* Latr., Isopoda, Crustacea) were collected in the litter layer of uncontaminated woodland near Nova Gorica (Slovenia) or near Ljubljana. In the laboratory, the animals were kept in a terrarium (20 × 35 × 20 cm) filled with a layer of moistened sand and soil (2–5 cm) and a thick layer of partly decomposed hazelnut tree leaves (*Corylus avellana*). The substratum in the terrarium was heated up to 80 °C for several hours to destroy predators (spiders) before the animals were placed in. Hazelnut tree leaves were collected in uncontaminated woodland and air-dried at room temperature. The culture was sprayed daily with water and kept at room temperature and high humidity up to three weeks before the experiments.

2.3. Food Preparation

Hazelnut tree leaves were collected in uncontaminated woodland and dried at room temperature. Dry leaves were cut up into pieces of approximately the same surface area and a weight of 100 mg. Solutions of different concentrations of diazinon, imidacloprid or acetamiprid were prepared from a standard solution diluted with distilled water. A volume of 150 μL of the solution was applied with a pipette to the lower leaf surfaces as small droplets and evenly spread over the surface with a paintbrush. The nominal concentrations of diazinon in the leaves were: 5, 10, 50, 100 or 150 μg diazinon/g dry food, while the concentrations in the case of imidacloprid or acetamiprid were 10 and 25 μg/g dry food. The animals in the control group were fed with hazelnut tree leaves prepared in the same way, yet with distilled water.

2.4. Experimental Set-up and Biomarker Analyses

Experimental set-up and biomarker analyses for diazinon (Stanek et al., 2006) and for imidacloprid (Blažič, 2006; Drobne et al., submitted) have been described previously.

For acetamiprid the same toxicity testing conditions were applied as for the other two pesticides, and the same set of biomarkers was analysed as in the tests for imidacloprid.

3. Results and Discussion

3.1. Comparative Toxicity of Diazinon, Imidacloprid and Acetamiprid to the Terrestrial Isopod Porcellio Scaber

Toxicity data for imidacloprid, diazinon and acetamiprid generated in the two-week laboratory tests with the terrestrial isopod *P. scaber* were compared. The results showed that both neonicotinoids have similar effects on *P. scaber*, while organophosphate diazinon proved to be less toxic when the LO(A)EC or LO(A)EL values are compared (Tables 2 and 3). However, when the hazard quotients (HQ) for the three pesticides are compared, the highest value was obtained for diazinon (HQ = 344), followed by acetamiprid (HQ = 13.3) and imidacloprid (HQ = 3.33). Since HQ is calculated by dividing the estimated daily dose with the reference dose, higher values indicate higher threat to *P. scaber*.

Environmental risk assessment of pesticides and other chemicals often uses the risk quotient (RQ) method to characterize risk quantitatively. An RQ is calculated by dividing an environmental exposure value by a toxicity end-point value. The numerical ranking of RQs for the purpose of comparing potential ecological risks is a valid approach because the rankings are significantly correlated regardless of the degree of exposure refinement (Peterson, 2006). For acute risk determination RQ is calculated with values for LD_{50} or LC_{50} and for chronic risk assessment NOEC or LOEC values are used. Higher RQ value corresponds to the higher potential risk (U.S. EPA, 2004).

3.2. Toxicity of Diazinon, Imidacloprid and Acetamiprid to the Terrestrial Isopod Porcellio Scaber Compared to Honeybees: Interspecies Comparison

Literature data on hazard indices (HI) for acetamiprid, imidacloprid and diazinon for honeybees show that acetamiprid is much less hazardous to honeybees than diazinon and imidacloprid, respectively (Table 4). When risk quotient (RQ) values for the honeybees are compared, again acetamiprid appears less acutely dangerous for honeybees than imidacloprid and diazinon, respectively. Similarly, when RQ values are determined for *P. scaber*, imidacloprid and acetamiprid pose similar risk, but diazinon is far more dangerous (Table 5).

Table 3. Risk assessment of diazinon, imidacloprid and acetamiprid for adult or juvenile isopods (*Porcellio scaber*)

a) Diazinon (Stanek et al., 2006)	
$PEC^{1}_{diazinon}$	7–80 µg /g
estimated daily dose, PEL/day	3.2 µg/g wet Wt./ day (consumption: 0.04 g/g wt. b w/day)
LOAEC isopods (juveniles and adults)	50 µg / g food
LOAEL isopods (juveniles and adults)	2.54 µg /mg wet Wt./ day
NOEL isopods	< 0.23 µg/g wet Wt./ day

(juveniles and adults)	
NOAEL isopods (juveniles and adults)	< 2.54 μg/g wet Wt./ day
$RQ_{diazinon}$	**PEC / LOAEC = 1.6**
$RfD_{diazinon}$	NOEL/100 < 0.0023 μg/g wet Wt./ day
$RfD_{diazinon}$	NOAEL/100 < 0.025 μg/g wet Wt./ day
$HQ_{diazinon}$	estimated daily dose/RfD = 3.2/0.0093= 344
b) Imidacloprid (IMI) (Drobne et al., submitted)	
$PEC^{2}_{imidacloprid}$	0.1–0.2 μg /g 0.14–1.6 μg /g (nectar)
estimated daily dose, PEL/day	0.008 μg/g wet Wt./ day (consumption: 0.04 g / g wt. b w/day)
$LOAEC_{imidacloprid}$ isopods (juveniles and adults)	10 μg/g food
$LOAEL_{imidacloprid}$ isopods (juveniles and adults)	0.24 μg/g wet Wt./ day
$NOEL_{imidacloprid}$ isopods (juveniles and adults)	< 0.24 μg/g wet Wt./ day
$NOAEL_{imidacloprid}$ isopods (juveniles and adults)	< 0.24 μg/g wet Wt./ day
$RQ_{imidacloprid}$	**PEC / LOAEC = 0.02**
$RfD_{imidacloprid}$	NOEL/100 < 0.0024 μg/g wet Wt./ day
$RfD_{imidacloprid}$	NOAEL/100 < 0.0024 μg/g wet Wt./ day
$HQ_{imidacloprid}$	estimated daily dose / RfD = 0.008 / 0.0024 = 3.33
c) Acetamiprid	
$EEC_{acetamiprid}^{3}$	0.1–0.8 μg /g
estimated daily dose, EEL/day	0.032 μg/g wet Wt./ day (consumption: 0.04 g / g wt. b w/day)
$LOAEC_{acetamiprid}$ isopods (adults)	10 μg/g food μg/g food
$LOAEL_{acetamiprid}$ isopods (adults)	0.24 μg/g wet Wt./ day
$NOEL_{acetamiprid}$ isopods (adults)	< 0.24 μg/g wet Wt./ day
$NOAEL_{acetamiprid}$ isopods (adults)	< 0.24 μg/g wet Wt./ day
$RQ_{acetamiprid}$	**EEC / LOAEC = 0.08**
$RfD_{acetamiprid}$	NOEL/100 < 0.0024 μg/g wet Wt./ day
$RfD_{acetamiprid}$	NOAEL/100 < 0.0024 μg/g wet Wt./ day
$HQ_{acetamiprid}$	estimated daily dose / RfD = 13.3

[1] U.S. EPA, 2004, [2] SERA, 2005, [3] estimation based on the data published by SANCO (2001).

LO(A)EC, LO(A)EL and NO(A)EL refer to Lowest or No-observable (Adverse) Effects Concentrations or Levels, with adverse being used to indicate reduction of higher-level endpoints (survival, weight gain, feeding rate); NOEL also includes lower-level biomarkers, like reduction of acetylcholinesterase (diazinon) or glutathione S-transferase (imidacloprid, acetamiprid) activity. RQ = risk quotient, RfD = reference dose, PEC (L) = predicted environmental concentration (level), EEC (L) = estimated environmental concentration (level), HQ = hazard quotient.

Table 4. Relationship between LD_{50} and Hazard Index for selected pesticides against honeybee. (California EPA, 2000; Council directive 91/414/EEC, 2006)

Pesticide	LD_{50} (μg/honeybee)	**Application Rate** (g/ha)	**Hazard Index** = (g/ha)/(LD_{50}*100)
Imidacloprid (topical)	**0.0179**	224–900	125.1
Imidacloprid (oral)	**0.057**	224–900	40.0
Chlorpyrifos (oral)	0.059	480	81.4
Malathion (oral)	0.27	1260	46.7
Diazinon (oral)	**0.37**	400	10.8
Carbaryl (oral)	1.3	850	6.5
Permethrin (oral)	0.075	40	5.3
Deltamethrin (oral)	0.035	7.5	2.1
Acetamiprid (contact)	7.07	168.2	0.24
Acetamiprid (oral)	**14.5**	168.2	0.116
2,4-D (oral)	100	908	0.091
Diuron (oral)	145	908	0.063

Table 5. Comparison of toxicity data for honeybees published in the literature and for *P. scaber* obtained in our work (SERA, 2005; California EPA, 2000; Council directive 91/414/EEC, 2006)

Pesticide (route of exposure)	RQ (PEC/ NOEC)$_{bees}$	**RQ** (PEC/ NOEL)$_{bees}$	**RQ** (PEC/ LD_{50})$_{bees}$	**RQ** (PEC/ NOEC)$_{P.scabe}$	**RQ** (PEC/ NOEL)$_{P.scaber}$
Imidacloprid (oral)	**10** (0.2 / 0.02)	**13.3** (0.2 / 0.015)	**3.5** (0.2 / 0.057)		
Imidacloprid (feeding)	**4.2** (0.2 / 0.48)			**0.02** (0.2 / 10)	**0.83** (0.2 / 0,24)
Diazinon (oral)	-	-	**216** (80 / 0.37)	**8.0** (80 / 10)	**320** (80 / 0.25)
Acetamiprid (oral)	**0.013** (0.8 / 62.5	-	**0.06** (0.8 /14.5)	**< 0.08** (0.8 / <10)	**< 3.2** (0.8/<0.25)

When the RQ values are compared between bees and isopods, imidacloprid has higher potential risk to bees compared to isopods, but acetamiprid has slightly higher potential risk to isopods than to bees (Table 5). Diazinon has much higher potential risk to both, bees and isopods.

Since at the moment there are not enough data on environmentally relevant concentrations of these pesticides, both RQ and HI values are of limited value for an accurate environmental risk assessment (Table 3). Besides, different pesticides are tested in different tests and different types of toxicity data are generated (NOEC, LOEC, LC_x, LD_x).

This problem could be avoided by comparative toxicity testing where in the same single-species toxicity test, toxicity endpoints at different levels of biological complexity are assessed and different types of toxicity data are generated. Terrestrial isopods are among those rare organisms in the terrestrial environment that fulfill all these requirements (Drobne, 1997; Stanek at al., 2006). Besides, terrestrial isopods were successfully used also in the field toxicity studies (Nolde et al., 2006; Lapanje et al., 2007).

3.3. Future Perspectives in Pesticide Toxicity Testing

In general, in risk assessment numerous sources of uncertainty are inherent. Therefore the extrapolation of toxicity data from the laboratory animal studies to the environment and human exposure scenarios is inexact. The assumptions that we made regarding the exposure events and durations are worst-case scenarios. Conversely, the potential for synergistic toxicity is not directly considered in the risk assessment process.

The current means of addressing these uncertainties are through extreme conservatism in all extrapolations and assumptions and by the use of large "uncertainty factors" that reduce the chemical dose considered safe. For example, determination of Reference Doses is typically done by finding the dose at which no effects are produced in selected test organisms and dividing that dose by a "safety factor" of 10 to 10,000. These safety factors are used to account for uncertainty and to be sure that even the most sensitive organisms would not be adversely affected at the Reference Dose. Use of such large safety factors may often result in RfDs that are actually far below a dose that could produce effects. This approach is prudent because the process of risk assessment is intended to support decision making that is protective of the environment rather than to accurately reflect the toxic potential of chemicals. But a question appears about the applicability and reliability of such risk assessments.

Without adequate data, our ability to identify true environmental hazards is reduced and the anxiety over chemicals increases. Costly errors can result when evaluations are made on the basis of inappropriate or poorly documented data. The more accurate the data we use to conduct risk assessments, the more confident we can be that our efforts to protect the environment are appropriate and effective.

The toxicity testing with isopods, though limited, illustrates how the methodology can be applied to produce reliable, comparative toxicity data. In order to expand and complete the pesticide risk assessment, we must verify, expand and optimize the database on pesticide fate, transport, and toxicity. To reach this goal for toxicity data, we suggest the extensive comparative toxicity studies with less test species but a battery of toxicity endpoints. This would substantially improve the quality of toxicity data and would help to set more realistic safety factors and subsequently a more relevant risk assessment.

4. References

Agrichemical and Environmental News, October 2001, Issue No. 186, http://aenews.wsu.edu/Oct01AENews/Oct01AENews.htm#anchor382974

Blažič, M; *Toxicity of imidacloprid to the terrestrial isopod Porcellio scaber (Isopoda, Crustacea): master's thesis.* Nova Gorica Polytechnic, School of Environmental Sciences, 2006.

Buckingham, SD; Lapied, B; LeCorronc, H; Grolleau, F; Sattelle, DB. 1997. Imidacloprid actions on insect neuronal acetylcholine receptors. *J. Exp. Biol.* 200, 2685-2692.

California EPA, *Department of pesticide regulation, 2000.* http://www.ent.uga.edu/bees/Pollination/Table_Insecticides_Miticides.htm

Council directive 91/414/EEC, 1991. http://www.furs.si/law/EU/ffs/eng/en_1991L0414_do_001.pdf

Council directive 91/414/EEC, 2006. http://www.efsa.europa.eu/etc/medialib/efsa/science/praper/conclusions/praper_concl_diazinon.Par.0008.File.dat/praper_diazinon_addendum_final.pdf)

Drobne, D; 1997. Terrestrial isopods – A good choice for toxicity testing of pollutants in the terrestrial environment. *Environ. Toxicol. Chem.* 6, 1159-64.

Drobne, D; Blažič, M; Van Gestel, CAM; Lešer, V; Zidar, P; Jemec, A; Trebše, P; submitted. *Toxicity of imidacloprid to the terrestrial isopod Porcellio scaber* (Isopoda, Crustacea).

Engenheiro, EL; Hankard, PK; Sousa, JP; Lemos, MF; Weeks, JM; Soares, AMVM; 2005. Influence of dimethoate on acetylcholinesterase activity and locomotor function in terrestrial isopods. *Environ. Toxicol. Chem.* 24, 603-609.

Food Quality Protection Act, 1996. http://www.epa.gov/pesticides/regulating/laws/fqpa/fqpa_implementation.htm

Jemec, A; Tišler, T; Drobne, D; Sepčić K; Fournier, D; Trebše, P; 2007. Comparative toxicity of imidacloprid, of its commercial liquid formulation and of diazinon to a non-target arthropod, the microcrustacean *Daphnia magna*. *Chemosphere* 68, 1408-1418.

Lapanje, A; Rupnik, M; Drobne, D; 2007. Gut bacterial community structure (*Porcellio scaber*, Isopoda, Crustacea) as a measure of community level response to long-term and short-term metal pollution. *Environ. Toxicol. Chem.* 26, 755-763.

Løkke, H; van Gestel, CAM; 1998. *Handbook of Soil Invertebrate Toxicity Tests,* John Wiley & Sons, Chichester, New York.

Nolde, N; Drobne, D; Valant, J; Padovan, I; Horvat, M; 2006. Lysosomal membrane stability in laboratory- and field-exposed terrestrial isopods *Porcellio scaber* (Isopoda, Crustacea). *Environ.Toxicol.Chem.* 25, 2114-2122.

Peterson RKD, 2006. Comparing ecological risks of pesticides: the utility of a Risk Quotient ranking approach across refinements of exposure. *Pest. manag. sci.* 62, 46-56.

Pope, CN, 1999. Organophosphorus pesticides: Do they all have the same mechanism of toxicity? *J. Toxicol. Env. Heal.* B, 2161-181.

Paoletti, MG; Hassall, M; 1999. Woodlice (Isopoda: Oniscidea): their potential for assessing sustainability and use as bioindicators. *Agriculture, Ecosystems & Environment,* 74, 157-165.

Ribeiro, S; Guilhermino, L; Sousa, JP; Soares, AMVM; 1999. Novel bioassay based on acetylcholinesterase and lactate dehydrogenase activities to evaluate the toxicity of chemicals to soil isopods. *Ecotoxicol. Environ. Saf.* 44, 287-293.

SANCO, 2001, 1392, final. 16 June 2004. http://ec.europa.eu/food/plant/protection/evaluation/newactive/acetamiprid.pdf

SERA (Syracuse Environmental Research Associates, Inc.), 2005. Imidacloprid-Human health and ecological risk assessment-final report; prepared for USDA, Forest Service, USA (SERA TR 05-43-24-03a). http://www.fs.fed.us/foresthealth/pesticide/risk_assessments/122805_Imidacloprid.pdf

Stanek, K; 2004. Sensitivity and specificity of acetylcholinesterase activity in *Porcellio scaber* (Isopoda, Crustacea) as a biomarker of diazinon exposure: *master's thesis.* Nova Gorica: Polytechnic, School of Environmental Sciences. Nova Gorica, Slovenia.

Stanek, K; Drobne, D; Trebše, P; 2006. Linkage of biomarkers along levels of biological complexity in juvenile and adult diazinon fed terrestrial isopod (*Porcellio scaber*, Isopoda, Crustacea). *Chemosphere* 64, 1745-1752.

TDC Environmental, 2003. Insecticide market trends and potential water quality implications. TDC Environmental, LLC, San Mateo, CA, USA.

Tomizawa, M; Casida, JE; 2005. Neonicotinoid insecticide toxicology: mechanisms of selective Action. *Annu. Rev. Pharmacol.* 45, 247-268.

U.S. EPA, 2001. http://www.epa.gov/oppbead1/pestsales/01pestsales/table_of_contents 2001_2.htm

U.S. EPA, 2004. *Office of prevention, pesticides and toxic substances.* EPA 738-R-04-006, Diazinon, interim reregistration eligibility decision. http://www.epa.gov/oppsrrd1/ REDs/diazinon_ired.pdf

Vink, K; van Straalen, NM; 1999. Effects of benomyl and diazinon on isopod-mediated leaf soil litter decomposition in microcosms. *Pedobiologia* 43, 345-359.

Ware, GW; Whitacre, DM; 2004. IPM World textbook. http://ipmworld.umn.edu/ chapters/ware.htm

In: Progress in Pesticides Research
Editor: Christopher M. Kanzantzakis

ISBN 978-1-60456-201-9

Short Communications B

DETERMINATION OF PESTICIDE RESIDUES IN FOOD USING LARGE VOLUME INJECTION WITH PROGRAMMABLE TEMPERATURE VAPORISER AND GAS CHROMATOGRAPHY – MASS SPECTROMETRY

Aniko Kende[1,2], Kornel Torkos[1], Izabella Csernyak[1,2], Krisztina Vind[1,2], Tamas Rikker[2]

[1]Institute of Chemistry, Eötvös University, 1117 Budapest, Hungary
[2]Wessling International Research and Education Center, 1047 Budapest, Hungary

ABSTRACT

The use of large volume injection is becoming widespread in the gas chromatographic analysis of pesticides in food since this technique allows the user to improve two major issues of the field: increase method sensitivity and simplify sample preparation. By increasing the amount of sample injected, the detection limit of the method can be decreased, which is necessary to comply with the new maximum residue levels of legislation. Large volume injection with solvent venting involves on-line sample concentration, which may replace the sample concentration step often used at the end of sample preparation procedures.

This chapter reports on large volume injection optimisation with temperature programmable vaporiser (PTV) discussing continuous slow injection. After method optimisation fortified food samples were analysed using multi residue sample preparation methods DFG S19d[1] QuEChERs[2]. Along with the evaluation of a number of false negatives and reproducibility, the impact of each method on instrument maintenance was examined to determine suitability for routine analytical screening.

For the analysis of vegetable and fruit samples a 100 μl continuous slow injection method is suggested in combination with QuEChERs sample preparation due to its low number of false negatives and acceptable matrix contamination on the GC-MS instrument.

1. Introduction

Pesticides are a chemically diverse group of compounds used extensively in modern agriculture to increase production. Some pesticides are known to cause birth defects [3, 4], known or probable carcinogens [5, 6] or highly toxic [7, 8], that is why their residues in food are to be controlled by several acts as the 396/2005/EC in the European Union and the 5/2002 EüM-FVM in Hungary. Legislation defines maximum residue levels (MRLs) for agricultural produce – active ingredient pairs, where the number of pesticides to be tested in a produce may result in over 400 target compounds. This and the diversity of food products available created the need for fast and sensitive multi residue analytical methods, as MRLs can be as low as 10 μg/kg.

1.1. Multi Residue Sample Preparation Methods

Multi residue methods are usually composed of sample clean-up, pre-concentration, liquid or gas chromatographic separation and detection by either a range of selective detectors such as FLD (LC) or ECD, NPD, AED (GC) and/or the both universal and selective mass spectrometer. Two widely accepted pesticide residue methods are the DFG S19 method of the German Food Safety Authority [9] and the QuEChERs method developed by S.J. Lehotay and M. Anastassiades [10]. After solvent extraction of the homogenized sample the DFG S19 applies liquid-liquid extraction and gel permeation chromatography (GPC) for sample clean up, while the QuEChERs method uses a special sorbent to remove matrix impurities. Both sample preparation methods result in an extract suitable for gas chromatographic analysis.

1.2. Large Volume Injection Techniques

One possible way in gas chromatography to increase method sensitivity and decrease detection limits is to increase the injected sample volume [11]. The analyzed sample volume in case of splitless injection is usually 1-2 μl, which can be extended up to 5 μl with the use of pulsed splitless injection [12, 13]. When the injected sample volume exceeds the capacity of traditional split/splitless injection, one can describe it as large volume injection. There are several techniques reported for large volume injection, for example the large volume on-column or the loop injection, however, because of availability the most widespread technique is the temperature programmable vaporiser (also called cooled inlet system). As this chapter will concentrate on the latter, the prior techniques are presented only very shortly.

In large volume on-column injection the injector's roll is reduced to being an interface between the sampler (autosampler or HPLC) and the gas chromatograph. When the sample enters the chromatograph, it is at room temperature and in liquid phase. The chromatograph is equipped with a special column system. First the sample enters the retention gap, which is a non-coated quartz capillary, connected to a coated pre-column or retention column. Between the pre-column and the analytical column there is a valve for venting the solvent vapour, called solvent-vapour exit (SVE). From the retention gap the sample is transferred to the pre-column by the carrier gas, where an initial separation takes place, in which the solvent overtakes the target compounds. The solvent is then vented through the SVE valve. Immediately after the solvent has left the system, the valve is closed, and the target compounds enter the analytical column [14].

Loop injection is used in HPLC – GC coupled systems, where the loop is filled by the HPLC elute. The solvent venting process can be very similar to the above described large volume on-column injection, applying a retention gap, pre-column and a solvent vapour exit. The main difference is that during the solvent elimination step the sample is transferred by the generated solvent vapour itself and not by the carrier gas [15-17].

1.3. Large Volume Injection with Temperature Programmable Vaporiser

Large volume injection with programmable temperature vaporiser was first reported by Vogt in 1979 [18, 19]. In his report he adjusted the injector temperature a few degrees below the boiling point of the solvent and executed a slow injection while venting the generated solvent vapour. Based on this idea Poy developed the first programmable temperature injector [20]. Today most instrument producers offer programmable temperature vaporisers for their gas chromatographs.

The inner diameter of the commercially available programmable temperature injectors is substantially smaller than thar of the traditional injectors to allow dynamic heating and cooling of the injector. The injector can be heated at a rate of up to 720 – 1000 °C/min. Sub-boiling point venting – where the initial injector temperature is set just merely below the boiling point of the applied solvent – can be accomplished without a cryogen if the target compounds have significantly higher boiling point than the solvent, as it has been described in several works [21-23]. For cold venting – where the initial injector temperature is set significantly lower than the solvent's boiling point the means of cooling can be selected from several options by the user. The lowest temperature can be achieved by liquid nitrogen as the cryogen (down to -150 °C). For this purpose the injector is equipped with a special entrance valve for the liquid nitrogen and an expansion shell. The nitrogen enters the injector in liquid phase, then expands during phase transition and cools the injector body. Another possible cryogen is liquid carbon dioxide, which works in the same way as liquid nitrogen, but requires a more durable entrance valve. When using carbon dioxide as cryogen the lowest achievable temperature is -50 °C. In case of applications which do not require extreme low temperatures, Peltier cooling is a possible choice. Peltier cooling applies a heat exchanger, where the heat exchanger liquid is circulated in a spiral around the injector body. The achievable temperature is about the laboratory temperature minus 20 °C. The gas system of the PTV injector is similar to the traditional injectors.

The programmable temperature injector has several modes of injection, such as the hot or cold split/splitless, on-column and solvent vent. Large volume injection is performed in solvent vent mode with the programmable temperature injector. When using solvent vent, there must be a sufficient boiling point difference between the solvent and the analytes to reduce the loss of the more volatile components [28].

Sample introduction into the injector can be performed fractionally, when the autosampler introduces the sample in multiple small volume portions, which technique is called multiple injections, or with a continuous slow flow injection.

The mechanism of the multiple injections is based on timing of the sample introductions. The scope of the optimisation process is to find the appropriate injection temperature and vent flow for a set portion size and interval of injections. Portion size has to be in accordance with the speed of solvent evaporation as it is avoidable that liquid sample enters the column, while if the interval between two injections is too large, analyte loss can be observed [24].

In case of the continuous slow injection the entire sample volume is introduced in one, slow injection to the injector by a controlled speed autosampler [25]. The aim of the optimisation here is the find a set of injection parameters when the sample's solvent content evaporates in the injector and leaves through the split valve, while the analytes are concentrated in a thin film of solvent on the surface of the inlet liner. Parameters are the head pressure, split flow, injector temperature and injection speed. Ideally the speed of sample injection equals the speed of evaporation of the solvent, and as such the sample film generating on the liner wall does not overflow or dry out [26, 27].

A good starting point for determining of the initial parameter values is Equation 1 [28]. For apolar solvents and smaller injection volumes (appr. up to 25 μl) it gives near optimal values.

$$I = \frac{M \cdot P_S}{\rho \cdot R \cdot T_a} \cdot \frac{P_a}{P_{inlet}} \cdot F_{split}, \text{ where} \qquad \text{Equation 1}$$

I	injection speed or solvent elimination rate [μl/min]
M	molecular weight of the solvent [g/mol]
P_s	vapour pressure of solvent at initial injector temperature [bar]
ρ	density of solvent [g/ml]
R	gas constant 83.12 $\frac{[ml][bar]}{[mol][K]}$
T_a	initial injector temperature [K]
P_a	ambient pressure (usually 1.013 bar)
P_{inlet}	head or inlet pressure (gauge + ambient pressure)
F_{split}	split flow [ml/min]

During sample introduction the injector is cold (10-40 °C), the oven holds its initial temperature, the split valve is open and the vent flow is started. Carrier gas pressure towards the column is reduced to 0 in order to reduce the amount of solvent entering the column during solvent vent. When the injection is done and the needle of the syringe is removed from the injector, head pressure is restored, the temperature program of the injector and the oven starts and the split valve is closed. From this point on the procedure is analogous to splitless injection: after the splitless time has elapsed the split valve opens to remove sample residues from the injector [28].

1.4. The Aim of the Study

This chapter reports on large volume injection optimisation with temperature programmable vaporiser applying cold venting with continuous slow injection. After method optimisation fortified food samples were analysed comparing multi residue sample preparation methods (DFG S19 and QuEChERs). Along with the evaluation of number of false negatives in the 10 μg/kg concentration range and reproducibility, the impact of each method on instrument maintenance was examined for suitability for routine analytical screening.

2. Experimental

2.1. Chemicals

Cyclohexane, dichloromethane, ethyl acetate and acetonitrile super purity solvents were purchased from Romil (Cambridge, UK), and the pesticide standards were purchased from Dr. Ehrenstorfer (Augsburg, Germany). A stock solution of each pesticide of 1 mg/ml was prepared by dilution with a) cyclohexane or dichloromethane according to solubility properties, b) acetonitrile. Two 25 μg/ml intermediate standard mix of pesticides were prepared with a) cyclohexane and b) acetonitrile (Table 1). Working solutions were prepared daily, with concentrations depending on injected sample volume. For fortification of food samples the acetonitrile solution was used. Sodium sulphate and magnesium sulphate were purchased from Reanal (Budapest, Hungary), Bondesil-PSA 40 μm sorbent was purchased from Varian (#12213024/100g).

Table 1.

Pesticide	RT (min)	Q1	Q2	Q3	Q4
Butylate	7.66	57	146	156	217
Propachlor	8.96	120	77	126	
Quintozene	10.19	237	249	295	297
Terbufos	10.13	57	231	103	153
2,4,5-T	10.19	233	235	268	270
Methyl-parathion	11.01	263	125	109	
Malathion	11.67	125	173	93	127
Chlorpyriphos	11.89	97	197	199	314
Isodrin	12.38	193	195	263	
Tetrachlorvinphos	13.16	109	329	331	
Prothiophos	13.53	113	162	267	309
Ethion	14.59	97	231	153	125
Captafol	15.77	79	77	80	
Bifenthrin	16.37	181	165	166	
Phosalon	17.18	182	121	154	
Leptophos	17.22	171	377	375	
Coumaphos	18.80	362	226	364	210
Cypermethrin α	19.98	163	181	165	209
Deltamethrin	22.83	181	253	251	255

2.2. Instrumentation

Experiments were conducted on an Agilent 6890 gas chromatograph – FID and 5973 mass selective detector (MSD) system was equipped with Gerstel MPS2 autosampler and a Gerstel CIS4 (cooled inlet system) injector with Peltier cooling. The analytical column was HP-5 MS+ 30 m x 0.25 mm x 0.25 μm, with 1 m same phase, same diameter pre-column attached to it with a Column Union Assembly (Agilent #5182-0744). In order to maintain the 30 m column length the first 1 m pre-column was trimmed from the analytical column [29]. The FID detector (300 °C, He 5.0 carrier gas, H_2 40 ml/min, N_2 45 ml/min and air 450 ml/min) was used in initial injection optimization, the MSD detector was applied for the development of the finalized method and residue analysis.

Chromatographic Conditions for 100 μl Large Volume Injection

Injection Parameters for Cyclohexane Extracts

Injected sample volume 100 μl, head pressure 7 psi, injection speed 0.6 μl/s (total injection time 2.8 min), injector temperature program: 30 °C (0.6min) 720 °C/min to 280 °C, vent flow 50 ml/min, vent pressure 0 psi, vent time 0.4 min, split vent time 2 min, pure flow to split valve 50 ml/min.

Injection Parameters for Acetonitrile Extracts

Injected sample volume 100 μl, head pressure 7 psi, injection speed 0.1 μl/s (total injection time 16.7 min), injector temperature program: 40 °C (0.6min) 720 °C/min to 280 °C, vent flow 50 ml/min, vent pressure 0 psi, vent time 0.4 min, split vent time 2 min, pure flow to split valve 50 ml/min.

Carrier gas He 5.0, carrier gas flow on column 1 ml/min, oven temperature program: 50 °C (1 min) 50 °C/min to 180 °C, 10 °C/min to 280 °C (2.25 min). Mass spectrometer data: transfer line 280 °C, solvent delay 3 min, and ion source temperature 230 ° C, quadruple 150 ° C using Selected Ion Monitoring (Table 1).

2.3. Injection Optimization

A reference chromatogram of 1 μl splitless injection was obtained in order to be able to evaluate any discrimination large volume injection may cause. Injected sample volume was then increased to 10 μl as well as the working standard solution was diluted 10 fold. The investigated parameters and their ranges are listed in Table 2, the calculated injection speeds for initial injector temperature and vent flow pairs are listed in Table 3 a) and b) for cyclohexane and acetonitrile respectively.

When the injected sample volume was increased to 100 μl, injection parameters had to be re-optimized for acetonitrile solution, as the settings optimized for 10 μl injections were not suitable. An additional f factor [28] was introduced to reduce the speed of injection.

$$I = \frac{M \cdot P_S}{\rho \cdot R \cdot T_a} \cdot \frac{P_a}{P_{inlet}} \cdot F_{split} \cdot f \qquad \text{Equation 2}$$

Table 2.

Parameter	Abbr.	Unit	Values
Initial injector temperature	$T_{inj\ ini}$	°C	20 / 40 / 60
Initial injector temperature hold time	$t_{inj\ ini}$	min	0.4 / 0.5 / 0.6 / 0.7
Injection speed	I	µl/s	Calculated (see Table 3)
Vent time	-	min	0.2 / 0.3 / 0.4 / 0.5
Vent flow	-	ml/min	50 / 75 / 100
Splitless time	SVT	min	1 / 1.5 / 2 / 2.5

Table 3 a).

Injection speed$_{cyclohexane}$ (µl/s)		Vent flow (ml/min)		
		50	75	100
$T_{inj\ ini}$ (°C)	20	0.32	0.47	0.62
	40	0.68	1.02	1.35
	60	1.35	2.03	2.70

Table 3 b).

Injection speed$_{acetonitrile}$ (µl/s)		Vent flow (ml/min)		
		50	75	100
$T_{inj\ ini}$ (°C)	20	0.15	0.22	0.28
	40	0.32	0.47	0.62
	60	0.62	0.93	1.23

The initial experimental value of the f factor was 0.8, which was decreased with an 0.1 increment to 0.5.

All parameter sets were run in triplicate and the average of the three runs were used for comparison. The optimized cyclohexane (DFG S19) and acetonitrile (QuEChERs) methods were externally calibrated at three concentration levels (10, 50 and 100 µg/kg).

2.4. Application on Food Samples

Both the DFG S19 (Module E1) and the QuEChERs extraction procedure can be found described elsewhere [9], [10]. Method protocols were strictly followed, except a few minor changes were introduced where necessary to fit available laboratory equipment. Gel Permeation Chromatography (GPC) was conducted on a LC Tech GPC 1122 solvent delivery system MI4-110-KC1 with Biorad Bio-Beads S-X3 (200-400 mesh).

The examined food samples (Chinese cabbage, carrot and apple) were purchased at the local market. Two set of samples were fortified and prepared using the two sample preparation procedures. Fortification was done at three concentration levels at 10, 50 and 100 µg/kg in three parallels.

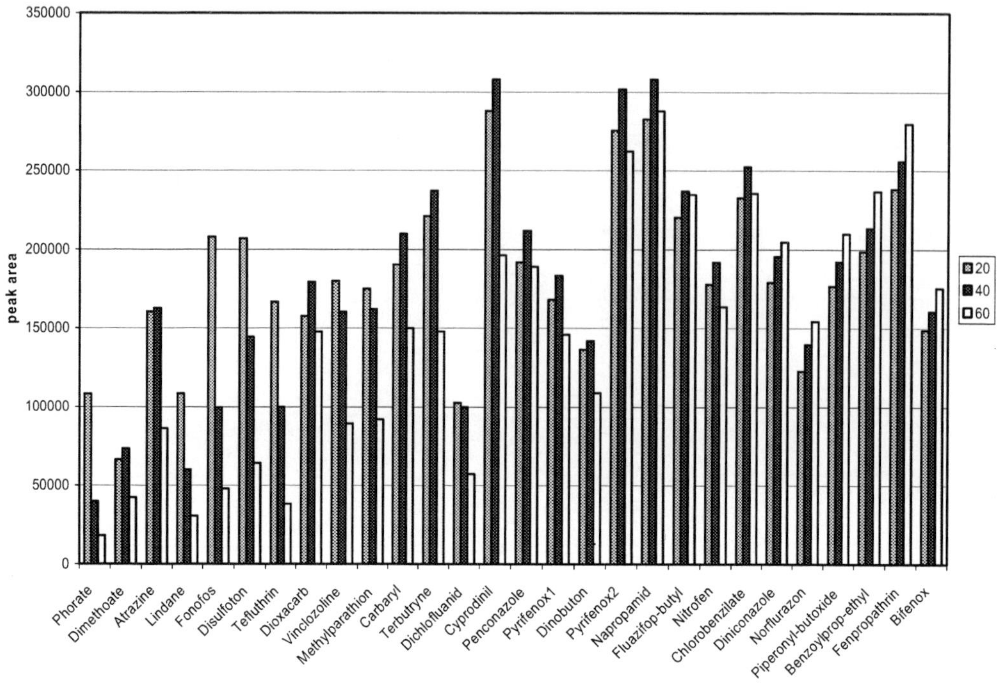

Figure 1 a). The effect of initial injector temperature on target peak areas in cyclohexane solution.

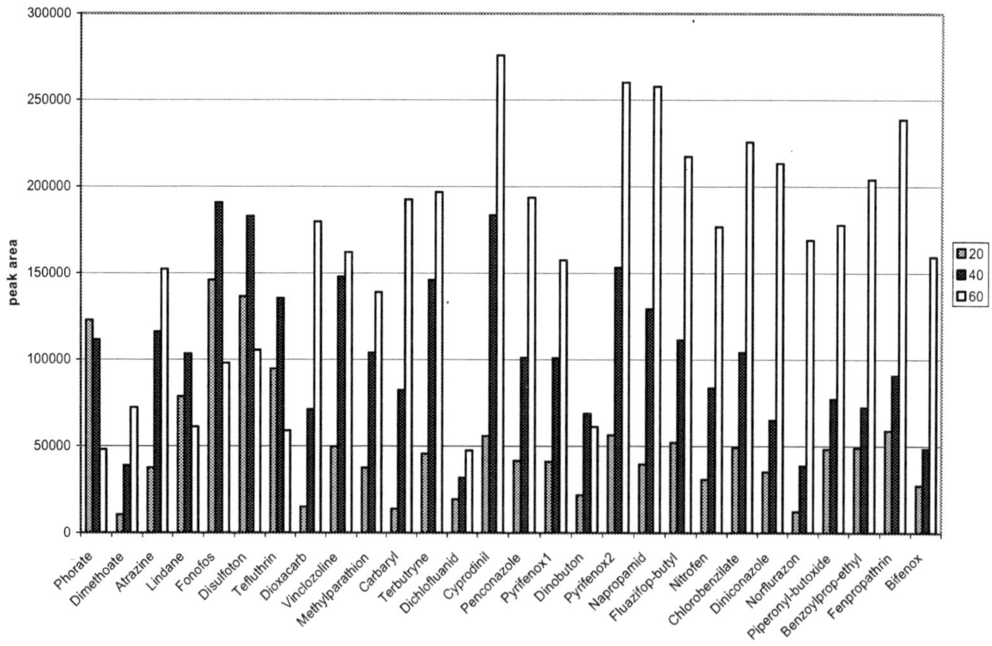

Figure 1 b). The effect of initial injector temperature on target peak areas in acetonitrile solution.

3. Results and Discussion

3.1. Initial Injector Temperature

Cyclohexane Solution

High initial injector temperature resulted in peak area loss of early eluting compounds regardless the applied vent flow (Figure 1a). While the high initial injector temperature was favorable for the late eluting compounds, it caused distortion of the chromatogram.

Acetonitrile Solution

The chromatographic distortion observed for cyclohexane was even more pronounced for acetonitrile (Figure 1b). At 20 and 40 °C the late eluting compounds suffered losses, at 60 °C the early eluting pesticides were partially lost.

We noticed that discrimination of late eluting compounds occurred in conjunction with large solvent peaks in the FID chromatograms. The loss of late eluting compounds may be explained by insufficient solvent removal, where excess solvent remains in the inlet liner during the temperature ramp, delaying the evaporation of higher boiling compounds, part of which stays in the injector until the end of the splitless time. When the split valve opens, the residues are flushed out of the injector by the carrier gas and thus lost for analysis.

3.2. Vent Time and Initial Injector Temperature Hold Time

Both the initial injector temperature hold time ($t_{inj\ ini}$) and the vent time starts when the syringe is removed from the injector following sample introduction, where $t_{inj\ ini}$ incorporates vent time. During vent time the split valve is open, allowing additional solvent venting. After the vent time the split valve is closed and the head pressure is restored. The time window between $t_{inj\ ini}$ and vent time allows the head pressure to stabilize. For both studied solvents the least amount of time between syringe removal and the temperature ramp gave the best results (Figure 2). If the vent time was increased, it resulted in loss of peak area of the early eluting compounds.

3.3. Splitless Time

In general splitless time has to be sufficiently long to allow the analytes to be transferred onto the column without loss before the oven temperature program starts. If splitless time is too long, peak shape is deteriorated [11]. The same was observable in case of large volume injection in case of both solvents; however optimal splitless time was significantly longer (2 min) than in case of splitless injection (0.7 min). To heat the injector from 20 °C to 280 °C at a 720 °C/min temperature ramp, takes 0.36 min which does not justify the 2 min optimal splitless time alone. However it might be caused by the analyte condensate in the cold injector. Thus when the injector is heated, the analytes first have to be desorbed, evaporated and then transferred onto the column.

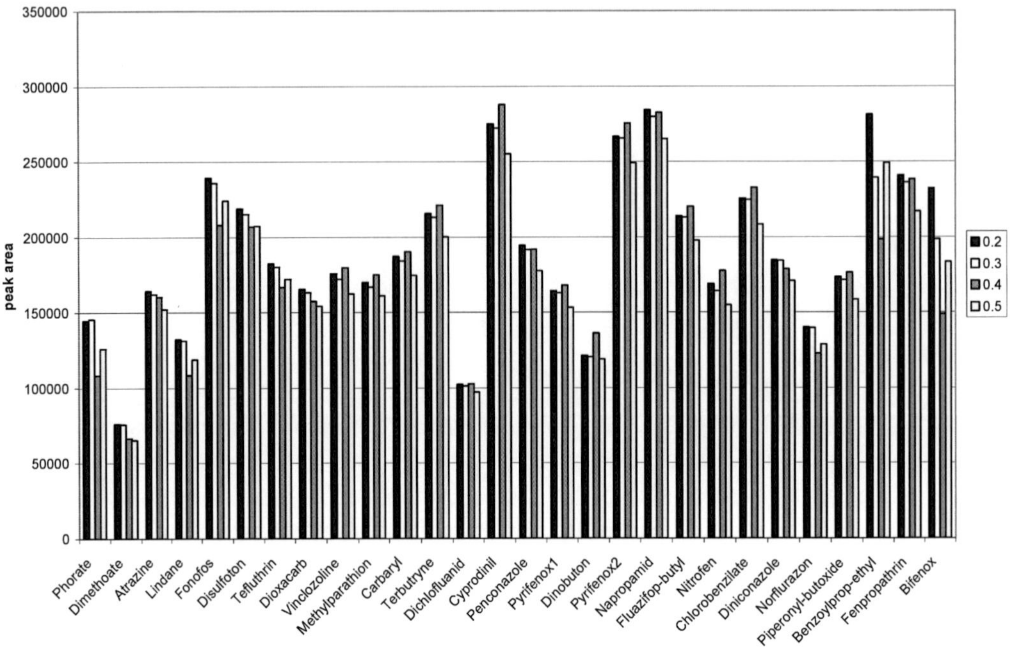

Figure 2. The effect of vent time on target peak areas in cyclohexane solution.

3.4. Vent Flow

Continuous slow flow injection allows the user to decide if a fast or a gas-saving method is preferred: fast sample introduction requires high vent flow, while if the required sample throughput allows for a more timely injection the average gas usage of a GC-MS system is sufficient.

In order to examine the effect of vent flow independently of other parameters, injection speed was matched to each studied vent flow value, calculated with Equation 1. Higher vent flow gave better results for both solvents (Figure 3 a and b), but the achieved gain in peak areas is more significant in case of the acetonitrile solution. A possible cause of this observation may be the polarity difference between of the two studied solvents.

3.5. Comparison of 1 μl Splitless and 10 μl Large Volume Injection

Throughout the method optimization those parameters were favored, which resulted in a chromatogram most similar to the one obtained with the original, 1 μl splitless injection. The amount of pesticides introduced into the gas chromatograph was the same, using a 1/10 dilution of the standard solution used in splitless experiments for large volume injection. The optimized parameter set was the following: head pressure 7 psi, injection speed 0.62 μl/s for cyclohexane, 0.28 μl/s for acetonitrile, injector temperature program: 20 °C (0.6 min) 720 °C/min to 280 °C, vent flow 100 ml/min, vent pressure 0 psi, vent time 0.4 min, split vent time 2 min, pure flow to split valve 50 ml/min. Average RSD of target peak areas in three parallel injections were 1.1 % and 1.3 % for cyclohexane and acetonitrile solutions respectively.

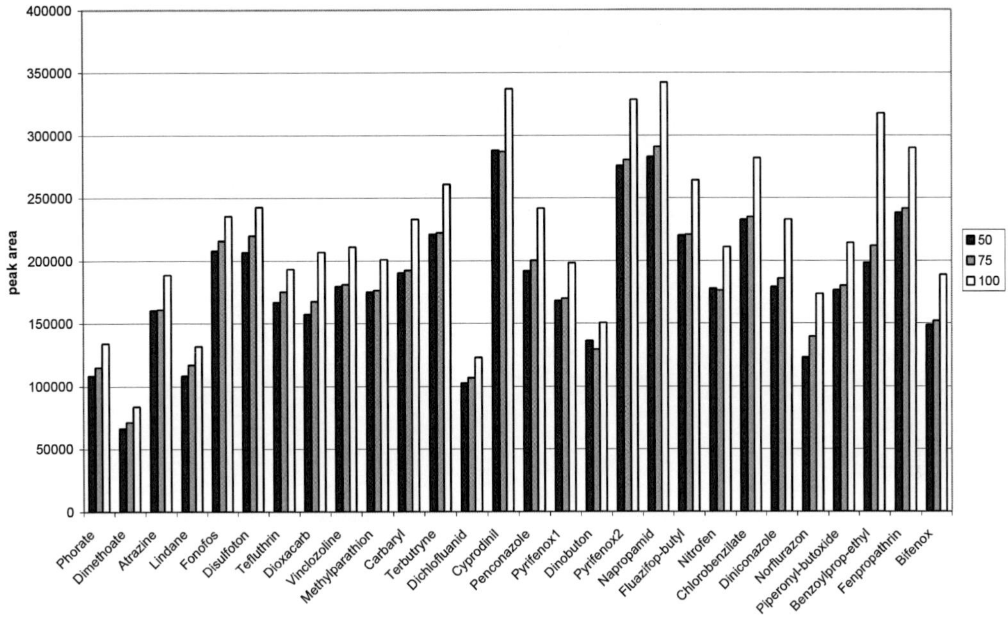

Figure 3 a). The effect of vent flow on target peak areas in cyclohexane solution.

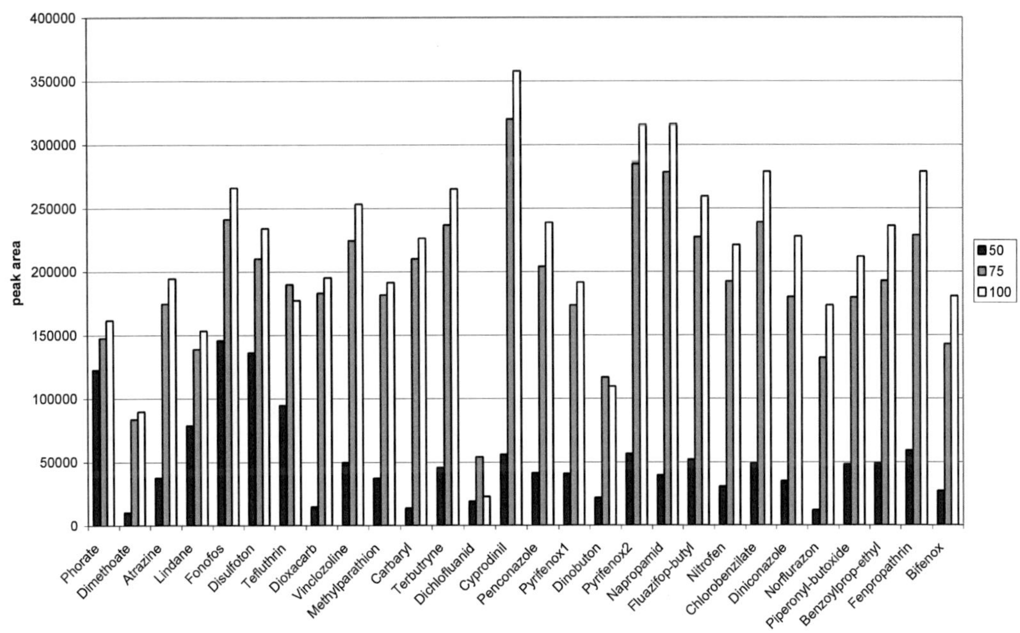

Figure 3 b). The effect of vent flow on target peak areas in acetonitrile solution.

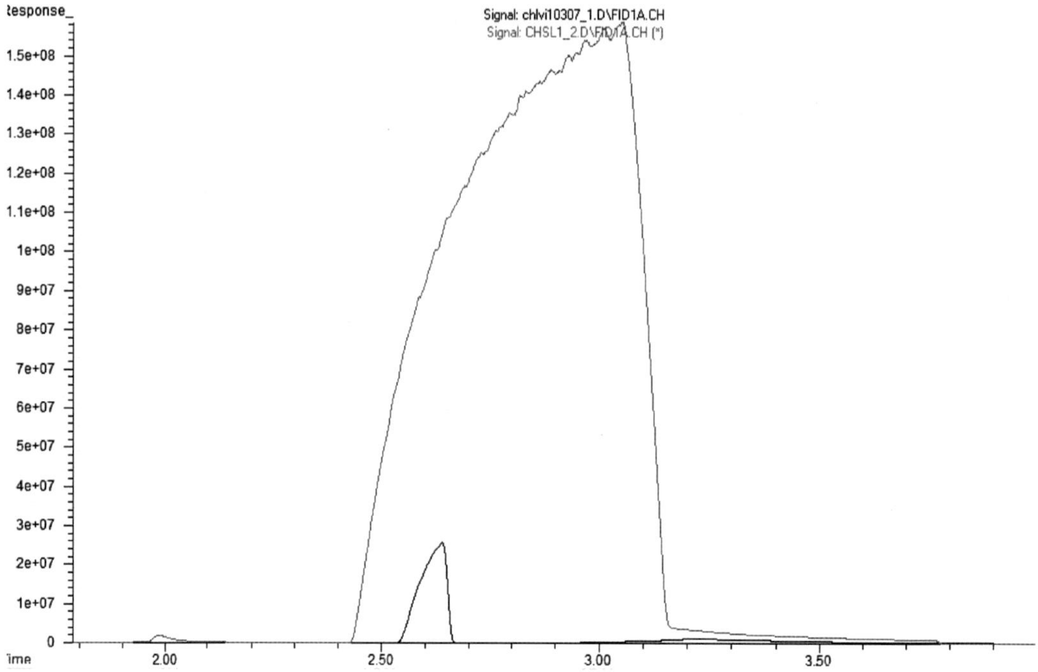

Figure 4 a). Solvent peaks in 1 µl splitless (grey) and 10 µl solvent venting large volume injection (black).

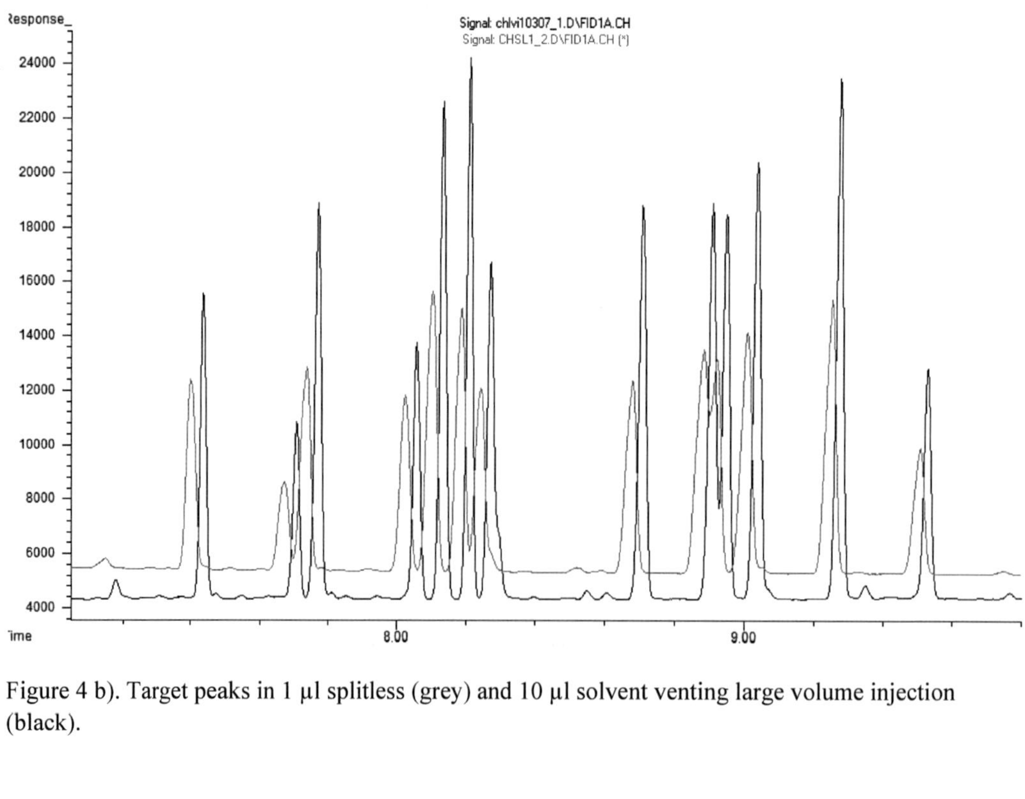

Figure 4 b). Target peaks in 1 µl splitless (grey) and 10 µl solvent venting large volume injection (black).

Table 4.

Solvent of standard solution	10 µg/l		50 µg/l		100 µg/l	
	$RT_{RSD\%}$	$c_{RSD\%}$	$RT_{RSD\%}$	$c_{RSD\%}$	$RT_{RSD\%}$	$c_{RSD\%}$
Cyclohexane	0.02	5.89	0.01	5.78	0.01	5.34
Acetonitrile	0.01	8.51	0.01	6.14	0.03	5.16

On Figure 4a the solvent peaks are compared, on Figure 4b a section of the target analytes' chromatogram (cyclohexane, FID). The efficiency of solvent venting was proved by the small solvent peak in the 10 µl large volume injection, which had a significantly smaller peak area than the 1 µl splitless injection. When target peak areas were compared, large volume injection gave larger areas. A possible explanation of it is that when the sample is injected in splitless injection, the majority of the solution is transferred onto the analytical column during the splitless time, out of which the majority is the solvent. In large volume injection the solvent is eliminated from the injector before the splitless time, which allows more of the analytes to reach the column before the splitting.

3.6. Increasing the Injected Sample Volume to 100 µl

Increasing the injected sample volume up to 100 µl was straightforward in case of cyclohexane, as the calculated injection speed (0.6 µl/s) gave good results. Vent flow was reduced to 50 ml/min. For acetonitrile the injection speed had to be further reduced using the additional f factor. The optimal value was found to be f=0.6, with an injection speed of 0.1 µl/s.

Reproducibility of the developed methods was determined from three parallels at three concentration levels, average RSD% of retention times and peak areas are listed in Table 4. Both methods were calibrated for three concentration levels (10, 50 and 100 µg/l), equations of calibration curve fits and R^2 values are listed in Table 5 for each compound. The calibration acceptance criterion was $R^2 \geq 0.99$.

3.7. Method Application on Vegetable and Fruit Samples

Reproducibility and recovery results for the fortified food samples are listed in Table 6. In both methods recovery and reproducibility values showed significant deviation, this might improve by practice of laboratory personal. As the study was conducted in order to decide between the two sample preparation procedures and their compatibility with large volume injection, their general performance was more important than direct compliance with validation requirements.

Table 5.

	Acetonitrile / QuEChERs		Cyclohexane / DFG S19	
	Curve fit	R^2	Curve fit	R^2
Butylate	R=2.0698e4*A	1.000	R=1.1915e4*A+5.7512e4	0.991
Propachlor	R=-4.0686e2*A^2+1.0428e5*A-6.8038e4	1.000	R=2.6703e4*A+1.5026e3	1.000
Quintozene	R=9.3238e3*A+3.1253e4	1.000	R=1.1011e4*A	1.000
Terbufos	R=5.7226e1*A^2-5.8049e2*A+1.1717e5	1.000	R=6.2829e3*A	1.000
2,4,5-T	R=8.981e3*A+1.8014e4	1.000	R=5.8115e3*A	0.994
Methyl parathion	R=1.553e4*A+7.085e3	1.000	R=6.3037e1* A^2+2.7782e3*A+2.2834e4	1.000
Malathion	R=3.6749e4*A+1.1049e5	0.992	R=1.336e2* A^2+1.1396e4*A+6.293e4	1.000
Chlorpyriphos	R=1.9944e4*A+5.3523e3	1.000	R=2.0188e4*A	0.990
Isodrin	R=1.1241e4*A+5.3097e4	0.998	R=1.0506e4*A	1.000
Tetrachlorvinphos	R=1.9673e4*A+5.4811e3	1.000	R=2.7501e4*A	1.000
Prothiophos	R=1.2615e4*A+3.9763e4	1.000	R=2.3629e4*A	0.997
Ethion	R=3.2446e4*A	0.999	R=3.2264e4*A	0.993
Captafol	R=1.6668e2* A^2+4.847e4*A-2.1325e5	1.000	R=3.6727e4*A	0.997
Bifenthrin	R=1.0095e5*A+3.4185e5	0.989	R=1.1281e5*A	1.000
Phosalon	R=8.0519e1* A^2+5.569e3*A+3.8274e4	1.000	R=2.5533e4*A	0.997
Leptophos	R=2.4283e4*A+1.6022e4	0.994	R=3.2505e4*A	0.990
Coumaphos	R=3.7535e3*A	0.991	R=3.0049e4*A	0.994
Cypermethrin a	R=-5.701e1* A^2+1.9382e4*A-8.8262e4	1.000	R=8.4462e1* A^2+7.6722e3*A+1.8016e4	1.000
Deltamethrin	R=7.9305e3*A	1.000	R=5.5637e1*A^2+4.9646e3*A+3.0659e4	1.000

Table 6 a).

	Chinese cabbage						Carrot						Apple					
	10 μg/kg		*50 μg/kg*		*100 μg/kg*		*10 μg/kg*		*50 μg/kg*		*100 μg/kg*		*10 μg/kg*		*50 μg/kg*		*100 μg/kg*	
	rec%	rsd%	rec%	rsd%	rec%	rsd%	rec%	rsd%	rec%	rsd%	rec%	rsd%	rec%	rsd%	rec%	rsd%	rec%	rsd%
2,4,5-T	103,0	10,0	38,9	51,6	33,2	70,4	31,6	3,2	63,2	36,0	35,0	6,3	127,8	5,5	71,0	12,6	73,9	5,0
Bentazone			55,0						83,0	24,2	33,3	6,7	158,1	6,9	124,1	6,2	80,3	17,6
Bifenthrin	58,7	23,6	52,1	7,4	41,3	50,0	27,0	16,0	23,3	23,1	12,9	7,3	58,5	26,8	31,4	12,5	44,1	10,4
Butylate			123,3	6,7	90,5	64,7			86,5	63,2	57,0	7,7			131,2	15,1	101,1	27,5
Captafol													145,8	14,1	63,9	17,3	41,5	17,0
Chlorpyriphos													135,2	20,5	63,5	27,1	67,1	5,5
Cypermethrin α	83,7	13,9	46,2	14,9	27,0	44,9	84,7	6,2	30,1	19,0	18,4	18,4	88,6	16,3	32,1	9,5	38,6	2,8
Deltamethrin	103,0		43,3	20,9	38,4	48,9	203,7	60,8	36,3	16,3	32,8	38,1	153,4	37,7	40,4	9,0	48,5	0,6
Dimethoate	294,1	1,1					152,3	0,9	47,1	7,0					182,7	7,7	157,9	10,2
Ethion			58,7	15,6	60,8	5,3			115,7				125,1	15,6	41,5	35,2	56,0	5,0
Isodrin	44,9	18,0	45,4	9,2	39,4	61,4			24,0	34,7	12,0	7,2	199,7	54,9	38,6	20,6	50,7	4,9
Leptophos	79,9	2,2	39,6	4,9	25,8	34,0	67,4	2,6	26,3	27,8	10,7	23,9	103,1	11,2	38,8	8,0	41,7	4,2
Malathion	134,6	1,3	76,7	9,3	70,7	16,5			86,9	38,1	53,7	4,0	161,8	6,0	79,1	10,1	85,6	6,8
Methyl parathion	136,1	1,8	55,8	10,2	24,6	39,9							165,6	5,8	76,9	11,3	77,6	6,4
Phosalon			39,9	2,5					33,3	33,0	15,4	7,5	111,4	19,0	52,7	7,6	58,5	9,1
Propachlor	73,0	26,7	79,4	7,9	60,0	61,4			80,9	36,5	47,9	4,4	101,7	8,7	78,9	15,9	89,6	6,9
Prothiophos	77,1	1,6	55,7	24,9	37,2	53,8	31,1	70,7	20,3	30,4	11,0	16,1	104,4	12,5	26,8	10,2	29,3	4,7
Quintozene	45,6	11,1	39,9	6,9	29,9	59,8	31,1	6,6	30,6	38,5	18,9	4,0	58,2	10,4	36,6	18,6	42,3	3,5
Tetrachlorvinphos	24,3	1,9											55,5	11,3	26,2	9,9	32,8	7,7
Average	***96,7***	*9,4*	***56,7***	*13,8*	***44,5***	*47,0*	***84,9***	*20,9*	***52,5***	*30,6*	***27,6***	*11,7*	***120,8***	*16,7*	***65,1***	*13,9*	***64,1***	*8,2*

Table 6 b).

	Chinese cabbage						Carrot						Apple					
	10 μg/kg		*50 μg/kg*		*100 μg/kg*		*10 μg/kg*		*50 μg/kg*		*100 μg/kg*		*10 μg/kg*		*50 μg/kg*		*100 μg/kg*	
	rec%	rsd%	rec%	rsd%	rec%	rsd%	rec%	rsd%	rec%	rsd%	rec%	rsd%	rec%	rsd%	rec%	rsd%	rec%	rsd%
2,4,5-T	122,8	43,0	96,9	3,1	19,1	12,1	11,5	42,5	29,7	40,7	26,5	27,6	77,9	89,0	125,1	33,2	74,9	62,2
Bentazone			1,1	9,6	0,3	60,2	13,1	0,0	2,6	0,0	1,3	0,0	32,4	81,4	1,3	35,1	1,0	77,6
Bifenthrin	110,0	11,9	56,3	15,9	84,0	35,0	19,0	32,7	16,9	53,9	19,2	31,1	86,3	41,1	98,0	44,9	58,3	72,7
Butylate	87,8	79,2	24,1	11,6	57,4	29,9	15,0	22,0	16,5	77,2	11,2	24,7	39,8	56,0	100,9	15,6	44,3	53,8
Captafol	33,0	57,6	2,4	13,0	0,0	60,6	28,1	119,2	19,1	46,1	7,6	63,4	20,9	51,3	474,9	40,2	19,8	47,8
Chlorpyriphos	116,9	12,8	90,3	9,3	107,8	14,0	83,8	80,3	36,4	30,3	33,1	22,4	116,9	83,6	109,5	38,1	59,9	63,8
Cypermethrin α	72,1	16,2	26,3	5,0	48,0	54,4	0,2	16,0	6,6	45,3	8,1	18,8	46,4	42,2	32,2	60,4	14,2	75,2
Deltamethrin	92,5	11,8	39,2	30,4	50,0	37,9	100,5	36,3	19,0	8,3	13,0	6,4	115,6	50,9	40,7	44,3	15,0	72,3
Ethion	196,8	5,2	63,0	10,3	74,5	27,8	35,5	11,6	23,4	27,7	19,9	27,0	87,3	68,4	69,7	43,3	39,6	62,6
Isodrin	109,3	36,6	95,8	7,3	16,2	16,3	23,8	24,2	22,1	43,4	28,6	18,3	62,3	124,4	124,1	30,1	74,6	61,8
Leptophos	17,4	29,5	10,5	6,7	18,5	46,6	3,9	25,2	3,1	68,2	4,0	36,8	14,0	56,0	16,5	64,7	7,1	46,2
Malathion	39,1	17,2	27,4	24,0	41,5	33,5	22,5	21,0	5,4	94,0	8,3	48,5	4,2	19,9	12,4	87,5	7,8	97,3
Methyl parathion	38,2	9,6	16,7	12,6	24,1	29,8	0,6	21,9	4,9	46,7	5,6	30,2	16,7	17,4	12,8	58,1	7,1	67,0
Phosalon	132,2	6,6	42,1	4,2	49,4	36,7	90,8	4,3	26,8	18,7	19,3	21,2	81,3	15,9	61,1	43,4	30,4	50,9
Propachlor	2,3	22,6	61,2	5,8	86,1	6,9	89,7	9,2	11,8	49,2	22,5	25,4	42,0	22,4	64,2	48,7	39,7	77,2
Prothiophos	161,5	12,0	72,0	8,7	87,8	21,1	16,8	27,8	17,3	48,0	20,2	26,7	48,8	137,1	92,3	38,8	51,5	68,5
Quintozene	178,5	19,8	123,0	4,8	56,1	14,8	10,3	69,9	29,5	33,0	36,9	17,7	55,1	19,2	122,1	36,7	76,7	59,6
Tetrachlorvinphos	130,9	14,5	18,8	11,6	19,3	42,4	0,5	13,4	3,5	62,9	3,1	40,6	3,8	22,6	9,9	53,1	6,7	38,3
Average	***96,5***	*23,0*	***45,9***	*10,5*	***45,8***	*32,6*	***28,9***	*34,3*	***15,6***	*43,4*	***15,4***	*27,0*	***51,7***	*57,9*	***81,4***	*46,3*	***32,8***	*64,7*

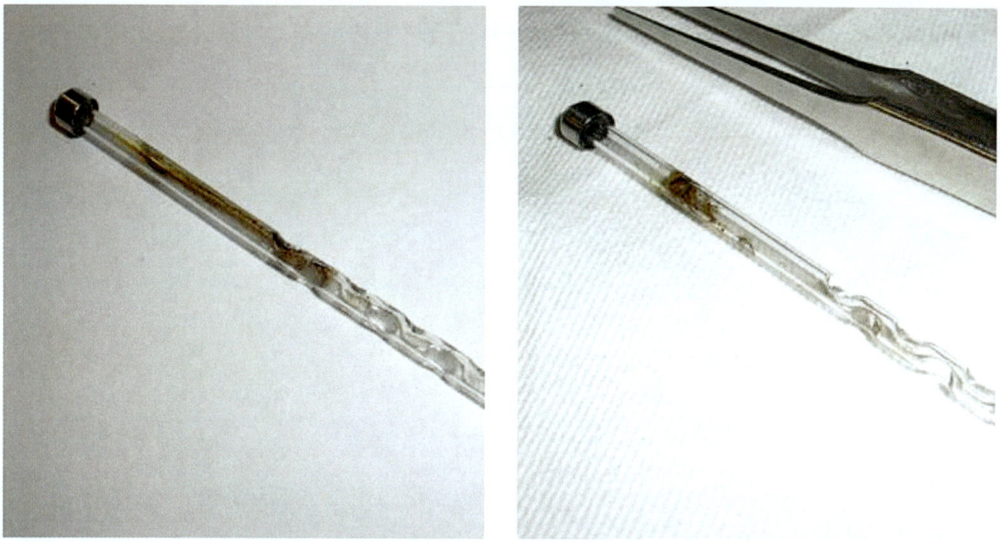

Figure 5. PTV injector liners after 40 sample extract injections a) DFG S19 sample preparation b) QuEChERs sample preparation.

Methods were evaluated on a number of false negatives, relative recovery and reproducibility, as well as their impact on the analytical instrument. Samples prepared with DFG S19 had better average reproducibility than the samples prepared with the QuEChERs method; however there were numerous compounds which were not recovered from DFG S19 extracts, which strongly reduces the method's reliability. This was caused not only by insufficient extraction, but also by the large amount of matrix contamination observable in the chromatograms. Comparison of the two sample preparation procedures' impact on the instrument was done by the examination of the liner after two identical sequences consisting of 40 sample extract injections (Figure 5). Liner contamination in case of the DFG S19 method was significantly more than after the QuEChERs sequence, raising the need for more frequent instrument maintenance both on injector and detector side.

4. Conclusion

Two 100 μl large volume injection – GC-MS methods were developed for cyclohexane and acetonitrile solutions of 19 pesticides using continuous slow injection and solvent venting. The optimised methods were calibrated and applied on fortified food samples prepared with DFG S19 and QuEChERs procedures, where the latter approach proved to be more suitable for routine residue screening based on the number of false negatives and impact on the analytical instrument.

5. ACKNOWLEDGMENTS

The authors wish to thank Denes Fabian (Kromat Ltd, Hungary) for his excellent help with Gerstel equipment, and thank Kromat Ltd and the Food Analysis Laboratory of the Wessling Hungary Ltd for their support.

6. REFERENCES

[1] Analysis of Foodstuffs - DFG S19 (1999) §35 LMBG 00.00.34
[2] M. Anastassiades, S.J. Lehotay, D. Stajnbaher, F.J. Schenck, J. AOAC Int., 86 (2003) 412-431
[3] Perera, F.P. *Neurotoxicology* 26 (2005) 573.
[4] Barr, D.B., Needham L.L. J. Chromatogr., B, *Biomed. Sci. Appl.* 778 (2002) 5.
[5] Stoppelli, I.M.; Crestana, S. *Environ. Int.* 31 (2005) 731.
[6] Dinham, B. *Public Health* 9 (2005) 203.
[7] Zillmer, E.A.; Montenegro, L.M.; Wiser, J.; Barth, J.T.; Spyker, D.A. *Arch. Clin. Neuropsychol.* 11 (1996) 77.
[8] Roberts, D.M.,; Dissanayake, W.; Sheriff, M.H.R.; Eddleston, M. *J. Clin. Neurosci.* 11 (2004) 760.
[9] Analysis of Foodstuffs – DFG S19 (1999) §35 LMBG 00.00.34.
[10] Anastassiades, M.; Lehotay, S.J.; Stajnbaher, D.; Schenck, F.J. *J. AOAC International,* 86 (2003) 412-431.
[11] Grob, K. (2001) Split and Splitless Injection for Quantitative Gas Chromatography: Concepts, Processes, Practical Guidelines, Sources of Error, Wiley, USA.
[12] Grob, K.; Biedermann, M. *J. Chromatogr.* A, 897 (2000) 237-246.
[13] Fattori M. *Hewlett-Packard Company*, FlowCalc V. A.02.07 software.
[14] Grob, K. *On-column Injection in Capillary Gas Chromatography,* Huthig, Heidelberg, Germany, 1987.
[15] Teske, J.; Engewald, W. *Trends in analytical chemistry* 21 (2002) 584-593
[16] Konig, S. de; Kurano, M.; Janssen, H.G.; Brinkman, U.A. *Th. J. Chromatogr.* A 1023 (2004) 165-174.
[17] Grob, K.; Biedermann, M. *J. Chromatogr.* A, 750 (1996) 11-23.
[18] Vogt, W.; Jacob, K.; Obwexer, H.W. *J. Chromatogr.* 174 (1979) 437.
[19] Vogt, W.; Jacob, K.; Ohnesorge A.B.; Obwexer, H.W. *J. Chromatogr.* 186 (1979) 197.
[20] Poy, E.S.; Visani, F.; Terrosi, F. *J. Chromatogr.* 217 (1981) 81.
[21] Agüera, A.; Contreras, M.; Crespo, J.; Fernández-Alba, A. *R. Analyst,* 2002, 127, 347–354.
[22] Leandro, C. C.; Fussell, R. J.; Keely, B. J. *J. Chromatography A,* 1085 (2005) 207–212.
[23] Ramos, L.; Vreuls, J.J.; Brinkman, U.A.Th. *J. Chromatography A,* 891 (2000) 275–286.
[24] Mol, H.G.J.; Janssen, H.G.; Cramers, C.A.; Brinkman, U.A.Th. *Trends in analytical chemistry,* 15 (1996) 4, 206-214.
[25] Staniewski, J.; Janssen, H.G.; Cramers, C.A.; Rijks, J.A. *J. Microcolumn Separations,* 4 (1992) 331.
[26] Engewald, W.; Teske, J.; Efer, J. *J. Chromatography A,* 842 (1999) 143-161.

[27] Stan, H.J.; Linkerhägner, M. *J. Chromatography A,* 727 (1996) 275-289.
[28] Inlets for capillary gas chromatography, Gerstel publication, www.gerstel.com
[29] Kende, A.; Csizmazia, Z.; Rikker, T.; Angyal, V.; Torkos, K. *Microchemical Journal* 84 (2006) 63–69.

In: Progress in Pesticides Research
Editor: Christopher M. Kanzantzakis

ISBN 978-1-60456-201-9

Chapter 1

THE EVALUATION MULTI-RESIDUES METHOD FOR DETERMINATION OF 858 PESTICIDES AND RELATED CHEMICAL IN GRAINS AND CEREALS BY GC-MS AND LC-MS-MS

Guo-Fang Pang**, Yong-ming Liu, Yan-Zhong Cao, Chun-Lin Fan, Yu-Jing Lian, Feng Zheng, Jin-Jie Zhang, Xue-Min Li, Zeng-Yin Li

Qinhuangdao Entry-Exit Inspection and Quarantine Bureau, 39 Haibin Rd, Qinhuangdao, Hebei, 066002, People's Republic of China

ABSTRACT

In the paper, a systematic study was conducted regarding the mass spectrometric characteristics of GC-MS and LC-MS-MS and GPC behaviors characteristics for 858 pesticides, and a GC-MS analytical database has been established for 541 pesticides, a LC-MS-MS analytical database for 629 pesticides and a GPC behavioral characteristic database for 751 pesticides.

Under the GC-MS and LC-MS-MS optimized analytical conditions based on these datebase, a series of the comparison testes were completed. 486 pesticides and related chemicals in 5 kinds of grain samples such as wheat, barley, rice, maize and oats were pretreated by ASE extraction and SPE cleanup as well as 465 pesticides were pretreated by homogeneous and shaking extraction and GPC cleanup. A GC-MS analytical method has been established for determination of 475 pesticides in grains and cereals, and same as LC-MS-MS analytical method for determination of 372 pesticides and related chemicals.

Of these, 263 pesticides can be analyzed by both GC-MS and LC-MS-MS methods, with 212 pesticides only suitable for determination by GC-MS and 109 pesticides only by LC-MS-MS. The total quantifying determination reached 584 pesticides, accounting for 68% of the pesticides studied, while the rest made up 32 % considered unfit for determination by the method described. With 475 pesticides analyled by GC-MS, grain samples were mixed with Celit 545, which was then extracted on accelerated solvent extractor with acetonitrile; the extracts were cleaned up via Envi-18 cartridge and then cleaned up via Envi-Carb and Sep-Pak NH_2 cartridge in series, with the pesticides eluted with acetonitrile +toluene (3+1) , and the eluates were concentrated and exchanged with hexane before being used for analysis. For 372 pesticides analyzed by LC-MS-MS, they were extracted with cyclohexane+ ethyl acetate (1+1) mixed solvent on homogenizer and shaker, cleaned up on GPC equipped with Bio-beads S-X3 column, collecting 22-44min fractions, concentrated and volumed with acetonitrile +water (3+2) before being used for analysis.

For GC-MS method, 5 kinds of grain samples were chosen for fortified recovery and precision tests at low, medium and high levels, with average recoveries of 475 pesticides and related chemicals falling between 60%-130% for 453 pesticides, accounting for 95%; 30%-60% for 22 pesticides, making up the total number of 5%; the relative standard deviation (RSD) falling within 30% involved 451 pesticides, accounting for 95%; those within 30%-40% involved 24 pesticides, making up the total number of 5%.

For LC-MS-MS, 5 kinds of grain samples were also selected for fortified recoveries and precision tests at two levels, with the average recoveries of 372 pesticides and related chemicals falling within 60% -130% for 322 pesticides, accounting for 86 %; Recoveries for 31 pesticides at low levels were unstable, with RSD% larger than 25% but high-levels recoveries were larger than 60% with RSD% less than 25%, accounting for 8%; recoveries at both low levels and high levels for 9 pesticides ranged between 40%~60%, with RSD% less than 30%, making up 3%; only low-level fortified tests were carried out for 10 pesticides because of their limited quantity, with satisfactory results obtained for both recoveries and RSD%, accounting for 3%. There were 50 pesticides of such kind in total, accounting for 14%. Recoveries and precision tests proved that such methods are capable of quantifying 584 pesticides in the above-mentioned 5 grain samples. Limit of detection of GC-MS method is between 0.0025mg/kg-0.8000mg/kg and that of LC-MS-MS method is 0.04 μg/kg –8.00mg/kg, which is subject to the nature of the pesticides.

INTRODUCTION

The wide-use of pesticides, veterinary drugs and other agricultural chemicals boosted the development of the agriculture and animal husbandry, but has accelerated the contamination of human survival environment, agricultural products and foodstuffs, which has drawn high attention by WHO and countries across the world. On May 29, 2006, Japan implemented "Positive List System for Agricultural Chemical Residues in Foods" which specified concretely the agricultural chemical residues in all foods. Among these chemicals, 15 pesticides and veterinary drugs were forbidden to use, with 53862 MRLs prescribed for 797 pesticides, veterinary drugs and feeds additives, while for those agricultural chemical unspecified, 0.01mg/kg Uniform Standard would be carried out, with no prescriptions of residue limits for only 65 varieties of natural and chemical synthetic matters treated as exempt. From May 29, 2006, foods and foodstuffs could enter the Japanese markets only when they were proved be in conformity with the "Positive List System for Agricultural

* E-mail: gfpang@heinfo.net; qhdciqgfpang@yahoo.com.cn

Chemical Residues in Foods". In view of the residue limit specifications across the world, Japan's "Positive List System for Agricultural Chemical Residues in Foods" is the most stringent residue limit standard system of all. At present, Canada, New Zealand, Germany, USA, EU and other countries and regions are also preparing to adopt the Uniform Standard, with the basic criteria being 0.1mg/kg-0.01mg/kg. With requirements for pesticide residue limit of different countries getting lower and lower and inspection items more and more, it is undoubtedly a severe challenge for residue analysts to churn out a quick analytical method of high sensitivity, high selectivity and high resolvability in order to meet a plurality of choosy requirements.

In recent years, quick and simultaneous analytical techniques of multiresidues have turned out to be a hot topic and a hard nut to crack for residue chemists of different countries worldwide. Presently, pesticides analytical methods adopted mostly in the world is mainly acetones [1~2,15~16], acetonitrile [3~11] or ethyl acetate [12~14,17~18] as extracting solvents, with the method generally by either shaking or homogenizing [4~7] or SFE or ASE[12~14] for extraction. For cleanup techniques, liquid-liquid partition [1~2], GPC[10~11,15~16] or SPE cartridge [4~14] is adopted. For analyses, GC or LC is equipped with one or a multitude of selective detectors[1~2, 8~16], GC-AED[15~16], GC-MS[3~9], LC-MS-MS[4~7], etc. In the U.S, Luke et al [1~2] proposed that acetone be the extracting solvent and liquid-liquid partitioning be the cleanup for determination of 79 pesticides by GC and LC. Liao et al [3] adopted acetonitrile as extracting solvent and used GC-MS for determination of 143 pesticides after salting-out. Lehotay et al [4~7] established a QuEChERS method(quick, easy, cheap, effective, rugged and safe), with acetonitrile as extracting solvent and cleaned up via dispersive-SPE containing PSA after salting-out with $MgSO_4$ and NaCl before analyzing approximately 200 pesticides by GC-MS and LC-MS-MS. Fillon et al from Canada [8~9] used acetonitrile as extracting solvent, rid of the co-extracting matters by C_{18}, adopted in-line rCarbon SPE column and Amion propyl SPE column for cleanup and analyzed 251 pesticides by GC-MS and LC-FLD. Ueno et al from Japan [10~11] used acetonitrile as extracting solvent, adopted GPC equipped with graphitized carbon column for pre-cleanup after salting-out, then used in-line silica-gel and Florisil column for further cleanup before analyzing 87 pesticides by GC-ECD/FPD/NPD equipped with double columns. Obana et al [12~14] adopted ASE (accelerated solvent extraction) or SFE (supercritical fluid extraction) technique for sample extraction, used high capacity absorbent polymer to absorb moisture with ethyl acetate as extracting solvent, used grphitized carbon column for cleanup before analyzing 110 pesticides with GC-FPD, GC-MS-NCI/EI and HPLC. Stan et al from Germany [15~16] adopted the modified DFG-mlutimethod S19 pre-treatment method with acetone as extracting solvent and GPC for cleanup and analyzed over 400 pesticides in foods with GC-AED, GC-ECD, GC-NPD, GC-MS.

For the purpose of establishing a quick and simultaneous analytical method of high sensitivity, high selectivity and high resolutions, the author's laboratory has conducted a systematic study of analytical residue techniques of over 800 pesticides commonly used in the world since 2001 and has now established 4 multiresidue analytical methods [19~22], which are respectively used for determination of 446 pesticide residues in fruits and vegetables, 405 in grains and cereals, 437 in animal tissues and 450 in honeys, fruit juices and fruit wines and have become China national standards. In these studies, LC-MS-MS only serves as the supplementary resort to GC-MS determination of residues with the pesticide varieties of 60-70 analyzed. Our present study is a continuation and deepening of our first-stage study during which LC-MS-MS and GC-MS were given equal status; a systematic study was conducted to GC-MS and LC-MS-MS mass spectrometric characteristics and GPC behaviors for 828 pesticides; a GC-MS data database was established for 541 pesticides, a LC-MS-MS data database for 629 and a GPC behavioral data database for 751. Under the optimized GC-MS and LC-MS-MS analytical conditions, the cleanup efficiency of 5 kinds of grains was evaluated, through which an analytical method was established for determination of 475 pesticides and related chemicals in grains by GC-MS, 372 pesticides by LC-MS-MS. Compared to our first-stage method, the analytical varieties by LC-MS-MS increased from 43 to the present 372 with the detected varieties increasing 9 times; those by GC-MS increased by 113 in total than before, both methods by GC-MS and LC-MS-MS were capable of detecting 263 varieties. Therefore, both GC-MS and LC-MS-MS methods were capable of detecting a total variety of 584, which increased by 179 in compared to our previous similar method, which has greatly boosted the analytical capability of the method.

EXPERIMENTAL

Reagents and Materials

Acetonitrile and methanol (HPLC grade), cyclohexane, hexane, and ethyl acetate (P.R. grade) were purchased from Dikma company. Toluene (Absolv grade), dichloromethane (P.R. grade) and acetone (HPLC grade) were purchased from Tedia company. Celit 545 (G.R.grade) were purchased from Fluka company. Sodium sulfate, anhydrous (A. R. grade) heated at $650^{0}C$ for 4h and kept in a desiccator. Envi-18 cartridge, 12mL, 2g, and Envi-Carb cartridge, 6mL, 0.5g were purchased from Supelco. Sep-Pak NH_2 cartridge, 3mL, 0.5g, were purchased from Waters. Pesticide standards, purity≥95%, were purchased from LGC promochem.

Stock standard solution: accurately weigh 5mg-10mg of individual pesticide standards (accurate to 0.1mg) into a 10mL volumetric flask, dissolve and dilute to volume with toluene, toluene-acetone, dichloromethane, methanol, acetonitrile, etc., depending upon each individual compound' solubility.

Mixed standard solution (mixed standard solution A, B, C, D, and E): depending upon properties and retention time of each pesticide, 475 pesticides of GC-MS, and 372 pesticides of LC-MS-MS are divided into five groups respectively, A, B, C, D, and E. The concentration of each pesticide are determined by its sensitivity on the instrument used for analysis. 475 pesticides name, grouping, solvent selected and concentration of mixed standard solution by GC-MS are listed at Table 1. 372 pesticides name, grouping, solvent selected and concentration of mixed standard solution by LC-MS-MS are listed at Table 2. Depending on group number, mixed standard solution concentration, and stock standard solution concentration, appropriate amounts of individual stock standard solutions were pipetted into 100-mL volumetric flasks, GC-MS pesticides being diluted to volume with toluene and LC-MS-MS pesticides being diluted with methanol. Mixed standard solutions were stored in dark below 4°C and were used for one month.

Table 1. 475 pesticides grouping, solvent selected and concentration of mixed standard solution by GC-MS

No.	Group	Compound	Solvent	Concentration of mixed standard solution /mg/L
ISTD		Heptachlor-epoxide	Toluene	
1	A	Allidochlor	Toluene	5.0
2	A	Dichlormid	Toluene	5.0
3	A	Etridiazol	Toluene	7.5
4	A	Chlormephos	Toluene	5.0
5	A	Propham	Toluene	2.5
6	A	Cycloate	Toluene	2.5
7	A	Diphenylamine	Toluene	2.5
8	A	Ethalfluralin	Toluene	10.0
9	A	Phorate	Toluene	2.5
10	A	Thiometon	Toluene	2.5
11	A	Quintozene	Toluene	5.0
12	A	Atrazine-desethyl	Toluene+Acetone (8+2)	2.5
13	A	Clomazone	Toluene	2.5
14	A	Diazinon	Toluene	2.5
15	A	Fonofos	Toluene	2.5
16	A	Etrimfos	Toluene	2.5
17	A	Simazine	Methanol	2.5
18	A	Propetamphos	Toluene	2.5
19	A	Secbumeton	Toluene	2.5
20	A	Dichlofenthion	Toluene	2.5
21	A	Pronamide	Toluene+Acetone (9+1)	2.5
22	A	Mexacarbate	Toluene	7.5
23	A	Aldrin	Toluene	5.0
24	A	Dinitramine	Toluene	10.0
25	A	Ronnel	Toluene	5.0
26	A	Prometryne	Toluene	2.5
27	A	Cyprazine	Toluene+Acetone (9+1)	20.0
28	A	Vinclozolin	Toluene	2.5
29	A	Beta-HCH	Toluene	2.5
30	A	Metalaxyl	Toluene	7.5
31	A	Chlorpyrifos (-ethyl)	Toluene	2.5
32	A	Methyl-Parathion	Toluene	10.0
33	A	Anthraquinone	Dichlormethane	2.5
34	A	Delta-HCH	Toluene	5.0
35	A	Fenthion	Toluene	2.5
36	A	Malathion	Toluene	10.0
37	A	Fenitrothion	Toluene	5.0
38	A	Paraoxon-ethyl	Toluene	10.0
39	A	Triadimefon	Toluene	5.0
40	A	Parathion	Toluene	10.0
41	A	Pendimethalin	Toluene	10.0
42	A	Linuron	Toluene+Acetone (9+1)	10.0
43	A	Chlorbenside	Toluene	5.0

Table 1. Continued

No.	Group	Compound	Solvent	Concentration of mixed standard solution /mg/L
44	A	Bromophos-ethyl	Toluene	2.5
45	A	Quinalphos	Toluene	2.5
46	A	trans-Chlordane	Toluene	2.5
47	A	Phenthoate	Toluene	5.0
48	A	Metazachlor	Toluene	7.5
49	A	Fenothiocarb	Acetone	5.0
50	A	Prothiophos	Toluene	2.5
51	A	Chlorfurenol	Toluene+Acetone (9+1)	7.5
52	A	Dieldrin	Toluene	5.0
53	A	Procymidone	Toluene	2.5
54	A	Methidathion	Toluene	5.0
55	A	Napropamide	Toluene	7.5
56	A	Oxadiazone	Toluene	2.5
57	A	Fenamiphos	Toluene	7.5
58	A	Tetrasul	Toluene	2.5
59	A	Aramite	Dichlormethane	2.5
60	A	Bupirimate	Toluene	2.5
61	A	Carboxin	Toluene	7.5
62	A	Flutolanil	Toluene	2.5
63	A	4,4'-DDD	Toluene	2.5
64	A	Ethion	Toluene	5.0
65	A	Sulprofos	Toluene	5.0
66	A	Etaconazole-1	Toluene	7.5
67	A	Etaconazole-2	Toluene	7.5
68	A	Myclobutanil	Toluene	2.5
69	A	Diclofop-methyl	Toluene	2.5
70	A	Propiconazole	Toluene	7.5
71	A	Fensulfothin	Toluene	5.0
72	A	Bifenthrin	Hexane	2.5
73	A	Mirex	Toluene	2.5
74	A	Benodanil	Toluene	7.5
75	A	Nuarimol	Toluene+Acetone (9+1)	5.0
76	A	Methoxychlor	Toluene	2.5
77	A	Oxadixyl	Toluene	2.5
78	A	Tetramethirn	Toluene	5.0
79	A	Tebuconazole	Toluene	7.5
80	A	Norflurazon	Toluene+Acetone (9+1)	2.5
81	A	Pyridaphenthion	Toluene	2.5
82	A	Phosmet	Toluene	5.0
83	A	Tetradifon	Toluene	2.5
84	A	Oxycarboxin	Toluene+Acetone (9+1)	15.0
85	A	cis-Permethrin	Toluene	2.5
86	A	trans-Permethrin	Toluene	2.5
87	A	Pyrazophos	Toluene	5.0
88	A	Cypermethrin	Toluene	7.5

Table 1. Continued

No.	Group	Compound	Solvent	Concentration of mixed standard solution /mg/L
89	A	Fenvalerate	Toluene	10.0
90	A	Deltamethrin	Toluene	15.0
91	B	EPTC	Toluene	7.5
92	B	Butylate	Toluene	7.5
93	B	Dichlobenil	Toluene	0.5
94	B	Pebulate	Toluene	7.5
95	B	Nitrapyrin	Toluene	7.5
96	B	Mevinphos	Toluene	5.0
97	B	Chloroneb	Toluene	2.5
98	B	Tecnazene	Toluene	5.0
99	B	Heptanophos	Toluene	7.5
100	B	Hexachlorobenzene	Toluene	2.5
101	B	Ethoprophos	Toluene	7.5
102	B	cis-Diallate	Toluene	5.0
103	B	Propachlor	Toluene	7.5
104	B	trans-Diallate	Toluene	5.0
105	B	Trifluralin	Toluene	5.0
106	B	Chlorpropham	Toluene	5.0
107	B	Sulfotep	Toluene	2.5
108	B	Sulfallate	Toluene	5.0
109	B	Alpha-HCH	Toluene	2.5
110	B	Terbufos	Toluene	5.0
111	B	Terbumeton	Toluene	7.5
112	B	Profluralin	Toluene	10.0
113	B	Dioxathion	Toluene	10.0
114	B	Propazine	Toluene	2.5
115	B	Chlorbufam	Toluene	5.0
116	B	Dicloran	Toluene+Acetone (9+1)	5.0
117	B	Terbuthylazine	Toluene	2.5
118	B	Monolinuron	Toluene	10.0
119	B	Flufenoxuron	Toluene+Acetone (8+2)	7.5
120	B	Cyanophos	Toluene	5.0
121	B	Chlorpyrifos-methyl	Toluene	2.5
122	B	Desmetryn	Toluene	2.5
123	B	Dimethachlor	Toluene	7.5
124	B	Alachlor	Toluene	7.5
125	B	Pirimiphos-methyl	Toluene	2.5
126	B	Terbutryn	Toluene	5.0
127	B	Thiobencarb	Toluene	5.0
128	B	Aspon	Toluene	5.0
129	B	Dicofol	Toluene	40.0
130	B	Metolachlor	Toluene	2.5
131	B	Oxy-chlordane	Toluene	2.5
132	B	Methoprene	Toluene	10.0
133	B	Bromofos	Toluene	5.0

Table 1. Continued

No.	Group	Compound	Solvent	Concentration of mixed standard solution /mg/L
134	B	Ethofumesate	Toluene	5.0
135	B	Isopropalin	Toluene	5.0
136	B	Endosulfan I	Toluene	15.0
137	B	Propanil	Toluene+Acetone (9+1)	5.0
138	B	Isofenphos	Toluene	5.0
139	B	Crufomate	Toluene	15.0
140	B	Chlorfenvinphos	Toluene	7.5
141	B	cis-Chlordane	Toluene	5.0
142	B	Tolylfluanide	Toluene	7.5
143	B	4,4'-DDE	Toluene	2.5
144	B	Butachlor	Toluene	5.0
145	B	Chlozolinate	Toluene	5.0
146	B	Crotoxyphos	Toluene	15.0
147	B	Iodofenphos	Toluene	5.0
148	B	Tetrachlorvinphos	Toluene	7.5
149	B	Chlorbromuron	Toluene	60.0
150	B	Profenofos	Toluene	15.0
151	B	Fluorochloridone	Toluene	5.0
152	B	2,4'-DDD	Toluene	2.5
153	B	Endrin	Toluene	30.0
154	B	Hexaconazole	Toluene	15.0
155	B	Chlorfenson	Toluene	5.0
156	B	2,4'-DDT	Toluene	5.0
157	B	Paclobutrazol	Toluene	7.5
158	B	Methoprotryne	Toluene	7.5
159	B	Chloropropylate	Toluene	2.5
160	B	Flamprop-methyl	Toluene	2.5
161	B	Nitrofen	Toluene	15.0
162	B	Oxyfluorfen	Toluene	10.0
163	B	Chlorthiophos	Toluene	7.5
164	B	Endosulfan II	Toluene	15.0
165	B	Flamprop-Isopropyl	Toluene	2.5
166	B	4,4'-DDT	Toluene	5.0
167	B	Carbofenothion	Toluene	5.0
168	B	Benalyxyl	Toluene	2.5
169	B	Edifenphos	Toluene	5.0
170	B	Triazophos	Toluene	7.5
171	B	Cyanofenphos	Toluene	2.5
172	B	Chlorbenside sulfone	Toluene	5.0
173	B	Endosulfan-Sulfate	Toluene	7.5
174	B	Bromopropylate	Toluene	5.0
175	B	Benzoylprop-ethyl	Toluene	7.5
176	B	Fenpropathrin	Toluene	5.0
177	B	Leptophos	Toluene	5.0
178	B	EPN	Toluene	10.0

Table 1. Continued

No.	Group	Compound	Solvent	Concentration of mixed standard solution /mg/L
179	B	Hexazinone	Toluene	7.5
180	B	Phosalone	Toluene	5.0
181	B	Azinphos-methyl	Toluene	15.0
182	B	Fenarimol	Toluene	5.0
183	B	Azinphos-ethyl	Toluene	5.0
184	B	Coumaphos	Toluene	15.0
185	B	Cyfluthrin	Toluene	30.0
186	B	Fluvalinate	Toluene	30.0
187	C	Dichlorvos	Methanol	120.0
188	C	Biphenyl	Toluene	2.5
189	C	Vernolate	Toluene	2.5
190	C	3,5-Dichloroaniline	Toluene	2.5
191	C	Molinate	Toluene	2.5
192	C	Methacrifos	Toluene	2.5
193	C	2-Phenylphenol	Toluene	2.5
194	C	Tetrahydrophthalimide	Toluene	7.5
195	C	Fenobucarb	Toluene	5.0
196	C	Benfluralin	Toluene	2.5
197	C	Hexaflumuron	Toluene	15.0
198	C	Triallate	Toluene	5.0
199	C	Pyrimethanil	Toluene	2.5
200	C	Gamma-HCH	Toluene	5.0
201	C	Disulfoton	Toluene	2.5
202	C	Atrizine	Toluene+Acetone (9+1)	2.5
203	C	Heptachlor	Toluene	7.5
204	C	Iprobenfos	Toluene	7.5
205	C	Isazofos	Toluene	5.0
206	C	Plifenate	Toluene	5.0
207	C	Transfluthrin	Toluene	2.5
208	C	Fluchloralin	Toluene	10.0
209	C	Tolclofos-methyl	Toluene	2.5
210	C	Propisochlor	Toluene	2.5
211	C	Metobromuron	Toluene	15.0
212	C	Metribuzin	Toluene	7.5
213	C	HCH, epsilon-	Methanol	5.0
214	C	Formothion	Toluene	5.0
215	C	Diethofencarb	Toluene	15.0
216	C	Dimepiperate	Ethyl Acetate	5.0
217	C	Bioallethrin-1	Toluene	10.0
218	C	Bioallethrin-2	Toluene	10.0
219	C	2,4'-DDE	Toluene	2.5
220	C	Fenson	Toluene	2.5
221	C	Diphenamid	Toluene	2.5
222	C	Chlorthion	Toluene	5.0
223	C	Prallethrin	Toluene	7.5

Table 1. Continued

No.	Group	Compound	Solvent	Concentration of mixed standard solution /mg/L
224	C	Penconazole	Toluene	7.5
225	C	Mecarbam	Toluene	10.0
226	C	Tetraconazole	Toluene	7.5
227	C	Propaphos	Toluene	5.0
228	C	Flumetralin	Toluene	5.0
229	C	Triadimenol	Toluene	7.5
230	C	Pretilachlor	Toluene	5.0
231	C	Kresoxim-methyl	Toluene	2.5
232	C	Fluazifop-butyl	Toluene	2.5
233	C	Chlorfluazuron	Toluene	7.5
234	C	Chlorobenzilate	Toluene	2.5
235	C	Uniconazole	Cyclohexane	5.0
236	C	Flusilazole	Toluene	7.5
237	C	Fluorodifen	Toluene	2.5
238	C	Diniconazole	Toluene	7.5
239	C	Piperonyl butoxide	Toluene	2.5
240	C	Propargite	Toluene	5.0
241	C	Mepronil	Toluene	2.5
242	C	Dimefuron	Toluene+Acetone (8+2)	10.0
243	C	Diflufenican	Toluene	2.5
244	C	Phenothrin	Toluene	2.5
245	C	Fludioxonil	Toluene+Acetone (8+2)	2.5
246	C	Fenoxycarb	Toluene	15.0
247	C	Sethoxydim	Toluene	22.5
248	C	Anilofos	Toluene	5.0
249	C	Acrinathrin	Toluene	5.0
250	C	Lambda-Cyhalothrin	Toluene	2.5
251	C	Mefenacet	Toluene	7.5
252	C	Permethrin	Toluene	5.0
253	C	Pyridaben	Toluene	2.5
254	C	Fluoroglycofen-ethyl	Toluene	30.0
255	C	Bitertanol	Toluene	7.5
256	C	Etofenprox	Toluene	2.5
257	C	Cycloxydim	Toluene	240.0
258	C	Alpha-Cypermethrin	Toluene	5.0
259	C	Flucythrinate	Cyclohexane	5.0
260	C	Esfenvalerate	Toluene	10.0
261	C	Difenconazole	Toluene	15.0
262	C	Flumioxazin	Cyclohexane	5.0
263	C	Flumiclorac-pentyl	Toluene	5.0
264	D	Dimefox	Toluene	60.0
265	D	Disulfoton-sulfoxide	Toluene	5.0
266	D	Pentachlorobenzene	Toluene	2.5
267	D	Tri-Iso-Butyl Phosphate	Toluene	2.5
268	D	Crimidine	Toluene	2.5

Table 1. Continued

No.	Group	Compound	Solvent	Concentration of mixed standard solution /mg/L
269	D	BDMC-1	Toluene	5.0
270	D	Chlorfenprop-methyl	Toluene	2.5
271	D	Thionazin	Toluene	2.5
272	D	2,3,5,6-Tetrachloroaniline	Toluene	2.5
273	D	Tri-N-Butyl Phosphate	Toluene	5.0
274	D	2,3,4,5-Tetrachloroanisole	Toluene	2.5
275	D	Pentachloroanisole	Toluene	2.5
276	D	Tebutam	Toluene	5.0
277	D	Dioxabenzofos	Methanol	25.0
278	D	Methabenzthiazuron	Toluene+Acetone (9+1)	25.0
279	D	Simetone	Toluene	5.0
280	D	Atratone	Toluene	2.5
281	D	Desisopropyl-atrazine	Toluene+Acetone (8+2)	20.0
282	D	Terbufos Sulfone	Toluene	2.5
283	D	Tefluthrin	Toluene	2.5
284	D	Bromocylen	Toluene	2.5
285	D	Trietazine	Toluene	2.5
286	D	Etrimfos oxon	Toluene	2.5
287	D	Cycluron	Toluene	7.5
288	D	2,6-Dichlorobenzamide	Toluene+Acetone (8+2)	5.0
289	D	DE-PCB 28	Toluene	2.5
290	D	DE-PCB 31	Toluene	2.5
291	D	Desethyl-sebuthylazine	Toluene+Acetone (8+2)	5.0
292	D	2,3,4,5-Tetrachloroaniline	Toluene	5.0
293	D	Musk ambrette	Toluene	2.5
294	D	Musk xylene	Toluene	2.5
295	D	Pentachloroaniline	Toluene	2.5
296	D	Aziprotryne	Toluene	20.0
297	D	Sebutylazine	Toluene+Acetone (8+2)	2.5
298	D	Isocarbamid	Toluene+Acetone (9+1)	12.5
299	D	DE-PCB 52	Toluene	2.5
300	D	Musk moskene	Toluene	2.5
301	D	Prosulfocarb	Toluene	2.5
302	D	Dimethenamid	Toluene	2.5
303	D	Fenchlorphos oxon	Toluene	5.0
304	D	BDMC-2	Toluene	5.0
305	D	Paraoxon-methyl	Toluene	40.0
306	D	Monalide	Toluene	5.0
307	D	Musk tibeten	Toluene	2.5
308	D	Isobenzan	Toluene	2.5
309	D	Octachlorostyrene	Toluene	2.5
310	D	Pyrimitate	Toluene	2.5
311	D	Isodrin	Toluene	2.5
312	D	Isomethiozin	Toluene	5.0
313	D	Trichloronat	Toluene	2.5

Table 1. Continued

No.	Group	Compound	Solvent	Concentration of mixed standard solution /mg/L
314	D	Dacthal	Toluene	2.5
315	D	4,4-Dichlorobenzophenone	Toluene	2.5
316	D	Nitrothal-isopropyl	Toluene	5.0
317	D	Musk ketone	Toluene	2.5
318	D	Rabenzazole	Toluene	2.5
319	D	Cyprodinil	Toluene	2.5
320	D	Isofenphos oxon	Toluene	5.0
321	D	Dicapthon	Toluene	12.5
322	D	DE-PCB 101	Toluene	2.5
323	D	MCPA-Butoxyethyl ester	Toluene	2.5
324	D	Isocarbophos	Toluene	5.0
325	D	Phorate sulfone	Toluene	2.5
326	D	Chlorfenethol	Toluene	2.5
327	D	Trans-Nonachlor	Toluene	2.5
328	D	Dinobuton	Toluene	25.0
329	D	DEF	Toluene	5.0
330	D	Flurochloridone	Methanol	5.0
331	D	Bromfenvinfos	Toluene+Acetone (8+2)	2.5
332	D	Perthane	Toluene	2.5
333	D	DE-PCB 118	Toluene	2.5
334	D	4,4-Dibromobenzophenone	Toluene	2.5
335	D	Flutriafol	Toluene+Acetone (9+1)	5.0
336	D	Mephosfolan	Toluene	5.0
337	D	Athidathion	Toluene	5.0
338	D	DE-PCB 153	Toluene	2.5
339	D	Diclobutrazole	Toluene+Acetone (8+2)	10.0
340	D	Disulfoton sulfone	Toluene	5.0
341	D	Hexythiazox	Toluene	20.0
342	D	DE-PCB 138	Toluene	2.5
343	D	Cyproconazole	Toluene	2.5
344	D	Clodinafop-propargyl	Toluene	5.0
345	D	Fenthion sulfoxide	Toluene	10.0
346	D	Fluotrimazole	Toluene	2.5
347	D	Fluroxypr-1-methylheptyl ester	Toluene	2.5
348	D	Fenthion sulfone	Toluene	10.0
349	D	Triphenyl phosphate	Toluene	2.5
350	D	Metamitron	Toluene+Acetone (8+2)	25.0
351	D	DE-PCB 180	Toluene	2.5
352	D	Tebufenpyrad	Toluene	2.5
353	D	Lenacil	Toluene+Acetone (8+2)	25.0
354	D	Bromuconazole-1	Toluene	5.0
355	D	Desbrom- leptophos	Toluene	2.5
356	D	Bromuconazole-2	Toluene	5.0
357	D	Nitralin	Toluene+Acetone (8+2)	25.0

Table 1. Continued

No.	Group	Compound	Solvent	Concentration of mixed standard solution /mg/L
358	D	Fenamiphos sulfoxide	Toluene	80.0
359	D	Fenamiphos sulfone	Toluene+Acetone (8+2)	10.0
360	D	Fenpiclonil	Toluene+Acetone (8+2)	10.0
361	D	Fluquinconazole	Toluene+Acetone (8+2)	2.5
362	D	Fenbuconazole	Toluene+Acetone (8+2)	5.0
363	E	Propoxur-1	Toluene	5.0
364	E	Isoprocarb -1	Toluene	5.0
365	E	Methamidophos	Toluene	10.0
366	E	Acenaphthene	Toluene	2.5
367	E	Dibutyl succinate	Toluene	5.0
368	E	Phthalimide	Toluene	5.0
369	E	Chlorethoxyfos	Toluene	5.0
370	E	Isoprocarb -2	Toluene	5.0
371	E	Pencycuron	Toluene	10.0
372	E	Tebuthiuron	Toluene	10.0
373	E	demeton-S-methyl	Toluene	10.0
374	E	Cadusafos	Toluene	10.0
375	E	Propoxur-2	Toluene	5.0
376	E	Phenanthrene	Toluene	2.5
377	E	Fenpyroximate	Toluene	20.0
378	E	Tebupirimfos	Toluene	5.0
379	E	Prohydrojasmon	Cyclohexane	10.0
380	E	Dichloran	Toluene	5.0
381	E	Pyroquilon	Toluene	2.5
382	E	Propyzamide	Toluene	5.0
383	E	Pirimicarb	Toluene	5.0
384	E	Phosphamidon -1	Toluene	20.0
385	E	Benoxacor	Toluene	5.0
386	E	Bromobutide	Cyclohexane	2.5
387	E	Acetochlor	Toluene	5.0
388	E	Tridiphane	Isooctane	10.0
389	E	Terbucarb	Toluene	5.0
390	E	Esprocarb	Toluene	5.0
391	E	Fenfuram	Toluene	5.0
392	E	Acibenzolar-S-Methyl	Cyclohexane	5.0
393	E	Benfuresate	Toluene	5.0
394	E	Dithiopyr	Toluene	2.5
395	E	Mefenoxam	Toluene	5.0
396	E	Malaoxon	Toluene	40.0
397	E	Phosphamidon -2	Toluene	20.0
398	E	Simeconazole	Toluene	5.0
399	E	Chlorthal-dimethyl	Toluene	5.0
400	E	Thiazopyr	Toluene	5.0
401	E	Dimethylvinphos	Toluene	5.0
402	E	Butralin	Toluene	10.0

Table 1. Continued

No.	Group	Compound	Solvent	Concentration of mixed standard solution /mg/L
403	E	Zoxamide	Toluene+Acetone (8+2)	5.0
404	E	Pyrifenox -1	Toluene	20.0
405	E	Allethrin	Toluene	10.0
406	E	Dimethametryn	Toluene	2.5
407	E	Quinoclamine	Toluene	10.0
408	E	Methothrin-1	Toluene	5.0
409	E	Flufenacet	Toluene	20.0
410	E	Methothrin-2	Toluene	5.0
411	E	Pyrifenox -2	Toluene	20.0
412	E	Fenoxanil	Toluene	5.0
413	E	Phthalide	Acetone	10.0
414	E	Furalaxyl	Toluene	5.0
415	E	Thiamethoxam	Toluene	10.0
416	E	Mepanipyrim	Toluene	2.5
417	E	Bromacil	Toluene	5.0
418	E	Picoxystrobin	Toluene	5.0
419	E	Butamifos	Cyclohexane	2.5
420	E	Imazamethabenz-methyl	Toluene	7.5
421	E	Metominostrobin-1	Acetonitrile	10.0
422	E	TCMTB	Toluene	40.0
423	E	Methiocarb Sulfone	Toluene+Acetone (8+2)	80.0
424	E	Imazalil	Toluene	10.0
425	E	Isoprothiolane	Toluene	5.0
426	E	Cyflufenamid	Cyclohexane	40.0
427	E	Pyriminobac-Methyl	Cyclohexane	10.0
428	E	Isoxathion	Cyclohexane	20.0
429	E	Metominostrobin-2	Acetonitrile	10.0
430	E	Diofenolan -1	Toluene	5.0
431	E	Thifluzamide	Acetonitrile	20.0
432	E	Diofenolan -2	Toluene	5.0
433	E	Quinoxyphen	Toluene	2.5
434	E	Chlorfenapyr	Toluene	20.0
435	E	Trifloxystrobin	Toluene	10.0
436	E	Imibenconazole-des-benzyl	Toluene+Acetone (8+2)	10.0
437	E	Isoxadifen-Ethyl	Toluene	5.0
438	E	Fipronil	Toluene	20.0
439	E	Imiprothrin-1	Toluene	5.0
440	E	Carfentrazone-Ethyl	Toluene	5.0
441	E	Imiprothrin-2	Toluene	5.0
442	E	Epoxiconazole -1	Toluene	20.0
443	E	Pyraflufen Ethyl	Toluene	5.0
444	E	Pyributicarb	Toluene	5.0
445	E	Thenylchlor	Toluene	5.0
446	E	Clethodim	Toluene	10.0
447	E	Mefenpyr-diethyl	Toluene	7.5

Table 1. Continued

No.	Group	Compound	Solvent	Concentration of mixed standard solution /mg/L
448	E	Famphur	Toluene	10.0
449	E	Etoxazole	Cyclohexane	15.0
450	E	Pyriproxyfen	Toluene	5.0
451	E	Epoxiconazole-2	Toluene	20.0
452	E	Picolinafen	Toluene	2.5
453	E	Iprodione	Toluene	10.0
454	E	Piperophos	Toluene	7.5
455	E	Ofurace	Toluene	7.5
456	E	Bifenazate	Toluene	20.0
457	E	Endrin ketone	Toluene	10.0
458	E	Clomeprop	Acetonitrile	2.5
459	E	Fenamidone	Toluene	2.5
460	E	Naproanilide	Acetone	2.5
461	E	Pyraclostrobin	Toluene	60.0
462	E	Lactofen	Toluene	20.0
463	E	Tralkoxydim	Toluene	20.0
464	E	Pyraclofos	Cyclohexane	20.0
465	E	Dialifos	Toluene	80.0
466	E	Spirodiclofen	Toluene	20.0
467	E	Halfenprox	Cyclohexane	5.0
468	E	Flurtamone	Toluene	5.0
469	E	Pyriftalid	Toluene	2.5
470	E	Silafluofen	Toluene	2.5
471	E	Pyrimidifen	Acetonitrile	5.0
472	E	Acetamiprid	Toluene	10.0
473	E	Butafenacil	Toluene	2.5
474	E	Cafenstrole	Acetonitrile	10.0
475	E	Fluridone	Toluene	5.0

Table 2. 372 pesticides grouping, solvent selected and concentration of mixed standard solution by LC-MS-MS

No.	Group	Compound	Solvent	Concentration of mixed standard solution /mg/L
1	A	Pirimicarb	Methanol	0.10
2	A	Mevinphos	Toluene	2.00
3	A	Ethidimuron	Methanol	2.00
4	A	Imidacloprid	Methanol	10.00
5	A	Thiofanox sufone	Methanol	10.00
6	A	Secbumeton	Methanol	0.05
7	A	Simetryn	Methanol	0.10
8	A	Thiacloprid	Methanol	0.20
9	A	Methoprotryne	Methanol	1.00
10	A	Paraoxon methyl	Methanol	0.50

Table 2. Continued

No.	Group	Compound	Solvent	Concentration of mixed standard solution /mg/L
11	A	Tepraloxydim	Methanol	8.00
12	A	Spiroxamine	Methanol	0.02
13	A	Cyanazine	Methanol	0.30
14	A	Malaoxon	Methanol	0.10
15	A	Fenthion sulfoxide	Methanol	0.10
16	A	Cycluron	Methanol	0.10
17	A	Prometryne	Methanol	0.05
18	A	Chlorpyrifos methyl	Methanol	1.00
19	A	Ethoxyquin	Methanol	6.00
20	A	Rabenzazole	Methanol	0.05
21	A	Monolinuron	Toluene	2.00
22	A	Isoprocarb	Methanol	1.00
23	A	Propachlor	Methanol	0.10
24	A	Dimefuron	Methanol	0.20
25	A	Propham	Toluene	10.00
26	A	Clomazone	Methanol	0.30
27	A	Triadimenol	Methanol	1.00
28	A	Paclobutrazol	Methanol	0.05
29	A	Cyprodinil	Methanol	0.10
30	A	Desmedipham	Methanol	0.50
31	A	Famphur	Methanol	0.60
32	A	Methidathion	Methanol	1.50
33	A	Fenarimol	Methanol	0.60
34	A	Iprovalicarb	Methanol	0.20
35	A	Dimethenamid	Methanol	0.10
36	A	Myclobutanil	Methanol	0.30
37	A	Isoxaflutole	Methanol	10.00
38	A	Etaconazole	Methanol	0.10
39	A	Diclobutrazole	Methanol	0.06
40	A	Napropamide	Methanol	0.10
41	A	Flusilazole	Methanol	0.10
42	A	Bitertanol	Methanol	1.00
43	A	Penconazole	Methanol	0.06
44	A	Flutolanil	Methanol	0.10
45	A	Acetochlor	Methanol	10.00
46	A	Fenitrothion	Methanol	8.00
47	A	Azinphos ethyl	Methanol	10.00
48	A	Edifenphos	Methanol	0.20
49	A	Isomethiozin	Methanol	0.06
50	A	Bromfenvinfos	Methanol	0.40
51	A	Benalyxyl	Methanol	0.06
52	A	Triflumuron	Methanol	0.60
53	A	Anilofos	Methanol	0.10
54	A	Pyraflufen ethyl	Methanol	0.60

Table 2. Continued

No.	Group	Compound	Solvent	Concentration of mixed standard solution /mg/L
55	A	Clodinafop propargyl	Methanol	0.20
56	A	Diazinon	Toluene	0.06
57	A	Dinitramine	Toluene	4.00
58	A	Pyrazophos	Methanol	0.04
59	A	Terbucarb	Methanol	1.00
60	A	Tolclofos methyl	Methanol	30.00
61	A	Haloxyfop-methyl	Methanol	0.03
62	A	Pretilachlor	Methanol	0.20
63	A	Diflufenican	Methanol	0.03
64	A	Indoxacarb	Methanol	0.60
65	A	Cloquintocet mexyl	Methanol	0.06
66	A	Quizalofop-ethyl	Methanol	0.30
67	A	Cis and trans diallate	Methanol	6.00
68	A	Butylate	Methanol	60.00
69	A	Tetramethirn	Methanol	0.60
70	A	Dioxathion	Methanol	2000.00
71	A	Dichlormid	Methanol	1.00
72	A	Allethrin	Methanol	3.00
73	A	Hexythiazox	Methanol	0.10
74	A	Methothrin	Methanol	200.00
75	A	Profluralin	Methanol	300.00
76	A	Triallate	Methanol	10.00
77	A	Phthalic acid, dicyclohexyl ester	Methanol	0.60
78	A	Prothiophos	Methanol	10.00
79	B	Pirmicarb-desmethyl	Methanol	0.05
80	B	Acephate	Methanol	2.00
81	B	Carbendazim	Methanol	0.05
82	B	Vamidothion sulfone	Methanol	1.20
83	B	Clothianidin	Methanol	6.00
84	B	3-hydroxy carbofuran	Methanol	0.20
85	B	Fenuron	Methanol	1.00
86	B	Difenzoquat-methyl sulfate	Methanol	0.20
87	B	TEPP	Methanol	1.00
88	B	Desethyl-sebuthylazine	Methanol	3.00
89	B	Phosphamidon	Methanol	0.30
90	B	Imibenzonazole-des-benzyl	Methanol	0.50
91	B	Mephosfolan	Methanol	0.10
92	B	Isouron	Methanol	0.10
93	B	Chlormephos	Methanol	300.00
94	B	Cythioate	Methanol	6.00
95	B	Propoxur	Methanol	1.00
96	B	Triasulfuron	Methanol	300.00
97	B	Bendiocarb	Methanol	0.20
98	B	Flutriafol	Methanol	0.30

Table 2. Continued

No.	Group	Compound	Solvent	Concentration of mixed standard solution /mg/L
99	B	Thiofanox	Methanol	30.00
100	B	Mefenoxam	Methanol	0.05
101	B	Diuron	Methanol	0.05
102	B	Difenoxuron	Methanol	0.20
103	B	Methobromuron	Toluene	0.40
104	B	Demeton-S	Methanol	5.00
105	B	Dimethachlor	Methanol	0.10
106	B	Prothoate	Methanol	0.20
107	B	Propazine	Methanol	0.10
108	B	Disulfoton sulfone	Methanol	0.30
109	B	Phenmedipham	Methanol	0.50
110	B	Furalaxyl	Methanol	0.10
111	B	Terbuthylazine	Methanol	0.05
112	B	Flurtamone	Methanol	0.10
113	B	Bromuconazole	Methanol	0.20
114	B	Chlorimuron ethyl	Methanol	1.00
115	B	Molinate	Methanol	0.50
116	B	Fenhexamid	Methanol	0.10
117	B	Pyriftalid	Methanol	0.05
118	B	Chlorbufam	Methanol	20.00
119	B	Pronamide	Methanol	0.30
120	B	Isoxaben	Methanol	0.05
121	B	Ethofume sate	Methanol	80.00
122	B	Bifenazate	Methanol	1.50
123	B	Diphenylamin	Methanol	0.20
124	B	Flamprop methyl	Methanol	0.50
125	B	Iprobenfos	Methanol	0.50
126	B	Metconazole	Methanol	0.10
127	B	Tetrachlorvinphos	Toluene	0.30
128	B	Triazophos	Toluene	0.05
129	B	Propiconazole	Methanol	0.10
130	B	Neburon	Methanol	0.30
131	B	Fluthiacet methyl	Methanol	0.30
132	B	Aclonifen	Methanol	1.00
133	B	Buprofezin	Methanol	0.05
134	B	Triphenyl phosphate	Methanol	0.05
135	B	Parathion	Methanol	60.00
136	B	Etrimfos	Methanol	0.10
137	B	Flamprop isopropyl	Methanol	0.10
138	B	Pirimiphos-methyl	Methanol	0.05
139	B	Cyanofenphos	Methanol	0.50
140	B	Phorate	Methanol	8.00
141	B	Cycloxydim	Methanol	1.00
142	B	Trifloxystrobin	Methanol	0.05

Table 2. Continued

No.	Group	Compound	Solvent	Concentration of mixed standard solution /mg/L
143	B	Clethodim	Methanol	1.00
144	B	Picolinafen	Methanol	0.10
145	B	Bioallethrin	Methanol	1.00
146	B	Pyriproxyfen	Methanol	0.05
147	B	Fenazaquin	Methanol	0.05
148	B	Trifluralin	Toluene	50.00
149	B	Propargite	Methanol	2.00
150	B	DEF	Methanol	0.10
151	B	Bioresmethrin	Methanol	1.00
152	C	Simeton	Methanol	0.20
153	C	Trichlorphon	Methanol	3.00
154	C	Phorate sulfoxide	Methanol	50.00
155	C	Brompyrazon	Methanol	1.50
156	C	Acetamiprid	Methanol	0.30
157	C	Prometon	Methanol	0.05
158	C	Imazamethabenz-methyl	Methanol	0.30
159	C	Fenamiphos sulfoxide	Methanol	0.20
160	C	Oxycarboxin	Methanol	0.20
161	C	Metoxuron	Methanol	0.20
162	C	Bromacil	Methanol	1.00
163	C	Monuron	Methanol	2.00
164	C	Thifensulfuron-methyl	Methanol	0.50
165	C	Fenamiphos sulfone	Methanol	0.10
166	C	Quinoclamine	Methanol	1.00
167	C	Disulfoton-sulfoxide	Methanol	0.20
168	C	Pyrimethanil	Methanol	0.20
169	C	Thiometon	Methanol	900.00
170	C	Metalaxyl	Methanol	0.10
171	C	Diethyltoluamide	Methanol	0.05
172	C	Ofurace	Methanol	0.30
173	C	Fensulfothin	Methanol	0.20
174	C	Dimethametryn	Methanol	0.05
175	C	Demeton(O+S)	Methanol	0.40
176	C	Naled	Methanol	285.00
177	C	Propanil	Methanol	0.40
178	C	Triticonazole	Isooctane	0.20
179	C	Fluridone	Methanol	0.10
180	C	Prochloraz	Methanol	0.20
181	C	Methiocarb	Methanol	0.20
182	C	Methfuroxam	Methanol	2.00
183	C	Benoxacor	Methanol	0.50
184	C	Ethoxysulfuron	Methanol	4.00
185	C	Epoxiconazole	Methanol	10.00
186	C	Fluquinconazole	Methanol	0.05

Table 2. Continued

No.	Group	Compound	Solvent	Concentration of mixed standard solution /mg/L
187	C	Mepanipyrim	Methanol	0.10
188	C	Methoxyfenozide	Methanol	2.00
189	C	Mepronil	Methanol	0.10
190	C	Fenoxycarb	Methanol	0.20
191	C	Thenylchlor	Methanol	0.40
192	C	Ditalimfos	Methanol	10.00
193	C	Fluorochloridone	Methanol	0.60
194	C	Pyrimitate	Methanol	0.05
195	C	Bromobutide	Methanol	0.30
196	C	Monalide	Methanol	0.10
197	C	Kresoxim-methyl	Methanol	10.00
198	C	Quinalphos	Methanol	0.20
199	C	Fenthion	Methanol	0.50
200	C	Benzoylprop-ethyl	Methanol	1.00
201	C	Fonofos	Methanol	0.30
202	C	Coumaphos	Methanol	0.10
203	C	Phosalone	Methanol	2.00
204	C	Phoxim	Methanol	2.00
205	C	Piperophos	Methanol	0.05
206	C	Chlorphoxim	Methanol	10.00
207	C	MCPA-butoxyethyl ester	Methanol	10.00
208	C	Piperonyl butoxide	Methanol	0.10
209	C	Dithiopyr	Methanol	0.30
210	C	Flumiclorac-pentyl	Methanol	0.50
211	C	Oxyflurofen	Methanol	2.00
212	C	Butachlor	Methanol	1.00
213	C	Iodofenphos	Methanol	3.00
214	C	Temephos	Methanol	1.00
215	C	Pendimethalin	Methanol	0.30
216	C	Fenpyroximate	Methanol	0.10
217	C	Pyrethrin	Methanol	20.00
218	C	Fenpropathrin	Methanol	10.00
219	C	Pyridaben	Methanol	0.10
220	C	Aspon	Methanol	0.20
221	C	Spirodiclofen	Methanol	1.00
222	C	Phenothrin	Methanol	10.00
223	D	Ethirimol	Methanol	0.20
224	D	Dicrotophos	Methanol	1.00
225	D	Metamitron	Methanol	0.10
226	D	Crimidine	Methanol	0.30
227	D	Dimethoate	Methanol	0.40
228	D	Terbumeton	Methanol	0.05
229	D	Desmetryn	Methanol	0.05
230	D	Imazalil	Methanol	0.30

Table 2. Continued

No.	Group	Compound	Solvent	Concentration of mixed standard solution /mg/L
231	D	Tebuthiuron	Methanol	0.05
232	D	Hexazinone	Methanol	0.10
233	D	Carbetamide	Methanol	0.40
234	D	Dodemorph	Methanol	0.05
235	D	Triamiphos	Methanol	0.20
236	D	Chlorotoluron	Methanol	0.10
237	D	Carbaryl	Methanol	0.50
238	D	Fluometuron	Methanol	1.50
239	D	Fosthiazate	Methanol	0.20
240	D	Isoproturon	Methanol	0.10
241	D	Carboxin	Methanol	0.10
242	D	Fenfuram	Methanol	0.10
243	D	Norflurazon	Methanol	0.20
244	D	3,4,5-trimethacarb	Methanol	0.05
245	D	Nuarimol	Methanol	0.50
246	D	Diphenamid	Methanol	0.05
247	D	Buturon	Methanol	0.30
248	D	Thionazin	Methanol	1.00
249	D	Benodanil	Methanol	0.10
250	D	Fenpiclonil	Methanol	0.05
251	D	Cyproconazole	Methanol	0.08
252	D	Fenobucarb	Methanol	0.20
253	D	Simeconazole	Methanol	0.10
254	D	Anilazine	Methanol	50.00
255	D	Crufomate	Methanol	8.00
256	D	Aziprotryne	Methanol	0.10
257	D	Azoxystrobin	Methanol	0.05
258	D	Triadimefon	Methanol	0.50
259	D	Ethoprophos	Methanol	0.20
260	D	Pyridaphenthion	Methanol	0.20
261	D	Coumatetralyl	Methanol	10.00
262	D	Fenbuconazole	Methanol	0.05
263	D	Trietazine	Methanol	0.10
264	D	Isoprothiolane	Methanol	0.15
265	D	Tebutam	Methanol	0.05
266	D	Flufenacet	Methanol	0.10
267	D	Propetamphos	Methanol	8.00
268	D	Tebufenozide	Methanol	2.00
269	D	Chlorfenvinphos	Toluene	0.30
270	D	Mecarbam	Methanol	1.00
271	D	Difenoconazole	Methanol	0.05
272	D	Tri-iso-butyl phosphate	Methanol	0.05
273	D	Pyraclofos	Methanol	0.30
274	D	Bensulide	Methanol	2.50

Table 2. Continued

No.	Group	Compound	Solvent	Concentration of mixed standard solution /mg/L
275	D	Sulfallate	Toluene	6.00
276	D	Dacthal	Methanol	300.00
277	D	Thiobencarb	Methanol	0.10
278	D	Mefenpyr-diethyl	Methanol	0.38
279	D	Pencycuron	Methanol	0.05
280	D	Sulfotep	Methanol	0.20
281	D	Profenefos	Methanol	0.20
282	D	4,4-dichlorobenzophenone	Methanol	0.30
283	D	Isofenphos	Methanol	20.00
284	D	Clomeprop	Methanol	0.40
285	D	Kadethrin	Methanol	0.20
286	D	Sethoxydim	Methanol	0.30
287	D	Terbufos	Methanol	18.00
288	D	Lactofen	Methanol	400.00
289	D	Tebupirimfos	Methanol	10.00
290	D	Chlorpyrifos (-ethyl)	Methanol	0.30
291	D	Ethion	Methanol	0.40
292	D	Sulprofos	Toluene	0.40
293	D	Butralin	Methanol	0.10
294	D	Bromophos-ethyl	Methanol	2.50
295	D	DE-PCB 52	Methanol	6.00
296	E	Thiabendazole	Methanol	0.30
297	E	2,6-dichlorobenzamide	Methanol	3.00
298	E	Atratone	Methanol	0.05
299	E	Isocarbamid	Methanol	0.20
300	E	Chloridazon	Methanol	1.00
301	E	Dioxacarb	Methanol	1.50
302	E	Tricyclazole	Methanol	0.10
303	E	Phosfolan	Cyclohexane	1.00
304	E	Triazoxide	Methanol	7.00
305	E	Ametryn	Methanol	0.10
306	E	Allidochlor	Methanol	10.00
307	E	Fenpropimorph	Methanol	0.04
308	E	Azamethiphos	Methanol	0.10
309	E	Lenacil	Methanol	2.00
310	E	Cinosulfuron	Methanol	0.20
311	E	Thiodicarb	Methanol	1.00
312	E	Metribuzin	Methanol	0.30
313	E	Pyrifenox	Methanol	0.20
314	E	Terbutryn	Methanol	0.04
315	E	DMST	Methanol	0.60
316	E	Azaconazole	Methanol	0.05
317	E	Atrazine	Methanol	0.03
318	E	Cyprazine	Methanol	0.03

Table 2. Continued

No.	Group	Compound	Solvent	Concentration of mixed standard solution /mg/L
319	E	Heptanophos	Methanol	1.00
320	E	Dipropetryn	Methanol	0.03
321	E	Metazachlor	Methanol	0.10
322	E	Dimethomorph	Methanol	0.10
323	E	Sebutylazine	Methanol	0.05
324	E	Bupirimate	Methanol	0.10
325	E	Phorate sulfone	Methanol	0.40
326	E	Pyrazosulfuron-ethyl	Methanol	0.10
327	E	Diethofencarb	Methanol	0.40
328	E	Azinphos-methyl	Methanol	10.00
329	E	Linuron	Methanol	1.00
330	E	Fenamiphos	Methanol	0.04
331	E	Uniconazole	Methanol	0.10
332	E	Phosmet	Methanol	0.30
333	E	Mefenacet	Methanol	0.10
334	E	Tetraconazole	Methanol	0.10
335	E	Tebuconazole	Methanol	0.05
336	E	Propyzamide	Methanol	0.20
337	E	Methyl parathion	Methanol	10.00
338	E	Hexaconazole	Methanol	0.06
339	E	Metolachlor	Methanol	0.05
340	E	Malathion	Methanol	0.30
341	E	Alachlor	Methanol	2.00
342	E	Diniconazole	Methanol	0.10
343	E	Furmecyclox	Methanol	10.00
344	E	Athidathion	Methanol	1.00
345	E	Isazofos	Methanol	0.10
346	E	Nitralin	Methanol	1.00
347	E	Dibutyl succinate	Methanol	4.00
348	E	Picoxystrobin	Methanol	10.00
349	E	Tri-n-butyl phosphate	Methanol	0.05
350	E	Cadusafos	Methanol	0.20
351	E	Phenthoate	Methanol	1.00
352	E	Cartap	Methanol	10.00
353	E	Alpha-HCH	Methanol	2.00
354	E	Pyraclostrobin	Methanol	0.20
355	E	Thiazopyr	Methanol	0.30
356	E	Pebulate	Methanol	10.00
357	E	Prohydrojasmon	Methanol	2.00
358	E	Cycloate	Methanol	0.50
359	E	Prallethrin	Methanol	0.40
360	E	Diofenolan	Methanol	190.00
361	E	Methiocarb sulfoxid	Methanol	0.60
362	E	EPN	Methanol	0.50

Table 2. Continued

No.	Group	Compound	Solvent	Concentration of mixed standard solution /mg/L
363	E	Prosulfocarb	Methanol	0.05
364	E	Benfuracarb	Methanol	10.00
365	E	Fluazifop-butyl	Methanol	0.10
366	E	Pirimiphos ethyl	Methanol	0.03
367	E	Tralkoxydim	Methanol	0.50
368	E	Pyributicarb	Methanol	0.10
369	E	Benfluralin	Toluene+Acetone (9+1)	40.00
370	E	Chlorthiophos	Methanol	0.50
371	E	Isopropalin	Methanol	1.00
372	E	Etofenprox	Methanol	10.00

Internal standard solution of GC-MS : accurately weigh 3.5 mg heptachlor epoxide (**epo**) into a 100-mL volumetric flask and dissolve and dilut to volume with toluene.

Working standard mixed solutions in matrix of GC-MS were prepared by diluting 40 μ L internal standard solution and an appropriate amount of mixed standard solution to 1.0 mL with blank extract which had been taken through the method with the rest of the samples, and mixing thoroughly.

Working standard mixed solutions in matrix of LC-MS-MS were prepared by mixing different concentrations of working mixed standard solutions with matrix. These solutions were used to construct calibration plots. Working standard mixed solutions in matrix must be prepared fresh.

Apparatus

GC-MS was performed with an Agilent 5973N instrument equipped with EI and a 30 m×0.25 mm capillary column coated with a 0.25 μ m film of DB-1701. Column temperature was maintained at 40°C for 1 min, then programmed at 30° /min to 130°C, then at 5° /min to 250°C, and finally at 10° /min to 300°C, which was held for 5 min. Helium, purity≥ 99.999%, was used as carrier gas at a flow rate of 1.2 mL /min. The injection port temperature was 290°C and 1μ L samples were injected splitless with the purge on after 1.5 min. The MS ionization energy was 70 eV, the ion-source temperature 230°C, and the GC-MS interface temperature 280°C. Selected ion monitoring was performed with one quantifying ion and two or three qualifying ions for each compound. All detected ions in each group were detected according to programmed-time and sequence of peaking. The retention time, quantifying ion, qualifying ions, and the abundance ratios of quantifying and qualifying ions for each compound are listed in Table 3. The programmed time and dwell time for the ions detected for each compound in each group are listed in Table 4.

Table 3. The retention time, quantifying ion, qualifying ions, and the abundance ratios of quantifying and qualifying ions for GC-MS

No.	Compound	Retention time /min	Quantifying ion	Qualifying ions 1	Qualifying ions 2	Qualifying ions 3
ISTD	Heptachlor-epoxide	22.10	353(100)	355(79)	351(52)	
1	Allidochlor	8.78	138(100)	158(10)	173(15)	
2	Dichlormid	9.74	172(100)	166(41)	124(79)	
3	Etridiazol	10.42	211(100)	183(73)	140(19)	
4	Chlormephos	10.53	121(100)	234(70)	154(70)	
5	Propham	11.36	179(100)	137(66)	120(51)	
6	Cycloate	13.56	154(100)	186(5)	215(12)	
7	Diphenylamine	14.55	169(100)	168(58)	167(29)	
8	Ethalfluralin	15.00	276(100)	316(81)	292(42)	
9	Phorate	15.46	260(100)	121(160)	231(56)	153(3)
10	Thiometon	16.20	88(100)	125(55)	246(9)	
11	Quintozene	16.75	295(100)	237(159)	249(114)	
12	Atrazine-desethyl	16.76	172(100)	187(32)	145(17)	
13	Clomazone	17.00	204(100)	138(4)	205(13)	
14	Diazinon	17.14	304(100)	179(192)	137(172)	
15	Fonofos	17.31	246(100)	137(141)	174(15)	202(6)
16	Etrimfos	17.92	292(100)	181(40)	277(31)	
17	Simazine	17.85	201(100)	186(62)	173(42)	
18	Propetamphos	17.97	138(100)	194(49)	236(30)	
19	Secbumeton	18.36	196(100)	210(38)	225(39)	
20	Dichlofenthion	18.80	279(100)	223(78)	251(38)	
21	Pronamide	18.72	173(100)	175(62)	255(22)	
22	Mexacarbate	18.83	165(100)	150(66)	222(27)	
23	Aldrin	19.67	263(100)	265(65)	293(40)	329(8)
24	Dinitramine	19.35	305(100)	307(38)	261(29)	
25	Ronnel	19.80	285(100)	287(67)	125(32)	
26	Prometryne	20.13	241(100)	184(78)	226(60)	
27	Cyprazine	20.18	212(100)	227(58)	170(29)	
28	Vinclozolin	20.29	285(100)	212(109)	198(96)	
29	Beta-HCH	20.31	219(100)	217(78)	181(94)	254(12)
30	Metalaxyl	20.67	206(100)	249(53)	234(38)	
31	Chlorpyrifos (-ethyl)	20.96	314(100)	258(57)	286(42)	
32	Methyl-Parathion	20.82	263(100)	233(66)	246(8)	200(6)
33	Anthraquinone	21.49	208(100)	180(84)	152(69)	
34	Delta-HCH	21.16	219(100)	217(80)	181(99)	254(10)
35	Fenthion	21.53	278(100)	169(16)	153(9)	
36	Malathion	21.54	173(100)	158(36)	143(15)	
37	Fenitrothion	21.62	277(100)	260(52)	247(60)	
38	Paraoxon-ethyl	21.57	275(100)	220(60)	247(58)	
39	Triadimefon	22.22	208(100)	210(50)	181(74)	
40	Parathion	22.32	291(100)	186(23)	235(35)	263(11)
41	Pendimethalin	22.59	252(100)	220(22)	162(12)	
42	Linuron	22.44	61(100)	248(30)	160(12)	
43	Chlorbenside	22.96	268(100)	270(41)	143(11)	
44	Bromophos-ethyl	23.06	359(100)	303(77)	357(74)	
45	Quinalphos	23.10	146(100)	298(28)	157(66)	
46	trans-Chlordane	23.29	373(100)	375(96)	377(51)	
47	Phenthoate	23.30	274(100)	246(24)	320(5)	

Table 3. Continued

No.	Compound	Retention time /min	Quantifying ion	Qualifying ions 1	Qualifying ions 2	Qualifying ions 3
48	Metazachlor	23.32	209(100)	133(120)	211(32)	
49	Fenothiocarb	23.79	72(100)	160(37)	253(15)	
50	Prothiophos	24.04	309(100)	267(88)	162(55)	
51	Chlorflurenol	24.15	215(100)	152(40)	274(11)	
52	Dieldrin	24.43	263(100)	277(82)	380(30)	345(35)
53	Procymidone	24.36	283(100)	285(70)	255(15)	
54	Methidathion	24.49	145(100)	157(2)	302(4)	
55	Napropamide	24.84	271(100)	128(111)	171(34)	
56	Oxadiazone	25.06	175(100)	258(62)	302(37)	
57	Fenamiphos	25.29	303(100)	154(56)	288(31)	217(22)
58	Tetrasul	25.85	252(100)	324(64)	254(68)	
59	Aramite	25.60	185(100)	319(37)	334(32)	
60	Bupirimate	26.00	273(100)	316(41)	208(83)	
61	Carboxin	26.25	235(100)	143(168)	87(52)	
62	Flutolanil	26.23	173(100)	145(25)	323(14)	
63	4,4'-DDD	26.59	235(100)	237(64)	199(12)	165(46)
64	Ethion	26.69	231(100)	384(13)	199(9)	
65	Sulprofos	26.87	322(100)	156(62)	280(11)	
66	Etaconazole-1	26.81	245(100)	173(85)	247(65)	
67	Etaconazole-2	26.89	245(100)	173(85)	247(65)	
68	Myclobutanil	27.19	179(100)	288(14)	150(45)	
69	Diclofop-methyl	28.08	253(100)	281(50)	342(82)	
70	Propiconazole	28.15	259(100)	173(97)	261(65)	
71	Fensulfothion	27.94	292(100)	308(22)	293(73)	
72	Bifenthrin	28.57	181(100)	166(25)	165(23)	
73	Mirex	28.72	272(100)	237(49)	274(80)	
74	Benodanil	29.14	231(100)	323(38)	203(22)	
75	Nuarimol	28.90	314(100)	235(155)	203(108)	
76	Methoxychlor	29.38	227(100)	228(16)	212(4)	
77	Oxadixyl	29.50	163(100)	233(18)	278(11)	
78	Tetramethirn	29.59	164(100)	135(3)	232(1)	
79	Tebuconazole	29.51	250(100)	163(55)	252(36)	
80	Norflurazon	29.99	303(100)	145(101)	102(47)	
81	Pyridaphenthion	30.17	340(100)	199(48)	188(51)	
82	Phosmet	30.46	160(100)	161(11)	317(4)	
83	Tetradifon	30.70	227(100)	356(70)	159(196)	
84	Oxycarboxin	31.00	175(100)	267(52)	250(3)	
85	cis-Permethrin	31.42	183(100)	184(15)	255(2)	
86	trans-Permethrin	31.68	183(100)	184(15)	255(2)	
87	Pyrazophos	31.60	221(100)	232(35)	373(19)	
88	Cypermethrin	33.19 33.38 33.46 33.56	181(100)	152(23)	180(16)	
89	Fenvalerate	34.45 34.79	167(100)	225(53)	419(37)	181(41)
90	Deltamethrin	35.77	181(100)	172(25)	174(25)	
91	EPTC	8.54	128(100)	189(30)	132(32)	
92	Butylate	9.49	156(100)	146(115)	217(27)	

Table 3. Continued

No.	Compound	Retention time /min	Quantifying ion	Qualifying ions 1	Qualifying ions 2	Qualifying ions 3
93	Dichlobenil	9.75	171(100)	173(68)	136(15)	
94	Pebulate	10.18	128(100)	161(21)	203(20)	
95	Nitrapyrin	10.89	194(100)	196(97)	198(23)	
96	Mevinphos	11.23	127(100)	192(39)	164(29)	
97	Chloroneb	11.85	191(100)	193(67)	206(66)	
98	Tecnazene	13.54	261(100)	203(135)	215(113)	
99	Heptanophos	13.78	124(100)	215(17)	250(14)	
100	Hexachlorobenzene	14.69	284(100)	286(81)	282(51)	
101	Ethoprophos	14.40	158(100)	200(40)	242(23)	168(15)
102	cis-Diallate	14.75	234(100)	236(37)	128(38)	
103	Propachlor	14.73	120(100)	176(45)	211(11)	
104	trans-Diallate	15.29	234(100)	236(37)	128(38)	
105	Trifluralin	15.23	306(100)	264(72)	335(7)	
106	Chlorpropham	15.49	213(100)	171(59)	153(24)	
107	Sulfotep	15.55	322(100)	202(43)	238(27)	266(24)
108	Sulfallate	15.75	188(100)	116(7)	148(4)	
109	Alpha-HCH	16.06	219(100)	183(98)	221(47)	254(6)
110	Terbufos	16.83	231(100)	153(25)	288(10)	186(13)
111	Terbumeton	17.20	210(100)	169(66)	225(32)	
112	Profluralin	17.36	318(100)	304(47)	347(13)	
113	Dioxathion	17.51	270(100)	197(43)	169(19)	
114	Propazine	17.67	214(100)	229(67)	172(51)	
115	Chlorbufam	17.85	223(100)	153(53)	164(64)	
116	Dicloran	17.89	206(100)	176(128)	160(52)	
117	Terbuthylazine	18.07	214(100)	229(33)	173(35)	
118	Monolinuron	18.15	61(100)	126(45)	214(51)	
119	Flufenoxuron	18.83	305(100)	126(67)	307(32)	
120	Cyanophos	18.73	243(100)	180(8)	148(3)	
121	Chlorpyrifos-methyl	19.38	286(100)	288(70)	197(5)	
122	Desmetryn	19.64	213(100)	198(60)	171(30)	
123	Dimethachlor	19.80	134(100)	197(47)	210(16)	
124	Alachlor	20.03	188(100)	237(35)	269(15)	
125	Pirimiphos-methyl	20.30	290(100)	276(86)	305(74)	
126	Terbutryn	20.61	226(100)	241(64)	185(73)	
127	Thiobencarb	20.63	100(100)	257(25)	259(9)	
128	Aspon	20.62	211(100)	253(52)	378(14)	
129	Dicofol	21.33	139(100)	141(72)	250(23)	251(4)
130	Metolachlor	21.34	238(100)	162(159)	240(33)	
131	Oxy-chlordane	21.63	387(100)	237(50)	185(68)	
132	Methoprene	21.71	73(100)	191(29)	153(29)	
133	Bromofos	21.75	331(100)	329(75)	213(7)	
134	Ethofumesate	21.84	207(100)	161(54)	286(27)	
135	Isopropalin	22.10	280(100)	238(40)	222(4)	
136	Endosulfan I	23.10	241(100)	265(66)	339(46)	
137	Propanil	22.68	161(100)	217(21)	163(62)	
138	Isofenphos	22.99	213(100)	255(44)	185(45)	
139	Crufomate	22.93	256(100)	182(154)	276(58)	
140	Chlorfenvinphos	23.19	323(100)	267(139)	269(92)	
141	cis-Chlordane	23.55	373(100)	375(96)	377(51)	

Table 3. Continued

No.	Compound	Retention time /min	Quantifying ion	Qualifying ions 1	Qualifying ions 2	Qualifying ions 3
142	Tolylfluanide	23.45	238(100)	240(71)	137(210)	
143	4,4'-DDE	23.92	318(100)	316(80)	246(139)	248(70)
144	Butachlor	23.82	176(100)	160(75)	188(46)	
145	Chlozolinate	23.83	259(100)	188(83)	331(91)	
146	Crotoxyphos	23.94	193(100)	194(16)	166(51)	
147	Iodofenphos	24.33	377(100)	379(37)	250(6)	
148	Tetrachlorvinphos	24.36	329(100)	331(96)	333(31)	
149	Chlorbromuron	24.37	61(100)	294(17)	292(13)	
150	Profenofos	24.65	339(100)	374(39)	297(37)	
151	Fluorochloridone	25.14	311(100)	313(64)	187(85)	
152	2,4'-DDD	24.94	235(100)	237(65)	165(39)	199(15)
153	Endrin	25.15	263(100)	317(30)	345(26)	
154	Hexaconazole	24.92	214(100)	231(62)	256(26)	
155	Chlorfenson	25.05	302(100)	175(282)	177(103)	
156	2,4'-DDT	25.56	235(100)	237(63)	165(37)	199(14)
157	Paclobutrazol	25.21	236(100)	238(37)	167(39)	
158	Methoprotryne	25.63	256(100)	213(24)	271(17)	
159	Chloropropylate	25.85	251(100)	253(64)	141(18)	
160	Flamprop-methyl	25.90	105(100)	77(26)	276(11)	
161	Nitrofcn	26.12	283(100)	253(90)	202(48)	139(15)
162	Oxyfluorfen	26.13	252(100)	361(35)	300(35)	
163	Chlorthiophos	26.52	325(100)	360(52)	297(54)	
164	Endosulfan II	26.72	241(100)	265(66)	339(46)	
165	Flamprop-Isopropyl	26.70	105(100)	276(19)	363(3)	
166	4,4'-DDT	27.22	235(100)	237(65)	246(7)	165(34)
167	Carbofenothion	27.19	157(100)	342(49)	199(28)	
168	Benalaxyl	27.54	148(100)	206(32)	325(8)	
169	Edifenphos	27.94	173(100)	310(76)	201(37)	
170	Triazophos	28.23	161(100)	172(47)	257(38)	
171	Cyanofenphos	28.43	157(100)	169(56)	303(20)	
172	Chlorbenside sulfone	28.88	127(100)	99(14)	89(33)	
173	Endosulfan-Sulfate	29.05	387(100)	272(165)	389(64)	
174	Bromopropylate	29.30	341(100)	183(34)	339(49)	
175	Benzoylprop-ethyl	29.40	292(100)	365(36)	260(37)	
176	Fenpropathrin	29.56	265(100)	181(237)	349(25)	
177	Leptophos	30.19	377(100)	375(73)	379(28)	
178	EPN	30.06	157(100)	169(53)	323(14)	
179	Hexazinone	30.14	171(100)	252(3)	128(12)	
180	Phosalone	31.22	182(100)	367(30)	154(20)	
181	Azinphos-methyl	31.41	160(100)	132(71)	77(58)	
182	Fenarimol	31.65	139(100)	219(70)	330(42)	
183	Azinphos-ethyl	32.01	160(100)	132(103)	77(51)	
184	Coumaphos	33.22	362(100)	226(56)	364(39)	334(15)
185	Cyfluthrin	32.94 33.12	206(100)	199(63)	226(72)	
186	Fluvalinate	34.94 35.02	250(100)	252(38)	181(18)	
187	Dichlorvos	7.80	109(100)	185(34)	220(7)	
188	Biphenyl	9.00	154(100)	153(40)	152(27)	

Table 3. Continued

No.	Compound	Retention time /min	Quantifying ion	Qualifying ions 1	Qualifying ions 2	Qualifying ions 3
189	Vernolate	9.82	128(100)	146(17)	203(9)	
190	3,5-Dichloroaniline	11.20	161(100)	163(62)	126(10)	
191	Molinate	11.92	126(100)	187(24)	158(2)	
192	Methacrifos	11.86	125(100)	208(74)	240(44)	
193	2-Phenylphenol	12.47	170(100)	169(72)	141(31)	
194	Tetrahydrophthalimide	13.39	151(100)	123(16)	122(16)	
195	Fenobucarb	14.60	121(100)	150(32)	107(8)	
196	Benfluralin	15.23	292(100)	264(20)	276(13)	
197	Hexaflumuron	16.20	176(100)	279(28)	277(43)	
198	Triallate	17.12	268(100)	270(73)	143(19)	
199	Pyrimethanil	17.28	198(100)	199(45)	200(5)	
200	Gamma-HCH	17.48	183(100)	219(93)	254(13)	221(40)
201	Disulfoton	17.61	88(100)	274(15)	186(18)	
202	Atrizine	17.64	200(100)	215(62)	173(29)	
203	Heptachlor	18.49	272(100)	237(40)	337(27)	
204	Iprobenfos	18.44	204(100)	246(18)	288(17)	
205	Isazofos	18.54	161(100)	257(53)	285(39)	313(14)
206	Plifenate	18.87	217(100)	175(96)	242(91)	
207	Transfluthrin	19.04	163(100)	165(23)	335(7)	
208	Fluchloraline	18.89	306(100)	326(87)	264(54)	
209	Tolclofos-methyl	19.69	265(100)	267(36)	250(10)	
210	Propisochlor	19.89	162(100)	223(200)	146(17)	
211	Metobromuron	20.07	61(100)	258(11)	170(16)	
212	Metribuzin	20.33	198(100)	199(21)	144(12)	
213	HCH, epsilon-	20.78	181(100)	219(76)	254(15)	217(40)
214	Formothion	21.42	170(100)	224(97)	257(63)	
215	Diethofencarb	21.43	267(100)	225(98)	151(31)	
216	Dimepiperate	22.28	119(100)	145(30)	263(8)	
217	Bioallethrin-1	22.29	123(100)	136(24)	107(29)	
218	Bioallethrin-2	22.34	123(100)	136(24)	107(29)	
219	2,4'-DDE	22.64	246(100)	318(34)	176(26)	248(65)
220	Fenson	22.54	141(100)	268(53)	77(104)	
221	Diphenamid	22.87	167(100)	239(30)	165(43)	
222	Chlorthion	22.86	297(100)	267(162)	299(45)	
223	Prallethrin	23.11	123(100)	105(17)	134(9)	
224	Penconazole	23.17	248(100)	250(33)	161(50)	
225	Mecarbam	23.46	131(100)	296(22)	329(40)	
226	Tetraconazole	23.35	336(100)	338(33)	171(10)	
227	Propaphos	23.92	304(100)	220(108)	262(34)	
228	Flumetralin	24.10	143(100)	157(25)	404(10)	
229	Triadimenol	24.22	112(100)	168(81)	130(15)	
230	Pretilachlor	24.67	162(100)	238(26)	262(8)	
231	Kresoxim-methyl	25.04	116(100)	206(25)	131(66)	
232	Fluazifop-butyl	25.21	282(100)	383(44)	254(49)	
233	Chlorfluazuron	25.27	321(100)	323(71)	356(8)	
234	Chlorobenzilate	25.90	251(100)	253(65)	152(5)	
235	Uniconazole	26.15	234(100)	236(40)	131(15)	
236	Flusilazole	26.19	233(100)	206(33)	315(9)	
237	Fluorodifen	26.59	190(100)	328(35)	162(34)	

Table 3. Continued

No.	Compound	Retention time /min	Quantifying ion	Qualifying ions 1	Qualifying ions 2	Qualifying ions 3
238	Diniconazole	27.03	268(100)	270(65)	232(13)	
239	Piperonyl butoxide	27.46	176(100)	177(33)	149(14)	
240	Propargite	27.87	135(100)	350(7)	173(16)	
241	Mepronil	27.91	119(100)	269(26)	120(9)	
242	Dimefuron	27.82	140(100)	105(75)	267(36)	
243	Diflufenican	28.45	266(100)	394(25)	267(14)	
244	Phenothrin	29.08 29.21	123(100)	183(74)	350(6)	
245	Fludioxonil	28.93	248(100)	127(24)	154(21)	
246	Fenoxycarb	29.57	255(100)	186(82)	116(93)	
247	Sethoxydim	29.63	178(100)	281(51)	219(36)	
248	Anilofos	30.68	226(100)	184(52)	334(10)	
249	Acrinathrin	31.07	181(100)	289(31)	247(12)	
250	Lambda-Cyhalothrin	31.11	181(100)	197(100)	141(20)	
251	Mefenacet	31.29	192(100)	120(35)	136(29)	
252	Permethrin	31.57	183(100)	184(14)	255(1)	
253	Pyridaben	31.86	147(100)	117(11)	364(7)	
254	Fluoroglycofen-ethyl	32.01	447(100)	428(20)	449(35)	
255	Bitertanol	32.25	170(100)	112(8)	141(6)	
256	Etofenprox	32.75	163(100)	376(4)	183(6)	
257	Cycloxydim	33.05	178(100)	279(7)	251(4)	
258	Alpha-Cypermethrin	33.35	163(100)	181(84)	165(63)	
259	Flucythrinate	33.58 33.85	199(100)	157(90)	451(22)	
260	Esfenvalerate	34.65	419(100)	225(158)	181(189)	
261	Difenoconazole	35.40	323(100)	325(66)	265(83)	
262	Flumioxazin	35.50	354(100)	287(24)	259(15)	
263	Flumiclorac-pentyl	36.34	423(100)	308(51)	318(29)	
264	Dimefox	5.62	110(100)	154(75)	153(17)	
265	Disulfoton-sulfoxide	8.41	212(100)	153(61)	184(20)	
266	Pentachlorobenzene	11.11	250(100)	252(64)	215(24)	
267	Tri-Iso-Butyl Phosphate	11.65	155(100)	139(67)	211(24)	
268	Crimidine	13.13	142(100)	156(90)	171(84)	
269	BDMC-1	13.25	200(100)	202(104)	201(13)	
270	Chlorfenprop-methyl	13.57	165(100)	196(87)	197(49)	
271	Thionazin	14.04	143(100)	192(39)	220(14)	
272	2,3,5,6-Tetrachloroaniline	14.22	231(100)	229(76)	158(25)	
273	Tri-N-Butyl Phosphate	14.33	155(100)	211(61)	167(8)	
274	2,3,4,5-Tetrachloroanisole	14.66	246(100)	203(70)	231(51)	
275	Pentachloroanisole	15.19	280(100)	265(100)	237(85)	
276	Tebutam	15.30	190(100)	106(38)	142(24)	
277	Dioxabenzofos	16.14	216(100)	201(26)	171(5)	
278	Methabenzthiazuron	16.34	164(100)	136(81)	108(27)	
279	Simetone	16.69	197(100)	196(40)	182(38)	
280	Atratone	16.70	196(100)	211(68)	197(105)	
281	Desisopropyl-atrazine	16.69	173(100)	158(84)	145(73)	
282	Terbufos Sulfone	16.79	231(100)	288(11)	186(15)	
283	Tefluthrin	17.24	177(100)	197(26)	161(5)	
284	Bromocylen	17.43	359(100)	357(99)	394(14)	

Table 3. Continued

No.	Compound	Retention time /min	Quantifying ion	Qualifying ions 1	Qualifying ions 2	Qualifying ions 3
285	Trietazine	17.53	200(100)	229(51)	214(45)	
286	Etrimfos oxon	17.83	292(100)	277(35)	263(12)	
287	Cycluron	17.95	89(100)	198(36)	114(9)	
288	2,6-Dichlorobenzamide	17.93	173(100)	189(36)	175(62	
289	DE-PCB 28	18.15	256(100)	186(53)	258(97)	
290	DE-PCB 31	18.19	256(100)	186(53)	258(97)	
291	Desethyl-sebuthylazine	18.32	172(100)	174(32)	186(11)	
292	2,3,4,5-Tetrachloroaniline	18.55	231(100)	229(76)	233(48)	
293	Musk ambrette	18.62	253(100)	268(35)	223(18)	
294	Musk xylene	18.66	282(100)	297(10)	128(20)	
295	Pentachloroaniline	18.91	265(100)	263(63)	230(8)	
296	Aziprotryne	19.11	199(100)	184(83)	157(31)	
297	Sebutylazine	19.26	200(100)	214(14)	229(13)	
298	Isocarbamide	19.24	142(100)	185(2)	143(6)	
299	DE-PCB 52	19.48	292(100)	220(88)	255(32)	
300	Musk moskene	19.46	263(100)	278(12)	264(15)	
301	Prosulfocarb	19.51	251(100)	252(14)	162(10)	
302	Dimethenamid	19.55	154(100)	230(43)	203(21)	
303	Fenchlorphos oxon	19.72	285(100)	287(70)	270(7)	
304	BDMC-2	19.74	200(100)	202(101)	201(12)	
305	Paraoxon-methyl	19.83	230(100)	247(93)	200(40)	
306	Monalide	20.02	197(100)	199(31)	239(45)	
307	Musk tibeten	20.40	251(100)	266(25)	252(14)	
308	Isobenzan	20.55	311(100)	375(31)	412(7)	
309	Octachlorostyrene	20.60	380(100)	343(94)	308(120)	
310	Pyrimitate	20.59	305(100)	153(116)	180(49)	
311	Isodrin	21.01	193(100)	263(46)	195(83)	
312	Isomethiozin	21.06	225(100)	198(86)	184(13)	
313	Trichloronat	21.10	297(100)	269(86)	196(16)	
314	Dacthal	21.25	301(100)	332(31)	221(16)	
315	4,4-Dichlorobenzophenone	21.29	250(100)	252(62)	215(26)	
316	Nitrothal-isopropyl	21.69	236(100)	254(54)	212(74)	
317	Musk ketone	21.70	279(100)	294(28)	128(16)	
318	Rabenzazole	21.73	212(100)	170(26)	195(19)	
319	Cyprodinil	21.94	224(100)	225(62)	210(9)	
320	Isofenphos oxon	22.04	229(100)	201(2)	314(12)	
321	Dicapthon	22.44	262(100)	263(10)	216(10)	
322	DE-PCB 101	22.62	326(100)	254(66)	291(18)	
323	MCPA-Butoxyethyl ester	22.61	300(100)	200(71)	182(41)	
324	Isocarbophos	22.87	136(100)	230(26)	289(22)	
325	Phorate sulfone	23.15	199(100)	171(30)	215(11)	
326	Chlorfenethol	23.29	251(100)	253(66)	266(12)	
327	Trans-Nonachlor	23.62	409(100)	407(89)	411(63)	
328	Dinobuton	23.88	211(100)	240(15)	223(15)	
329	DEF	24.08	202(100)	226(51)	258(55)	
330	Flurochloridone	24.31	311(100)	187(74)	313(66)	
331	Bromfenvinfos	24.62	267(100)	323(56)	295(18)	
332	Perthane	24.81	223(100)	224(20)	178(9)	
333	DE-PCB 118	25.08	326(100)	254(38)	184(16)	

Table 3. Continued

No.	Compound	Retention time /min	Quantifying ion	Qualifying ions 1	Qualifying ions 2	Qualifying ions 3
334	4,4-Dibromobenzophenone	25.30	340(100)	259(30)	185(179)	
335	Flutriafol	25.31	219(100)	164(96)	201(7)	
336	Mephosfolan	25.29	196(100)	227(49)	168(60)	
337	Athidathion	25.63	145(100)	330(1)	129(12)	
338	DE-PCB 153	25.64	360(100)	290(62)	218(24)	
339	Diclobutrazole	25.95	270(100)	272(68)	159(42)	
340	Disulfoton sulfone	26.16	213(100)	229(4)	185(11)	
341	Hexythiazox	26.48	227(100)	156(158)	184(93)	
342	DE-PCB 138	26.84	360(100)	290(68)	218(26)	
343	Cyproconazole	27.23	222(100)	224(35)	223(11)	
344	Clodinafop-propargyl	27.74	349(100)	238(96)	266(83)	
345	Fenthion sulfoxide	28.06	278(100)	279(290)	294(145)	
346	Fluotrimazole	28.39	311(100)	379((60)	233(36)	
347	Fluroxypr-1-methylheptyl ester	28.45	366(100)	254(67)	237(60)	
348	Fenthion sulfone	28.55	310(100)	136(25)	231(10)	
349	Triphenyl phosphate	28.65	326(100)	233(16)	215(20)	
350	Metamitron	28.63	202(100)	174(52)	186(12)	
351	DE-PCB 180	29.05	394(100)	324(70)	359(20)	
352	Tebufenpyrad	29.06	318(100)	333(78)	276(44)	
353	Lenacil	29.70	153(100)	136(6)	234(2)	
354	Bromuconazole-1	29.90	173(100)	175(65)	214(15)	
355	Desbrom- leptophos	30.15	377(100)	171(97)	375(72)	
356	Bromuconazole-2	30.72	173(100)	175(67)	214(14)	
357	Nitralin	30.92	316(100)	274(58)	300(15)	
358	Fenamiphos sulfoxide	31.03	304(100)	319(29)	196(22)	
359	Fenamiphos sulfone	31.34	320(100)	292(57)	335(7)	
360	Fenpiclonil	32.37	236(100)	238(66)	174(36)	
361	Fluquinconazole	32.62	340(100)	342(37)	341(20)	
362	Fenbuconazole	34.02	129(100)	198(51)	125(31)	
363	Propoxur-1	6.58	110(100)	152(16)	111(9)	
364	Isoprocarb -1	7.56	121(100)	136(34)	103(20)	
365	Methamidophos	9.37	94(100)	95(112)	141(52)	
366	Acenaphthene	10.79	164(100)	162(84)	160(38)	
367	Dibutyl succinate	12.20	101(100)	157(19)	175(5)	
368	Phthalimide	13.21	147(100)	104(61)	103(35)	
369	Chlorethoxyfos	13.43	153(100)	125(67)	301(19)	
370	Isoprocarb -2	13.69	121(100)	136(34)	103(20)	
371	Pencycuron	14.30	125(100)	180(65)	209(20)	
372	Tebuthiuron	14.25	156(100)	171(30)	157(9)	
373	demeton-S-methyl	15.19	109(100)	142(43)	230(5)	
374	Cadusafos	15.13	159(100)	213(14)	270(12)	
375	Propoxur-2	15.48	110(100)	152(19)	111(8)	
376	Phenanthrene	16.97	188(100)	160(9)	189(16)	
377	Fenpyroximate	17.49	213(100)	142(21)	198(9)	
378	Tebupirimfos	17.61	318(100)	261(107)	234(100)	
379	prohydrojasmon	17.80	153(100)	184(41)	254(7)	
380	Dichloran	18.10	176(100)	206(87)	124(101)	
381	Pyroquilon	18.28	173(100)	130(69)	144(38)	

Table 3. Continued

No.	Compound	Retention time /min	Quantifying ion	Qualifying ions 1	Qualifying ions 2	Qualifying ions 3
382	Propyzamide	19.01	173(100)	255(23)	240(9)	
383	Pirimicarb	19.08	166(100)	238(23)	138(8)	
384	Phosphamidon -1	19.66	264(100)	138(62)	227(25)	
385	Benoxacor	19.62	120(100)	259(38)	176(19)	
386	Bromobutide	19.70	119(100)	232(27)	296(6)	
387	Acetochlor	19.84	146(100)	162(59)	223(59)	
388	Tridiphane	19.90	173(100)	187(90)	219(46)	
389	Terbucarb	20.06	205(100)	220(52)	206(16)	
390	Esprocarb	20.01	222(100)	265(10)	162(61)	
391	Fenfuram	20.35	109(100)	201(29)	202(5)	
392	Acibenzolar-S-Methyl	20.42	182(100)	135(64)	153(34)	
393	Benfuresate	20.68	163(100)	256(17)	121(18)	
394	Dithiopyr	20.78	354(100)	306(72)	286(74)	
395	Mefenoxam	20.91	206(100)	249(46)	279(11)	
396	Malaoxon	21.17	127(100)	268(11)	195(15)	
397	Phosphamidon -2	21.36	264(100)	138(54)	227(17)	
398	Simeconazole	21.41	121(100)	278(14)	211(34)	
399	Chlorthal-dimethyl	21.39	301(100)	332(27)	221(17)	
400	Thiazopyr	21.91	327(100)	363(73)	381(34)	
401	Dimethylvinphos	22.21	295(100)	297(56)	109(74)	
402	Butralin	22.24	266(100)	224(16)	295(60)	
403	Zoxamide	22.30	187(100)	242(68)	299(9)	
404	Pyrifenox -1	22.50	262(100)	294(15)	227(15)	
405	Allethrin	22.60	123(100)	107(24)	136(20)	
406	Dimethametryn	22.83	212(100)	255(9)	240(5)	
407	Quinoclamine	22.89	207(100)	172(259)	144(64)	
408	Methothrin-1	22.92	123(100)	135(89)	104(41)	
409	Flufenacet	23.09	151(100)	211(61)	363(6)	
410	Methothrin-2	23.19	123(100)	135(73)	104(12)	
411	Pyrifenox -2	23.50	262(100)	294(17)	227(16)	
412	Fenoxanil	23.58	140(100)	189(14)	301(6)	
413	Phthalide	23.51	243(100)	272(28)	215(20)	
414	Furalaxyl	23.97	242(100)	301(24)	152(40)	
415	Thiamethoxam	24.38	182(100)	212(92)	247(124)	
416	Mepanipyrim	24.29	222(100)	223(53)	221(9)	
417	Bromacil	24.73	205(100)	207(46)	231(5)	
418	Picoxystrobin	24.97	335(100)	303(43)	367(9)	
419	Butamifos	25.41	286(100)	200(57)	232(37)	
420	Imazamethabenz-methyl	25.50	144(100)	187(117)	256(95)	
421	Metominostrobin-1	25.61	191(100)	238(56)	196(75)	
422	TCMTB	25.59	180(100)	238(108)	136(30)	
423	Methiocarb Sulfone	25.56	200(100)	185(40)	137(16)	
424	Imazalil	25.72	215(100)	173(66)	296(5)	
425	Isoprothiolane	25.87	290(100)	231(82)	204(88)	
426	Cyflufenamid	26.02	91(100)	412(11)	294(11)	
427	Pyriminobac-Methyl	26.34	302(100)	330(107)	361(86)	
428	Isoxathion	26.51	313(100)	105(341)	177(208)	
429	Metominostrobin-2	26.76	196(100)	191(36)	238(89)	
430	Diofenolan -1	26.81	186(100)	300(57)	225(25)	

Table 3. Continued

No.	Compound	Retention time /min	Quantifying ion	Qualifying ions 1	Qualifying ions 2	Qualifying ions 3
431	Thifluzamide	27.26	449(100)	447(97)	194(308)	
432	Diofenolan -2	27.14	186(100)	300(58)	225(31)	
433	Quinoxyphen	27.14	237(100)	272(37)	307(29)	
434	Chlorfenapyr	27.60	247(100)	328(47)	408(42)	
435	Trifloxystrobin	27.71	116(100)	131(40)	222(30)	
436	Imibenconazole-des-benzyl	27.86	235(100)	270(35)	272(35)	
437	Isoxadifen-Ethyl	27.90	204(100)	222(76)	294(44)	
438	Fipronil	28.34	367(100)	369(69)	351(15)	
439	Imiprothrin-1	28.31	123(100)	151(55)	107(54)	
440	Carfentrazone-Ethyl	28.29	312(100)	340(135)	376(32)	
441	Imiprothrin-2	28.50	123(100)	151(21)	107(17)	
442	Epoxiconazole -1	28.58	192(100)	183(24)	138(35)	
443	Pyraflufen Ethyl	28.91	412(100)	349(41)	339(34)	
444	Pyributicarb	28.87	165(100)	181(23)	108(64)	
445	Thenylchlor	29.12	127(100)	288(25)	141(17)	
446	Clethodim	29.21	164(100)	205(50)	267(15)	
447	Mefenpyr-diethyl	29.55	227(100)	299(131)	372(18)	
448	Famphur	29.80	218(100)	125(27)	217(22)	
449	Etoxazole	29.64	300(100)	330(69)	359(65)	
450	Pyriproxyfen	30.06	136(100)	226(8)	185(10)	
451	Epoxiconazole-2	29.73	192(100)	183(13)	138(30)	
452	Picolinafen	30.27	238(100)	376(77)	266(11)	
453	Iprodione	30.24	187(100)	244(65)	246(42)	
454	Piperophos	30.42	320(100)	140(123)	122(114)	
455	Ofurace	30.36	160(100)	232(83)	204(35)	
456	Bifenazate	30.38	300(100)	258(99)	199(100)	
457	Endrin ketone	30.45	317(100)	250(31)	281(58)	
458	Clomeprop	30.48	290(100)	288(279)	148(206)	
459	Fenamidone	30.66	268(100)	238(111)	206(32)	
460	Naproanilide	31.89	291(100)	171(96)	144(100)	
461	Pyraclostrobin	31.98	132(100)	325(14)	283(21)	
462	Lactofen	32.06	442(100)	461(25)	346(12)	
463	Tralkoxydim	32.14	283(100)	226(7)	268(8)	
464	Pyraclofos	32.18	360(100)	194(79)	362(38)	
465	Dialifos	32.27	186(100)	357(143)	210(397)	
466	Spirodiclofen	32.50	312(100)	259(48)	277(28)	
467	Halfenprox	32.62	263(100)	237(6)	476(5)	
468	Flurtamone	32.78	333(100)	199(63)	247(25)	
469	Pyriftalid	32.94	318(100)	274(71)	303(44)	
470	Silafluofen	33.18	287(100)	286(274)	258(289)	
471	Pyrimidifen	33.63	184(100)	186(32)	185(10)	
472	Acetamiprid	33.87	126(100)	152(99)	166(58)	
473	Butafenacil	33.85	331(100)	333(34)	180(35)	
474	Cafenstrole	34.36	100(100)	188(69)	119(25)	
475	Fluridone	37.61	328(100)	329(100)	330(100)	

Table 4. The programmed- time and dwell time for the ions detected for each compound in each group for GC-MS

No.	Time /min	Ion /amu	Dwell time /ms
Group A			
1	8.30	138, 158, 173	200
2	9.60	124, 140, 166, 172, 183, 211	90
3	10.50	121, 154, 234	200
4	10.75	120, 137, 179	200
5	11.70	154, 186, 215	200
6	14.40	167, 168, 169	200
7	14.90	121, 142, 143, 153, 183, 195, 196, 198, 230, 231, 260, 276, 292, 316	30
8	16.20	88, 125, 246	200
9	16.70	137, 138, 145, 172, 174, 179, 187, 202, 204, 205, 237, 246, 249, 295, 304	30
10	17.80	138, 173, 175, 181, 186, 194, 196, 201, 210, 225, 236, 255, 277, 292	30
11	18.80	150, 165, 173, 175, 222, 223, 251, 255, 279	50
12	19.20	125, 143, 229, 261, 263, 265, 293, 305, 307, 329	50
13	19.80	125, 261, 263, 265, 285, 287, 293, 305, 307, 329	50
14	20.10	170, 181, 184, 198, 200, 206, 212, 217, 219, 226, 227, 233, 234, 241, 246, 249, 254, 258, 263, 264, 266, 268, 285, 286, 314	10
15	21.40	143, 152, 153, 158, 169, 173, 180, 181, 208, 217, 219, 220, 247, 254, 256, 260, 275, 277, 278, 351, 353, 355	10
16	22.30	61, 143, 160, 162, 181, 186, 208, 210, 220, 235, 248, 252, 263, 268, 270, 291, 351, 353, 355	20
17	23.00	133, 143, 146, 157, 209, 211, 246, 268, 270, 274, 298, 303, 320, 357, 359, 373, 375, 377	20
18	23.70	72, 104, 133, 145, 152, 157, 160, 162, 209, 211, 215, 253, 255, 260, 263, 267, 274, 277, 283, 285, 297, 302, 309, 345, 380	10
19	24.80	128, 145, 154, 157, 171, 175, 198, 217, 225, 240, 255, 258, 271, 283, 285, 288, 302, 303	20
20	25.50	154, 185, 217, 252, 253, 254, 288, 303, 319, 324, 334	50
21	26.00	87, 139, 143, 145, 165, 173, 199, 208, 231, 235, 237, 251, 253, 273, 316, 323, 384	20
22	26.80	145, 150, 156, 165, 173, 179, 199, 231, 235, 237, 245, 247, 280, 288, 322, 323, 384	20
23	27.90	165, 166, 173, 181, 253, 259, 261, 281, 292, 293, 308, 342	40
24	28.60	118, 160, 165, 166, 181, 203, 212, 227, 228, 231, 235, 237, 272, 274, 314, 323	30
25	29.30	135, 163, 164, 212, 227, 228, 232, 233, 250, 252, 278	40
26	30.00	102, 145, 159, 160, 161, 188, 199, 227, 303, 317, 340, 356	40
27	31.00	175, 183, 184, 220, 221, 223, 232, 250, 255, 267, 373	40
28	33.00	127, 180, 181	200
29	34.40	167, 181, 225, 419	150
30	35.70	172, 174, 181	200
Group B			
1	7.80	128, 132, 189	200
2	8.80	146, 156, 217	200
3	9.70	128, 136, 161, 171, 173, 203	90
4	10.70	127, 164, 192, 194, 196, 198	90
5	11.70	191, 193, 206	200
6	13.40	124, 203, 215, 250, 261	100

Table 4. Continued

No.	Time /min	Ion /amu	Dwell time /ms
7	14.40	158, 168, 200, 242, 282, 284, 286	80
8	14.70	116, 120, 128, 148, 153, 171, 176, 188, 202, 211, 213, 234, 236, 238, 264, 266, 282, 284, 286, 306, 322, 335	10
9	16.00	116, 148, 183, 188, 219, 221, 254	80
10	16.80	153, 186, 231, 288	150
11	17.10	153, 160, 164, 169, 172, 173, 176, 197, 206, 210, 214, 223, 225, 229, 270, 318, 330, 347	20
12	18.20	61, 126, 160, 173, 176, 206, 214, 229	60
13	18.70	126, 127, 134, 148, 164, 171, 172, 180, 192, 197, 198, 210, 213, 223, 243, 286, 288, 305, 307	20
14	19.90	134, 171, 188, 197, 198, 210, 213, 237, 269, 276, 290, 305	40
15	20.60	100, 185, 211, 226, 241, 253, 257, 259, 378	50
16	21.20	73, 139, 141, 153, 161, 162, 167, 185, 191, 207, 213, 224, 226, 237, 238, 240, 250, 251, 286, 304, 318, 329, 331, 333, 351, 353, 355, 387	10
17	22.00	161, 167, 207, 222, 224, 226, 238, 264, 280, 286, 351, 353, 355	40
18	22.70	161, 163, 170, 171, 182, 185, 205, 213, 217, 241, 255, 256, 265, 267, 269, 276, 323, 339	20
19	23.40	137, 160, 176, 188, 238, 240, 246, 248, 259, 267, 269, 316, 318, 323, 331, 373, 375, 377	20
20	23.90	61, 160, 166, 176, 188, 193, 194, 246, 248, 250, 259, 292, 294, 297, 316, 318, 329, 331, 333, 339, 374, 377, 379	20
21	24.90	61, 105, 165, 167, 172, 175, 177, 187, 199, 214, 231, 235, 236, 237, 238, 256, 263, 292, 294, 297, 302, 305, 311, 313, 317, 339, 345, 374	10
22	25.60	77, 105, 139, 141, 165, 169, 171, 199, 202, 213, 223, 235, 237, 251, 252, 253, 256, 271, 276, 283, 297, 300, 325, 360, 361	10
23	26.70	105, 157, 165, 195, 199, 235, 237, 246, 276, 297, 325, 339, 342, 360, 363	30
24	27.60	148, 157, 161, 169, 172, 173, 201, 206, 257, 303, 310, 325	40
25	28.90	89, 99, 126, 127, 157, 161, 169, 172, 181, 183, 257, 260, 265, 272, 292, 303, 339, 341, 349, 365, 387, 389	10
26	29.80	79, 181, 183, 265, 311, 349	90
27	30.00	128, 157, 169, 171, 189, 252, 310, 323, 341, 375, 377, 379	40
28	31.20	132, 139, 154, 160, 161, 182, 189, 251, 310, 330, 341, 367	40
29	32.90	180, 199, 206, 226, 266, 308, 334, 362, 364	50
30	34.00	181, 250, 252	200
Group C			
1	7.30	109, 185, 220	200
2	8.70	152, 153, 154	200
3	9.30	58, 128, 129, 146, 188, 203	90
4	11.20	126, 161, 163	200
5	11.75	125, 126, 141, 158, 169, 170, 187, 208, 240	50
6	13.50	122, 123, 124, 151, 215, 250	90
7	14.70	107, 121, 150, 264, 276, 292	90
8	16.00	174, 202, 217	200
9	16.50	126, 141, 143, 156, 168, 176, 198, 199, 200, 210, 225, 268, 270, 277, 279	30
10	17.60	88, 173, 183, 186, 200, 215, 219, 254, 274	50
11	18.40	104, 130, 159, 161, 204, 237, 246, 257, 272, 285, 288, 313, 337	40
12	18.90	128, 129, 161, 163, 165, 175, 204, 217, 242, 246, 257, 264, 285, 288,	20

Table 4. Continued

No.	Time /min	Ion /amu	Dwell time /ms
		303, 306, 313, 326, 335	
13	19.80	73, 89, 146, 162, 185, 212, 223, 227, 250, 265, 267	50
14	20.30	61, 144, 146, 162, 170, 185, 198, 199, 212, 213, 223, 227, 258	40
15	20.70	61, 103, 118, 144, 170, 181, 198, 199, 210, 217, 219, 222, 240, 254, 255	30
16	21.35	108, 117, 151, 160, 161, 170, 219, 221, 224, 225, 257, 267, 351, 353, 355	30
17	22.20	107, 108, 119, 123, 136, 145, 176, 219, 221, 246, 248, 263, 318, 351, 353, 355	20
18	22.70	77, 141, 165, 167, 174, 176, 206, 234, 239, 246, 248, 267, 268, 297, 299, 318	20
19	23.20	105, 123, 134, 161, 248, 250, 267, 297, 299	50
20	23.50	131, 143, 157, 161, 171, 220, 248, 250, 262, 296, 304, 329, 336, 338, 404	30
21	24.30	112, 130, 162, 168, 238, 262	90
22	25.10	112, 116, 130, 131, 162, 168, 206, 233, 234, 235, 238, 262	40
23	25.30	254, 282, 321, 323, 356, 383	90
24	26.00	131, 152, 206, 233, 234, 236, 251, 253, 315	50
25	26.90	149, 162, 176, 177, 190, 232, 268, 270, 328	50
26	27.90	105, 119, 120, 135, 140, 173, 266, 267, 269, 350, 394	50
27	28.80	105, 117, 123, 140, 145, 160, 183, 266, 267, 350, 394	50
28	29.00	117, 123, 127, 145, 154, 160, 183, 248, 350	50
29	29.60	116, 178, 186, 191, 219, 255	90
30	30.30	132, 162, 178, 184, 219, 226, 281, 293, 334	50
31	31.10	120, 136, 141, 147, 181, 183, 184, 192, 197, 247, 255, 289, 309, 364	30
32	32.00	112, 141, 147, 170, 183, 184, 255, 309, 364, 428, 447, 449	40
33	32.60	112, 141, 163, 170, 183, 376, 428, 447, 449	50
34	33.10	163, 165, 178, 181, 251, 279	90
35	33.80	157, 199, 451	200
36	34.70	181, 225, 250, 252, 419	100
37	35.40	259, 265, 287, 323, 325, 354	90
38	36.40	308, 318, 423	200
Group D			
1	5.50	110, 153, 154	200
2	8.00	153, 184, 212	200
3	11.00	139, 155, 211, 215, 250, 252	90
4	13.00	142, 156, 165, 171, 196, 197, 200, 201, 202	50
5	14.00	143, 155, 158, 167, 192, 203, 211, 220, 229, 231, 246	40
6	15.00	106, 142, 190, 237, 265, 280	90
7	16.00	108, 136, 145, 158, 164, 171, 173, 182, 186, 196, 197, 201, 211, 216, 213, 288	20
8	17.20	161, 174, 177, 197, 200, 202, 214, 229, 246, 357, 359, 394	40
9	17.90	89, 114, 128, 172, 173, 174, 175, 186, 189, 198, 223, 229, 230, 231, 233, 253, 256, 258, 263, 265, 268, 277, 282, 292, 297	10
10	19.20	142, 143, 154, 157, 162, 184, 185, 199, 200, 201, 202, 203, 214, 220, 229, 230, 247, 251, 252, 255, 263, 264, 270, 278, 285, 287, 292	10
11	20.00	153, 180, 197, 199, 200, 201, 202, 230, 239, 247, 251, 252, 266, 305, 308, 311, 343, 375, 380, 412	15
12	21.00	115, 184, 193, 195, 196, 198, 215, 221, 225, 250, 252, 263, 269, 276,	20

Table 4. Continued

No.	Time /min	Ion /amu	Dwell time /ms
		285, 297, 301, 332	
13	21.60	128, 170, 194, 195, 210, 212, 224, 225, 236, 254, 279, 294	40
14	22.10	129, 155, 182, 184, 200, 201, 210, 212, 216, 224, 225, 229, 230, 254, 262, 263, 291, 300, 314, 326, 351, 353, 355	10
15	23.00	136, 171, 199, 215, 230, 251, 253, 266, 289, 407, 409, 411	40
16	23.90	130, 148, 178, 187, 202, 211, 223, 224, 226, 240, 258, 267, 295, 299, 311, 313, 323	20
17	25.00	129, 130, 145, 148, 164, 168, 184, 185, 196, 201, 218, 219, 227, 254, 259, 290, 299, 326, 330, 340, 360	15
18	26.00	156, 159, 184, 185, 213, 218, 227, 229, 270, 272, 290, 360	40
19	27.10	143, 160, 171, 206, 222, 223, 224, 230, 238, 251, 266, 294, 312, 338, 349	30
20	28.00	136, 174, 186, 202, 215, 231, 233, 237, 254, 278, 279, 294, 310, 311, 326, 366, 379	20
21	29.00	136, 153, 192, 194, 220, 234, 276, 318, 324, 333, 359, 394	40
22	30.00	160, 161, 171, 173, 175, 214, 317, 375, 377	50
23	30.80	173, 175, 196, 213, 230, 274, 292, 300, 304, 316, 319, 320, 335, 373	30
24	32.40	147, 236, 238, 340, 341, 342	90
25	34.00	125, 129, 198	200
Group E			
1	5.50	110, 111, 152	200
2	7.00	103, 107, 121, 122, 136	100
3	9.00	94, 95, 141	200
4	10.40	160, 162, 164, 205, 206, 220	100
5	12.00	101, 157, 175	200
6	12.90	103, 104, 121, 125, 130, 136, 147, 153, 301	60
7	13.90	125, 156, 157, 171, 180, 209	100
8	14.80	109, 110, 111, 142, 145, 152, 159, 185, 213, 230, 370	50
9	16.80	98, 100, 126, 142, 145, 153, 160, 184, 187, 188, 189, 198, 213, 232, 234, 254, 261, 273, 318	30
10	17.95	98, 100, 124, 126, 130, 144, 145, 173, 176, 177, 187, 198, 206, 213, 225, 232, 240, 273	30
11	18.70	138, 166, 173, 238, 240, 255	100
12	19.20	109, 119, 120, 135, 138, 146, 153, 162, 173, 176, 182, 187, 201, 202, 205, 206, 219, 220, 222, 223, 227, 232, 259, 264, 265, 296	20
13	20.30	109, 121, 127, 135, 153, 163, 182, 195, 201, 202, 206, 249, 256, 268, 279, 286, 306, 354	30
14	20.90	121, 127, 138, 195, 206, 211, 221, 227, 249, 264, 268, 278, 279, 301, 327, 332, 363, 381	30
15	21.95	109, 187, 224, 242, 266, 295, 297, 299, 351, 353, 355	50
16	22.30	104, 107, 123, 135, 136, 144, 151, 172, 187, 209, 211, 212, 227, 240, 242, 255, 262, 294, 299, 363	35
17	23.30	140, 152, 189, 215, 227, 272, 243, 262, 272	50
18	24.00	112, 128, 149, 168, 182, 205, 207, 212, 221, 222, 223, 231, 236, 247, 264, 303, 335, 367	30
19	25.00	91, 112, 128, 136, 137, 144, 168, 173, 180, 185, 187, 191, 196, 200, 204, 215, 231, 232, 238, 256, 286, 290, 294, 296, 412	20
20	26.05	105, 125, 157, 177, 186, 191, 196, 225, 238, 300, 302, 313, 314, 330, 361	40

Table 4. Continued

No.	Time /min	Ion /amu	Dwell time /ms
21	26.90	116, 131, 186, 194, 204, 222, 225, 235, 237, 247, 270, 272, 294, 300, 307, 328, 351, 367, 369, 408, 447, 449	30
22	28.00	107, 123, 138, 151, 183, 192, 235, 260, 270, 272, 295, 312, 327, 340, 351, 367, 369, 376	30
23	28.60	108, 127, 141, 164, 165, 181, 205, 267, 288, 339, 349, 412	50
24	29.20	120, 125, 136, 137, 138, 164, 183, 185, 187, 192, 205, 206, 217, 218, 226, 227, 236, 240, 244, 246, 249, 299, 300, 330, 359, 372,	20
25	30.05	122, 136, 140, 148, 160, 185, 187, 199, 204, 206, 214, 226, 229, 232, 238, 244, 246, 250, 258, 266, 268, 285, 288, 290, 300, 317, 319, 320, 376	20
26	31.60	111, 132, 137, 144, 171, 186, 194, 199, 210, 226, 237, 247, 259, 263, 268, 274, 277, 291, 303, 312, 318, 325, 333, 346, 357, 360, 362, 442, 461, 476	20
27	33.00	126, 152, 166, 180, 184, 185, 186, 258, 286, 287, 331, 333	50
28	34.00	100, 119, 188	200
29	37.00	328, 329, 330	200

LC-MS-MS was performed with an AB API3200Q TRAP instrument equipped with ESI and a 150 mm×2.1 mm, 3 μm particle, Atlantis dC18 Column. Details of the mobile phase and flow rate are given in Table 5.

Table 5. Mobile phase and flow rate

Step	Time (min)	Flow rate /μL/min	Water /%	Acetonitrile /%
0	0.00	200	90.0	10.0
1	4.00	200	50.0	50.0
2	15.00	200	40.0	60.0
3	20.00	200	20.0	80.0
4	25.00	200	5.0	95.0
5	32.00	200	5.0	95.0
6	32.01	200	90.0	10.0
7	40.00	200	90.0	10.0

The column temperature was 40°C and the injection volume 20 μL. Positive ion scan mode was used with multiple reaction monitoring (MRM). The ion spray potential was 5,500 V, the nebulizer ,curtain and auxiliary warm gas pressures were 0.483 ,0.138 and 0.379 MPa, and the ion source temperature was 725°C. MRM transitions for precursor/product ion, Quantifying for precursor/product ion, Declustering potential, Collision energy and Collision cell exit potential see Table 6.

Table 6. MRM transitions for precursor/product ion, Quantifying for precursor/product ion, Declustering potential, Collision energy and Collision cell exit potential

No.	Compound	Retention time /min	Quantifyin g for precursor/ product ion	MRM transitions for precursor /product ion	Declus-tering potential /V	Collision energy /V	Collision cell exit potential /V
1	Pirimicarb	9.54	239.2/72.0	239.2/72.0; 239.2/182.2	36	35; 22	2; 2
2	Mevinphos	10.11 10.86	225.0/127.0	225.0/127.0; 225.0/193.1	18	23; 9	3; 2
3	Ethidimuron	10.23	265.1/208.1	265.1/208.1; 265.1/162.1	50	17; 30	3; 3
4	Imidacloprid	10.27	256.1/175.1	256.1/175.1; 256.1/209.1	25	26; 24	6; 5
5	Thiofanox sufone	10.52	251.2/ 57.1	251.2/ 57.1; 251.2/76.1	39	17; 17	3; 3
6	Secbumeton	10.63	226.2/170.1	226.2/170.1; 226.2/142.1	28	25; 32	6; 6
7	Simetryn	10.77	214.2/124.1	214.2/124.1; 214.2/96.1	39	27; 34	2; 3
8	Thiacloprid	11.41	253.1/126.1	253.1/126.1; 253.1/186.1	52	27; 18	2.5; 2.5
9	Methoprotryne	11.86	272.2/198.2	272.2/198.2; 272.2/170.1	42	30; 24	6; 5
10	Paraoxon methyl	12.04	248.0/202.1	248.0/202.1; 248.0/90.0	37	25; 35	2; 2
11	Tepraloxydim	12.17 18.29	342.2/250.2	342.2/250.2; 342.2/166.1	32	19; 31	3; 2
12	Spiroxamine	12.19	298.2/144.2	298.2/144.2; 298.2/100.1	38	45; 30	2; 2
13	Cyanazine	12.29	241.1/214.1	241.1/214.1; 241.1/104.0	33	22; 44	2; 2
14	Malaoxon	12.62	315.1/99.0	315.1/99.0; 315.1/127.0	19	36; 17	2; 2
15	Fenthion sulfoxide	12.77	295.1/280.0	295.1/280.0; 295.1/127.0	37	23; 40	5; 2
16	Cycluron	13.34	199.2/89.0	199.2/89.0; 199.2/72.0	43	20; 30	2; 2 .
17	Prometryne	13.37	242.2/158.1	242.2/158.1; 242.2/200.2	48	33; 25	2; 2
18	Chlorprifos methyl	13.53	339.3/283.1	339.3/283.1; 339.3/121.1	50	24; 41	4; 2
19	Ethoxyquin	13.64	218.3/174.2	218.3/174.2; 218.3/160.2	57	30; 39	2; 2
20	Rabenzazole	13.75	213.2/118.0	213.2/118.0; 213.2/157.1	40	50; 40	3; 2
21	Monolinuron	14.08	215.1/126.0	215.1/126.0; 215.1/148.1	31	25; 20	3; 2
22	Isoprocarb	14.53	194.1/95.0	194.1/95.0; 194.1/137.1	27	25; 13	3; 2
23	Propachlor	14.93	212.1/170.1	212.1/170.1; 212.1/94.1	31	19; 38	2; 2
24	Dimefuron	15.16	339.1/72.0	339.1/72.0; 339.1/167.1	35	21; 29	2; 6
25	Propham	15.23	180.1/138.0	180.1/138.0; 180.1/120.0	23	13; 22	4; 3
26	Clomazone dimethazone	15.46	240.1/125.0	240.1/125.0; 240.1/89.1	39	28; 61	3; 2
27	Triadimenol	15.56	296.1/70.0	296.1/70.0; 296.1/99.1	22	30; 21	3; 3
28	Paclobutrazol	15.79	294.2/70.0	294.2/70.0; 294.2/125.0	34	49; 50	2; 4
29	Cyprodinil	15.92	226.2/93.0	226.2/93.0; 226.2/118.0	56	52; 46	2; 2
30	Desmedipham	16.35	301.2/182.1	301.2/182.1; 301.2/136.1	27	12; 31	2; 2
31	Famphur	16.53	326.0/93.0	326.0/93.0; 326.0/217.0	42	45; 30	2; 2
32	Methidathion	16.75	303.0/145.1	303.0/145.1; 303.0/85.0	18	12; 29	4; 2
33	Fenarimol	16.99	331.0/268.1	331.0/268.1; 331.0/259.1	43	30; 31	9; 3
34	Iprovalicarb	17.05	321.1/119.0	321.1/119.0; 321.1/203.2	57	32; 12	3; 2
35	Dimethenamid	17.09	276.1/244.1	276.1/244.1; 276.1/168.1	22	25; 22	2; 2
36	Myclobutanil	17.31	289.1/70.0	289.1/70.0; 289.1/125.0	33	38; 45	2; 2
37	Isoxaflutole	17.54	360.0/251.1	360.0/251.1; 360.0/220.1	38	22; 51	2; 2
38	Etaconazole	17.57	328.1/159.1	328.1/159.1; 328.1/205.1	48	42; 20	2; 2
39	Diclobutrazole	18.51	328.0/70.0	328.0/70.0; 328.0/159.0	35	60; 54	2; 2
40	Napropamide	18.64	272.2/129.1	272.2/129.1; 272.2/171.2	29	21; 25	5; 2
41	Flusilazole	18.98	316.1/165.1	316.1/165.1; 316.1/247.1	31	40; 23	5; 8
42	Bitertanol	19.34 19.90	338.2/269.2	338.2/269.2; 338.2/99.1	38	12; 21	4; 3

Table 6. Continued

No.	Compound	Retention time /min	Quantifying for precursor/ product ion	MRM transitions for precursor /product ion	Declustering potential /V	Collision energy /V	Collision cell exit potential /V
43	Penconazole	19.39	284.1/159.0	284.1/159.0; 284.1/70.0	35	38; 37	2; 2
44	Flutolanil	20.13	324.2/262.1	324.2/262.1; 324.2/282.1	30	25; 17	3; 3
45	Acetochlor	20.26	270.2/148.2	270.2/148.2; 270.2/133.1	24	29; 45	2; 2
46	Fenitrothion	20.45	278.0/127.0	278.0/127.0; 278.0/264.0	48	28; 21	3; 2
47	Azinphos ethyl	20.93	346.0/261.1	346.0/261.1; 346.0/233.0	19	12; 18	4; 5
48	Edifenphos	21.25	311.1/109.0	311.1/109.0; 311.1/283.0	40	47; 17	3; 3
49	Isomethiozin	21.63	269.1/200.0	269.1/200.0; 269.1/172.1	33	30; 20	3; 2
50	Bromfenvinfos	21.97	402.9/155.0	402.9/155.0; 402.9/170.0	26	17; 55	2; 2
51	Benalyxyl	22.09	326.2/208.2	326.2/208.2; 326.2/148.1	28	20; 32	6; 3
52	Triflumuron	22.79	359.0/156.1	359.0/156.1; 359.0/139.0	24	25; 47	5; 2
53	Anilofos	23.12	368.0/199.0	368.0/199.0; 368.0/125.0	27	20; 45	3; 3
54	Pyraflufen ethyl	23.39	413.0/339.0	413.0/339.0; 413.0/289.1	59	28; 41	4; 4
55	Clodinafop propargyl	23.49	350.1/266.1	350.1/266.1; 350.1/238.1	36	21; 32	4; 2
56	Diazinon	23.70	305.1/169.1	305.1/169.1; 305.1/153.2	31	30; 29	2; 2
57	Dinitramine	23.75	323.0/305.0	323.0/305.0; 323.0/247.0	56	22; 28	2; 2
58	Prazophos	23.87	374.1/222.1	374.1/222.1; 374.1/194.2	59	31; 47	2; 3
59	Terbucarb	24.77	278.2/222.1	278.2/222.1; 278.2/166.1	32	11; 19	3; 2
60	Tolclofos methyl	24.80	301.2/175.1	301.2/175.1; 301.2/125.2	33	35; 25	7; 4
61	Haloxyfop-methyl	25.17	376.0/316.0	376.0/316.0; 376.0/288.0	45	24; 35	5; 3
62	Pretilachlor	25.32	312.1/252.1	312.1/252.1; 312.1/176.2	50	24; 39	3; 2
63	Diflufenican	25.42	395.0/266.2	395.0/266.2; 395.0/375.0	30	34; 20	3; 3
64	Indoxacarb	25.52	528.0/218.0	528.0/218.0; 528.0/249.0	45	32; 24	2; 4
65	Cloquintocet mexyl	25.55	336.1/238.1	336.1/238.1; 336.1/192.1	29	21; 39	3; 3
66	Quizalofop-ethyl	25.85	373.0/299.1	373.0/299.1; 373.0/271.0	50	25; 35	2; 2
67	Cis and trans diallate	26.01	270.0/109.0	270.0/109.0; 270.1/228.0	30	39; 17	2; 2
68	Butylate	26.26	218.1/57.0	218.1/57.0; 218.1/190.1	26	30, 12	2; 2
69	Tetramethirn	26.50 26.79	332.2/164.1	332.2/164.1; 332.2/135.1	18	30; 23	2; 2
70	Dioxathion	27.37	457.0/97.0	457.0/97.0; 457.0/115.0	60	66; 51	4; 4
71	Dichlormid	27.53 27.80	225.1/69.0	225.1/69.0; 225.1/105.0	72	32; 35	2; 18
72	Allethrin	27.01 27.53	303.2/135.1	303.2/135.1; 303.2/123.2	40	19; 15	2; 2
73	Hexythiazox	27.73	353.1/228.1	353.1/228.1; 353.1/271.1	38	21; 19	2; 10
74	Methothrin	27.83	320.2/123.1	320.2/123.1; 320.2/135.1	23	20; 26	2; 2
75	Profluralin	28.18	348.1/276.1	348.1/276.1; 318.1/330.1	27	14; 12	4; 4
76	Triallate	28.43	304.0/143.0	304.0/143.0; 304.0/86.0	50	39; 24	2; 2
77	Phthalic acid,dicyclobexyl ester	29.56	331.2/167.0	331.2/167.0; 331.2/149.1	20	18; 36	2; 2
78	Prothiophos	30.22	345.0/241.0	345.0/241.0; 345.0/259.0	21	25; 20	6; 3
79	Pirmicarb-desmethyl	2.29	225.1/72.0	225.1/72.0; 225.1/168.1	36	31; 19	2; 2
80	Acephate	2.30	184.1/143.0	184.1/143.0; 184.1/113.0	15	11; 30	2; 2
81	Carbendazim	2.31	192.1/160.1	192.1/160.1; 192.1/132.1	27	25; 41	2; 2
82	Vamidothion sulfone	2.43	320.1/178.1	320.1/178.1; 320.1/169.0	37	22; 27	2; 2
83	Clothianidin	10.14	250.2/132.0	250.2/132.0; 250.2/169.1	35	26; 19	3; 3
84	3-hydroxy carbofuran	10.17	238.2/163.1	238.2/163.1; 238.2/181.1	22	19; 14	2; 2

Table 6. Continued

No.	Compound	Retention time /min	Quantifying for precursor/ product ion	MRM transitions for precursor /product ion	Declustering potential /V	Collision energy /V	Collision cell exit potential /V
85	Fenuron	10.25	165.1/72.0	165.1/72.0; 165.1/120.0	31	31; 24	2; 2
86	Difenzoquat-methyl sulfate	10.10 10.45	249.1/130.0	249.1/130.0; 249.1/193.1	52	55; 39	3; 2
87	TEPP	11.27	291.1/179.0	291.1/179.0; 291.1/207.0	23	30; 20	2; 2
88	Desethyl-sebuthylazine	11.37	202.0/146.0	202.0/146.0; 202.0/104.1	43	25; 39	3; 2
89	Phosphamidon	11.39	300.1/174.1	300.1/174.1; 300.1/127.0	27	18; 31	3; 2
90	Imibenzonazole-des-benzyl	11.64	271.0/174.0	271.0/174.0; 271.0/70.0	55	38; 39	2; 3
91	Mephosfolan	11.67	270.1/140.1	270.1/140.1; 270.1/168.1	40	36; 26	2; 5
92	Isouron	11.97	212.2/72.0	212.2/72.0; 212.2/167.1	28	40; 22	2; 3
93	Chlormephos	12.35	235.0/153.0	235.0/153.0; 235.0/125.0	27	15; 20	5; 2
94	Cythioate	12.46	298.0/125.0	298.0/125.0; 298.0/217.1	58	33; 24	2; 5
95	Propoxur	12.85	210.1/168.1	210.1/168.1; 210.1/111.0	21	10; 21	4; 4
96	Triasulfuron	12.85	402.1/167.1	402.1/167.1; 402.1/141.1	32	26; 30	4; 2
97	Bendiocarb	12.92	224.1/109.0	224.1/109.0; 224.1/167.1	20	25; 12	3; 5
98	Flutriafol	13.19	302.1/123.0	302.1/123.0; 302.1/109.0	33	41; 49	2; 3
99	Thiofanox	13.59	219.3/57.1	219.3/57.1; 219.3/76.1	24	16; 16	3; 3
100	Mefenoxam	13.74	280.1/220.0	280.1/220.0; 280.1/192.1	40	18; 24	2; 2
101	Diuron	13.95	233.1/72.0	233.1/72.0; 233.1/160.1	32	39; 35	3; 2
102	Difenoxuron	14.07	287.1/123.1	287.1/123.1; 287.1/214.1	46	30; 32	2; 2
103	Methobromuron	14.60	259.0/170.1	259.0/170.1; 259.0/148.0	32	25; 19	4; 2
104	Demeton-s	14.78	259.1/89.0	259.1/89.0; 259.1/61.0	12	20; 25	2; 2
105	Dimethachloro	15.12	256.1/224.2	256.1/224.2; 256.1/148.2	27	18; 35	4; 2
106	Prothoate	15.12	286.1/199.0	286.1/199.0; 286.1/227.1	23	21; 14	3; 3
107	Propazine	15.82	230.1/188.1	230.1/188.1; 230.1/146.1	40	23; 30	2; 6
108	Disulfoton sulfone	15.85	307.0/97.0	307.0/97.0; 307.0/115.0	28	56; 43	2; 3
109	Phenmedipham	16.24	301.1/136.0	301.1/136.0; 301.1/168.1	30	27; 11	3; 6
110	Furalaxyl	16.40	302.2/242.2	302.2/242.2; 302.2/270.2	31	21; 15	4; 4
111	Terbuthylazine	16.48	230.1/174.1	230.1/174.1; 230.1/110.0	34	21; 42	2; 2
112	Flurtamone	16.67	334.1/247.1	334.1/247.1; 334.1/105.1	55	37; 45	3; 2
113	Bromuconazole	17.17 18.30	376.0/159.0	376.0/159.0; 376.0/173.0	37	38; 42	2; 2
114	Chlorimuron ethyl	17.44	415.0/186.1	415.0/186.1; 415.0/213.1	30	30; 21	2; 4
115	Molinate	17.82	188.1/126.1	188.1/126.1; 188.1/98.1	25	19; 26	2; 3
116	Fenhexamid	17.90	302.0/97.1	302.0/97.1; 302.0/142.2	48	31; 32	2; 3
117	Pyriftalid	17.92	319.0/139.1	319.0/139.1; 319.0/157.2	56	40; 34	4; 4
118	Chlorbufam	18.20	224.1/154.1	224.1/154.1; 224.1/172.1	16	24; 12	5; 3
119	Pronamide	18.47	256.1/190.1	256.1/190.1; 256.1/173.0	40	20; 28	2; 7
120	Isoxaben	18.88	333.1/165.0	333.1/165.0; 331.0/150.1	43	29; 55	2; 2
121	Ethofume sate	18.98	287.0/259.0	287.0/259.0; 287.0/121.0	50	15; 22	2; 2
122	Bifenazate	19.03	301.2/198.1	301.2/198.1; 301.2/170.1	25	14; 28	6; 3
123	Diphenylamin	19.08	170.0/152.0	170.0/152.0; 170.0/93.0		37; 37	6; 4
124	Flamprop methyl	19.10	336.1/105.1	336.1/105.1; 336.1/77.0	22	25; 73	2; 3
125	Iprobenfos	19.39	289.1/205.1	289.1/205.1; 289.1/91.0	20	12; 39	2; 3
126	Metconazole	19.74	320.2/70.0	320.2/70.0; 320.2/125.0	37	50; 55	2; 2
127	Tetrachlorvinphos	19.91	365.0/127.0	365.0/127.0; 365.0/204.0	32	18; 51	5; 8

Table 6. Continued

No.	Compound	Retention time /min	Quantifying for precursor/ product ion	MRM transitions for precursor /product ion	Declus-tering potential /V	Collision energy /V	Collision cell exit potential /V
128	Triazophos	20.15	314.1/162.1	314.1/162.1; 314.1/119.1	42	27; 48	4; 4
129	Propiconazole	20.46	342.1/159.1	342.1/159.1; 342.1/205.1	52	38; 20	6; 2
130	Neburon	21.10	275.1/88.1	275.1/88.1; 275.1/160.1	33	23; 41	2; 2
131	Fluthiacet methyl	21.81	404.0/274.0	404.0/274.0; 404.0/215.0	64	40; 55	2; 2
132	Aclonifen	22.07	265.1/248.0	265.1/248.0; 265.1/182.1	43	25; 40	4; 2
133	Buprofezin	22.72	306.2/116.1	306.2/116.1; 306.2/106.1	18	21; 41	2; 3
134	Triphenyl phosphate	23.39	327.1/152.1	327.1/152.1; 327.1/215.0	52	53; 35	4; 4
135	Parathion	23.41	292.1/236.2	292.1/236.2; 292.1/94.0	50	23; 30	8; 2
136	Etrimfos	23.47	293.1/125.0	293.1/125.0; 293.1/265.1	29	35; 21	2; 4
137	Flamprop isopropyl	23.61	364.1/105.1	364.1/105.1; 364.1/304.1	22	33; 13	2; 5
138	Pirimiphos methyl	23.92	306.2/108.1	306.2/108.1; 306.2/95.0	42	42; 50	3; 3
139	Cyanofenphos	24.37	304.0/157.0	304.0/157.0; 304.0/276.0	38	30; 16	2; 2
140	Phorate	24.82	261.0/75.0	261.0/75.0; 261.0/143.0	21	22; 25	2; 2
141	Cycloxydim	25.71	326.2/280.2	326.2/280.2; 326.2/180.2	21	17; 28	4; 2
142	Trifloxystrobin	25.82	409.3/206.2	409.3/206.2; 409.3/186.1	35	18; 25	3; 3
143	Clethodim	26.31	360.1/164.1	360.1/164.1; 360.1/206.2	25	28; 20	2; 2
144	Picolinafen	26.68	377.0/238.0	377.0/238.0; 377.0/145.0	56	35; 69	8; 6
145	Bioallethrin	27.14	303.1/135.1	303.1/135.1; 303.1/123.1	35	20; 25	2; 2
146	Pyriproxyfen	27.28	322.1/96.0	322.1/96.0; 322.1/227.1	25	25; 18	2; 3
147	Fenazaquin	28.10	307.2/161.2	307.2/161.2; 307.2/147.1	37	24; 29	2; 2
148	Trifluralin	28.10	336.0/252.1	336.0/252.1; 336.0/294.1	48	27; 18	2; 3
149	Propargite	28.64	368.1/175.1	368.1/175.1; 368.1/81.0	33	23; 46	2; 2
150	DEF	29.93	315.1/169.0	315.1/169.0; 315.1/113.0	50	21; 34	2; 2
151	Bioresmethrin	30.12	339.2/171.1	339.2/171.1; 339.2/143.1	33	21; 35	2; 2
152	Simeton	9.29	198.1/128.1	198.1/128.1; 198.1/100.1	33	28; 37	2; 2
153	Trichlorphon	9.78	257.0/127.1	257.0/127.1; 257.0/109.0	28	23; 25	2; 2
154	Phorate sulfoxide	9.87	261.1/111.0	261.1/111.0; 261.1/129.0	27	34; 27	3; 2
155	Brompyrazon	10.53	266.0/92.0	266.0/92.0; 266.0/104.0	46	41; 35	3; 3
156	Acetamiprid	10.65	223.2/126.0	223.2/126.0; 223.2/90.0	28	30; 48	2; 3
157	Prometon	10.68	226.2/184.1	226.2/184.1; 226.2/142.0	35	25; 31	2; 2
158	Imazamethabenz-methyl	10.87	289.1/86.0	289.1/86.0; 289.1/229.0	33	35; 25	3; 1
159	Fenamiphos sulfoxide	11.06	320.1/292.1	320.1/292.1; 320.1/171.1	37	20; 30	3; 5
160	Oxycarboxin	11.37	268.0/175.0	268.0/175.0; 268.0/147.1	35	21; 32	2; 2
161	Metoxuron	11.55	229.1/72.0	229.1/72.0; 229.1/156.1	41	39; 34	2; 2
162	Bromacil	11.78	261.0/205.0	261.0/205; 261.0/188.0	30	23; 39	2; 6
163	Monuron	12.00	199.0/126.0	199.0/126.0; 199.0/99.0	25	39; 55	2; 2
164	Thifensulfuron-methyl	12.21	388.1/167.1	388.1/167.1; 388.0/205.1	24	25; 35	2; 3
165	Fenamiphos sulfone	12.30	336.1/266.2	336.1/266.2; 336.1/188.2	45	26; 36	4; 2
166	Quinoclamine	12.36	208.1/105.0	208.1/105.0; 208.1/154.1	49	36; 35	3; 2
167	Disulfoton-sulfoxide	13.12	291.0/129.0	291.0/129.0; 291.0/97.0	34	41; 49	2; 2
168	Pyrimethanil	13.27	200.2/107.0	200.2/107.0; 200.2/168.1	41	34; 40	2; 2
169	Thiometon	13.50	247.1/89.0	247.1/89.0; 247.1/61.0	8	19; 20	3; 2
170	Metalaxyl	13.76	280.1/220.2	280.1/220.2; 280.1/192.2	35	19; 25	2; 2
171	Diethyltoluamide	13.82	192.2/119.0	192.2/119.0; 192.2/91.0	28	25; 42	4; 2
172	Ofurace	13.91	282.1/254.2	282.1/254.2; 282.1/160.2	26	15; 31	3; 2

Table 6. Continued

No.	Compound	Retention time /min	Quantifying for precursor/ product ion	MRM transitions for precursor /product ion	Declustering potential /V	Collision energy /V	Collision cell exit potential /V
173	Fensulfothin	14.20	309.0/157.1	309.0/157.1; 309.0/173.0	35	33; 31	3; 2
174	Dimethametryn	14.69	256.2/186.1	256.2/186.1; 256.2/96.1	36	28; 20	2; 2
175	Demeton(O+S)	14.76	259.1/89.0	259.1/89.0; 259.1/61.0	14	26; 30	2; 3
176	Naled	14.80	378.9/127.1	378.9/127.1; 378.9/109.0	32	16; 55	3; 4
177	Propanil	15.77	218.0/162.1	218.0/162.1; 218.0/127.0	45	23; 41	2; 2
178	Triticonazole	16.10	318.2/70.0	318.2/70.0; 318.2/125.1	28	42; 50	2; 3
179	Fluridone	16.22	330.1/309.1	330.1/309.1; 330.1/259.2	54	47; 61	3; 5
180	Prochloraz	16.31	376.1/308.0	376.1/308.0; 376.1/266.0	31	17; 22	5; 4
181	Methiocarb	16.34	226.2/121.1	226.2/121.1; 226.2/169.1	19	25; 12	2; 2
182	Methfuroxam	16.79	230.2/137.1	230.2/137.1; 230.2/111.1	32	27; 20	3; 2
183	Benoxacor	17.35	260.0/149.2	260.0/149.2; 260.0/134.1	40	23; 41	4; 2
184	Ethoxysulfuron	17.42	399.2/261.1	399.2/261.1; 399.2/218.1	60	21; 33	3; 3
185	Epoxiconazole	14.82 17.84	330.1/121.1	330.1/121.1; 330.1/141.1	32	32; 25	2; 2
186	Fluquinconazole	18.39	376.0/349.1	376.0/349.1; 376.0/307.0	53	26; 32	9; 4
187	Mepanipyrim	18.86	224.2/77.0	224.2/77.0; 224.2/106.0	40	55; 36	2; 2
188	Methoxyfenozide	19.31	369.2/149.1	369.2/149.1; 369.2/313.1	19	25; 10	5; 5
189	Mepronil	19.48	270.2/119.1	270.2/119.1; 270.2/228.2	28	31; 20	2; 3
190	Fenoxycarb	20.03	302.1/116.0	302.1/116.0; 302.1/88.0	26	17; 31	2; 2
191	Thenylchlor	20.23	324.1/127.0	324.1/127.0; 324.1/99.0	30	24; 45	2; 2
192	Ditalimfos	20.34	300.0/148.1	300.0/148.1; 300.0/130.1	40	19; 42	2; 2
193	Fluorochloridone	20.52	312.1/292.1	312.1/292.1; 312.1/212.2	49	28; 43	5; 3
194	Pyrimitate	21.75	306.1/170.2	306.1/170.2; 306.1/154.2	35	27; 28	3; 2
195	Bromobutide	21.83	312.2/194.1	312.2/194.1; 312.2/119.1	35	18; 26	2; 2
196	Monalide	20.15 21.98	240.1/85.1	240.1/85.1; 240.1/93.0	39	27; 53	2; 3
197	Kresoxim-methyl	22.29	314.1/235.1	314.1/235.1; 314.1/222.2	15	20; 18	3; 2
198	Quinalphos	22.37	299.1/163.1	299.1/163.1; 299.1/147.1	34	33; 31	2; 2
199	Fenthion	23.25	279.0/169.1	279.0/169.1; 279.0/105.1	30	24; 36	3; 4
200	Benzoylprop-ethyl	23.71	366.0/105.0	366.0/105.0; 366.0/320.0	25	30; 13	2; 4
201	Fonofos	24.31	247.1/109.0	247.1/109.0; 247.1/137.1	15	28; 15	2; 4
202	Coumaphos	24.39	363.1/227.2	363.1/227.2; 363.1/307.1	63	34; 25	4; 4
203	Phosalone	24.93	368.1/182.0	368.1/182.0; 368.1/111.0	31	23; 59	2; 2
204	Phoxim	24.95	299.0/97.0	299.0/97.0; 299.0/129.0	36	35; 18	2; 3
205	Piperophos	25.17	354.1/171.0	354.1/171.0; 354.1/143.0	26	30; 43	3; 2
206	Chlorphoxim	25.50	333.0/97.0	333.0/97.0; 333.0/163.1	22	34; 20	2; 2
207	MCPA-butoxyethyl ester	25.95	301.0/227.1	301.0/227.1; 301.0/155.1	43	8; 30	5; 2
208	Piperonyl butoxide	26.55	356.2/177.1	356.2/177.1; 356.2/119.0	15	19; 50	2; 2
209	Dithiopyr	26.62	402.0/354.0	402.0/354.0; 402.0/296.0	38	24; 35	5; 2
210	Flumiclorac-pentyl	27.00	441.1/308.0	441.1/308.0; 441.1/252.1	45	36; 51	3; 3
211	Oxyflurofen	27.21	362.0/316.1	362.0/316.1; 362.0/237.1	32	18; 30	5; 2
212	Butachlor	27.25	312.2/238.1	312.2/238.1; 312.2/147.2	22	13; 46	2; 2
213	Iodofenphos	27.38	412.9/125.0	412.9/125.0; 412.9/287.0	43	30; 37	2; 10
214	Temephos	27.61	467.0/419.0	467.0/419.0; 467.0/341.0	40	27; 41	5; 4
215	Pendimethalin	27.85	282.0/212.0	282.0/212.0; 282.0/194.1	23	17; 25	2; 2
216	Fenpyroximate	28.38	422.2/366.2	422.2/366.2; 422.2/214.2	33	23; 38	6; 5

Table 6. Continued

No.	Compound	Retention time /min	Quantifying for precursor/ product ion	MRM transitions for precursor /product ion	Declus-tering potential /V	Collision energy /V	Collision cell exit potential /V
217	Pyrethrin	28.70	329.1/133.1	329.1/133.1; 329.1/143.1	39	25; 26	2; 5
218	Fenpropathrin	29.20	350.2/125.1	350.2/125.1; 350.2/97.1	35	23; 42	2; 2
219	Pyridaben	29.62	365.1/147.1	365.1/147.1; 365.1/309.1	65	37; 19	2; 3
220	Aspon	29.74	379.1/115.0	379.1/115.0; 379.1/97.0	27	46; 82	4; 2
221	Spirodiclofen	29.89	411.1/71.0	411.1/71.0; 411.1/313.1	48	29; 17	2; 5
222	Phenothrin	30.78 30.98	351.1/183.2	351.1/183.2; 351.1/333.2	45	30; 14	3; 5
223	Ethirimol	2.36	210.2/140.1	210.2/140.1; 210.2/98.0	46	30; 39	2; 2
224	Dicrotophos	9.23	238.1/112.1	238.1/112.1; 238.1/127.0	18	18; 24	2; 2
225	Metamitron	10.02	203.1/175.1	203.1/175.1; 203.1/104.0	33	22; 35	5; 3
226	Crimidine	10.04	172.1/136.2	172.1/136.2; 172.1/107.1	37	27; 30	4; 5
227	Dimethoate	10.55	230.0/199.0	230.0/199.0; 230.0/125.0	19	11; 28	3; 1.5
228	Terbumeton	10.73	226.2/170.1	226.2/170.1; 226.2/114.0	34	25; 34	3; 4
229	Desmetryn	10.79	214.1/172.1	214.1/172.1; 214.1/82.1	48	23; 45	4; 3
230	Imazalil	11.03	297.0/159.0	297.0/159.0; 297.0/255.0	45	36; 25	2; 3
231	Tebuthiuron	11.32	229.2/172.2	229.2/172.2; 229.2/116.0	37	25; 37	2; 3
232	Hexazinone	11.43	253.2/171.1	253.2/171.1; 253.2/85.1	32	23; 45	3; 3
233	Carbetamide	11.50	237.1/192.1	237.1/192.1; 237.1/118.1	18	11; 18	2; 4
234	Dodemorph	11.58 11.76	282.3/116.1	282.3/116.1; 282.3/98.1	29	30; 40	2; 2
235	Triamiphos	11.90	295.1/135.1	295.1/135.1; 295.1/110.0	47	34; 55	2; 5
236	Chlorotoluron	13.30	213.1/72.0	213.1/72.0; 213.1/140.1	29	38; 33	3; 3
237	Carbaryl	13.50	202.1/145.1	202.1/145.1; 202.1/127.1	20	15; 40	2; 2
238	Fluometuron	13.51	233.1/72.0	233.1/72.0; 233.1/160.0	29	40; 39	2; 4
239	Fosthiazate	13.51	284.1/104.0	284.1/104.0; 284.1/228.1	22	34; 13	2; 3
240	Isoproturon	13.84	207.2/72.0	207.2/72.0; 207.2/165.1	27	35; 19	2; 2
241	Carboxin	13.90	236.1/143.1	236.1/143.1; 236.1/124.0	37	22; 31	2; 2
242	Fenfuram	14.24	202.1/109.0	202.1/109.0; 202.1/120.0	38	34; 21	2; 2
243	Norflurazon	14.39	304.0/284.1	304.0/284.1; 304.0/160.1	66	27; 41	6; 5
244	3,4,5-Trimethacarb	14.65	194.2/137.2	194.2/137.2; 194.2/122.2	18	16; 37	2; 4
245	Nuarimol	14.99	315.1/81.0	315.1/81.0; 315.1/252.1	50	47; 31	2; 3
246	Diphenamid	15.22	240.1/134.1	240.1/134.1; 240.1/167.1	35	32; 31	2; 6
247	Buturon	15.52	237.1/84.1	237.1/84.1; 237.0/126.1	39	24; 40	2; 2
248	Thionazin	15.74	249.1/97.0	249.1/97.0; 249.1/221.2	22	43; 12	3; 2
249	Benodanil	15.82	324.1/231.0	324.1/231.0; 324.1/203.0	50	30; 47	6; 2
250	Fenpiclonil	16.09	237.0/202.1	237.0/202.1; 237.0/140.2	44	28; 53	3; 2
251	Cyproconazole	16.22	292.1/70.0	292.1/70.0; 292.1/125.0	38	45; 42	2; 4
252	Fenobucarb	16.34	208.2/95.0	208.2/95.0; 208.2/152.1	36	26; 12	2; 2
253	Simeconazole	16.96	294.2/70.1	294.2/70.1; 294.2/135.1	59	39; 30	2; 4
254	Anilazine	16.98	275.1/155.2	275.1/155.2; 275.1/169.0	33	34; 33	3; 6
255	Crufomate	17.26	292.1/236.0	292.1/236.0; 292.1/108.1	46	24; 40	4; 4
256	Aziprotryne	17.53	226.1/156.1	226.1/156.1; 226.1/198.1	20	19; 13	3; 2
257	Azoxystrobin	17.70	404.0/372.0	404.0/372.0; 404.0/344.1	32	20; 33	6; 3
258	Triadimefon	18.00	294.2/197.1	294.2/197.1; 294.2/225.2	29	20; 17	2; 2
259	Ethoprophos	18.08	243.1/131.0	243.1/131.0; 243.1/97.0	33	27; 46	3; 3
260	Pyridaphenthion	18.32	341.1/189.2	341.1/189.2; 341.1/205.2	46	28; 26	2; 3
261	Coumatetralyl	18.63	293.2/175.1	293.2/175.1; 293.2/91.2	58	29; 56	3; 3

Table 6. Continued

No.	Compound	Retention time /min	Quantifying for precursor/ product ion	MRM transitions for precursor /product ion	Declus-tering potential /V	Collision energy /V	Collision cell exit potential /V
262	Fenbuconazole	19.16	337.1/125.0	337.1/125.0; 337.1/70.0	36	45; 43	2; 3
263	Trietazine	19.22	230.1/99.1	230.1/99.1; 230.1/132.1	40	32; 30	3; 2
264	Isoprothiolane	19.82	291.1/189.1	291.1/189.1; 291.0/231.1	13	28; 15	2; 3
265	Tebutam	19.91	234.2/91.1	234.2/91.1; 234.2/192.2	54	40; 21	1.5; 3
266	Flufenacet	20.53	364.0/194.0	364.0/194.0; 364.0/152.0	11	17; 27	7; 6
267	Propetamphos	20.66	282.1/156.1	282.1/156.1; 282.1/110.0	30	15; 40	2; 3
268	Tebufenozide	21.36	353.1/133.1	353.1/133.1; 353.1/297.1	46	30; 11	2; 2
269	Chlorfenvinphos	21.36	359.0/99.0	359.0/99.0; 359.0/155.1	27	45; 18	3; 2
270	Mecarbam	21.70	330.0/199.0	330.0/199.0; 330.0/171.0	19	20; 24	7; 7
271	Difenoconazole	22.50	406.1/251.0	406.1/251.0; 406.1/337.0	40	35; 23	4; 5
272	Tri-iso-butyl phosphate	22.60 23.01	267.1/155.1	267.1/155.1; 267.1/211.2	24	12; 9	2; 2
273	Pyraclofos	22.98	361.1/257.0	361.1/257.0; 361.0/275.0	43	30; 28	5; 2
274	Bensulide	23.45	398.0/158.1	398.0/158.1; 398.0/314.0	21	33; 14	2; 5
275	Sulfallate	23.55	224.1/88.2	224.1/88.2; 224.1/116.1	25	30; 17	2; 1.5
276	Dacthal	23.87	331.0/299.0	331.0/299.0; 331.0/221.0	57	25; 45	8; 1.5
277	Thiobencarb	24.08	258.1/125.0	258.1/125.0; 258.1/89.0	31	31; 71	3; 2
278	Mefenpyr-diethyl	24.35	373.0/160.0	373.0/160.0; 373.0/186.0	25	45; 50	3; 1.5
279	Pencycuron	24.49	329.2/125.0	329.2/125.0; 329.2/218.1	35	40; 20	1.5; 3
280	Sulfotep	24.66	323.0/115.0	323.0/115.0; 323.0/171.1	25	41; 20	2; 2
281	Profenefos	25.28	373.0/302.9	373.0/302.9; 373.0/345.0	34	24; 17	3; 5
282	4,4-dichloro benzophenone	25.43 26.42	251.1/139.0	251.1/139.0; 251.1/111.1	32	25; 52	2; 4
283	Isofenphos	25.69	346.1/121.0	346.1/121.0; 346.1/245.0	18	54; 18	2; 3
284	Clomeprop	25.88	324.1/120.1	324.1/120.1; 324.1/203.1	47	27; 21	4; 2
285	Kadethrin	26.93	397.1/171.1	397.1/171.1; 397.1/143.1	33	19; 35	3; 1.5
286	Sethoxydim	27.09	328.2/178.1	328.2/178.1; 328.2/282.2	40	28; 16	2; 3
287	Terbufos	27.42	289.0/103.1	289.0/103.1; 289.0/187.0	20	14; 19	4; 2
288	Lactofen	27.59	462.1/344.0	462.1/344.0; 462.1/223.1	64	17; 42	14; 9
289	Tebupirimfos	27.66	319.1/153.2	319.1/153.2; 319.1/277.1	34	42; 20	1.5; 4
290	Chlorsulfuron	13.06	358.0/141.1	358.0/141.1; 358.0/167.0	23	28; 27	2; 3
291	Ethion	28.17	385.0/143.0	385.0/143.0; 385.0/199.1	19	35; 12	3; 2
292	Sulprofos	28.31	323.0/219.1	323.0/219.1; 323.0/155.0	42	24; 32	2; 1.5
293	Butralin	28.62	296.1/240.1	296.1/240.1; 296.1/222.1	20	19; 30	4; 4
294	Bromophos-ethyl	29.86	393.0/337.0	393.0/337.0; 393.0/162.1	35	25; 52	3; 2
295	DE-PCB 52	32.22	308.2/72.0	308.2/72.0; 308.2/114.0	49	49; 30	3; 3
296	Thiabendazole	2.32	202.1/175.1	202.1/175.1; 202.1/131.1	43	35; 45	3; 2
297	2,6-dichlorobenzamide	9.55	190.0/173.0	190.0/173.0; 190.0/145.0	44	31; 36	2; 6
298	Atratone	10.03	212.2/170.2	212.2/170.2; 212.2/100.1	30	28; 36	6; 2
299	Isocarbamid	10.11	186.2/87.0	186.2/87.0; 186.2/130.1	25	25; 16	3; 2
300	Chloridazon	10.33	222.1/104.0	222.1/104.0; 222.1/92.0	43	31; 37	2; 3
301	Dioxacarb	10.54	224.1/123.1	224.1/123.1; 224.1/167.1	15	22; 11	3; 3
302	Tricyclazole	10.71	190.1/136.1	190.1/136.1; 190.1/163.1	42	38; 31	2; 2
303	Phosfolan	10.87	256.1/140.0	256.1/140.0; 256.1/168.0	33	33; 20	5; 2
304	Triazoxide	11.54	248.0/150.1	248.0/150.1; 248.0/124.0	47	48; 45	2; 2
305	Ametryn	11.90	228.2/96.1	228.2/96.1; 228.2/158.1	35	36; 32	2; 5

Table 6. Continued

No.	Compound	Retention time /min	Quantifying for precursor/ product ion	MRM transitions for precursor /product ion	Declustering potential /V	Collision energy /V	Collision cell exit potential /V
306	Allidochlor	12.09	174.1/98.1	174.1/98.1; 174.1/106.0	27	14; 15	3; 2
307	Fenpropimorph	12.09	304.3/147.2	304.3/147.2; 304.3/119.1	45	39; 53	2; 2
308	Azamethiphos	12.22	325.0/183.0	325.0/183.0; 325.0/112.0	26	21; 51	3; 2
309	Lenacil	12.35	235.1/153.1	235.1/153.1; 235.1/136.0	26	25; 46	2; 2
310	Cinosulfuron	12.50	414.1/183.1	414.1/183.1; 414.1/157.1	39	28; 34	2; 3
311	Thiodicarb	12.53	355.1/108.0	355.1/108.0; 355.1/88.0	18	24; 30	2; 2
312	Metribuzin	12.56	215.1/187.2	215.1/187.2; 215.1/171.1	37	23; 27	3; 5
313	Pyrifenox	12.97	295.0/93.1	295.0/93.1; 295.0/263.0	43	42; 25	2; 2
314	Terbutryn	13.45	242.2/186.1	242.2/186.1; 242.2/91.0	35	25; 37	2; 2
315	DMST	13.61	215.3/106.1	215.3/106.1; 215.3/151.2	35	19; 11	3; 3
316	Azaconazole	13.72	300.1/159.0	300.1/159.0; 300.1/231.1	35	40; 23	2; 4
317	Atrazine	13.79	216.2/174.2	216.2/174.2; 216.2/104.0	33	24; 40	2; 3
318	Cyprazine	13.80	228.2/186.1	228.2/186.1; 228.2/108.1	48	24; 37	2; 3
319	Heptanophos	14.31	251.0/127.0	251.0/127.0; 251.0/89.0	30	20; 49	2; 2
320	Dipropetryn	14.90	256.1/144.1	256.1/144.1; 256.1/172.1	39	37; 30	4; 2
321	Metazachlor	14.91	278.1/134.1	278.1/134.1; 278.1/210.1	15	21; 14	3; 2
322	Dimethomorph	15.14 15.56	388.1/301.1	388.1/301.1; 388.1/165.1	40	28; 45	3; 2
323	Sebutylazine	15.60	230.1/174.1	230.1/174.1; 230.1/104.0	37	25; 42	2; 2
324	Bupirimate	15.92	317.2/108.0	317.2/108.0; 317.2/210.2	37	36; 30	4; 2
325	Phorate sulfone	16.13	293.0/143.1	293.0/143.1; 293.0/171.0	31	24; 14	2; 4
326	Pyrazosulfuron-ethyl	16.83	415.1/182.1	415.1/182.1; 415.1/369.1	26	31; 17	2; 3
327	Diethofencarb	16.89	268.1/152.1	268.1/152.1; 268.1/180.2	17	29; 22	3; 3
328	Azinphos-methyl	16.90	318.1/132.1	318.1/132.1; 318.1/261.1	19	21; 10	2; 10
329	Linuron(a)	16.92	249.0/160.1	249.0/160.1; 249.0/182.1	32	25; 21	2; 2
330	Fenamiphos	17.10	304.1/217.1	304.1/217.1; 304.1/202.0	30	29; 46	7; 2
331	Uniconazole	17.32	292.1/70.0	292.1/70.0; 292.1/125.0	44	47; 46	2; 2
332	Phosmet	17.51	318.0/160.1	318.0/160.1; 318.0/133.0	22	20; 48	5; 4
333	Mefenacet	17.85	299.1/120.1	299.1/120.1; 299.1/148.1	23	36; 23	2; 2
334	Tetraconazole	18.35	372.0/159.0	372.0/159.0; 372.0/150.1	41	44; 46	6; 4
335	Tebuconazole	18.46	308.2/70.0	308.2/70.0; 308.2/125.0	57	45; 55	2; 3
336	Propyzamide	18.50	256.0/190.0	256.0/190.0; 256.0/173.0	36	20; 32	2; 2
337	Methyl-parathion	18.67	264.1/232.1	264.1/232.1; 264.1/250.1	33	20; 18	3; 5
338	Hexaconazole	19.06	314.1/70.1	314.1/70.1; 314.1/159.1	34	45; 45	3; 6
339	Metolachlor	19.93	284.1/252.1	284.1/252.1; 284.1/176.2	34	22; 36	3; 2
340	Malathion	19.93	331.0/127.1	331.0/127.1; 331.0/99.0	30	18; 35	2; 2
341	Alachlor	20.17	270.2/238.2	270.2/238.2; 270.2/162.2	29	14; 25	3; 2
342	Diniconazole	20.31	326.1/70.0	326.1/70.0; 326.1/159.0	34	50; 45	3; 3
343	Furmecyclox	20.99	252.2/170.1	252.2/170.1; 252.2/110.1	23	17; 31	3; 2
344	Athidathion	21.30	331.0/145.1	331.0/145.1; 331.0/85.0	17	9; 35	6; 3
345	Isazofos	21.72	314.1/120.1	314.1/120.1; 314.1/162.1	32	40; 22	2; 2
346	Nitralin	22.39	346.1/304.1	346.1/304.1; 346.1/262.0	27	20; 25	4; 2
347	Dibutyl succinate	22.60	231.1/157.1	231.1/157.1; 231.1/101.0	25	10; 21	2; 2
348	Picoxystrobin	22.97	368.2/145.1	368.2/145.1; 368.2/205.2	21	27; 12	3; 3
349	Tri-n-butyl phosphate	23.07	267.2/99.0	267.2/99.0; 267.2/155.1	41	24; 14	3; 2
350	Cadusafos	23.29	271.1/159.1	271.1/159.1; 271.1/97.0	27	14; 36	2; 3
351	Phenthoate	23.56	321.1/135.1	321.1/135.1; 321.1/163.1	20	26; 15	2; 2

Table 6. Continued

No.	Compound	Retention time /min	Quantifying for precursor/ product ion	MRM transitions for precursor /product ion	Declus-tering potential /V	Collision energy /V	Collision cell exit potential /V
352	Cartap hydrochloride	23.87	238.0/210.1	238.0/210.1; 238.0/117.0	30	25; 30	2; 3
353	Alpha-hch	23.93	306.1/164.1	306.1/164.1; 306.1/108.1	40	28; 41	2; 2
354	Pyraclostrobin	24.10	388.0/194.0	388.0/194.0; 388.0/163.0	6	19; 29	7; 7
355	Thiazopyr	24.23	397.1/335.1	397.1/335.1; 397.1/275.1	48	42; 51	5; 9
356	Pebulate	24.54	204.2/128.1	204.2/128.1; 204.2/176.2	27	15; 15	4; 4
357	Prohydrojasmon	24.55	255.1/97.1	255.1/97.1; 255.1/195.1	35	36; 17	3; 2
358	Cycloate	24.60	216.2/83.0	216.2/83.0; 216.2/134.1	28	24; 18	2; 2
359	Prallethrin	25.24	301.1/105.1	301.1/105.1; 301.1/133.1	27	29; 17	2; 2
360	Diofenolan	25.27	301.2/253.2	301.2/253.2; 301.2/265.2	35	10; 13	2; 7
361	Methiocarb sulfoxid	25.59 35.15	259.2/101.0	259.2/101.0; 259.2/111.1	20	30; 26	2; 2
362	Epn	25.76	324.0/157.1	324.0/157.1; 324.0/296.0	41	34; 19	4; 2
363	Prosulfocarb	25.88	252.1/91.0	252.1/91.0; 252.1/128.1	29	38; 15	2; 2
364	Benfuracarb	26.81	411.1/195.2	411.1/195.2; 411.1/252.1	21	34; 19	3; 4
365	Fluazifop butyl	27.34	384.1/282.1	384.1/282.1; 384.1/328.1	43	28; 22	11; 5
366	Flubenzimine	27.18	417.1/187.3	417.1/187.3; 417.1/167.3	57	36; 33	3; 7
367	Tralkoxydim	27.66	330.2/138.1	330.2/138.1; 330.2/284.2	38	28; 15	5; 9
368	Pyributicarb	27.86	331.1/181.1	331.1/181.1; 331.1/190.1	40	23; 24	2; 2
369	Benfluralin	28.12	336.0/57.0	336.0/57.0; 336.0/236.0	45	30; 20	3; 3
370	Chlorthiophos	28.57	361.0/305.0	361.0/305.0; 361.0/225.1	41	21; 29	4; 3
371	Isopropalin	29.67	310.2/226.0	310.2/226.0; 310.2/208.1	30	27; 31	2; 2
372	Etofenprox	31.11	394.2/135.2	394.2/135.2; 394.2/359.2	11	35; 14	2; 6

Gel Permeation Chromatography was performed with a Gilson GPC instrument equipped with Bio-Beads S-X3 (400mm×25mm) cleanup column. Mobil phase: cyclohexane+ ethyl acetate (1+1); flow rate: 5 mL / min; detection wavelengh: 254 nm; injection volume: 5 mL; commenced collecting time: 22 min; stopped collecting time: 40 min.

Accelerated solvent extraction was performed with a Dionex ASE 300 equipped with 34-mL cell. Other equipment included homogenizer, rotary evaporator, analytical balances capable of weighing to 0.1 mg or to 0.01 g, 200-mL pear-shaped flasks, a 1-mL pipette, and a grinder.

Extraction

1. For GC-MS Pesticides

Test sample **10 g** (accurate to 0.01 g) and 10 g Celite 545 were weighed, mixed thoroughly, placed in a 34-mL cell of the accelerated solvent extractor, and extracted with acetonitrile by heating for 5 min, then in the static state for 3 min, then by cycling twice at 1,500 psi and 80°C. The sample was then rinsed with acetonitrile (60% of cell volume) and purged with nitrogen for 100 s. The extracts were then mixed thoroughly. Half of the extracts (corresponding to 5 g test sample) were concentrated to **ca 2mL** for cleanup.

2. For LC-MS-MS Pesticides

Weigh 10 g (accurate to 0.01 g) of the test sample into a centrifuge tube filled with 10 g anhydrous sodium sulfate. Add 30 mL mixed solvent of cyclohexane + ethyl acetate (1+1), and homogenize in homogenizer for 1 min, shake on shaker for 10min, and then centrifuge for 3 min. The supernatants are passed through glass funnel containing anhydrous sodium sulfate and collected in pear–shaped flask. Residues with 30 mL cyclohexane–ethyl acetate (1+1) were extracted twice, the extracts are combined in pear–shaped flask, then placed in a waterbath of 40^{0}C and evaporated to ca 2mL on a rotary evaporator for cleanup.

Cleanup

1. For GC-MS Pesticides

The Envi-18 cartridge was fixed to a support and conditioned with 10 mL acetonitrile before addition of the sample. A pear-shaped flask was connected to the cartridge and the extracts obtained as described above were decanted on to the cartridge. The cartridge was eluted with 15 mL acetonitrile and the eluate was concentrated to ca 1 mL by rotary evaporation at 40°C. Sodium sulfate (ca 2 cm) was placed on top of an Envi- Carb cartridge and this was connected in series to the top of **a** Sep-Pak NH_2 cartridge. The cartridges were fixed to a support and conditioned with 4 mL acetonitrile+toluene (3+1). When the conditioning solution reached the top of the sodium sulfate the cartridges were connected to a pearshaped flask and the concentrated sample obtained as described above was added to the cartridges. The pearshaped flask was rinsed with 3×2 mL acetonitrile+toluene(3+1), and the washings were also applied to the cartridges. A reservoir was attached to the cartridges and the pesticides were eluted with 25 mL **acetonitrile+toluene (3+1)**. The eluate was concentrated to ca 0.5 mL by rotary evaporation at 40°C, exchanged with hexane (2×5 mL) and diluted to ca 1 mL. Internal standard solution (40 μL) was added in the solution, after thorough mixing, was ready for GC-MS analysis.

2. For LC-MS-MS Pesticides

Transfer the concentrated extracts to a 10 mL volumetric flask with cychexane+ethyl acetate (1+1). Rinse pear-shaped flask twice with 5 mL cychexane +ethyl acetate (1+1) and transfer them to the above mentioned 10mL volumetric flask. Dilute to volume with cychexane +ethyl acetate (1+1) and mix thoroughly. Filter the solution to 10 mL test tube with 0.20 μm filter membrane for GPC cleanup. The fraction of 22 min ~ 40 min are collected in a 100 mL pear-shaped flask and concentrated to ca 0.5 mL on ratary evaporator at 40^{0}C. Evaporate it to dryness using nitrogen evaporator. Dissolve residues with 1.0mL acetonitrile + water (3+2) and pass the 0.20 μm filter membrane. The solution is ready for LC-MS-MS determination.

Determination

1. Qualitative and Quantitative GC-MS

For the samples determined, five injections were required to analyze all the pesticides by GC-MS under the operating conditions given above. If the retention time of the peaks of the sample solution were the same as those of the peaks from working standard mixed solution, and the selected ions appeared in the background-subtracted mass spectrum, and the abundance ratios of the selected ions were within the expected limits, the sample was confirmed to contain the pesticide. If the results were not definitive, the sample was re-injected with acquisition in scan mode (if sensitivity was sufficient), or with additional confirmatory ions, or using other instruments of higher sensitivity.

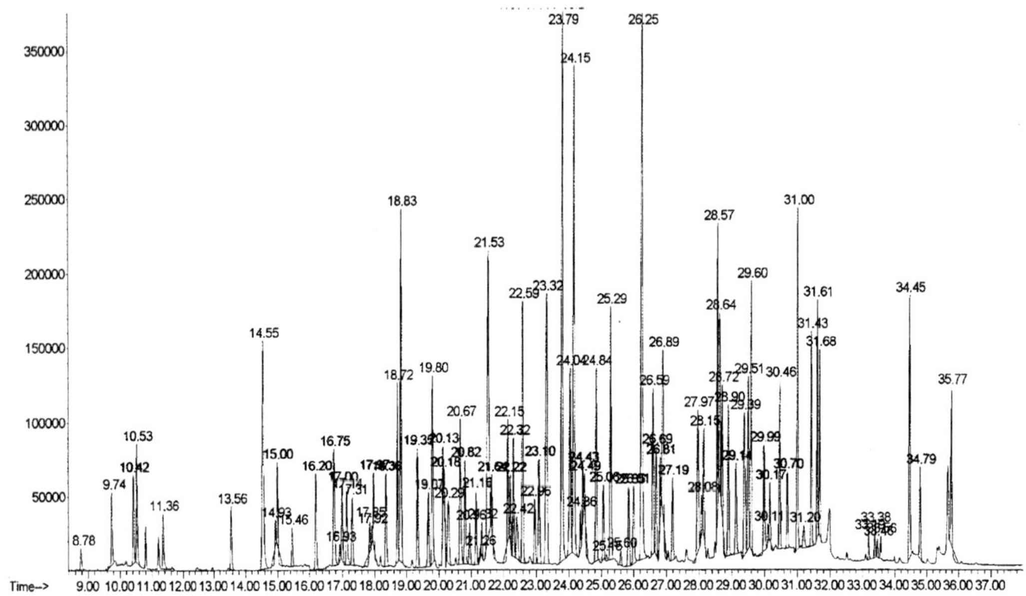

Figure 1. GC-MS chromatogram of A group standards in wheat blank matrix extracts.

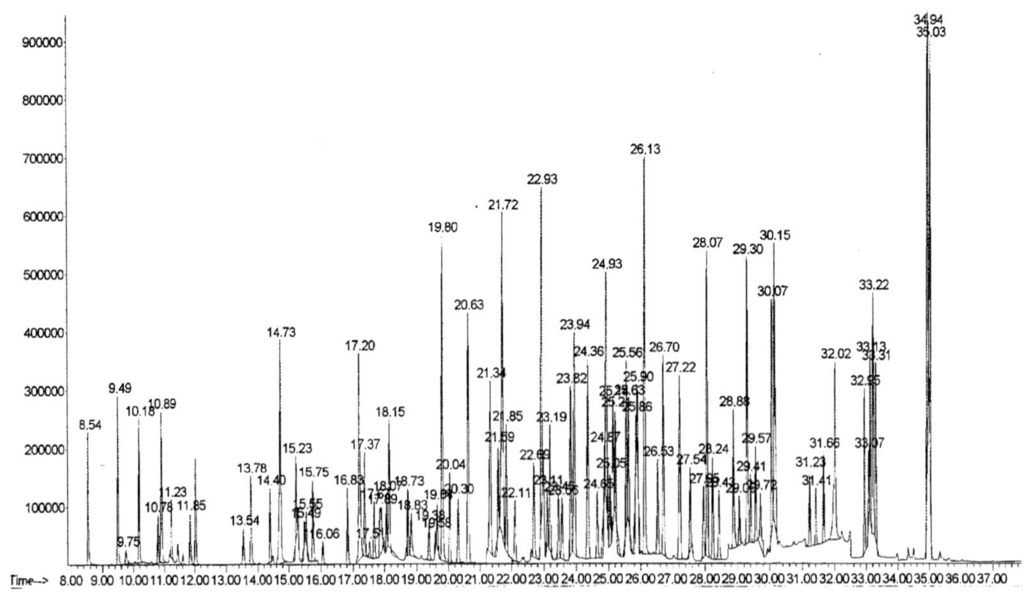

Figure 2. GC-MS chromatogram of B group standards in wheat blank matrix extracts.

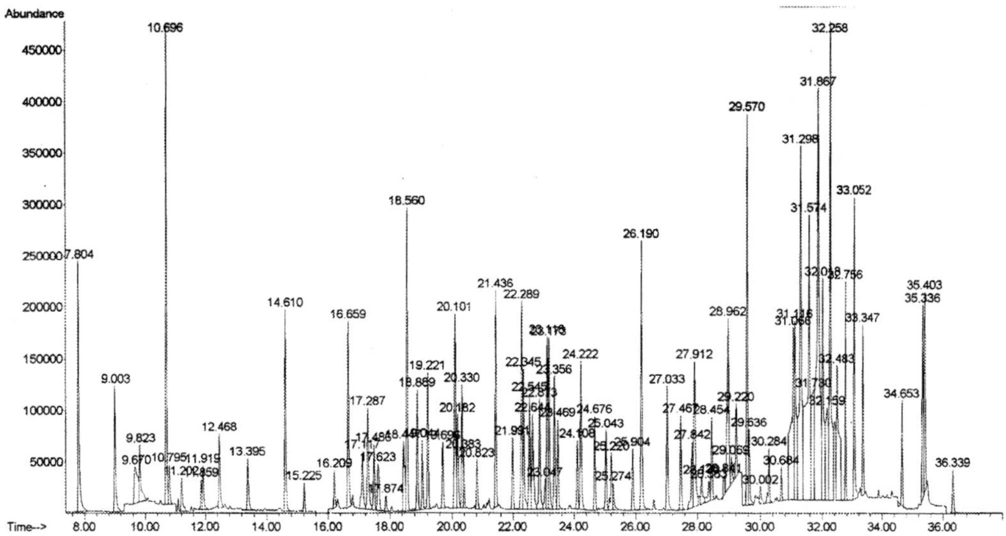

Figure 3. GC-MS chromatogram of C group standards in wheat blank matrix extracts.

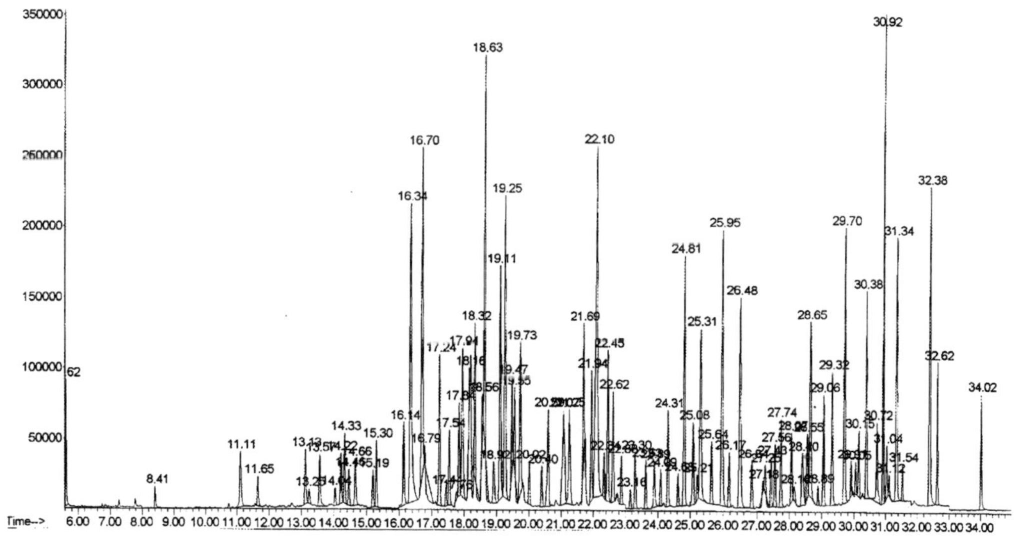

Figure 4. GC-MS chromatogram of D group standards in wheat blank matrix extracts.

Quantification was performed using the internal standard (heptachlor epoxide-exo) and the quantifying ion for GC-MS. To compensate for matrix effects, quantitation was based on a mixed standard prepared in blank matrix extract. The internal standard was also used to check retention times and to adjust the carrier gas head pressure so heptachlor epoxide-exo eluted at exactly 22.10 min. The concentrations in the standard solution and in the sample solution analyzed must be close. GC-MS chromatograms obtained from grain samples for the five groups A, B, C, D and E are given in Figure 1~5.

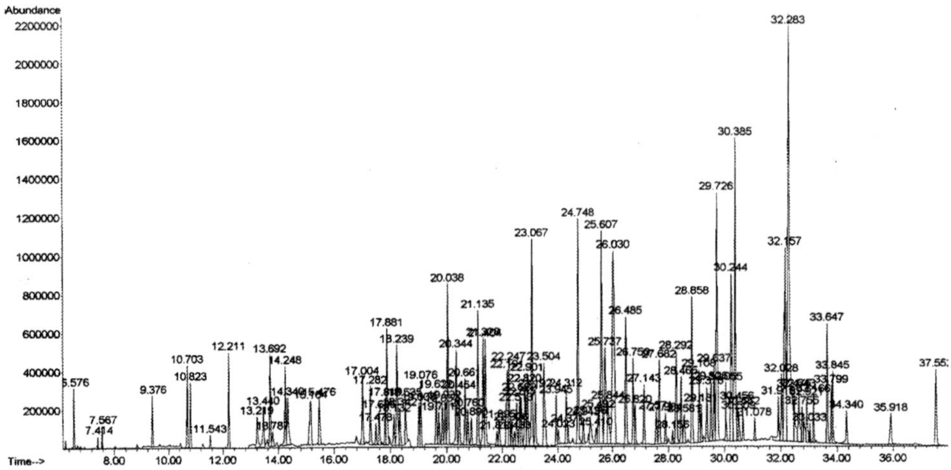

Figure 5. GC-MS chromatogram of E group standards in wheat blank matrix extracts.

2. Qualitative and Quantitative LC-MS-MS

Sample solutions were analyzed using the LC-MS-MS operating conditions given above. For the samples determined, identification of the target analytes in unknown samples was based on the LC retention time compared with that of a standard and the unique combination of a precursor-product ion. External calibration with standard curves was used for quantitation by LC-MS-MS. To compensate for matrix effects, quantitation was based on a series of working standard solutions prepared in blank matrix extract. The standard plots were established by separate injection of working standard mixed solutions of different concentration in matrix. The responses of the pesticides in sample solutions should be within the linear range of instrumental detection. MRM chromatograms of 372 standards and related chemicals are shown in Figure 6.

RESULTS AND DISCUSSION

Selection of Pesticide Varieties

In accordance with the pesticide varieties of maximum residue limit prescribed for various foods by International Alimentarius Codex, China, USA, EU, Germany, Canada and Japan's "Positive List System", 858 world commonly used pesticide standards were purchased and their GC-MS and LC-MS-MS spectrometric characteristics were evaluated.

1. GC-MS Characteristics

After analysis of their physical and chemical properties of 858 pesticides such as their solubility etc., 19 pesticides were found unfit for instrumental analysis by GC-MS, which is shown in Table 7.

Under optimized GC-MS instrumental conditions, over 839 pesticides were respectively perfomed GC-MS in scanning mode, from which full scan spectra and retention time for standards were obtained, among which 298 had no peaks or extremely low sensitivity, being unable to reach the analytical requirements for pesticides. See Table 8.

A group

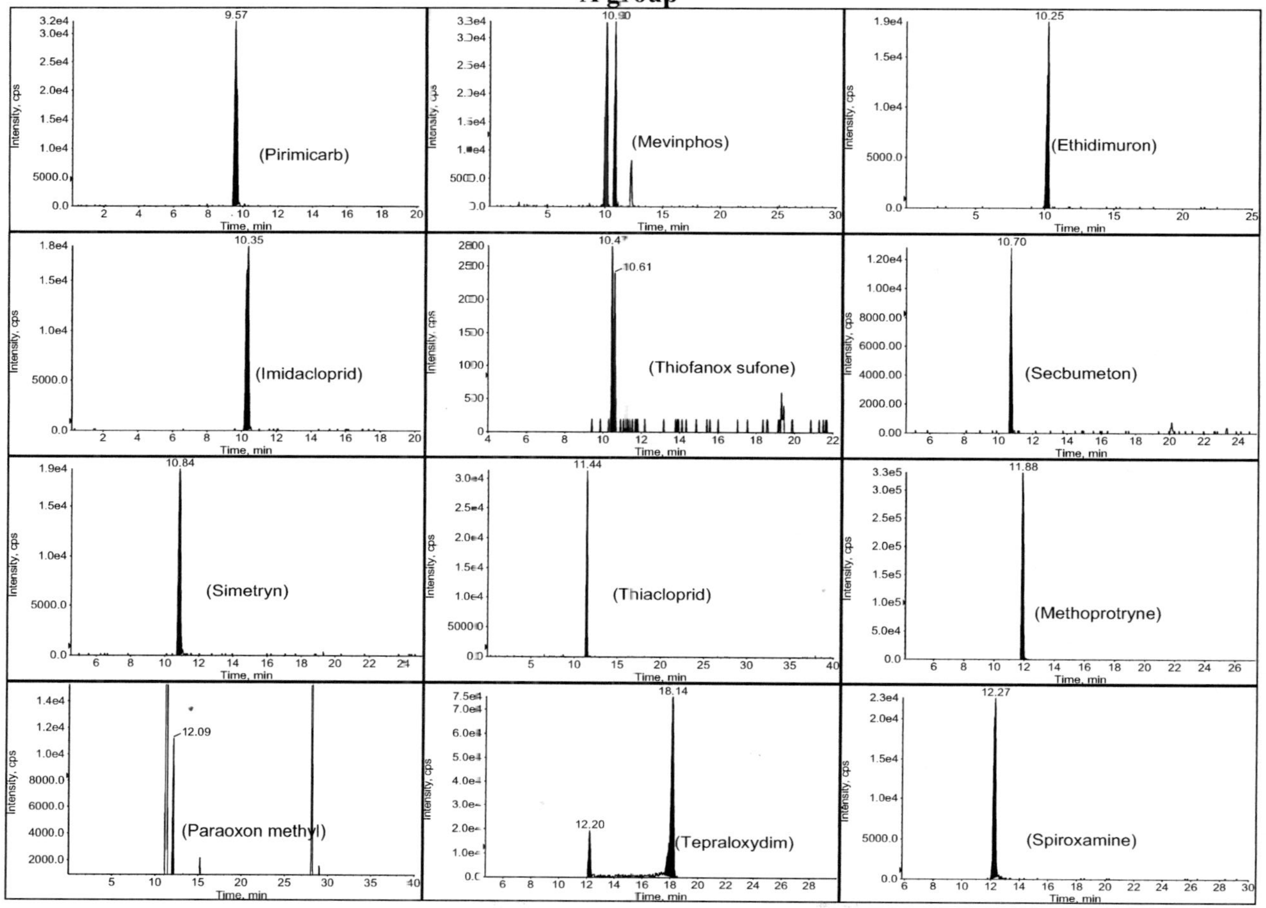
(Pirimicarb)
9.57
(Mevinphos)
(Ethidimuron)
10.25
(Imidacloprid)
10.35
(Thiofanox sufone)
10.61
(Secbumeton)
10.70
(Simetryn)
10.84
(Thiacloprid)
11.44
(Methoprotryne)
11.88
(Paraoxon methyl)
12.09
(Tepraloxydim)
18.14
12.20
(Spiroxamine)
12.27
Intensity, cps
Time, min

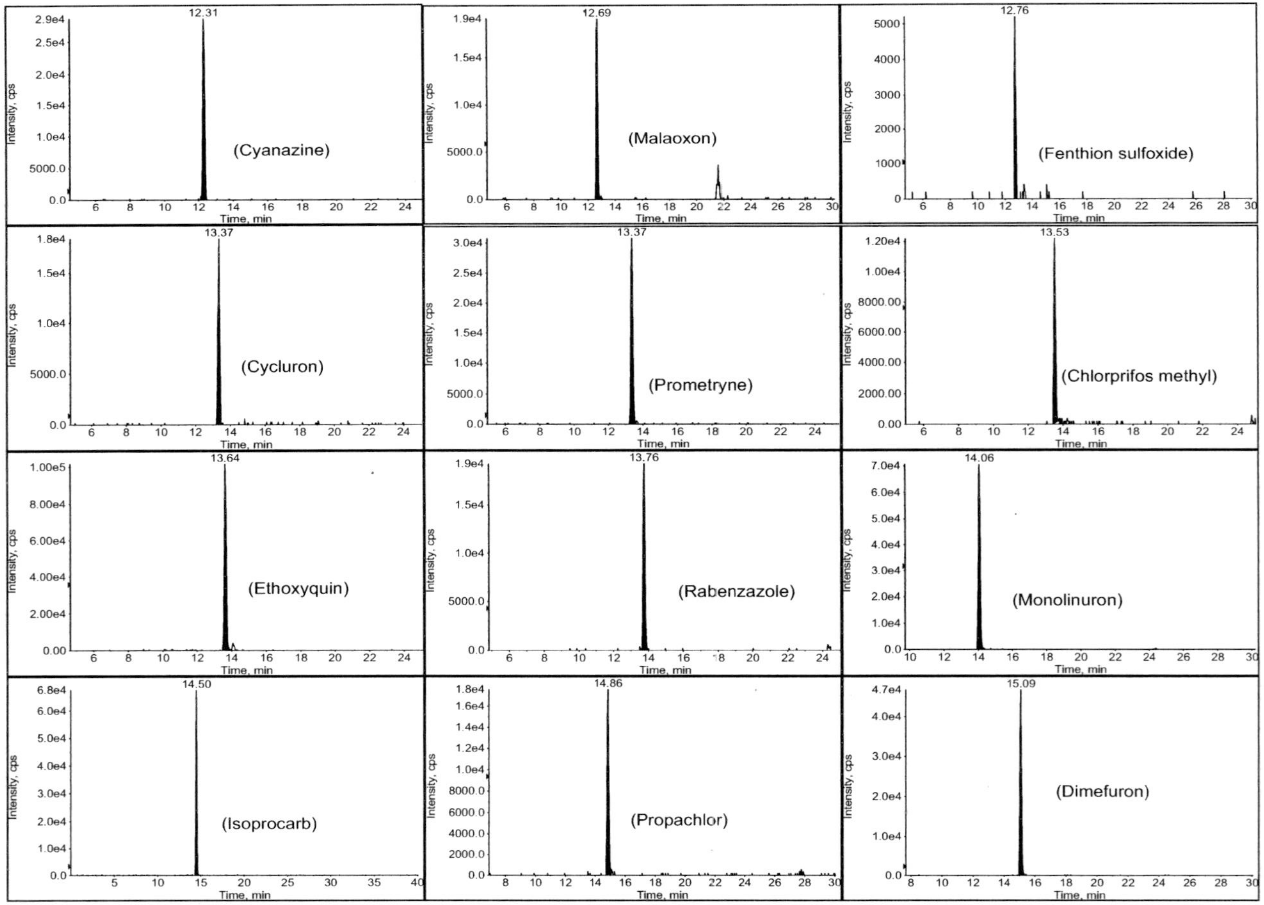
12.31
(Cyanazine)
12.69
(Malaoxon)
12.76
(Fenthion sulfoxide)
13.37
(Cycluron)
13.37
(Prometryne)
13.53
(Chlorprifos methyl)
13.64
(Ethoxyquin)
13.76
(Rabenzazole)
14.06
(Monolinuron)
14.50
(Isoprocarb)
14.86
(Propachlor)
15.09
(Dimefuron)
Intensity, cps
Time, min

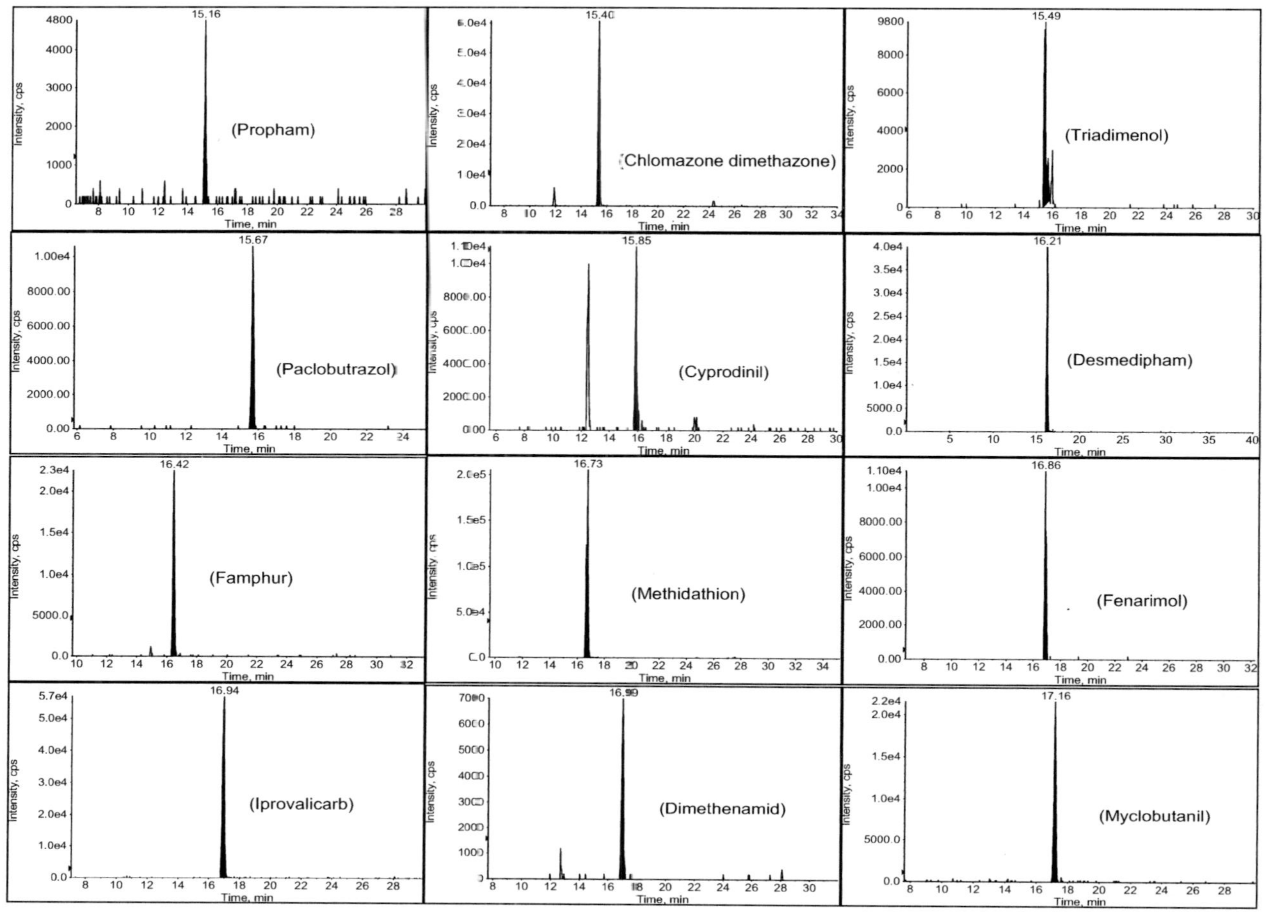
(Propham)
(Chlomazone dimethazone)
(Triadimenol)
(Paclobutrazol)
(Cyprodinil)
(Desmedipham)
(Famphur)
(Methidathion)
(Fenarimol)
(Iprovalicarb)
(Dimethenamid)
(Myclobutanil)
Intensity, cps
Time, min

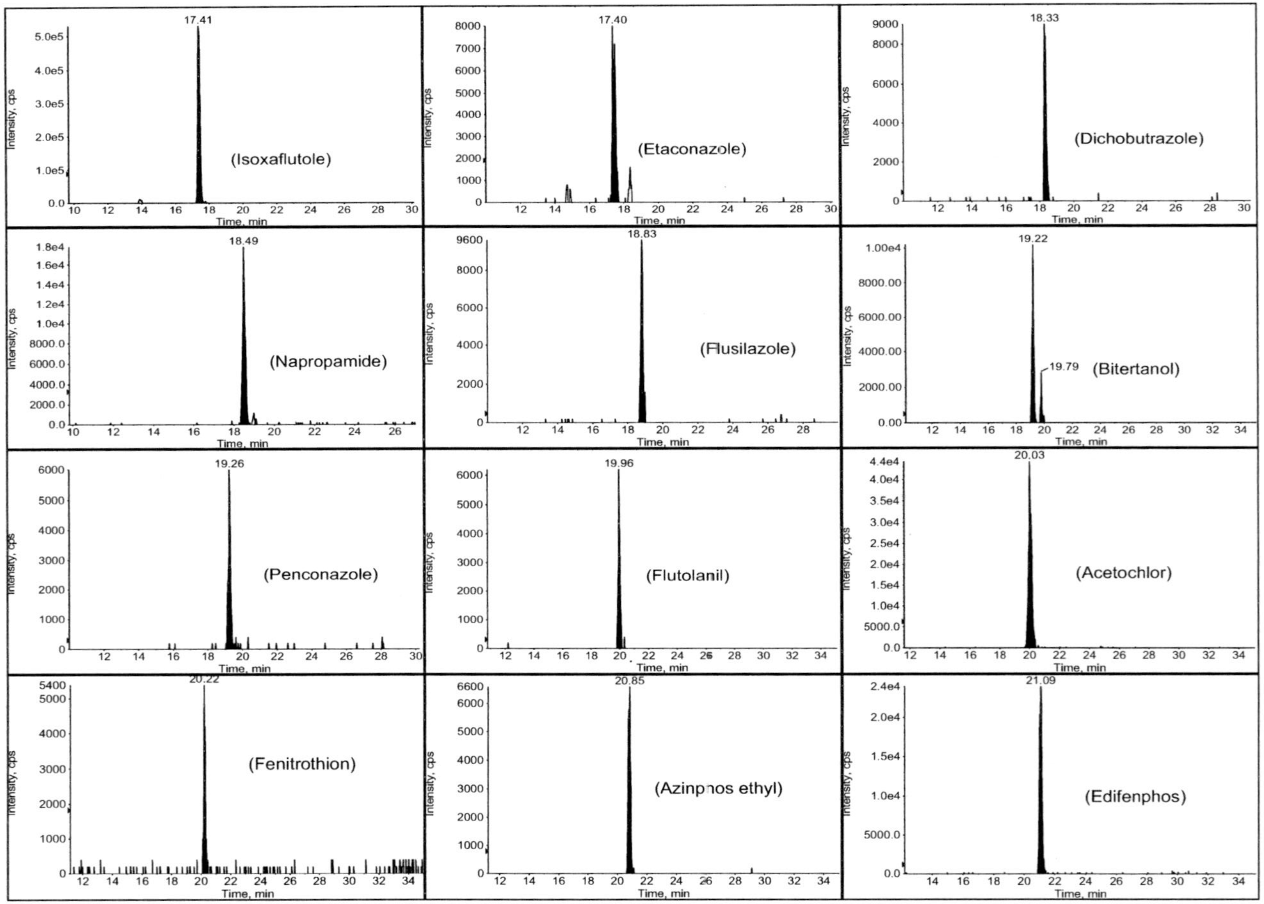

(Isoxaflutole)
17.41
(Etaconazole)
17.40
(Dichobutrazole)
18.33
(Napropamide)
18.49
(Flusilazole)
18.83
(Bitertanol)
19.22
19.79
(Penconazole)
19.26
(Flutolanil)
19.96
(Acetochlor)
20.03
(Fenitrothion)
20.22
(Azinphos ethyl)
20.85
(Edifenphos)
21.09
Intensity, cps
Time, min

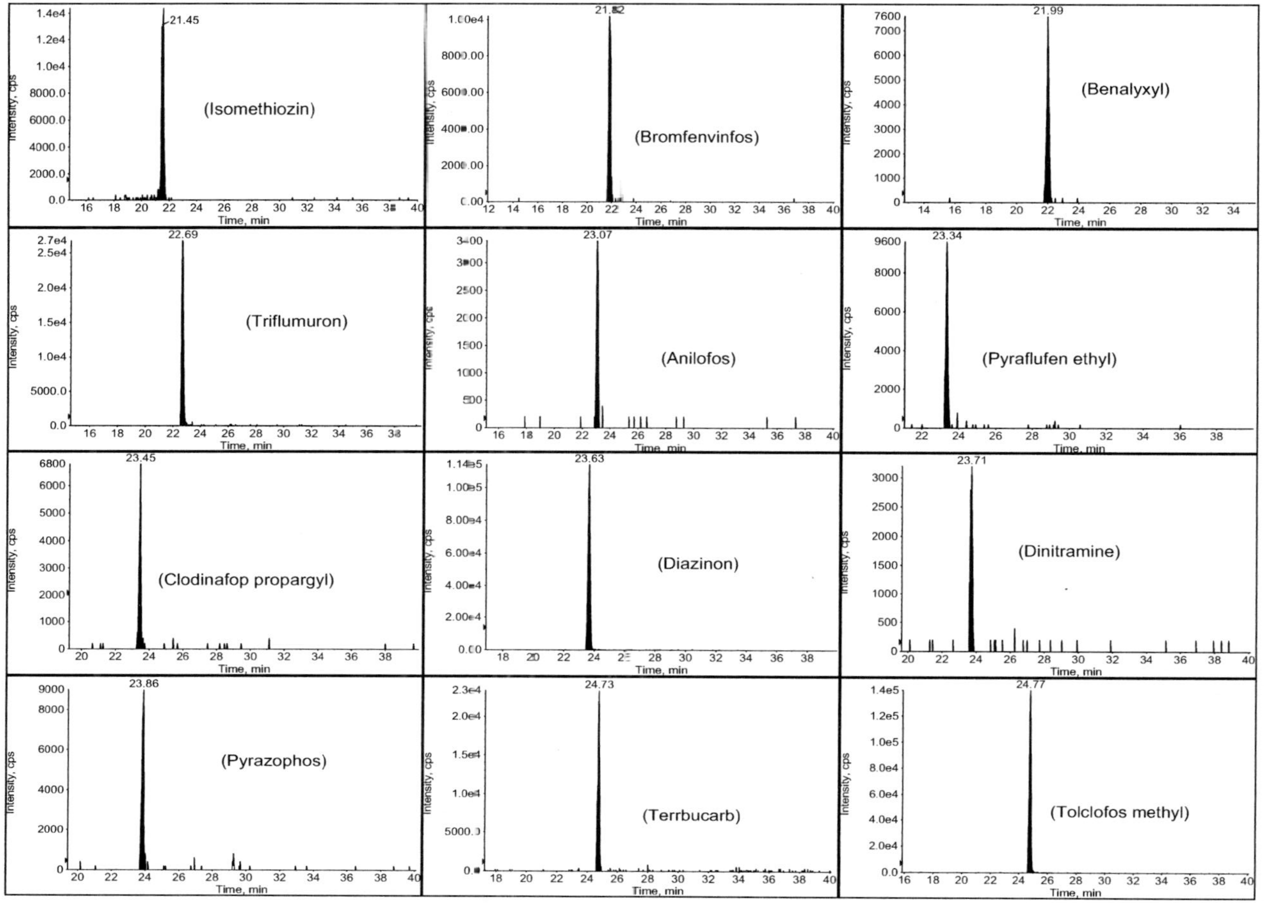
(Isomethiozin)
21.45
(Bromfenvinfos)
(Benalyxyl)
21.99
(Triflumuron)
22.69
(Anilofos)
23.07
(Pyraflufen ethyl)
23.34
(Clodinafop propargyl)
23.45
(Diazinon)
23.63
(Dinitramine)
23.71
(Pyrazophos)
23.86
(Terrbucarb)
24.73
(Tolclofos methyl)
24.77
Intensity, cps
Time, min

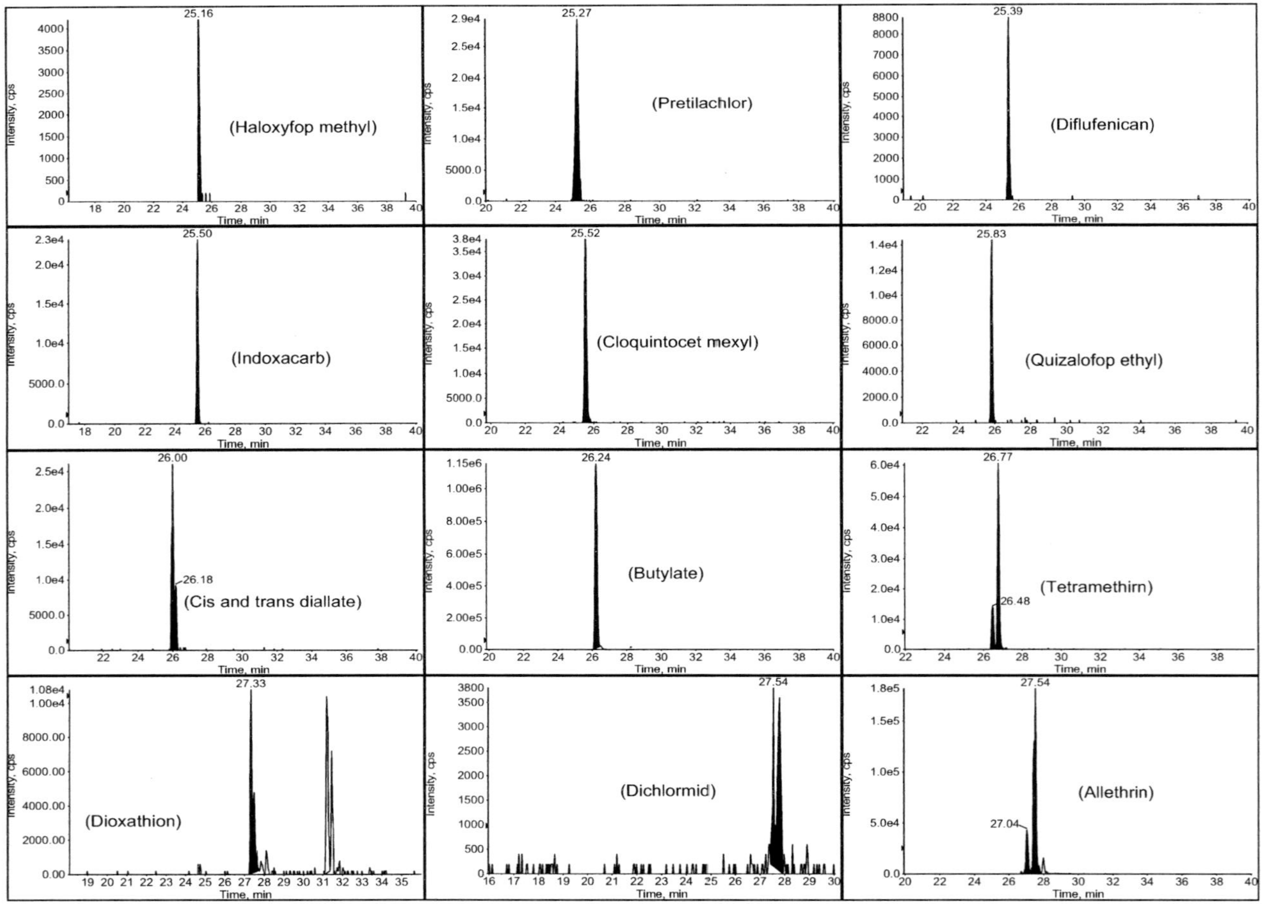
(Haloxyfop methyl)
25.16
(Pretilachlor)
25.27
(Diflufenican)
25.39
(Indoxacarb)
25.50
(Cloquintocet mexyl)
25.52
(Quizalofop ethyl)
25.83
(Cis and trans diallate)
26.00
26.18
(Butylate)
26.24
(Tetramethirn)
26.77
26.48
(Dioxathion)
27.33
(Dichlormid)
27.54
(Allethrin)
27.54
27.04
Intensity, cps
Time, min

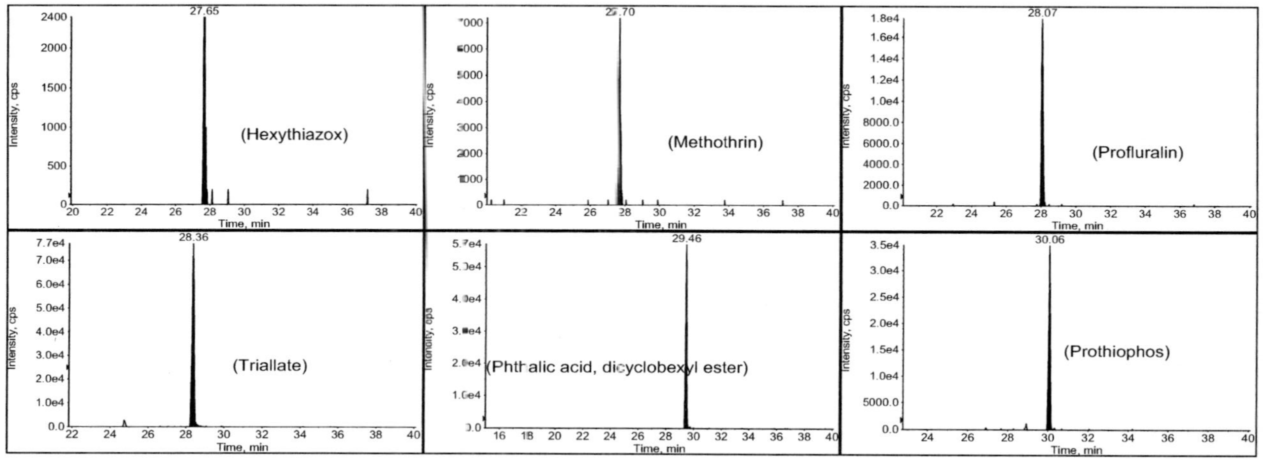

27.65
(Hexythiazox)
(Methothrin)
28.07
(Profluralin)
28.36
(Triallate)
29.46
(Phthalic acid, dicyclobexyl ester)
30.06
(Prothiophos)

B group

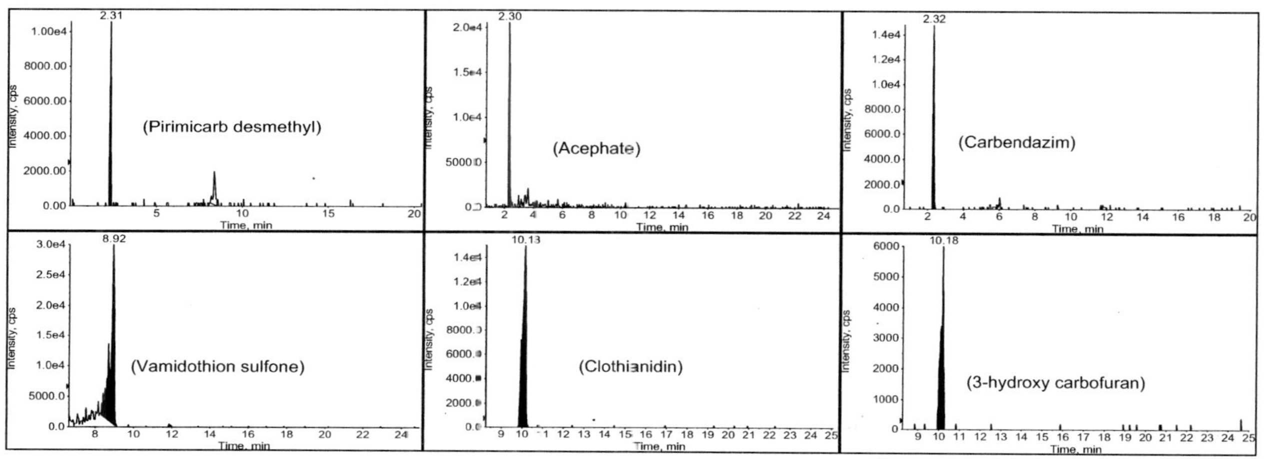

2.31
(Pirimicarb desmethyl)
2.30
(Acephate)
2.32
(Carbendazim)
8.92
(Vamidothion sulfone)
10.13
(Clothianidin)
10.18
(3-hydroxy carbofuran)

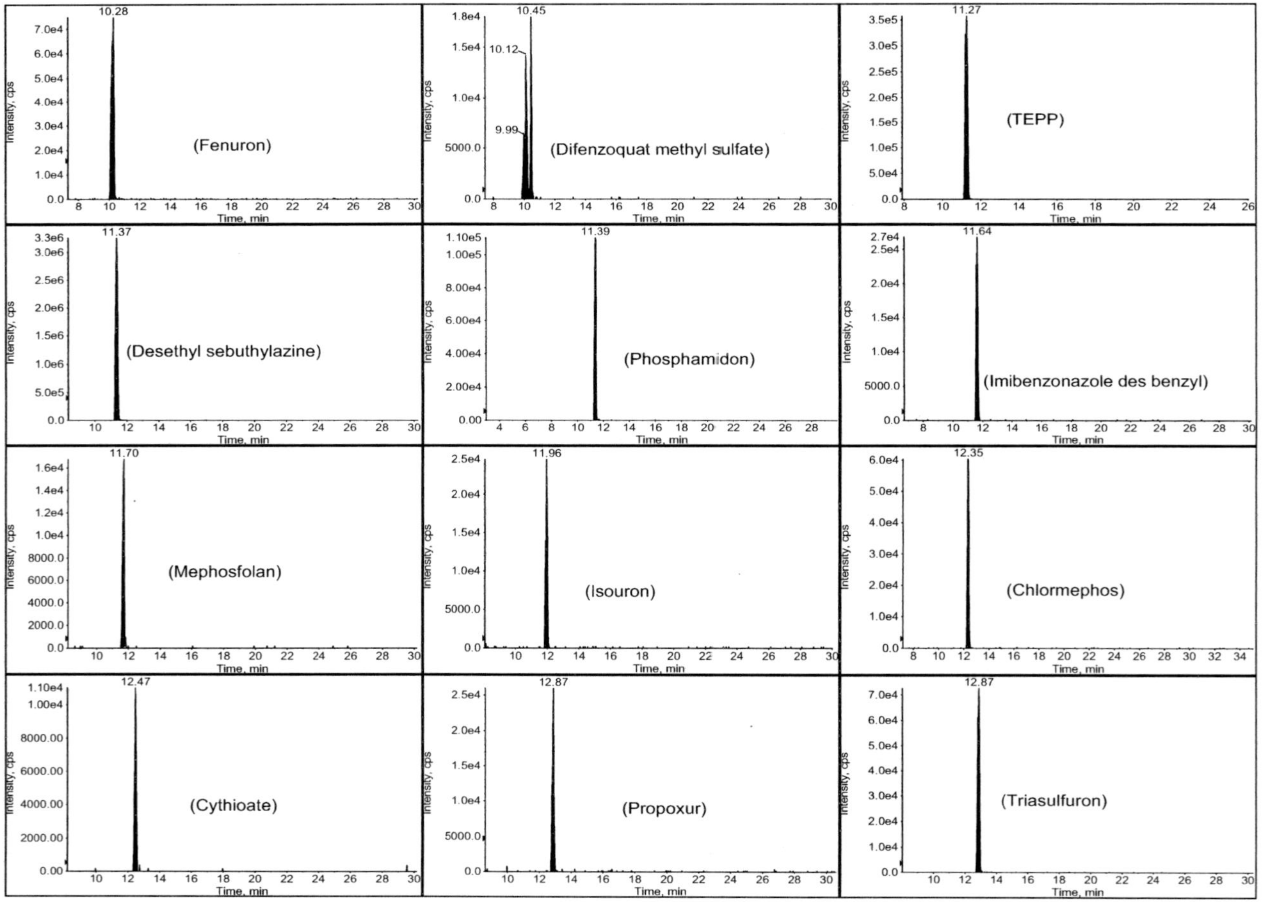
(Fenuron)
10.28
(Difenzoquat methyl sulfate)
10.45
10.12
9.99
(TEPP)
11.27
(Desethyl sebuthylazine)
11.37
(Phosphamidon)
11.39
(Imibenzonazole des benzyl)
11.64
(Mephosfolan)
11.70
(Isouron)
11.96
(Chlormephos)
12.35
(Cythioate)
12.47
(Propoxur)
12.87
(Triasulfuron)
12.87
Intensity, cps
Time, min

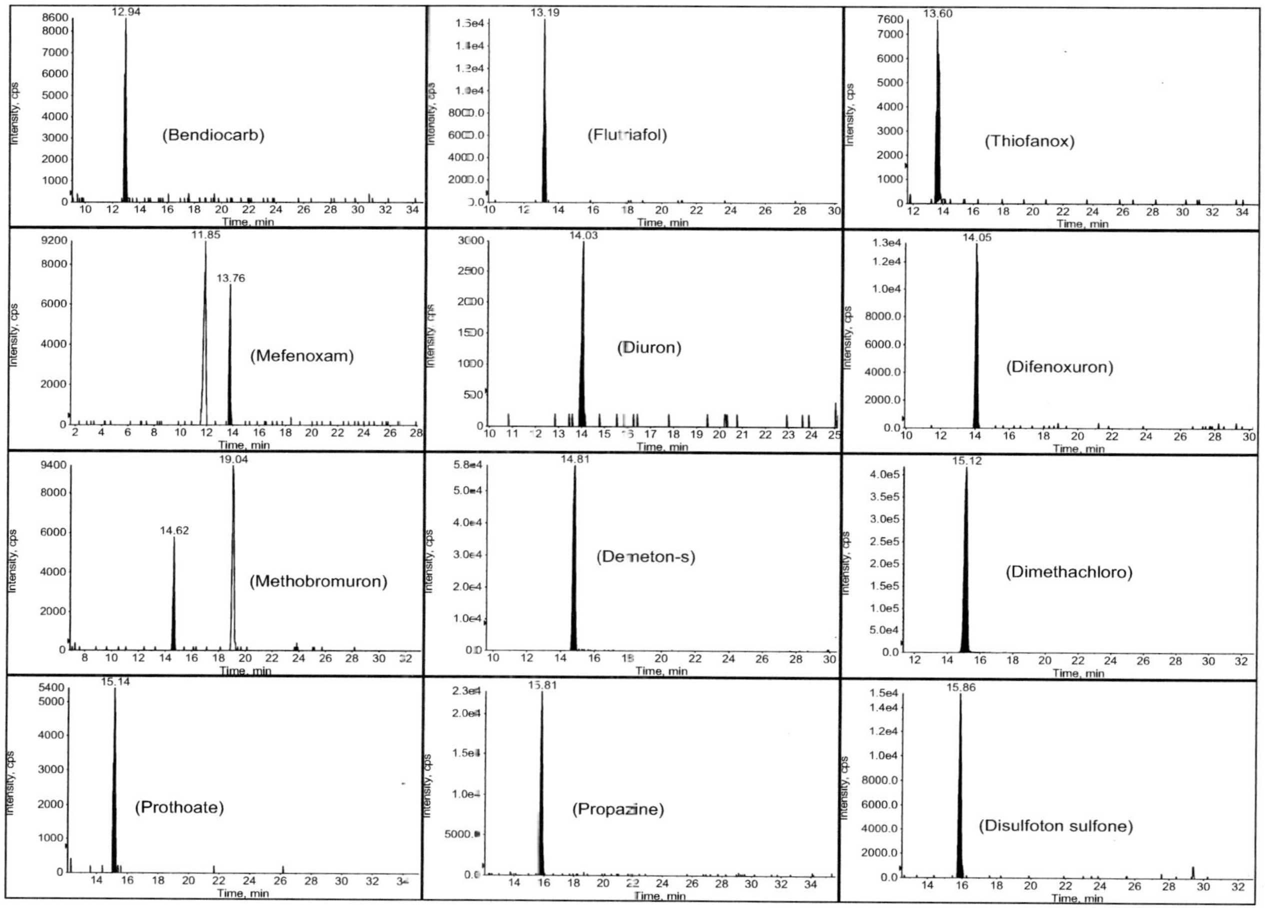
(Bendiocarb)
(Flutriafol)
(Thiofanox)
(Mefenoxam)
(Diuron)
(Difenoxuron)
(Methobromuron)
(Demeton-s)
(Dimethachloro)
(Prothoate)
(Propazine)
(Disulfoton sulfone)
Intensity, cps
Time, min

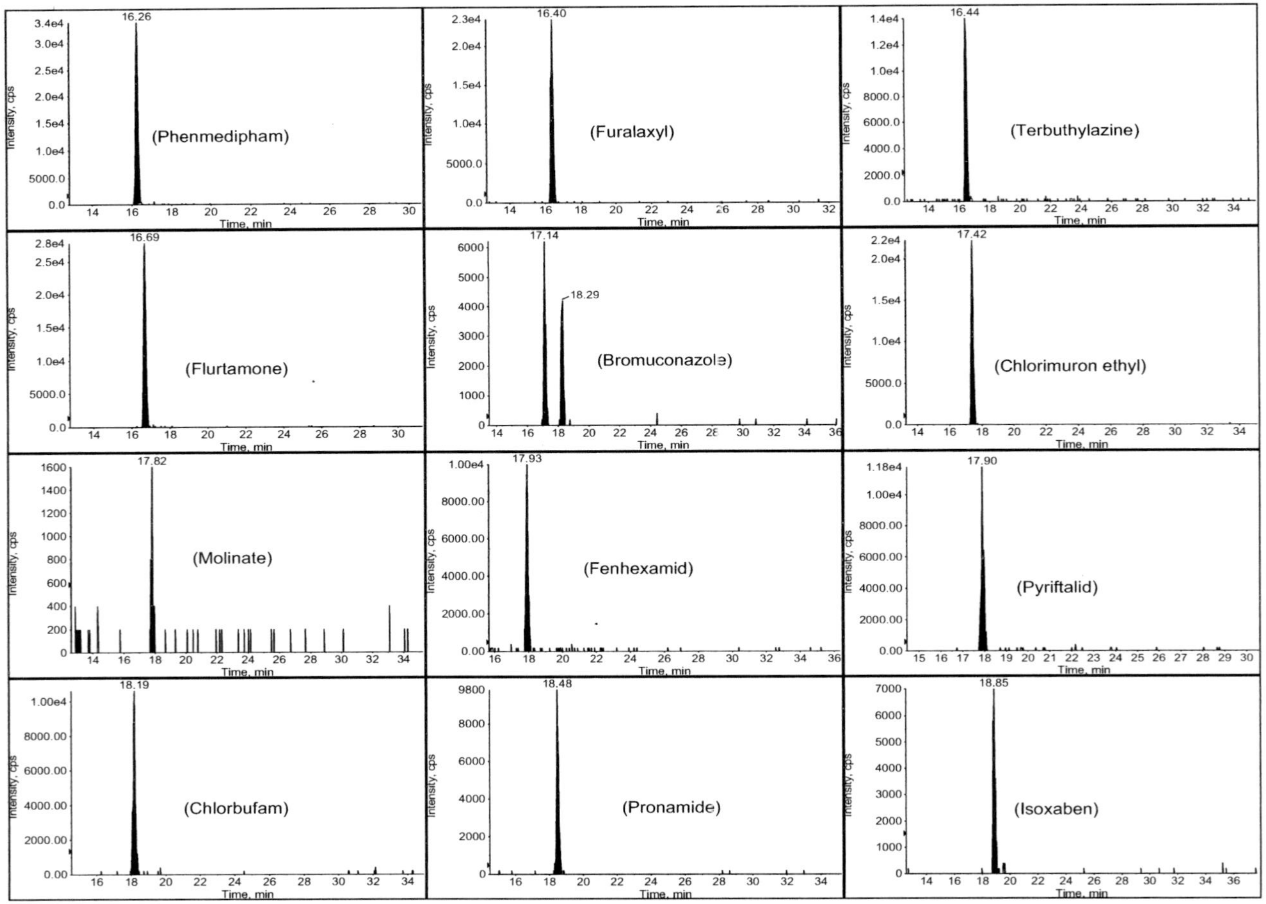
(Phenmedipham)
16.26
(Furalaxyl)
16.40
(Terbuthylazine)
16.44
(Flurtamone)
16.69
(Bromuconazole)
17.14
18.29
(Chlorimuron ethyl)
17.42
(Molinate)
17.82
(Fenhexamid)
17.93
(Pyriftalid)
17.90
(Chlorbufam)
18.19
(Pronamide)
18.48
(Isoxaben)
18.85
Intensity, cps
Time, min

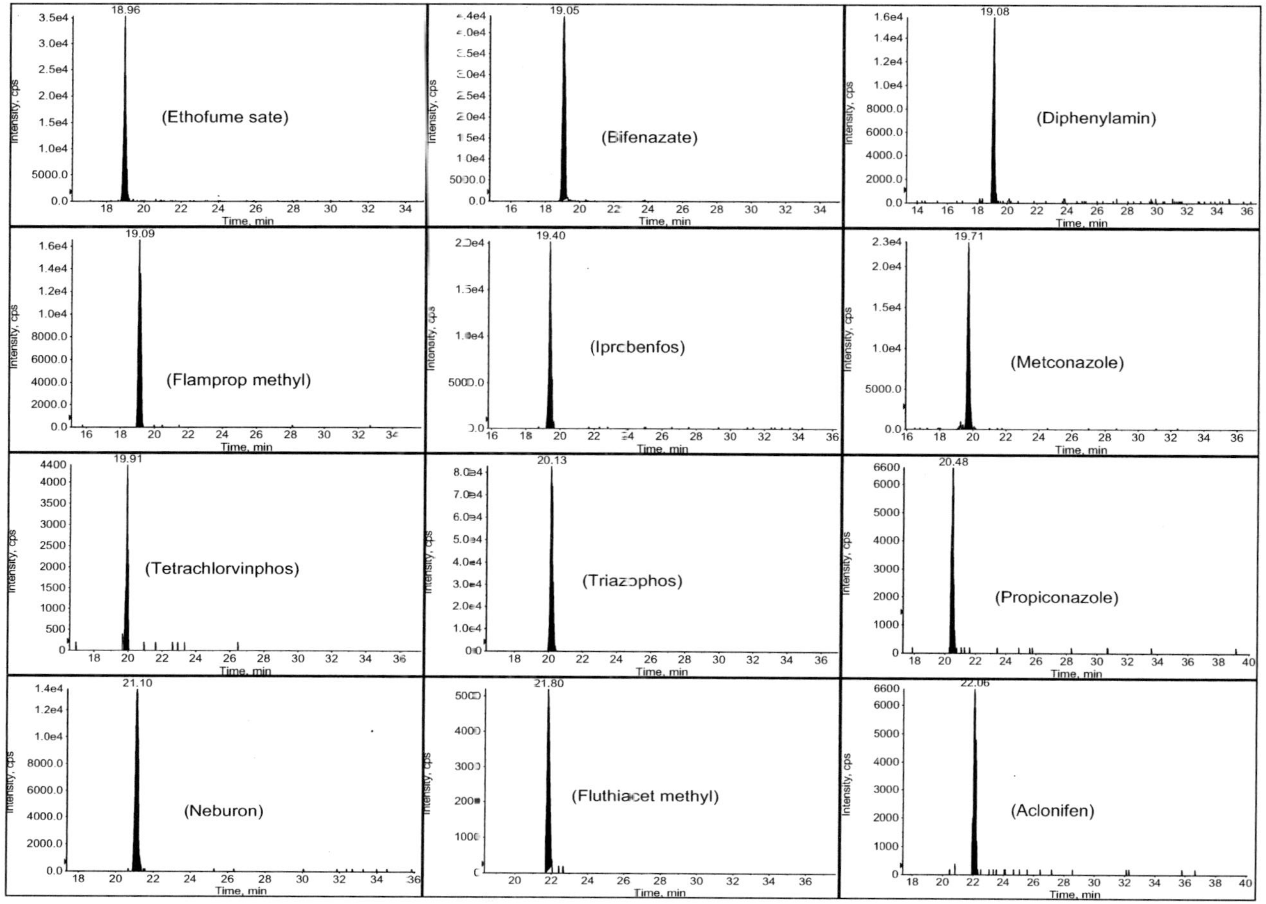

(Ethofume sate)
(Bifenazate)
(Diphenylamin)
(Flamprop methyl)
(Iprobenfos)
(Metconazole)
(Tetrachlorvinphos)
(Triazophos)
(Propiconazole)
(Neburon)
(Fluthiacet methyl)
(Aclonifen)
Intensity, cps
Time, min
18.96
19.05
19.08
19.09
19.40
19.71
19.91
20.13
20.48
21.10
21.80
22.06

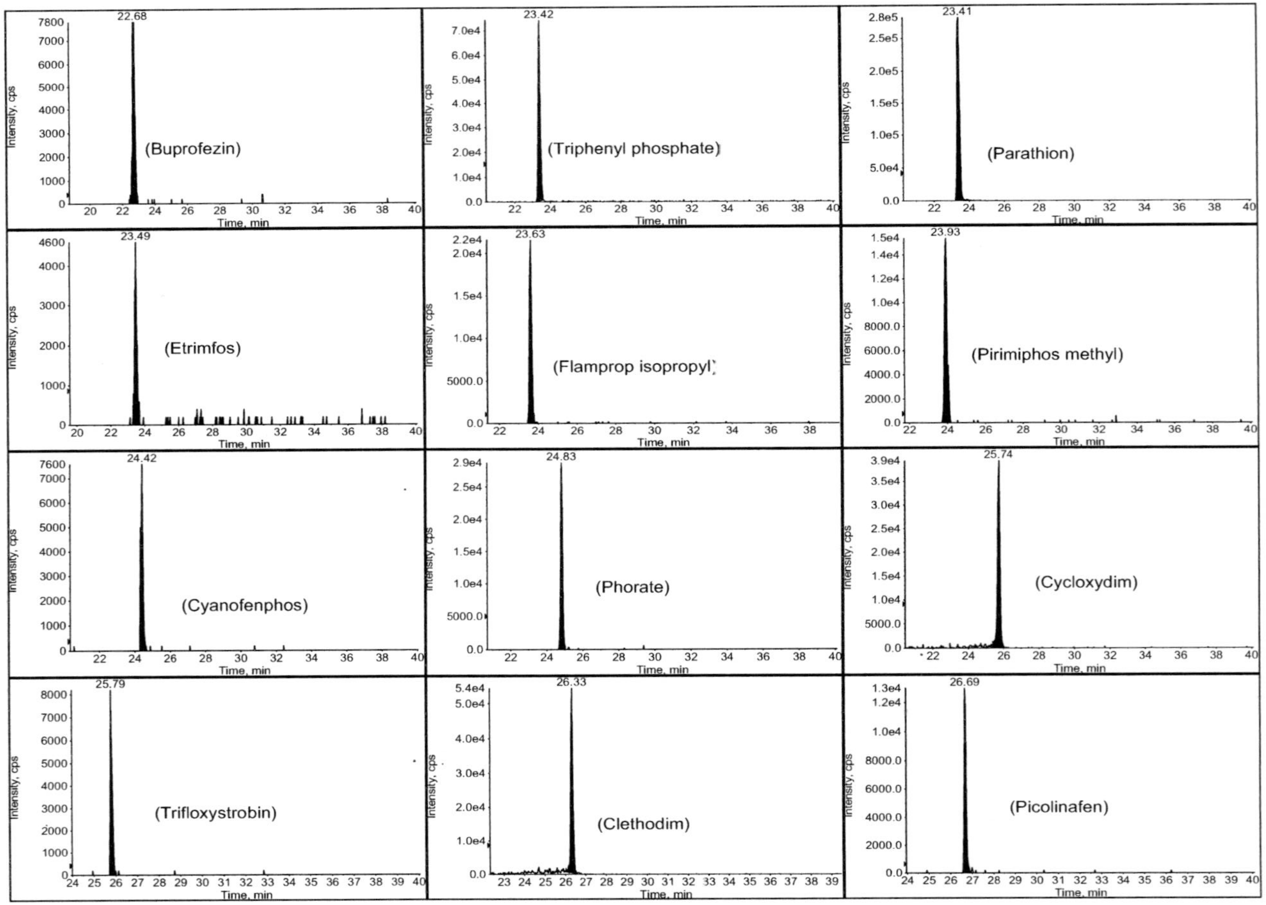

(Buprofezin)
22.68
(Triphenyl phosphate)
23.42
(Parathion)
23.41
(Etrimfos)
23.49
(Flamprop isopropyl)
23.63
(Pirimiphos methyl)
23.93
(Cyanofenphos)
24.42
(Phorate)
24.83
(Cycloxydim)
25.74
(Trifloxystrobin)
25.79
(Clethodim)
26.33
(Picolinafen)
26.69
Intensity, cps
Time, min

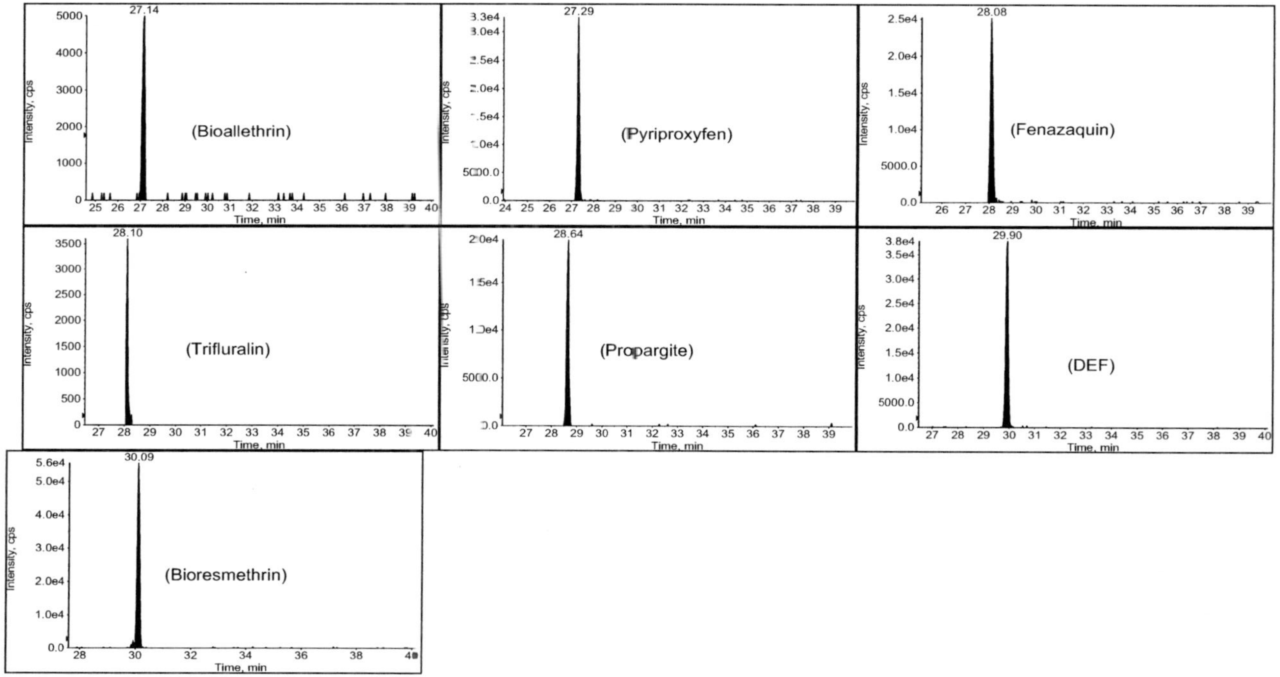
27.14
(Bioallethrin)
27.29
(Pyriproxyfen)
28.08
(Fenazaquin)
28.10
(Trifluralin)
28.64
(Propargite)
29.90
(DEF)
30.09
(Bioresmethrin)
Intensity, cps
Time, min

C group

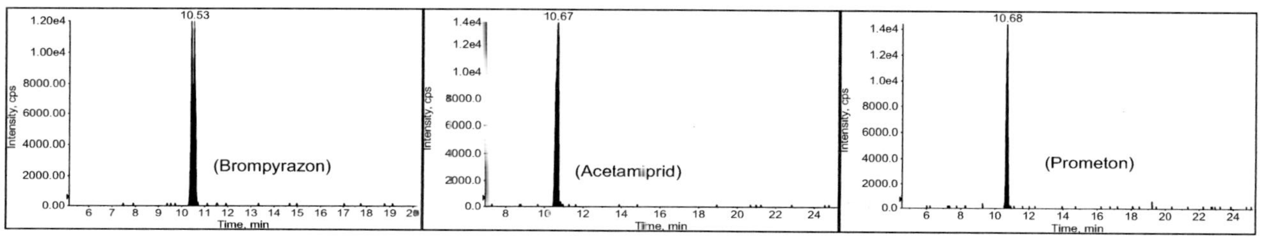
10.53
(Brompyrazon)
10.67
(Acetamiprid)
10.68
(Prometon)
Intensity, cps
Time, min

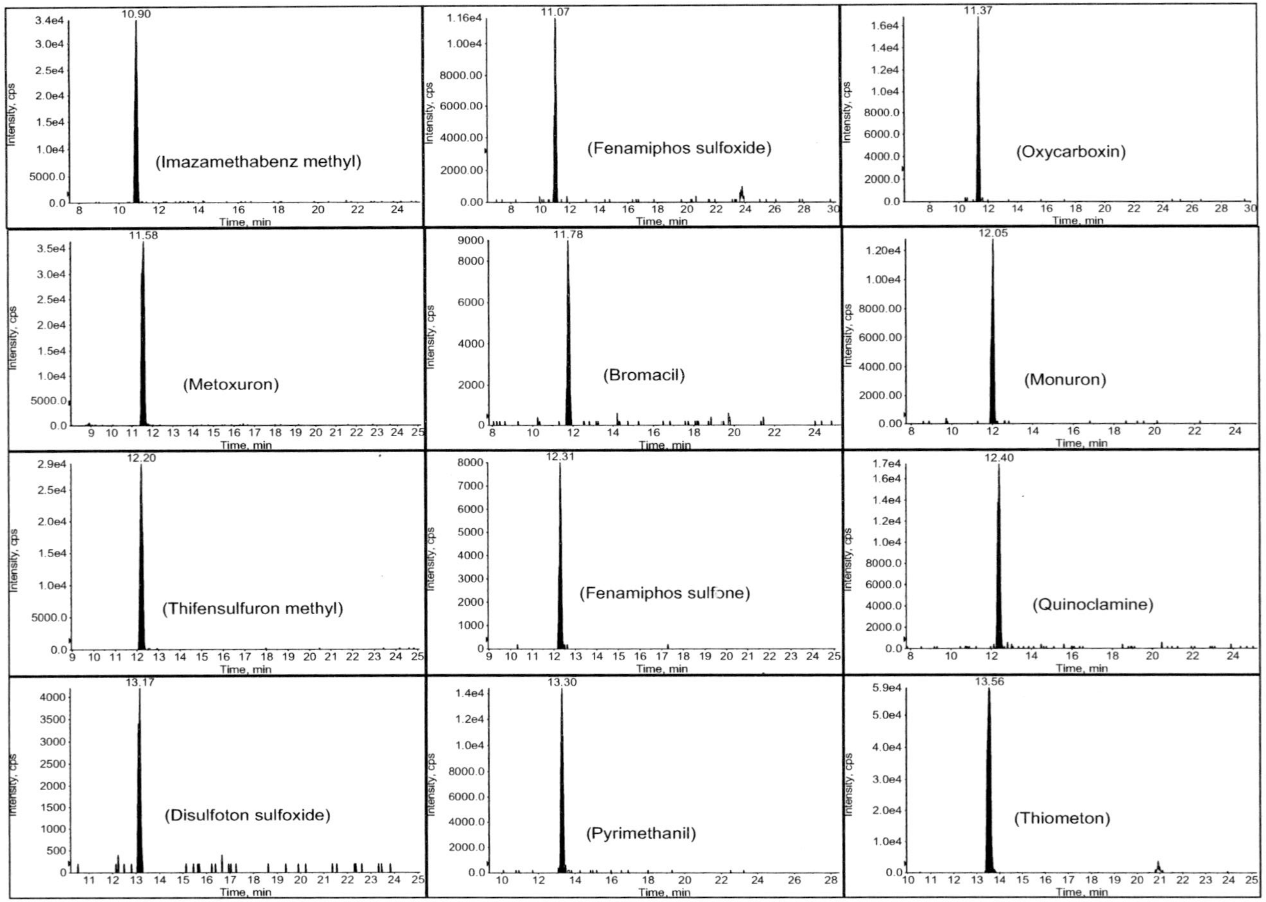

(Imazamethabenz methyl)
(Fenamiphos sulfoxide)
(Oxycarboxin)
(Metoxuron)
(Bromacil)
(Monuron)
(Thifensulfuron methyl)
(Fenamiphos sulfone)
(Quinoclamine)
(Disulfoton sulfoxide)
(Pyrimethanil)
(Thiometon)
Intensity, cps
Time, min

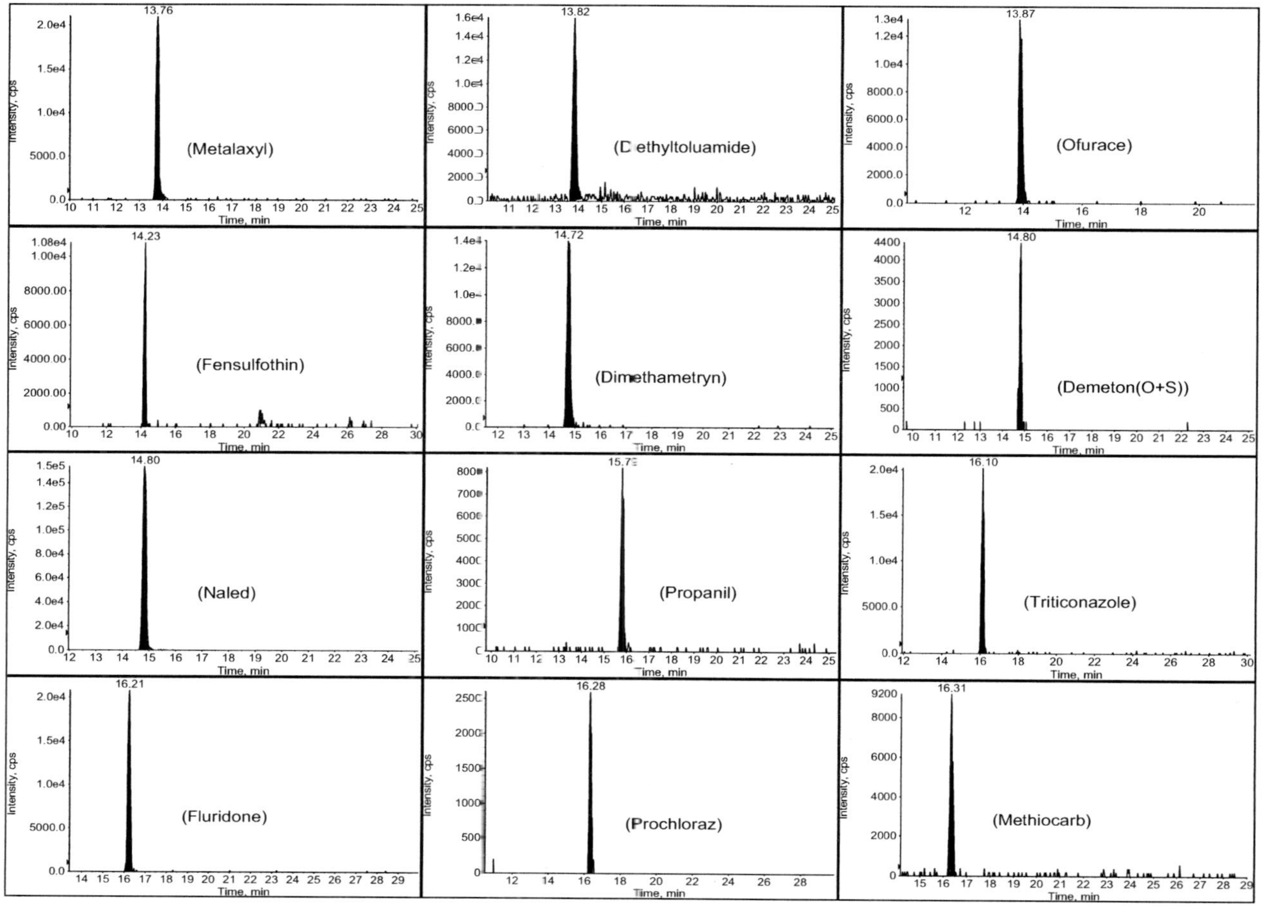
(Metalaxyl)
13.76
(Diethyltoluamide)
13.82
(Ofurace)
13.87
(Fensulfothin)
14.23
(Dimethametryn)
14.72
(Demeton(O+S))
14.80
(Naled)
14.80
(Propanil)
15.73
(Triticonazole)
16.10
(Fluridone)
16.21
(Prochloraz)
16.28
(Methiocarb)
16.31
Intensity, cps
Time, min

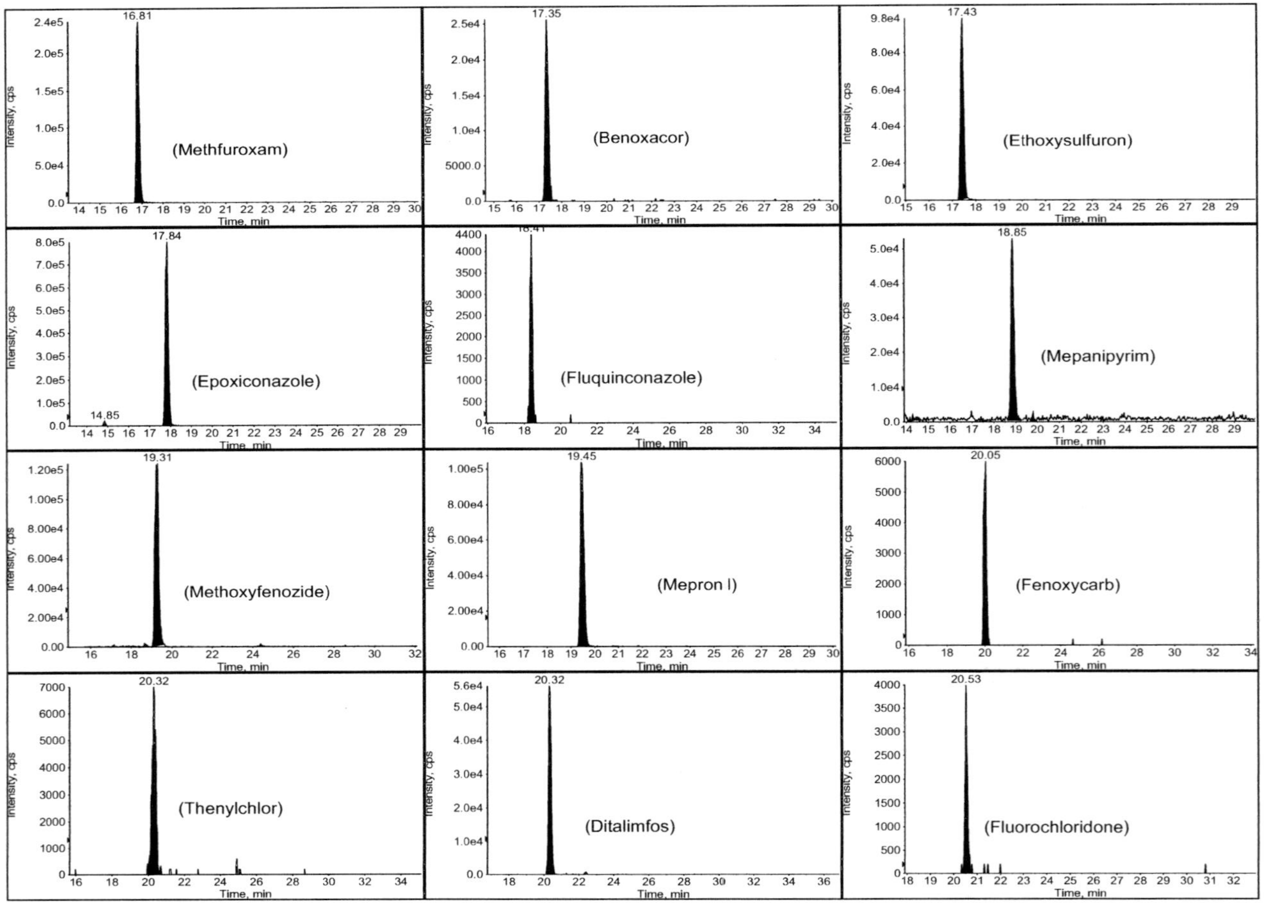

(Methfuroxam)
16.81
(Benoxacor)
17.35
(Ethoxysulfuron)
17.43
(Epoxiconazole)
17.84
14.85
(Fluquinconazole)
(Mepanipyrim)
18.85
(Methoxyfenozide)
19.31
(Mepron I)
19.45
(Fenoxycarb)
20.05
(Thenylchlor)
20.32
(Ditalimfos)
20.32
(Fluorochloridone)
20.53
Intensity, cps
Time, min

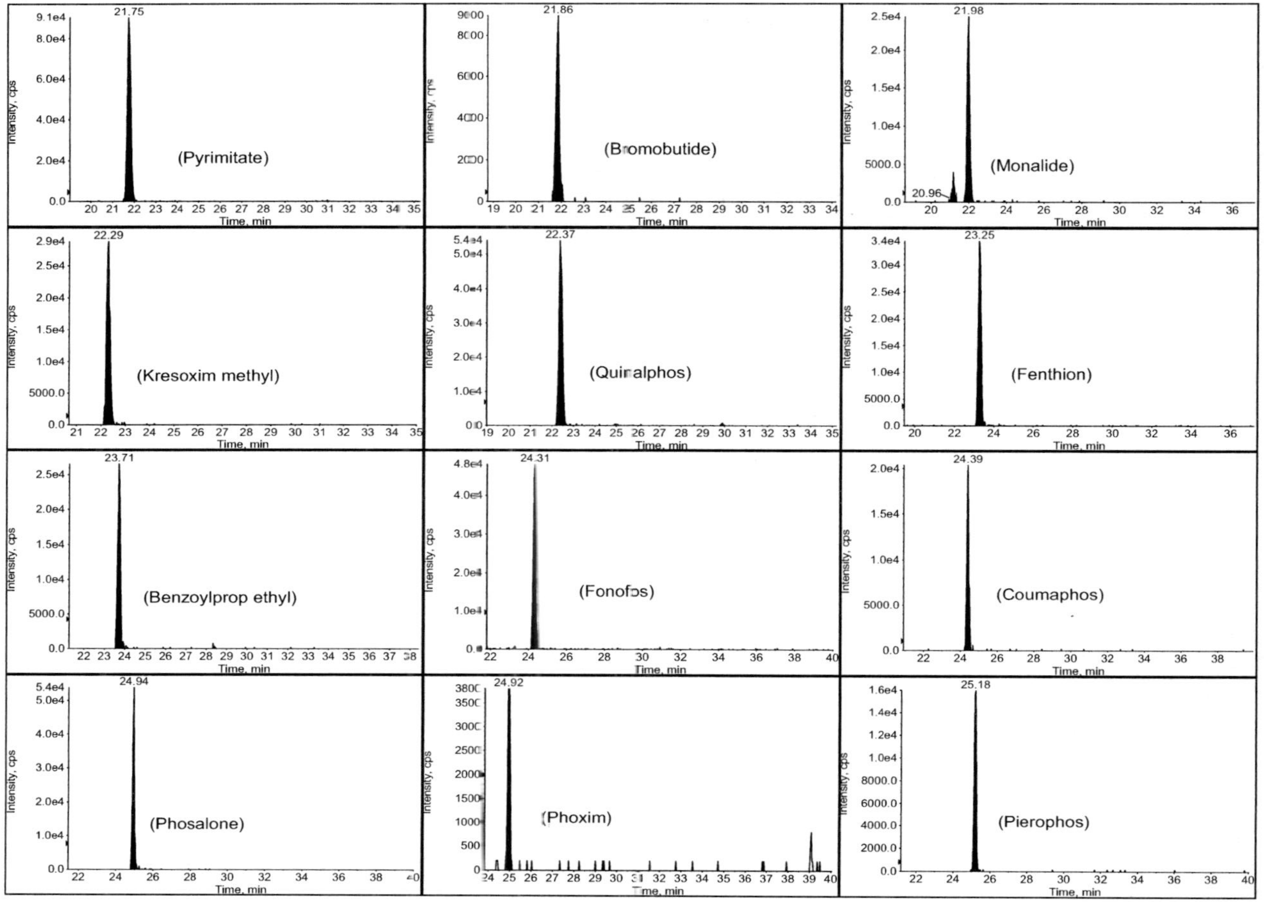

(Pyrimitate)
21.75
(Bromobutide)
21.86
(Monalide)
21.98
20.96
(Kresoxim methyl)
22.29
(Quinalphos)
22.37
(Fenthion)
23.25
(Benzoylprop ethyl)
23.71
(Fonofos)
24.31
(Coumaphos)
24.39
(Phosalone)
24.94
(Phoxim)
24.92
(Pierophos)
25.18
Intensity, cps
Time, min

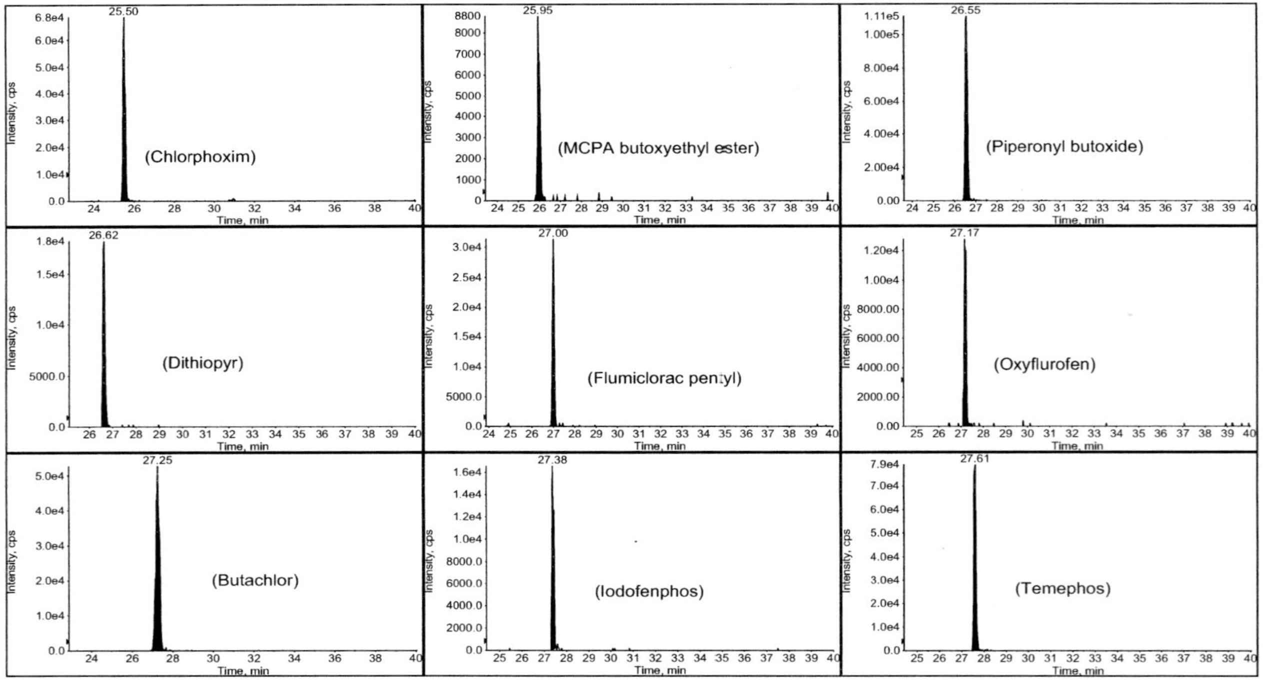
(Chlorphoxim)
25.50
(MCPA butoxyethyl ester)
25.95
(Piperonyl butoxide)
26.55
(Dithiopyr)
26.62
(Flumiclorac pentyl)
27.00
(Oxyflurofen)
27.17
(Butachlor)
27.25
(Iodofenphos)
27.38
(Temephos)
27.61
Intensity, cps
Time, min

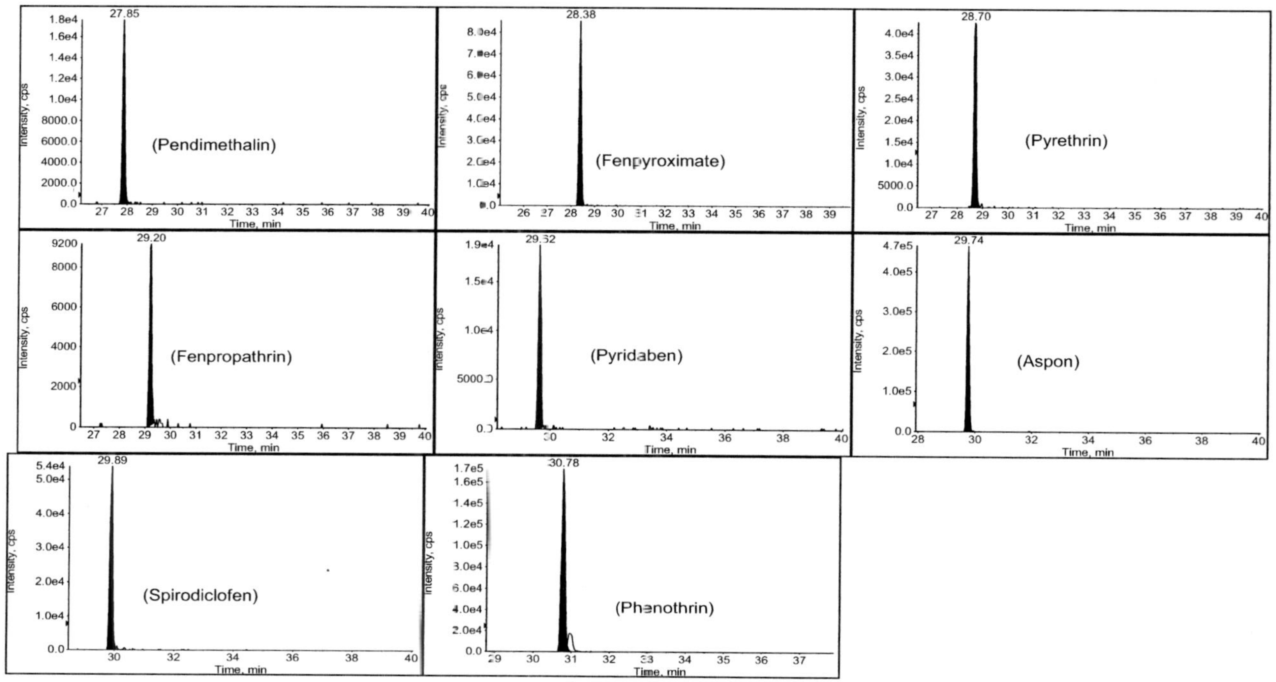
(Pendimethalin)
27.85
(Fenpyroximate)
28.38
(Pyrethrin)
28.70
(Fenpropathrin)
29.20
(Pyridaben)
29.32
(Aspon)
29.74
(Spirodiclofen)
29.89
(Phenothrin)
30.78
Intensity, cps
Time, min

D group

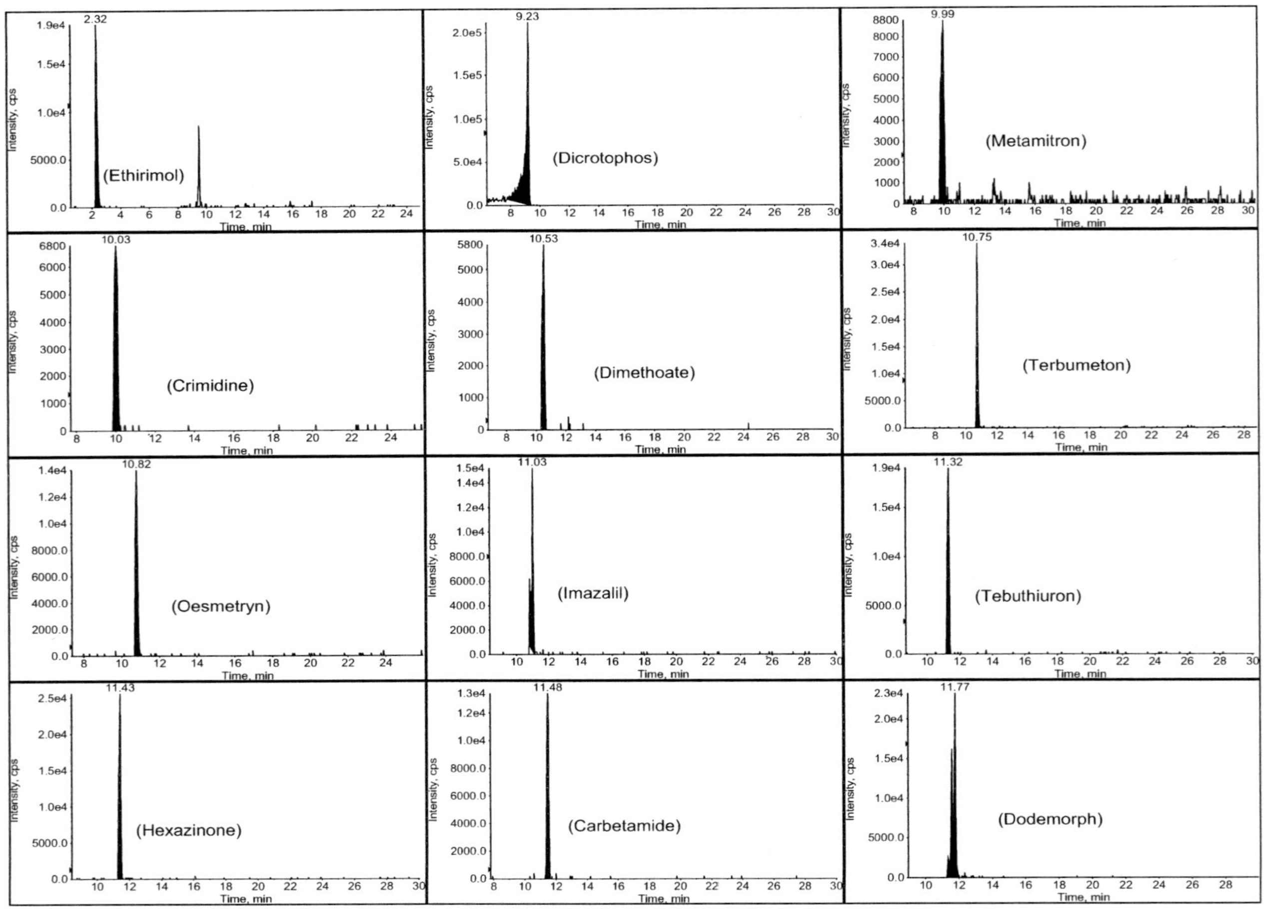
(Ethirimol)
(Dicrotophos)
(Metamitron)
(Crimidine)
(Dimethoate)
(Terbumeton)
(Oesmetryn)
(Imazalil)
(Tebuthiuron)
(Hexazinone)
(Carbetamide)
(Dodemorph)
Intensity, cps
Time, min

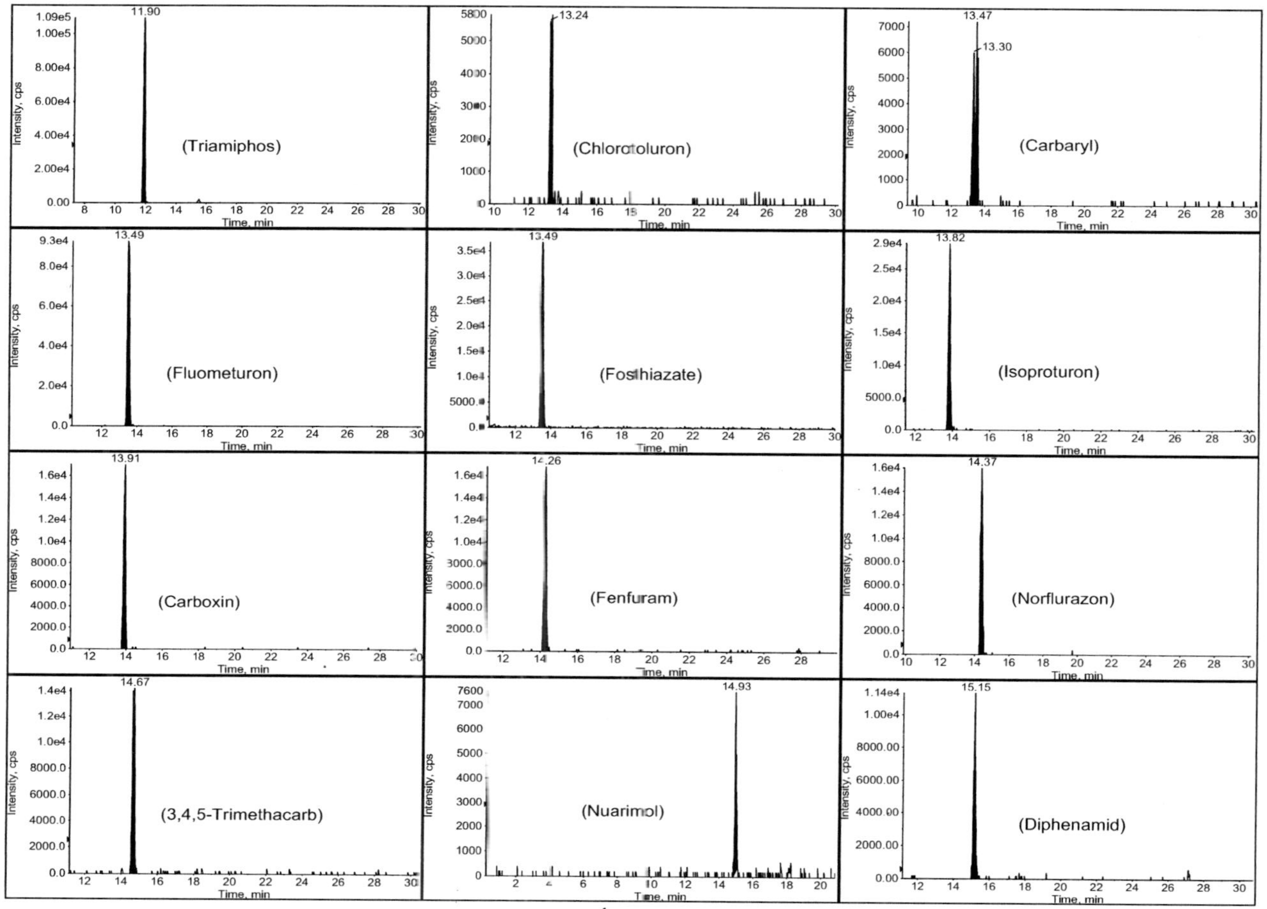
(Triamiphos)
11.90
(Chlorotoluron)
13.24
(Carbaryl)
13.47
13.30
(Fluometuron)
13.49
(Fosthiazate)
13.49
(Isoproturon)
13.82
(Carboxin)
13.91
(Fenfuram)
14.26
(Norflurazon)
14.37
(3,4,5-Trimethacarb)
14.67
(Nuarimol)
14.93
(Diphenamid)
15.15
Intensity, cps
Time, min

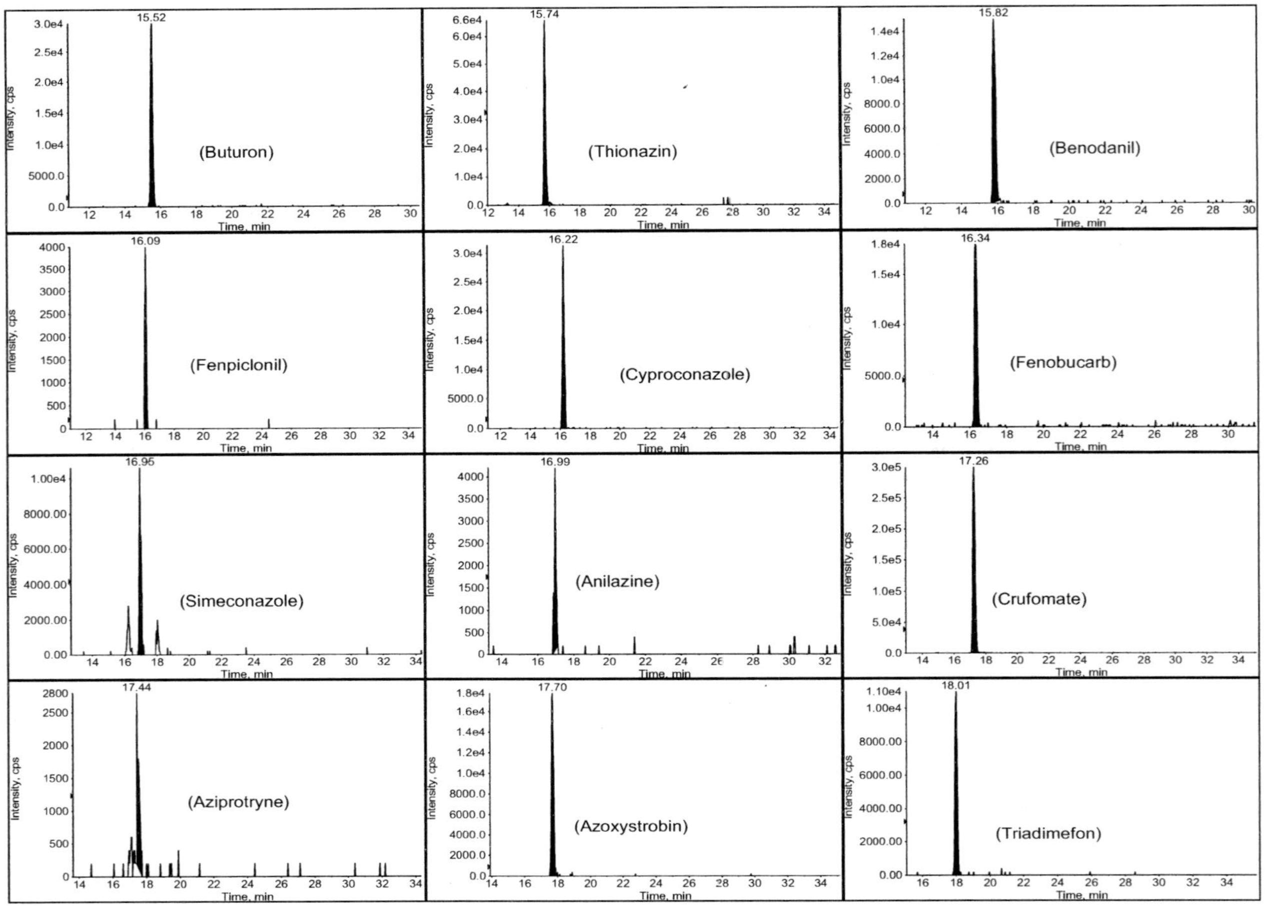

(Buturon)
15.52
(Thionazin)
15.74
(Benodanil)
15.82
(Fenpiclonil)
16.09
(Cyproconazole)
16.22
(Fenobucarb)
16.34
(Simeconazole)
16.95
(Anilazine)
16.99
(Crufomate)
17.26
(Aziprotryne)
17.44
(Azoxystrobin)
17.70
(Triadimefon)
18.01
Intensity, cps
Time, min

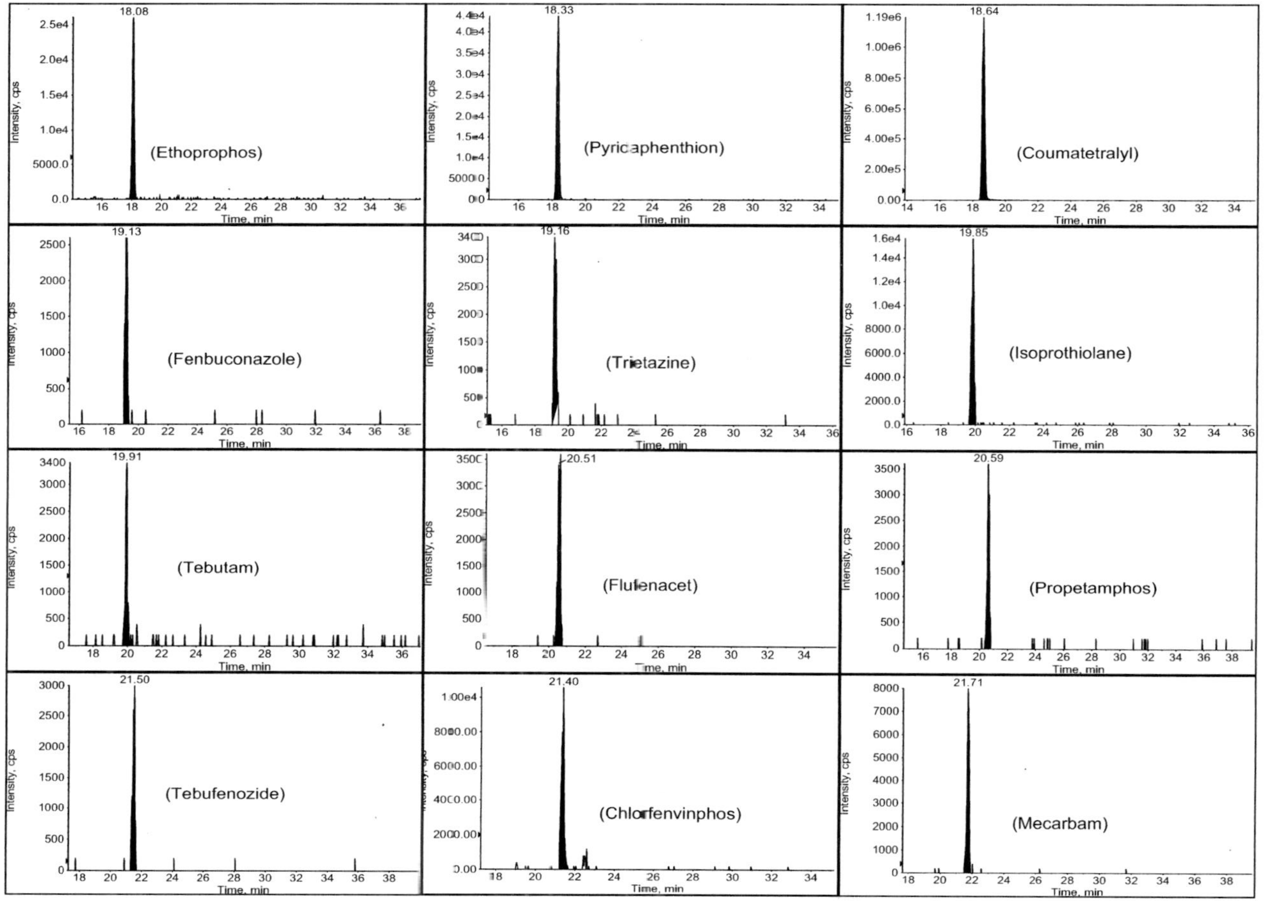

18.08
(Ethoprophos)
18.33
(Pyricaphenthion)
18.64
(Coumatetralyl)
19.13
(Fenbuconazole)
19.16
(Trietazine)
19.85
(Isoprothiolane)
19.91
(Tebutam)
20.51
(Flufenacet)
20.59
(Propetamphos)
21.50
(Tebufenozide)
21.40
(Chlorfenvinphos)
21.71
(Mecarbam)
Intensity, cps
Time, min

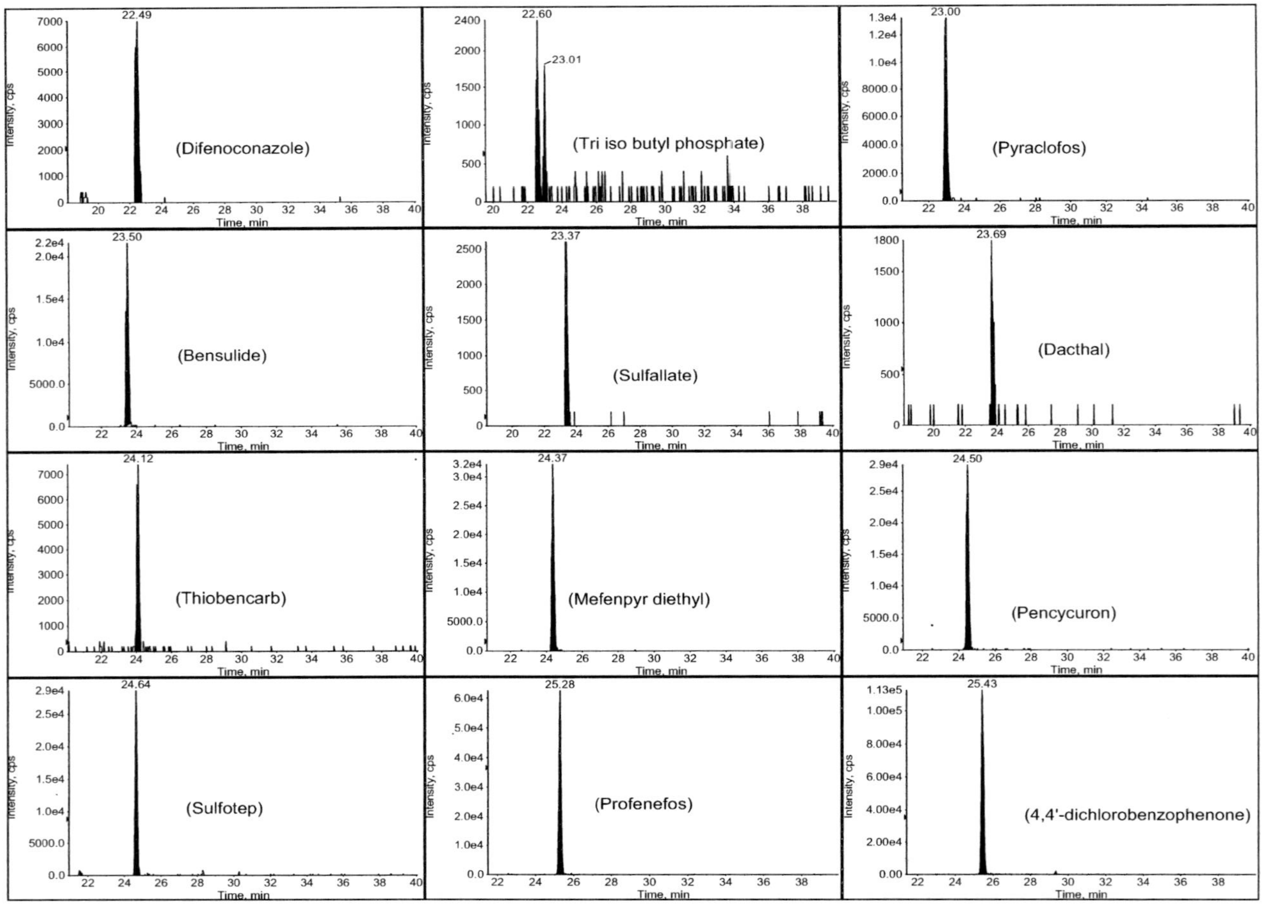

22.49
(Difenoconazole)
22.60
23.01
(Tri iso butyl phosphate)
23.00
(Pyraclofos)
23.50
(Bensulide)
23.37
(Sulfallate)
23.69
(Dacthal)
24.12
(Thiobencarb)
24.37
(Mefenpyr diethyl)
24.50
(Pencycuron)
24.64
(Sulfotep)
25.28
(Profenefos)
25.43
(4,4'-dichlorobenzophenone)
Intensity, cps
Time, min

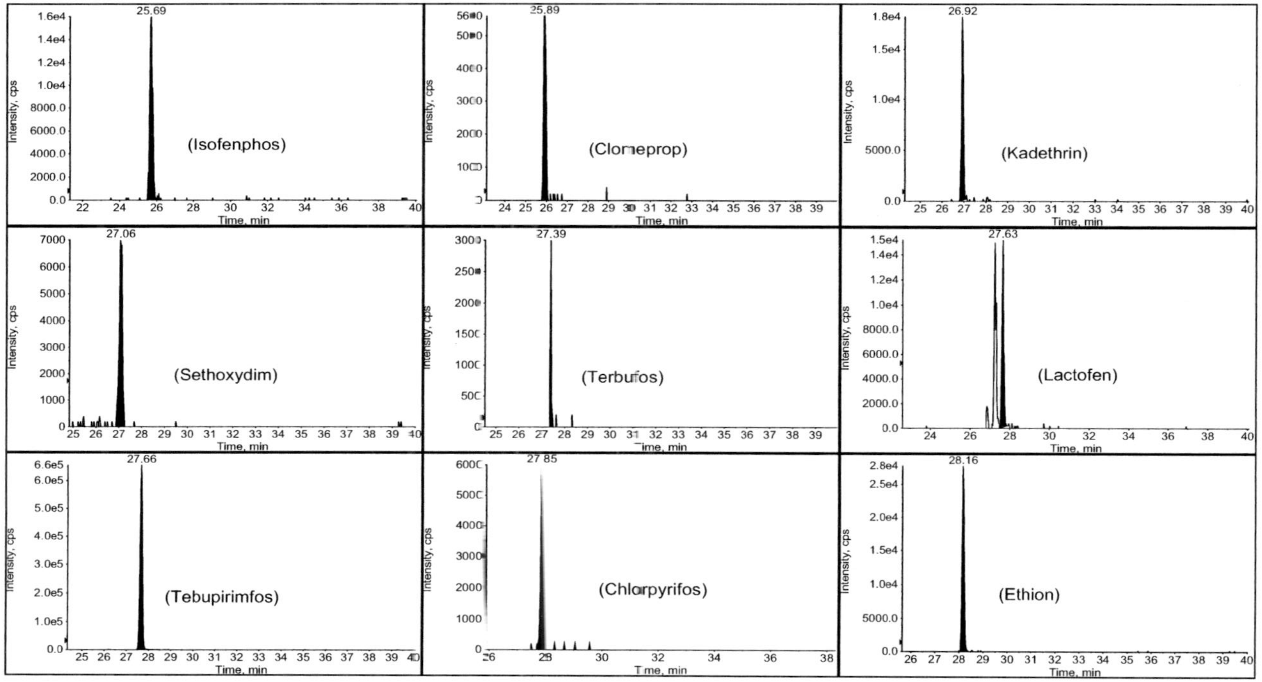
25.69
(Isofenphos)
25.89
(Clomeprop)
26.92
(Kadethrin)
27.06
(Sethoxydim)
27.39
(Terbufos)
27.63
(Lactofen)
27.66
(Tebupirimfos)
27.85
(Chlorpyrifos)
28.16
(Ethion)
Intensity, cps
Time, min

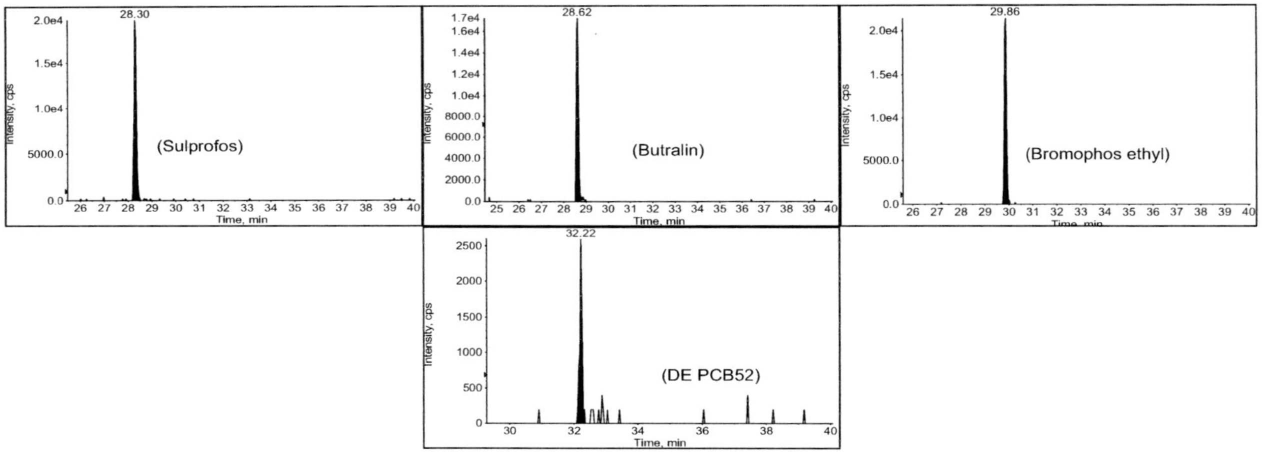
28.30
(Sulprofos)
28.62
(Butralin)
29.86
(Bromophos ethyl)
32.22
(DE PCB52)
Intensity, cps
Time, min

E group

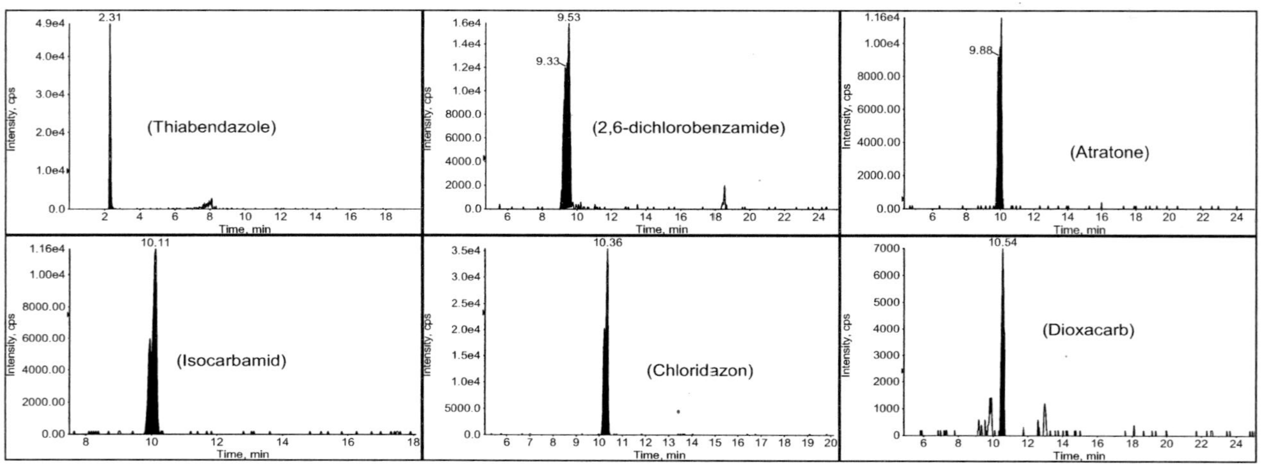
2.31
(Thiabendazole)
9.53
9.33
(2,6-dichlorobenzamide)
9.88
(Atratone)
10.11
(Isocarbamid)
10.36
(Chloridazon)
10.54
(Dioxacarb)
Intensity, cps
Time, min

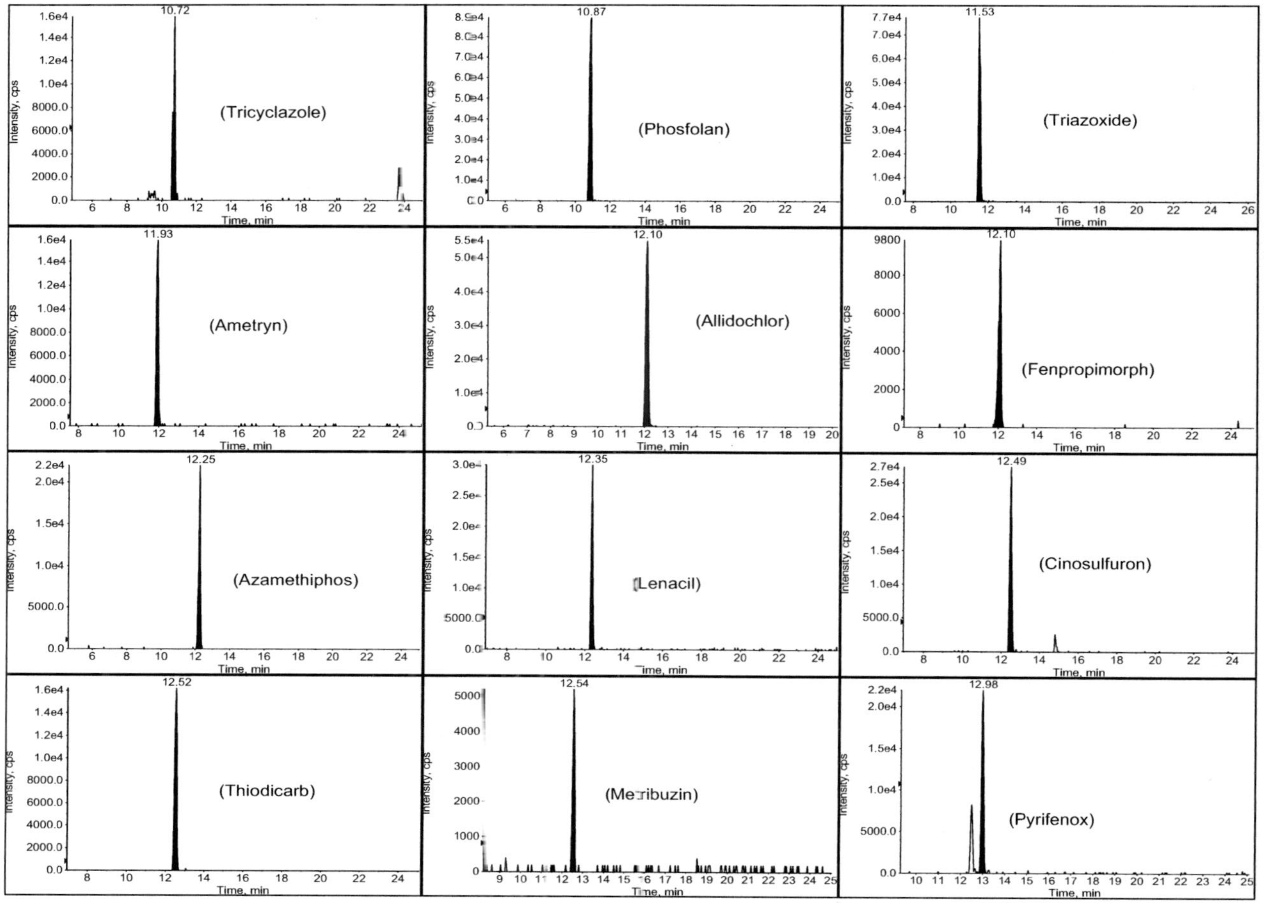

(Tricyclazole)
10.72
(Phosfolan)
10.87
(Triazoxide)
11.53
(Ametryn)
11.93
(Allidochlor)
12.10
(Fenpropimorph)
12.10
(Azamethiphos)
12.25
(Lenacil)
12.35
(Cinosulfuron)
12.49
(Thiodicarb)
12.52
(Metribuzin)
12.54
(Pyrifenox)
12.98
Intensity, cps
Time, min

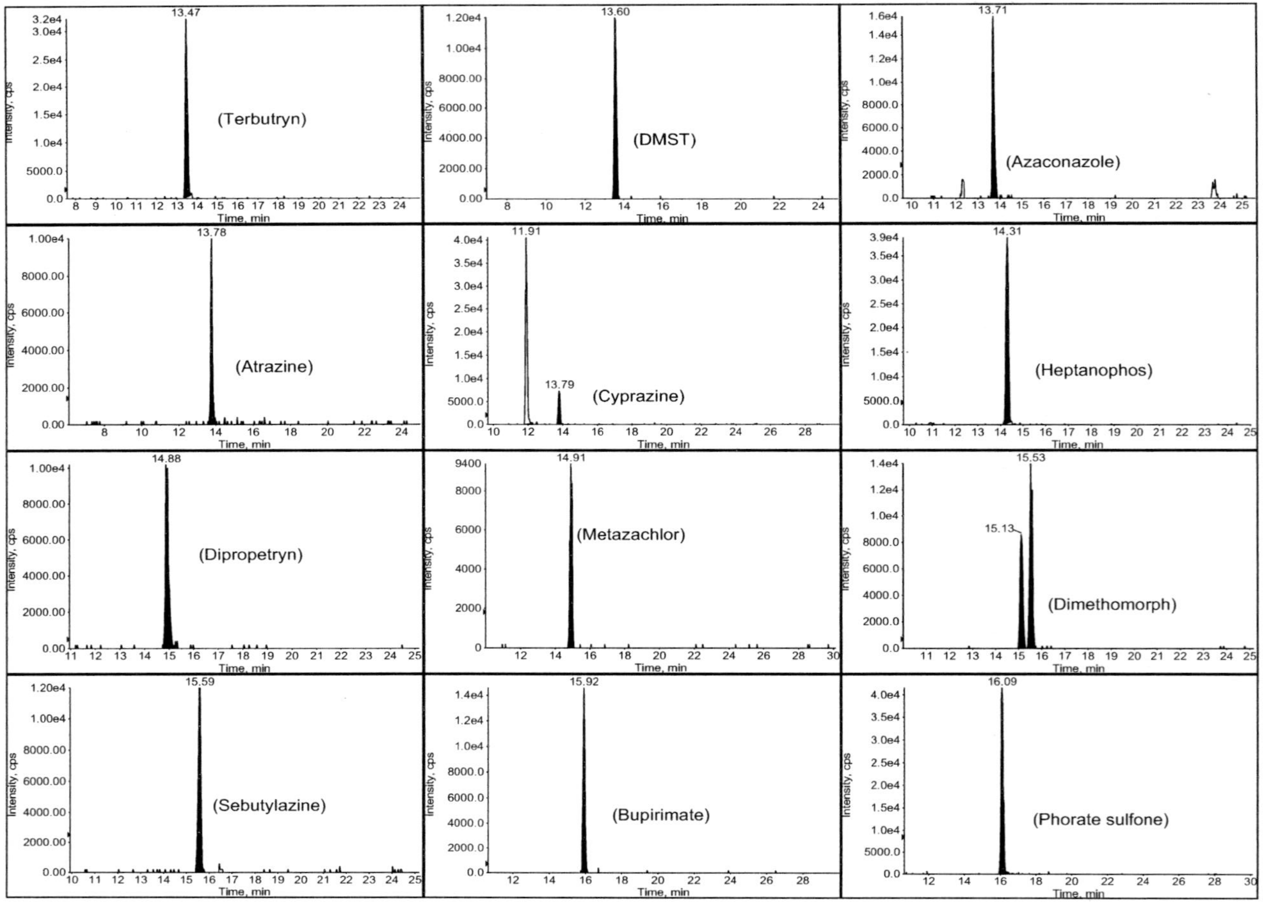
(Terbutryn)
13.47
(DMST)
13.60
(Azaconazole)
13.71
(Atrazine)
13.78
(Cyprazine)
11.91
13.79
(Heptanophos)
14.31
(Dipropetryn)
14.88
(Metazachlor)
14.91
(Dimethomorph)
15.13
15.53
(Sebutylazine)
15.59
(Bupirimate)
15.92
(Phorate sulfone)
16.09
Intensity, cps
Time, min

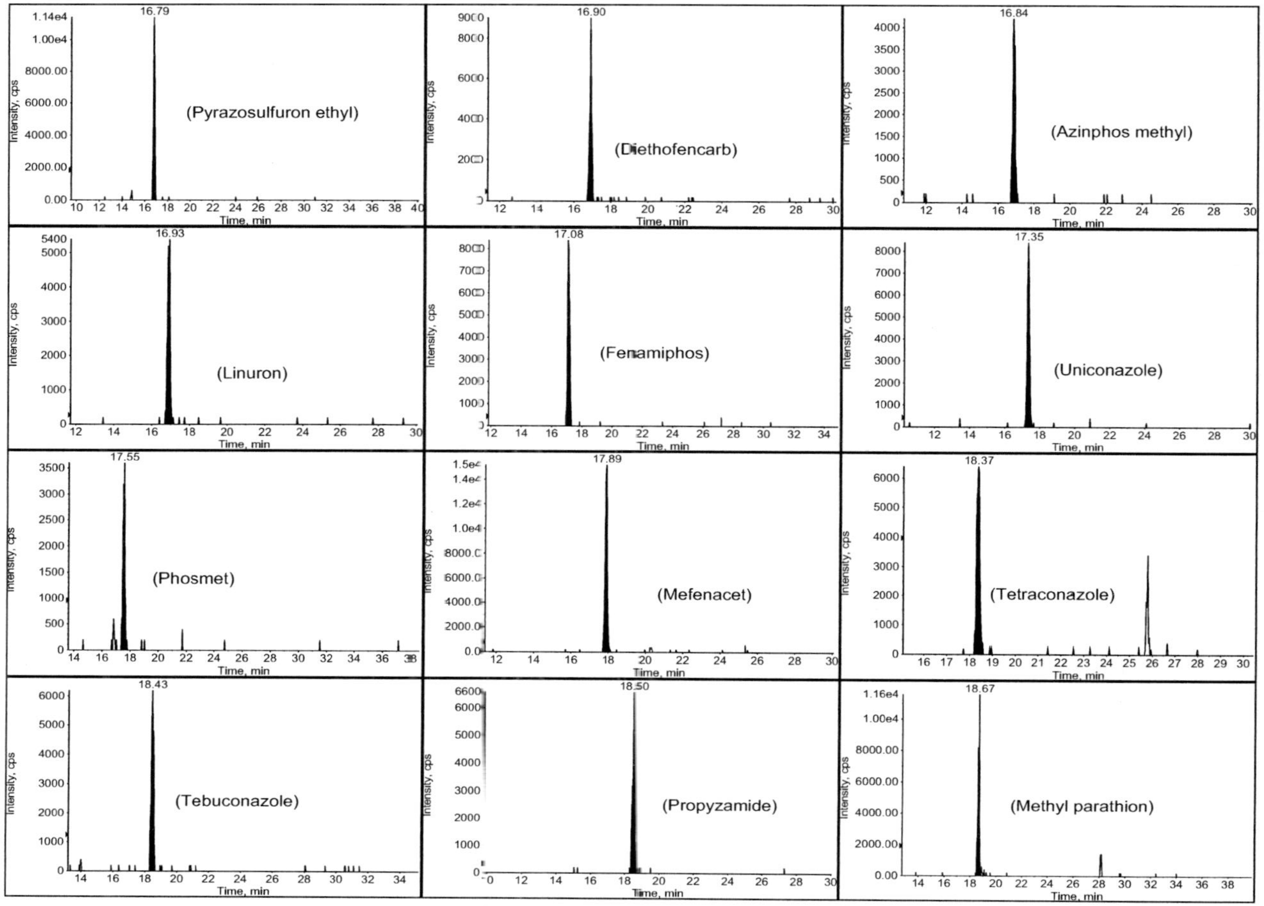
(Pyrazosulfuron ethyl)
16.79
(Diethofencarb)
16.90
(Azinphos methyl)
16.84
(Linuron)
16.93
(Fenamiphos)
17.08
(Uniconazole)
17.35
(Phosmet)
17.55
(Mefenacet)
17.89
(Tetraconazole)
18.37
(Tebuconazole)
18.43
(Propyzamide)
18.50
(Methyl parathion)
18.67
Intensity, cps
Time, min

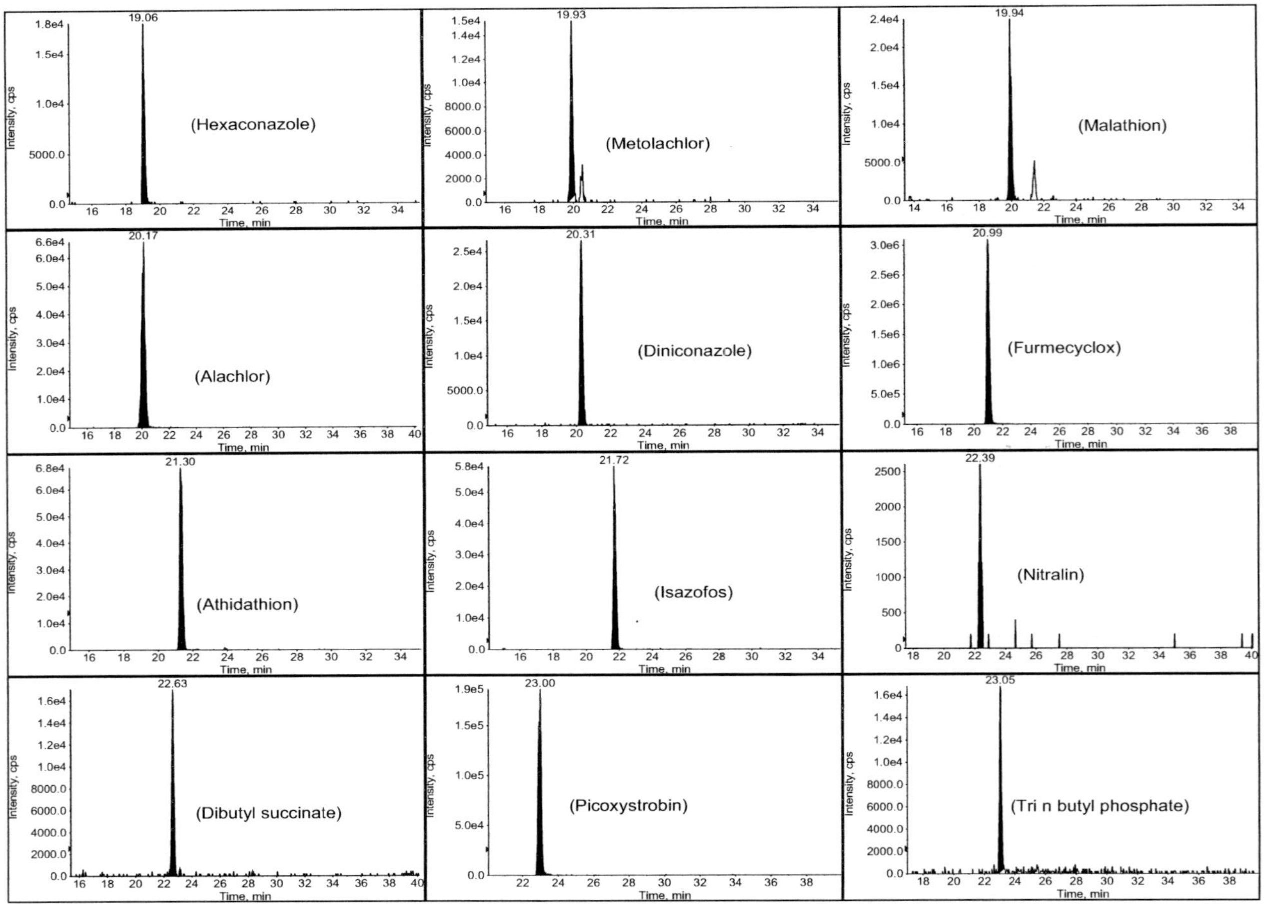
(Hexaconazole)
19.06
(Metolachlor)
19.93
(Malathion)
19.94
(Alachlor)
20.17
(Diniconazole)
20.31
(Furmecyclox)
20.99
(Athidathion)
21.30
(Isazofos)
21.72
(Nitralin)
22.39
(Dibutyl succinate)
22.63
(Picoxystrobin)
23.00
(Tri n butyl phosphate)
23.05
Intensity, cps
Time, min

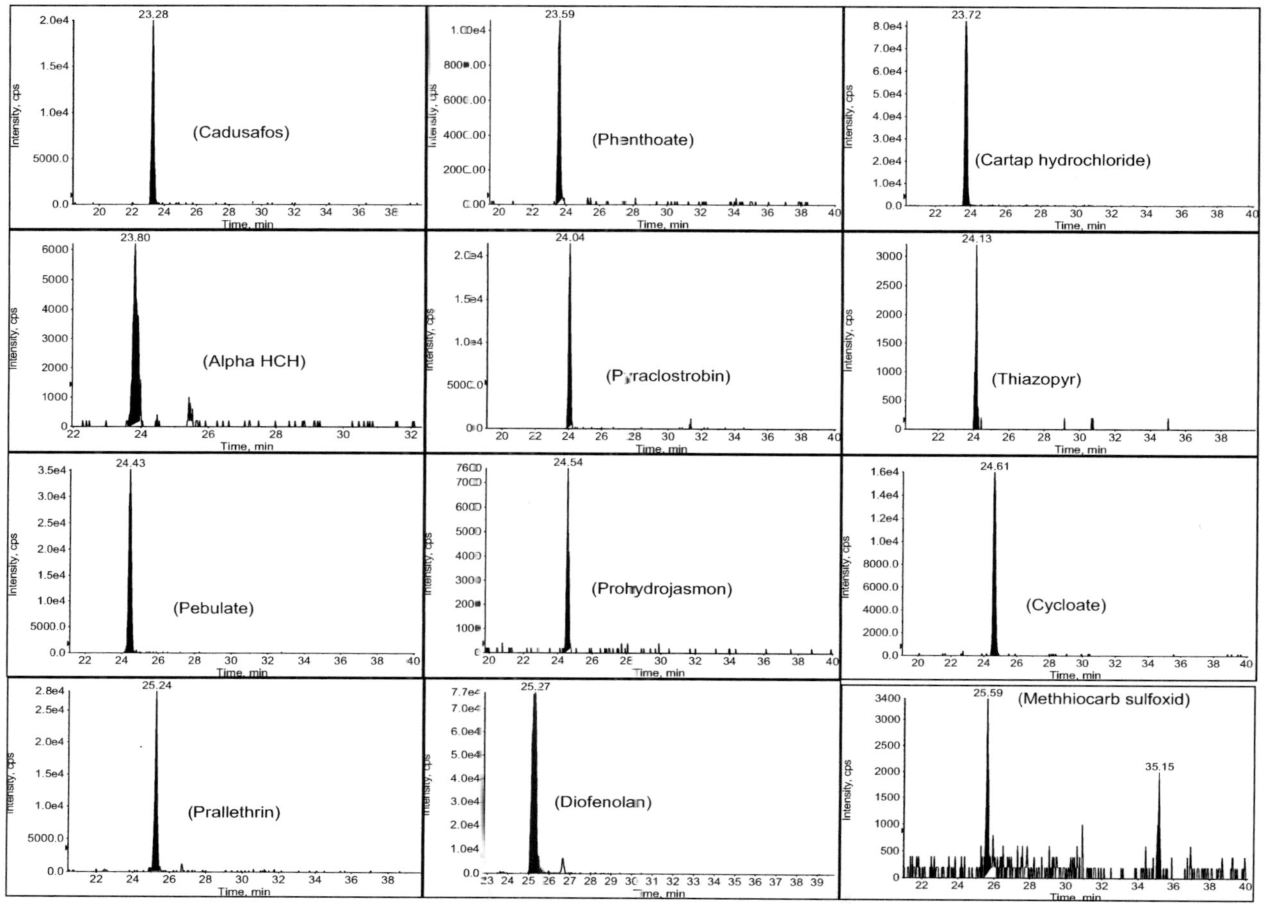

(Cadusafos)
(Phenthoate)
(Cartap hydrochloride)
(Alpha HCH)
(Pyraclostrobin)
(Thiazopyr)
(Pebulate)
(Prohydrojasmon)
(Cycloate)
(Prallethrin)
(Diofenolan)
(Methhiocarb sulfoxid)

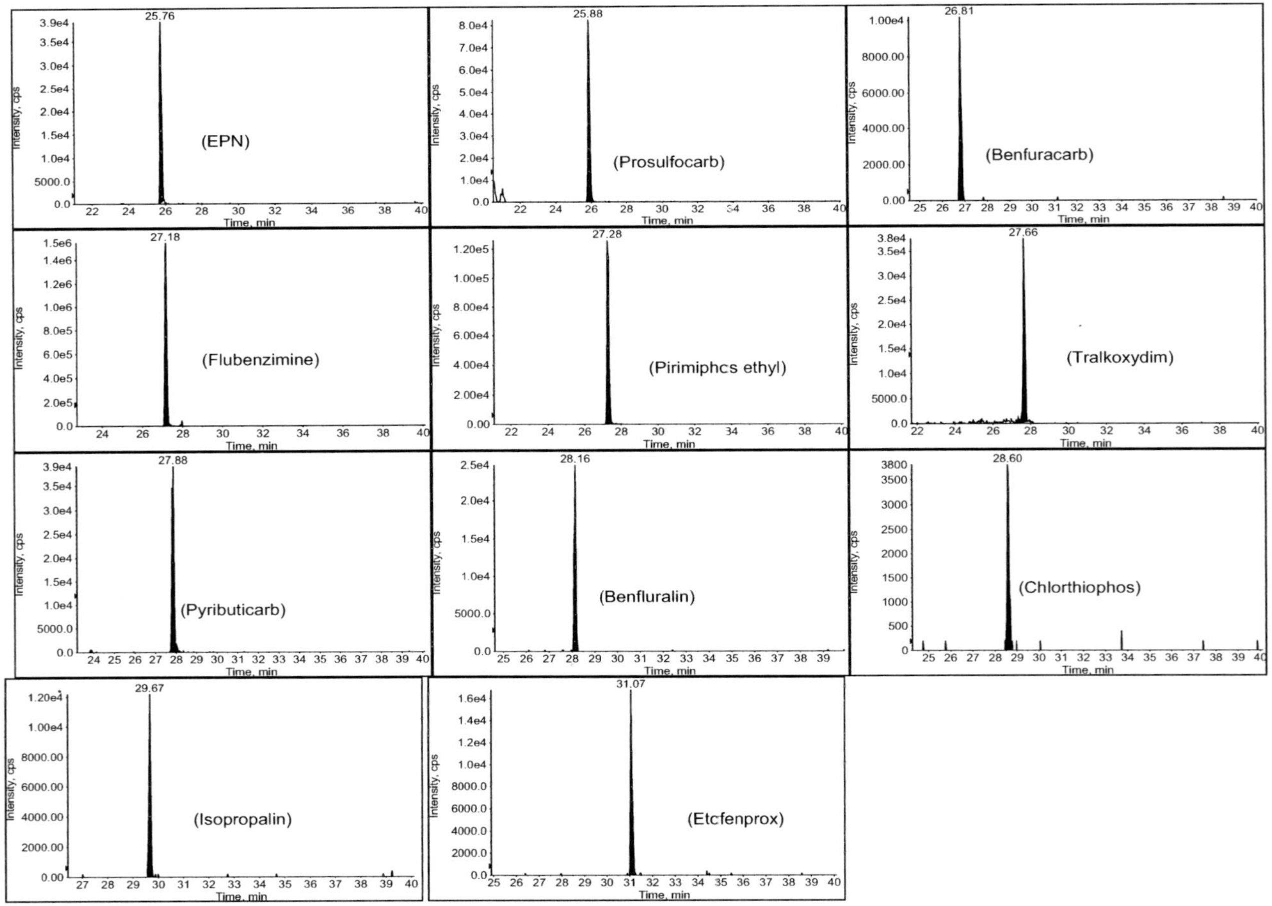

Figure 6. A, B, C, D and E five groups MRM chromatograms of 372 standards and related chemicals.

Table 7. Pesticides unfit for instrumental analysis by GC-MS

No.	Compound	No.	Compound
1	1,2-dibromo ethane	11	Methyl bromide
2	1,2-dichloro ethane	12	Nicotine
3	1-naphthyl acetic acid	13	Oxine-Copper
4	Bromide	14	Propineb
5	Carbon disulphide	15	Quinmerac
6	Carbon tetrachloride	16	S 421(Octachlorodipropyl ether)
7	Guazatine triacetate	17	Sulfur
8	Inorganic bromide	18	Zineb
9	Mancozeb	19	Ziram
10	Maneb		

Table 8. 298 Pesticides with no peaks or extremely low sensitivity by GC-MS

No.	Compound	No.	Compound
1	2,4-DB	33	Asulam
2	2,4,5-T	34	Atrazine
3	2,4-D	35	Azaconazole
4	2,6-Difluorobenzoic acid	36	Azamethiphos
5	3,4,5-Trimethacarb	37	Azocyclotin
6	3-hydroxy carbofuran	38	Azoxystrobin
7	3-phenylphenol	39	Benazolin
8	4-Chlorophenoxyacctic acid	40	Bendiocarb
9	6-chloro-4-hydroxy-3-phenyl-pyridazin	41	Benomyl
10	Abamectin	42	Bensulfuron-methyl
11	Acephate	43	Bensulide
12	Acifluorfen	44	Bensultap
13	Aclonifen	45	Bentazone
14	Aldicarb	46	Benzoximate
15	Aldicarb sulfone	47	Binapacryl
16	Aldicarb- sulfoxide	48	Bioresmethrin
17	Aldimorph	49	Brompyrazon
18	Aldoxycarb	50	Butocarboxim
19	Alloxydim-sodium	51	Butocarboxim-sulfoxide
20	Amidosulfuron	52	Butoxycarboxim
21	Aminocarb	53	Butoxycarboxim-sulfoxid
22	Amitrole	54	Buturon
23	Amobam	55	Camphechlor
24	Anilazine	56	Carbaryl
25	Aroclor 1221	57	Carbendazim
26	Aroclor 1232	58	Carbofuran
27	Aroclor 1242	59	Cartap hydrochloride
28	Aroclor 1248	60	Chloramphenicolum
29	Aroclor 1254	61	Chlorbenzuron
30	Aroclor 1260	62	Chlordecone (Kepone)
31	Aroclor 1262	63	Chlormequat
32	Aroclor 1268	64	Chloropicrin

Table 8. Continued

No.	Compound	No.	Compound
65	Chlorotoluron	112	Dimethoate
66	Chlorphoxim	113	Dimethomorph
67	Chlorpropylate	114	Dimethyl phthalate
68	Chlorpyrifos-methyl-oxon	115	Dimethylaminosulfanilide (DMSA)
69	Chlorsulfuron	116	Dimethylaminosulfotoluidide (DMST)
70	Chlortoluron	117	Dinocap technical mixture of isomers
71	Cinosulfuron	118	Dinoseb
72	Clophen A 30	119	Dinoseb acetate
73	Clophen A 40	120	Dinotefuran
74	Clophen A 50	121	Dioxacarb
75	Clophen A 60	122	Diquat dibromide hydrate
76	Cloprop	123	Dithianon
77	Clopyralld	124	Diuron
78	Cloransulam-methyl	125	DNOC
79	Clothianidin	126	Dodemorph
80	Coumatetralyl	127	Dodine
81	Cyazofamid	128	Emamectin benzoate
82	Cyclanilide	129	Endothal
83	Cycloprothrin	130	Endrin aldehyde
84	Cyhexatin	131	Ethephon
85	Cymoxanil	132	Ethidimuron
86	Cyprodinil	133	Ethiofencarb
87	Cythioate	134	Ethiofencarb-sulfone
88	Dalapon	135	Ethiofencarb-sulfoxide
89	Daminozide	136	Ethirimol
90	Demephion solution	137	Ethylene thiourea
91	Demephion(Tinox)	138	Famoxadone
92	Demeton-S sulfoxide	139	Fenaminosulf
93	Demeton-s-methyl sulfone	140	Fenazaflor
94	Demeton-s-methyl sulfoxide	141	Fenbutatin oxide
95	Desamino-metamitron	142	Fenhexamid
96	Desmedipham	143	Fenoprop (Silvex,2,4,5-TP)
97	Desmethyl-norflurazon	144	Fenoxaprop-ethyl
98	Desmethyl-pirimicarb	145	Fenpropimorph
99	Diafenthiuron	146	Fenthion oxon
100	Dicamba	147	Fenthion oxon sulfone
101	Dichlone	148	Fenthion oxon sulfoxide
102	Dichlorofop-methyl	149	Fentin acetate
103	Dichlorprop	150	Fentin-chloride
104	Diclosulam	151	Fenuron
105	Dienochlor	152	Flazasulfuron
106	Diethyltoluamide	153	Fluazinam
107	Difenoxuron	154	Fluazuron
108	Diflubenzuron	155	Flubenzimine
109	Diflufenzopyr-sodium	156	Flucarbazone-sodium
110	Dimehypo	157	Flumethrin
111	Dimethirimol	158	Flumetsulam

Table 8. Continued

No.	Compound	No.	Compound
159	Fluometuron	206	Methobromuron
160	Fluoroimide	207	Methomyl
161	Fluroxypyr	208	Methoxyfenozide
162	Flusulfamide	209	Methyl isothiocyanate
163	Fluthiacet-Methyl	210	Metiram
164	Fomesafen	211	Metolcarb
165	Forchlorfenuron	212	Metosulam
166	Formetanate hydrochloride	213	Metoxuron
167	Fosetyl-aluminium	214	Metsulfuron-methyl
168	Fosthiazate	215	Milbemectin
169	Fosthietan	216	Monuron
170	Furathiocarb	217	Neburon
171	Furmecyclox	218	Nicosulfuron
172	Gibberellic acid	219	Nitenpyram
173	Glufosinate ammonium	220	Novaluron
174	Glyphosate	221	Octachlorodipropyl ether
175	Haloxyfop-2-ethoxyethyl	222	Olaquindox
176	Haloxyfop-methyl	223	Omethoate
177	Hydramethylnon	224	Ortho-phenylphenol
178	Imazamox	225	Oryzalin
179	Imazapic	226	Oxamyl
180	Imazapyr	227	Oxolinic acid
181	Imazaquin	228	Oxy-Demeton methyl
182	Imazethapyr	229	Paraquat Dichloride
183	Imibenconazole	230	Pentachlorophenol
184	Imidacloprid	231	Phenkapton
185	Iminoctadine triacetate	232	Phenmedipham
186	Indoxacarb	233	Phorate sulfoxide
187	Iodosulfuron-methyl	234	Phosfolan
188	Ioxynil	235	Phosmet-oxon
189	Iprovalicarb	236	Phoxim
190	Isoproturon	237	Phthalic acid, di-(2-ethylhexyl) ester
191	Isouron	238	Phthalic acid, dibutyl ester
192	Isoxaben	239	Phthalic acid, dicyclohexyl ester
193	Isoxaflutole	240	Picloram
194	Kadethrin	241	Pirmicarb-desmethyl
195	Kelevan	242	Primisulfuron-methyl
196	Lufenuron	243	Prohexadione-calcium
197	Maleic hydrazide	244	Promecarb
198	MCPB	245	Propamocarb
199	Mepiquat chloride	246	Propoxycarbazone-sodium
200	Merphos	247	Propylene thiourea
201	Mesosuifuron-methyl	248	Prosulfuron
202	Mesotrion	249	Prothoate
203	Metaldehyde	250	Pymetrozin
204	Methiocarb	251	Pyrazolynate
205	Methiocarb sulfoxide	252	Pyrazosulfuron-ethyl

Table 8. Continued

No.	Compound	No.	Compound
253	Pyrethrins	276	Thiofanox
254	Pyridate	277	Thiofanox-sulfone
255	Pyridinitril	278	Thiofanox-sulfoxide
256	Pyrifenox	279	Thiophanate-methyl
257	Pyrithiobac sodium	280	Thiophanat-ethyl
258	Quizalofop-ethyl	281	Tiamulin-fumerate
259	Rimsulfuron	282	Tolfenpyrad
260	Rotenone	283	Triacloprid
261	Spinosad	284	Triasulfuron
262	Sulfanitran	285	Triazoxide
263	Sulfentrazone	286	Tribenuron-methyl
264	Sulfosulfuron	287	Trichlorfon
265	TCA-sodium	288	Trichlorphon
266	Tebufenozide	289	Triclopyr
267	Teflubenzuron	290	Tridemorph
268	Temephos	291	Triflumuron
269	Tert-Butylamine	292	Triflusulfuron-methyl
270	Thiabendazole	293	Triforine
271	Thiacloprid	294	Trimethylsulfonium iodide
272	Thidiazuron	295	Vamidothion
273	Thifensulfuron-methyl	296	Vamidothion sulfone
274	Thiocyclam hydrogenoxalate	297	Vamidothion sulfoxide
275	Thiodicarb	298	Zeta cypermethrin

Tests found that 32 pesticides had very poor stability, and the prepared mixed standard solutions were quickly decomposed, being incapable of added into a series of tests like fortification recoveries and so on, which were declared unsuitable for analysis by the said method as well. See Table 9.

The remaining 509 pesticides were prepared into five kinds of standard mixed solutions such as A, B, C, D and E, which were then added into practical samples to observe their stabilities in the samples and their suitability for extraction and cleanup conditions. Results showed that another 34 pesticides were unfit for analysis by GC-MS, with the main cause being the low recoveries 30%. See Table 10.

Table 9. 32 pesticides with poor stabilities unfit for analysis by GC-MS

No.	Compound	No.	Compound
1	Amidithion	9	Chloridazon
2	Barban	10	Chlorimuron-ethyl
3	Benfuracarb	11	Chlorthiamid
4	bifenox	12	Clofentezine
5	Bromoxynil octanoate	13	Cyphenothrin
6	Captan	14	Cyromazine
7	Carbetamide	15	Dazomet
8	Chinomethionat	16	Demeton –O

Table 9. Continued

No.	Compound	No.	Compound
17	Demeton -S	25	TEPP
18	Dicrotophos	26	terbacil
19	Difenzoquat-methyl sulfate	27	Thiram
20	Ethoxyquin	28	Tralomethrin
21	Leptophos oxon	29	Tricyclazole
22	Metconazole	30	Triflumizole
23	monocrotophos	31	Triticonazole
24	Probenazole	32	XMC

Table 10. 34 Pesticides with recoveries lower than 30% unfit for analysis by GC-MS

No.	Compound	No.	Compound
1	Ametryn	18	Fenazaquin
2	Amitraz	19	Fenpropidin
3	Buprofezin	20	Folpet
4	Captafol	21	Fuberidazole
5	Carbosulfan	22	Halosulfuran-methyl
6	Chlordimeform	23	Methfuroxam
7	Chlorothalonil	24	Methyl trithion
8	Chromafenozide	25	Naled (Dibrom)
9	Chrysene	26	Phthalic acid, benzyl butyl ester
10	Cloquintocet-mexyl	27	Pirimiphos-ethyl
11	Cyanazine	28	Prochloraz
12	Dichlofluanid	29	Prometon
13	Dimethipin	30	Resmethrin
14	Dinoterb	31	Simetryn
15	Dipropetryn	32	Spiroxamine
16	Ditalimfos	33	Tepraloxydim
17	Erbon	34	Triamiphos

Table 11. 43 pesticides unsuitable for analysis by LC-MS-MS with ESI source

No.	Compound	No.	Compound
1	1,2-dibromo ethane	15	Chinomethionat
2	Aroclor 1221	16	Clofentezine
3	Aroclor 1232	17	Clophen A 30
4	Aroclor 1242	18	Clophen A 40
5	Aroclor 1248	19	Clophen A 50
6	Aroclor 1254	20	Clophen A 60
7	Aroclor 1260	21	Desamino-metamitron
8	Aroclor 1262	22	Desmethyl-pirimicarb
9	Aroclor 1268	23	Diclosulam
10	Atrazine-Desethyl	24	Diflubenzuron
11	BDMC-1	25	Dithianon
12	BDMC-2	26	Fluroxypr-1-methylheptyl ester
13	Benomyl	27	Fosetyl-aluminium
14	Carbon disulphide	28	Inorganic bromide

Table 11. Continued

No.	Compound	No.	Compound
29	Mancozeb	37	Oxolinic acid
30	Maneb	38	Paraoxon-ethyl
31	Mesosuifuron-methyl	39	Sulfosulfuron
32	Metaldehyde	40	Sulfur
33	Methyl bromide	41	Thidiazuron
34	Metiram	42	Zineb
35	Olaquindox	43	Ziram
36	Oxine-Copper		

2. LC-MS-MS Characteristics

After analysis of 810 pesticides, 43 pesticides were found to have difficulty of ionization with low ion ratio, which was unsuitable for analysis by LC-MS-MS with ESI source. See Table 11.

Syringe pump was adopted to directly infuse analyte, precursor ion of the remaining 767 pesticides and related chemicals were found by Q1 scan mode of ESI source of LC-MS-MS before product ion scan, through which two transations were selected for each compounds. After analysis of the information obtained, 137 pesticides were found unable to locate the precursor ion, which was unfit for analysis by LC-MS-MS. See Table 12.

Table 12. 137 pesticides unable to locate the precursor ions

No.	Compound	No.	Compound
1	1-naphthyl acetic acid	25	Carbofenothion
2	2,3,4,5-tetrachloroaniline	26	Chlorbenside
3	2,3,4,5-tetrachloroanisole	27	Chlorbenside sulfone
4	2,3,5,6-tetrachloroaniline	28	Chlordecone
5	2,4,5-T	29	Chlorethoxyfos
6	2,4'-DDT	30	Chlorfenprop-methyl
7	3,5-Dichloroaniline	31	Chlorfurenol
8	Acenaphthene	32	Chlorobenzilate
9	Acequinocyl	33	Chloroneb
10	Alpha-Cypermethrin	34	Chlozolinate
11	Amidosulfuron	35	Chrysene-d12
12	Amitraz	36	Cis-1,2,3,6-Tetrahydrophthalimide
13	Anthraquinone	37	Cis-Chlordanc
14	Aramite	38	Cyanohos
15	Barban	39	Cyflufenamid
16	Benzoximate	40	Cyhexatin
17	Beta-HCH	41	Cypermethin
18	Binapacryl	42	DDE, 2.4'-
19	bromocylen	43	Delta-HCH
20	Bromofos	44	DE-PCB 101
21	Bromoxynil octanoate	45	DE-PCB 138
22	Camphechlor	46	DE-PCB 153
23	Captafol	47	DE-PCB 28
24	Captan	48	Desbrom- leptophos

Table 11. Continued

No.	Compound	No.	Compound
49	Dicamba	94	Musk ambrette
50	Dichlobenil	95	Musk xylene
51	Dichlone	96	Naproanilide
52	Dichloran	97	Nitrofen
53	Dicofol	98	Oxychlordane
54	Dimehypo	99	Oxy-Demeton methyl
55	Dimethipin	100	P,p'-DDD
56	Dinocap	101	Pentachloroaniline
57	Diquat dibromide hydrate	102	Perthane
58	Dodine	103	Phenanthrene
59	Endosulfan	104	Phthalide
60	Endrin	105	Phthalimide
61	Endrin aldehyde	106	Plifenate
62	Ethalfluralin	107	p'p-DDE
63	Fenaminosulf	108	p'p-DDT
64	Fenazaflor	109	Primisulfuron-methyl
65	Fenoprop (Silvex,2,4,5-TP)	110	Prohexadione-calcium
66	Fenthion oxon	111	Propaphos
67	Fentin acetate	112	Propineb
68	Flazasulfuron	113	Prosulfuron
69	Flumetralin	114	Pyridate
70	Fluoroimide	115	Pyriminobac-methyl(z)
71	fluotrimazole	116	Quinclorac
72	Formetanate hydrochloride	117	Quintozene
73	Gamma-Cyhalothrin	118	Rimsulfuron
74	Guazatine triacetate	119	Ronnel
75	Halfenprox	120	Rotenone
76	Halosulfuran-methyl	121	Octachlorodipropyl ether
77	HCH, epsilon-	122	Silafluofen
78	Imazapic	123	Tecnazene
79	Iminoctadine triacetate	124	Tefluthrin
80	Iprodione	125	Terbufos sulfone
81	Isobenzan	126	Tert-butylamine
82	Isodrin	127	Tetrasul
83	Isoxadifen-ethyl	128	Thifluzamide
84	Leptophos oxon	129	Thiocyclam hydrogenoxalate
85	Methiocarb Sulfone	130	Tralomethrin
86	Methomyl	131	Trans-Chlodane
87	Methoxychlor	132	Trans-nonachlor
88	Methyl isothiocyanate	133	Trichloronat
89	Methyl trithion	134	Triclopyr
90	Metsulfuron-methyl	135	Tridiphane
91	Milbemectin	136	Vinclozolin
92	Milbemectin	137	Zeta cypermethrin
93	Mirex		

For the remaining 629 pesticides and related chemicals, mass spectrum data database were established under 3 different Collision Energy (CE), according to the database, 629 pesticides and related chemicals were optimized under LC-MS-MS conditions. The ion source of mass spectrum was ESI, and the scanning mode was positive ions.

Tests found that 42 pesticides and related chemicals gave a low ESI response or no signal at all, which was unfit for analytical requirement. See Table 13. 59 pesticides required the negative ion scanning mode for analysis of pesticides and related chemicals. See Table 14. Only 1 product ion was located, and 16 pesticides and related chemicals were found being incapable of qualification. See Table 15. 47 pesticides and related were unable to identify due to other causes. See Table 16.

For the remaining 465 pesticides and related chemicals, they were divided into give groups of A, B, C, D and E according to retention time and were prepared into mixed standard solutions, which were then added into oats, barley, maize, rice and wheat samples in order to investigate their stability and their suitability for extraction and cleanup conditions. Tests found that 93 pesticides and related chemicals had less than 30% fortification recoveries at two levels; some even had no recoveries, with coefficients of variations more than 30%, which were declared to be unsuitable for analysis by the said method. See Table 17.

Eventually, 475 pesticides were chosen by gradual selection to be suitable for determination by GC-MS; 372 were suitable for determination by LC-MS-MS.

Table 13. 42 pesticides with too much low sensitivity unfit for analytical requirements

No.	Compound	No.	Compound
1	Abamectin	22	Endrin ketone
2	Aldrin-R	23	Esfenvalerate
3	Benfuresate	24	Ethiofencarb-sulfone
4	Benzofenap	25	Famoxadone
5	Bifenthrin	26	Fenamidone
6	Biphenyl	27	Fenbutatin oxide
7	Butocarboxim	28	Flumethrin
8	Carbosulfan	29	Haloxyfop-2-ethoxyethyl
9	Carfentrazone-ethyl	30	Methomyl-oxime
10	Chlorfenapyr	31	Methoprene
11	Chloropicrin	32	Musk ketone
12	Chlorothalonil	33	Nitrothal-isopropyl
13	Chlorthiamid	34	Oxadiazone
14	Cycloprothrin	35	Phenkapton
15	Cymexanil	36	Phthalic acid, di-(2-ethylhexyl) ester
16	Cyphenothrin	37	Procymidone
17	DE-PCB 118	38	Pyroquilon
18	Desmethyl-norflurazon	39	TCA-sodium
19	Dieldrin	40	TCMTB
20	Dimepiperate	41	Tribenuron-methyl
21	DMSA	42	XMC

Table 14. 59 pesticides required the negative ion scanning for analysis of pesticides and related chemicals

No.	Compound	No.	Compound
1	2,4-D	31	Flumioxazin
2	2,4-DB	32	Fluorodifen
3	2,6-Difluorobenzoic acid	33	Fluroxypyr
4	3-phenylphenol	34	Fomesafen
5	4-Chlorophenoxyacetic acid	35	Forchlorfenuron
6	Acifluorfen	36	Gibberellic acid
7	Bentazone	37	Imibenconazole
8	Bifenox	38	imibenconazole
9	Bromoxynil	39	Ioxynil
10	chlorfenethol	40	Isoxathion
11	Chlorpropham	41	Kelevan
12	Chromafenozide	42	Lambda-Cyhalothrin
13	Cis-1,2,3,6-Tetrahydrophthalimide	43	Lufenuron
14	Clopyralld	44	MCPB
15	Cymoxanil	45	Mesotrion
16	Dalapon	46	Metominostrobin-(E)
17	Dichlorofop-methyl	47	metominostrobin-(Z)
18	Dichlorprop	48	musk tibeten
19	Dicloran	49	octachlorostyrene
20	Dienochlor	50	pentachlorobenzene
21	Diflufenzopyr-sodium	51	Phthalic acid, dibutyl ester
22	dinobuton	52	Picloram
23	Dinoterb	53	Propoxycarbazone-sodium
24	DNOC	54	Pyrithlobac sodium
25	Endosulfan-Sulfate	55	Sulfanitran
26	Endothal	56	Sulfentrazone
27	Ethephon	57	Teflubenzuron
28	Fenoxaprop-ethyl	58	Tiamulin-fumerate
29	Fluazinam	59	Tolfenpyrad
30	Flucarbazone-sodium		

Table 15. Only 1 product ion was located, and 16 pesticides and related chemicals were incapable of scientific identification

No.	Compound	No.	Compound
1	2-Phenylphenol	9	IBiphenyl
2	Amitrole	10	Mesotrion
3	Chlorfenson	11	Metolcarb
4	Chlorpropylate	12	Nitrapyrin
5	Chrysene	13	Pentachloroanisole
6	Etoxazole	14	Resmethrin
7	Halosulfuran methyl	15	Triforine
8	Hymexazol	16	Zoxamide

Table 16. 47 pesticides and related chemicals were unable to identify due to other causes

No.	Compound	No.	Compound
1	4,4,-DDD	25	*Dimethylvinphos*
2	Acibenzolar-S-methyl	26	Dinoseb
3	Acrinathrin	27	Esprocarb
4	Acrylamide	28	Fenson
5	Aldicarb	29	Fenthion oxon sulfoxide
6	Alloxydim-sodium	30	Fenvalerate
7	amidithion	31	Fluazifop-butyl
8	Amobam	32	Flucythrinate
9	Benzyladenine	33	Flufenoxuron
10	Bromide	34	Flusulfamide
11	Bromopropylate	35	Glyphosate
12	Butoxycarboxim	36	Iodosulfuron-methyl sodium
13	Butoxycarboxim-sulfoxid	37	isocarbophos
14	Cafenstrol	38	Meleic Hydrazide
15	Carbofuran	39	Omethoate
16	Carbonyl sulfide	40	Promecarb
17	Chlorphenapyr	41	Pyrimidifen
18	Cloprop	42	Quinmerac
19	Cyfluthrin	43	Quinoxyphen
20	Dalapon acid	44	Schradan/OMPA
21	Demephion(Tinox)	45	Tetradifon
22	DE-PCB 180	46	Triacloprid
23	dicapthon	47	Uniconazole
24	dicapthon		

Selection of ASE Conditions for GC-MS

On the basis of ASE operating conditions reported by Richter et al. [23], in US Environment Protection Agency (USEPA) standard SW-846 [24], and in German Standard L00.00-34 [25], the extraction conditions selected for the ASE instrument were: pressure 1,500 psig, sample cell temperature 80°C, heating time 5 min, static extraction time 3 min, number of cycles 2, and purge time for nitrogen gas 100 s. To increase extraction efficiency Celite 545 was chosen as desiccant and dispersive during sample preparation to reduce moisture and increase permeation of solvents into the grain matrixes. When the extraction conditions had been fixed, extractions were conducted to compare the extraction efficiency of the three solvents acetonitrile, hexane+acetone (2+1), and cyclohexane+ethyl acetate (1+1). It was found that the efficiency of extraction of most of the pesticides by these three solvents did not vary much, with some exceptions. Less vegetable fat was obtained with acetonitrile as extracting solvent than with the other two solvents. In the proposed method the extracts needed evaporating to dryness and redissolving in acetonitrile before three-column cleanup after extraction with hexane+acetone or cyclohexane+ethyl acetate. So not only was less fat extracted but evaporation was also omitted when acetonitrile was used as the extracting solvent. The study by Mastovska et al. [26] also proved the advantage of using acetonitrile as extraction solvent. Acetonitrile was therefore chosen as extraction solvent in this method.

Table 17. 93 pesticides and related chemicals were unsuitable for analysis by the said method

No.	Compound	No.	Compound
1	4,4'-dibromobenzophenone	48	Fludioxonil
2	6-chloro-4-hydroxy-3-phenyl-pyridazin	49	Flufenoxuron
3	Alanycarb	50	Flumetsulam
4	Aldicarb sulfone	51	Fluoroglycofen-ethyl
5	Aldoxycarb	52	Fluoazuron
6	Asulam	53	Formothion
7	Bensulfuron-methyl	54	Heptachlor
8	Bensultap	55	Hexachlorobenzene
9	Butafenacil	56	Hexaflumuron
10	Butocarboxim sulfoxide	57	Hydramethylnon
11	Chloransulam methyl l	58	Imazamox
12	Chlorfluazuron	59	Imazapyr
13	Chlormequat	60	Imazaquin
14	Chlorsulfuron	61	Imazethapyr
15	Chlorthal-dimethyl	62	Imiprothrin
16	Chlorthion	63	Iodosulfuron-methyl
17	Cyazofamid	64	Leptophos
18	Cyromazine	65	Maleic hydrazide
19	Daminozide	66	Mepiquat chloride
20	Dazomet	67	Methacrifos
21	Demeton s methl sulfoxide	68	Methamidophos
22	Demeton s methyl sulfone	69	Metosulam
23	Demeton-s-methyl	70	Monocrotophos
24	De-pcb 31	71	Nicotine
25	Diafenthiuron	72	Nitenpyram
26	Dialifos	73	Novaluron
27	Dichlofenthion	74	Oryzalin
28	Dichlofluanid	75	Oxamyl
29	Dichlorvos	76	Pentachlorophenol
30	Dimefox	77	Phosmet-oxon
31	Dimethirimol	78	Propoxycarbazone-sodium
32	Dinoseb acetate	79	Propamocarb
33	Dinotefuran	80	Propylene thiourea
34	Disulfoton	81	Pymetrozin
35	Emamectin benzoate	82	Spinosad
36	EPTC	83	Tau fluvalinate
37	Ethiofencarb	84	Thiamethoxam
38	Ethiofencarb sulfoxide	85	Thiofanox-sulfoxide
39	Ethylene thiourea	86	Thiophanat ethyl
40	Etridiazol	87	Thiophanate methyl
41	Fenoxanil	88	Thiram
42	Fenpropidin	89	Tolyfluanid
43	Fentin-chloride	90	Transfluthrin
44	Fipronil	91	Tridemorph
45	Florasulam	92	Triflumizole
46	Flubenzimine	93	Vernolate
47	Fluchloralin		

Selection of GPC Conditions for LC-MS-MS

In our previous studies, ASE acetonitrile extraction and SPE cleanup were also adopted for LC-MS-MS analysis. Tests found that after ASE extraction and three-cartridge cleanup some pesticides tended to have very low recoveries, making the pesticide varieties for instrumental analysis decrease apparently. Therefore, in our present study, extraction and cleanup procedures were improved. ASE extraction was replaced with homogenous and shaking method, three - cartridge cleanup with GPC. Tests found acetonitrile was still used as extraction solvent, and the solvent exchange was needed during GPC cleanup, and acetonitrile was changed to cyclohexane+ethyl acetate (1+1), through which test steps were added and recoveries were lowered. If cyclohexane+ethyl acetate(1+1) was adopted as extracting solvent directly, test efficiency was increased. Tests found that cyclohexane+ethyl acetate(1+1) had very good extraction effect on pesticide varieties analyzed by LC-MS-MS as well., but during preparation of grain samples 30mL cyclohexane+ethyl acetate (1+1) was used for extraction three times to ensure extracting efficiency due to the relatively big difference of sizes and matrixes.

Three- cartridge cleanup method was replaced with GPC method. A gel permeation chromatography behavior appraisal was conducted to751 pesticides and related chemical solutions suitable for GPC analysis. See Table 18 for the beginning of collection time (peaking time), stoppage of collection time (stoppage of peaking time) and collection volumes, etc for each pesticide and related chemicals. It could be seen from Table 18 that 81 out of these 751 pesticides and related chemicals started to peak in the first 20 minutes, accounting for 10.8% of the analyzed. 670 pesticides and related chemicals started to peak 20 minutes later, accounting for 89.2%, among which 549 pesticide and related chemicals began to peak 22 later, making up 73% of the pesticides and related chemicals analyzed. It could be seen from the data in Table 14 that all the pesticides and chemicals finished peaking within 40 minutes. Based on the above-mentioned conditions, three different collection times of 20 min, 21min and 22 were selected respectively, and the ending of the collection time was chosen at 40min. The cleanup tests were carried out with rice, wheat and soybean samples. Tests found that at the beginning collection time of 22 min the GPC cleaned-up samples were basically removed of the interferences of the fat in grains on the determination, and the varieties and recoveries of the analyzed pesticides and related chemicals were both satisfactory.

Table 18. 751 pesticides and related chemical of gel permeation chromatography behavior

No.	Compound	Beginning collection time /min	Stoppage collection time /min	Beginning collection volume/mL	Stoppage collection volume/mL	Collection volume/mL
1	2,3,4,5-Tetrachloroaniline	24	30	120	150	30
2	2,3,4,5-Tetrachloroanisole	23	35	115	175	60
3	2,3,5,6-Tetrachloroaniline	23	33	115	165	50
4	2,4-D	22	33	110	165	55
5	2,4-DB	22	33	110	165	55
6	2,6-Dichlorobenzamide	22	34	110	170	60
7	2,6-Difluorobenzoic acid	23	30	115	150	35

Table 18. Continued

No.	Compound	Beginning collection time /min	Stoppage collection time /min	Beginning collection volume/mL	Stoppage collection volume/mL	Collection volume/mL
8	2-Phenylphenol	22	33	110	165	55
9	3,5-Dichloroaniline	22	30	110	150	40
10	3,5-xylyl methylcarbamate	23	27	115	135	20
11	3.4-Trimethacarb	22	33	110	165	55
12	3-Hydroxycarbofuran	22	33	110	165	55
13	3-phenylphenol	24	31	120	155	35
14	4,4-Dibromobenzophenone	29	38	145	190	45
15	4,4-Dichlorobenzophenone	27	36	135	180	45
16	4-Bromo-3-dimethylphenyl N-methylcarbamate	22	33	110	165	55
17	4-Chlorophenoxyacetic acid	24	35	120	175	55
18	6-chloro-4-hydroxy-3-phenyl-pyridazin	25	32	125	160	35
19	Acephate	22	33	110	165	55
20	Acequinocyl	20	30	100	150	50
21	Acetamiprid	26	34	130	170	40
22	Acetochlor	23	29	115	145	30
23	Acifluorfen	17	30	85	150	65
24	Aclonifen	26	35	130	175	45
25	Acrinathrin	15	21	75	105	30
26	Alachlor	24	31	120	155	35
27	Aldicarb	23	32	115	160	45
28	Aldicarb sulfone	23	33	115	165	50
29	Aldicarb sulfoxide	24	32	120	160	40
30	Aldoxycarb	24	30	120	150	30
31	Aldrin	22	33	110	165	55
32	Allethrin mixture	20	29	100	145	45
33	Allidochlor	23	31	115	155	40
34	Alloxydim-sodium	21	29	105	145	40
35	Alpha-cypermethrin	21	27	105	135	30
36	Ametryn	22	29	110	145	35
37	Amidithion	22	36	110	180	70
38	Amidosulfuron	22	33	110	165	55
39	Aminocarb	24	30	120	150	30
40	Amitraz	23	36	115	180	65
41	Amitrole	23	32	115	160	45
42	Amobam	22	35	110	175	65
43	Anilazine	22	30	110	150	40
44	Anilofos	24	31	120	155	35
45	Anthraquinone	20	42	100	210	110
46	Aramite	18	27	90	135	45
47	Aroclor 1221	26	35	130	175	45
48	Aroclor 1232	26	36	130	180	50
49	Aroclor 1242	26	35	130	175	45
50	Aroclor 1254	25	35	125	175	50
51	Aroclor 1260	25	35	125	175	50
52	Aroclor 1262	25	35	125	175	50
53	Aroclor 1268	26	32	130	160	30

Table 18. Continued

No.	Compound	Beginning collection time /min	Stoppage collection time /min	Beginning collection volume/mL	Stoppage collection volume/mL	Collection volume/mL
54	Aspon	23	34	115	170	55
55	Asulam	21	29	105	145	40
56	Athidathion	19	33	95	165	70
57	Atratone	22	33	110	165	55
58	Atrazine	20	27	100	135	35
59	atrazine Des-ethyl	20	30	100	150	50
60	Atrazine-desethyl	20	26	100	130	30
61	Azaconazole	23	34	115	170	55
62	Azamethiphos	33	39	165	195	30
63	Azinphos-ethyl	25	35	125	175	50
64	Azinphos-methyl	24	38	120	190	70
65	Aziprotryne	23	30	115	150	35
66	Azoxystrobin	24	35	120	175	55
67	Benalaxyl	23	33	115	165	50
68	Benazolin	23	40	115	200	85
69	Bendiocarb	24	33	120	165	45
70	Benfluralin	18	28	90	140	50
71	Benfuracarb	20	30	100	150	50
72	Benfuresate	22	31	110	155	45
73	Benodanil	24	33	120	165	45
74	Benoxacor	26	33	130	165	35
75	Bensulfuron-methyl	25	32	125	160	35
76	Bensulide	20	33	100	165	65
77	Bensultap	24	33	120	165	45
78	Bentazone	23	31	115	155	40
79	Benzofenap	23	33	115	165	50
80	Benzoximate	25	33	125	165	40
81	Benzoylprop-ethyl	23	30	115	150	35
82	Bifenazate	22	30	110	150	40
83	Bifenox	22	34	110	170	60
84	Bifenthrin	21	34	105	170	65
85	Bioallethrin	19	33	95	165	70
86	Bioresmethrin	21	33	105	165	60
87	Biphenyl	23	36	115	180	65
88	Bitertanol	19	27	95	135	40
89	Bromacil	22	29	110	145	35
90	Bromfenvinfos	23	31	115	155	40
91	Bromobutide	22	36	110	180	70
92	Bromocylen	22	32	110	160	50
93	Bromophos (-methyl)	23	33	115	165	50
94	Bromophos-ethyl	24	31	120	155	35
95	Bromopropylate	22	33	110	165	55
96	Bromoxynil	25	32	125	160	35
97	Brompyrazon	22	32	110	160	50
98	Bromuconazole	22	33	110	165	55
99	Bupirimate	21	28	105	140	35
100	Buprofezin	22	31	110	155	45
101	Butachlor	22	37	110	185	75

Table 18. Continued

No.	Compound	Beginning collection time /min	Stoppage collection time /min	Beginning collection volume/mL	Stoppage collection volume/mL	Collection volume/mL
102	Butafenacil	19	25	95	125	30
103	Butamifos	22	38	110	190	80
104	Butocarboxim	26	33	130	165	35
105	Butocarboxim-sulfoxide	25	32	125	160	35
106	Butoxycarboxim	24	33	120	165	45
107	Butoxycarboxim-sulfoxid	25	32	125	160	35
108	Butralin	21	28	105	140	35
109	Buturon	21	27	105	135	30
110	Butylate	22	30	110	150	40
111	Cadusafos	22	34	110	170	60
112	Cafenstrol	23	32	115	160	45
113	Captafol	22	33	110	165	55
114	Captan	23	35	115	175	60
115	Carbaryl	26	33	130	165	35
116	Carbendazim	16	33	80	165	85
117	Carbetamide	20	34	100	170	70
118	Carbofuran	24	31	120	155	35
119	Carbophenothion	23	35	115	175	60
120	Carbosulfan	19	28	95	140	45
121	Carboxin	28	37	140	185	45
122	carfentrazone—ethyl	20	29	100	145	45
123	Cartap	22	37	110	185	75
124	Chlorbenside	27	34	135	170	35
125	Chlorbenside sulfone	23	33	115	165	50
126	Chlorbenside sulfone	25	33	125	165	40
127	Chlorbromuron	23	30	115	150	35
128	Chlorbufam	21	28	105	140	35
129	Chlordecone (Kepone)	22	32	110	160	50
130	Chlordimeform	26	34	130	170	40
131	Chlorfenapyr	20	26	100	130	30
132	Chlorfenethol (Dimite)	22	34	110	170	60
133	Chlorfenprop-methyl	23	34	115	170	55
134	Chlorfenson	23	33	115	165	50
135	Chlorfenvinphos	22	32	110	160	50
136	Chlorfluazuron	17	23	85	115	30
137	Chlorfluazuron	16	23	80	115	35
138	Chlorflurenol-methyl	25	34	125	170	45
139	Chlorfurenol	25	34	125	170	45
140	Chloridazon	23	33	115	165	50
141	Chlorimuron-ethyl	20	31	100	155	55
142	Chlormephos	24	32	120	160	40
143	Chlormequat	20	28	100	140	40
144	Chlorobenzilate	22	28	110	140	30
145	Chloroneb	23	35	115	175	60
146	Chloropicrin	23	33	115	165	50
147	Chloropropylate	21	33	105	165	60
148	Chlorothalonil	27	35	135	175	40
149	Chlorotoluron	23	31	115	155	40

Table 18. Continued

No.	Compound	Beginning collection time /min	Stoppage collection time /min	Beginning collection volume/mL	Stoppage collection volume/mL	Collection volume/mL
150	Chlorphoxim	23	29	115	145	30
151	Chlorpropham	21	33	105	165	60
152	Chlorpyrifos (-ethyl)	22	31	110	155	45
153	Chlorpyrifos-methyl	23	33	115	165	50
154	Chlorsulfuron	23	32	115	160	45
155	Chlorthal-dimethyl	23	33	115	165	50
156	Chlorthiamid	22	32	110	160	50
157	Chlorthion	24	32	120	160	40
158	Chlorthion	25	33	125	165	40
159	Chlorthiophos	24	33	120	165	45
160	Chlozolinate	20	28	100	140	40
161	Chromafenozide	22	37	110	185	75
162	Chrysene	37	47	185	235	50
163	Cinosulfuron	17	33	85	165	80
164	Cis Permethrin	22	33	110	165	55
165	Cis-1,2,3,6-Tetrahydrophthalimide	18	33	90	165	75
166	Cis-Chlordane	22	32	110	160	50
167	clethodim	25	32	125	160	35
168	Clodinafop-propargyl	21	28	105	140	35
169	Clomazone	24	33	120	165	45
170	Clomeprop	22	35	110	175	65
171	Cloprop	22	33	110	165	55
172	Clopyralld	23	34	115	170	55
173	Cloquintocet-mexyl	23	30	115	150	35
174	Cloransulam-methyl	23	31	115	155	40
175	Clothianidin	22	29	110	145	35
176	Coumaphos	23	33	115	165	50
177	Coumatetralyl	26	34	130	170	40
178	Crimidine	28	35	140	175	35
179	Crotoxyphos	22	31	110	155	45
180	Crufomate	22	33	110	165	55
181	Cyanazine	19	33	95	165	70
182	Cyanofenphos	22	32	110	160	50
183	Cyanophos	20	28	100	140	40
184	Cyclanilide	23	33	115	165	50
185	Cycloate	23	34	115	170	55
186	Cycloprothrin	23	36	115	180	65
187	Cycloxydim	20	35	100	175	75
188	Cycluron	23	33	115	165	50
189	Cyfluthrin	19	26	95	130	35
190	Cymoxanil	22	30	110	150	40
191	Cypermethrin	20	33	100	165	65
192	Cyprazine	21	27	105	135	30
193	Cyproconazole	24	30	120	150	30
194	Cyprodinil	26	33	130	165	35
195	Cyromazine	23	34	115	170	55
196	Cythioate	22	30	110	150	40

Table 18. Continued

No.	Compound	Beginning collection time /min	Stoppage collection time /min	Beginning collection volume/mL	Stoppage collection volume/mL	Collection volume/mL
197	Dacthal (DCPA)	27	33	135	165	30
198	Dalapon	22	33	110	165	55
199	Daminozide	21	35	105	175	70
200	Dazomet	27	41	135	205	70
201	DDD, o.p'-	23	33	115	165	50
202	DDD, p.p'-	22	32	110	160	50
203	DDE, o.p'-	25	32	125	160	35
204	DDE, p.p'-	24	33	120	165	45
205	DDT, o.p'-	26	32	130	160	30
206	DDT, p.p'-	23	33	115	165	50
207	DEF	22	29	110	145	35
208	Deltamethrin	19	33	95	165	70
209	Demephion(Tinox)	20	33	100	165	65
210	Demeton (O+S)	23	32	115	160	45
211	Demeton-S	23	30	115	150	35
212	Demeton-S-methyl	24	33	120	165	45
213	Demeton-S-methyl sulfoxide	22	30	110	150	40
214	Demeton-S-methyl suphone	23	33	115	165	50
215	deneton-s-methyl	23	34	115	170	55
216	DE-PCB 101 2,2',4,5,5'-Pentachlorobiphenyl	23	34	115	170	55
217	DE-PCB 1182,3',4,4',5-Pentachlorobiphenyl	27	35	135	175	40
218	DE-PCB 138 2,2',3,4,4',5'-Hexachlorobiphenyl	23	34	115	170	55
219	DE-PCB 153 2,2',4,4',5,5'-Hexachlorobiphenyl	23	34	115	170	55
220	DE-PCB 180 2,2',3,4,4',5,5'-Heptachlorobiphenyl	22	35	110	175	65
221	DE-PCB 28 2,4,4'-Trichlorobiphenyl	23	34	115	170	55
222	DE-PCB 31 2,4',5-Trichlorobiphenyl	23	35	115	175	60
223	DE-PCB 52 2,2',5,5'-Tetrachlorobiphenyl	23	34	115	170	55
224	Desethyl-sebuthylazine	20	25	100	125	25
225	Desisopropyl-atrazine	21	27	105	135	30
226	desmedipham	21	27	105	135	30
227	desmethylformamido-pirimicarb	23	37	115	185	70
228	Desmethyl-norflurazon	19	33	95	165	70
229	Desmethyl-pirimicarb	22	29	110	145	35
230	Desmetryn	23	32	115	160	45
231	Di-allate-mixture of cis and trans isomers	22	34	110	170	60
232	Diafenthiuron	19	26	95	130	35
233	Dialifos	24	31	120	155	35
234	Diazinon	21	32	105	160	55

Table 18. Continued

No.	Compound	Beginning collection time /min	Stoppage collection time /min	Beginning collection volume/mL	Stoppage collection volume/mL	Collection volume/mL
235	Dibutyl succinate	23	30	115	150	35
236	Dicamba	23	29	115	145	30
237	Dicapthon	26	33	130	165	35
238	Dichlobenil	23	34	115	170	55
239	Dichlofluanid	22	31	110	155	45
240	Dichlone	29	38	145	190	45
241	Dichlormid	23	32	115	160	45
242	Dichlorprop	20	33	100	165	65
243	Dichlorvos	22	33	110	165	55
244	Diclobutrazole	22	36	110	180	70
245	Diclofenthion	23	32	115	160	45
246	Diclofluanid	24	32	120	160	40
247	Diclofop-methyl	22	34	110	170	60
248	Dicloran	24	32	120	160	40
249	Dicofol	23	33	115	165	50
250	Dicrotophos	26	35	130	175	45
251	Dieldrin	23	34	115	170	55
252	Dienochlor	23	33	115	165	50
253	Diethofencarb	20	30	100	150	50
254	Diethyltoluamide	26	34	130	170	40
255	Difenconazole	22	33	110	165	55
256	Difenoxuron	25	33	125	165	40
257	Difenzoquat-methyl sulfate	22	37	110	185	75
258	Diflufenican	19	28	95	140	45
259	Diflufenzopyr-sodium	21	35	105	175	70
260	Dimefox	23	33	115	165	50
261	Dimefuron	19	27	95	135	40
262	Dimepiperate	22	33	110	165	55
263	Dimethachlor	23	32	115	160	45
264	Dimethametryn	19	32	95	160	65
265	Dimethenamid	24	33	120	165	45
266	Dimethipin	21	33	105	165	60
267	Dimethirimol	23	29	115	145	30
268	Dimethoate	23	34	115	170	55
269	Dimethomorph	29	37	145	185	40
270	Dimethylaminosulfanilide	22	33	110	165	55
271	Dimethylaminosulfotoluidide	22	33	110	165	55
272	Diniconazole	19	30	95	150	55
273	Dinitramine	20	33	100	165	65
274	Dinobuton	21	28	105	140	35
275	Dinoseb	23	30	115	150	35
276	Dinoseb acetate	22	29	110	145	35
277	Dinotefuran, 10ng/ul	22	33	110	165	55
278	Dinoterb	23	30	115	150	35
279	Diofenolan	24	32	120	160	40
280	Dioxabenzofos (Salithion)	28	35	140	175	35
281	Dioxacarb	22	32	110	160	50
282	Dioxathion	22	33	110	165	55

Table 18. Continued

No.	Compound	Beginning collection time /min	Stoppage collection time /min	Beginning collection volume/mL	Stoppage collection volume/mL	Collection volume/mL
283	Diphenamid	17	22	85	110	25
		28	35	140	175	35
284	Diphenylamine	27	33	135	165	30
285	Dipropetryn	20	27	100	135	35
286	Diquat dibromide hydrate	23	34	115	170	55
287	Disufoton	26	31	130	155	25
288	Disulfoton sulfone	22	34	110	170	60
289	Disulfoton sulfoxide	22	34	110	170	60
290	Ditalimfos	23	34	115	170	55
291	Dithiopyr	19	26	95	130	35
292	Diuron	22	31	110	155	45
293	DMSA	23	30	115	150	35
294	DMST	22	29	110	145	35
295	DNOC	27	34	135	170	35
296	Dodemorph	19	29	95	145	50
297	Edifenphos	27	37	135	185	50
298	Endosulfan sulfate	22	32	110	160	50
299	Endosulfan, (alpha beta)	22	34	110	170	60
300	Endothal	26	32	130	160	30
301	Endrin	22	34	110	170	60
302	Endrin ketone	18	33	90	165	75
303	EPN	24	33	120	165	45
304	epoxiconazole	23	30	115	150	35
305	EPTC	19	28	95	140	45
306	Erbon	22	31	110	155	45
307	Esfenvalerate	20	33	100	165	65
308	Etaconazole	23	34	115	170	55
309	Ethalfluralin	19	33	95	165	70
310	Ethephon	21	33	105	165	60
311	Ethidimuron	24	31	120	155	35
312	Ethiofencarb	24	32	120	160	40
313	Ethiofencarbsulfon	23	34	115	170	55
314	Ethiofencarbsulfoxid	23	34	115	170	55
315	Ethion	23	31	115	155	40
316	Ethirimol	22	30	110	150	40
317	Ethofumesate	22	32	110	160	50
318	Ethoprophos	24	32	120	160	40
319	Ethoxysulfuron	22	33	110	165	55
320	Ethylene thiourea	14	37	70	185	115
321	Etofenprox	23	29	115	145	30
322	Etridiazole	27	33	135	165	30
323	Etrimfos	22	33	110	165	55
324	Etrimfos oxon (PO-etrimfos)	24	30	120	150	30
325	Famoxadone	22	29	110	145	35
326	Famphur	26	33	130	165	35
327	Fenamidone	22	30	110	150	40
328	Fenamiphos	22	29	110	145	35
329	Fenamiphos sulfone	21	34	105	170	65

Table 18. Continued

No.	Compound	Beginning collection time /min	Stoppage collection time /min	Beginning collection volume/mL	Stoppage collection volume/mL	Collection volume/mL
330	Fenamiphos sulfoxide	22	34	110	170	60
331	Fenarimol	23	33	115	165	50
332	Fenazaflor	20	34	100	170	70
333	Fenazaquin	22	33	110	165	55
334	Fenbuconazole	22	34	110	170	60
335	Fenbutatin oxide	22	31	110	155	45
336	Fenchlorphos oxon	22	34	110	170	60
337	Fenfuram	24	30	120	150	30
338	Fenhexamid	21	33	105	165	60
339	Fenitorthion	26	33	130	165	35
340	Fenitrothion	24	33	120	165	45
341	Fenobucarb	22	28	110	140	30
342	fenothiocarb	23	33	115	165	50
343	Fenoxanil	22	31	110	155	45
344	Fenoxaprop-ethyl	23	33	115	165	50
345	Fenoxycarb	22	30	110	150	40
346	Fenpiclonil	21	33	105	165	60
347	Fenpropathrin	20	28	100	140	40
348	Fenpropidin	20	25	100	125	25
349	Fenpropimorph	18	33	90	165	75
350	fenpyroximate	21	29	105	145	40
351	Fenson	21	33	105	165	60
352	Fensulfothion	23	33	115	165	50
353	Fenthion	27	37	135	185	50
354	Fenthion PO-sulfone	22	34	110	170	60
355	Fenthion PO-sulfoxide	23	37	115	185	70
356	Fenthion PS-sulfone	23	33	115	165	50
357	Fentin-chloride	21	34	105	170	65
358	Fenuron	26	33	130	165	35
359	Fenvalerate	16	27	80	135	55
360	Fipronil	16	21	80	105	25
361	Flamprop-isopropyl	20	33	100	165	65
362	Flamprop-methyl	23	31	115	155	40
363	Flazasulfuron	21	33	105	165	60
364	florasulam	23	33	115	165	50
365	Fluazifop-butyl	18	26	90	130	40
366	Fluazinam	17	23	85	115	30
367	Fluazuron	17	23	85	115	30
368	Flubenzimine	18	24	90	120	30
369	Fluchloralin	21	27	105	135	30
370	Flucythrinate	20	30	100	150	50
371	Fludioxonil	20	26	100	130	30
372	Flufenacet	20	26	100	130	30
373	Flufenoxuron	16	24	80	120	40
374	Flumethrin	18	26	90	130	40
375	Flumetralin	20	26	100	130	30
376	Flumetsulam	23	31	115	155	40
377	Flumiclorac-pentyl	22	31	110	155	45

Table 18. Continued

No.	Compound	Beginning collection time /min	Stoppage collection time /min	Beginning collection volume/mL	Stoppage collection volume/mL	Collection volume/mL
378	Flumioxazin	23	35	115	175	60
379	Fluometuron	21	26	105	130	25
380	Fluorochloridone	20	27	100	135	35
381	Fluorodifen	26	34	130	170	40
382	Fluoroglycofen-ethyl	20	26	100	130	30
383	Fluoroimide	19	30	95	150	55
384	Fluotrimazole	22	33	110	165	55
385	Fluquinconazole	24	32	120	160	40
386	Fluridone	23	30	115	150	35
387	Flurochloridone	20	27	100	135	35
388	Fluroxypr 1-methylheptyl	18	24	90	120	30
389	Fluroxypyr	21	36	105	180	75
390	Flurtamone	21	28	105	140	35
391	Flusilazole	22	28	110	140	30
392	Flusulfamide	21	27	105	135	30
393	Fluthiacet-Methyl	28	36	140	180	40
394	Flutolanil	19	26	95	130	35
395	Flutriafol	22	33	110	165	55
396	Folpet	28	36	140	180	40
397	Fomesafen	18	28	90	140	50
398	Fonofos	26	34	130	170	40
399	Fonofos, O-analogue	26	33	130	165	35
400	Forchlorfenuron	23	35	115	175	60
401	Formothion	22	37	110	185	75
402	fosthiazate	22	36	110	160	50
403	Fuberidazole	25	32	125	160	35
404	Furalaxyl	26	33	130	165	35
405	Furathiocarb	22	32	110	160	50
406	Furmecyclox	25	32	125	160	35
407	gamma-Cyhalothrin	23	38	115	190	75
408	Gibberellic acid	18	28	90	140	50
409	Halosulfuran-methyl	22	40	110	200	90
410	Haloxyfop-ehyoxyethyl	19	25	95	125	30
411	Haloxyfop-methyl	19	33	95	165	70
412	HCH, alpha-	23	34	115	170	55
413	HCH, beta-	23	34	115	170	55
414	HCH, delta-	23	33	115	165	50
415	HCH-ε	22	30	110	150	40
416	Heptachlor	23	34	115	170	55
417	Heptachlor epoxide, cis-	23	33	115	165	50
418	Heptachlor epoxide, trans-	20	34	100	170	70
419	Heptanophos	22	34	110	170	60
420	Hexachlorobenzene	25	37	125	185	60
421	Hexaconazole	22	33	110	165	55
422	Hexaflumuron	16	33	80	165	85
423	Hexazinone	28	36	140	180	40
424	Hexythiazox	25	33	125	165	40
425	Hydramethylnon	16	30	80	150	70

Table 18. Continued

No.	Compound	Beginning collection time /min	Stoppage collection time /min	Beginning collection volume/mL	Stoppage collection volume/mL	Collection volume/mL
426	Imazalil	24	33	120	165	45
427	Imazamethabenz-methyl	22	28	110	140	30
428	Imazamox	19	31	95	155	60
429	Imazapyr	23	32	115	160	45
430	Imazaquin	24	31	120	155	35
431	Imazethapyr	23	29	115	145	30
432	Imibenconazole	27	33	135	165	30
433	Imidacloprid	25	35	125	175	50
434	Imiprothrin	22	31	110	155	45
435	indoxacarb	20	26	100	130	30
436	Iodofenphos	23	34	115	170	55
437	iodosulfuron-methyl	23	32	115	160	45
438	Ioxynil	26	37	130	185	55
439	Iprobenfos	22	29	110	145	35
440	Iprodione	18	23	90	115	25
441	iprovalicarb	19	24	95	120	25
442	Isazofos	22	33	110	165	55
443	Isobenzan	23	33	115	165	50
444	Isocarbamid	23	34	115	170	55
445	Isocarbophos	26	33	130	165	35
446	Isodrin	23	34	115	170	55
447	Isofenphos	20	33	100	165	65
448	Isomethiozin	26	36	130	180	50
449	Isoprocarb	22	29	110	145	35
450	Isopropalin	19	31	95	155	60
451	Isoprothiolane	27	34	135	170	35
452	Isoproturon	22	31	110	155	45
453	Isouron	20	33	100	165	65
454	Isoxaben	21	33	105	165	60
455	Isoxaflutole	21	33	105	165	60
456	Isoxathion	23	33	115	165	50
457	Kadethrin	23	33	115	165	50
458	Kelevan	23	34	115	170	55
459	Kresoxim-methyl	22	32	110	160	50
460	Lactofen	18	27	90	135	45
461	Lambda-Cyhalothrin	18	33	90	165	75
462	Lenacil	23	34	115	170	55
463	Leptophos	27	34	135	170	35
464	Leptophos desbrom	27	34	135	170	35
465	Leptophos oxon	17	27	85	135	50
466	Lindane	22	31	110	155	45
467	Linuron	22	30	110	150	40
468	Lufenuron	16	19	80	95	15
469	Malaoxon	22	30	110	150	40
470	Malathion	22	30	110	150	40
471	Maleic hydrazide	23	31	115	155	40
472	MCPA butoxyethyl ester	22	34	110	170	60
473	MCPB	19	31	95	155	60

Table 18. Continued

No.	Compound	Beginning collection time /min	Stoppage collection time /min	Beginning collection volume/mL	Stoppage collection volume/mL	Collection volume/mL
474	Mecarbam	22	33	110	165	55
475	Mefenacet	23	38	115	190	75
476	Mefenoxam	23	33	115	165	50
477	Mefenpyr-diethyl	23	29	115	145	30
478	Mepanipyrim	26	34	130	170	40
479	Mephosfolan	27	36	135	180	45
480	Mepiquat chloride	23	33	115	165	50
481	Mepronil	21	30	105	150	45
482	Merphos	22	29	110	145	35
483	Mesotrion	26	40	130	200	70
484	Metalaxyl	25	33	125	165	40
485	Metamitron	27	34	135	170	35
486	Metamitron	23	33	115	165	50
487	Metazachlor	23	33	115	165	50
488	Metconazole	22	33	110	165	55
489	Methabenzthiazuron	28	39	140	195	55
490	Methacrifos	24	33	120	165	45
491	Methamidophos	25	34	125	170	45
492	Methfuroxam	23	33	115	165	50
493	Methidathion	28	36	140	180	40
494	Methiocarb	23	32	115	160	45
495	Methiocarb sulfoxide	22	32	110	160	50
496	Methomyl	23	36	115	180	65
497	Methoprene	18	28	90	140	50
498	Methoprotryne	21	34	105	170	65
499	Methothrin	20	32	100	160	60
500	Methoxychlor	26	34	130	170	40
501	Methoxyfenozide	21	30	105	150	45
502	Metobromuron	24	33	120	165	45
503	Metolachlor	23	33	115	165	50
504	Metolcarb	22	32	110	160	50
505	Metominostrobin-(E)	23	32	115	160	45
506	metominostrobin-(Z)	20	32	100	160	60
507	Metosulam	24	33	120	165	45
508	Metoxuron	25	33	125	165	40
509	Metribuzin	23	34	115	170	55
510	metsulfuron-methyl	23	32	115	160	45
511	Mevinphos	23	34	115	170	55
512	Mexacarbate	22	34	110	170	60
513	Mirex	22	32	110	160	50
514	Molinate	21	36	105	180	75
515	Monalide	20	33	100	165	65
516	Monocrotophos	22	33	110	165	55
517	Monolinuron	23	31	115	155	40
518	Monuron	20	27	100	135	35
519	Musk ambrette	23	32	115	160	45
520	Musk Ketone	22	30	110	150	40
521	Musk moskene	22	32	110	160	50

Table 18. Continued

No.	Compound	Beginning collection time /min	Stoppage collection time /min	Beginning collection volume/mL	Stoppage collection volume/mL	Collection volume/mL
522	Musk tibetene	23	33	115	165	50
523	Musk xylene	23	33	115	165	50
524	Myclobutanil	22	34	110	170	60
525	Naled (Dibrom)	23	34	115	170	55
526	Napropamide	22	32	110	160	50
527	Neburon	21	27	105	135	30
528	Nicosulfuron	24	34	120	170	50
529	Nicotine	28	34	140	170	30
530	Nitenpyram	26	35	130	175	45
531	Nitralin	20	34	100	170	70
532	Nitrapyrin	24	33	120	165	45
533	Nitrofen	27	33	135	165	30
534	Nitrothal-isopropyl	21	33	105	165	60
535	Nonachlor, trans	22	34	110	170	60
536	Norflurazon	21	33	105	165	60
537	Novaluron	15	20	75	100	25
538	Nuarimol	23	30	115	150	35
539	Octachlorostyrene	23	34	115	170	55
540	Ofurace	24	34	120	170	50
541	Omethoate	22	33	110	165	55
542	Oryzalin	17	34	85	170	85
543	Oxadiazon	21	28	105	140	35
544	Oxadixyl	23	35	115	175	60
545	Oxamyl	23	35	115	175	60
		29	37			
546	Oxycarboxin	26	34	130	170	40
547	Oxy-chlordane	22	30	110	150	40
548	Oxyflurofen	21	33	105	165	60
549	Paclobutrazol	21	33	105	165	60
550	Paraoxon	22	33	110	165	55
551	Paraoxon	23	31	115	155	40
552	Paraoxon-methyl	23	34	115	170	55
553	Paraquat Dichloride	23	40	115	200	85
554	Parathion-ethyl	22	33	110	165	55
555	Parathion-methyl	25	32	125	160	35
556	Pebulate	22	33	110	165	55
557	Penconazole	21	33	105	165	60
558	pencycuron	24	32	120	160	40
559	Pendimethalin	23	33	115	165	50
560	Pentachloroaniline	25	33	125	165	40
561	Pentachloroanisole	28	35	140	.175	35
562	Pentachlorobenzene	23	35	115	175	60
563	Pentachlorophenol	23	34	115	170	55
564	Permethrin	22	29	110	145	35
565	Permethrin	23	32	115	160	45
566	Perthane	21	34	105	170	65
567	Phenkapton	23	33	115	165	50
568	phenmedipham	21	28	105	140	35

Table 18. Continued

No.	Compound	Beginning collection time /min	Stoppage collection time /min	Beginning collection volume/mL	Stoppage collection volume/mL	Collection volume/mL
569	Phenothrin	20	30	100	150	50
570	Phenthoate	24	33	120	165	45
571	Phorate	22	33	110	165	55
572	Phorate sulfone	22	34	110	170	60
573	Phorate sulfoxide	22	35	110	175	65
574	Phosalone	23	34	115	170	55
575	Phosmet	23	28	115	140	25
		29	37	145	185	40
576	Phosmet, O-analogue	23	36	115	180	65
577	Phosmet-oxon	23	33	115	165	50
578	Phosphamidon	24	31	120	155	35
579	phoxim	23	30	115	150	35
580	Phthalic acid, benzyl butyl ester	23	33	115	165	50
581	Phthalic acid, di-(2-ethylhexyl) ester	18	24	90	120	30
582	Phthalic acid, dibutyl ester	22	33	110	165	55
583	Phthalic acid, dicyclohexyl ester	22	34	110	170	60
584	Picloram	23	30	115	150	35
585	Picolinafen	20	29	100	145	45
586	Picoxystrobin	22	28	110	140	30
587	Piperonyl butoxide	22	28	110	140	30
588	Piperophos	23	29	115	145	30
589	Pirimicarb	23	36	115	180	65
590	Pirimiphos-ethyl	21	33	105	165	60
591	Pirimiphos-methyl	22	33	110	165	55
592	Pirmicarb-desmethyl	23	24	115	120	5
593	Plifenate	22	33	110	165	55
594	Prallethrin	21	30	105	150	45
595	Pretilachlor	22	29	110	145	35
596	primisulfuron-methyl	18	24	90	120	30
597	Probenazole	22	30	110	150	40
598	Prochloraz	23	33	115	165	50
599	Procymidone	23	29	115	145	30
600	Profenofos	23	33	115	165	50
601	Profluralin	19	33	95	165	70
602	Prohydrojasmon	23	32	115	160	45
603	Promecarb	21	33	105	165	60
604	Prometon	22	28	110	140	30
605	Prometryne	21	34	105	170	65
606	Pronamide	20	33	100	165	65
607	Propachlor	23	32	115	160	45
608	Propamocarb	22	34	110	170	60
609	Propanil	22	28	110	140	30
610	Propaphos	22	33	110	165	55
611	Propargite	20	34	100	170	70
612	Propazine	19	33	95	165	70
613	Propetamphos	20	27	100	135	35

Table 18. Continued

No.	Compound	Beginning collection time /min	Stoppage collection time /min	Beginning collection volume/mL	Stoppage collection volume/mL	Collection volume/mL
614	Propham	24	34	120	170	50
615	Propiconazole	22	32	110	160	50
616	Propisochlor	22	30	110	150	40
617	Propoxur	23	33	115	165	50
618	Propylene thiourea	26	34	130	170	40
619	Propyzamide	21	30	105	150	45
620	Prosulfocarb	25	33	125	165	40
621	Prosulfuron	19	33	95	165	70
622	Prothiofos	23	33	115	165	50
623	Prothoate	28	33	140	165	25
624	Pymetrozin	26	34	130	170	40
625	Pyraclofos	29	40	145	200	55
626	Pyraclostrobin	25	33	125	165	40
627	Pyraflufen Ethyl	23	30	115	150	35
628	Pyrazolynate (Pyrazolate)	22	27	110	135	25
629	Pyrazophos	23	33	115	165	50
630	Pyrazosulfuron-ethyl	21	32	105	160	55
631	pyrethrin	20	27	100	135	35
632	Pyributicarb	23	31	115	155	40
633	Pyridaben	18	32	90	160	70
634	Pyridaphenthion	23	33	115	165	50
635	Pyridate	23	30	115	150	35
636	Pyrifenox	26	33	130	165	35
637	Pyriftalid	27	35	135	175	40
638	Pyrimethanil	24	33	120	165	45
639	Pyrimidifen	26	32	130	160	30
640	Pyrimitate	22	33	110	165	55
641	Pyriproxyfen	24	32	120	160	40
642	Pyrithlobac sodium	23	32	115	160	45
643	Pyroquilon	35	44	175	220	45
644	Quinalphos	25	33	125	165	40
645	Quinmerac	27	36	135	180	45
646	Quinoclamine	27	35	135	175	40
647	Quinoxyphen	28	35	140	175	35
648	Quintozene	30	36	150	180	30
649	quizalofop-ethyl	24	31	120	155	35
650	Rabenzazole	25	32	125	160	35
651	Resmethrin	22	33	110	165	55
652	Rimsulfuron	25	36	125	180	55
653	Ronnel	23	33	115	165	50
654	S421(Octachlorodipropyl ether)	26	34	130	170	40
655	Sebutylazine	20	26	100	130	30
656	Secbumeton	22	33	110	165	55
657	Sethoxydim	15	32	75	160	85
658	Simazine	22	34	110	170	60
659	Simeconazole	20	31	100	155	55
660	Simeton	24	33	120	165	45
661	Simetryn	24	33	120	165	45

Table 18. Continued

No.	Compound	Beginning collection time /min	Stoppage collection time /min	Beginning collection volume/mL	Stoppage collection volume/mL	Collection volume/mL
662	Spinosad	17	24	85	120	35
663	Spirodiclofen	21	29	105	145	40
664	Spiroxamine	21	34	105	170	65
665	Sulfallate	25	37	125	185	60
666	Sulfanitran	18	26	90	130	40
667	Sulfentrazone	21	33	105	165	60
668	Sulfotep	22	33	110	165	55
669	Sulprofos	24	35	120	175	55
670	Tau-fluvalinate	14	25	70	125	55
671	TCA-sodium	22	33	110	165	55
672	TCMTB	26	37	130	185	55
673	Tebuconazole	21	33	105	165	60
674	Tebufenozide	19	29	95	145	50
675	Tebufenpyrad	21	28	105	140	35
676	Tebupirimfos	20	27	100	135	35
677	Tebutam	23	33	115	165	50
678	Tebuthiuron	24	32	120	160	40
679	Tecnazene	23	33	115	165	50
680	Teflubenzuron	20	26	100	130	30
681	Tefluthrin	18	23	90	115	25
		24	33	120	165	45
682	Temephos (Abate)	24	33	120	165	45
683	TEPP	17	32	85	160	75
684	Tepraloxydim	24	32	120	160	40
685	Terbacil	22	33	110	165	55
686	Terbucarb	20	31	100	155	55
687	Terbufos	23	29	115	145	30
688	Terbufos sulfone O-analogue	23	33	115	165	50
689	Terbumeton	22	33	110	165	55
690	Terbutryne	22	33	110	165	55
691	Terbutylazine	21	32	105	160	55
692	Tetrachlorvinphos	25	32	125	160	35
693	Tetraconazole	19	30	95	150	55
694	Tetradifon	23	33	115	165	50
695	Tetrahydrophthalimide	23	33	115	165	50
696	Tetramethrin	23	34	115	170	55
697	Tetrasul	27	34	135	170	35
698	Thenylchlor	26	34	130	170	40
699	Thiabendazole	27	35	135	175	40
700	Thiacloprid	27	35	135	175	40
701	Thiacloprid	28	36	140	180	40
702	Thiamethoxam	27	37	135	185	50
703	Thiazopyr	20	27	100	135	35
704	Thifensulfuron-methyl	23	35	115	175	60
705	Thiobencarb	26	33	130	165	35
706	Thiodicarb	28	38	140	190	50
707	thiofanox	24	30	120	150	30
708	Thiofanox sulfone	23	34	115	170	55

Table 18. Continued

No.	Compound	Beginning collection time /min	Stoppage collection time /min	Beginning collection volume/mL	Stoppage collection volume/mL	Collection volume/mL
709	Thiofanox-sulfoxid	23	34	115	170	55
710	Thiometon	18	34	90	170	80
711	Thionazin	26	33	130	165	35
712	thiophanat-methyl	22	30	110	150	40
713	Thiram	38	50	190	250	60
714	Tolclofos-methyl	21	31	105	155	50
715	Tolfenpyrad	22	31	110	155	45
716	Tolylfluanide	19	33	95	165	70
717	Tralkoxydim	23	30	115	150	35
718	trans-Chlordane	25	34	125	170	45
719	Transfluthrin	19	28	95	140	45
720	trans-permethrin	22	29	110	145	35
721	Triadimefon	22	32	110	160	50
722	Triadimenol	20	34	100	170	70
723	Triallate	23	30	115	150	35
724	Triamiphos	22	33	110	165	55
725	Triasulfuron	22	33	110	165	55
726	Triazophos	23	33	115	165	50
727	Triazoxide	32	40	160	200	40
728	Tribenuron-methyl	26	33	130	165	35
729	Trichlorfon	23	30	115	150	35
730	Trichloronat	22	31	110	155	45
731	Tricyclazole	33	43	165	215	50
732	Tridemorph	22	30	110	150	40
733	Trietazine	22	29	110	145	35
734	Trifloxystrobin	21	27	105	135	30
735	Triflumizole	20	26	100	130	30
736	Triflumuron	18	25	90	125	35
737	Trifluralin	18	34	90	170	80
738	Triflusulfuron-methyl	18	33	90	165	75
739	Triforine	18	26	90	130	40
740	Tri-isobutyl phosphate	22	31	110	155	45
741	Trimethylsulfonium iodide	22	31	110	155	45
742	Tri-n-butyl phosphate	23	34	115	170	55
743	Triphenyl phosphate	23	32	115	160	45
744	Triticonazole	22	33	110	165	55
745	Uniconazole	22	28	110	140	30
746	Vamidothion	22	35	110	175	65
747	Vamidothion sulfone	23	33	115	165	50
748	Vamidothion sulfoxide	25	33	125	165	40
749	Vernolate	22	32	110	160	50
750	Vinclozolin	22	30	110	150	40
751	Zoxamide	20	27	100	135	35

Selection of GC-MS Conditions

In GC-MS analysis correct choice of the qualifying and quantifying ions for each target compound is of great importance. Because of the great variety of pesticides determined and the overall retention time, which was relatively short, selection of ions was based on four principles: choice of ions with greater than 10% abundance compared with the base peak, trying to use the molecular ions, characteristic adducts of the molecular ion, characteristic fragment ions, and all their isotope ions as detecting ions; choice of fragment ions with high abundance, for example base peaks;choice of ions not observed for other pesticides and effluents, to try to minimize interference between different pesticides, especially interference from pesticides with retention times close to that of the target compound; and choice of ions for which the signal-to-noise ratio was greater than 3 after background subtraction. On the basis of these principles, tests of the standard solutions to be analyzed were conducted initially to acquire the full-scan mass spectra and to measure retention times. A quantitative ion and two qualitative ions were selected for each compound, and three qualitative ions were selected for the forbidden pesticides. On the basis of the retention times of the 475 GC-compatible pesticides, groups A, B, C, D andE were formed, each containing approximately 100 pesticides. To ensure sensitivity for each pesticide, all the ions to be determined for each group were monitored by programmed time based on the sequence of the compound peaks. The number of ions within any programmed time was fixed appropriately and attention was paid to the peaks eluting near the end of one programmed period and those eluting near the beginning of the next of programmed period. The dwell time for each ion was adjusted to ensure each peak was of a cycled scan in the constant cycling scan time and the number of dataacquisition points was sufficient for all the compounds monitored. Changing the dwell time did not affect the results of integration.

Selection of LC-MS-MS Conditions

1. Optimization of Mass Spectrometric Conditions

The syringe pump was adopted to inject respectively the pesticides and related chemical standard solutions at a flow rate of 5μL/min into the ion source and Q1 scan of each pesticide and relate chemical was carried out in the positive ion mode to obtain the molecule ion peaks for each pesticide and related chemical. Product ion scan was implemented for each pesticide and related chemical to obtain fragmented ion messages. Qualitative and quantitative precusor/product ions were selected for pesticides before making DP to each precusor/product ions and optimizing CE in order to have the analytical response reach the optimum. Based on the properties of pesticides, each ten pesticides formed one group and were submitted for monitoring by in-line LC-MS-MS to acquire the retention time under the said condition. In accordance with the sequence of the peaking, 465 pesticides divided into five groups, and several representative pesticides of different chemical properties were selected from each group, which were then optimized with ion spray potential, ion source temperature, nebulizer ,curtain and auxiliary heated gas pressures to make the intensity of precusor/product ions of each pesticide and related chemicals reach the maximum. The mass spectrometric conditions in the above mentioned were finally avilable to said method.

2. Optimization of Separation Conditions of LC

Regarding analysis of pesticide and related chemicals multiresidues of very different polarities, LC colums C18 and C8 type columns were commonly adopted for LC separation, which was mainly listed as: Atlantis dC18, 3μm, 2.1mm×150mm□Inertsil, C8-3, 5μm, 2.1mm×150mm□Luna, C18, 5μm, 2mm×50mm and Luna, C18, 3μm, 2mm×50mm. No remarkble differences were found after comparing the above-mentioned LC columns with those in the method. Of these columns, Atlantis dC18 turned out to be a little better, so the said column was adopted as separation column. On the selection and optimization of mobile phase, four optimized tests were carried out on the basis of the past priority gradient procedures, with the results shown in Table 19. As for the separation effect of peaks, monitoring time and the sensitivity of the pesticides and related chemicals, Test Method 3 was superior to others. Therefore, the third LC mobile phase gradient procedure was adopted in this method.

Table 19. Selection and optimization of mobile phase

Test 1				Test 2			
Time (min)	Flow rate (μl/min)	acetonitrile (%)	water (%)	Time (min)	Flow rate (μl/min)	acetonitrile (%)	water (%)
0	200	30	70	0	200	10	90
10	200	40	60	10	200	60	40
15	200	60	40	15	200	60	40
22	200	80	20	22	200	80	20
40	200	95	5	40	200	95	5
45	200	95	5	45	200	95	5
45.01	200	30	70	45.01	200	10	90
55	200	30	70	55	200	10	90
Test 3				Test 4			
0	200	10	90	0	200	20	80
4	200	50	50	4	200	50	50
15	200	60	40	15	200	60	40
20	200	80	20	20	200	80	20
25	200	95	5	25	200	95	5
32	200	95	5	35	200	95	5
32.01	200	10	90	35.01	200	20	80
40	200	10	90	40	200	20	80

Linear Range, LOD, LOQ

(1) Under the GC-MS conditions selected, analysis was performed to determine linear range for the 505 pesticides. The fortification concentration giving a signal-to-noise ratio ≥ 5 for each pesticide was regarded as the LOD of the method and the fortification concentration giving a signalto- noise ratio ≥ 10 was fixed as LOQ of the method. The LOD, LOQ, linear

ranges, correlation equations, and correlation coefficients of 505 pesticides are shown in Tables 20.

(2) Under the LC-MS-MS conditions selected, analysis was performed to determine linear range for the 465 pesticides. The fortification concentration giving a signal-to-noise ratio ≥ 5 for each pesticide was regarded as the LOD of the method and the fortification concentration giving a signalto- noise ratio ≥ 10 was fixed as LOQ of the method. The LOD, LOQ, linear ranges, correlation equations, and correlation coefficients of 465 pesticides are shown in Tables 21.

Recovery and Precision

1. For GC-MS Pesticides

The different grain samples free from pesticide residues were used for fortification precision tests. The samples were held for 30min after adding pesticides so that the pesticides were thoroughly absorbed before proceeding to extraction, cleanup and determination according to the said GC-MS method. Table 22a and 22b reflected the data of recoveries and precision at three and two fortification levels for 5 grains, from which it could be seen that the average recovery range of the pesticides determined by GC-MS fell within 42%-132%, among which recoveries for 454 pesticides were found to be within the range of 60%-120%, accounting for 95%, the recoveries for 21 pesticides such as sethoxydim, cycluron, cycloxydim, maxacarbate, vamidothion, dichlobenil, EPTC, isofenphos oxon, secbumeton, oxycarboxin, dimefox, tolyfluanide, methamidophos, pencycuron, pirimicarb, pyrifenox, methiocarb sulfone, imazalil, clethodim, tralkoxydim, pyrimidifen belonged to the range of 42%-60%, accounting for 4 %, the recoveries for 11pesticides such as mecarbam, aminocarb, isazofos, dimefuron, fluoroglycofen-ethyl, triazophos, phosalon, tetradifon, oxamyl, aziprotryne, 2,4'-DDD fell within 120%-132%, making up 1 %. RSD for all the pesticides fell below 38%, of which 456 pesticides were found to be below 30%, making up 97%. 14 pesticides such as isodrin, mexacarbate, fenamiphos sulfoxide, indoxacarb, anthraquinone, flufenoxuron, fluorochloridone, secbumeton, dichlorvos, mevinphos, methomyl, dimethoate, demeton-s-methyl, vamidothion belonged to the range of 30%-38%, accounting for3 %. All this demonstrated fully that the method has relatively good reproducibility.

Table 20. The LOD, LOQ, linear ranges, correlation equations, and correlation coefficients of 505 pesticides by GC-MS

No.	Compound	Retention time/min	LOQ/ mg/kg	LOD /mg/kg	Linear ranges	Correlation equation	Correlation coefficient
1	Allidochlor	8.82	0.0500	0.0250	0.1250 ~ 5.000	$Y=4.83\times10^5 X - 5.48\times10^4$	0.9986
2	Dichlormid	9.80	0.0250	0.0125	0.0625 ~ 2.500	$Y=3.26\times10^7 X+1.03\times10^6$	0.9981
3	Etridiazol	10.47	0.0750	0.0375	0.1875 ~ 7.500	$Y=8.12\times10^5 X - 2.65\times10^5$	0.9945
4	Chlormephos	10.59	0.0500	0.0250	0.1250 ~ 5.000	$Y=1.19\times10^6 X + 1.14\times10^5$	0.9928
5	Propham	11.43	0.0250	0.0125	0.0625 ~ 2.500	$Y=1.01\times10^6 X - 5.76\times10^4$	0.9975
6	Cycloate	13.61	0.0250	0.0125	0.0625 ~ 2.500	$Y=2.26\times10^6 X - 1.16\times10^5$	0.9986
7	Diphenylamin	14.59	0.0250	0.0125	0.0625 ~ 2.500	$Y=5.81\times10^6 X - 3.97\times10^5$	0.9975
8	Chlordimeform	15.00	0.0500	0.0250	0.1250 ~ 2.500	$Y=3.39\times10^5 X - 4.28\times10^4$	0.9966
9	Ethalfluralin	15.07	0.1000	0.0500	0.2500 ~ 10.00	$Y=5.28\times10^5 X - 2.60\times10^5$	0.9926
10	Phorate	15.54	0.0250	0.0125	0.0625 ~ 2.500	$Y=5.32\times10^5 X - 4.50\times10^4$	0.9969
11	Thiometon	16.28	0.0250	0.0125	0.0625 ~ 2.500	$Y=3.00\times10^6 X - 2.40\times10^5$	0.9967
12	Quintozene	16.80	0.0500	0.0250	0.1250 ~ 5.000	$Y=2.98\times10^5 X - 5.32\times10^4$	0.9958
13	Atrazine-Desethyl	16.85	0.0250	0.0125	0.0625 ~ 2.500	$Y=2.26\times10^6 X - 1.93\times10^5$	0.9956
14	Clomazone	17.08	0.0250	0.0125	0.0625 ~ 2.500	$Y=3.04\times10^6 X - 2.30\times10^5$	0.9976
15	Diazinon	17.21	0.0250	0.0125	0.0625 ~ 2.500	$Y=6.86\times10^5 X - 4.99\times10^4$	0.9980
16	Fonofos	17.38	0.0250	0.0125	0.0625 ~ 2.500	$Y=1.43\times10^6 X - 1.10\times10^5$	0.9975
17	Etrimfos	17.98	0.0250	0.0125	0.0625 ~ 2.500	$Y=8.55\times10^5 X - 6.25\times10^4$	0.9976
18	Simazine	17.99	0.0250	0.0125	0.0625 ~ 2.500	$Y=1.19\times10^6 X - 9.71\times10^4$	0.9965

Table 20. Continued

No.	Compound	Retention time/min	LOQ/ mg/kg	LOD /mg/kg	Linear ranges	Correlation equation	Correlation coefficient
19	Propetamphos	18.06	0.0250	0.0125	0.0625 ~ 2.500	$Y=2.44\times10^6 X - 2.24\times10^5$	0.9968
20	Secbumeton	18.43	0.0250	0.0125	0.0625 ~ 2.500	$Y=3.16\times10^6 X - 3.21\times10^5$	0.9952
21	Dichlofenthion	18.87	0.0250	0.0125	0.0625 ~ 2.500	$Y=1.45\times10^6 X - 8.71\times10^4$	0.9986
22	Pronamide	18.89	0.0250	0.0125	0.0625 ~ 2.500	$Y=2.33\times10^6 X - 2.70\times10^5$	0.9905
23	Mexacarbate	18.91	0.0750	0.0375	0.1875 ~ 7.500	$Y=2.74\times10^6 X - 9.75\times10^5$	0.9924
24	Aldrin	19.71	0.0500	0.0250	0.1250 ~ 5.000	$Y=6.16\times10^5X - 6.01\times10^4$	0.9990
25	Dinitramine	19.74	0.1000	0.0500	0.2500 ~ 10.00	$Y=8.42\times10^5X - 3.98\times10^5$	0.9956
26	Ronnel	19.84	0.0500	0.0250	0.1250 ~ 5.000	$Y=2.37\times10^6 X - 3.41\times10^5$	0.9978
27	Prometrye	20.20	0.0250	0.0125	0.0625 ~ 2.500	$Y=1.97\times10^6 X - 1.82\times10^5$	0.9961
28	Cyprazine	20.25	0.0250	0.0125	0.0625 ~ 2.500	$Y=1.81\times10^6 X - 1.38\times10^5$	0.9978
29	Chlorothalonil	20.36	1.0000	0.5000	2.5000 ~ 5.000	$Y=5.41\times10^5X - 1.03\times10^6$	1.0000
30	Vinclozolin	20.37	0.0250	0.0125	0.0625 ~ 2.500	$Y=4.84\times10^5X - 3.88\times10^4$	0.9973
31	beta-HCH	20.38	0.0250	0.0125	0.0625 ~ 2.500	$Y=6.87\times10^5X - 3.02\times10^4$	0.9992
32	Metalaxyl	20.76	0.0750	0.0375	0.1875 ~ 7.500	$Y=1.23\times10^6 X - 2.59\times10^5$	0.9982
33	Chlorpyifos(Ethyl)	21.02	0.0250	0.0125	0.0625 ~ 2.500	$Y=6.84\times10^5X - 4.74\times10^4$	0.9985
34	Methyl-Parathion	21.12	0.1000	0.0500	0.2500 ~ 10.00	$Y=1.06\times10^6 X - 6.40\times10^5$	0.9892
35	Anthraquinone	21.55	0.0250	0.0125	0.0625 ~ 2.500	$Y=2.27\times10^6 X - 3.41\times10^5$	0.9862
36	Delta-HCH	21.24	0.0500	0.0250	0.1250 ~ 5.000	$Y=6.21\times10^5X - 7.51\times10^4$	0.9986

37	Fenthion	21.59	0.0250	0.0125	0.0625 ~ 2.500	$Y=2.91\times10^{6} X - 2.33\times10^{5}$	0.9973
38	Malathion	21.63	0.1000	0.0500	0.2500 ~ 10.00	$Y=1.43\times10^{6} X - 5.23\times10^{5}$	0.9974
39	Fenitrothion	21.73	0.0500	0.0250	0.1250 ~ 5.000	$Y=1.13\times10^{6} X - 2.61\times10^{5}$	0.9957
40	Paraoxon-Ethyl	21.68	0.1000	0.0500	0.2500 ~ 10.00	$Y=3.39\times10^{5} X - 2.09\times10^{5}$	0.9892
41	Triadimefon	22.30	0.0500	0.0250	0.1250 ~ 5.000	$Y=1.20\times10^{6} X - 2.03\times10^{5}$	0.9976
42	Parathion	22.42	0.1000	0.0500	0.2500 ~ 10.00	$Y=1.08\times10^{6} X - 5.62\times10^{5}$	0.9935
43	Pendimethalin	22.66	0.1000	0.0500	0.2500 ~ 10.00	$Y=2.46\times10^{6} X - 1.12\times10^{6}$	0.9960
44	Linuron	22.54	0.1000	0.0500	0.2500 ~ 10.00	$Y=4.30\times10^{5} X - 3.08\times10^{5}$	0.9730
45	Chlorbenside	23.01	0.0500	0.0250	0.1250 ~ 5.00	$Y=6.55\times10^{5} X - 1.37\times10^{5}$	0.9946
46	Bromophos-Ethyl	23.17	0.0250	0.0125	0.0625 ~ 2.500	$Y=8.39\times10^{5} X - 6.63\times10^{4}$	0.9973
47	Quinalphos	23.18	0.0250	0.0125	0.0625 ~ 2.500	$Y=1.93\times10^{6} X - 1.91\times10^{5}$	0.9959
48	Trans-Chlodane	23.35	0.0250	0.0125	0.0625 ~ 2.500	$Y=1.09\times10^{6} X - 5.33\times10^{4}$	0.9989
49	Phenthoate	23.38	0.0500	0.0250	0.1250 ~ 5.000	$Y=1.37\times10^{6} X - 2.52\times10^{5}$	0.9970
50	Metazachlor	23.42	0.0750	0.0375	0.1875 ~ 7.500	$Y=1.18\times10^{6} X - 2.26\times10^{5}$	0.9984
51	Fenothiocarb	23.90	0.4000	0.2000	1.0000 ~ 5.000	$Y=2.27\times10^{4} X - 1.21\times10^{4}$	0.9955
52	Prothiophos	24.09	0.0250	0.0125	0.0625 ~ 2.500	$Y=9.17\times10^{5} X - 9.14\times10^{4}$	0.9957
53	Folpet	24.11	0.3000	0.1500	0.7500 ~ 30.00	$Y=4.67\times10^{5} X - 1.14\times10^{6}$	0.9739
54	Chlorfurenol	24.23	0.0750	0.0375	0.1875 ~ 7.500	$Y=4.48\times10^{6} X - 1.29\times10^{6}$	0.9960
55	Dieldrin	24.49	0.0500	0.0250	0.1250 ~ 5.000	$Y=3.00\times10^{5} X - 2.58\times10^{4}$	0.9994
56	Procymidone	24.44	0.0250	0.0125	0.0625 ~ 2.500	$Y=1.18\times10^{6} X - 6.98\times10^{4}$	0.9979
57	Methidathion	24.58	0.0500	0.0250	0.1250 ~ 5.000	$Y=2.58\times10^{6} X - 5.55\times10^{5}$	0.9948

58	Cyanazine	24.97	0.0375	0.0000	0.1875 ~ 7.500	$Y=1.26\times10^{6} X - 4.10\times10^{5}$	0.9952
59	Napropamide	24.98	0.0750	0.0375	0.1875 ~ 7.500	$Y=9.05\times10^{5} X - 2.34\times10^{5}$	0.9963
60	Oxadiazone	25.13	0.0250	0.0125	0.0625 ~ 2.500	$Y=1.56\times10^{6}X - 1.01\times10^{5}$	0.9980
61	Fenamiphos	25.39	0.0750	0.0375	0.1875 ~ 7.500	$Y=1.80\times10^{6} X - 8.73\times10^{5}$	0.9864
62	Tetrasul	25.87	0.0250	0.0125	0.0625 ~ 2.500	$Y=1.60\times10^{6} X - 1.21\times10^{5}$	0.9958
63	Aramite	25.88	0.0250	0.0125	0.0625 ~ 2.500	$Y=3.89\times10^{5} X - 3.66\times10^{4}$	0.9927
64	Bupirimate	26.10	0.0250	0.0125	0.0625 ~ 2.500	$Y=1.73\times10^{6} X - 1.57\times10^{5}$	0.9954
65	Carboxin	26.35	0.0750	0.0375	0.1875 ~ 7.500	$Y=1.49\times10^{6} X - 5.15\times10^{5}$	0.9930
66	Flutolanil	26.36	0.0250	0.0125	0.0625 ~ 2.500	$Y=6.85\times10^{6} X - 7.11\times10^{5}$	0.9951
67	p,p'-DDD	26.37	0.0250	0.0125	0.0625 ~ 2.500	$Y=4.46\times10^{6} X - 5.15\times10^{5}$	0.9930
68	Ethion	26.72	0.0500	0.0250	0.1250 ~ 5.000	$Y=1.91\times10^{6} X - 4.27\times10^{5}$	0.9944
69	Sulprofos	26.94	0.0500	0.0250	0.1250 ~ 5.000	$Y=1.45\times10^{6} X - 2.67\times10^{5}$	0.9959
70	Etaconazole-1	26.97	0.0750	0.0375	0.1875 ~ 7.500	$Y=5.57\times10^{5} X - 1.69\times10^{5}$	0.9948
71	Etaconazole-2	27.18	0.0750	0.0375	0.1875 ~ 7.500	$Y=8.08\times10^{5} X - 1.78\times10^{5}$	0.9976
72	Myclobutanil	27.29	0.0250	0.0125	0.0625 ~ 2.500	$Y=1.90\times10^{6} X - 1.74\times10^{5}$	0.9958
73	Dichlorofop-Methyl	28.14	0.0250	0.0125	0.0625 ~ 2.500	$Y=1.23\times10^{6} X - 4.28\times10^{4}$	0.9958
74	Propiconazole	28.22	0.0750	0.0375	0.1875 ~ 7.500	$Y=6.40\times10^{5} X - 1.94\times10^{5}$	0.9942
75	Fensulfothion	28.24	0.0500	0.0250	0.1250 ~ 5.000	$Y=3.84\times10^{5} X - 5.31\times10^{4}$	0.9900
76	Bifenthrin	28.62	0.0250	0.0125	0.0625 ~ 2.500	$Y=7.25\times10^{6} X - 7.29\times10^{5}$	0.9936
77	Carbosulfan	28.80	0.0750	0.0375	0.1875 ~ 7.500	$Y=1.06\times10^{6} X - 2.53\times10^{5}$	0.9987
78	Mirex	28.81	0.0250	0.0125	0.0625 ~ 2.500	$Y=1.29\times10^{6} X - 6.90\times10^{4}$	0.9974

79	Benodanil	28.82	0.0750	0.0375	0.1875 ~ 7.500	$Y=5.34\times10^{6} X - 2.7\times10^{6}$	0.9899
80	Nuarimol	28.97	0.0500	0.0250	0.1250 ~ 5.000	$Y=6.42\times10^{5}X - 1.13\times10^{5}$	0.9962
81	Methoxychlor	29.43	0.0250	0.0125	0.0625 ~ 2.500	$Y=5.40\times10^{6}X - 6.40\times10^{5}$	0.9921
82	Oxadxyl	29.57	0.0250	0.0125	0.0625 ~ 2.500	$Y=7.29\times10^{5}X - 6.65\times10^{4}$	0.9897
83	Tetramethrin	29.66	0.0500	0.0250	0.1250 ~ 5.000	$Y=4.35\times10^{6}X - 9.91\times10^{5}$	0.9928
84	Tebuconazole	29.67	0.0750	0.0375	0.1875 ~ 7.500	$Y=1.39\times10^{6}X - 4.05\times10^{5}$	0.9941
85	Norflurazon	30.10	0.0250	0.0125	0.0625 ~ 2.500	$Y=1.16\times10^{6}X - 1.33\times10^{5}$	0.9879
86	Pyridaphenthion	30.25	0.0250	0.0125	0.0625 ~ 2.500	$Y=1.04\times10^{6} X - 1.47\times10^{5}$	0.9841
87	Phosmet	30.54	0.0500	0.0250	0.1250 ~ 5.000	$Y=2.04\times10^{6} X - 6.08\times10^{5}$	0.9861
88	Tetradifon	30.75	0.0250	0.0125	0.0625 ~ 2.500	$Y=4.97\times10^{5} X - 2.54\times10^{4}$	0.9966
89	Oxycarboxin	31.04	0.1500	0.0750	0.3750 ~ 15.00	$Y=1.34\times10^{6} X - 1.23\times10^{6}$	0.9850
90	Cis-Permethrin	31.45	0.0250	0.0125	0.0625 ~ 2.500	$Y=4.92\times10^{6} X - 4.33\times10^{5}$	0.9935
91	Chloridazon	31.59	0.0250	0.0125	0.0625 ~ 2.500	$Y=4.46\times10^{6} X - 5.32\times10^{5}$	0.9904
92	Trans-Permethrin	31.72	0.0250	0.0125	0.0625 ~ 2.500	$Y=4.19\times10^{6} X - 4.17\times10^{5}$	0.9911
93	Pyrazophos	31.65	0.0500	0.0250	0.1250 ~ 5.000	$Y=2.23\times10^{6} X - 5.32\times10^{5}$	0.9904
94	Cypermethrin	33.20	0.0750	0.0375	0.1875 ~ 7.500	$Y=6.63\times10^{5} X - 1.77\times10^{5}$	0.9919
		33.30					
		33.47					
		33.57					
95	Fenvalerate	34.02	0.1000	0.0500	0.2500 ~ 10.00	$Y=1.54\times10^{6} X - 4.39\times10^{5}$	0.9928
		34.62					
96	Deltamethrin	35.82	0.1500	0.0750	0.3750 ~ 15.00	$Y=9.80\times10^{5} X - 6.51\times10^{5}$	0.9921
97	EPTC	8.60	0.0750	0.0375	0.1875 ~ 7.500	$Y=1.57\times10^{6} X - 1.31\times10^{5}$	0.9993
98	Butylate	9.56	0.0750	0.0375	0.1875 ~ 7.500	$Y=1.10\times10^{6} X - 1.13\times10^{5}$	0.9994
99	Dichlobenil	9.84	0.0100	0.0050	0.0250 ~ 0.500	$Y=2.99\times10^{6} X - 1.58\times10^{4}$	0.9998
100	Pebulate	10.25	0.0750	0.0375	0.1875 ~ 7.500	$Y=2.11\times10^{6} X - 2.26\times10^{5}$	0.9994

101	Nitrapyrin	10.97	0.0750	0.0375	0.1875 ~ 7.500	$Y=1.41\times10^{6}X - 4.43\times10^{5}$	0.9957
102	Mevinphos	11.36	0.0500	0.0250	0.1250 ~ 5.000	$Y=1.77\times10^{6}X - 3.53\times10^{5}$	0.9957
103	Chloroneb	11.94	0.0250	0.0125	0.0625 ~ 2.500	$Y=2.04\times10^{6}X - 9.36\times10^{4}$	0.9990
104	Tecnazene	13.63	0.1000	0.0500	0.2500 ~ 5.000	$Y=3.92\times10^{5}X - 6.99\times10^{4}$	0.9985
105	Heptanophos	13.91	0.0750	0.0375	0.1875 ~ 7.500	$Y=1.71\times10^{6}X - 4.01\times10^{5}$	0.9979
106	Hexachlorobenzene	14.60	0.0250	0.0125	0.0625 ~ 2.500	$Y=1.80\times10^{6}X - 5.09\times10^{4}$	0.9996
107	Ethoprophos	14.63	0.0750	0.0375	0.1875 ~ 7.500	$Y=1.04\times10^{6}X - 2.32\times10^{5}$	0.9982
108	Cis-Diallate	14.76	0.0500	0.0250	0.1250 ~ 5.000	$Y=2.70\times10^{5}X - 1.65\times10^{4}$	0.9998
109	Propachlor	14.85	0.0750	0.0375	0.1875 ~ 7.500	$Y=2.89\times10^{6}X - 3.00\times10^{5}$	0.9998
110	Trans-Diallate	14.84	0.0500	0.0250	0.1250 ~ 5.000	$Y=1.02\times10^{6}X - 9.85\times10^{4}$	0.9994
		15.39					
111	Trifluralin	15.34	0.0500	0.0250	0.1250 ~ 5.000	$Y=1.83\times10^{6}X - 4.47\times10^{5}$	0.9930
112	Chlorpropham	15.61	0.0500	0.0250	0.1250 ~ 5.000	$Y=7.24\times10^{5}X - 1.22\times10^{5}$	0.9971
113	Sulfotep	15.67	0.0250	0.0125	0.0625 ~ 2.500	$Y=1.61\times10^{6}X - 8.57\times10^{4}$	0.9991
114	Sulfallate	15.86	0.0500	0.0250	0.1250 ~ 5.000	$Y=2.84\times10^{6}X - 6.05\times10^{5}$	0.9958
115	alpha-HCH	16.17	0.0250	0.0125	0.0625 ~ 2.500	$Y=6.51\times10^{5}X - 2.25\times10^{4}$	0.9994
116	Terbufos	16.94	0.0500	0.0250	0.1250 ~ 5.000	$Y=1.77\times10^{6}X - 2.82\times10^{5}$	0.9977
117	Terbumeton	17.32	0.0750	0.0375	0.1875 ~ 7.500	$Y=2.47\times10^{6}X - 5.48\times10^{5}$	0.9983
118	Profluralin	17.48	0.1000	0.0500	0.2500 ~ 10.00	$Y=9.72\times10^{5}X - 4.42\times10^{5}$	0.9955
119	Dioxathion	17.64	0.1000	0.0500	0.2500 ~ 10.00	$Y=1.58\times10^{5}X - 2.19\times10^{4}$	0.9993
120	Propazine	17.78	0.0250	0.0125	0.0625 ~ 2.500	$Y=1.63\times10^{6}X - 9.61\times10^{4}$	0.9988
121	Chlorbufam	17.97	0.1000	0.0500	0.2500 ~ 10.00	$Y=5.75\times10^{5}X - 2.01\times10^{5}$	0.9921
122	Dicloran	18.01	0.1000	0.0500	0.2500 ~ 10.00	$Y=6.59\times10^{5}X - 1.86\times10^{5}$	0.9963
123	Terbuthylazine	18.18	0.0250	0.0125	0.0625 ~ 2.500	$Y=9.62\times10^{5}X - 1.21\times10^{5}$	0.9886

124	Monolinuron	18.32	0.1000	0.0500	0.2500 ~ 10.00	$Y=1.78\times10^6 X - 9.80\times10^5$	0.9888
125	Flufenoxuron	18.95	0.0750	0.0375	0.1875 ~ 7.500	$Y=1.40\times10^5 X + 5.53\times10^4$	0.9929
126	Monocrotophos	18.96	0.1000	0.0500	0.2500 ~ 2.500	$Y=3.15\times10^4 X + 1.27\times10^4$	0.9941
127	Cyanohos	18.97	0.0500	0.0250	0.1250 ~ 5.000	$Y=2.01\times10^6 X - 2.99\times10^5$	0.9980
128	Chlorprifos-Methyl	19.48	0.0250	0.0125	0.0625 ~ 2.500	$Y=1.55\times10^6 X - 1.25\times10^5$	0.9972
129	Desmetryn	19.75	0.0250	0.0125	0.0625 ~ 2.500	$Y=1.92\times10^6 X - 1.86\times10^5$	0.9964
130	Dimethachloro	19.90	0.0750	0.0375	0.1875 ~ 7.500	$Y=3.98\times10^6 X - 7.53\times10^5$	0.9989
131	Alachlor	20.15	0.0750	0.0375	0.1875 ~ 7.500	$Y=1.27\times10^6 X - 2.26\times10^5$	0.9990
132	Pirimiphos-Methyl	20.40	0.0250	0.0125	0.0625 ~ 2.500	$Y=1.36\times10^6 X - 1.07\times10^5$	0.9977
133	Terbutryn	20.72	0.0500	0.0250	0.1250 ~ 5.000	$Y=1.95\times10^6 X - 3.07\times10^5$	0.9980
134	Thiobencarb	20.73	0.0500	0.0250	0.1250 ~ 5.000	$Y=4.00\times10^6 X - 5.32\times10^5$	0.9987
135	Aspon	20.84	0.0500	0.0250	0.1250 ~ 5.000	$Y=3.58\times10^6 X - 1.49\times10^5$	0.9994
136	Dicofol	21.42	0.0500	0.0250	0.1250 ~ 5.000	$Y=1.00\times10^6 X - 1.25\times10^5$	0.9980
137	Metolachlor	21.45	0.0250	0.0125	0.0625 ~ 2.500	$Y=2.13\times10^6 X - 1.69\times10^5$	0.9977
138	Oxychlordane	21.72	0.1000	0.0500	0.2500 ~ 2.500	$Y=1.97\times10^5 X - 1.23\times10^4$	0.9997
139	Pirimiphos-Ethyl	21.74	0.0500	0.0250	0.1250 ~ 5.000	$Y=1.16\times10^6 X - 1.83\times10^5$	0.9980
140	Methoprene	21.79	0.1000	0.0500	0.2500 ~ 10.00	$Y=3.30\times10^6 X - 1.31\times10^6$	0.9975
141	Bromofos	21.85	0.0500	0.0250	0.1250 ~ 5.000	$Y=1.50\times10^6 X - 1.98\times10^5$	0.9987
142	Dichlofluanid	21.86	0.1500	0.0750	0.3750 ~ 15.00	$Y=5.53\times10^5 X - 3.40\times10^5$	0.9977
143	Ethofumesate	22.00	0.0500	0.0250	0.1250 ~ 5.000	$Y=2.00\times10^6 X - 2.03\times10^5$	0.9992
144	Endosulfan-1	23.07	0.1500	0.0750	0.3750 ~ 15.00	$Y=1.47\times10^5 X - 3.22\times10^4$	0.9991

145	Propanil	23.08	0.0500	0.0250	0.1250 ~ 5.000	$Y=1.99\times10^{6} X - 4.75\times10^{5}$	0.9930
146	Isofenphos	23.11	0.0500	0.0250	0.1250 ~ 5.000	$Y=1.54\times10^{6} X - 2.40\times10^{5}$	0.9982
147	Crufomate	23.22	0.1500	0.0750	0.3750 ~ 15.00	$Y=1.85\times10^{6} X - 1.63\times10^{6}$	0.9914
148	Chlorfenvinphos	23.32	0.0750	0.0375	0.1875 ~ 7.500	$Y=8.38\times10^{5} X - 2.30\times10^{5}$	0.9976
149	Cis-Chlordane	23.65	0.0500	0.0250	0.1250 ~ 5.000	$Y=9.13\times10^{5} X - 7.13\times10^{4}$	0.9994
150	Tolyfluanide	23.68	0.0750	0.0375	0.1875 ~ 7.500	$Y=5.19\times10^{5} X - 1.68\times10^{5}$	0.9981
151	p,p'-DDE	24.00	0.0250	0.0125	0.0625 ~ 2.500	$Y=1.41\times10^{6} X - 5.29\times10^{4}$	0.9994
152	Butachlor	24.02	0.0500	0.0250	0.1250 ~ 5.000	$Y=1.84\times10^{6} X - 3.26\times10^{5}$	0.9970
153	Chlozolinate	24.04	0.0500	0.0250	0.1250 ~ 5.000	$Y=6.77\times10^{5} X - 6.41\times10^{4}$	0.9994
154	Crotoxyphos	24.10	0.1500	0.0750	0.3750 ~ 15.00	$Y=4.98\times10^{5} X - 4.39\times10^{5}$	0.9904
155	Iodofenphos	24.44	0.0500	0.0250	0.1250 ~ 5.000	$Y=1.32\times10^{6} X - 3.31\times10^{5}$	0.9940
156	Tetrachlorvinphos	24.48	0.0750	0.0375	0.1875 ~ 7.500	$Y=1.61\times10^{6} X - 4.88\times10^{5}$	0.9975
157	Chlorbromuron	24.49	0.6000	0.3000	1.5000 ~ 60.00	$Y=1.63\times10^{5} X - 6.33\times10^{5}$	0.9843
158	Profenofos	24.76	0.1500	0.0750	0.3750 ~ 15.00	$Y=4.53\times10^{5} X - 2.76\times10^{5}$	0.9969
159	Fluorochloridone	25.14	0.2000	0.1000	0.5000 ~ 5.000	$Y=1.14\times10^{5} X - 1.75\times10^{4}$	0.9994
160	Buprofezin	25.16	0.0500	0.0250	0.1250 ~ 5.000	$Y=7.95\times10^{6} X - 8.95\times10^{5}$	0.9987
161	o,p'-DDD	25.17	0.0500	0.0250	0.1250 ~ 2.500	$Y=1.16\times10^{5} X + 2.92\times10^{4}$	0.9997
162	Endrin	25.18	0.3000	0.1500	0.7500 ~ 30.00	$Y=3.08\times10^{5} X - 2.37\times10^{5}$	0.9982
163	Hexaconazole	25.20	0.3000	0.1500	0.7500 ~ 15.00	$Y=5.28\times10^{4} X - 2.04\times10^{4}$	0.9989
164	Chlorfenson	25.24	0.0500	0.0250	0.1250 ~ 5.000	$Y=5.95\times10^{5}X - 8.13\times10^{4}$	0.9980
165	o,p'-DDT	25.34	0.0500	0.0250	0.1250 ~ 5.000	$Y=2.71\times10^{6} X - 4.24\times10^{5}$	0.9978

166	Paclobutrazol	25.44	0.0750	0.0375	0.1875 ~ 7.500	$Y=1.76\times10^6 X - 6.51\times10^5$	0.9941
167	Methoprotryne	25.75	0.0750	0.0375	0.1875 ~ 7.500	$Y=1.80\times10^6 X - 5.13\times10^5$	0.9973
168	Erbon	25.83	0.0500	0.0250	0.1250 ~ 2.500	$Y=4.86\times10^5 X - 4.10\times10^4$	0.9994
169	Chlorpropylate	25.96	0.0250	0.0125	0.0625 ~ 2.500	$Y=2.99\times10^6 X - 2.62\times10^5$	0.9970
170	Flamprop-Methyl	26.03	0.0250	0.0125	0.0625 ~ 2.500	$Y=6.68\times10^6 X - 4.53\times10^5$	0.9983
171	Nitrofen	26.24	0.1500	0.0750	0.3750 ~ 15.00	$Y=1.20\times10^6 X - 1.04\times10^6$	0.9905
172	Oxyflurofen	26.25	0.1000	0.0500	0.2500 ~ 10.00	$Y=1.54\times10^6 X - 7.81\times10^5$	0.9945
173	Chlorthiophos	26.63	0.0750	0.0375	0.1875 ~ 7.500	$Y=6.89\times10^5 X - 1.58\times10^5$	0.9978
174	Endosulfan-2	26.75	0.6000	0.3000	1.5000 ~ 15.00	$Y=2.45\times10^4 X - 1.10\times10^4$	0.9994
175	Flamprop-Isopropyl	26.83	0.0250	0.0125	0.0625 ~ 2.500	$Y=7.37\times10^6 X - 5.64\times10^5$	0.9974
176	p,p'-DDT	27.31	0.0500	0.0250	0.1250 ~ 5.000	$Y=2.76\times10^6 X - 5.05\times10^5$	0.9971
177	Carbofenothion	27.32	0.0500	0.0250	0.1250 ~ 5.000	$Y=1.32\times10^6 X - 2.79\times10^5$	0.9955
178	Benalaxyl	27.66	0.0250	0.0125	0.0625 ~ 2.500	$Y=3.38\times10^6 X - 2.55\times10^5$	0.9972
179	Edifenphos	28.07	0.0500	0.0250	0.1250 ~ 5.000	$Y=1.25\times10^6 X - 3.66\times10^5$	0.9912
180	Triazophos	28.37	0.0750	0.0375	0.1875 ~ 7.500	$Y=1.05\times10^6 X - 3.63\times10^5$	0.9944
181	Cyanofenphos	28.56	0.0250	0.0125	0.0625 ~ 2.500	$Y=2.34\times10^6 X - 1.98\times10^5$	0.9967
182	Chlorbenside Sulfone	29.01	0.0500	0.0250	0.1250 ~ 5.000	$Y=1.89\times10^6 X - 2.89\times10^5$	0.9966
183	Endosulfan-Sulfate	29.18	0.0750	0.0375	0.1875 ~ 7.500	$Y=2.46\times10^5 X - 3.43\times10^4$	0.9985
184	Bromopropylate	29.39	0.0500	0.0250	0.1250 ~ 5.000	$Y=3.62\times10^6 X - 6.59\times10^5$	0.9963
185	Benzoylprop-Ethyl	29.51	0.0750	0.0375	0.1875 ~ 7.500	$Y=5.20\times10^5 X - 1.00\times10^5$	0.9983
186	Fenpropathrin	29.66	0.0500	0.0250	0.1250 ~ 5.000	$Y=4.35\times10^5 X - 7.38\times10^4$	0.9966

187	Captafol	30.18	0.4500	0.2250	1.1250 ~ 45.00	$Y=2.33\times10^5 X - 9.85\times10^5$	0.9564
188	EPN	30.28	0.1000	0.0500	0.2500 ~ 10.00	$Y=2.08\times10^6 X - 1.17\times10^6$	0.9913
189	Hexazinone	30.30	0.0750	0.0375	0.1875 ~ 7.500	$Y=5.21\times10^6 X - 1.27\times10^6$	0.9970
190	Bifenox	30.95	0.0125	0.0000	0.2500 ~ 10.00	$Y=1.24\times10^6 X - 5.52\times10^5$	0.9826
191	Phosalone	31.32	0.0500	0.0250	0.1250 ~ 5.000	$Y=1.19\times10^6 X - 2.42\times10^5$	0.9955
192	Azinphos-Methyl	31.52	0.3000	0.1500	0.7500 ~ 15.00	$Y=1.82\times10^5 X - 1.18\times10^5$	0.9971
193	Fenarimol	31.74	0.0500	0.0250	0.1250 ~ 5.000	$Y=1.19\times10^6 X - 1.33\times10^5$	0.9993
194	Azinphos-Ethyl	32.11	0.0500	0.0250	0.1250 ~ 5.000	$Y=1.15\times10^6 X - 2.82\times10^5$	0.9930
195	Prochloraz	33.15	0.1500	0.0750	0.3750 ~ 15.00	$Y=5.01\times10^5 X - 5.72\times10^4$	0.9919
196	Cyfluthrin	32.99	0.3000	0.1500	0.7500 ~ 30.00	$Y=2.38\times10^5 X - 1.95\times10^5$	0.9966
197	Coumaphos	33.31	0.1500	0.0750	0.3750 ~ 15.00	$Y=7.20\times10^5 X - 4.57\times10^5$	0.9955
198	Fluvalinate	35.04	0.3000	0.1500	0.7500 ~ 30.00	$Y=3.44\times10^6 X - 3.89\times10^6$	0.9954
		35.13					
199	Dichlorvos	7.93	0.1500	0.0750	0.3750 ~ 15.00	$Y=2.58\times10^6 X - 9.22\times10^5$	0.9990
200	Biphenyl	9.14	0.0250	0.0125	0.0625 ~ 2.500	$Y=6.61\times10^6 X - 1.66\times10^5$	0.9998
201	Propamocarb	9.75	0.0750	0.0375	0.1875 ~ 7.500	$Y=5.73\times10^6 X - 2.71\times10^6$	0.9912
202	Vernolate	9.98	0.0250	0.0125	0.0625 ~ 2.500	$Y=2.12\times10^6 X - 6.55\times10^4$	0.9997
203	3,5-Dichloroaniline	11.37	0.0250	0.0125	0.0625 ~ 2.500	$Y=9.29\times10^5 X + 1.10\times10^5$	0.9905
204	Molinate	12.09	0.0250	0.0125	0.0625 ~ 2.500	$Y=2.55\times10^6 X - 8.80\times10^4$	0.9996
205	Methacrifos	12.10	0.0250	0.0125	0.0625 ~ 2.500	$Y=1.10\times10^6 X - 5.10\times10^4$	0.9994
206	2-Phenylphenol	12.67	0.0250	0.0125	0.0625 ~ 1.2500	$Y=1.70\times10^6 X + 3.54\times10^4$	0.9946

207	Cis1,2,3,6Tetrahydrophthalimide	13.60	0.0750	0.0375	0.1875 ~ 7.500	$Y=1.06\times10^{6} X - 2.31\times10^{5}$	0.9985
208	Fenobucarb	14.80	0.0500	0.0250	0.1250 ~ 5.000	$Y=5.44\times10^{6} X - 7.85\times10^{5}$	0.9982
209	Benfluralin	15.40	0.0250	0.0125	0.0625 ~ 2.500	$Y=2.07\times10^{6} X - 2.74\times10^{5}$	0.9888
210	Hexaflumuron	16.47	0.0250	0.0125			
211	Prometon	16.85	0.0750	0.0375	0.1875 ~ 7.500	$Y=1.47\times10^{6} X - 3.19\times10^{5}$	0.9988
212	Triallate	17.31	0.0500	0.0250	0.1250 ~ 5.000	$Y=8.94\times10^{5} X - 9.82\times10^{4}$	0.9993
213	Pyrimethanil	17.48	0.0250	0.0125	0.0625 ~ 2.500	$Y=5.84\times10^{6} X - 4.82\times10^{5}$	0.9980
214	gamma-HCH	17.70	0.0500	0.0250	0.1250 ~ 5.000	$Y=7.39\times10^{5} X - 3.27\times10^{4}$	0.9998
215	Disulfoton	17.80	0.0250	0.0125	0.0625 ~ 2.500	$Y=2.37\times10^{6} X - 1.88\times10^{5}$	0.9983
216	Atrizine	17.85	0.0250	0.0125	0.0625 ~ 2.500	$Y=1.60\times10^{6} X - 1.05\times10^{5}$	0.9991
217	Heptachlor	18.69	0.0750	0.0375	0.1875 ~ 7.500	$Y=6.01\times10^{5} X - 1.16\times10^{5}$	0.9990
218	Probenazole	18.70	0.1000	0.0500	0.2500 ~ 2.500	$Y=1.26\times10^{5} X - 1.20\times10^{4}$	0.9997
219	Iprobenfos	18.70	0.0750	0.0375	0.1875 ~ 7.500	$Y=1.47\times10^{6} X - 6.69\times10^{5}$	0.9895
220	Isazofos	18.78	0.0500	0.0250	0.1250 ~ 5.000	$Y=8.48\times10^{5} X - 9.19\times10^{4}$	0.9994
221	Plifenate	19.08	0.1000	0.0500	0.2500 ~ 5.000	$Y=4.07\times10^{5} X - 7.16\times10^{4}$	0.9992
222	Fenpropimorph	19.28	0.0250	0.0125	0.0625 ~ 2.500	$Y=8.00\times10^{6} X - 5.62\times10^{5}$	0.9989
223	Transfluthrin	19.30	0.0250	0.0125	0.0625 ~ 2.500	$Y=3.03\times10^{6} X - 1.80\times10^{5}$	0.9991
224	Fluchloralin	19.32	0.1000	0.0500	0.2500 ~ 10.00	$Y=9.56\times10^{5} X - 4.69\times10^{5}$	0.9952
225	Tolclofos-methyl	19.90	0.0250	0.0125	0.0625 ~ 2.500	$Y=3.46\times10^{6} X - 1.95\times10^{5}$	0.9991
226	Ametryn	20.30	0.0750	0.0375	0.1875 ~ 7.500	$Y=1.96\times10^{6} X - 4.43\times10^{5}$	0.9989
227	Simetryn	20.40	0.0500	0.0250	0.1250 ~ 5.000	$Y=2.26\times10^{6} X - 2.94\times10^{5}$	0.9990

228	Methobromuron	20.54	0.1500	0.0750	0.3750 ~ 15.00	$Y=1.67\times10^5 X - 6.17\times10^4$	0.9986
229	Metribuzin	20.59	0.0750	0.0375	0.1875 ~ 7.500	$Y=1.61\times10^6 X - 3.48\times10^5$	0.9986
230	Dipropetryn	21.04	0.0250	0.0125	0.0625~2.500	$Y=1.67\times10^6 X - 1.52\times10^5$	0.9976
231	Formothion	21.65	0.2000	0.1000	0.5000~5.000	$Y=1.12\times10^5 X - 3.13\times10^4$	0.9993
232	Diethofencarb	21.75	0.1500	0.0750	0.3750~15.00	$Y=1.15\times10^6X - 6.62\times10^5$	0.9977
233	Dimepiperate	21.77	1.0000	0.5000	2.500~5.000	$Y=2.12\times10^4X + 4.02\times10^4$	1.0000
234	Bioallethrin-1	22.54	0.1000	0.0500	0.2500~10.00	$Y=1.37\times10^6X - 7.09\times10^5$	0.9941
235	Bioallethrin-2	22.74	0.1000	0.0500	0.2500~10.00	$Y=1.49\times10^6X - 5.77\times10^5$	0.9983
236	O,p'-DDE	22.84	0.0250	0.0125	0.0625~2.500	$Y=2.54\times10^6X - 7.65\times10^4$	0.9998
237	Fenson	22.86	0.0250	0.0125	0.0625~2.500	$Y=2.26\times10^6X - 1.16\times10^5$	0.9993
238	Chinomethionat	22.96	0.5000	0.2500	1.250~2.500	$Y=5.69\times10^4X - 3.15\times10^4$	1.0000
239	Diphenamid	23.00	0.0250	0.0125	0.0625~2.500	$Y=4.01\times10^6X - 2.59\times10^5$	0.9989
240	Chlorthion	23.10	0.2000	0.1000	0.5000~5.000	$Y=3.72\times10^5X - 2.23\times10^5$	0.9857
241	Prallethrin	23.32	0.0750	0.0375	0.1875~7.500	$Y=3.52\times10^6X - 1.22\times10^6$	0.9964
242	Penconazole	23.40	0.0750	0.0375	0.1875~7.500	$Y=2.32\times10^6X - 4.76\times10^5$	0.9992
243	Mecarbam	23.67	0.1000	0.0500	0.2500~10.00	$Y=2.87\times10^5X - 2.80\times10^5$	0.9986
244	Tetraconazole	23.68	0.0750	0.0375	0.1875~7.500	$Y=2.72\times10^6X - 5.43\times10^5$	0.9992
245	Flumetralin	24.30	0.0500	0.0250	0.1250~5.000	$Y=2.51\times10^6X - 7.70\times10^5$	0.9873
246	Triadimenol	24.44	0.0750	0.0375	0.1875~7.500	$Y=1.90\times10^6X - 5.96\times10^5$	0.9972
247	Pretilachlor	24.87	0.0500	0.0250	0.1250~5.000	$Y=1.40\times10^6X - 2.48\times10^5$	0.9980
248	Kresoxim-methyl	25.27	0.0250	0.0125	0.0625~2.500	$Y=2.45\times10^6X - 2.13\times10^5$	0.9976

249	Fluazifop-butyl	25.41	0.0250	0.0125	0.0625 ~ 2.500	$Y=2.18\times10^6X - 2.27\times10^5$	0.9961
250	Chlorfluazuron	25.50	0.1500	0.0750	0.3750 ~ 7.500	$Y=1.55\times10^5X - 9.15\times10^3$	0.9980
251	Chlorobenzilate	26.13	0.0250	0.0125	0.0625 ~ 2.500	$Y=2.62\times10^6X - 2.29\times10^5$	0.9977
252	Uniconazole	26.34	0.0500	0.0250	0.1250 ~ 5.000	$Y=1.23\times10^6X - 2.13\times10^5$	0.9983
253	Flusilazole	26.40	0.0750	0.0375	0.1875 ~ 7.500	$Y=4.23\times10^6X - 1.07\times10^6$	0.9983
254	Diniconazole	27.27	0.0750	0.0375	0.1875 ~ 7.500	$Y=1.62\times10^6X - 6.37\times10^5$	0.9936
255	Piperonyl butoxide	27.67	0.0250	0.0125	0.0625 ~ 2.500	$Y=3.50\times10^6X - 4.69\times10^5$	0.9919
256	Propargite	28.10	0.0500	0.0250	0.1250 ~ 5.000	$Y=1.43\times10^6X - 1.30\times10^5$	0.9995
257	Mepronil	28.12	0.0250	0.0125	0.0625 ~ 2.500	$Y=5.96\times10^6X - 6.83\times10^5$	0.9941
258	Dimefuron	28.14	0.1000	0.0500	0.2500 ~ 10.00	$Y=2.79\times10^5X - 1.10\times10^4$	0.9885
259	Diflufenican	29.63	0.0250	0.0125	0.0625 ~ 2.500	$Y=3.75\times10^6X - 4.25\times10^5$	0.9950
260	Fenazaquin	29.14	0.0250	0.0125	0.0625 ~ 2.500	$Y=5.25\times10^6X - 5.73\times10^5$	0.9950
261	Phenothrin	29.34	0.0250	0.0125	0.0625 ~ 2.500	$Y=2.13\times10^6X - 2.47\times10^5$	0.9950
262	Fludioxonil	29.18	0.0250	0.0125	0.0625 ~ 2.500	$Y=2.12\times10^6X - 2.08\times10^5$	0.9963
263	Fenoxycarb	29.75	0.1500	0.0750	0.3750 ~ 15.00	$Y=1.81\times10^5X + 1.11\times10^5$	0.9902
264	Sethoxydim	29.82	0.2250	0.1125	0.5625 ~ 22.50	$Y=2.11\times10^5X + 1.26\times10^5$	0.9868
265	Amitraz	30.44	0.0750	0.0375	0.1875 ~ 7.500	$Y=4.35\times10^5X - 8.43\times10^4$	0.9993
266	Anilofos	30.88	0.1000	0.0500	0.2500 ~ 5.000	$Y=5.65\times10^5X - 1.83\times10^5$	0.9954
267	Acrinathrin	31.22	0.0500	0.0250	0.1250 ~ 5.000	$Y=1.21\times10^6X - 2.74\times10^5$	0.9939
268	Lambda-cyhalothrin	31.29	0.0250	0.0125	0.0625 ~ 2.500	$Y=1.83\times10^6X - 1.43\times10^5$	0.9982
269	Mefenacet	31.48	0.0750	0.0375	0.1875 ~ 7.500	$Y=2.02\times10^6X - 7.14\times10^5$	0.9928

270	Permethrin	31.72	0.0500	0.0250	0.1250 ~ 5.000	$Y=2.92\times10^{6}X - 5.21\times10^{5}$	0.9967
271	Pyridaben	32.00	0.0250	0.0125	0.0625 ~ 2.500	$Y=5.18\times10^{6}X - 4.92\times10^{5}$	0.9958
272	Fluoroglycofen-ethyl	32.18	0.3000	0.1500	0.7500 ~ 30.00	$Y=4.13\times10^{5}X - 8.77\times10^{5}$	0.9826
273	Bitertanol	32.40	0.0750	0.0375	0.1875 ~ 7.500	$Y=4.06\times10^{6}X - 1.42\times10^{6}$	0.9931
274	Etofenprox	32.88	0.0250	0.0125	0.0625 ~ 2.500	$Y=6.57\times10^{6}X - 3.60\times10^{5}$	0.9961
275	Cycloxydim	32.20	0.3000	0.1500	0.7500 ~ 30.00	$Y=7.64\times10^{5}X + 8.13\times10^{5}$	0.9907
276	Esfenvalerate	33.00	0.1000	0.0500	0.2500 ~ 10.00	$Y=3.68\times10^{5}X - 7.15\times10^{4}$	0.9983
277	Alpha-cypermethrin	33.48	0.0500	0.0250	0.1250 ~ 5.000	$Y=1.42\times10^{6}X - 2.51\times10^{5}$	0.9974
278	Difenconazole	34.32	0.1500	0.0750	0.3750 ~ 15.00	$Y=1.08\times10^{6}X - 3.85\times10^{5}$	0.9860
279	Flumiclorac-pentyl	36.57	0.0500	0.0250	0.1250 ~ 5.000	$Y=9.93\times10^{5}X - 2.33\times10^{5}$	0.9903
280	Dimefox	5.70	0.0750	0.0375	0.1875 ~ 7.500	$Y=1.34\times10^{6}X - 1.07\times10^{5}$	0.9992
281	Disulfoton-sulfoxide	8.50	0.0500	0.0250	0.1250 ~ 5.000	$Y=2.77\times10^{5}X - 2.21\times10^{4}$	0.9993
282	Pentachlorobenzene	11.18	0.0250	0.0125	0.0625 ~ 2.500	$Y=2.14\times10^{6}X - 3.01\times10^{4}$	0.9998
283	Crimidine	13.30	0.0250	0.0125	0.0625 ~ 2.500	$Y=1.32\times10^{6}X - 5.33\times10^{4}$	0.9987
284	BDMC-1	13.50	0.1000	0.0500	0.2500 ~ 5.000	$Y=2.00\times10^{5}X + 2.44\times10^{4}$	0.9978
285	Chlorfenprop-methyl	13.70	0.0250	0.0125	0.0625 ~ 2.500	$Y=1.34\times10^{6}X - 4.92\times10^{4}$	0.9989
286	Thionazin	14.19	0.0250	0.0125	0.0625 ~ 2.500	$Y=6.15\times10^{5}X - 3.47\times10^{4}$	0.9969
287	2,3,5,6-tetrachloroaniline	14.31	0.0250	0.0125	0.0625 ~ 2.500	$Y=1.72\times10^{6}X - 5.14\times10^{4}$	0.9994
288	Tri-n-butyl phosphate	14.46	0.0500	0.0250	0.1250 ~ 5.000	$Y=1.40\times10^{6}X - 2.15\times10^{5}$	0.9972
289	2,3,4,5-tetrachloroanisole	14.76	0.0250	0.0125	0.0625 ~ 2.500	$Y=1.11\times10^{6}X - 2.36\times10^{4}$	0.9997
290	Pentachloroanisole	15.28	0.0250	0.0125	0.0625 ~ 2.500	$Y=8.25\times10^{5}X - 1.64\times10^{4}$	0.9996

291	Tebutam	15.43	0.0500	0.0250	0.1250 ~ 5.000	$Y=1.22\times10^6X - 8.63\times10^4$	0.9990
292	Dioxabenzofos	16.31	0.2500	0.1250	0.6250 ~ 25.00	$Y=4.35\times10^5X - 2.36\times10^5$	0.9980
293	Methabenzthiazuron	16.53	0.2500	0.1250	0.6250 ~ 25.00	$Y=1.30\times10^6X - 1.65\times10^6$	0.9930
294	Simeton	16.86	0.0500	0.0250	0.1250 ~ 5.000	$Y=1.93\times10^6X - 2.26\times10^5$	0.9984
295	Atratone	16.89	0.0250	0.0125	0.0625 ~ 2.500	$Y=2.38\times10^6X - 1.24\times10^5$	0.9988
296	Desisopropyl-atrazine	16.91	0.2000	0.1000	0.5000 ~ 20.00	$Y=8.37\times10^5X - 4.16\times10^5$	0.9975
297	Terbufos sulfone	16.94	0.0250	0.0125	0.0625 ~ 2.500	$Y=2.08\times10^6X - 1.20\times10^5$	0.9980
298	Tefluthrin	17.37	0.0250	0.0125	0.0625 ~ 2.500	$Y=5.47\times10^6X - 2.89\times10^5$	0.9980
299	Fonofos	17.37	0.0250	0.0125	0.0625 ~ 2.500	$Y=1.45\times10^6X - 7.52\times10^4$	0.9987
300	Bromocylen	17.54	0.0250	0.0125	0.0625 ~ 2.500	$Y=5.19\times10^5X - 2.05\times10^4$	0.9990
301	Trietazine	17.70	0.0250	0.0125	0.0625 ~ 2.500	$Y=2.17\times10^6X - 9.66\times10^4$	0.9987
302	Etrimfos oxon	17.97	0.0250	0.0125	0.0625 ~ 2.500	$Y=4.07\times10^6X - 1.85\times10^5$	0.9990
303	Cycluron	18.11	0.0750	0.0375	0.1875 ~ 7.500	$Y=6.73\times10^5X - 1.27\times10^5$	0.9982
304	2,6-dichlorobenzamide	18.16	0.0500	0.0250	0.1250 ~ 5.000	$Y=1.57\times10^6X - 1.74\times10^5$	0.9983
305	DE-PCB 28	18.26	0.0250	0.0125	0.0625 ~ 1.250	$Y=5.60\times10^6X - 5.83\times10^4$	0.9997
306	DE-PCB 31	18.29	0.0250	0.0125	0.0625 ~ 1.250	$Y=5.45\times10^6X - 5.64\times10^4$	0.9997
307	Desethyl-sebuthylazine	18.52	0.0500	0.0250	0.1250 ~ 5.000	$Y=3.58\times10^6X - 3.93\times10^5$	0.9985
308	2,3,4,5-tetrachloroaniline	18.71	0.0500	0.0250	0.1250 ~ 5.000	$Y=1.61\times10^6X - 1.25\times10^5$	0.9989
309	Pentachloroaniline	19.01	0.0250	0.0125	0.0625 ~ 2.500	$Y=1.63\times10^6X - 5.38\times10^4$	0.9990
310	Aziprotryne	19.30	0.2000	0.1000	0.5000 ~ 20.00	$Y=4.43\times10^5X - 1.80\times10^5$	0.9979
311	Sebutylazine	19.45	0.0250	0.0125	0.0625 ~ 2.500	$Y=3.74\times10^6X - 1.70\times10^5$	0.9989

312	Isocarbamid	19.46	0.1250	0.0625	0.3125 ~ 12.50	$Y=2.55\times10^{6}X - 9.72\times10^{5}$	0.9965
313	DE-PCB 52	19.59	0.0250	0.0125	0.0625 ~ 2.500	$Y=1.92\times10^{6}X - 3.05\times10^{4}$	0.9997
314	Prosulfocarb	19.65	0.0250	0.0125	0.0625 ~ 2.500	$Y=1.04\times10^{6}X - 5.13\times10^{4}$	0.9988
315	Dimethenamid	19.74	0.0250	0.0125	0.0625 ~ 2.500	$Y=3.66\times10^{6}X - 1.48\times10^{5}$	0.9990
316	Fenchlorphos oxon	19.85	0.0500	0.0250	0.1250 ~ 5.000	$Y=2.18\times10^{6}X - 1.77\times10^{5}$	0.9991
317	BDMC-2	19.91	0.0500	0.0250	0.1250 ~ 5.000	$Y=1.46\times10^{6}X - 3.49\times10^{5}$	0.9912
318	Paraoxon-methyl	20.04	0.2000	0.1000	0.5000 ~ 5.000	$Y=3.57\times10^{5}X - 2.04\times10^{5}$	0.9852
319	Monalide	20.19	0.0500	0.0250	0.1250 ~ 5.000	$Y=7.65\times10^{5}X - 5.97\times10^{4}$	0.9988
320	Isobenzan	20.68	0.0250	0.0125	0.0625 ~ 2.500	$Y=5.29\times10^{5}X - 1.51\times10^{4}$	0.9995
321	Octachlorostyrene	20.70	0.0250	0.0125	0.0625 ~ 2.500	$Y=6.84\times10^{5}X - 1.55\times10^{4}$	0.9997
322	Isodrin	21.14	0.0250	0.0125	0.0625 ~ 2.500	$Y=7.73\times10^{5}X - 9.06\times10^{3}$	0.9982
323	Isomethiozin	21.22	0.0500	0.0250	0.1250 ~ 5.000	$Y=1.60\times10^{6}X - 2.21\times10^{5}$	0.9970
324	Trichloronat	21.24	0.0250	0.0125	0.0625 ~ 2.500	$Y=1.49\times10^{6}X - 6.67\times10^{4}$	0.9989
325	Dacthal	21.41	0.0250	0.0125	0.0625 ~ 2.500	$Y=4.04\times10^{6}X - 1.08\times10^{5}$	0.9995
326	dichlorobenzophenone	21.43	0.0250	0.0125	0.0625 ~ 2.500	$Y=1.13\times10^{6}X - 9.71\times10^{4}$	0.9959
327	Nitrothal-isopropyl	21.86	0.0500	0.0250	0.1250 ~ 5.000	$Y=1.80\times10^{6}X - 4.91\times10^{5}$	0.9911
328	Rabenzazole	21.91	0.0250	0.0125	0.0625 ~ 2.500	$Y=5.20\times10^{6}X - 6.05\times10^{5}$	0.9931
329	Cyprodinil	22.07	0.0250	0.0125	0.0625 ~ 2.500	$Y=5.21\times10^{6}X - 3.06\times10^{5}$	0.9977
330	Isofenphos oxon	22.49	0.0250	0.0125			
331	Methfuroxam	22.59	0.0250	0.0125	0.0625 ~ 2.500	$Y=1.52\times10^{6}X - 1.33\times10^{5}$	0.9954
332	Dicapthon	22.62	0.1250	0.0625	0.3125 ~ 12.50	$Y=1.75\times10^{6}X - 1.05\times10^{6}$	0.9946
333	DE-PCB 101	22.73	0.0250	0.0125	0.0625 ~ 2.500	$Y=1.74\times10^{6}X - 3.67\times10^{4}$	0.9994

334	MCPA-butoxyethyl ester	22.75	0.0250	0.0125	0.0625 ~ 2.500	$Y=6.47\times10^{5}X - 4.71\times10^{4}$	0.9966
335	Phorate sulfone	23.38	0.0250	0.0125	0.0625 ~ 2.500	$Y=8.24\times10^{5}X - 5.08\times10^{4}$	0.9946
336	Chlorfenethol	23.46	0.0250	0.0125	0.0625 ~ 2.500	$Y=2.04\times10^{6}X - 1.30\times10^{5}$	0.9975
337	Trans-nonachlor	23.76	0.0250	0.0125	0.0625 ~ 2.500	$Y=1.13\times10^{6}X - 3.81\times10^{4}$	0.9990
338	Dinobuton	24.06	0.5000	0.2500	1.2500 ~ 25.00	$Y=5.32\times10^{5}X - 1.34\times10^{6}$	0.9788
339	DEF	24.21	0.0500	0.0250	0.1250 ~ 5.000	$Y=7.30\times10^{5}X - 1.05\times10^{5}$	0.9971
340	Flurochloridone	24.54	0.0500	0.0250	0.1250 ~ 5.000	$Y=1.26\times10^{6}X - 1.66\times10^{5}$	0.9979
341	Bromfenvinfos	24.80	0.0250	0.0125	0.0625 ~ 2.500	$Y=1.20\times10^{6}X - 1.00\times10^{5}$	0.9937
342	Perthane	24.92	0.0250	0.0125	0.0625 ~ 2.500	$Y=7.03\times10^{6}X - 3.66\times10^{5}$	0.9974
343	Ditalimfos	25.01	0.0250	0.0125	0.0625 ~ 2.500	$Y=2.75\times10^{6}X - 1.67\times10^{5}$	0.9976
344	DE-PCB 118	25.18	0.0250	0.0125	0.0625 ~ 2.500	$Y=2.10\times10^{6}X - 6.14\times10^{4}$	0.9985
345	Dibromobenzo-phenone	25.43	0.0500	0.0250	0.1250 ~ 2.500	$Y=5.97\times10^{5}X - 7.56\times10^{4}$	0.9945
346	Flutriafol	25.51	0.0500	0.0250	0.1250 ~ 5.000	$Y=1.40\times10^{6}X - 1.75\times10^{5}$	0.9985
347	Mephosfolan	25.56	0.0500	0.0250	0.1250 ~ 5.000	$Y=9.51\times10^{5}X - 2.57\times10^{5}$	0.9890
348	Athidathion	25.75	0.0500	0.0250	0.1250 ~ 5.000	$Y=1.88\times10^{5}X - 5.69\times10^{3}$	0.9995
349	DE-PCB 153	25.76	0.0250	0.0125	0,0625 ~ 2.500	$Y=1.60\times10^{6}X - 4.19\times10^{4}$	0.9992
350	Diclobutrazole	26.14	0.1000	0.0500	0.2500 ~ 10.00	$Y=1.90\times10^{6}X - 5.90\times10^{5}$	0.9957
351	Disulfoton sulfone	26.40	0.0500	0.0250	0.1250 ~ 5.000	$Y=1.39\times10^{6}X - 1.94\times10^{5}$	0.9966
352	Hexythiazox	26.73	0.2000	0.1000	0.500 ~ 20.00	$Y=3.00\times10^{5}X - 1.65\times10^{5}$	0.9981
353	DE-PCB 138	26.96	0.0250	0.0125	0.0625 ~ 2.500	$Y=1.31\times10^{6}X - 3.38\times10^{4}$	0.9991
354	Resmethrin-1	27.38	0.0500	0.0250	0.1250 ~ 5.000	$Y=2.98\times10^{5}X - 1.11\times10^{4}$	0.9823

355	Cyproconazole	27.43	0.0250	0.0125	0.0625 ~ 2.500	$Y=1.31\times10^6X - 6.22\times10^4$	0.9983
356	Resmethrin-2	27.55	0.0500	0.0250	0.1250 ~ 5.000	$Y=7.98\times10^5X - 1.33\times10^5$	0.9953
357	Phthalic acid,benzyl butyl ester	27.70	0.0250	0.0125	0.0625 ~ 2.500	$Y=8.67\times10^5X - 4.54\times10^4$	0.9973
358	Clodinafop propargyl	27.92	0.0500	0.0250	0.1250 ~ 5.000	$Y=9.81\times10^5X - 3.07\times10^5$	0.9820
359	Fenthion sulfoxide	28.28	0.1000	0.0500	0.2500 ~ 5.000	$Y=2.45\times10^5X - 1.08\times10^5$	0.9863
360	Fluotrimazole	28.56	0.0250	0.0125	0.0625 ~ 2.500	$Y=1.55\times10^6X - 5.73\times10^4$	0.9981
361	Fluroxypr-1-methylheptyl ester	28.62	0.0500	0.0250	0.1250 ~ 2.500	$Y=3.14\times10^5X - 2.20\times10^4$	0.9974
362	Fenthion sulfone	28.77	0.1000	0.0500	0.2500 ~ 10.00	$Y=4.68\times10^5X + 1.45\times10^5$	0.9839
363	Triphenyl phosphate	28.82	0.0250	0.0125	0.0625 ~ 2.500	$Y=2.59\times10^6X - 6.86\times10^4$	0.9990
364	Metamitron	28.84	1.0000	0.5000	2.500 ~ 25.00	$Y=4.59\times10^4X - 4.78\times10^4$	0.9994
365	DE-PCB 180	29.15	0.0250	0.0125	0.0625 ~ 2.500	$Y=1.16\times10^6X - 3.07\times10^4$	0.9989
366	Tebufenpyrad	29.18	0.0250	0.0125	0.0625 ~ 2.500	$Y=1.69\times10^6X - 8.49\times10^4$	0.9965
367	Cloquintocet-mexyl	29.44	0.0250	0.0125	0.0625 ~ 2.500	$Y=4.90\times10^6X - 4.73\times10^5$	0.9915
368	Lenacil	29.86	0.0250	0.0125	0.6250 ~ 25.00	$Y=1.31\times10^6X - 7.83\times10^5$	0.9967
369	Bromuconazole-1	30.06	0.0500	0.0250	0.1250 ~ 5.000	$Y=4.45\times10^5X + 7.05\times10^3$	0.9916
370	Leptophos	30.26	0.0250	0.0125	0.0625 ~ 2.500	$Y=9.66\times10^5X - 3.34\times10^4$	0.9990
371	Phosmet	30.54	0.1000	0.0500	0.2500 ~ 10.00	$Y=1.57\times10^6X + 3.93\times10^4$	0.9950
372	Bromuconazole-2	30.86	0.0500	0.0250	0.1250 ~ 2.500	$Y=7.98\times10^5X - 5.29\times10^4$	0.9994
373	Nitralin	31.11	0.2500	0.1250	0.6250 ~ 25.00	$Y=1.35\times10^6X - 1.87\times10^6$	0.9907
374	Fenamiphos sulfoxide	31.19	0.1000	0.0500	0.2500 ~ 10.00	$Y=3.38\times10^5X - 2.60\times10^3$	0.9561
375	Fenamiphos sulfone	31.53	0.1000	0.0500	0.2500 ~ 10.00	$Y=2.06\times10^6X - 7.52\times10^5$	0.9954

376	Pyrazophos	31.66	0.0250	0.0125	0.0625 ~ 2.500	$Y=4.90\times10^{5}X - 2.73\times10^{4}$	0.9977
377	Fenpiclonil	32.56	0.1000	0.0500	0.2500 ~ 10.00	$Y=1.72\times10^{6}X - 2.58\times10^{5}$	0.9981
378	Fluquinconazole	32.71	0.0250	0.0125	0.0625 ~ 2.500	$Y=3.40\times10^{6}X - 1.22\times10^{4}$	0.9991
379	Fenbuconazole	34.21	0.0500	0.0250	0.1250 ~ 5.000	$Y=2.96\times10^{6}X + 9.45\times10^{5}$	0.9896
380	Propoxur-1	6.58	0.0500	0.0250	0.1250 ~ 5.000	$Y=1.45X+5.56\times10^{-1}$	0.9570
381	Isoprocarb -1	7.56	0.0500	0.0250	0.1250 ~ 5.000	$Y=5.44\times10^{-1}X+1.86\times10^{-1}$	0.9670
382	Methamidophos	9.37	0.1000	0.0500	0.2500 ~ 10.00	$Y=1.71X - 6.58\times10^{-1}$	0.9940
383	Acenaphthene	10.79	0.0250	0.0125	0.0625 ~ 2.500	$Y=5.7X+7.57\times10^{-2}$	0.9980
384	Dibutyl succinate	12.20	0.0500	0.0250	0.1250 ~ 5.000	$Y=6.82X - 2.14\times10^{-1}$	1.0000
385	Phthalimide	13.21	0.0500	0.0250	0.1250 ~ 5.000	$Y=1.27X - 1.34\times10^{-1}$	0.9980
386	Chlorethoxyfos	13.43	0.0500	0.0250	0.1250 ~ 5.000	$Y=1.67X - 4.32\times10^{-2}$	1.0000
387	Isoprocarb -2	13.69	0.0500	0.0250	0.1250 ~ 5.000	$Y=6.46X - 7.58\times10^{-1}$	0.9670
388	Pencycuron	14.30	0.0500	0.0250	0.1250 ~ 5.000	$Y=3.35X - 2.21\times10^{-1}$	0.9980
389	Tebuthiuron	14.25	0.1000	0.0500	0.2500 ~ 10.00	$Y=2.9X - 7.67\times10^{-1}$	0.9960
390	Demeton-S-methyl	15.19	0.1000	0.0500	0.2500 ~ 10.00	$Y=5.71\times10^{-1}X - 5.10\times10^{-2}$	0.9990
391	Cadusafos	15.13	0.1000	0.0500	0.2500 ~ 10.00	$Y=2.26X - 2.17\times10^{-1}$	0.9990
392	Propoxur-2	15.48	0.0500	0.0250	0.1250 ~ 5.000	$Y=6.32X - 1.34$	0.9890
393	Naled	15.50	0.4000	0.2000	1.0000 ~ 40.00	$Y=4.46\times10^{-1}X - 6.33\times10^{-1}$	0.9960
394	Phenanthrene	16.97	0.0250	0.0125	0.0625 ~ 2.500	$Y=8.74X - 1.65\times10^{-2}$	0.9990
395	Spiroxamine -1	17.26	0.0500	0.0250	0.1250 ~ 5.000	$Y=5.27X - 5.65\times10^{-1}$	0.9950
396	Fenpyroximate	17.49	0.2000	0.1000	0.5000 ~ 20.00	$Y=1.97\times10^{-1}X - 3.44\times10^{-2}$	0.9960

397	Tebupirimfos	17.61	0.0500	0.0250	0.1250 ~ 5.000	$Y=8.44\times10^{-1}X - 1.53\times10^{-2}$	0.9980
398	Prohydrojasmon	17.80	0.1000	0.0500			
399	Fenpropidin	17.85	0.0500	0.0250	0.1250 ~ 5.000	$Y=1.21\times10^{1}X - 1.26$	0.9970
400	Dichloran	18.10	0.0500	0.0250	0.1250 ~ 5.000	$Y=8.33\times10^{-1}X - 6.01\times10^{-2}$	0.9990
401	Pyroquilon	18.28	0.0250	0.0125	0.0625 ~ 2.500	$Y=3.8X - 9.6\times10^{-2}$	1.0000
402	Spiroxamine -2	18.23	0.0500	0.0250	0.1250 ~ 5.000	$Y=1.25\times10^{1}X - 1.23$	0.9980
403	Dinoterb	18.52	0.4000	0.2000			
404	Propyzamide	19.01	0.0500	0.0250	0.1250 ~ 5.000	$Y=2.94X - 2.44\times10^{-1}$	0.9990
405	Pirimicarb	19.08	0.0500	0.0250	0.1250 ~ 5.000	$Y=5.18X - 1.7\times10^{-1}$	1.0000
406	Phosphamidon -1	19.66	0.2000	0.1000		$Y=1.83\times10^{-1}X - 2\times10^{-1}$	0.9770
407	Benoxacor	19.62	0.0500	0.0250	0.1250 ~ 5.000	$Y=2.74X - 2.12\times10^{-1}$	0.9990
408	Bromobutide	19.70	0.0250	0.0125			
409	Acetochlor	19.84	0.0500	0.0250	0.1250 ~ 5.000	$Y=1.29X - 4.33\times10^{-2}$	1.0000
410	Tridiphane	19.90	0.1000	0.0500			
411	Terbucarb-2	20.06	0.0500	0.0250	0.1250 ~ 5.000	$Y=6.76X - 3.79\times10^{-2}$	1.0000
412	Esprocarb	20.01	0.0500	0.0250			
413	Fenfuram	20.35	0.0500	0.0250	0.1250 ~ 5.000	$Y=6.91X - 6.72\times10^{-1}$	0.9990
414	Acibenzolar-s-methyl	20.42	0.0500	0.0250			
415	Benfuresate	20.68	0.0500	0.0250	0.1250 ~ 5.000	$Y=4.25X - 1.95\times10^{-2}$	1.0000
416	Dithiopyr	20.78	0.0250	0.0125	0.0625 ~ 2.500	$Y=2.28X - 3.3\times10^{-2}$	0.9990
417	Mefenoxam	20.91	0.0500	0.0250	0.1250 ~ 5.000	$Y=1.24X - 5.51\times10^{-2}$	
418	Malaoxon	21.17	0.4000	0.2000	1.0000 ~ 40.00	$Y=1.65X - 3.28$	0.9880
419	Phosphamidon -2	21.36	0.2000	0.1000	0.5000 ~ 20.00	$Y=2.44X - 4.03\times10^{-1}$	0.9980
420	Simeconazole	21.41	0.0500	0.0250	0.1250 ~ 5.000	$Y=3.05X - 2.44\times10^{-1}$	1.0000
421	Chlorthal-dimethyl	21.39	0.0500	0.0250	0.1250 ~ 5.000	$Y=2.61X+6.8\times10^{-2}$	0.9980
422	Thiazopyr	21.91	0.0500	0.0250	0.1250 ~ 5.000	$Y=7.9\times10^{-1}X - 3.26\times10^{-2}$	0.9990
423	Dimethylvinphos	22.21	0.0500	0.0250	0.1250 ~ 5.000	$Y=3.18X - 3.31\times10^{-1}$	0.9980

424	Butralin	22.24	0.1000	0.0500	0.2500 ~ 10.00	$Y=2.77X - 4.35\times10^{-1}$	0.9940
425	Zoxamide	22.30	0.0500	0.0250	0.1250 ~ 5.00	$Y=9.78\times10^{-1}X+2.6\times10^{-2}$	0.9990
426	Pyrifenox -1	22.50	0.2000	0.1000	0.5000 ~ 20.00	$Y=4.07\times10^{-1} - 8.48\times10^{-2}$	0.9990
427	Allethrin	22.60	0.1000	0.0500	0.2500 ~ 10.00	$Y=1.88X - 3.8\times10^{-1}$	0.9980
428	Dimethametryn	22.83	0.0250	0.0125	0.0625 ~ 2.500	$Y=9.04X - 3.08\times10^{-1}$	1.0000
429	Quinoclamine	22.89	0.1000	0.0500	0.2500 ~ 10.00	$Y=6.31\times10^{-1} X - 1.57\times10^{-1}$	0.9980
430	Methothrin-1	22.92	0.0500	0.0250	0.1250 ~ 5.000	$Y=8.75\times10^{-1}X - 8.41\times10^{-2}$	0.9990
431	Flufenacet	23.09	0.2000	0.1000	0.5000 ~ 20.00	$Y=1.95X - 6.19\times10^{-1}$	0.9990
432	Methothrin-2	23.19	0.0500	0.0250	0.1250 ~ 5.000	$Y=2.61X - 2.31\times10^{-1}$	0.9990
433	Pyrifenox -2	23.50	0.2000	0.1000	0.5000 ~ 20.00	$Y=9.74\times10^{-1}X - 1.01\times10^{-1}$	0.9980
434	Fenoxanil	23.58	0.0500	0.0250	0.1250 ~ 2.500	$Y=9.69\times10^{-1}X+3.46\times10^{-2}$	0.9960
435	Phthalide	23.51	0.1000	0.0500			
436	Furalaxyl	23.97	0.0500	0.0250	0.1250 ~ 2.500	$Y=2.64X - 1.12\times10^{-1}$	1.0000
437	Thiamethoxam	24.38	0.1000	0.0500	0.2500 ~ 5.000	$Y=3.98\times10^{-1}X - 9.13\times10^{-2}$	0.9960
438	Mepanipyrim	24.29	0.0250	0.0125	0.0625 ~ 2.500	$Y=7.2X - 5.17\times10^{-1}$	0.9970
439	Captan	24.55	0.4000	0.2000	1.0000 ~ 40.00	$Y=1.61\times10^{-1}X - 2.67\times10^{-1}$	0.9940
440	Bromacil	24.73	0.2000	0.1000	0.5000 ~ 20.00	$Y=1.91X - 3.65\times10^{-1}$	0.9990
441	Triadimenol-1	24.75	0.0500	0.0250	0.1250 ~ 5.000	$Y=2.22X - 1.88\times10^{-1}$	0.9990
442	Picoxystrobin	24.97	0.0500	0.0250	0.1250 ~ 5.000	$Y=1.11X - 3.65\times10^{-2}$	0.9990
443	Triadimenol-2	25.25	0.0500	0.0250	0.1250 ~ 5.000	$Y=4.32\times10^{-1}X - 6.77\times10^{-1}$	0.9920
444	Butamifos	25.41	0.0250	0.0125			
445	Imazamethabenz-methyl	25.50	0.0750	0.0375	0.1875 ~ 7.500	$Y=5.47\times10^{-1}X+2.14\times10^{-4}$	0.9960
446	Metominostrobin-1	25.61	0.1000	0.0500			
447	TCMTB	25.59	0.4000	0.2000			
448	Methiocarb sulfone	25.56	0.2000	0.1000	0.5000 ~ 5.000	$Y=1.69\times10^{-2}X+8.96\times10^{-4}$	0.9930
449	Imazalil	25.72	0.1000	0.0500	0.2500 ~ 10.00	$Y=9.69\times10^{-1}X - 1.86\times10^{-1}$	0.9980
450	Isoprothiolane	25.87	0.0500	0.0250	0.1250 ~ 5.000	$Y=7.2\times10^{-1}X - 2.98\times10^{-2}$	0.9990
451	Cyflufenamid	26.02	0.4000	0.2000			
452	Methyl trithion	26.36	0.4000	0.2000			
453	Pyriminobac-methyl	26.34	0.1000	0.0500			

454	Isoxathion	26.51	0.2000	0.1000			
455	Metominostrobin-2	26.76	0.1000	0.0500			
456	Diofenolan -1	26.81	0.0500	0.0250	0.1250 ~ 5.000	$Y=2.13X - 2.6\times10^{-1}$	0.9970
457	Thifluzamide	27.26	0.2000	0.1000			
458	Diofenolan -2	27.14	0.0500	0.0250	0.1250 ~ 5.000	$Y=1.1X - 1.04\times10^{-1}$	0.9990
459	Quinoxyphen	27.14	0.0250	0.0125	0.0625 ~ 2.500	$Y=3.6X - 1.47\times10^{-1}$	0.9990
460	Chlorfenapyr	27.60	0.2000	0.1000	0.5000 ~ 20.00	$Y=2.29\times10^{-1}X - 1.84\times10^{-2}$	0.9990
461	Trifloxystrobin	27.71	0.1000	0.0500	0.2500 ~ 10.00	$Y=2.69X - 3.93\times10^{-1}$	0.9990
462	Imibenconazole-des-benzyl	27.86	0.1000	0.0500			
463	Isoxadifen-ethyl	27.90	0.0500	0.0250			
464	Fipronil	28.34	0.2000	0.1000	0.5000 ~ 20.00	$Y=8.2\times10^{-1}X - 2.91\times10^{-1}$	0.9960
465	Imiprothrin-1	28.31	0.0500	0.0250	0.1250 ~ 5.000	$Y=5.69\times10^{-1}X - 8.96\times10^{-2}$	0.9850
466	Carfentrazone-ethyl	28.29	0.0500	0.0250	0.1250 ~ 5.000	$Y=9.73\times10^{-1}X - 8.03\times10^{-2}$	0.9990
467	Imiprothrin-2	28.50	0.0500	0.0250	0.1250 ~ 5.000	$Y=2.72X - 5.56\times10^{-1}$	0.9900
468	Halosulfuran-methyl	28.42	0.1000	0.0500			
469	Epoxiconazole -1	28.58	0.2000	0.1000	0.5000 ~ 20.00	$Y=6.79\times10^{-2}X+3.91\times10^{-2}$	0.9790
470	Pyraflufen ethyl	28.91	0.0500	0.0250	0.1250 ~ 5.000	$Y=1.18X - 4.03\times10^{-2}$	0.9990
471	Pyributicarb	28.87	0.0500	0.0250	0.1250 ~ 5.000	$Y=6.7X - 6.22\times10^{-1}$	0.9990
472	Thenylchlor	29.12	0.0500	0.0250	0.1250 ~ 5.000	$Y=4.53X - 2.81\times10^{-1}$	1.0000
473	Clethodim	29.21	0.1000	0.0500	0.2500 ~ 10.00	$Y=8.75\times10^{-1}X - 8.65\times10^{-2}$	0.9970
474	Chrysene	29.35	0.0250	0.0125	0.0625 ~ 2.500	$Y=7.15X - 3.13\times10^{-1}$	0.9990
475	Mefenpyr-diethyl	29.55	0.0750	0.0375	0.1875 ~ 7.500	$Y=8.69\times10^{-1}X - 8.09\times10^{-1}$	1.0000
476	Famphur	29.80	0.1000	0.0500	0.2500 ~ 10.00	$Y=1.78X+9.96\times10^{-2}$	0.9910
477	Etoxazole	29.64	0.1500	0.0750			
478	Pyriproxyfen	30.06	0.0250	0.0125	0.0625 ~ 2.500	$Y=2.75X - 1.92\times10^{-1}$	0.9960
479	Epoxiconazole-2	29.73	0.2000	0.1000	0.5000 ~ 20.00	$Y=2.75X - 5.32\times10^{-1}$	0.9990
480	Tepraloxydim	29.76	0.0500	0.0250	0.1250 ~ 5.000	$Y=1.02\times10^{-1}X+9.14\times10^{-3}$	0.9950
481	Picolinafen	30.27	0.0250	0.0125	0.0625 ~ 2.500	$Y=3.35X - 2.06\times10^{-1}$	0.9980
482	Iprodione	30.24	0.1000	0.0500	0.2500 ~ 10.00	$Y=1.5X - 4.14\times10^{-1}$	0.9980
483	Piperophos	30.42	0.0750	0.0375	0.1875 ~ 7.500	$Y=3.04X - 5.31\times10^{-1}$	0.9990

484	Ofurace	30.36	0.0750	0.0375	0.1875 ~ 7.500	$Y=9.18\times10^{-1}X - 9.29\times10^{-2}$	0.9990
485	Bifenazate	30.38	0.2000	0.1000			
486	Chromafenozide	30.42	0.2000	0.1000			
487	Endrin ketone	30.45	0.1000	0.0500	0.2500 ~ 10.00	$Y=4.12\times10^{-1}X+9.06\times10^{-3}$	0.9980
488	Clomeprop	30.48	0.0250	0.0125			
489	Fenamidone	30.66	0.0250	0.0125			
490	Naproanilide	31.89	0.0250	0.0125			
491	Pyraclostrobin	31.98	0.6000	0.3000	1.5000 ~ 60.00	$Y=4.97\times10^{-1}X - 1.15$	0.9910
492	Lactofen	32.06	0.2000	0.1000	0.5000 ~ 20.00	$Y=3.3\times10^{-1}X - 1.88\times10^{-1}$	0.9930
493	Tralkoxydim	32.14	0.2000	0.1000	0.5000 ~ 20.00	$Y=1.92X+2.21\times10^{-1}$	0.9930
494	Pyraclofos	32.18	0.2000	0.1000			
495	Dialifos	32.27	0.2000	0.1000			
496	Spirodiclofen	32.50	0.2000	0.1000	0.5000 ~ 20.00	$Y=1.95\times10^{-1}X - 1.04\times10^{-1}$	0.9950
497	Halfenprox	32.62	0.1000	0.0500			
498	Flurtamone	32.78	0.1000	0.0500	0.2500 ~ 5.000	$Y=4.89\times10^{-1}X - 2.29\times10^{-1}$	0.9330
499	Pyriftalid	32.94	0.0250	0.0125	0.0625 ~ 2.500	$Y=1.73X - 9.39\times10^{-1}$	0.9980
500	Silafluofen	33.18	0.0250	0.0125	0.0625 ~ 2.500	$Y=6.53\times10^{-1}X - 3.35\times10^{-2}$	0.9960
501	Pyrimidifen	33.63	0.1000	0.0500			
502	Acetamiprid	33.87	0.3000	0.1500	0.7500 ~ 30.00	$Y=3.28\times10^{-1}X - 1.54\times10^{-1}$	0.9930
503	Butafenacil	33.85	0.0250	0.0125	0.0625 ~ 2.500	$Y=4.09X - 3.83\times10^{-1}$	0.9860
504	Cafenstrole	34.36	0.3000	0.1500			
505	Fluridone	37.61	0.0500	0.0250	0.125 ~ 5.000	$Y=5.33X - 1.01$	0.9920

Tables 21. The LOD, LOQ, linear ranges, correlation equations, and correlation coefficients of 465 pesticides by LC-MS-MS

No.	Compound	Retention time/min	LOQ/ μg/kg	LOD/ μg/kg	Linear ranges	Correlation equation	Correlation coefficient
1	Acetochlor	20.26	40.00	20.00	1.0000 ~ 40.000	$Y=9.37\times10^3X+3.39\times10^4$	0.9903
2	Allethrin	27.01 27.53	12.00	6.00	0.3000 ~ 6.0000	$Y=6.12\times10^3X+1.48\times10^4$	0.9917
3	Anilofos	23.12	0.40	0.20	0.0100 ~ 0.2000	$Y=2.88\times10^3X+4.12\times10^3$	0.9909
4	Azinphos ethyl	20.93	40.00	20.00	1.0000 ~ 20.000	$Y=1.06\times10^3X+9.96\times10^3$	0.9932
5	Benalyxyl	22.09	0.24	0.12	0.0060 ~ 0.2400	$Y=1.47\times10^3X-1.17\times10^3$	0.9931
6	Bitertanol	19.34 19.90	4.00	2.00	0.1000 ~ 2.0000	$Y=1.19\times10^3X+7.94$	0.9923
7	Bromfenvinfos	21.97	1.60	0.80	0.0400 ~ 1.6000	$Y=2.17\times10^3X+6.54\times10^3$	0.9969
8	Butocarboxim sulfoxide	2.31	40.00	20.00	1.0000 ~ 4.0000	$Y=2.82\times10^4X-3.02\times10^4$	0.9969
9	Butylate	26.26	240	120	6.0000 ~ 120.00	$Y=2.03\times10^5X+3.86\times10^5$	0.9992
10	Chlorprifos methyl	13.53	20.00	10.00	0.1000 ~ 4.0000	$Y=7.97X+2.2\times10^4$	0.0976
11	Cis&trans diallate	26.01	24.00	12.00	0.6000 ~ 24.000	$Y=4.37\times10^3X+1.76\times10^4$	0.9950
12	Clodinafop propargyl	23.49	0.80	0.40	0.0200 ~ 0.8000	$Y=2.88\times10^3X+3.46\times10^3$	0.9934
13	Clomazone dimethazone	15.46	1.20	0.60	0.0300 ~ 1.2000	$Y=8.99\times10^3X+1.51\times10^4$	0.9961
14	Cloquintocet mexyl	25.55	0.24	0.12	0.0060 ~ 0.2400	$Y=6.53\times10^3X+3.32\times10^3$	0.9996
15	Cyanazine	12.29	1.20	0.60	0.0300 ~ 1.2000	$Y=4.38\times10^3X+1.01\times10^4$	0.9989
16	Cycluron	13.34	0.40	0.20	0.0100 ~ 0.4000	$Y=2.36\times10^3X+1.3\times10^3$	0.9990
17	Cyprodinil	15.92	0.40	0.20	0.0200 ~ 0.4000	$Y=1.96\times10^3X-1.06\times10^3$	0.9976

18	Demeton s methyl sulfone	9.27	8.00	4.00	0.2000 ~ 8.0000	$Y=1.26\times10^4X+2.9\times10^4$	0.9926
19	Demeton s methly sulfoxide	2.31	16.00	8.00	0.4000 ~ 16.000	Y=199X+288	0.9945
20	DE-PCB 31	12.28	40.00	20.00	1.0000 ~ 40.000	$Y=11.7X+8.52\times10^3$	0.9925
21	Desmedipham	16.35	2.00	1.00	0.0500 ~ 2.0000	$Y=9.52\times10^3X+1.13\times10^4$	0.9973
22	Dialifos	25.68	8.00	4.00	0.2000 ~ 4.0000	$Y=985X+2.88\times10^3$	0.9926
23	Diazinon	23.70	0.24	0.12	0.0060 ~ 0.2400	$Y=1.89\times10^4X+3.19\times10^4$	0.9970
24	Dichlofluanid	22.23	40.00	20.00	1.0000 ~ 40.000	$Y=1.27\times10^3X-2.56\times10^3$	0.9960
25	Dichlormid	27.53 27.80	4.00	2.00	0.1000 ~ 4.0000	$Y=233X+7.27\times10^3$	0.9919
26	Dichlorvos	17.8	20.00	10.00	0.5000 ~ 20.000	$Y=7.22\times10^3X+2.75\times10^4$	0.9983
27	Diclobutrazole	18.51	0.24	0.12	0.0060 ~ 0.2400	$Y=1.68\times10^3X-6.46\times10^3$	0.9911
28	Diflufenican	25.42	0.12	0.06	0.0060 ~ 0.1200	$Y=1.01\times10^3X+2.1\times10^3$	0.9981
29	Dimefuron	15.16	0.80	0.40	0.0200 ~ 0.8000	$Y=5.63\times10^3X+1.72\times10^4$	0.9940
30	Dimethenamid	17.09	0.40	0.20	0.0200 ~ 0.4000	$Y=1.09\times10^3X+2.68\times10^3$	0.9946
31	Dinitramine	23.75	80.00	40.00	0.4000 ~ 16.000	Y=474X+767	0.9911
32	Dioxathion	27.37	8000	4000	200.00 ~ 2000.0	$Y=2.05\times10^3X+1.38\times10^4$	0.9921
33	Edifenphos	21.25	0.80	0.40	0.0200 ~ 0.8000	$Y=4.38\times10^3X+1.94\times10^4$	0.9930
34	Emamectin benzoate	16.32	0.20	0.10	0.0100 ~ 0.2000	$Y=2.42\times10^3X-606$	0.9978
35	Etaconazole	17.57	0.40	0.20	0.0100 ~ 0.4000	$Y=1.88\times10^3X-1.6\times10^3$	0.9960
36	Ethidimuron	10.23	40.00	20.00	0.4000 ~ 4.0000	$Y=1.19\times10^4X+5.05\times10^4$	0.999
37	Ethiofencarb sulfoxide	9.42	4.00	2.00	0.1000 ~ 2.0000	$Y=1.09\times10^4X-2.6\times10^4$	0.9943
38	Ethoxyquin	19.1	120.00	60.00	0.6000 ~ 24.000	$Y=1.45\times10^4X-2.94\times10^4$	0.9947

39	Famphur	17.53	2.40	1.20	0.0600 ~ 2.4000	$Y=3.36\times10^3X+5.79\times10^3$	0.9985
40	Fenarimol	16.99	2.40	1.20	0.0600 ~ 2.4000	$Y=1.73\times10^3X+3.64\times10^3$	0.9976
41	Fenitrothion	20.45	160.00	80.00	0.8000 ~ 16.000	$Y=611X+2.54\times10^3$	0.9959
42	Fenpropidin	18.70	0.08	0.04	0.0020 ~ 0.0800	$Y=3.67\times10^3X+2.43\times10^3$	0.9981
43	Fenthion sulfoxide	12.77	0.40	0.20	0.0100 ~ 0.2000	$Y=657X+2.31\times10^3$	0.9948
44	Fluazifop butyl	27.34	0.40	0.20	0.0100 ~ 0.4000	$Y=2.62\times10^3X+8.78\times10^3$	0.9942
45	Flufenoxuron	27.66	1.20	0.60	0.0300 ~ 1.2000	$Y=950X+2.35\times10^4$	0.9980
46	Flusilazole	18.98	0.40	0.20	0.0200 ~ 0.4000	$Y=1.66\times10^3X-376$	0.9982
47	Flutolanil	20.13	0.40	0.20	0.0100 ~ 0.4000	$Y=735X+4.08\times10^3$	0.9908
48	Haloxyfop-methyl	25.17	0.12	0.06	0.0060 ~ 0.1200	$Y=626X+5.87\times10^3$	0.9931
49	Heptachlor	34.92	40.00	20.00	2.0000 ~ 40.000	$Y=732X+2.47\times10^4$	0.9945
50	Hexythiazox	27.73	0.40	0.20	0.0200 ~ 0.4000	$Y=337X+1.45\times10^3$	0.9915
51	imazethapyr	16.40	480	240	12.000 ~ 120.00	$Y=6.65\times10^5X+3.98\times10^6$	0.9926
52	Imidacloprid	10.27	200 00	100.00	1.0000 ~ 40.000	$Y=1.47\times10^4X+3.81\times10^4$	0.9931
53	Indoxacarb	25.52	2.40	1.20	0.0600 ~ 1.2000	$Y=2.41\times10^3X+9.72\times10^3$	0.9944
54	Iprovalicarb	17.05	4.00	2.00	0.0200 ~ 0.4000	$Y=1.07\times10^4X+2.94\times10^4$	0.9975
55	Isomethiozin	21.63	0.24	0.12	0.0060 ~ 0.2400	$Y=2.95\times10^3X+2.79\times10^3$	0.9970
56	Isoprocarb	14.53	4.00	2.00	0.1000 ~ 4.0000	$Y=9.27\times10^3X+2.01\times10^4$	0.9987
57	Isoxaflutole	17.54	40.00	20.00	1.0000 ~ 40.000	$Y=8.09\times10^4X+3.23\times10^5$	0.9915
58	Malaoxon	12.62	0.40	0.20	0.0100 ~ 0.2000	$Y=2.91\times10^3X+9.38\times10^3$	0.9908
59	Methidathion	16.75	6.00	3.00	0.1500 ~ 3.0000	$Y=3.17\times10^4X+1.37\times10^5$	0.9981

60	Methoprotryne	11.86	4.00	2.00	0.1000 ~ 4.0000	$Y=3.94\times10^4X+1\times10^5$	0.9950
61	Methothrin	27.83	800	400	20.000 ~ 800.00	$Y=261X+1.51\times10^4$	0.9962
62	Mevinphos	10.11 10.86	8.00	4.00	0.2000 ~ 8.0000	$Y=5.11\times10^3X-3.98\times10^3$	0.9996
63	Monocrotophos	2.45	0.80	0.40	0.0200 ~ 0.4000	$Y=944X-2.29\times10^3$	0.9902
64	Monolinuron	14.08	8.00	4.00	0.2000 ~ 8.0000	$Y=9\times10^3X+2.5\times10^4$	0.9958
65	Myclobutanil	17.31	6.00	3.00	0.0300 ~ 1.2000	$Y=3.69\times10^3X+4.2\times10^3$	0.9994
66	Napropamide	18.64	0.40	0.20	0.0100 ~ 0.4000	$Y=2.92\times10^3X+9.11\times10^3$	0.9959
67	Nitenpyram	2.30	8.00	4.00	0.2000 ~ 8.0000	$Y=595X-337$	0.9990
68	Paclobutrazol	15.79	0.20	0.10	0.0050 ~ 0.2000	$Y=1.47\times10^3X-2.3\times10^3$	0.9913
69	Paraoxon methyl	12.04	2.00	1.00	0.0500 ~ 1.0000	$Y=1.28\times10^4X+8.79\times10^4$	0.9918
70	Penconazole	19.39	0.24	0.12	0.0120 ~ 0.2400	$Y=1.03\times10^3X+976$	0.9957
71	Pentachlorophenol	10.20	80.00	40.00	8.0000 ~ 80.000	$Y=202X+3.06\times10^3$	0.9901
72	Phthalic acid, dicyclobexyl ester	29.56	2.40	1.20	0.0600 ~ 2.4000	$Y=5.77\times10^3X+4.34\times10^4$	0.9907
73	Pirimicarb	9.54	0.40	0.20	0.0100 ~ 0.4000	$Y=5.92\times10^3X+7.18\times10^3$	0.9969
74	Prazophos	23.87	0.16	0.08	0.0040 ~ 0.0800	$Y=1.44\times10^3X+2.02\times10^3$	0.9998
75	Pretilachlor	25.32	0.80	0.40	0.0200 ~ 0.8000	$Y=5.92\times10^3X+1.33\times10^4$	0.9984
76	Profluralin	28.18	6000	3000	30.000 ~ 300.00	$Y=2.51\times10^3X-718$	0.9947
77	Prometryne	13.37	0.20	0.10	0.0050 ~ 0.2000	$Y=4.35\times10^3X+4.03\times10^3$	0.9981
78	Propachlor	14.93	0.40	0.20	0.0100 ~ 0.4000	$Y=2.7\times10^3X+5.86\times10^3$	0.9996
79	Propham	15.23	40.00	20.00	1.0000 ~ 40.000	$Y=625X-13.1$	0.9958
80	Prothiophos	30.22	200.00	100.00	1.0000 ~ 40.000	$Y=3.59\times10^3X+8.95\times10^3$	0.994

81	Pymetrozin	2.29	20.00	10.00	0.5000 ~ 20.000	$Y=5.24\times10^4X+9.27\times10^4$	0.9931
82	Pyraflufen ethyl	23.39	2.40	1.20	0.0600 ~ 2.4000	$Y=5.7\times10^3X+5.62\times10^3$	0.9978
83	Quizalofop-ethyl	25.85	1.20	0.60	0.0300 ~ 1.2000	$Y=1.26\times10^3X+4.85\times10^3$	0.9957
84	Rabenzazole	13.75	0.20	0.10	0.0050 ~ 0.2000	$Y=2.5\times110^3X+1.03\times10^4$	0.9962
85	Secbumeton	10.63	0.20	0.10	0.0050 ~ 0.2000	$Y=1.54\times10^3X+5.35\times10^3$	0.9971
86	Simetryn	10.77	0.40	0.20	0.0100 ~ 0.4000	$Y=3.05\times10^3X+2.43\times10^3$	0.9982
87	Spiroxamine	12.19	0.08	0.04	0.0020 ~ 0.0800	$Y=4.22\times10^3X+1.11\times10^3$	0.9983
88	Tepraloxydim	12.17 18.29	32.00	16.00	0.8000 ~ 32.000	$Y=1.82\times10^4X+3.78\times10^4$	0.9955
89	Terbucarb	24.77	20.00	10.00	0.1000 ~ 4.0000	$Y=3.54\times10^3X+1.06\times10^4$	0.9988
90	Tetramethirn	26.50 26.79	2.40	1.20	0.0600 ~ 2.4000	$Y=6.89\times10^3X+8.21\times10^3$	0.9975
91	Thiacloprid	11.41	0.80	0.40	0.0200 ~ 0.4000	$Y=4.38\times10^3X+6.28\times10^3$	0.9965
92	Thiofanox sufone	24.2	40.00	20.00	1.0000 ~ 40.000	$Y=269X-1.36\times10^3$	0.9928
93	Thiophanate-methyl	18.5	40.00	20.00	2.0000 ~ 40.000	$Y=5.2\times10^4X+1.08\times10^6$	0.9978
94	Thiophanat-ethyl	21.3	40.00	20.00	1.0000 ~ 20.000	$Y=7.53\times10^4X+3.14\times10^5$	0.9957
95	Tolclofos methyl	24.80	120	60.00	3.0000 ~ 120.00	$Y=1.69\times10^4X+6.41\times10^4$	0.9913
96	Tolylfluanid	21.6	2000	1000	50.000 ~ 2000.0	$Y=1.78\times10^4X+1.45\times10^5$	0.9939
97	Triadimenol	15.56	20.00	10.00	0.1000 ~ 4.0000	$Y=1.94\times10^3X-2.68\times10^3$	0.9974
98	Triallate	28.43	40.00	20.00	1.0000 ~ 4.0000	$Y=1.64\times10^4X+2.66\times10^4$	0.9997
99	Triflumuron	22.79	2.40	1.20	0.0600 ~ 2.4000	$Y=4.61\times10^3X+3.68\times10^3$	0.9972
100	3-hydroxy carbofuran	10.17	4.00	2.00	0.0200 ~ 0.8000	$Y=881X+2.83\times10^3$	0.9936
101	6-chloro-4-hydroxy-3-phenyl-pyridazin	5.53	1.20	0.60	0.0300 ~ 1.2000	$Y=1.28\times10^3X+3.24\times10^3$	0.9977

No.	Compound						
102	Acephate	2.30	8.00	4.00	0.2000 ~ 8.0000	$Y=2.92\times10^3X+1.93\times10^3$	0.9942
103	Aclonifen	22.07	4.00	2.00	0.1000 ~ 4.0000	$Y=1.22\times10^3 X +1.37\times10^3$	0.9961
104	Bendiocarb	12.92	0.80	0.40	0.0200 ~ 0.8000	$Y=1.17\times10^3X+912$	0.9935
105	bensultap	14.32	200	100	5.0000 ~ 200.00	$y=1.52\times10^5 X -1.95\times10^3$	0.9936
106	Bifenazate	19.03	30.00	15.00	0.1500 ~ 6.0000	$Y=8.77\times10^3X-1.18\times10^4$	0.9962
107	Bioallethrin	27.14	4.00	2.00	0.1000 ~ 4.0000	$Y=202X-284$	0.9910
108	Bioresmethrin	30.12	4.00	2.00	0.1000 ~ 4.0000	$Y=7.64\times10^3X-1.5\times10^4$	0.9968
109	Bromuconazole	17.17 18.30	0.80	0.40	0.0200 ~ 0.8000	$Y=983X-1.46\times10^3$	0.9985
110	Buprofezin	22.72	0.20	0.10	0.0100 ~ 0.2000	$Y=1.35\times10^3X-430$	0.9975
111	Carbendazim	2.31	0.20	0.10	0.0100 ~ 0.2000	$Y=1.37\times10^3X-96.6$	0.9947
112	chlorarosulam-methyl	19.44	20.00	10.00			
113	Chlorbufam	18.20	80.00	40.00	2.0000 ~ 80.000	$Y=1.84\times10^3X+2.81\times10^3$	0.9992
114	Chlorimuron-ethyl	22.17	20.00	10.00	0.1000 ~ 4.0000	$Y=3.27\times10^3X+8.94\times10^3$	0.9950
115	Chlormephos	12.35	1200	600	30.000 ~ 1200.0	$Y=8.93\times10^3X+2.42\times10^4$	0.9965
116	Chlormequat	1.95	200	100			
117	Chlorsulfuron	14.52	4.00	2.00	0.1000 ~ 4.0000	$Y=6.38\times10^3X+1.54\times10^4$	0.9912
118	Clethodim	26.31	20.00	10.00	0.1000 ~ 4.0000	$Y=7.63\times10^3X-2.68\times10^3$	0.9993
119	Clothianidin	10.14	120.00	60.00	0.6000 ~ 24.000	$Y=3.37\times10^3X+7.91\times10^3$	0.9978
120	Cyanofenphos	24.37	2.00	1.00	0.0500 ~ 2.0000	$Y=1.24\times10^3X+506$	0.9974
121	Cyazofamid	22.40	2.00	1.00	0.0500 ~ 2.0000	$Y=1.03\times10^3X-1.41\times10^3$	0.9981
122	Cycloxydim	25.71	20.00	10.00	0.1000 ~ 4.0000	$Y=7.34\times10^3X+1.3\times10^4$	0.9969
123	Cythioate	12.46	24.00	12.00	0.6000 ~ 24.000	$Y=1.3\ 7\times10^3X+958$	0.9961

124	Daminozide	1.72	1.20	0.60	0.0300 ~ 1.2000	$Y=1.14\times10^3X+2.07\times10^3$	0.9986
125	DEF	29.93	0.40	0.20	0.0100 ~ 0.4000	$Y=4.75\times10^3X-6.81\times10^3$	0.9963
126	Demeton-s	14.78	20.00	10.00	0.5000 ~ 20.000	$Y=7.51\times10^3X+7.57\times10^4$	0.9668
127	Desethyl-sebuthylazine	11.37	12 00	6.00	0.3000 ~ 12.000	$Y=1.6\times10^5X+4.35\times10^5$	0.9929
128	diafenthiuron	32.29	1.20	0.60	0.0300 ~ 1.2000	Y=122X+227	0.9935
129	Difenoxuron	14.07	0.80	0.40	0.0200 ~ 0.8000	$Y=1.97\times10^3X+8.05\times10^3$	0.9954
130	Difenzoquat-methyl sulfate	10.1 10.45	0.80	0.40	0.0200 ~ 0.8000	$Y=1.54\times10^3X+4.1\times10^3$	0.9972
131	dimefox	3.44	8.00	4.00	0.2000 ~ 8.0000	Y=587X-194	0.9182
132	Dimethachloro	19.04	0.40	0.20	0.0100 ~ 0.4000	$Y=9.81\times10^4X+2.57\times10^5$	0.9952
133	Dimethachloro	15.12	160	80.00	4.0000 ~ 160.00	Y=-1.35X+238	0.9866
134	Diphenylamin	19.08	4.00	2.00	0.0200 ~ 0.8000	$Y=2.63\times10^3X-8.02\times10^3$	0.9994
135	Disulfoton sulfone	15.85	1.20	0.60	0.0300 ~ 1.2000	$Y=2.66\times10^3X+2.78\times10^3$	0.9998
136	Diuron	13.95	0.20	0.10	0.0100 ~ 0.2000	Y=414X-158	0.9962
137	Ethofume sate	18.98	320	160	8.0000 ~ 320.00	$Y=4.82\times10^3X+5.05\times10^3$	0.9963
138	Ethylene thiourea	2.29	24.00	12.00			
139	Etrimfos	23.47	0.40	0.20	0.0100 ~ 0.4000	$Y=960X-3.21\times10^3$	0.9924
140	Fenazaquin	28.10	0.20	0.10	0.0100 ~ 0.2000	$Y=4.86\times10^3X-1.06\times10^4$	0.9952
141	Fenhexamid	20.14	0.40	0.20	0.0100 ~ 0.4000	$Y=1.61\times10^3X-2.27\times10^3$	0.9963
142	Fenuron	10.25	4.00	2.00	0.1000 ~ 4.0000	$Y=1.52\times10^4X+1.26\times10^4$	0.9953
143	Flamprop isopropyl	23.61	0.40	0.20	0.0100 ~ 0.4000	$Y=3.43\times10^3X+3.63\times10^3$	0.9997
144	Flamprop methyl	19.10	2.00	1.00	0.0500 ~ 2.0000	$Y=2.69\times10^3X+6.67\times10^3$	0.9977
145	Fluchloralin	26.39	80.00	40.00	2.0000 ~ 80.000	Y=656X-766	0.9970

146	Fluoazuron	26.97	0.80	0.40	0.0200 ~ 0.8000	$Y=2.66\times10^3X+1.67\times10^3X$	0.9951
147	Flurtamone	16.67	2.00	1.00	0.0100 ~ 0.4000	$Y=4.39\times10^3X+7.05\times10^3$	0.9970
148	Fluthiacet methyl	21.81	1.20	0.60	0.0300 ~ 1.2000	$Y=1.4\times10^3X+1.82\times10^3$	0.9940
149	Flutriafol	13.19	1.20	0.60	0.0300 ~ 1.2000	$Y=1.72\times10^3X+5.33\times10^3$	0.9966
150	Formothion	10.45	8000	4000	200.00 ~ 8000.0	$Y=1.64X-18.8$	0.9880
151	Furalaxyl	16.40	0.40	0.20	0.0100 ~ 0.4000	$Y=3.89\times10^3X+1.13\times10^4$	0.9935
152	Hexaflumuron	25.03	1.20	0.60	0.0300 ~ 1.2000	$Y=1.12\times10^3X+666$	0.9973
153	Hydramethylnon	22.70	2.00	1.00	0.0500 ~ 2.0000	$Y=1.16\times10^4X+3.58\times10^4$	0.9947
154	Imibenzonazole-des-benzyl	11.64	10.00	5.00	0.0500 ~ 2.0000	$Y=1.12\times10^3X+91.5$	0.9988
155	Iprobenfos	19.39	2.00	1.00	0.0500 ~ 2.0000	$Y=3.06\times10^3X+2.26\times10^3$	0.9985
156	Isouron	11.97	0.40	0.20	0.0100 ~ 0.4000	$Y=3.03\times10^3X+9.05\times10^3$	0.9939
157	Isoxaben	18.88	0.20	0.10	0.0100 ~ 0.2000	$Y=1.67\times10^3X+7.5\times10^3$	0.9948
158	Mefenoxam	13.74	0.20	0.10	0.0200 ~ 0.2000	$Y=628X+1.67\times10^4$	0.9938
159	Mephosfolan	11.67	2.00	1.00	0.0100 ~ 0.4000	$Y=2.61\times10^3X+3.22\times10^3$	0.9938
160	Metconazole	19.74	0.40	0.20	0.0100 ~ 0.4000	$Y=3.9\times10^3X+2.67\times10^3$	0.9971
161	Methobromuron	14.60	1.60	0.80	0.0400 ~ 1.6000	$Y=679X+63.5$	0.9949
162	Molinate	17.82	2.00	1.00	0.0500 ~ 2.0000	$Y=1.58\times10^3X-1.59\times10^4$	0.9024
163	Neburon	21.10	1.20	0.60	0.0300 ~ 1.2000	$Y=2.32\times10^3X+7.01\times10^3$	0.9911
164	Nicotine	1.50	8.00	4.00	0.2000 ~ 8.0000	$Y=1.05\times10^4X+1.21\times10^3$	0.9974
165	Novaluron	26.83	0.40	0.20	0.0100 ~ 0.4000	$Y=715X+320$	0.9980
166	Parathion	23.41	240	120	6.0000 ~ 240.00	$Y=3.98\times10^4X+1.05\times10^5$	0.9940

167	Phenmedipham	16.24	2.00	1.00	0.0500 ~ 2.0000	$Y=5.4\times10^3X+9.17\times10^3$	0.9988
168	Phorate	24.82	32.00	16.00	0.8000 ~ 32.000	$Y=8..09\times10^3X-1.56\times10^5$	0.9922
169	Phosphamidon	11.39	1.20	0.60	0.0300 ~ 1.2000	$Y=5.52\times10^3X+7.43\times10^3$	0.9914
170	Picolinafen	26.68	0.40	0.20	0.0100 ~ 0.4000	$Y=2.07\times10^3X-4.37\times10^3$	0.9934
171	Pirimiphos methyl	23.92	0.20	0.10	0.0100 ~ 0.2000	$Y=3.24+10^3X-3.41\times10^3$	0.9994
172	Pirmicarb-desmethyl	2.29	0.20	0.10	0.0100 ~ 0.2000	$Y=944X+3.15\times10^3$	0.9981
173	Pronamide	18.47	1.20	0.60	0.0300 ~ 1.2000	$Y=1.43\times10^3X-446$	0.9952
174	Propargite	28.64	8.00	4.00	0.2000 ~ 8.0000	$Y=2.19\times10^3X+2.03\times10^3$	0.9930
175	Propazine	15.82	0.40	0.20	0.0100 ~ 0.4000	$Y=3.67\times10^3X-1.08\times10^3$	0.9982
176	Propiconazole	20.46	0.40	0.20	0.0100 ~ 0.4000	$Y=1.28\times10^3 X-131$	0.9986
177	Propoxur	12.85	4.00	2.00	0.1000 ~ 4.0000	$Y=3.32\times10^3X-2.69\times10^3$	0.9966
178	Prothoate	15.12	0.80	0.40	0.0200 ~ 0.8000	$Y=565X+1.67\times10^3$	0.9960
179	Pyriftalid	17.92	0.20	0.10	0.0100 ~ 0.2000	$Y=1.62\times10^3X+2.51\times10^3$	0.9983
180	Pyriproxyfen	27.28	0.20	0.10	0.0100 ~ 0.2000	$Y=4.05\times10^3X-5.23\times10^3$	0.9991
181	Tau fluvalinate	30.74	8.00	4.00	0.2000 ~ 8.0000	$Y=307X-1.01\times10^3$	0.9935
182	TEPP	11.27	4.00	2.00	0.1000 ~ 4.0000	$Y=1.77\times10^4X-7.03\times10^4$	0.9986
183	Terbuthylazine	16.48	0.20	0.10	0.0100 ~ 0.2000	$Y=2.8\times10^3X-2.76\times10^3$	0.9991
184	Tetrachlorvinphos	19.91	1.20	0.60	0.0300 ~ 1.2000	$Y=618X+2.79\times10^3$	0.9965
185	Thiamethoxam	9.50	40.00	20.00	1.0000 ~ 20.000	$Y=6.87\times10^3X-6.02e\times10^3$	0.9904
186	Thiofanox	13.59	120	60.00	3.0000 ~ 120.00	$Y=787X+4.05\times10^4$	0.9916
187	Transfluthrin	34.92	120	60.00	3.0000 ~ 120.00	$Y=1.28X+9.32\times10^3$	0.0317

188	Triasulfuron	13.89	6000	3000	30.000 ~ 300.00	$Y=1.25\times10^4X+2.44\times10^4$	0.9960
189	Triazophos	20.15	0.20	0.10	0.0100 ~ 0.2000	$Y=1.31\times10^4X+4\times10^4$	0.9952
190	Trifloxystrobin	25.82	0.20	0.10	0.0100 ~ 0.2000	$Y=923X+1.47\times10^3$	0.9955
191	Trifluralin	28.10	200	100	20.000 ~ 200.00	$Y=466X-4.77\times10^3$	0.9900
192	Triphenyl phosphate	23.39	0.20	0.10	0.0100 ~ 0.2000	$Y=1.32\times10^4X-2.93\times10^3$	0.9971
193	Vamidothion sulfone	2.43	24.00	12.00	0.1200 ~ 4.8000	$Y=8.5\times10^3X+1.73\times10^4$	0.9960
194	Acetamiprid	10.65	1.20	0.60	0.0300 ~ 1.2000	$Y=2.71\times10^3X+110$	0.9993
195	Aspon	29.74	0.80	0.40	0.0200 ~ 0.8000	$Y=1.3\times10^4X-2.48\times10^3$	0.9933
196	Asulam Pestanal	22.80	40.00	20.00	1.0000 ~ 40.000	$Y=9.21\times10^3X+6.4\times10^3$	0.9967
197	Benoxacor	17.35	2.00	1.00	0.0500 ~ 2.0000	$Y=1.31\times10^3X+1.13\times10^3$	0.9963
198	Benzoylprop-ethyl	23.71	4.00	2.00	0.1000 ~ 4.0000	$Y=1.3\times10^3X-39.5$	0.9962
199	Bromacil	11.78	2.00	1.00	0.0500 ~ 2.0000	$Y=1.21\times10^3X-1.53\times10^3$	0.9978
200	Bromobutide	21.83	6.00	3.00	0.0300 ~ 1.2000	$Y=1.75\times10^3X+3.35\times10^3$	0.9987
201	Brompyrazon	10.53	6.00	3.00	0.1500 ~ 6.0000	$Y=2.03\times10^3X+2.8\times10^3$	0.9914
202	Butachlor	27.25	4.00	2.00	0.1000 ~ 4.0000	$Y=3.33\times10^3X+2.46\times10^3$	0.9983
203	Butafenacil	21.65	2.00	1.00	0.0500 ~ 2.0000	$Y=874X+2.51\times10^3$	0.9912
204	Chlorphoxim	25.50	40.00	20.00	1.0000 ~ 40.000	$Y=9.42\times10^3X+3.46\times10^4$	0.9952
205	Chlorthal-dimethyl	23.87	2400	1200	60.000 ~ 2400.0	$Y=106X+523$	0.9128
206	Chlorthion	24.32	600	300	15.000 ~ 600.00	$Y=1.09\times10^3X+5.07\times10^4$	0.9402
207	Coumaphos	24.39	0.40	0.20	0.0100 ~ 0.4000	$Y=785X-1.71\times10^3$	0.9933
208	Demeton(O+S)	14.76	1.60	0.80	0.0400 ~ 1.6000	$Y=-198X+4.36\times10^4$	0.9655

209	Dichlofenthion	29.80	2.40	1.20	0.0600 ~ 2.4000	$Y=814X-1.93\times10^3$	0.9961
210	Diethyltoluamide	13.82	0.20	0.10	0.0200 ~ 0.4000	$Y=1.81\times10^3X+8.36\times10^4$	0.9974
211	Dimethametryn	14.69	0.20	0.10	0.0100 ~ 0.4000	$Y=2.52\times10^3X+3.67\times10^3$	0.9984
212	Disulfoton-sulfoxide	13.12	0.80	0.40	0.0200 ~ 0.8000	$Y=410X+4.75\times10^3$	0.9923
213	Ditalimfos	20.34	40.00	20.00	1.0000 ~ 40.000	$Y=1.59\times10^4X-6.24\times10^4$	0.9962
214	Dithiopyr	26.62	1.20	0.60	0.0300 ~ 1.2000	$Y=585X-963$	0.9957
215	Epoxiconazole	14.82 17.84	40.00	20.00	1.0000 ~ 40.000	$Y=1.32\times10^5X+2.97\times10^5$	0.9964
216	ethoxysulfuron	23.10	80.00	40.00	0.8000 ~ 16.000	$Y=1.15\times10^4X+1.32\times10^5$	0.9998
217	Etridiazol	21.58	160.00	80.00	4.0000 ~ 160.00	$Y=505X-4.79\times10^3$	0.9114
218	Fenamiphos sulfone	12.30	0.40	0.20	0.0100 ~ 0.4000	$Y=903X+643$	0.9992
219	Fenamiphos sulfoxide	11.06	0.80	0.40	0.0200 ~ 0.8000	$Y=1.43\times10^3X+1.88\times10^3$	0.9951
220	Fenoxanil	21.83 22.05	1.20	0.60	0.0300 ~ 1.2000	$Y=415X+4.93\times10^3$	0.9885
221	Fenoxycarb	20.03	0.80	0.40	0.0200 ~ 0.8000	$Y=1.12\times10^3X-505$	0.9911
222	Fenpropathrin	29.20	40.00	20.00	1.0000 ~ 40.000	$Y=388X+1.44\times10^3$	0.9368
223	Fenpyroximate	28.38	0.40	0.20	0.0100 ~ 0.4000	$Y=3.01\times10^3X+5.82\times10^3$	0.9953
224	Fensulfothin	14.20	0.80	0.40	0.0200 ~ 0.8000	$Y=1.32\times10^3X+3.7\times10^3$	0.9936
225	Fenthion	23.25	2.00	1.00	0.0500 ~ 2.0000	$Y=1.88\times10^3X+720$	0.9974
226	Fludioxonil	16.97	8160	4080	204.00 ~ 8160.0	$Y=163X+3.14\times10^3$	0.9952
227	Flumiclorac-pentyl	27.00	2.00	1.00	0.0500 ~ 2.0000	$Y=887X+2.33\times10^3$	0.9987
228	Fluorochloridone	20.52	2.40	1.20	0.0600 ~ 2.4000	$Y=571X-394$	0.9938
229	Fluoroglycofen-ethyl	26.41	40.00	20.00	1.0000 ~ 40.000	$Y=79.1X+1.32\times10^3$	0.9724

230	Fluquinconazole	18.39	0.20	0.10	0.0100 ~ 0.4000	$Y=195X-748$	0.9909
231	Fluridone	16.22	0.40	0.20	0.0100 ~ 0.4000	$Y=3.13\times10^3X+3.54\times10^3$	0.9912
232	Fonofos	24.31	1.20	0.60	0.0300 ~ 1.2000	$Y=2.1\times10^3X+-5.21\times10^3$	0.9973
233	Hexachlorobenzene	13.98	1600	800	40.000 ~ 1600.0	$Y=398X+1.1\times10^3$	0.9972
234	Imazamethabenz-methyl	10.87	1.20	0.60	0.0300 ~ 1.2000	$Y=4.74\times10^3X+391$	0.9963
235	imazamox	15.26	10.00	5.00	0.2500 ~ 10.000	$Y=8.17\times10^3X+2.9310^4$	0.9901
236	Imazaquin	16.32	40.00	20.00	1.0000 ~ 40.000	$Y=6.01\times10^4X+2.72\times10^5$	0.9929
237	Iodofenphos	27.38	12.00	6.00	0.3000 ~ 12.000	$Y=544X-300$	0.9971
238	Kresoxim-methyl	22.29	40.00	20.00	1.0000 ~ 40.000	$Y=1.5\times10^3X+7.18\times10^3$	0.9946
239	Leptophos	30.09	20.00	10.00	0.5000 ~ 20.000	$Y=330X+1.5\times10^3$	0.9650
240	Maleic hydrazide	2.23	20.00	10.00	0.5000 ~ 20.000	$Y=919X+2.77\times10^3$	0.9909
241	MCPA-butoxyethyl ester	25.95	40.00	20.00	1.0000 ~ 40.000	$Y=345X+72.5$	0.9954
242	Mepanipyrim	18.86	0.40	0.20	0.0100 ~ 0.4000	$Y=2.45\times10^3X+8.18\times10^3$	0.9962
243	Mepronil	19.48	0.40	0.20	0.0100 ~ 0.4000	$Y=5.65\times10^3X+1.87\times10^4$	0.9985
244	Metalaxyl	13.76	0.40	0.20	0.0100 ~ 0.4000	$Y=3.49\times10^3X+8.13\times10^3$	0.9987
245	Methamidophos	2.30	20.00	10.00	0.5000 ~ 10.000	$Y=8.36\times10^3X+1.83\times10^4$	0.9908
246	Methfuroxam	16.79	8.00	4.00	0.2000 ~ 8.0000	$Y=4.29\times10^4X-8.52\times10^4$	0.9951
247	Methiocarb	16.34	0.80	0.40	0.0200 ~ 0.8000	$Y=1.44\times10^3X-1.68\times10^3$	0.9968
248	Methoxyfenozide	19.31	8.00	4.00	0.2000 ~ 8.0000	$Y=8.48\times10^3X+9.52\times10^4$	0.9694
249	Metoxuron	11.55	0.80	0.40	0.0200 ~ 0.4000	$Y=6.05\times10^3X-3.42\times10^3$	0.9931
250	Monalide	20.15 21.98	0.40	0.20	0.0100 ~ 0.4000	$Y=1.17\times10^3X+1.38\times10^3$	0.9966

251	Monuron	12.00	8.00	4.00	0.4000 ~ 8.0000	$Y=1.41\times10^3X+1.52\times10^4$	0.9934
252	Naled	21.20	4000	2000	20.000 ~ 800.00	$Y=3.79\times10^4X+5.81\times10^4$	0.9920
253	Ofurace	13.91	1.20	0.60	0.0300 ~ 0.3000	$Y=2.08\times10^3X+2.16\times10^3$	0.9984
254	Oryzalin	19.84	20 00	10.00	0.5000 ~ 20.000	$Y=1.3\times10^3X+2.13\times10^3$	0.9940
255	Oxamyl	8.06	480	240	24.000 ~ 480.00	$Y=1.09\times10^3X+1.74\times10^4$	0.9934
256	Oxycarboxin	11.37	0.80	0.40	0.0200 ~ 0.8000	$Y=2.24\times10^3X+1.45\times10^3$	0.9963
257	Oxyflurofen	27.21	8.00	4.00	0.2000 ~ 8.0000	Y=478X+41.2	0.9927
258	Pendimethalin	27.85	1.20	0.60	0.0300 ~ 1.2000	Y=581X-47.6	0.9967
259	Phenothrin	30.78 30.98	40.00	20.00	1.0000 ~ 40.000	$Y=7.04\times10^3X-1.38\times10^3$	0.9941
260	Phorate sulfoxide	9.87	120.00	60.00	3.0000 ~ 120.00	$Y=1.24\times10^3X-120$	0.9946
261	Phosalone	24.93	8.00	4.00	0.2000 ~ 8.0000	$Y=6.03\times10^3X+1.76\times10^4$	0.9929
262	Phoxim	24.95	8.00	4.00	0.2000 ~ 8.0000	$Y=609X+1.41\times10^3$	0.9940
263	Piperonyl butoxide	26.55	0.40	0.20	0.0100 ~ 0.4000	$Y=4.46\times10^3X-859$	0.9978
264	Piperophos	25.17	0.20	0.10	0.0100 ~ 0.4000	$Y=2.56\times10^3-1.43\times10^3$	0.9994
265	Prochloraz	16.31	0.80	0.40	0.0400 ~ 0.8000	Y=405X-203	0.9964
266	Prometon	10.68	0.20	0.10	0.0100 ~ 0.4000	$Y=1.96\times10^3X+4.19\times10^3$	0.9977
267	Propanil	15.77	1.60	0.80	0.0400 ~ 1.6000	$Y=1.2\times10^3X+233$	0.9991
268	Propylene thiourea	2.45	160	80.00	4.0000 ~ 160.00	$Y=1.71\times10^4X+2.03\times10^4$	0.9958
269	Pyrethrin	28.70	80.00	40.00	2.0000 ~ 80.000	$Y=1.63\times10^3X-1.1\times10^3$	0.9967
270	Pyridaben	29.62	0.40	0.20	0.0100 ~ 0.2000	Y=784X-279	0.9953
271	Pyrimethanil	13.27	0.80	0.40	0.0200 ~ 0.8000	$Y=2.1\times10^3X+4.31\times10^3$	0.9988

272	Pyrimitate	21.75	0.20	0.10	0.0100 ~ 0.4000	$Y=5.42\times10^{3}X-4.16\times10^{3}$	0.9972
273	Quinalphos	22.37	0.80	0.40	0.0200 ~ 0.8000	$Y=3.12\times10^{3}X+4.76\times10^{3}$	0.9971
274	Quinoclamine	12.36	4.00	2.00	0.1000 ~ 4.0000	$Y=2.61\times10^{3}X+4.39\times10^{3}$	0.9978
275	Simeton	9.29	0.80	0.40	0.0200 ~ 0.8000	$Y=1.78\times10^{3}X-351$	0.9806
276	Spinosad	14.14	0.20	0.10	0.0100 ~ 0.4000	$Y=3.86\times10^{3}X-1.21\times10^{4}$	0.9997
277	Spirodiclofen	29.89	4.00	2.00	0.2000 ~ 4.0000	$Y=1.76\times10^{3}X+1.55\times10^{4}$	0.9961
278	Temephos	27.61	4.00	2.00	0.1000 ~ 2.0000	$Y=4.01\times10^{3}X+1.17\times10^{3}$	0.9982
279	Thenylchlor	20.23	1.60	0.80	0.0400 ~ 1.6000	$Y=1.06\times10^{3}X+8.07\times10^{3}$	0.9942
280	Thifensulfuron-methyl	18.30	10.00	5.00	0.0500 ~ 2.0000	$Y=3.66\times10^{3}X+2.03\times10^{4}$	0.9921
281	Thiometon	13.50	3600	1800	90.000 ~ 3600.0	$Y=1.16\times10^{4}X+7.22\times10^{4}$	0.9915
282	Thiram	2.48	400	200	10.000 ~ 200.00	$Y=3.15\times10^{4}X+2.18\times10^{5}$	0.9930
283	Trichlorphon	9.78	12.00	6.00	0.3000 ~ 12.000	$Y=6.39\times10^{3}X-3.11\times10^{4}$	0.9924
284	Triticonazole	16.10	0.80	0.40	0.0200 ~ 0.8000	$Y=3.26\times10^{3}X+1.83\times10^{3}$	0.9978
285	3,4,5-Trimethacarb	14.65	0.40	0.20	0.0050 ~ 0.2000	$Y=4.88\times10^{3}X+3.86\times10^{3}$	0.9994
286	dibromobenzophenone	13.55	1.60	0.80	0.0200 ~ 0.8000	$Y=384X+2.24\times10^{5}$	0.9167
287	dichlorobenzophenone	25.43; 26.42	2.40	1.20	0.0300 ~ 1.2000	$Y=5.49\times10^{3} X +1.11\times10^{4}$	0.9957
288	Aldicarb sulfone	7.83	14.40	7.20	0.3600 ~ 7.2000	$Y=251X-1.12\times10^{3}$	0.9996
289	Anilazine	16.98	400	200	5.000 ~ 200.000	$Y=683X+482$	0.9978
290	Aziprotryne	17.53	0.80	0.40	0.0100 ~ 0.4000	$Y=1.33\times10^{3}X-883$	0.9993
291	Azoxystrobin	17.70	0.40	0.20	0.0050 ~ 0.2000	$Y=2.5\times10^{3}X+1.6\times10^{4}$	0.9991
292	Benodanil	15.82	0.80	0.40	0.0100 ~ 0.2000	$Y=1.21\times10^{3}X+5.4\times10^{3}$	0.9965

293	Bensulide	23.45	20.00	10.00	0.2500 ~ 10.000	$Y=3.62\times10^3X+2.1\times10^4$	0.9987
294	Bromophos-ethyl	29.86	20.00	10.00	0.2500 ~ 10.000	$Y=755X+1.63\times10^3$	0.9982
295	Butralin	28.62	2.00	1.00	0.0100 ~ 0.4000	$Y=757X-391$	0.9991
296	Buturon	15.52	2.40	1.20	0.0300 ~ 1.2000	$Y=1.04\times10^3X+1.61\times10^3$	0.9973
297	Carbaryl	13.50	4.00	2.00	0.0500 ~ 2.0000	$Y=873X+8.38\times10^3$	0.9951
298	Carbetamide	11.50	3.20	1.60	0.0400 ~ 1.6000	$Y=1.72\times10^3X+6.89\times10^3$	0.9992
299	Carboxin	13.90	0.80	0.40	0.0100 ~ 0.4000	$Y=2.23\times10^3X+2.08\times10^4$	0.9900
300	cis and trans -Chlorfenvinphos	21.36	2.40	1.20	0.0300 ~ 1.2000	$Y=1.67\times10^3X+9.04\times10^3$	0.9972
301	Chlorotoluron	13.30	0.80	0.40	0.0100 ~ 0.4000	$Y=622X+6.57\times10^3$	0.9981
302	Chlorpyrifos	27.87	2.40	1.20	0.0300 ~ 1.2000	$Y=1.05\times10^3X-560$	0.9997
303	Clomeprop	25.88	3.20	1.60	0.0400 ~ 1.6000	$Y=880X+7.95\times10^3$	0.9938
304	Coumatetralyl	23.8	200.00	100.00	1.0000 ~ 40.000	$Y=1.58\times105X+2.7\times10^6$	0.9919
305	Crimidine	10.04	2.40	1.20	0.0300 ~ 1.2000	$Y=3.5\times10^3X+5.08\times10^3$	0.9979
306	Crufomate	17.26	64.00	32.00	0.8000 ~ 32.000	$Y=3.83\times10^4X+4.71\times10^5$	0.9950
307	Cyproconazole	16.22	0.64	0.32	0.0080 ~ 0.3200	$Y=1.55\times10^3X-4.71\times10^3$	0.9995
308	Cyromazine	2.2	80.00	40.00	1.0000 ~ 40.000	$Y=6.52\times10^3X+6.8\times10^4$	0.9942
309	Dacthal	23.87	2400	1200	30.000 ~ 1200.0	$Y=-2.4X+657$	0.9980
310	Dazomet	8.86	320.00	160.00	4.0000 ~ 160.00	$Y=9.64\times10^3-1.74\times10^4$	0.9962
311	Demeton-s-methyl	12.27	32000	16000			
312	DE-PCB 52	32.22	120.00	60.00	0.6000 ~ 24.000	$Y=541X+7.82\times10^4$	0.9409
313	Dicrotophos	9.23	3.20	1.60	0.0400 ~ 0.8000	$Y=1.46\times10^4X+6.17\times10^3$	0.9991
314	Difenoconazole	22.50	0.40	0.20	0.0050 ~ 0.2000	$Y=1.65\times10^3X+6.36\times10^3$	0.9996

315	Dimethoate	10.55	3.20	1.60	0.0400 ~ 1.6000	$Y=1.14\times10^{3}X+4.23\times10^{3}$	0.9993
316	Diphenamid	15.22	0.40	0.20	0.0050 ~ 0.2000	$Y=6.72\times10^{3}X+3.07\times10^{4}$	0.9972
317	Disulfoton	25.19	80.00	40.00	1.0000 ~ 40.000	$Y=151X-1.05\times10^{3}$	0.9986
318	Dodemorph	11.58 11.76	0.40	0.20	0.0050 ~ 0.2000	$Y=4.6\times10^{3}X+66$	0.9996
319	EPTC	21.32	240	120	3.0000 ~ 120.00	$Y=9.27\times10^{4} X +5.29\times10^{5}$	0.9976
320	Ethion	28.17	3.20	1.60	0.0400 ~ 1.6000	$Y=3.08\times10^{3}X+9.24\times10^{3}$	0.9995
321	Ethirimol	2.36	1.60	0.80	0.0200 ~ 0.8000	$Y=1.13\times10^{3}X+434$	0.9934
322	Ethoprophos	18.08	1.60	0.80	0.0200 ~ 0.8000	$Y=3.24\times10^{3}X+1.15\times10^{4}$	0.9980
323	Fenbuconazole	19.16	0.40	0.20	0.0050 ~ 0.2000	$Y=545X+1.66\times10^{3}$	0.9983
324	Fenfuram	14.24	0.80	0.40	0.0100 ~ 0.4000	$Y=2.8\times10^{3}X+8.13\times10^{3}$	0.9994
325	Fenobucarb	16.34	1.60	0.80	0.0200 ~ 0.8000	$Y=1.6\times10^{3}X+4.34\times10^{3}$	0.9998
326	Fenpiclonil	16.09	1.00	0.50	0.0050 ~ 0.0500	$Y=1.31\times10^{4}X+7.86\times10^{4}$	0.9976
327	Fentin-chloride	14.26	80.00	40.00	1.0000 ~ 40.000	$Y=3.21\times10^{4}X+2.03\times10^{5}$	0.9977
328	Fipronil	22.51	16.00	8.00	0.2000 ~ 8.0000	$Y=251X+639$	0.9985
329	Flufenacet	20.53	0.80	0.40	0.0100 ~ 0.4000	$Y=618X+7.71\times10^{3}$	0.9926
330	Fluometuron	13.51	12.00	6.00	0.1500 ~ 6.0000	$Y=1.63\times10^{4}X+1.12\times10^{5}$	0.9975
331	Fosthiazate	13.51	1.60	0.80	0.0200 ~ 0.8000	$Y=3.64\times10^{3} X +4.44\times10^{4}$	0.9972
332	Hexazinone	11.43	0.80	0.40	0.0100 ~ 0.4000	$Y=3.58\times10^{3}X+1.47\times10^{3}$	0.9998
333	Imazalil	11.03	2.40	1.20	0.0300 ~ 1.2000	$Y=1.67\times10^{3}X-504$	0.9978
334	Imiprothrin	19.40 19.64	8.00	4.00	0.1000 ~ 4.0000	$Y=-5.25X+1.07\times10^{4}$	0.9330
335	Isofenphos	25.69	160.00	80.00	2.0000 ~ 40.000	$Y=2.38\times10^{3}X+1.59\times10^{4}$	0.9928

336	Isoprothiolane	19.82	0.80	0.40	0.0100 ~ 0.4000	$Y=2.32\times10^{3}X+2.21\times10^{4}$	0.9920
337	Isoproturon	13.84	0.80	0.40	0.0100 ~ 0.4000	$Y=3.02\times10^{3}X+2.81\times10^{4}$	0.9926
338	Kadethrin	26.93	1.60	0.80	0.0200 ~ 0.8000	$Y=2.15\times10^{3}X+8.25\times10^{3}$	0.9990
339	Lactofen	27.59	8000	4000	40.000 ~ 1600.0		
340	Mecarbam	21.70	8.00	4.00	0.1000 ~ 4.0000	$Y=1.27\times10^{3}X+9.7\times10^{3}$	0.9970
341	Mefenpyr-diethyl	24.35	3.00	1.50	0.0375 ~ 1.5000	$Y=3.83\times10^{3}X+1.3\times10^{4}$	0.9981
342	Metamitron	10.02	0.80	0.40	0.0100 ~ 0.4000	$Y=5.17\times10^{4}X+1.6\times10^{5}$	0.9988
343	Methacrifos	16.88	80.00	40.00	1.0000 ~ 40.000	$Y=2.73\times10^{3}X+4.45\times10^{3}$	0.9987
344	Norflurazon	14.39	1.60	0.80	0.0200 ~ 0.8000	$Y=1.6\times10^{3}X+1.56\times10^{4}$	0.9917
345	Nuarimol	14.99	4.00	2.00	0.0500 ~ 2.0000	$Y=1.98\times10^{3}X+9.03\times10^{3}$	0.9958
346	Desmetryn	10.79	0.40	0.20	0.0050 ~ 0.2000	$Y=2.31\times10^{3}X-5.23\times10^{3}$	0.9994
347	Pencycuron	24.49	0.40	0.20	0.0050 ~ 0.2000	$Y=4.16\times10^{3}X+3.06\times10^{3}$	0.9983
348	Prepoxycarbazone -sodium	14.81	16.00	8.00			
349	Profenefos	25.28	1.60	0.80	0.0200 ~ 0.8000	$Y=2.21\times10^{3}X+3.83\times10^{3}$	0.9992
350	Propamocarb	1.82 2.33	4.00	2.00	0.0500 ~ 2.0000	$Y=1.92\times10^{4}X+1.01\times10^{5}$	0.9994
351	Propetamphos	20.66	64.00	32.00	0.8000 ~ 32.000	$Y=861X+2.06\times10^{4}$	0.9912
352	Pyraclofos	22.98	2.40	1.20	0.0300 ~ 1.2000	$Y=2.1\times10^{3}X+5.61\times10^{3}$	0.9993
353	Pyridaphenthion	18.32	1.60	0.80	0.0200 ~ 0.8000	$Y=6.07\times10^{3}X+1.84\times10^{4}$	0.9981
354	Sethoxydim	27.09	2.40	1.20	0.0300 ~ 1.2000	$Y=1.44\times10^{3}X-5.81\times10^{3}$	0.9955
355	Simeconazole	16.96	0.80	0.40	0.0100 ~ 0.4000	$Y=1.56\times10^{3}X+942$	0.9993
356	Sulfallate	23.55	48.00	24.00	0.6000 ~ 24.000	$Y=5.27\times10^{3}X+3.56\times10^{3}$	0.9998

357	Sulfotep	24.66	1.60	0.80	0.0200 ~ 0.8000	$Y=2.59\times10^3X-2.5\times10^3$	0.9988
358	Sulprofos	28.31	3.20	1.60	0.0400 ~ 1.6000	$Y=2.88\times10^3X+4.65\times10^3$	0.9991
359	Tebufenozide	21.36	16.00	8.00	0.4000 ~ 4.0000	$Y=394X+5.87\times10^3$	0.9934
360	Tebupirimfos	27.66	80.00	40.00	1.0000 ~ 20.000	$Y=3\times10^5X+2.08\times10^6$	0.9946
361	Tebutam	19.91	0.40	0.20	0.0050 ~ 0.2000	$Y=3.64\times10^3+1.48\times10^4$	0.9959
362	Tebuthiuron	11.32	0.40	0.20	0.0050 ~ 0.2000	$Y=3.04\times10^3X+2.54\times10^3$	0.9998
363	Terbufos	27.42	144	72.00	1.8000 ~ 36.000	$Y=2.51\times10^3X+7.79\times10^3$	0.9987
364	Terbumeton	10.73	0.40	0.20	0.0050 ~ 0.2000	$Y=3.08\times10^3X+4.49\times10^3$	0.9992
365	Thiobencarb	24.08	0.80	0.40	0.0100 ~ 0.4000	$Y=2.19\times10^3X+2.35\times10^3$	0.9988
366	Thiofanox-sulfoxide	2.31	48.00	24.00	0.6000 ~ 24.000	$Y=2.22\times10^3 X+9.12\times10^4$	0.9593
367	Thionazin	15.74	8.00	4.00	0.1000 ~ 4.0000	$Y=8.03\times10^3X+4.29\times10^4$	0.9972
368	Triadimefon	18.00	4.00	2.00	0.0500 ~ 2.0000	$Y=2.05\times10^3X+6.61\times10^3$	0.9977
369	Triamiphos	11.90	1.60	0.80	0.0200 ~ 0.8000	$Y=1.45\times10^4X+3.41\times10^4$	0.9997
370	Tridemorph	14.32	8.00	4.00	0.1000 ~ 4.0000	$Y=1.59\times10^4X+1.01\times10^4$	0.9996
371	Trietazine	19.22	0.80	0.40	0.0100 ~ 0.4000	$Y=1.58\times10^3X-1.18\times10^3$	0.9992
372	Triflumizole	20.12	2.40	1.20	0.0300 ~ 1.2000	$Y=258X-954$	0.9985
373	Tri-iso-butyl phosphate	22.60; 23.01	0.40	0.20	0.0100 ~ 0.1000	$Y=563X+39.8$	0.9993
374	Vernolate	24.51	80.00	40.00			
375	2,6dichlorobenzamide	9.55	60.00	30.00	0.3000 ~ 12.000	$Y=0.69\times10^3X+298$	0.9957
376	Alachlor	20.17	8.00	4.00	0.2000 ~ 8.0000	$Y=5.39\times10^3X+1.37\times10^4$	0.9956
377	Alanycarb	13.57	2.40	1.20	0.0600 ~ 2.4000	$Y=115X+3.04\times10^4$	0.9318
378	Aldoxycarb	8.10	40.00	20.00	1.0000 ~ 40.000	$Y=963X -1.39\times10^3$	0.9947

379	Allidochlor	12.09	40.00	20.00	2.0000 ~ 20.000	$Y=1.01\times10^{4}X-4.08\times10^{4}$	0.9968
380	Alpha-hch	23.93	8.00	4.00	0.2000 ~ 8.0000	$Y=2.7\times10^{3}X-751$	0.9980
381	Ametryn	11.90	0.40	0.20	0.0100 ~ 0.4000	$Y=2.43\times10^{3}X-1.28\times10^{3}$	0.9974
382	Athidathion	21.30	20.00	10.00	0.1000 ~ 4.0000	$Y=3.91\times10^{3}X+9.37\times10^{3}$	0.9989
383	Atratone	10.03	0.20	0.10	0.0050 ~ 0.2000	$Y=3.07\times10^{3}X-50.8$	0.9994
384	Atrazine	13.79	0.12	0.06	0.0030 ~ 0.1200	$Y=1.3\times10^{3}X-1.3\times10^{3}$	0.9977
385	Azaconazole	13.72	0.20	0.10	0.0050 ~ 0.2000	$Y=1.97\times10^{3}X+1.02\times10^{3}$	0.9973
386	Azamethiphos	12.22	0.40	0.20	0.0100 ~ 0.4000	$Y=2.47\times10^{3}X+3.75\times10^{3}$	0.9921
387	Azinphos-methyl	16.90	40.00	20.00	1.0000 ~ 40.000	$Y=791X+7.49\times10^{3}$	0.9986
388	Benfluralin	28.12	800.00	400.00	4.0000 ~ 80.000	$Y=2.75\times10^{3}X+2.5\times10^{3}$	0.9904
389	Benfuracarb	24.20	40.00	20.00	2.0000 ~ 20.000	$Y=394X+6.71\times10^{3}$	0.9976
390	Bensulfuron methyl	15.62	40.00	20.00	1.0000 ~ 40.000	$Y=7.97\times10^{4}X+4.02\times10^{5}$	0.9931
391	Bupirimate	15.92	0.40	0.20	0.0100 ~ 0.4000	$Y=1.94\times10^{3}X+669$	0.9936
392	Cadusafos	23.29	0.80	0.40	0.0200 ~ 0.8000	$Y=3.52\times10^{3}X-4.84\times10^{3}$	0.9992
393	Cartap	23.87	40.00	20.00	1.0000 ~ 40.000	$Y=1.83\times10^{4}X+2.53\times10^{5}$	0.9940
394	Chlorfluazuron	28.38	4.00	2.00	0.1000 ~ 4.0000	$Y=1.71\times10^{3}X+1.29\times10^{4}$	0.9803
395	Chloridazon	10.33	4.00	2.00	0.1000 ~ 4.0000	$Y=6.04\times10^{3}X+7.52\times10^{3}$	0.9953
396	Chlorthiophos	28.57	2.00	1.00	0.0500 ~ 2.0000	$Y=595X-507$	0.9983
397	Cinosulfuron	18.70	4.00	2.00	0.0200 ~ 0.8000	$Y=2.75\times10^{3}X+2.23\times10^{4}$	0.9743
398	Cycloate	24.60	2.00	1.00	0.0500 ~ 2.0000	$Y=3.67\times10^{3}X-9.53\times10^{3}$	0.9975
399	Cyprazine	13.80	0.12	0.06	0.0030 ~ 0.1200	$Y=921X+233$	0.9918

400	Dibutyl succinate	22.60	16.00	8.00	0.4000 ~ 16.000	$Y=3.62\times10^3X+1.62\times10^3$	0.9993
401	Diethofencarb	16.89	1.60	0.80	0.0400 ~ 1.6000	$Y=1.32\times10^3X+463$	0.9940
402	Dimethirimol	2.37	2.00	1.00	0.0500 ~ 2.0000	$Y=4.65\times10^3X+2.01\times10^4$	0.9908
403	Dimethomorph	9.39 15.14; 15.56	0.40	0.20	0.0100 ~ 0.4000	$Y=1.78\times10^3X+3.2\times10^3$	0.9808
404	Diniconazole	20.31	0.40	0.20	0.0100 ~ 0.4000	$Y=1.39\times10^3X-1.77\times10^3$	0.9999
405	Dinoseb acetate	13.58	20.00	10.00	0.5000 ~ 20.000	$Y=6.03X+1.56\times10^4$	0.9275
406	Diofenolan	25.27	760	380	19.000 ~ 380.00	$Y=8.66\times10^3X+4.71\times10^4$	0.9958
407	Dioxacarb	10.54	30.00	15.00	0.1500 ~ 6.0000	$Y=1.29\times10^3X+3.28\times10^3$	0.9943
408	Dipropetryn	14.90	0.12	0.06	0.0060 ~ 0.0600	$Y=1.54\times10^3+1.06\times10^4$	0.9938
409	Dmst	13.61	2.40	1.20	0.0600 ~ 2.4000	$Y=1.84\times10^3X+52.6$	0.9994
410	EPN	25.76	2.00	1.00	0.0500 ~ 2.0000	$Y=1.4\times10^3X+2.8\times10^3$	0.9924
411	Ethiofencarb	20.1	2.00	1.00	0.0500 ~ 2.0000	$Y=4.32\times10^3X-757$	0.9957
412	Etofenprox	31.11	40.00	20.00	1.0000 ~ 40.000	$Y=1.12\times10^3X+1.08\times10^4$	0.9455
413	Fenamiphos	17.10	0.80	0.40	0.0040 ~ 0.1600	$Y=1.75\times10^3X+599$	0.9993
414	Fenpropimorph	12.09	0.16	0.08	0.0040 ~ 0.1600	$Y=1.98\times10^3X-6.98\times10^3$	0.9967
415	Florasulam	12.95	1.60	0.80	0.0400 ~ 1.6000	$Y=5.43\times10^4X+2.06\times10^5$	0.9956
416	flubenzimine	27.4	40.00	20.00	1.0000 ~ 40.000	$Y=5.39\times10^4X-2.09\times105$	0.9985
417	Flumetsulam	7.56	0.40	0.20	0.0100 ~ 0.4000	$Y=1.79\times10^3X+3.82\times10^3$	0.9929
418	furmecyclox	20.99	200.00	100.00	1.0000 ~ 40.000	$Y=2.08\times10^5X+8.11\times10^5$	0.9905
419	Heptanophos	14.31	4.00	2.00	0.1000 ~ 4.0000	$Y=6.67\times10^3 X -626$	0.9993
420	Hexaconazole	19.06	0.24	0.12	0.0060 ~ 0.2400	$Y=831X+185$	0.9927

421	Imazapyr	9.68	12.00	6.00	0.3000 ~ 6.0000	$Y=5.61\times10^{3}X-1.04\times10^{4}$	0.9971
422	Iodosulfuron-methyl	15.41	16.00	8.00	0.4000 ~ 8.0000	$Y=923X-171$	0.9912
423	Isazofos	21.72	0.40	0.20	0.0100 ~ 0.4000	$Y=2.9\times10^{3}X-243$	0.9987
424	Isocarbamid	10.11	0.80	0.40	0.0200 ~ 0.8000	$Y=2.65\times10^{3}X+3.92\times10^{3}$	0.9979
425	Isopropalin	29.67	4.00	2.00	0.1000 ~ 4.0000	$Y=1.24\times10^{3}X-1.12\times10^{3}$	0.9992
426	Lenacil	12.35	8.00	4.00	0.2000 ~ 8.0000	$Y=1.71\times104X+1.47\times10^{4}$	0.9976
427	Linuron	16.92	4.00	2.00	0.1000 ~ 4.0000	$Y=1.12\times10^{3}X+2.89\times10^{3}$	0.9966
428	Malathion	19.93	1.20	0.60	0.0300 ~ 0.6000	$Y=1.46\times10^{3}X-1.03\times10^{3}$	0.9970
429	Mefenacet	17.85	0.40	0.20	0.0100 ~ 0.4000	$Y=2.57\times10^{3}X+3.32\times10^{3}$	0.9963
430	Mepiquat chloride	2.20	1.20	0.60	0.0300 ~ 1.2000	$Y=1.19\times10^{4}X+1.15\times10^{5}$	0.9742
431	Metazachlor	14.91	0.40	0.20	0.0100 ~ 0.2000	$Y=1.55\times10^{3}X+3.65\times10^{3}$	0.9913
432	Methiocarb sulfoxid	25.59; 35.15	2.40	1.20	0.0600 ~ 2.4000	$Y=43X+1.83\times10^{4}$	0.9410
433	Methyl-parathion	18.67	200.00	100.00	1.0000 ~ 20.000	$Y=286X+127$	0.9972
434	Metolachlor	19.93	0.20	0.10	0.0050 ~ 0.2000	$Y=692X-260$	0.9953
435	metosulam	20.20	2.00	1.00	0.0500 ~ 2.0000	$Y=6.1\times10^{3}X+1.09\times10^{3}$	0.9969
436	Metribuzin	12.56	1.20	0.60	0.0300 ~ 1.2000	$Y=780X+610$	0.9990
437	Nitralin	22.39	4.00	2.00	0.1000 ~ 4.0000	$Y=640X-34.4$	0.9903
438	Pebulate	24.54	40.00	20.00	1.0000 ~ 40.000	$Y=2.34\times10^{4}X-1.01\times10^{5}$	0.9995
439	Phenthoate	23.56	4.00	2.00	0.4000 ~ 4.0000	$Y=1.22\times10^{3}X+3\times10^{4}$	0.9938
440	Phorate sulfone	16.13	1.60	0.80	0.0400 ~ 1.6000	$Y=6.34\times10^{3}X+2.97\times10^{4}$	0.9948
441	Phosfolan	10.87	4.00	2.00	0.1000 ~ 4.0000	$Y=1.58\times10^{4}X+1.31\times10^{4}$	0.9982
442	Phosmet	17.51	1.20	0.60	0.0300 ~ 1.2000	$Y=800X-2.1\times10^{3}$	0.9993
443	Phosmet-oxon	25.23	16.00	8.00	0.8000 ~ 8.0000	$Y=112X+2.57\times10^{3}$	0.9976
444	Picoxystrobin	22.97	40.00	20.00	1.0000 ~ 40.000	$Y=3.93\times10^{4}X+2.88\times10^{5}$	0.9913
445	Pirimiphos-ethyl	27.27	0.12	0.06	0.0030 ~ 0.1200	$Y=2.33\times10^{4}X-1.59\times10^{4}$	0.9990
446	Prallethrin	25.24	1.60	0.80	0.0800 ~ 0.8000	$Y=1.33\times10^{3}X+8.16\times10^{3}$	0.9982
447	Prohydrojasmon	24.55	8.00	4.00	0.2000 ~ 8.0000	$Y=933X+1.04\times10^{3}$	0.9917
448	Propyzamide	18.50	0.80	0.40	0.0200 ~ 0.8000	$Y=945X-1.5\times10^{3}$	0.9956
449	Prosulfocarb	25.88	0.20	0.10	0.0050 ~ 0.2000	$Y=2.99\times10^{3}X-1.29\times10^{3}$	0.9995
450	Pyraclostrobin	24.10	0.80	0.40	0.0200 ~ 0.8000	$Y=2.55\times10^{3}X+2.67\times10^{3}$	0.9969

451	Pyrazosulfuron-ethyl	16.83	2.00	1.00	0.0100 ~ 0.4000	$Y=2.11\times10^3X+1.02\times10^3$	0.9954
452	Pyributicarb	27.86	0.40	0.20	0.0100 ~ 0.4000	$Y=7.25\times10^3X-2.2\times10^4$	0.9982
453	Pyrifenox	12.97	0.40	0.20	0.0100 ~ 0.4000	$Y=2.62\times10^3X+856$	0.9950
454	Sebutylazine	15.60	0.20	0.10	0.0050 ~ 0.2000	$Y=1.85\times10^3X+2.88\times10^3$	0.9976
455	Tebuconazole	18.46	0.20	0.10	0.0005 ~ 0.2000	$Y=1.1\times10^3X+1.47\times10^3$	0.9925
456	Terbutryn	13.45	0.16	0.08	0.0040 ~ 0.1600	$Y=4.69\times10^3X+4.57\times10^3$	0.9980
457	Tetraconazole	18.35	2.00	1.00	0.0100 ~ 0.4000	$Y=1.38\times10^3X-767$	0.9989
458	Thiabendazole	2.32	6.00	3.00	0.0300 ~ 1.2000	$Y=4.39\times10^3X+2.15\times10^3$	0.9955
459	Thiazopyr	24.23	1.20	0.60	0.0300 ~ 1.2000	$Y=733X+3.09\times10^3$	0.9946
460	Thiodicarb	12.53	20.00	10.00	0.1000 ~ 4.0000	$Y=2.24\times10^3X+5.02\times10^3$	0.9901
461	Tralkoxydim	27.66	2.00	1.00	0.0500 ~ 2.0000	$Y=5.58\times10^3X+38$	0.9980
462	Triazoxide	11.54	28.00	14.00	0.7000 ~ 28.000	$Y=1.19\times10^4X+2.22\times10^4$	0.9979
463	Tricyclazole	10.71	2.00	1.00	0.0100 ~ 0.4000	$Y=2.4\times10^3X+283$	0.9999
464	Tri-n-butyl phosphate	23.07	0.20	0.10	0.0050 ~ 0.2000	$Y=2.1\times10^3X+4.22\times10^4$	0.9956
465	Uniconazole	17.32	0.40	0.20	0.0100 ~ 0.4000	$Y=1.44\times10^3X-488$	0.9980

Table 22a. Recovery and precision data for 362 pesticide residues in grains by GC-MS

No.	Compound	Level /mg/kg	Rice	Wheat	Maize	Barley	Oats	Ave.	RSD %	Level /mg/kg	Rice	Wheat	Maize	Barley	Oats	Ave.	RSD %	Level /mg/kg	Rice	Wheat	Maize	Barley	Oats	Ave.	RSD %
A Group																									
1	Allidochlor	0.0250	70.5	130.8	97.6	127.0	125.0	110.2	23.4	0.0500	74.0	120.7	100.2	132.4	111.2	107.7	20.7	0.2500	73.3	88.9	63.4	111.7	103.4	88.2	22.8
2	Dichlormid	0.0125	73.1	70.9	73.1	86.9	95.2	79.9	13.4	0.0250	85.1	90.3	79.3	92.7	90.9	87.6	6.2	0.1250	77.9	85.6	63.5	97.6	96.1	84.1	16.7
3	Etridiazol	0.0375	70.6	65.6	110.8	97.2	108.2	90.5	23.3	0.0750	63.9	83.4	80.6	91.7	94.2	82.8	14.5	0.3750	71.6	79.0	54.1	76.3	90.7	74.3	17.9
4	Chlormephos	0.0250	75.7	63.0	90.1	60.8	62.7	70.5	17.7	0.0500	89.1	81.7	74.3	67.4	62.7	75.0	14.2	0.2500	80.4	73.0	59.8	91.9	85.5	78.1	15.8
5	Propham	0.0125	98.9	82.1	89.1	128.8	116.7	103.1	18.8	0.0250	115.8	92.6	100.6	143.0	141.0	118.6	19.4	0.1250	79.8	145.4	117.4	93.5	92.9	105.8	24.5
6	Cycloate	0.0125	86.1	84.8	85.8	89.4	59.4	81.1	15.1	0.0250	90.5	92.9	93.3	93.2	90.2	92.0	1.6	0.1250	85.3	95.4	79.0	102.0	98.5	92.0	10.4
7	Diphenylamin	0.0125	74.3	72.1	51.3	85.9	74.6	71.7	17.5	0.0250	89.7	79.5	79.8	79.8	86.2	83.0	5.6	0.1250	85.6	93.7	84.3	113.8	101.6	95.8	12.8
8	Ethalfluralin	0.0500	83.9	92.3	82.2	104.2	113.9	95.3	14.2	0.1000	77.9	87.5	93.3	103.6	94.4	91.3	10.4	0.5000	80.4	95.8	82.9	101.4	94.0	90.9	9.8
9	Phorate	0.0125	86.3	90.1	92.8	118.2	121.8	101.8	16.5	0.0250	79.8	100.8	96.4	112.6	103.5	98.6	12.2	0.1250	85.1	98.9	83.0	106.2	100.9	94.8	10.8
10	Thiometon	0.0125	84.8	88.9	79.1	96.4	92.1	88.3	7.5	0.0250	84.4	97.1	89.4	95.9	81.1	89.6	7.8	0.1250	79.2	92.6	78.5	99.4	95.0	88.9	10.7
11	Quintozene	0.0250	88.7	120.7	94.3	130.7	129.1	112.7	17.6	0.0500	89.7	93.4	103.0	106.3	95.7	97.6	7.1	0.2500	83.8	98.3	81.5	108.9	102.4	95.0	12.5
12	Atrazine-desethyl	0.0125	84.0	102.8	91.3	111.3	131.6	104.2	17.8	0.0250	95.4	108.9	94.4	96.1	103.0	99.6	6.2	0.1250	94.6	102.1	78.2	101.4	105.2	96.3	11.2
13	Clomazone	0.0125	90.3	104.8	92.7	107.4	108.5	100.8	8.5	0.0250	98.0	110.3	105.6	112.9	103.0	106.0	5.6	0.1250	90.2	103.7	86.5	105.4	101.2	97.4	8.7
14	Diazinon	0.0125	89.1	98.5	91.1	97.1	103.0	95.7	5.9	0.0250	89.1	99.5	105.2	100.9	88.5	96.6	7.8	0.1250	85.8	101.5	86.1	105.1	96.1	94.9	9.3
15	Fonofos	0.0125	93.6	98.6	96.2	114.8	120.5	104.8	11.6	0.0250	91.8	104.0	97.2	110.6	101.4	101.0	7.0	0.1250	87.0	100.9	85.0	107.5	101.1	96.3	10.2
16	Etrimfos	0.0125	84.8	102.2	93.5	119.4	130.2	106.0	17.6	0.0250	96.7	100.6	100.5	100.3	92.5	98.1	3.6	0.1250	90.2	101.9	86.2	105.6	98.5	96.5	8.4
17	Simazine	0.0125	100.3	112.0	103.4	102.6	112.5	106.1	5.3	0.0250	96.3	105.8	103.8	97.7	93.6	99.4	5.2	0.1250	88.4	102.4	86.4	102.3	98.4	95.6	8.0
18	Propetamphos	0.0125	87.3	98.8	99.4	124.3	124.7	106.9	15.7	0.0250	100.4	106.9	110.8	139.8	106.3	112.8	13.8	0.1250	89.7	99.6	90.3	108.5	103.7	98.4	8.4
19	Secbumeton	0.0125	56.9	62.5	77.0	37.6	84.2	63.6	28.6	0.0250	60.5	69.5	80.0	33.9	45.8	57.9	31.7	0.1250	62.5	74.7	67.6	65.3	63.0	66.6	7.5
20	Dichlofenthion	0.0125	90.4	100.5	89.2	98.6	101.6	96.1	6.1	0.0250	95.3	102.8	103.7	103.3	96.0	100.2	4.2	0.1250	90.7	100.3	85.9	102.3	98.2	95.5	7.3
21	Pronamide	0.0125	95.6	104.9	113.6	118.3	154.4	117.4	19.1	0.0250	90.8	108.0	119.1	109.6	120.6	109.6	10.9	0.1250	90.2	105.8	90.7	108.3	106.3	100.3	9.0
22	Mexacarbate	0.0375	70.7	72.1	75.6	38.6	33.6	58.1	34.8	0.0750	58.6	69.2	54.0	33.9	33.9	49.9	31.3	0.3750	56.5	68.8	67.7	50.4	48.1	58.3	16.5
23	Aldrin	0.0250	93.5	96.3	86.7	92.4	97.3	93.2	4.5	0.0500	95.2	101.1	98.1	101.2	94.6	98.1	3.2	0.2500	89.8	98.6	80.8	102.3	93.2	92.9	8.9
24	Dinitramine	0.0500	62.2	87.3	81.8	70.6	80.0	76.4	13.1	0.1000	56.8	80.5	83.2	95.0	84.9	80.1	17.6	0.5000	71.9	84.2	79.0	88.3	77.4	80.1	7.9

Table 22a. Continued

No.	Compound	Level /mg/kg	Rice	Whea	Maize	Barley	Oats	Ave.	RSD %	Level /mg/kg	Rice	Whea	Maize	Barley	Oats	Ave.	RSD %	Level /mg/kg	Rice	Whea	Maize	Barley	Oats	Ave.	RSD %
25	Ronnel	0.0250	95.1	104.3	95.0	101.7	104.6	100.1	4.8	0.0500	94.1	103.6	102.7	101.6	86.8	97.8	7.4	0.2500	90.5	102.1	86.3	104.5	101.1	96.9	8.2
26	Prometrye	0.0125	64.7	83.9	88.5	57.6	56.6	70.3	21.4	0.0250	69.3	77.4	89.7	56.5	63.1	71.2	18.1	0.1250	67.6	80.5	73.7	83.1	42.5	69.5	23.4
27	Cyprazine	0.0125	93.7	109.1	111.3	112.8	118.5	109.1	8.5	0.0250	91.8	111.8	113.2	121.1	105.9	108.7	10.0	0.1250	92.2	106.4	90.9	108.8	94.0	98.5	8.6
28	Vinclozolin	0.0125	85.2	97.2	94.4	93.6	100.5	94.2	6.0	0.0250	91.1	107.5	110.3	104.9	98.3	102.4	7.5	0.1250	89.4	101.8	85.9	105.7	98.3	96.2	8.7
29	Beta-HCH	0.0125	92.3	124.0	108.2	127.7	109.8	112.4	12.5	0.0250	99.8	120.8	114.3	121.8	108.8	113.1	8.1	0.1250	95.3	107.6	90.0	106.6	100.3	100.0	7.5
30	Metalaxyl	0.0375	91.6	99.3	93.4	97.3	126.6	101.6	14.1	0.0750	96.1	104.6	102.7	111.2	115.3	106.0	7.1	0.3750	93.6	103.7	88.5	104.0	102.2	98.4	7.1
31	Chlorpyifos (ethyl)	0.0125	91.6	98.2	93.5	101.6	127.4	102.4	14.1	0.0250	95.3	101.5	102.6	108.5	106.5	102.9	5.0	0.1250	90.8	104.4	88.6	106.3	100.5	98.1	8.1
32	Methyl-parathion	0.0500	83.1	103.4	103.4	106.0	122.0	103.6	13.3	0.1000	80.2	113.0	109.2	116.4	107.7	105.3	13.7	0.5000	84.8	112.3	93.7	122.4	114.9	105.6	14.9
33	Anthraquinone	0.0125	51.5	77.8	98.2	56.9	122.3	81.3	36.2	0.0250	50.8	99.2	105.3	81.4	101.0	87.5	25.7	0.1250	72.0	87.2	87.7	101.0	82.4	86.1	12.2
34	Delta-HCH	0.0250	94.3	102.1	94.5	91.1	94.3	95.3	4.3	0.0500	98.0	104.7	102.2	106.9	58.4	94.0	21.5	0.2500	93.7	98.0	85.3	105.2	94.5	95.3	7.6
35	Fenthion	0.0125	95.5	109.3	122.8	117.8	128.4	114.8	11.2	0.0250	94.0	106.3	134.9	117.1	117.2	113.9	13.3	0.1250	88.5	103.0	94.5	108.6	101.0	99.1	7.9
36	Malathion	0.0500	90.2	99.3	109.8	102.0	119.3	104.1	10.6	0.1000	91.3	106.2	113.5	107.4	98.3	103.4	8.3	0.5000	92.5	103.2	90.6	109.4	103.0	99.7	7.9
37	Fenitrothion	0.0250	83.1	95.9	97.5	130.2	134.9	108.3	21.1	0.0500	82.4	105.0	103.6	120.3	105.9	103.4	13.1	0.2500	88.5	104.5	89.6	113.0	107.7	100.6	11.0
38	Paraoxon-ethyl	0.0500	81.1	106.0	109.4	126.1	126.8	119.9	21.7	0.1000	72.4	112.1	111.6	120.9	125.4	108.4	19.4	0.5000	88.0	115.6	95.7	131.1	126.0	111.3	16.9
39	Triadimefon	0.0250	93.6	93.1	94.0	106.2	108.7	99.1	7.7	0.0500	93.3	102.4	105.0	109.3	102.7	102.5	5.7	0.2500	91.6	103.6	90.2	107.1	98.8	98.3	7.5
40	Parathion	0.0500	82.1	90.8	94.7	97.9	122.1	97.5	15.3	0.1000	78.5	101.7	104.6	115.1	119.6	103.9	15.4	0.5000	87.2	106.0	90.4	115.1	111.0	101.9	12.2
41	Pendimethalin	0.0500	82.6	95.1	98.9	108.7	123.5	101.7	15.1	0.1000	78.2	103.9	101.0	114.6	113.5	102.3	14.4	0.5000	87.9	105.9	87.4	114.3	107.3	100.6	12.2
42	Linuron	0.0500	103.5	60.0	72.4	110.4	71.6	83.6	26.3	0.1000	79.4	80.8	101.3	43.5	64.9	74.0	28.9	0.5000	93.2	98.6	81.8	73.8	66.2	82.7	16.2
43	Chlorbenside	0.0250	84.1	89.3	88.1	106.4	101.3	93.8	10.1	0.0500	91.4	107.2	104.0	110.4	84.6	99.5	11.1	0.2500	90.6	107.6	88.7	109.8	99.5	99.2	9.6
44	Bromophos-ethyl	0.0125	95.9	102.2	85.1	103.9	104.0	98.2	8.2	0.0250	99.5	101.9	107.9	109.1	98.9	103.4	4.6	0.1250	90.7	105.4	88.6	107.0	97.8	97.9	8.5
45	Quinalphos	0.0125	91.6	107.0	105.6	139.7	115.9	112.0	15.9	0.0250	93.5	111.3	111.5	112.1	105.5	106.8	7.4	0.1250	92.1	109.2	92.0	113.2	101.3	101.5	9.5
46	Trans-Chlodane	0.0125	96.7	98.9	91.5	97.4	101.1	97.1	3.7	0.0250	98.9	103.7	104.4	99.6	90.6	99.4	5.5	0.1250	92.8	100.8	86.0	103.9	95.6	95.8	7.3
47	Phenthoate	0.0250	90.6	83.8	92.7	100.8	113.0	96.2	11.6	0.0500	91.6	96.9	105.2	103.8	96.7	98.8	5.7	0.2500	91.1	102.5	89.4	102.0	100.5	97.1	6.5
48	Metazachlor	0.0375	94.0	101.0	97.6	103.7	107.8	100.8	5.3	0.0750	94.5	104.7	104.2	105.8	101.4	102.1	4.5	0.3750	92.2	103.1	88.7	106.4	99.3	97.9	7.6
49	Fenothiocarb	0.2000	91.7	95.2	100.7	91.5	99.7	95.8	4.5	0.4000	86.1	102.5	105.3	106.0	101.6	100.3	8.1	2.0000	106.2	133.0	107.4	107.5	102.0	111.2	11.1
50	Prothiophos	0.0125	90.7	102.8	96.1	103.6	106.7	100.0	6.5	0.0250	87.9	104.7	100.4	107.6	102.0	100.5	7.5	0.1250	86.1	101.1	84.4	104.6	99.8	95.2	9.7

Table 22a. Continued

No.	Compound	Level /mg/kg	Rice	Wheat	Maize	Barley	Oats	Ave.	RSD %	Level /mg/kg	Rice	Wheat	Maize	Barley	Oats	Ave.	RSD %	Level /mg/kg	Rice	Wheat	Maize	Barley	Oats	Ave.	RSD %
51	Chlorfurenol	0.0375	90.1	97.5	95.8	102.4	117.2	100.6	10.2	0.0750	89.6	103.0	103.8	108.8	104.1	101.9	7.1	0.3750	91.4	104.2	89.9	106.5	101.4	98.7	7.7
52	Dieldrin	0.0250	87.4	116.4	90.4	94.6	101.4	98.0	11.8	0.0500	101.4	114.4	106.6	110.5	98.1	106.2	6.2	0.2500	93.3	104.4	86.2	103.0	96.7	96.7	7.7
53	Procymidone	0.0125	115.7	106.7	119.6	97.2	108.0	109.4	7.9	0.0250	82.2	109.1	107.5	111.8	102.9	102.7	11.6	0.1250	95.1	92.4	80.0	102.8	115.9	97.2	13.6
54	Methidathion	0.0250	92.2	99.1	103.7	103.2	123.1	104.2	11.0	0.0500	88.4	106.9	104.7	112.2	86.1	99.6	11.7	0.2500	91.7	108.4	92.6	91.7	102.3	97.3	7.8
55	Napropamide	0.0375	90.2	104.6	98.7	92.3	102.6	97.7	6.5	0.0750	90.7	108.4	105.9	109.1	102.8	103.4	7.3	0.3750	92.0	106.9	91.1	107.3	102.4	100.0	7.9
56	Oxadiazone	0.0125	95.2	94.5	100.1	82.0	84.4	91.2	8.4	0.0250	109.7	104.9	108.3	120.7	98.5	108.4	7.5	0.1250	93.6	105.1	89.7	103.6	97.4	97.9	6.6
57	Fenamiphos	0.0375	73.3	87.6	97.7	92.0	119.5	94.0	17.9	0.0750	73.5	97.9	102.5	102.7	103.9	96.1	13.4	0.3750	81.0	109.5	94.1	110.0	105.1	99.9	12.4
58	Tetrasul	0.0125	103.4	116.1	84.6	76.5	78.1	91.7	18.9	0.0250	95.5	94.1	88.7	93.4	76.9	89.7	8.5	0.1250	92.1	101.7	81.9	106.2	95.1	95.4	9.8
59	Aramite	0.0125	108.0	93.8	108.3	81.8	96.2	97.6	11.3	0.0250	101.5	104.2	102.1	97.2	96.8	100.4	3.2	0.1250	91.5	94.0	77.6	85.5	83.6	86.4	7.5
60	Bupirimate	0.0125	63.5	74.5	83.1	59.0	37.8	63.6	27.1	0.0250	67.1	75.3	89.2	62.3	69.6	72.7	14.2	0.1250	67.1	85.5	79.0	81.9	52.5	73.2	18.4
61	Carboxin	0.0375	64.9	80.8	66.6	74.5	60.8	69.5	11.6	0.0750	59.6	83.4	65.6	84.5	57.7	70.1	18.5	0.3750	64.6	84.1	77.3	89.8	78.5	78.9	11.9
62	Flutolanil	0.0125	91.7	92.1	97.6	91.7	103.2	95.3	5.3	0.0250	95.0	103.2	106.7	106.1	102.1	102.6	4.5	0.1250	91.9	106.8	92.1	105.5	100.6	99.4	7.2
63	p,p'-DDD	0.0125	64.9	80.8	66.6	74.5	60.8	69.5	11.6	0.0250	59.6	83.4	65.6	84.5	57.7	70.1	18.5	0.1250	64.6	84.1	77.3	89.8	78.5	78.9	11.9
64	Ethion	0.0250	91.4	97.5	97.3	105.8	130.6	104.5	14.8	0.0500	91.4	106.2	108.5	109.1	105.4	104.1	7.0	0.2500	91.2	106.6	92.2	109.5	106.3	101.2	8.6
65	Sulprofos	0.0250	89.3	98.4	95.4	98.7	105.8	97.5	6.1	0.0500	88.8	101.7	100.5	105.2	90.3	97.3	7.5	0.2500	87.4	102.2	87.9	106.9	97.5	96.4	8.9
66	Etaconazole-1	0.0375	79.6	81.9	122.6	95.9	111.0	98.2	18.9	0.0750	98.0	94.6	105.9	93.8	87.1	95.9	7.2	0.3750	82.5	108.2	100.8	105.9	100.5	99.6	10.1
67	Etaconazole-2	0.0375	88.3	93.9	102.5	88.6	94.5	93.6	6.1	0.0750	96.5	130.8	133.4	106.1	84.7	110.3	19.3	0.3750	78.3	107.9	91.3	112.6	84.3	94.9	15.7
68	Myclobutanil	0.0125	89.6	80.7	103.6	68.2	72.9	83.0	17.0	0.0250	95.8	102.6	101.1	89.0	97.9	97.7	10.2	0.1250	97.3	106.1	88.8	99.3	95.9	97.5	6.4
69	Dichlorofop-methyl	0.0125	91.1	105.7	89.6	94.3	66.9	89.5	15.8	0.0250	99.3	113.1	102.8	119.0	87.7	104.4	11.7	0.1250	79.8	100.8	75.9	107.3	92.5	91.3	14.7
70	Propiconazole	0.0375	82.2	85.3	108.2	82.1	92.0	89.9	12.2	0.0750	107.4	117.1	103.4	96.3	90.5	103.0	9.9	0.3750	83.3	105.7	90.9	102.7	97.9	96.1	9.5
71	Fensulfothion	0.0250	96.6	85.3	121.3	90.6	96.6	98.1	14.1	0.0500	120.7	94.0	94.9	96.7	80.3	97.3	15.0	0.2500	108.9	90.3	84.5	73.9	62.7	84.0	20.7
72	Bifenthrin	0.0125	94.2	96.2	94.7	95.0	103.3	96.7	3.9	0.0250	96.9	105.8	105.3	107.5	100.3	103.2	4.3	0.1250	92.3	103.2	89.2	107.5	100.8	98.6	7.7
73	Mirex	0.0125	96.6	93.8	88.1	94.2	92.3	93.0	3.4	0.0250	101.9	102.0	99.3	100.7	82.7	97.3	8.5	0.1250	95.1	99.2	80.7	96.8	88.6	92.1	8.1
74	Benodanil	0.0375	88.4	103.1	101.9	105.0	121.3	103.9	11.2	0.0750	88.0	109.1	109.0	112.7	113.4	106.5	9.9	0.3750	92.0	107.5	92.0	109.9	106.4	101.5	8.7
75	Nuarimol	0.0250	89.0	101.1	98.5	96.3	98.1	96.6	4.7	0.0500	97.5	99.9	105.7	131.8	139.5	114.9	16.9	0.2500	89.0	103.1	88.9	113.2	101.3	99.1	10.4
76	Methoxychlor	0.0125	88.6	93.4	101.6	86.5	109.6	95.9	10.0	0.0250	86.3	101.1	103.0	106.4	102.3	99.8	7.8	0.1250	89.3	100.3	88.1	66.6	81.8	85.2	14.5

Table 22a. Continued

No.	Compound	Level /mg/kg	Rice	Whea	Maize	Barley	Oats	Ave.	RSD %	Level /mg/kg	Rice	Whea	Maize	Barley	Oats	Ave.	RSD %	Level /mg/kg	Rice	Whea	Maize	Barley	Oats	Ave.	RSD %
77	Oxadxyl	0.0125	75.8	65.5	79.7	81.9	91.8	78.9	12.1	0.0250	93.5	99.3	103.2	102.9	82.2	96.2	9.1	0.1250	81.5	107.7	92.4	103.8	86.8	94.4	11.7
78	Tetramethrin	0.0250	90.2	97.8	100.3	102.7	119.5	102.1	10.6	0.0500	91.3	108.1	108.5	109.6	106.0	104.7	7.2	0.2500	91.1	106.0	92.5	108.0	108.3	101.2	8.5
79	Tebuconazole	0.0375	75.2	78.4	89.4	73.9	86.9	80.8	8.7	0.0750	89.8	94.4	101.1	83.7	77.2	89.2	10.4	0.3750	82.4	107.3	91.6	88.2	84.0	90.7	11.0
80	Norflurazon	0.0125	89.7	101.1	98.8	96.5	120.3	101.3	11.3	0.0250	93.7	111.8	102.5	100.5	129.4	107.6	12.8	0.1250	90.9	105.9	84.6	109.0	114.8	101.1	12.6
81	Pyridaphenthion	0.0125	92.4	117.6	124.0	131.8	113.1	115.8	12.8	0.0250	89.6	135.4	122.6	120.7	117.3	117.1	14.4	0.1250	90.0	120.1	100.4	123.1	125.0	111.7	14.0
82	Phosmet	0.0250	90.0	88.8	100.4	106.5	135.9	104.3	18.4	0.0500	89.5	108.2	106.1	104.9	67.3	95.2	18.1	0.2500	88.8	109.5	99.5	124.5	120.8	108.6	13.7
83	Tetradifon	0.0125	95.6	92.0	92.5	121.0	109.1	102.0	12.4	0.0250	109.4	119.2	114.2	143.0	123.3	121.8	10.6	0.1250	95.8	103.5	92.5	116.3	124.3	106.5	12.7
84	Oxycarboxin	0.0750	76.0	47.3	80.3	67.1	52.2	64.6	22.4	0.1500	70.2	44.1	60.9	49.1	39.5	56.8	12.4	0.7500	69.8	44.0	64.8	65.8	50.8	60.0	14.1
85	cis-permethrin	0.0125	95.6	98.0	81.1	103.0	136.3	102.8	19.9	0.0250	99.3	107.9	108.8	91.9	103.4	102.3	6.8	0.1250	93.6	85.4	91.0	109.9	105.6	97.1	10.6
86	Trans-permethrin	0.0125	93.9	119.5	100.0	106.7	135.5	111.1	15.0	0.0250	96.9	108.8	108.4	113.3	111.2	107.7	5.9	0.1250	93.6	106.7	94.2	110.8	108.3	102.7	8.0
87	Pyrazophos	0.0250	92.0	95.8	111.7	112.9	124.9	107.4	12.6	0.0500	96.6	113.2	113.3	112.8	71.1	101.4	18.1	0.2500	90.6	110.1	95.7	115.1	108.2	103.9	9.9
88	Cypermethrin	0.0375	95.6	120.0	118.3	122.7	125.0	118.3	13.7	0.0750	98.4	111.8	115.2	125.7	114.4	113.1	8.6	0.3750	92.7	118.2	97.3	112.8	110.3	106.3	10.2
89	Fenvalerate	0.0500	95.1	83.9	97.5	99.4	117.6	98.7	10.8	0.1000	96.6	92.6	99.9	113.2	148.0	110.1	20.5	0.5000	94.2	110.8	100.3	134.9	126.3	113.3	15.1
90	Deltamethrin	0.0750	104.6	78.7	90.6	88.0	81.6	88.7	11.4	0.1500	95.3	74.1	109.1	73.9	47.7	80.0	29.3	0.7500	98.9	66.1	92.4	74.9	50.4	76.6	25.7
B group																									
91	EPTC	0.0375	54.8	62.3	54.7	56.0	70.9	59.7	11.7	0.0750	60.4	54.3	60.2	72.7	84.0	66.3	18.0	0.3750	67.6	71.9	76.8	79.5	84.5	76.1	8.6
92	Butylate	0.0375	76.7	81.6	108.9	86.7	83.0	87.4	14.4	0.0750	73.6	68.0	83.2	89.9	132.2	89.4	28.4	0.3750	74.2	79.0	85.6	86.7	101.1	85.3	11.9
93	Dichlobenil	0.0050	34.3	45.9	42.3	52.0	69.2	48.7	26.9	0.0100	68.7	44.7	50.2	58.4	91.2	62.6	29.3	0.0500	64.4	68.1	74.2	81.5	87.9	75.2	12.8
94	Pebulate	0.0375	68.8	77.6	96.6	88.6	131.5	92.6	26.1	0.0750	69.8	70.5	70.3	99.7	108.3	83.7	22.4	0.3750	78.8	88.2	89.0	88.6	95.7	88.1	6.9
95	Nitrapyrin	0.0375	60.2	63.8	76.5	74.4	90.1	73.0	16.1	0.0750	58.4	57.6	55.2	74.7	81.3	65.4	18.0	0.3750	68.7	69.8	75.7	67.8	76.4	71.7	5.7
96	Mevinphos	0.0250	78.7	79.4	98.3	101.5	116.0	94.8	16.7	0.0500	63.7	43.6	65.2	105.6	104.2	76.4	35.8	0.2500	87.3	102.5	101.8	96.9	99.2	97.5	6.3
97	Chloroneb	0.0125	68.7	72.3	82.2	84.2	99.0	81.3	14.6	0.0250	76.4	69.2	66.5	93.2	96.5	80.3	17.1	0.1250	80.3	86.2	89.3	91.2	94.0	88.2	6.0
98	Tecnazene	0.0500	58.1	59.5	92.1	88.3	96.9	79.0	23.7	0.1000	74.0	84.8	75.1	105.9	95.1	87.0	15.6	0.5000	84.3	91.4	91.0	87.2	91.2	89.0	3.6
99	Heptanophos	0.0375	79.1	60.5	94.3	68.0	110.0	82.4	24.3	0.0750	78.3	65.6	71.9	105.2	112.5	86.7	24.1	0.3750	87.5	95.2	97.3	92.1	97.9	94.0	4.5
100	Hexachlorobenzene	0.0125	61.7	76.7	82.5	88.4	88.5	79.6	14.0	0.0250	64.0	64.3	57.0	89.0	80.7	71.0	18.7	0.1250	68.5	80.9	77.3	86.0	81.7	78.9	8.3
101	Ethoprophos	0.0375	85.4	71.6	83.6	92.1	104.3	87.4	13.7	0.0750	88.5	82.4	74.8	103.1	102.7	90.3	13.8	0.3750	88.8	95.3	97.1	93.2	96.9	94.3	3.6

Table 22a. Continued

No.	Compound	Level /mg/kg	Rice	Whea	Maize	Barley	Oats	Ave.	RSD %	Level /mg/kg	Rice	Whea	Maize	Barley	Oats	Ave.	RSD %	Level /mg/kg	Rice	Whea	Maize	Barley	Oats	Ave.	RSD %
102	cis-diallate	0.0250	95.6	89.6	101.4	96.2	107.7	98.1	6.9	0.0500	85.9	74.6	67.3	96.3	99.3	84.7	16.2	0.2500	86.7	92.2	95.8	96.9	98.5	94.0	5.0
103	Propachlor	0.0375	87.5	80.5	92.5	83.4	90.7	86.9	5.8	0.0750	96.2	74.0	72.0	99.2	101.7	88.6	16.3	0.3750	87.5	93.4	94.1	92.6	94.6	92.4	3.1
104	Trans-diallate	0.0250	87.2	81.8	92.0	92.8	94.3	89.6	5.7	0.0500	87.2	78.6	70.5	100.1	97.1	86.7	14.3	0.2500	87.4	92.3	94.9	92.2	94.6	92.3	3.2
105	Trifluralin	0.0250	79.7	81.1	93.4	89.4	95.1	87.7	8.0	0.0500	81.3	77.1	69.3	104.4	102.7	87.0	18.1	0.2500	85.5	95.3	99.3	87.8	94.4	92.5	6.1
106	Chlorpropham	0.0250	94.9	95.5	98.4	115.8	114.8	103.9	10.1	0.0500	89.8	88.5	79.1	113.3	108.5	95.8	15.1	0.2500	87.6	95.0	97.2	98.8	99.8	95.7	5.1
107	Sulfotep	0.0125	88.4	83.8	97.7	92.1	100.0	92.4	7.2	0.0250	89.8	79.8	73.0	102.8	99.7	89.0	14.3	0.1250	89.5	95.0	95.8	94.2	96.3	94.2	2.9
108	Sulfallate	0.0250	68.1	68.7	85.4	81.4	93.0	79.3	13.6	0.0500	54.0	68.8	63.3	98.0	96.2	76.1	26.2	0.2500	76.0	87.8	89.4	87.0	89.9	86.0	6.7
109	alpha-HCH	0.0125	109.6	80.8	92.6	84.7	112.4	96.0	15.0	0.0250	91.6	75.9	68.7	68.5	102.6	75.4	22.5	0.1250	87.1	101.8	93.5	97.3	95.4	95.0	5.7
110	Terbufos	0.0250	87.8	81.9	94.3	90.9	96.6	90.3	6.4	0.0500	67.9	81.0	74.2	102.5	101.0	85.3	18.4	0.2500	87.7	95.1	95.6	97.4	94.3	94.0	3.9
111	Terbumeton	0.0375	70.6	66.1	84.0	62.9	66.1	69.9	11.9	0.0750	75.7	59.9	59.4	71.7	38.4	61.0	23.9	0.3750	78.3	83.2	85.4	66.8	39.0	70.5	27.0
112	Profluralin	0.0500	83.9	82.8	98.6	90.8	95.6	90.3	7.7	0.1000	82.9	76.0	68.0	104.6	100.3	86.4	18.2	0.5000	84.8	93.1	96.6	86.3	93.7	90.9	5.6
113	Dioxathion	0.0500	86.7	70.8	74.7	81.1	77.0	78.1	7.8	0.1000	104.4	75.3	65.4	85.6	77.7	81.7	17.9	0.5000	89.2	79.8	70.6	78.5	79.2	79.5	8.3
114	Propazine	0.0125	98.5	91.2	99.8	92.1	98.0	95.9	4.1	0.0250	95.5	83.5	73.5	104.0	104.8	92.3	14.7	0.1250	91.3	95.2	98.0	94.2	97.7	95.3	2.9
115	Chlorbufam	0.0500	86.5	87.1	92.8	99.3	102.6	93.6	7.7	0.1000	83.3	81.9	68.7	112.7	107.9	90.9	20.6	0.5000	90.6	97.7	99.0	96.1	101.6	97.0	4.2
116	Dicloran	0.0500	89.5	87.1	94.3	113.5	96.1	96.1	10.8	0.1000	43.8	68.1	53.4	91.8	77.5	66.9	28.5	0.5000	75.7	92.0	90.0	90.3	81.9	86.0	8.1
117	Terbuthylazine	0.0125	101.4	90.8	108.8	69.5	98.1	93.7	16.0	0.0250	90.2	84.5	75.1	85.8	88.9	84.9	7.0	0.1250	90.3	84.7	100.6	87.0	91.3	90.8	6.7
118	Monolinuron	0.0500	99.2	92.6	112.1	80.2	97.3	96.3	12.0	0.1000	91.8	80.7	75.3	101.2	97.9	89.4	12.4	0.5000	89.9	84.3	98.7	81.1	79.3	86.7	9.0
119	Flufenoxuron	0.0375	99.5	62.5	92.5	40.7	49.3	68.9	37.3	0.0750	98.6	63.8	88.0	112.2	116.2	95.7	22.0	0.3750	104.0	71.6	120.2	106.9	98.2	100.2	17.9
120	Cyanohos	0.0250	92.6	87.8	95.4	94.8	103.1	94.7	5.8	0.0500	93.8	79.2	73.6	104.3	104.2	91.0	15.6	0.2500	89.7	92.8	95.4	94.9	97.7	94.1	3.2
121	Chlorprifos-methyl	0.0125	92.0	83.5	96.9	94.0	98.5	93.0	6.3	0.0250	93.0	85.0	74.1	104.8	104.9	92.4	14.3	0.1250	88.9	94.1	97.3	95.6	98.7	94.9	4.0
122	Desmetryn	0.0125	76.0	65.9	84.3	59.1	57.5	72.6	14.9	0.0250	61.0	64.4	59.6	64.2	53.3	61.5	13.2	0.1250	78.3	91.9	88.0	73.3	50.2	76.3	21.5
123	Dimethachloro	0.0375	94.1	89.1	95.7	92.3	100.9	94.4	4.6	0.0750	95.6	84.4	74.1	101.9	100.2	91.2	12.9	0.3750	89.8	94.3	94.6	93.0	96.3	93.6	2.6
124	Alachlor	0.0375	93.2	85.4	95.0	95.5	101.4	94.1	6.1	0.0750	95.3	83.5	73.8	101.9	100.7	91.1	13.2	0.3750	90.2	94.3	96.1	93.6	97.0	94.3	2.8
125	Pirimiphos-methyl	0.0125	76.8	65.7	86.1	75.7	58.2	72.5	14.9	0.0250	75.0	66.9	66.4	79.8	69.8	71.6	8.0	0.1250	81.6	88.9	94.3	79.6	71.2	83.1	10.7
126	Terbutryn	0.0250	77.4	71.8	86.7	70.3	67.0	76.6	14.5	0.0500	64.2	65.2	63.7	76.6	45.6	63.1	17.6	0.2500	79.7	86.7	88.5	75.4	48.8	75.8	21.1
127	Thiobencarb	0.0250	94.0	87.7	96.5	90.3	106.1	94.9	7.5	0.0500	92.7	84.6	72.5	99.1	108.8	91.5	15.1	0.2500	89.8	94.1	94.0	95.4	100.1	94.7	3.9

Table 22a. Continued

No. Compound	Level /mg/kg	Rice	Whea	Maize	Barley	Oats	Ave.	RSD %	Level /mg/kg	Rice	Whea	Maize	Barle	Oats	Ave.	RSD %	Level /mg/kg	Rice	Whea	Maiz	Barle	Oats	Ave.	RSD %
128 Aspon	Not add	0.0	0.0	0.0	0.0	0.0	0.0	#DIV	Not add	0.0	0.0	0.0	0.0	0.0	0.0	#DIV	Not add	0.0	0.0	0.0	0.0	0.0	0.0	#DIV
129 Dicofol	Not add	0.0	0.0	0.0	0.0	0.0	0.0	#DIV	Not add	0.0	0.0	0.0	0.0	0.0	0.0	#DIV	Not add	0.0	0.0	0.0	0.0	0.0	0.0	#DIV
130 Metolachlor	0.0125	95.3	114.9	113.3	117.8	127.9	113.8	10.4	0.0250	97.5	84.7	79.1	111.0	106.2	95.7	14.2	0.1250	90.3	96.9	99.1	97.5	100.5	96.9	4.1
131 Oxychlordane	0.0500	0.0	0.0	0.0	0.0	0.0	0.0	#DIV	0.1000	91.3	79.9	90.0	96.9	112.8	94.2	12.8	0.5000	88.4	90.5	90.8	91.2	97.6	91.7	3.8
132 Methoprene	0.0250	94.1	88.4	101.7	96.6	88.7	93.9	5.9	0.0500	90.3	79.4	70.1	105.1	95.8	88.2	15.6	0.2500	89.4	96.7	96.1	89.5	91.6	92.7	3.8
133 Bromofos	0.0500	100.1	83.8	98.3	86.7	91.0	92.0	7.7	0.1000	95.4	82.6	73.5	105.7	103.5	92.1	15.0	0.5000	88.7	91.6	94.0	93.1	97.3	93.0	3.4
134 Ethofumesate	0.0250	96.1	88.6	109.0	96.0	145.5	107.0	21.2	0.0500	98.4	85.1	80.2	103.3	121.1	97.6	16.5	0.2500	89.3	93.4	96.1	94.0	101.5	94.9	4.7
135 Isopropalin	0.0750	105.9	87.6	100.7	96.1	95.2	97.1	7.0	0.1500	105.1	98.3	104.1	102.6	101.0	102.2	2.6	0.7500	100.4	100.5	102.3	105.0	82.8	98.2	9.0
136 Endosulfan-1	0.0250	92.8	81.4	81.7	101.1	103.4	92.1	11.3	0.0500	104.8	89.6	89.2	102.7	103.4	98.0	8.0	0.2500	88.8	95.2	97.4	93.8	94.7	94.0	3.4
137 Propanil	0.0750	101.5	112.7	107.2	106.9	99.1	105.5	5.0	0.1500	94.5	97.2	77.1	100.6	99.7	93.8	10.3	0.7500	88.7	98.0	96.8	94.2	98.0	95.1	4.1
138 Isofenphos	0.0250	97.1	88.6	91.3	88.0	102.4	93.5	6.6	0.0500	93.4	83.1	69.4	103.7	104.3	90.8	16.3	0.2500	90.4	94.3	97.1	94.1	98.8	95.0	3.4
139 Crufomate	0.0250	92.1	91.4	112.9	99.1	107.3	100.6	9.4	0.0500	81.6	82.7	70.6	107.0	105.5	89.5	17.9	0.2500	87.5	99.8	103.5	92.2	98.6	96.3	6.6
140 Chlorfenvinphos	0.0750	96.3	91.0	103.6	95.9	103.6	98.1	5.6	0.1500	91.5	84.4	74.3	109.1	106.5	93.2	15.8	0.7500	90.1	97.5	100.2	94.7	100.5	96.6	4.5
141 Cis-chlordane	0.0250	93.2	84.4	91.9	91.3	96.6	91.5	4.9	0.0500	96.5	83.0	72.4	99.4	96.2	89.5	12.8	0.2500	90.1	92.5	92.5	92.4	94.4	92.4	1.6
142 Tolyfluanide	0.0375	62.0	64.6	75.9	71.6	73.2	69.5	8.5	0.0750	74.5	87.2	103.3	89.7	81.0	87.2	12.4	0.3750	60.6	70.3	46.6	47.4	45.7	54.1	20.2
143 p,p'-DDE	0.0125	96.9	85.2	91.9	90.7	96.7	92.3	5.3	0.0250	98.6	82.3	70.0	100.5	92.1	88.7	14.2	0.1250	90.7	93.7	91.6	93.6	92.3	92.4	1.4
144 Butachlor	0.0250	92.9	86.5	102.7	94.0	102.1	95.7	7.1	0.0500	94.3	81.9	74.7	101.6	99.9	90.5	13.0	0.2500	91.5	95.6	97.2	94.3	98.4	95.4	2.8
145 Chlozolinate	0.0250	87.6	73.3	90.4	82.8	90.0	84.8	8.4	0.0500	73.2	70.6	66.0	92.2	91.5	78.7	15.6	0.2500	79.4	83.2	86.1	85.9	89.9	84.9	4.6
146 Crotoxyphos	0.0750	92.2	101.0	121.7	96.0	112.2	104.6	11.6	0.1500	83.6	89.9	76.0	127.8	132.6	102.0	25.7	0.7500	91.0	101.1	111.6	114.4	123.5	108.3	11.6
147 Iodofenphos	0.0250	92.2	88.0	105.0	96.8	103.9	97.2	7.6	0.0500	87.1	83.5	70.9	106.2	104.4	90.4	16.4	0.2500	88.1	94.8	100.8	91.5	98.9	94.8	5.5
148 Tetrachlorvinphos	0.0375	98.2	92.4	112.1	97.8	108.1	101.7	8.0	0.0750	89.5	82.5	75.1	113.6	112.4	94.6	18.5	0.3750	90.4	92.7	101.3	96.0	103.6	96.8	5.8
149 Chlorbromuron	0.3000	98.5	71.4	86.2	68.0	70.0	78.8	16.7	0.6000	91.8	76.1	68.6	88.5	80.2	81.1	11.6	3.0000	100.7	69.4	93.6	90.9	77.7	86.5	14.6
150 Profenofos	0.0750	91.2	81.0	100.6	94.9	96.8	92.9	8.0	0.1500	84.3	77.7	70.8	103.3	101.9	87.6	16.6	0.7500	87.9	94.5	97.2	92.1	96.9	93.7	4.1
151 Fluorochloridone	0.1000	0.0	0.0	0.0	0.0	0.0	0.0	#DIV	0.2000	87.5	82.5	89.0	47.5	47.4	70.8	30.2	1.0000	94.8	96.7	102.0	91.8	97.5	96.6	3.9
152 o,p'-DDD	0.0250	100.4	96.8	104.5	95.3	103.6	100.1	4.1	0.0500	126.4	108.3	101.0	101.8	113.9	110.3	9.5	0.2500	99.9	146.8	121.9	142.5	130.5	128.3	14.6
153 Endrin	0.0250	99.5	107.4	104.7	116.9	147.5	115.2	16.6	0.0500	95.7	94.4	77.0	117.3	123.0	101.5	18.4	0.2500	88.9	99.0	98.6	97.9	104.2	97.7	5.7

Table 22a. Continued

No.	Compound	Level /mg/kg	Rice	Whea	Maize	Barley	Oats	Ave.	RSD %	Level /mg/kg	Rice	Whea	Maize	Barley	Oats	Ave.	RSD %	Level /mg/kg	Rice	Whea	Maize	Barley	Oats	Ave.	RSD %
154	Hexaconazole	0.1500	0.0	0.0	0.0	0.0	0.0	0.0	#DIV	0.3000	93.5	84.0	63.2	100.7	105.1	89.3	18.6	1.5000	89.0	98.9	97.9	96.8	100.3	96.6	4.6
155	Chlorfenson	0.1500	104.9	93.3	110.8	87.1	109.7	101.2	10.4	0.3000	88.6	72.6	81.6	93.9	111.2	89.6	16.2	1.5000	87.9	93.9	93.1	103.6	102.0	96.1	6.8
156	o,p'-DDT	0.0250	95.0	86.7	92.6	89.5	101.4	93.0	6.1	0.0500	92.2	77.8	67.4	95.4	92.5	85.0	14.1	0.2500	88.2	87.6	88.3	78.2	85.4	85.5	5.0
157	Paclobutrazol	0.0250	89.7	89.7	98.8	90.9	103.6	94.5	6.7	0.0500	89.0	82.0	72.3	93.9	96.6	86.8	11.3	0.2500	83.4	94.7	95.6	88.4	96.3	91.7	6.1
158	Methoprotryne	0.0375	66.4	74.6	77.0	65.1	45.0	67.6	18.2	0.0750	57.8	57.7	55.9	70.7	37.3	55.9	21.4	0.3750	76.8	85.5	87.8	68.1	45.5	72.7	23.5
159	Chlorpropylate	0.0125	95.0	89.7	100.4	94.9	103.5	96.7	5.6	0.0250	98.8	84.9	75.3	102.8	100.0	92.4	12.8	0.1250	89.5	94.2	96.7	94.4	98.4	94.6	3.5
160	Flamprop-methyl	0.0125	95.7	92.7	98.6	90.7	102.9	96.1	5.0	0.0250	99.4	89.7	76.9	101.3	101.8	93.8	11.4	0.1250	91.7	95.2	96.9	95.3	98.6	95.5	2.7
161	Nitrofen	0.0750	92.2	86.4	106.3	96.1	116.8	99.5	12.1	0.1500	81.5	77.4	64.7	104.2	103.9	86.3	20.1	0.7500	87.2	95.6	98.0	88.2	98.7	93.5	5.8
162	Oxyflurofen	0.0500	88.6	84.6	103.7	94.8	108.2	96.0	10.3	0.1000	84.3	76.5	64.5	103.9	103.1	86.5	19.8	0.5000	88.2	96.3	99.2	87.2	98.7	93.9	6.2
163	Chlorthiophos	0.0375	93.2	85.7	98.0	90.9	94.6	92.5	5.0	0.0750	82.2	82.2	74.5	99.9	100.5	87.9	13.3	0.3750	89.1	93.8	94.4	93.6	95.9	93.4	2.7
164	Endosulfan-2	0.3000	0.0	0.0	0.0	0.0	0.0	0.0	#DIV	0.6000	0.0	0.0	0.0	0.0	0.0	0.0	#DIV	3.0000	90.0	92.3	91.0	98.4	97.2	93.8	4.0
165	Flamprop-isopropyl	0.0125	96.8	89.2	99.8	95.0	103.5	96.9	5.5	0.0250	97.8	84.0	77.0	99.5	98.8	91.4	11.3	0.1250	90.3	95.2	97.7	95.3	99.2	95.6	3.5
166	p,p'-DDT	0.0250	94.2	88.3	93.5	94.0	112.7	96.5	9.7	0.0500	91.1	79.8	65.7	94.4	92.8	84.8	14.3	0.2500	86.2	83.7	85.8	72.4	80.9	81.8	6.9
167	Carbofenothion	0.0250	94.5	98.0	103.1	76.6	89.4	92.3	11.0	0.0500	81.6	88.4	77.8	107.0	112.2	93.4	16.5	0.2500	88.7	96.4	98.0	95.4	102.7	96.3	5.3
168	Benalaxyl	0.0125	97.8	88.4	94.7	80.3	115.9	95.4	13.9	0.0250	103.0	83.5	76.8	74.8	124.7	91.6	20.2	0.1250	90.0	94.9	96.2	89.8	106.2	95.4	7.0
169	Edifenphos	0.0250	96.3	94.0	117.0	89.3	114.2	102.1	12.3	0.0500	82.3	84.2	76.6	107.3	122.1	94.5	20.5	0.2500	100.1	108.6	122.3	94.7	107.7	106.7	9.8
170	Triazophos	0.0375	100.6	124.6	157.9	144.7	107.6	127.1	19.0	0.0750	96.0	101.4	119.6	143.1	161.0	124.2	22.2	0.3750	93.4	103.1	110.4	104.4	116.5	105.6	8.2
171	Cyanofenphos	0.0125	99.6	101.8	136.7	97.9	105.0	108.2	14.9	0.0250	102.7	92.2	93.0	105.9	106.0	100.0	6.9	0.1250	92.9	98.6	105.7	96.9	102.9	99.4	5.1
172	Chlorbenside sulfone	0.0250	98.4	106.2	105.4	80.7	86.0	95.3	12.1	0.0500	103.9	94.1	86.9	127.7	86.2	99.8	17.2	0.2500	89.8	95.1	96.3	101.8	113.6	99.3	9.1
173	Endosulfan-sulfate	0.0375	97.2	90.9	105.9	120.5	101.3	103.1	10.8	0.0750	100.7	87.0	78.9	114.4	101.4	96.5	14.3	0.3750	90.4	88.7	96.2	94.8	103.3	94.7	6.1
174	Bromopropylate	0.0250	95.6	89.2	97.6	93.4	103.6	95.9	5.6	0.0500	97.1	86.0	74.6	102.3	105.0	93.0	13.6	0.2500	91.0	96.2	98.3	95.6	100.2	96.2	3.6
175	Benzoylprop-ethyl	0.0375	95.7	104.4	101.5	81.6	91.1	94.9	9.5	0.0750	103.0	89.4	77.5	105.1	104.8	96.0	12.7	0.3750	92.3	96.5	95.9	94.0	97.0	95.1	2.1
176	Fenpropathrin	0.0250	99.3	114.1	95.7	87.0	87.8	96.8	11.4	0.0500	99.5	98.1	83.5	105.5	97.3	96.8	8.4	0.2500	91.2	99.3	98.5	96.1	119.3	100.9	10.7
177	Leptophos	0.0500	95.1	90.5	99.1	93.4	133.3	102.3	17.2	0.1000	98.1	84.3	74.7	100.4	103.7	92.2	13.3	0.5000	89.2	92.7	96.0	96.4	107.7	96.4	7.2
178	EPN	0.0375	95.5	86.2	107.9	129.5	86.0	101.0	18.1	0.0750	90.3	91.8	78.6	124.7	84.8	94.0	19.1	0.3750	88.5	97.6	113.4	95.4	91.6	97.3	9.9
179	Hexazinone	0.0000	92.0	85.3	93.7	82.8	87.9	88.3	5.2	0.0125	94.5	82.7	72.8	89.2	96.8	87.2	11.1	0.0625	88.1	90.5	92.0	85.5	92.8	89.8	3.3

Table 22a. Continued

No.	Compound	Level /mg/kg	Rice	Whea	Maize	Barley	Oats	Ave.	RSD %	Level /mg/kg	Rice	Whea	Maize	Barley	Oats	Ave.	RSD %	Level /mg/kg	Rice	Whea	Maiz	Barley	Oats	Ave.	RSD %
180	Phosalone	0.0250	108.0	115.0	150.4	124.9	155.4	130.7	16.2	0.0500	98.2	84.5	108.7	126.2	98.2	103.2	15.0	0.2500	91.0	96.2	105.2	103.6	131.2	105.4	14.7
181	Azinphos-methyl	0.1500	97.2	100.6	130.8	110.0	123.1	112.3	12.8	0.3000	94.0	98.1	87.4	132.3	142.1	110.8	22.3	1.5000	92.4	90.6	118.6	104.6	122.8	105.8	13.9
182	Fenarimol	0.0250	98.7	102.5	78.7	90.7	73.1	88.7	14.2	0.0500	96.0	85.9	69.9	108.5	94.2	90.9	15.7	0.2500	87.7	94.1	94.7	91.8	94.7	92.6	3.2
183	Azinphos-ethyl	0.0250	98.9	109.2	98.5	95.1	102.0	100.7	5.3	0.0500	94.3	93.3	155.1	90.4	87.2	104.1	27.6	0.2500	92.3	104.3	134.4	111.0	91.6	106.7	16.4
184	Coumaphos	0.0750	99.6	116.4	112.3	99.3	115.7	108.7	7.9	0.1500	97.2	90.5	80.6	123.5	116.3	101.6	17.6	0.7500	90.9	96.7	105.0	100.5	109.8	100.6	7.3
185	Cyfluthrin	0.1500	93.6	108.1	119.4	67.9	90.5	95.9	20.4	0.3000	104.6	94.3	96.5	115.0	91.8	100.4	9.4	1.5000	90.4	96.8	102.2	102.1	115.8	101.5	9.2
186	Fluvalinate	0.1500	93.7	85.8	101.5	92.0	99.5	94.5	6.6	0.3000	97.8	84.7	77.9	101.6	102.1	92.8	11.7	1.5000	90.1	90.5	101.1	94.7	100.5	95.4	5.5
C group																									
187	Dichlorvos	0.0750	51.3	67.5	102.6	84.9	84.9	78.2	25.0	0.1500	42.2	55.6	98.7	78.4	61.8	67.3	32.4	0.7500	56.8	66.4	71.7	75.0	76.5	69.3	11.5
188	Biphenyl	0.0125	65.3	90.9	102.8	83.0	118.9	92.2	21.9	0.0250	52.1	62.2	91.6	79.6	104.1	77.9	27.1	0.1250	52.2	62.0	63.9	70.9	89.6	67.7	20.6
189	Vernolate	0.0125	61.7	83.5	91.5	85.7	94.8	83.4	15.5	0.0250	60.3	70.8	92.1	85.4	95.9	80.9	18.5	0.1250	66.8	75.9	73.8	79.8	93.6	78.0	12.7
190	3,5-Dichloroaniline	0.0125	53.6	77.1	72.9	41.6	80.1	65.1	25.6	0.0250	51.3	62.6	58.0	68.2	73.9	62.8	13.9	0.1250	59.3	55.7	60.1	68.5	82.6	65.2	16.5
191	Molinate	0.0125	56.6	71.7	84.1	87.9	94.3	78.9	19.0	0.0250	71.6	66.3	84.4	89.7	99.4	82.3	16.4	0.1250	77.1	81.0	81.8	81.8	93.0	83.0	7.2
192	Methacrifos	0.0125	94.3	112.9	104.0	110.4	102.0	104.7	7.0	0.0250	70.1	83.8	112.3	104.6	104.0	95.0	18.4	0.1250	77.1	85.3	84.7	95.2	105.8	89.6	12.4
193	2-Phenylphenol	0.0125	76.6	78.1	68.9	106.6	113.6	88.8	22.5	0.0250	79.0	84.3	104.1	98.4	102.8	93.7	12.2	0.1250	84.4	87.9	88.1	94.3	103.4	91.6	8.2
194	cis-1,2,3,6 Tetrahydrophthalimide	0.0375	55.0	68.2	72.7	91.5	104.9	78.5	25.1	0.0750	65.4	84.1	79.1	84.4	107.4	84.1	18.0	0.3750	65.3	72.2	56.7	68.0	97.8	72.0	21.5
195	Fenobucarb	0.0250	91.0	112.8	115.5	108.9	106.9	107.0	8.9	0.0500	82.3	85.9	102.6	99.7	104.9	95.1	10.8	0.2500	88.3	91.6	91.6	94.4	104.2	94.0	6.5
196	Benfluralin	0.0125	82.6	98.0	106.7	108.2	112.0	101.5	11.5	0.0250	80.8	82.1	108.3	107.4	111.7	98.1	15.6	0.1250	83.9	89.5	91.7	99.9	106.2	94.2	9.4
197	Hexaflumuron	Not add	0.0	0.0	0.0	0.0	0.0	0.0	#DIV	Not add	0.0	0.0	0.0	0.0	0.0	0.0	#DIV	Not add	0.0	0.0	0.0	0.0	0.0	0.0	#DIV
198	Triallate	0.0250	85.3	97.0	102.1	99.3	101.5	97.0	7.1	0.0500	87.2	83.3	101.4	98.4	103.1	94.7	9.4	0.2500	89.6	90.2	90.4	90.9	98.0	91.8	3.8
199	Pyrimethanil	0.0125	76.1	96.8	100.6	72.5	71.4	83.5	16.8	0.0250	78.7	75.3	97.1	80.7	80.8	82.5	10.2	0.1250	85.1	86.2	89.4	91.6	92.2	88.9	3.6
200	gamma-HCH	0.0250	89.5	107.7	120.9	99.7	109.6	105.5	11.1	0.0500	86.4	82.8	104.3	95.8	108.1	95.5	11.5	0.2500	89.1	89.5	90.5	91.3	101.4	92.4	5.6
201	Disulfoton	0.0125	74.8	96.3	101.5	112.7	91.5	95.4	14.6	0.0250	78.1	84.9	100.3	96.1	97.4	91.4	10.3	0.1250	84.9	88.7	89.5	88.8	93.8	89.2	3.5
202	Atrizine	0.0125	88.0	100.3	105.7	95.9	102.6	98.5	7.0	0.0250	88.0	87.9	104.3	97.5	104.9	96.5	8.6	0.1250	92.3	92.0	93.9	93.2	102.2	94.7	4.5
203	Heptachlor	0.0375	81.8	98.3	115.2	98.6	104.3	99.6	12.1	0.0750	86.2	81.5	115.9	98.5	96.0	95.6	13.9	0.3750	87.3	85.4	88.0	91.3	98.4	90.1	5.7

Table 22a. Continued

No.	Compound	Level /mg/kg	Rice	Whea	Maize	Barley	Oats	Ave.	RSD %	Level /mg/kg	Rice	Whea	Maize	Barley	Oats	Ave.	RSD %	Level /mg/kg	Rice	Whea	Maiz	Barley	Oats	Ave.	RSD %
204	Iprobenfos	0.0375	71.4	85.0	85.1	121.4	121.8	97.0	23.9	0.0750	89.9	94.6	95.9	112.4	110.7	100.7	10.1	0.3750	91.1	98.2	86.2	111.9	106.6	98.8	10.7
205	Isazofos	0.0250	95.6	121.8	121.2	144.2	129.7	122.5	14.4	0.0500	91.2	80.6	88.7	102.1	102.1	93.0	9.9	0.2500	92.4	92.5	113.5	92.7	110.4	100.3	10.7
206	Plifenate	0.0500	111.4	104.2	98.5	107.2	113.0	106.8	5.4	0.1000	86.9	90.4	111.0	103.1	109.8	100.2	11.0	0.5000	86.9	88.4	90.3	92.7	102.1	92.1	6.5
207	Transfluthrin	0.0125	86.9	106.5	112.9	116.0	117.5	108.0	11.6	0.0250	87.7	85.9	107.0	108.7	111.4	100.1	12.3	0.1250	92.7	93.7	96.9	97.5	104.6	97.1	4.8
208	Fluchloralin	0.0500	79.7	100.8	110.0	103.8	110.2	100.9	12.4	0.1000	80.0	84.6	105.8	100.3	107.3	95.6	13.1	0.5000	82.4	89.3	90.5	96.8	100.6	91.9	7.7
209	Tolclofos-methyl	0.0125	88.2	103.9	96.8	100.9	102.8	98.5	6.5	0.0250	89.3	87.8	105.6	99.4	103.0	97.0	8.3	0.1250	90.7	91.8	93.5	92.7	100.8	93.9	4.2
210	Propisochlor	0.0125	97.6	121.8	109.2	98.4	98.7	105.1	9.9	0.0250	88.3	88.1	111.0	113.6	108.4	101.9	12.4	0.1250	97.5	99.2	100.0	106.6	108.9	102.4	4.9
211	Methobromuron	0.0750	87.9	100.8	102.6	90.1	103.7	97.0	7.7	0.1500	83.7	83.6	93.4	93.5	91.1	89.1	5.6	0.7500	87.6	86.6	90.5	88.5	97.4	90.1	4.8
212	Metribuzin	0.0375	83.7	95.8	98.8	95.3	108.8	96.5	9.3	0.0750	81.2	85.3	96.8	94.5	95.1	90.6	7.6	0.3750	87.7	87.1	90.3	89.1	96.1	90.1	4.0
213	epsilon-HCH	Not add	0.0	0.0	0.0	0.0	0.0	0.0	#DIV	Not add	0.0	0.0	0.0	0.0	0.0	0.0	#DIV	Not add	0.0	0.0	0.0	0.0	0.0	0.0	#DIV
214	Formothion	Not add	0.0	0.0	0.0	0.0	0.0	0.0	#DIV	Not add	0.0	0.0	0.0	0.0	0.0	0.0	#DIV	Not add	0.0	0.0	0.0	0.0	0.0	0.0	#DIV
215	Diethofencarb	0.0750	79.6	107.9	109.3	97.5	130.3	104.9	17.6	0.1500	85.7	94.1	122.2	104.0	117.6	104.7	14.7	0.7500	89.0	95.3	100.1	96.7	105.5	97.3	6.2
216	Dimepiperate	0.5000	80.5	103.2	115.7	101.0	100.9	100.2	12.6	1.0000	94.1	91.2	117.0	101.3	105.4	101.8	10.0	5.0000	92.2	92.2	99.4	98.3	105.5	97.5	5.7
217	Bioallethrin-1	0.0500	86.0	147.1	107.7	112.9	86.3	108.0	23 2	0.1000	87.6	114.3	107.7	128.8	104.0	108.5	13.8	0.5000	85.4	104.2	91.9	103.8	96.7	96.4	8.3
218	Bioallethrin-2	0.0500	80.8	103.2	118.9	105.5	82.1	98.1	16 7	0.1000	88.2	92.1	121.6	98.7	60.6	92.2	23.8	0.5000	90.3	94.4	97.6	93.5	94.7	94.1	2.8
219	o,p'-DDE	0.0125	86.5	103.0	100.9	99.8	95.4	97.1	6.7	0.0250	89.8	87.7	101.9	96.6	97.8	94.7	6.2	0.1250	91.2	91.8	90.4	91.3	96.0	92.2	2.4
220	Fenson	0.0125	88.0	96.4	125.0	95.3	105.5	102.0	14 0	0.0250	88.2	77.3	116.8	85.4	94.8	92.5	16.1	0.1250	90.9	82.0	83.3	93.4	104.9	90.9	10.2
221	Diphenamid	0.2500	84.1	121.3	102.5	97.3	105.5	102.1	13 2	0.5000	88.5	86.4	102.8	95.4	106.1	95.8	9.0	2.5000	93.7	93.1	94.4	92.6	101.8	95.1	4.0
222	Chlorthion	0.0125	88.2	104.5	115.6	98.0	86.0	98.5	12.4	0.0250	82.7	93.5	121.0	109.7	124.7	106.3	16.9	0.1250	89.1	100.4	102.6	113.1	117.6	104.6	10.7
223	Prallethrin	0.1000	81.0	94.1	101.6	108.1	88.9	94.7	11.2	0.2000	81.1	79.3	107.0	99.8	87.3	90.9	13.3	1.0000	83.3	83.3	85.2	100.9	84.9	87.5	8.6
224	Penconazole	0.0375	78.1	89.7	102.0	81.0	93.4	88.8	10.9	0.0750	72.7	75.8	98.0	81.3	94.9	84.5	13.4	0.3750	85.6	86.0	90.9	90.0	94.9	89.5	4.3
225	Mecarbam	0.0375	129.6	110.1	123.3	114.5	126.6	120.8	6.8	0.0750	107.0	87.9	113.5	131.6	106.3	109.3	14.4	0.3750	91.0	91.3	94.5	94.0	102.1	94.6	4.8
226	Tetraconazole	0.0500	84.8	98.6	105.4	101.3	107.6	99.5	9.0	0.1000	83.2	85.9	104.8	98.1	107.1	95.8	11.3	0.5000	91.0	91.8	95.0	90.4	98.5	93.3	3.6
227	Propaphos	Not add	0.0	0.0	0.0	0.0	0.0	0.0	#DIV	Not add	0.0	0.0	0.0	0.0	0.0	0.0	#DIV	Not add	0.0	0.0	0.0	0.0	0.0	0.0	#DIV
228	Flumetralin	0.0250	90.2	124.7	99.2	128.8	147.4	118.0	19.6	0.0500	88.9	89.0	109.4	101.3	121.2	102.0	13.6	0.2500	84.6	88.9	95.7	100.1	107.5	95.4	9.5
229	Triadimenol	0.0375	83.5	87.3	108.8	111.7	102.3	98.7	12.9	0.0750	78.7	77.5	105.1	100.1	114.5	95.2	17.3	0.3750	89.1	91.1	92.9	91.9	102.0	93.4	5.4

Table 22a. Continued

No.	Compound	Level /mg/kg	Rice	Whea	Maize	Barley	Oats	Ave.	RSD %	Level /mg/kg	Rice	Whea	Maize	Barley	Oats	Ave.	RSD %	Level /mg/kg	Rice	Whea	Maiz	Barley	Oats	Ave.	RSD %
230	Pretilachlor	0.0250	83.9	102.0	110.4	100.5	106.0	100.6	10.0	0.0500	87.3	87.6	106.8	96.7	106.7	97.0	9.9	0.2500	90.9	93.2	95.1	93.7	102.1	95.0	4.4
231	Kresoxim-methyl	0.0125	82.9	97.1	99.3	100.7	98.9	95.8	7.6	0.0250	86.9	87.0	99.7	97.3	102.4	94.7	7.7	0.1250	91.1	93.6	93.9	94.8	101.4	94.9	4.0
232	Fluazifop-butyl	0.0125	92.4	106.3	109.5	101.6	110.9	104.2	7.2	0.0250	87.6	87.8	109.4	100.6	112.9	99.6	11.9	0.1250	91.8	93.5	97.4	95.4	105.4	96.7	5.5
233	Chlorfluazuron	0.0750	94.3	92.0	58.1	108.2	80.0	86.5	21.7	0.1500	99.8	103.1	117.5	136.0	80.0	107.3	19.5	0.7500	71.7	118.4	74.7	80.1	80.5	85.1	22.3
234	Chlorobenzilate	0.0125	85.2	109.9	116.3	100.3	105.7	103.5	11.4	0.0250	86.8	88.9	100.4	96.2	109.3	96.3	9.4	0.1250	90.3	91.7	94.5	93.9	102.8	94.6	5.2
235	Uniconazole	0.0250	78.9	79.1	96.8	85.4	90.6	86.1	8.9	0.0500	84.6	81.9	96.7	91.7	106.6	92.3	10.7	0.2500	86.3	88.2	90.9	93.3	104.1	92.6	7.5
236	Flusilazole	0.0375	86.3	94.4	102.6	88.8	103.1	95.0	8.1	0.0750	77.7	82.5	104.4	88.9	102.3	91.2	13.0	0.3750	88.8	90.8	92.7	89.9	97.4	91.9	3.7
237	Fluorodifen	Not add	0.0	0.0	0.0	0.0	0.0	0.0	#DIV	Not add	0.0	0.0	0.0	0.0	0.0	0.0	#DIV	Not add	0.0	0.0	0.0	0.0	0.0	0.0	#DIV
238	Diniconazole	0.0375	85.3	100.7	105.7	89.0	108.1	97.8	10.4	0.0750	74.9	92.0	101.7	93.7	111.5	94.8	14.3	0.3750	89.2	91.6	92.4	93.9	105.7	94.6	6.8
239	Piperonyl butoxide	0.0125	87.4	102.9	106.8	104.7	118.8	104.1	10.8	0.0250	86.2	86.9	107.1	100.4	114.5	99.0	12.6	0.1250	91.1	92.7	96.5	94.2	107.2	96.3	6.6
240	Propargite	0.0250	80.0	75.0	72.0	70.2	69.1	73.3	6.0	0.0500	82.0	71.0	83.0	88.7	79.0	80.8	8.0	0.2500	83.5	76.8	88.0	76.5	85.3	85.3	6.1
241	Mepronil	0.0125	84.7	112.1	111.7	106.5	104.9	104.0	10.8	0.0250	84.5	95.4	113.1	98.2	109.1	100.1	11.4	0.1250	91.7	97.0	97.0	95.9	102.3	96.8	3.9
242	Dimefuron	0.0500	104.8	127.1	130.8	123.8	130.2	123.3	8.7	0.1000	96.6	118.4	86.0	73.9	90.7	93.1	17.6	0.5000	91.7	93.9	75.0	134.5	82.7	95.6	24.1
243	Diflufenican	0.0125	85.3	98.4	107.7	101.2	108.1	100.1	9.3	0.0250	83.8	86.2	108.7	98.4	110.3	97.5	12.6	0.1250	92.5	92.5	98.9	94.3	106.5	96.9	6.2
244	Phenothrin	0.0125	83.4	108.3	114.8	156.0	122.3	116.9	22.4	0.0250	89.6	91.2	112.2	137.7	124.7	111.1	18.8	0.1250	92.9	94.2	96.5	91.2	101.8	95.3	4.3
245	Fludioxonil	0.0125	88.0	93.5	99.9	84.6	96.3	92.4	6.7	0.0250	84.8	83.8	102.6	84.8	108.1	92.8	12.5	0.1250	92.5	91.3	90.9	82.6	99.8	91.4	6.7
246	Fenoxycarb	0.0750	95.5	122.1	118.2	100.5	100.7	107.4	11.1	0.1500	100.8	91.8	89.1	96.7	101.6	96.0	5.7	0.7500	91.7	76.1	72.5	78.3	87.9	81.3	10.0
247	Sethoxydim	0.1125	46.4	56.4	48.4	42.4	42.3	47.2	12.2	0.2250	56.6	68.1	54.5	56.3	40.7	55.2	17.7	1.1250	68.3	58.4	56.4	54.1	52.7	58.0	10.7
248	Anilofos	0.0500	87.1	114.6	124.7	120.0	120.5	113.4	13.3	0.1000	87.4	91.2	115.8	110.8	125.3	106.1	15.3	0.5000	94.0	95.2	103.9	102.4	118.6	102.8	9.6
249	Acrinathrin	0.0250	79.1	81.6	102.2	110.2	98.0	94.2	14.3	0.0500	84.0	66.9	117.1	108.4	102.6	95.8	21.1	0.2500	91.1	76.2	107.0	88.7	109.6	94.5	14.6
250	lambda-cyhalothrin	0.0125	81.8	85.3	105.2	115.2	134.0	104.3	20.7	0.0250	95.4	79.7	126.2	96.5	139.7	107.5	22.9	0.1250	91.7	74.0	94.2	87.0	119.3	93.2	17.7
251	Mefenacet	0.0375	89.3	103.8	118.2	111.1	123.6	109.2	12.2	0.0750	88.8	91.5	109.8	105.8	118.4	102.8	12.1	0.3750	93.7	96.8	100.3	99.3	114.0	100.8	7.7
252	Permethrin	0.0250	89.1	103.7	111.5	103.2	105.4	102.6	8.0	0.0500	88.0	86.8	107.4	101.7	109.3	98.6	10.8	0.2500	93.4	94.7	98.2	97.1	109.4	98.6	6.4
253	Pyridaben	0.0125	0.0	0.0	0.0	0.0	0.0	0.0	#DIV	0.0250	0.0	0.0	0.0	0.0	0.0	0.0	#DIV	0.1250	88.8	100.9	86.5	83.3	66.8	85.2	14.4
254	Fluoroglycofen-ethyl	0.1500	88.0	119.2	122.6	121.2	169.6	124.1	23.5	0.3000	88.4	89.0	124.1	88.9	134.4	104.9	21.4	1.5000	94.6	98.7	112.1	109.3	126.2	108.2	11.4
255	Bitertanol	0.0375	89.7	108.1	116.1	106.2	120.6	108.1	11.0	0.0750	81.1	88.9	104.1	100.6	118.1	98.6	14.5	0.3750	92.9	95.7	100.2	96.5	111.3	99.3	7.2

Table 22a. Continued

No.	Compound	Level /mg/kg	Rice	Whea	Maize	Barley	Oats	Ave.	RSD %	Level /mg/kg	Rice	Whea	Maize	Barley	Oats	Ave.	RSD %	Level /mg/kg	Rice	Whea	Maize	Barley	Oats	Ave.	RSD %
256	Etofenprox	0.0125	91.0	103.7	115.9	102.9	118.8	106.5	10.5	0.0250	88.2	87.6	109.2	101.4	115.9	100.5	12.5	0.1250	93.3	94.4	98.5	96.5	112.3	99.0	7.8
257	Cycloxydim	0.1500	45.6	47.8	44.8	48.9	59.3	49.3	11.9	0.3000	61.1	60.1	53.0	34.5	44.4	50.6	22.2	1.5000	67.5	64.5	56.5	58.3	61.4	61.6	7.3
258	alpha-cypermethrin	0.0250	80.9	82.6	69.0	95.9	82.0	82.1	11.6	0.0500	82.4	68.6	87.5	88.8	96.3	84.7	12.1	0.2500	86.9	75.1	83.1	85.1	95.4	85.1	8.6
259	Flucythrinate	Not add	0.0	0.0	0.0	0.0	0.0	0.0	#DIV	Not add	0.0	0.0	0.0	0.0	0.0	0.0	#DIV	Not add	0.0	0.0	0.0	0.0	0.0	0.0	#DI\
260	Esfenvalerate	0.0500	124.6	123.1	104.1	97.0	95.6	108.9	12.9	0.1000	70.1	52.7	79.9	71.2	92.7	73.3	20.0	0.5000	87.6	59.7	87.9	83.3	91.4	82.0	15.6
261	Difenconazole	0.0750	61.4	86.4	79.9	99.9	105.4	86.6	20.1	0.1500	70.1	79.5	83.0	91.4	124.2	89.6	23.2	0.7500	88.0	91.5	95.2	89.1	106.2	94.0	7.9
262	Flumioxazin	Not add	0.0	0.0	0.0	0.0	0.0	0.0	#DIV	Not add	0.0	0.0	0.0	0.0	0.0	0.0	#DIV	Not add	0.0	0.0	0.0	0.0	0.0	0.0	#DI\
263	Flumiclorac-pentyl	0.0250	83.8	104.3	116.3	102.0	111.9	103.7	12.1	0.0500	84.0	99.5	126.5	133.4	96.0	107.9	19.5	0.2500	93.3	98.7	104.9	119.0	135.0	110.2	15.3
D group																									
264	Dimefox	0.0375	50.2	61.4	57.7	67.8	81.2	63.7	18.4	0.0750	51.7	62.1	58.7	51.0	60.9	56.9	9.1	0.3750	71.0	76.1	92.4	59.1	61.5	72.0	18.5
265	Disulfoton-sulfoxide	0.0250	92.8	96.3	95.1	91.3	93.7	93.8	2.[illegible]	0.0500	94.3	106.1	103.8	88.2	103.0	99.1	7.6	0.2500	96.4	95.9	101.3	94.0	93.6	96.2	3.2
266	Pentachlorobenzene	0.0125	61.8	75.9	81.5	71.2	75.1	73.1	10.0	0.0250	69.4	77.8	79.2	72.6	83.0	76.4	7.1	0.1250	82.3	84.8	97.4	79.4	78.6	84.5	9.0
267	Tri-iso-butyl Phosphate	Not add	0.0	0.0	0.0	0.0	0.0	0.0	#DIV	Not add	0.0	0.0	0.0	0.0	0.0	0.0	#DIV	Not add	0.0	0.0	0.0	0.0	0.0	0.0	#DI\
268	Crimidine	0.0125	65.7	79.5	99.0	95.8	119.2	91.8	22 1	0.0250	72.9	87.4	96.2	69.5	57.7	76.7	19.8	0.1250	77.6	82.5	101.5	70.7	42.7	75.0	28.5
269	BDMC-1	0.0500	0.0	0.0	0.0	0.0	0.0	0.0	#DIV	0.1000	0.0	0.0	0.0	0.0	0.0	0.0	#DIV	0.5000	91.7	101.4	78.8	85.7	114.9	94.5	14.9
270	Chlorfenprop-methyl	0.0125	82.9	84.2	83.2	76.0	98.7	85.0	9.8	0.0250	87.8	89.2	104.7	84.1	84.5	90.1	9.4	0.1250	90.5	122.3	112.5	97.4	93.9	103.3	13.1
271	Thionazin	0.0125	98.0	127.4	92.6	93.2	86.1	99.5	16.3	0.0250	84.9	119.2	153.5	130.1	80.8	113.7	27.1	0.1250	90.6	102.6	121.7	98.8	88.7	100.5	13.1
272	2,3,5,6 -Tetrachloroaniline	0.0125	77.1	86.6	92.2	82.0	80.2	83.6	7.1	0.0250	83.6	97.3	94.5	84.3	94.0	90.8	7.0	0.1250	92.9	90.9	105.4	96.3	90.7	95.2	6.4
273	Tri-n-butyl phosphate	0.0250	71.7	102.6	93.0	102.6	112.7	96.5	16.1	0.0500	87.5	114.5	117.2	86.8	105.5	102.3	14.2	0.2500	91.6	105.9	114.9	95.8	95.3	100.7	9.5
274	2,3,4,5 -Tetrachloroanisole	0.0125	82.8	88.9	93.7	84.2	86.2	87.2	4.9	0.0250	88.7	104.2	101.8	84.3	97.5	95.3	8.9	0.1250	93.0	101.8	115.4	92.5	90.5	98.6	10.5
275	Pentachloroanisole	0.0125	86.3	88.1	94.1	89.3	83.9	88.3	4.3	0.0250	84.6	95.8	90.9	85.3	93.7	90.0	5.5	0.1250	91.5	91.0	103.9	88.3	87.7	92.5	7.1
276	Tebutam	0.0250	90.2	95.0	99.3	88.3	86.6	91 9	5.6	0.0500	90.3	99.5	99.9	86.5	97.6	94.7	6.4	0.2500	94.9	94.2	109.1	95.1	94.1	97.5	6.7
277	Dioxabenzofos	0.1250	86.6	99.1	100.7	100.9	89.5	95 4	7.1	0.2500	87.3	102.6	104.0	90.4	101.0	97.1	7.9	1.2500	95.1	96.4	110.4	95.2	94.3	98.3	6.9

Table 22a. Continued

No.	Compound	Level /mg/kg	Rice	Whea	Maize	Barley	Oats	Ave.	RSD %	Level /mg/kg	Rice	Whea	Maize	Barle	Oats	Ave.	RSD %	Level /mg/kg	Rice	Whea	Maiz	Barle	Oats	Ave.	RSD %
278	Methabenzthiazuron	0.1250	87.6	106.8	100.0	103.6	121.3	103.9	11.7	0.2500	72.2	95.9	98.7	82.6	93.5	88.6	12.5	1.2500	90.9	99.3	104.6	92.7	94.9	96.5	5.7
279	Simeton	0.0250	67.0	81.4	88.4	88.6	96.5	84.4	13.1	0.0500	58.1	75.1	83.1	53.8	66.0	67.2	17.9	0.2500	63.3	68.6	84.9	64.5	65.4	69.3	12.9
280	Atratone	0.0125	52.9	73.0	75.7	91.0	103.4	79.2	24.2	0.0250	58.0	80.7	87.2	57.0	68.1	70.2	19.2	0.1250	64.9	70.8	88.9	63.8	68.0	71.3	14.4
281	Desisopropyl-atrazin	0.1000	97.2	90.7	73.6	61.3	79.3	80.4	17.6	0.2000	76.0	82.6	64.1	66.2	96.1	77.0	16.9	1.0000	76.0	63.4	61.0	69.6	85.1	71.0	13.8
282	Terbufos sulfone	0.0125	87.6	102.0	99.8	93.1	92.7	95.0	6.1	0.0250	81.9	104.9	103.1	87.4	96.1	94.7	10.4	0.1250	92.2	97.6	110.0	94.1	90.9	96.9	8.0
283	Tefluthrin	0.0125	89.8	97.4	110.2	95.6	103.4	99.3	7.8	0.0250	88.1	101.8	96.5	88.9	101.3	95.3	6.9	0.1250	96.3	97.0	123.0	97.7	96.5	102.1	11.4
284	Bromocylen	0.0125	80.0	93.4	90.0	85.0	88.1	87.3	5.8	0.0250	86.1	97.5	95.3	86.9	96.4	92.4	5.9	0.1250	94.7	95.6	107.1	94.1	89.6	96.2	6.8
285	Trietazine	0.0125	92.5	94.8	97.1	95.1	94.0	94.7	1.8	0.0250	92.5	104.9	105.2	89.5	98.8	98.2	7.3	0.1250	95.7	98.7	110.6	97.3	94.7	99.4	6.5
286	Etrimfos oxon	0.0125	89.1	96.7	102.2	89.8	86.4	92.8	7.0	0.0250	88.6	102.6	102.0	89.0	99.3	96.3	7.2	0.1250	93.8	95.5	108.1	95.7	94.2	97.5	6.1
287	Cycluron	0.0375	30.0	65.6	35.8	69.0	70.2	48.1	21.0	0.0750	62.6	39.2	53.5	39.7	35.3	42.1	25.1	0.3750	63.0	61.7	62.0	48.7	46.1	60.3	19.8
288	2,6-dichlorobenzamide	0.0250	85.9	94.9	69.5	76.6	119.0	89.2	21.5	0.0500	90.4	94.0	64.4	64.7	101.0	82.9	20.7	0.2500	79.8	80.5	59.4	72.2	82.7	74.9	12.7
289	DE-PCB 28	0.0125	87.2	91.5	110.4	46.0	81.8	83.4	28.2	0.0250	91.1	99.2	91.3	88.7	74.2	88.9	10.3	0.1250	83.4	84.6	103.6	93.7	94.0	91.9	8.9
290	DE-PCB 31	0.0125	85.0	106.9	104.1	75.9	133.9	101.2	22.2	0.0250	92.6	128.4	121.4	79.1	108.9	106.1	19.2	0.1250	94.8	89.9	96.6	88.0	86.4	91.1	4.8
291	Desethyl-sebuthylazine	0.0250	91.6	97.8	90.9	91.5	88.7	92.1	3.7	0.0500	90.1	104.2	98.6	83.8	90.0	93.3	8.6	0.2500	86.3	83.2	93.3	91.3	91.9	89.2	4.8
292	2,3,4,5 -Tetrachloroaniline	0.0250	93.8	93.8	100.3	86.7	84.9	91.9	6.8	0.0500	89.3	101.0	102.6	86.7	96.3	95.2	7.4	0.2500	89.3	89.1	104.5	91.5	88.9	92.7	7.2
293	Musk ambrette	Not add	0.0	0.0	0.0	0.0	0.0	0.0	#DIV	Not add	0.0	0.0	0.0	0.0	0.0	0.0	#DIV	Not add	0.0	0.0	0.0	0.0	0.0	0.0	#DIV
294	Musk xylene	Not add	0.0	0.0	0.0	0.0	0.0	0.0	#DIV	Not add	0.0	0.0	0.0	0.0	0.0	0.0	#DIV	Not add	0.0	0.0	0.0	0.0	0.0	0.0	#DIV
295	Pentachloroaniline	0.0125	106.9	97.0	98.5	97.1	89.8	97.9	6.2	0.0250	90.0	102.6	100.4	90.4	99.2	96.5	6.1	0.1250	96.5	92.5	104.7	94.6	90.4	95.7	5.7
296	Aziprotryne	0.1000	87.2	120.9	135.3	123.9	103.8	114.2	16.5	0.2000	98.2	143.9	142.4	116.3	156.9	131.5	18.1	1.0000	99.1	115.7	138.0	117.6	123.2	118.7	11.8
297	Sebutylazine	0.0125	94.9	99.7	99.8	91.5	90.3	95.3	4.6	0.0250	92.1	104.6	104.3	88.6	86.6	95.2	9.1	0.1250	95.5	94.7	106.8	97.6	92.9	97.5	5.6
298	Isocarbamid	0.0625	91.3	97.0	83.4	83.1	99.9	90.9	8.4	0.1250	86.5	105.3	87.9	79.2	101.4	92.1	11.8	0.6250	87.4	81.9	81.7	81.2	89.9	84.4	4.7
299	DE-PCB 52	0.0125	84.4	89.0	100.0	100.0	109.1	96.5	10.2	0.0250	91.1	103.6	98.3	93.7	120.0	101.3	11.3	0.1250	92.7	89.1	104.1	97.7	95.1	95.7	5.9
300	Musk moskene	Not add	0.0	0.0	0.0	0.0	0.0	0.0	#DIV	Not add	0.0	0.0	0.0	0.0	0.0	0.0	#DIV	Not add	0.0	0.0	0.0	0.0	0.0	0.0	#DIV

Table 22a. Continued

No.	Compound	Level /mg/kg	Rice	Whea	Maize	Barley	Oats	Ave.	RSD %	Level /mg/kg	Rice	Whea	Maize	Barley	Oats	Ave.	RSD %	Level /mg/kg	Rice	Whea	Maiz	Barley	Oats	Ave.	RSD %
301	Prosulfocarb	0.0125	91.5	96.4	96.5	92.5	93.3	94.0	2.4	0.0250	85.4	110.8	107.2	90.6	100.1	98.8	10.9	0.1250	90.4	94.1	105.9	97.3	94.9	96.5	6.0
302	Dimethenamid	0.0125	93.5	98.4	99.2	93.2	99.6	96.8	3.3	0.0250	89.8	103.9	100.2	90.0	101.9	97.1	7.0	0.1250	96.5	97.5	109.9	97.3	95.0	99.2	6.1
303	Fenchlorphos oxon	0.0250	94.9	99.2	102.6	95.7	90.3	96.5	4.3	0.0500	86.3	102.5	100.0	91.9	99.4	96.0	7.0	0.2500	95.5	95.9	110.2	95.7	93.6	98.2	6.9
304	BDMC-2	0.0250	89.7	95.5	99.3	108.2	102.9	99.1	7.1	0.0500	80.4	93.1	120.6	94.7	100.7	97.9	15.0	0.2500	93.8	86.5	117.8	98.3	87.8	96.8	13.1
305	Paraoxon-methyl	0.1000	0.0	0.0	0.0	0.0	0.0	0.0	#DIV	0.2000	0.0	0.0	0.0	0.0	0.0	0.0	#DIV	1.0000	92.1	114.3	115.5	100.0	104.6	105.3	9.3
306	Monalide	0.0250	90.7	103.6	103.7	94.7	103.9	99.3	6.2	0.0500	97.0	106.2	103.6	88.9	119.7	103.0	11.1	0.2500	98.9	96.3	109.6	96.8	95.4	99.4	5.9
307	Musk tibeten	Not add	0.0	0.0	0.0	0.0	0.0	0.0	#DIV	Not add	0.0	0.0	0.0	0.0	0.0	0.0	#DIV	Not add	0.0	0.0	0.0	0.0	0.0	0.0	#DIV
308	Isobenzan	0.0125	0.0	0.0	0.0	0.0	0.0	0.0	#DIV	0.0250	92.2	100.6	95.3	84.5	94.3	93.4	6.2	0.1250	94.1	93.6	105.8	93.8	88.6	95.2	6.7
309	Octachlorostyrene	0.0125	93.1	91.1	89.8	83.3	78.2	87.1	7.[illegible]	0.0250	90.9	99.9	87.7	82.0	86.5	89.4	7.5	0.1250	94.1	89.2	97.9	90.6	81.0	90.6	7.0
310	Pyrimitate	Not add	0.0	0.0	0.0	0.0	0.0	0.0	#DIV	Not add	0.0	0.0	0.0	0.0	0.0	0.0	#DIV	Not add	0.0	0.0	0.0	0.0	0.0	0.0	#DIV
311	Isodrin	0.0125	79.4	50.3	74.9	108.4	119.7	86.6	32.1	0.0250	116.8	101.6	127.8	85.0	118.6	110.0	15.3	0.1250	106.2	96.7	110.8	95.5	95.4	100.9	7.1
312	Isomethiozin	0.0250	92.2	84.4	83.3	93.8	85.2	87.8	5.5	0.0500	77.8	94.5	93.8	83.3	90.4	88.0	8.2	0.2500	83.8	86.2	98.5	89.4	86.2	88.8	6.5
313	Trichloronat	0.0125	85.9	88.1	80.0	92.6	73.7	84.0	8.8	0.0250	88.4	97.2	85.1	87.9	88.8	89.5	5.1	0.1250	95.6	87.6	95.8	92.7	84.3	91.2	5.6
314	Dacthal	0.0125	93.2	96.2	101.4	92.5	85.4	93.7	6.2	0.0250	94.1	102.8	105.2	90.5	100.9	98.7	6.3	0.1250	101.9	102.8	117.9	98.7	96.2	103.5	8.2
315	4,4-Dichloro benzophenone	0.0125	84.8	96.6	100.2	105.4	104.4	98.3	8.5	0.0250	86.0	107.6	107.9	87.9	95.8	97.0	10.7	0.1250	94.7	102.0	114.3	94.9	93.4	99.8	8.8
316	Nitrothal-isopropyl	0.0250	85.3	104.0	102.4	109.2	105.7	101.3	9.2	0.0500	79.9	112.2	121.9	84.9	108.0	101.4	17.9	0.2500	92.8	108.5	116.9	91.0	93.5	100.5	11.4
317	Musk ketone	Not add	0.0	0.0	0.0	0.0	0.0	0.0	#DIV	Not add	0.0	0.0	0.0	0.0	0.0	0.0	#DIV	Not add	0.0	0.0	0.0	0.0	0.0	0.0	#DIV
318	Rabenzazole	0.0125	88.0	101.9	87.0	71.6	139.5	97.6	26.4	0.0250	83.2	110.5	101.3	76.7	116.6	97.7	17.6	0.1250	90.2	95.7	85.0	86.8	90.9	89.7	4.6
319	Cyprodinil	0.0125	72.3	81.3	88.9	95.0	98.7	87.2	12.2	0.0250	72.4	85.9	89.4	64.7	60.5	74.6	17.1	0.1250	70.6	69.6	85.9	77.3	60.2	72.7	13.1
320	Isofenphos oxon	0.0250	60.1	67.2	56.8	88.6	83.2	71.2	19.8	0.0500	48.1	69.5	56.1	36.7	45.8	51.2	24.0	0.2500	42.1	45.9	56.3	40.2	33.5	43.6	19.3
321	Dicapthon	0.0625	85.0	114.4	112.3	103.8	109.7	105.1	11.3	0.1250	78.3	119.6	133.0	88.0	107.5	105.3	21.3	0.6250	89.7	110.4	122.5	93.8	97.6	102.8	13.1
322	DE-PCB 101	0.0125	94.4	92.4	91.5	92.5	86.1	91.4	3.4	0.0250	92.5	99.5	93.2	88.0	94.0	93.4	4.4	0.1250	95.0	91.3	101.7	94.2	86.9	93.8	5.8
323	MCPA-butoxyethyl Ester	0.0125	87.9	100.0	101.1	105.7	99.9	98.9	6.7	0.0250	86.8	108.0	108.2	91.3	102.4	99.3	9.9	0.1250	96.5	101.8	115.7	97.3	96.8	101.6	8.0
324	Isocarbophos	Not add	0.0	0.0	0.0	0.0	0.0	0.0	#DIV	Not add	0.0	0.0	0.0	0.0	0.0	0.0	#DIV	Not add	0.0	0.0	0.0	0.0	0.0	0.0	#DIV

Table 22a. Continued

No.	Compound	Level /mg/kg	Rice	Whea	Maize	Barley	Oats	Ave.	RSD %	Level /mg/kg	Rice	Whea	Maize	Barley	Oats	Ave.	RSD %	Level /mg/kg	Rice	Whea	Maiz	Barley	Oats	Ave.	RSD %
325	Phorate sulfone	0.0125	90.9	102.8	99.8	95.0	92.0	96.1	5.3	0.0250	88.8	107.5	111.0	89.7	100.4	99.5	10.1	0.1250	94.6	99.4	112.7	97.0	96.0	99.9	7.4
326	Chlorfenethol	0.0125	91.9	95.9	100.6	89.1	94.2	94.4	4.6	0.0250	91.0	109.8	111.3	90.8	100.8	100.8	9.8	0.1250	96.9	96.9	111.2	95.4	94.0	98.9	7.1
327	trans-nonachlor	0.0125	91.7	92.8	96.6	90.9	85.7	91.5	4.3	0.0250	90.9	103.3	100.8	88.1	98.7	96.4	6.8	0.1250	95.4	93.7	106.4	94.7	90.3	96.1	6.3
328	Dinobuton	Not add	0.0	0.0	0.0	0.0	0.0	0.0	#DIV	Not add	0.0	0.0	0.0	0.0	0.0	0.0	#DIV	Not add	0.0	0.0	0.0	0.0	0.0	0.0	#DI\
329	DEF	0.0250	69.5	78.8	76.2	91.4	114.4	86.1	20.6	0.0500	99.3	131.7	142.5	68.6	103.2	109.1	26.7	0.2500	96.7	103.3	113.5	93.9	96.6	100.8	7.8
330	Flurochloridone	0.0250	92.9	97.3	101.6	93.6	94.6	96.0	3.7	0.0500	87.9	103.4	110.8	85.4	109.9	99.5	12.2	0.2500	94.6	96.0	109.9	95.7	95.8	98.4	6.6
331	Bromfenvinfos	0.0125	91.0	93.4	96.7	99.4	103.7	96.8	5.1	0.0250	84.6	111.1	120.9	84.4	117.2	103.6	17.2	0.1250	93.4	102.9	113.1	95.4	97.4	100.4	7.9
332	Perthane	0.0125	90.4	96.9	99.4	90.2	83.7	92.1	6.7	0.0250	90.8	108.1	104.8	84.8	111.7	100.0	11.6	0.1250	95.3	97.0	109.6	94.6	94.3	98.2	6.6
333	DE-PCB 118	0.0125	95.3	91.0	93.1	81.4	83.2	88.8	6.9	0.0250	93.9	104.4	93.3	90.0	107.1	97.8	7.7	0.1250	95.6	92.5	99.4	92.8	83.2	92.7	6.5
334	4,4-Dibromobenzo-phenone	0.0250	75.7	78.6	79.0	91.0	89.0	82.6	8.3	0.0500	84.9	104.9	105.8	85.2	108.4	97.8	12.0	0.2500	98.2	98.0	108.3	93.1	90.8	97.7	6.9
335	Flutriafol	0.0250	83.4	89.5	85.3	90.7	95.6	88.9	5.4	0.0500	93.0	116.5	109.2	84.2	108.1	102.2	12.9	0.2500	92.4	94.2	97.4	91.9	89.2	93.0	3.2
336	Mephosfolan	0.0250	81.7	103.0	95.0	105.4	127.8	102.6	16.4	0.0500	78.3	120.2	120.0	86.7	135.8	108.2	22.6	0.2500	93.7	107.1	104.9	89.3	96.6	98.3	7.6
337	Athidathion	0.0250	0.0	0.0	0.0	0.0	0.0	0.0	#DIV	0.0500	94.7	90.4	79.9	81.2	76.2	84.5	9.1	0.2500	97.8	87.7	95.9	91.9	82.2	91.1	6.9
338	DE-PCB 153	0.0125	93.6	88.7	85.5	89.9	78.5	87.2	6.5	0.0250	92.9	99.9	87.4	83.5	93.9	91.5	6.9	0.1250	95.0	90.9	97.7	91.0	82.0	91.3	6.5
339	Diclobutrazole	0.0500	88.2	96.5	93.1	95.4	99.1	94.5	4.3	0.1000	84.5	110.2	110.7	81.9	92.7	96.0	14.4	0.5000	89.5	96.3	105.3	89.3	85.0	93.1	8.5
340	Disulfoton sulfone	0.0250	91.0	82.5	90.4	83.6	80.8	85.7	5.5	0.0500	93.7	92.1	105.6	81.4	102.5	95.1	10.0	0.2500	95.1	87.7	107.4	94.3	94.7	95.8	7.5
341	Hexythiazox	0.1000	88.9	99.9	96.2	98.6	96.8	96.1	4.4	0.2000	90.2	117.0	116.6	88.1	109.2	104.2	13.5	1.0000	95.0	103.1	113.5	97.2	97.6	101.3	7.4
342	DE-PCB 138	0.0125	94.3	90.9	89.0	88.2	78.0	88.1	6.9	0.0250	94.4	102.3	90.3	84.3	100.7	94.4	7.9	0.1250	95.4	92.8	99.9	92.5	83.8	92.9	6.3
343	Cyproconazole	0.0125	63.9	90.7	95.3	87.2	108.6	89.1	18.3	0.0250	124.3	118.8	105.2	77.0	102.3	105.5	17.4	0.1250	90.3	97.9	105.4	87.0	85.2	93.1	9.0
344	Clodinafop-propargy	0.0250	74.6	100.0	103.1	103.4	129.4	102.1	19.0	0.0500	84.7	116.1	145.0	80.7	132.9	111.9	25.6	0.2500	89.2	110.2	117.2	90.9	102.8	102.1	11.9
345	Fenthion sulfoxide	0.0500	118.3	94.2	108.1	98.7	82.7	100.4	13.5	0.1000	92.9	86.6	75.9	90.3	70.3	83.2	11.7	0.5000	109.1	73.5	93.2	105.9	83.8	93.1	16.0
346	Fluotrimazole	0.0125	85.9	91.5	100.5	74.6	83.7	87.2	11.0	0.0250	91.2	105.6	109.7	85.0	99.4	98.2	10.3	0.1250	92.7	92.9	113.0	89.2	90.6	95.7	10.3
347	Fluroxypr-1 -methylheptyl ester	0.0250	0.0	0.0	0.0	0.0	0.0	0.0	#DIV	0.0500	98.5	121.1	97.4	86.9	108.3	102.4	12.6	0.2500	96.1	102.9	115.3	96.9	100.6	102.4	7.5
348	Fenthion sulfone	0.0500	91.7	100.3	101.1	99.9	98.5	98.3	3.9	0.1000	95.3	114.7	120.1	88.2	115.7	106.8	13.2	0.5000	98.5	96.4	112.3	99.5	105.0	102.3	6.3

Table 22a. Continued

No.	Compound	Level /mg/kg	Rice	Whea	Maize	Barley	Oats	Ave.	RSD %	Level /mg/kg	Rice	Whea	Maize	Barley	Oats	Ave.	RSD %	Level /mg/kg	Rice	Whea	Maize	Barley	Oats	Ave.	RSD %
349	Triphenyl phosphate	0.0125	95.5	98.1	100.0	95.5	117.3	101.3	9.0	0.0250	98.0	112.9	111.2	87.0	120.5	105.9	12.6	0.1250	96.8	98.5	110.2	96.9	101.3	100.7	5.5
350	Metamitron	0.5000	95.0	116.1	80.4	110.2	91.2	98.6	14.7	1.0000	98.5	136.9	98.1	113.3	106.7	110.7	14.4	5.0000	53.4	74.4	53.0	84.9	99.6	73.1	27.7
351	DE-PCB 180	0.0125	96.1	88.2	88.6	92.1	62.9	85.6	15.3	0.0250	95.9	103.7	88.6	83.1	91.8	92.6	8.4	0.1250	92.9	89.0	94.4	90.9	79.7	89.4	6.4
352	Tebufenpyrad	0.0125	92.8	95.5	100.8	92.5	93.0	94.9	3.7	0.0250	93.1	107.1	110.7	86.7	110.6	101.6	10.9	0.1250	95.3	99.2	113.8	94.7	98.0	100.2	7.8
353	Lenacil	0.0125	102.0	99.2	93.0	96.7	88.6	95.9	5.5	0.0250	96.0	119.5	118.6	87.3	106.3	105.6	13.3	0.1250	93.2	96.5	102.0	95.0	101.4	97.6	4.0
354	Bromuconazole-1	0.0250	81.7	52.0	60.8	62.5	57.1	62.8	18.0	0.0500	101.4	94.3	73.2	44.3	78.8	78.4	28.3	0.2500	90.0	98.0	115.9	71.0	65.5	88.1	23.2
355	Desbrom- leptophos	0.0125	94.8	106.9	98.9	90.8	86.8	95.6	8.1	0.0250	102.4	109.4	125.4	89.2	136.4	112.6	16.6	0.1250	96.0	99.6	112.5	97.3	101.2	101.3	6.4
356	Bromuconazole-2	0.0250	88.5	103.9	107.8	108.2	58.3	93.4	22.7	0.0500	92.9	118.0	121.2	91.0	91.4	102.9	14.9	0.2500	92.1	99.3	112.1	97.0	100.7	100.2	7.4
357	Nitralin	0.1250	85.7	86.2	79.6	94.2	88.8	86.9	6.1	0.2500	78.8	116.8	133.8	75.6	72.6	95.5	29.2	1.2500	82.9	119.3	126.4	88.9	87.4	101.0	20.0
358	Fenamiphos sulfoxid	0.0500	68.5	105.6	68.5	52.1	121.8	83.3	34.9	0.1000	56.9	122.8	103.3	57.0	137.8	101.6	23.8	0.5000	93.5	107.1	67.0	65.9	86.1	79.9	19.9
359	Fenamiphos sulfone	0.0500	85.2	103.4	74.5	87.1	118.5	93.8	18.4	0.1000	87.4	119.6	99.8	87.4	130.3	104.9	18.4	0.5000	109.3	108.8	90.7	86.5	108.4	100.7	11.1
360	Fenpiclonil	0.0500	91.7	111.4	81.4	110.3	65.5	92.0	21.2	0.1000	103.6	129.6	103.0	88.0	88.1	102.5	16.6	0.5000	100.7	101.6	103.8	93.5	121.9	104.3	10.1
361	Fluquinconazole	0.0125	140.1	96.8	103.4	98.0	80.3	103.7	21.3	0.0250	81.4	93.2	96.9	91.3	111.1	94.8	11.4	0.1250	94.7	95.6	112.0	98.3	103.8	100.9	7.1
362	Fenbuconazole	0.0250	91.7	119.7	99.2	67.3	67.5	89.1	25.0	0.0500	95.2	134.0	109.7	92.4	117.3	109.7	15.5	0.2500	84.0	84.6	85.9	98.4	103.5	91.3	9.9

Table 22b. Recovery and precision data for 124 pesticide residues (E group) in grains by GC-MS

No.	Compound	Level /mg/kg	Rice	Wheat	Oats	Barley	Maize	Ave.	RSD%	Level /mg/kg	Rice	Wheat	Oats	Barley	Maize	Ave.	RSD%
E group																	
1	Propoxur-1	0.0500	92.4	88.6	77.5	101.9	92.1	90.5	9.7	0.2000	101.9	102.7	86.0	86.0	105.3	96.4	9.9
2	Isoprocarb-1	0.0500	94.3	87.3	75.5	105.0	96.4	91.7	12.0	0.2000	117.0	115.5	84.7	82.8	91.2	98.2	17.0
3	Methamidophos	0.1000	90.8	94.7	51.3	105.2	84.0	85.2	24.0	0.4000	63.3	50.8	42.9	66.7	48.6	54.5	18.6
4	Acenaphthene	0.0250	96.5	82.9	73.1	72.4	53.2	75.6	21.0	0.1000	82.5	87.9	77.4	66.9	94.5	81.8	12.8
5	Dibutyl succinate	0.0500	87.1	82.8	80.1	94.7	79.0	84.8	7.5	0.2000	91.5	95.8	87.9	81.5	97.7	90.9	7.1
6	Phthalimide	0.0500	83.4	86.7	76.4	85.3	74.4	81.2	6.8	0.2000	94.9	98.2	85.6	83.3	125.5	97.5	17.3
7	Chlorethoxyfos	0.0500	105.8	88.4	81.2	95.2	80.9	90.3	11.6	0.2000	91.0	96.2	91.3	92.7	102.1	94.7	4.9

Table 22b.Continued

No.	Compound	Level /mg/kg	Rice	Wheat	Oats	Barley	Maize	Ave.	RSD%	Level /mg/kg	Rice	Wheat	Oats	Barley	Maize	Ave.	RSD%
8	Isoprocarb-2	0.0500	86.8	83.7	84.1	88.5	73.5	83.3	7.0	0.2000	86.6	90.9	94.6	91.1	111.1	94.9	10.0
9	Pencycuron	0.1000	208.7	237.3	101.9	121.4	256.3	185.1	37.6	0.4000	91.3	109.6	78.2	86.4	104.5	94.0	13.7
10	Tebuthiuron	0.1000	83.1	89.2	82.8	105.2	94.5	91.0	10.3	0.4000	95.8	99.1	84.0	75.2	87.9	88.4	10.8
11	Demeton-S-methyl	0.1000	83.5	58.2	82.3	109.7	98.9	86.5	22.5	0.4000	93.3	96.9	85.6	77.4	96.2	89.9	9.2
12	Cadusafos	0.1000	85.6	85.3	92.8	99.5	85.1	89.7	7.1	0.4000	92.5	98.1	88.9	82.2	103.8	93.1	8.9
13	Propoxur-2	0.0500	75.9	84.7	88.1	117.1	104.9	94.2	17.6	0.2000	72.8	84.2	94.1	86.2	87.9	85.0	9.1
14	Naled	0.4000	16.7	51.1	41.8	23.0	23.0	31.1	47.0	1.6000	24.4	8.2	36.2	41.2	12.1	24.4	59.2
15	Phenanthrene	0.0250	87.6	81.3	80.9	95.0	87.1	86.4	6.7	0.1000	92.6	97.2	85.4	81.2	95.5	90.4	7.6
16	Spiroxamine-1	0.0500	9.7	9.8	7.8	4.4	4.3	7.2	38.0	0.2000	4.3	4.6	1.0	0.7	3.3	2.8	66.6
17	Fenpyroximate	0.2000	88.6	80.0	82.8	80.1	72.5	80.8	7.2	0.8000	103.4	113.3	97.5	77.7	127.8	104.0	17.9
18	Tebupirimfos	0.0500	84.8	83.7	83.0	101.1	85.5	87.6	8.7	0.2000	92.0	97.3	91.4	82.1	94.2	91.4	6.2
19	Prohydrojamon	0.1000	96.9	81.7	82.7	110.8	107.2	95.8	14.1	0.4000	100.1	103.0	114.5	131.1	152.0	120.1	18.0
20	Fenpropidin	0.0500	25.9	20.6	20.5	3.4	5.2	15.1	67.2	0.2000	13.9	14.0	0.6	1.3	0.8	6.1	117.1
21	Dichloran	0.0500	103.7	78.3	90.9	102.9	91.0	93.3	11.2	0.2000	91.2	96.5	87.2	76.6	91.5	88.6	8.4
22	Pyroquilon	0.0250	87.4	86.3	80.9	101.4	90.3	89.3	8.5	0.1000	95.1	98.2	87.7	75.6	93.1	90.0	9.9
23	Spiroxamine-2	0.0500	7.6	9.8	7.7	10.0	2.0	7.4	43.5	0.2000	3.9	5.0	0.5	0.3	2.4	2.4	85.1
24	Dinoterb	0.4000	33.4	38.7	23.2	76.0	85.7	51.4	53.8	1.6000	89.2	77.3	81.2	81.9	99.7	85.8	10.3
25	propyzamide	0.0500	82.4	84.2	85.8	102.8	89.2	88.9	9.2	0.2000	94.0	98.1	90.9	85.1	96.9	93.0	5.6
26	Pirimicicarb	0.0500	45.5	46.5	47.7	37.7	43.3	44.1	8.9	0.2000	63.7	68.5	29.8	16.5	35.5	42.8	52.4
27	Phosphamidon-1	0.2000	75.0	107.9	101.4	103.4	126.2	102.8	17.9	0.8000	83.6	79.5	77.3	78.3	82.5	80.2	3.4
28	Benoxacor	0.0500	83.3	82.6	90.0	98.3	78.6	86.6	8.9	0.2000	89.9	91.9	93.8	86.5	93.8	91.2	3.3
29	Bromobutide	0.0250	84.1	85.4	90.6	106.0	103.0	93.8	10.8	0.1000	103.0	105.8	86.7	Not add	Not add	98.5	10.5
30	Acetochlor	0.0500	83.0	84.8	84.4	101.2	88.4	88.4	8.4	0.2000	94.6	98.8	92.4	86.3	105.2	95.5	7.4
31	Tridiphane	0.1000	78.3	81.5	92.5	Not add	Not add	84.1	8.9	0.4000	96.1	91.9	95.9	Not add	Not add	94.6	2.5
32	Terbucarb-2	0.0500	84.9	85.2	85.4	97.7	83.9	87.4	6.6	0.2000	93.0	97.4	88.7	90.3	131.2	100.1	17.7
33	Esprocarb	0.0500	101.8	101.1	109.3	117.1	98.3	105.5	7.2	0.2000	76.8	94.7	86.1	91.3	134.5	96.7	23.0
34	Fenfuram	0.0500	80.1	81.2	82.6	106.2	92.0	88.4	12.4	0.2000	88.0	93.7	85.5	73.7	73.7	82.9	10.8
35	Acibenzolar-S-methyl	0.0500	83.0	82.0	88.2	Not add	Not add	84.4	3.9	0.2000	103.2	105.6	Not add	Not add	Not add	104.4	1.6
36	Benfuresate	0.0500	89.0	86.4	85.6	86.0	74.2	84.2	6.8	0.2000	93.4	97.7	93.8	1.3	1.6	57.6	89.1
37	Dithiopyr	0.0250	85.4	84.8	85.0	99.2	86.4	88.2	7.0	0.1000	94.1	98.0	90.0	83.6	93.4	91.8	5.9

Table 22b.Continued

No.	Compound	Level /mg/kg	Rice	Wheat	Oats	Barley	Maize	Ave.	RSD%	Level /mg/kg	Rice	Wheat	Oats	Barley	Maize	Ave.	RSD%
38	Mefenoxam	0.0500	85.9	86.3	84.9	100.4	84.2	88.3	7.7	0.2000	94.5	99.1	91.2	87.0	100.7	94.5	6.0
39	Malaoxon	0.4000	71.0	80.1	103.1	103.9	78.5	87.3	17.4	1.6000	80.7	78.7	91.4	86.4	81.3	83.7	6.1
40	Phosphamidon-2	0.2000	81.4	90.9	96.4	88.3	72.6	85.9	10.7	0.8000	91.2	87.5	124.4	74.8	82.9	92.2	20.6
41	Simeconazole	0.0500	80.0	81.1	86.4	127.0	110.1	96.9	21.4	0.2000	92.0	95.4	87.3	86.9	124.3	97.2	16.0
42	Chlorthal-dimethyl	0.0500	85.5	84.9	84.3	96.8	86.9	87.7	5.9	0.2000	93.6	99.5	89.8	84.9	97.3	93.0	6.3
43	Thiazopyr	0.0500	87.8	86.8	89.5	101.6	88.3	90.8	6.7	0.2000	93.3	98.8	90.6	81.9	94.2	91.8	6.8
44	Dimethylvinphos	0.0500	80.2	80.6	90.0	105.4	85.3	88.3	11.7	0.2000	91.0	91.8	91.4	84.1	93.7	90.4	4.1
45	Butralin	0.1000	73.5	81.4	88.6	116.7	92.1	90.5	18.1	0.4000	91.1	96.5	93.6	84.4	92.9	91.7	4.9
46	Zoxamide	0.0500	89.4	110.5	84.6	97.7	87.9	94.0	11.1	0.2000	104.9	109.3	91.7	84.8	97.7	97.7	10.1
47	Pyrifenox -1	0.2000	47.6	50.7	47.3	44.5	48.2	47.7	4.6	0.8000	61.2	66.3	64.7	70.1	82.4	68.9	11.9
48	Allethrin	0.1000	80.8	80.0	86.6	103.4	88.1	87.8	10.7	0.4000	94.1	99.2	90.3	82.1	98.3	92.8	7.5
49	Dimethametryn	0.0250	72.2	73.9	76.2	79.1	77.9	75.9	3.7	0.1000	78.4	84.3	58.9	45.9	67.7	67.0	22.9
50	Quinoclamine	0.1000	74.6	76.8	82.5	117.6	104.2	91.1	20.7	0.4000	92.8	96.3	86.0	85.1	99.9	92.0	7.0
51	Methothrin-1	0.0500	69.0	69.5	73.5	117.3	103.1	86.5	25.8	0.2000	93.9	97.5	89.8	83.7	98.0	92.6	6.5
52	Flufenacet	0.2000	80.7	80.2	90.6	100.3	79.8	86.3	10.5	0.8000	93.5	90.0	93.2	92.1	122.4	98.2	13.8
53	Methothrin-2	0.0500	66.2	66.6	69.0	115.1	98.4	83.0	27.0	0.2000	92.5	97.6	86.3	81.9	89.1	89.5	6.7
54	Pyrifenox-2	0.2000	43.6	46.3	45.2	41.6	43.9	44.1	4.1	0.8000	60.4	65.3	75.4	63.8	73.2	67.6	9.4
55	Fenoxanil	0.0500	98.2	96.0	90.1	106.1	104.4	99.0	6.6	0.2000	123.2	137.9	88.5	78.5	88.3	103.3	24.9
56	Phthalide	0.1000	98.4	100.3	93.2	Not add	Not add	97.3	3.8	0.4000	102.4	100.9	Not add	Not add	Not add	101.7	1.1
57	Furalaxyl	0.0500	83.9	85.7	85.4	98.1	85.3	87.7	6.7	0.2000	95.4	100.2	90.9	84.7	97.0	93.6	6.4
58	Thiamethoxam	0.1000	113.5	117.6	89.0	110.0	129.1	111.9	13.1	0.4000	80.3	58.9	46.5	13.7	12.1	42.3	69.7
59	Mepanipyrim	0.0250	77.7	81.7	86.5	118.3	98.1	92.5	17.7	0.1000	93.0	99.2	82.2	77.5	58.8	82.1	19.0
60	Captan	0.4000	74.2	84.4	86.6	108.5	80.3	86.8	15.0	1.6000	49.4	39.5	91.5	80.1	94.7	71.0	35.4
61	Bromacil	0.0500	94.2	97.4	94.0	76.5	61.3	84.7	18.2	0.2000	95.2	92.7	79.4	64.9	97.1	85.9	15.9
62	Picoxystrobin	0.0500	84.1	84.5	85.2	102.9	88.9	89.1	8.9	0.2000	94.3	99.1	91.1	84.0	95.6	92.8	6.2
63	Butamifos	0.0250	80.7	77.6	87.9	Not add	Not add	82.0	6.4	0.1000	106.7	110.9	Not add	Not add	Not add	108.8	2.7
64	Imazamethabenz-methyl	0.0750	66.4	68.3	70.2	41.2	60.6	61.4	19.3	0.3000	79.4	76.8	55.9	45.3	65.6	64.6	22.1
65	Metominostrobin-1	0.1000	83.6	86.9	92.6	Not add	Not add	87.7	5.2	0.4000	105.8	106.2	Not add	Not add	Not add	106.0	0.3
66	TCMTB	0.4000	69.2	78.4	112.7	52.4	68.0	76.1	29.5	1.6000	92.9	71.1	54.0	88.7	70.8	75.5	20.7

Table 22b.Continued

No.	Compound	Level /mg/kg	Rice	Wheat	Oats	Barley	Maize	Ave.	RSD%	Level /mg/kg	Rice	Wheat	Oats	Barley	Maize	Ave.	RSD%
67	Imazalil	0.1000	33.3	35.1	30.7	37.5	46.0	36.5	16.0	0.4000	41.3	49.8	8.4	7.8	20.6	25.6	74.9
68	Isoprothiolane	0.0500	81.8	84.6	87.5	105.6	91.7	90.2	10.3	0.2000	94.1	99.3	88.9	92.3	119.6	98.8	12.4
69	Methiocarb sulfone	0.8000	39.4	40.2	79.3	111.3	90.9	72.2	44.0	3.2000	111.9	95.7	81.9	82.8	96.3	93.7	13.1
70	Cyflufenamid	0.4000	97.5	98.4	104.3	Not add	Not add	100.1	3.7	1.6000	96.8	97.5	65.6	61.8	83.3	81.0	20.8
71	Methyl trithion	0.4000	76.7	81.8	103.2	128.8	117.7	101.6	22.1	1.6000	90.6	87.1	101.8	123.1	145.5	109.6	22.3
72	Pyriminobac-methyl	0.1000	83.7	83.1	89.8	Not add	Not add	85.6	4.3	0.4000	104.0	105.1	Not add	Not add	Not add	104.5	0.8
73	Isoxathion	0.2000	68.8	70.5	96.8	86.8	77.8	80.1	14.6	0.8000	85.2	77.4	104.7	86.9	105.2	91.9	13.6
74	Metominostrobin-2	0.1000	58.2	66.0	76.4	Not add	Not add	66.8	13.7	0.4000	101.7	99.3	Not add	Not add	Not add	100.5	1.7
75	Diofenolan-1	0.0500	82.1	83.9	85.7	106.7	94.0	90.5	11.2	0.2000	94.4	99.5	88.2	83.6	93.9	91.9	6.7
76	Thifluzamide	0.2000	92.7	79.8	106.7	Not add	Not add	93.1	14.5	0.8000	Not add	Not adc	Not add	Not add	Not add	#DIV!	#DIV/0!
77	Diofenolan-2	0.0500	82.1	82.8	85.2	91.3	81.2	84.5	4.8	0.2000	95.0	98.4	95.1	82.7	100.0	94.2	7.2
78	Quinoxyphen	0.0250	67.4	71.2	72.2	71.1	62.0	68.8	6.1	0.1000	80.5	84.9	62.2	46.9	68.3	68.6	22.1
79	Chlorfenapyr	0.2000	84.4	80.8	85.6	100.5	87.2	87.7	8.6	0.8000	93.3	97.2	90.2	84.0	96.1	92.2	5.8
80	Trifloxystrobin	0.1000	80.3	83.8	89.3	93.9	71.3	83.7	10.3	0.4000	97.0	100.5	106.6	108.5	110.6	104.6	5.4
81	Imibenconazole-des-benzyl	0.1000	134.4	134.9	140.2	105.4	86.9	120.3	19.2	0.4000	124.7	123.0	84.9	72.2	101.9	101.4	22.8
82	Isoxadifen-ethyl	0.0500	83.7	84.3	91.4	89.7	69.8	83.8	10.1	0.2000	99.8	103.1	Not add	Not add	Not add	101.5	2.3
83	Fipronil	0.2000	79.0	61.7	85.4	99.9	84.8	82.1	16.8	0.8000	92.0	89.4	93.0	97.7	141.4	102.7	21.3
84	Imiprothrin-1	0.0500	98.6	66.9	95.5	84.7	72.9	83.7	16.5	0.2000	89.4	112.6	80.6	90.5	90.9	92.8	12.8
85	Carfentrazone-ethyl	0.0500	82.6	81.9	92.6	103.2	83.5	88.8	10.3	0.2000	93.9	98.2	105.9	84.9	111.5	98.9	10.5
86	Imiprothrin-2	0.0500	95.2	69.8	107.2	131.2	98.5	100.4	22.1	0.2000	96.5	102.5	104.4	87.0	125.9	103.2	13.9
87	Halosulfuran-methyl	0.1000	31.2	28.0	32.0	Not add	Not add	30.4	7.0	0.4000	68.3	29.1	61.1	Not add	Not add	52.8	39.5
88	Epoxiconazole-1	0.2000	99.5	91.6	82.2	84.8	81.4	87.9	8.7	0.8000	111.6	134.7	86.1	81.1	104.6	103.6	20.8
89	Pyraflufen ethyl	0.0500	84.8	84.2	85.9	99.9	91.2	89.2	7.4	0.2000	95.3	100.0	90.9	86.6	96.6	93.9	5.5
90	Pyributicarb	0.0500	84.4	82.4	83.7	106.4	92.1	89.8	11.1	0.2000	92.9	97.5	92.1	83.2	99.9	93.1	6.9
91	Thenylchlor	0.0500	75.7	77.2	80.1	100.6	87.5	84.2	12.1	0.2000	94.1	97.2	91.6	102.7	105.5	98.2	5.9
92	Clethodim	0.1000	21.7	27.3	27.2	52.8	37.3	33.2	36.9	0.4000	67.9	68.4	40.6	60.6	75.7	62.7	21.4
93	Chrysene	0.0250	0.0	0.0	1.9	2.2	6.3	2.1	124.7	0.1000	1.1	1.3	0.3	0.5	2.6	1.2	76.1
94	Mefenpyr-diethyl	0.0750	83.7	83.1	85.3	97.2	87.4	87.3	6.6	0.3000	91.8	96.4	88.8	84.1	97.4	91.7	6.0
95	Famphur	0.1000	83.2	67.6	91.2	103.6	86.8	86.5	15.4	0.4000	94.1	85.3	88.8	84.4	102.2	91.0	8.1

Table 22b.Continued

No.	Compound	Level /mg/kg	Rice	Wheat	Oats	Barley	Maize	Ave.	RSD%	Level /mg/kg	Rice	Wheat	Oats	Barley	Maize	Ave.	RSD%
96	Etoxazole	0.1500	91.9	95.8	97.0	104.3	94.5	96.7	4.8	0.6000	88.7	84.7	88.5	82.8	96.2	88.2	5.8
97	Pyriproxyfen	0.0500	75.1	76.2	79.8	99.3	84.6	83.0	11.9	0.2000	87.1	93.0	76.1	71.3	89.9	83.5	11.2
98	Epoxiconazole-2	0.2000	78.6	82.1	85.9	102.5	85.2	86.9	10.6	0.8000	90.2	93.4	87.1	79.4	98.9	89.8	8.1
99	Tepraloxydim	0.0500	103.0	95.4	83.0	100.4	33.3	83.0	34.8	0.2000	105.4	104.1	97.4	42.1	103.2	90.4	30.1
100	Picolinafen	0.0250	84.1	86.7	94.0	116.2	100.7	96.3	13.4	0.1000	94.3	97.8	86.4	83.0	98.9	92.1	7.6
101	Iprodione	0.1000	79.1	79.2	86.7	88.8	75.3	81.8	7.0	0.4000	98.9	100.9	82.0	93.0	149.7	104.9	24.9
102	Piperophos	0.0750	76.9	82.9	89.1	109.9	89.0	89.6	13.9	0.3000	94.5	97.1	87.3	90.1	122.6	98.3	14.4
103	Ofurace	0.0750	78.1	79.8	88.2	98.8	85.7	86.1	9.6	0.3000	93.3	91.4	88.4	98.2	88.4	92.0	4.4
104	Bifenazate	0.2000	100.5	121.2	122.2	Not add	Not add	114.6	10.7	0.8000	91.1	99.9	Not add	Not add	Not add	95.5	6.5
105	Chromafenozide	0.2000	103.6	75.0	69.8	82.1	69.6	80.0	17.6	0.8000	80.1	88.3	Not add	Not add	Not add	84.2	6.9
106	Endrin ketone	0.1000	78.6	76.7	86.8	99.3	78.7	84.0	11.2	0.4000	92.9	96.9	95.3	86.7	95.1	93.4	4.3
107	Clomeprop	0.0250	79.8	71.8	84.5	Not add	Not add	78.7	8.2	0.1000	103.4	106.4	Not add	Not add	Not add	104.9	2.0
108	Fenamidone	0.0250	82.9	64.9	93.7	110.4	96.1	89.6	18.9	0.1000	105.5	107.0	Not add	Not add	Not add	106.2	1.0
109	Naproanilide	0.0250	82.5	83.3	98.4	Not add	Not add	88.0	10.2	0.1000	106.0	108.5	Not add	Not add	Not add	107.3	1.7
110	Pyraclostrobin	0.6000	103.0	105.1	85.2	96.5	102.4	98.5	8.2	2.4000	88.2	111.0	130.0	113.4	130.8	114.7	15.2
111	Lactofen	0.2000	95.9	73.5	104.3	107.1	78.6	91.9	16.5	0.8000	90.4	97.6	100.2	95.4	112.9	99.3	8.5
112	Tralkoxydim	0.2000	19.2	24.1	30.0	54.6	34.8	32.6	42.0	0.8000	75.1	76.7	37.7	9.9	20.5	44.0	70.0
113	Pyraclofos	0.2000	76.8	84.8	111.8	Not add	Not add	91.1	20.1	0.8000	101.5	98.8	Not add	Not add	Not add	100.1	1.9
114	Dialifos	0.8000	78.3	71.0	103.1	Not add	Not add	84.1	20.1	3.2000	99.0	96.6	102.9	79.4	112.5	98.1	12.3
115	Spirodiclofen	0.2000	79.1	75.1	83.9	96.1	81.9	83.2	9.5	0.8000	96.2	94.7	90.8	77.2	109.1	93.6	12.2
116	Halfenprox	0.0500	69.8	77.9	93.5	Not add	Not add	80.4	15.0	0.2000	107.6	108.7	Not add	Not add	Not add	108.1	0.7
117	Flurtamone	0.0500	83.7	71.8	86.4	103.4	91.5	87.3	13.2	0.2000	101.8	91.8	79.5	87.7	137.2	99.6	22.6
118	Pyriftalid	0.0250	82.2	85.2	88.3	95.7	94.9	89.3	6.6	0.1000	95.8	99.2	88.8	86.1	101.4	94.3	7.0
119	Silafluofen	0.0250	82.9	75.2	78.2	113.5	84.2	86.8	17.7	0.1000	92.6	95.1	91.8	80.3	98.0	91.6	7.4
120	Pyrimidifen	0.0500	25.6	29.7	27.1	Not add	Not add	27.4	7.6	0.2000	54.1	49.2	61.0	65.2	62.3	58.4	11.2
121	Acetamiprid	0.1000	110.5	94.1	93.1	106.3	104.5	101.7	7.6	0.4000	95.0	73.1	85.3	97.3	88.6	87.9	10.9
122	Butafenacil	0.0250	75.9	86.8	93.4	108.6	91.2	91.2	13.0	0.1000	95.6	97.2	88.9	85.1	98.7	93.1	6.3
123	Cafenstrole	0.1000	76.7	77.5	100.2	Not add	Not add	84.8	15.8	0.4000	86.2	80.0	Not add	Not add	Not add	83.1	5.3
124	Fluridone	0.0500	69.3	82.0	77.8	80.4	55.0	74.9	9.9	0.2000	89.3	77.2	68.8	65.4	101.3	80.4	18.5

Table 23. Recovery and precision data for 465 pesticide residues in grains by LC-MS-MS

No.	Compound	Level /μg /kg	Oats	Barley	Maize	Rice	Wheat	Ave.	RSD%	Level /μg /kg	Oats	Barley	Maize	Rice	Wheat	Ave.	RSD%
A group																	
1	Propham	40.0	68.5	101.0	92.0	67.3	56.0	77.0	24.4	200.0	85.5	99.5	86.9	83.2	106.0	92.2	9.7
2	Isoprocarb	4.0	44.4	86.2	64.8	71.1	61.5	65.6	23.1	20.0	92.9	105.0	87.3	95.3	88.0	93.7	6.8
3	Cycluron	0.4	35.8	86.3	70.6	79.3	64.4	67.3	28.9	2.0	103.0	110.0	97.2	115.0	91.4	103.3	8.2
4	Butocarboxim sulfoxide	40.0	23.6	47.9	44.6	18.2	27.2	32.3	40.8	200.0	67.1	71.7	58.7	43.7	30.5	54.3	28.1
5	Propachlor	0.4	58.3	121.0	81.7	105.0	70.1	87.2	29.3	2.0	109.0	120.0	87.5	94.1	90.4	100.2	12.3
6	Rabenzazole	0.2	57.7	94.8	75.1	85.0	55.9	73.7	23.0	1.0	99.4	106.0	93.7	124.0	79.7	100.6	14.5
7	Simetryn	0.4	46.1	80.5	67.6	76.3	79.6	70.0	20.4	2.0	83.5	103.0	91.1	111.0	80.5	93.8	12.3
8	Monolinuron	8.0	68.1	104.0	81.1	79.1	59.2	78.3	21.5	40.0	95.2	112.0	91.0	128.0	92.2	103.7	13.8
9	Pymetrozin	20.0	N/A	N/A	N/A	N/A	N/A	#DIV/0!	#DIV/0!	100.0	N/A	N/A	N/A	N/A	N/A	#DIV/0!	#DIV/0!
10	Butylate	240.0	101.0	119.0	118.0	74.8	86.4	99.8	19.4	1200.0	101.0	102.0	103.0	66.4	105.0	95.5	15.3
11	Ethoxyquin	24.0	45.8	49.0	41.1	32.9	39.3	440.0	15.8	120.0	43.2	43.0	45.4	39.7	46.2	43.5	5.2
12	Dichlorvos	20.0	56.7	123.0	84.0	69.7	51.5	77.0	37.2	100.0	81.3	82.0	79.7	45.3	97.4	77.1	22.2
13	Monocrotophos	0.8	94.5	179.0	81.1	65.4	64.5	105.0	48.3	4.0	84.8	103.0	45.4	69.9	85.9	77.8	24.8
14	Mevinphos	8.0	41.9	91.6	54.7	72.8	49.8	62.2	32.2	40.0	75.6	92.7	81.4	102.0	82.6	86.9	10.8
15	Dichlormid	4.0	70.1	75.0	66.8	82.9	66.8	72.3	9.4	20.0	84.5	94.1	63.1	128.0	108.0	95.5	22.9
16	Secbumeton	0.2	57.7	86.7	44.1	68.1	68.1	64.9	24.1	1.0	41.8	104.0	85.2	111.0	72.1	82.8	29.8
17	Cyprodinil	0.4	86.3	103.0	78.0	91.9	84.9	88.8	10.5	2.0	86.9	122.0	93.7	148.0	113.0	112.7	19.3
18	Pirimicarb	0.4	42.6	41.0	33.6	46.3	45.1	41.7	12.0	2.0	97.4	129.0	86.1	109.0	78.0	99.9	17.9
19	Clomazone	1.2	58.1	104.0	74.2	80.7	64.4	76.3	23.3	6.0	89.1	101.0	87.1	120.0	80.9	95.6	14.5
20	Cyanazine	1.2	32.2	73.7	53.6	66.6	50.6	55.3	28.9	6.0	59.4	76.9	64.9	91.0	66.4	71.7	15.6
21	Ethiofencarb sulfoxide	4.0	32.8	87.4	67.6	23.9	25.1	47.4	60.4	20.0	73.0	89.3	79.8	37.1	54.8	66.8	27.9
22	Prometryne	0.2	74.8	111.0	78.2	101.0	73.7	87.7	19.5	1.0	79.5	101.0	83.6	124.0	91.4	95.9	16.5
23	Demeton s methl sulfoxide	16.0	41.0	55.6	43.4	30.5	28.5	39.8	27.5	80.0	74.2	73.6	66.7	63.5	46.6	64.9	15.4
24	Paraoxon methyl	2.0	48.6	101.0	72.2	52.2	49.8	64.8	34.6	10.0	82.5	102.0	79.8	108.0	73.3	89.1	15.1
25	Thiofanox sufone	40.0	63.2	71.0	58.3	51.0	93.5	67.4	24.2	200.0	72.3	68.5	75.0	66.7	60.2	68.5	7.4
26	Thiacloprid	0.8	61.1	87.7	47.0	52.6	53.7	60.4	26.6	4.0	81.3	92.2	78.2	116.0	91.5	91.8	14.5
27	Imidacloprid	40.0	24.1	74.2	42.8	39.4	32.8	42.7	44.6	200.0	78.8	89.6	75.0	75.6	58.5	75.5	13.2
28	Demeton s methyl sulfone	8.0	35.1	90.0	47.6	44.2	34.6	50.3	45.5	40.0	80.0	102.0	6.0	164.0	58.3	82.1	63.3
29	Ethidimuron	8.0	26.2	76.5	40.5	27.5	33.4	40.8	50.8	40.0	79.3	91.4	74.5	75.4	59.9	76.1	13.3

Table 23. Continued

No.	Compound	Level /µg /kg	Oats	Barley	Maize	Rice	Wheat	Ave.	RSD%	Level /µg /kg	Oats	Barley	Maize	Rice	Wheat	Ave.	RSD%
30	Pentachlorophenol	80.0	N/A	N/A	N/A	N/A	N/A	#DIV/0!	#DIV/0!	400.0	N/A	N/A	N/A	N/A	N/A	#DIV/0!	#DIV/0!
31	Isomethiozin	0.2	95.4	110.0	76.6	97.9	61.8	88.3	21.6	1.2	89.2	103.0	95.5	126.0	84.6	99.7	14.6
32	Cis and trans diallate	24.0	90.3	66.8	82.9	78.8	77.2	79.2	10.8	120.0	93.3	106.0	93.5	141.0	87.9	104.3	18.5
33	Acetochlor	40.0	84.3	95.3	96.5	99.2	70.8	89.2	13.2	200.0	98.8	114.0	95.1	114.0	88.2	102.0	10.2
34	Nitenpyram	8.0	20.7	11.7	19.0	17.4	21.7	18.1	21.7	40.0	39.6	42.9	29.8	28.0	28.8	33.8	18.3
35	Napropamide	0.4	78.6	112.0	93.9	75.1	70.5	86.0	19.7	2.0	105.0	116.0	96.0	109.0	85.4	102.3	10.4
36	Methoprotryne	4.0	65.4	92.6	79.5	87.2	79.1	80.8	12.7	20.0	92.9	110.0	88.3	123.0	81.5	99.1	15.3
37	DE-PCB 31	40.0	177.0	212.0	113.0	185.0	99.3	157.3	30.9	200.0	172.0	138.0	120.0	111.2	67.5	99.7	29.9
38	Fenpropidin	0.1	15.5	12.9	14.0	103.0	88.5	46.8	96.2	0.4	91.7	180.0	116.0	100.0	72.8	112.1	32.7
39	Dimethenamid	0.4	42.0	98.0	62.0	85.3	64.4	70.3	31.0	2.0	103.0	109.0	102.0	102.0	80.3	99.3	9.9
40	Fenitrothion	32.0	194.0	122.0	83.6	186.0	75.2	132.2	42.2	160.0	81.5	120.0	94.7	120.0	73.4	97.9	19.7
41	Terrbucarb	4.0	47.0	85.7	108.0	56.0	136.0	86.5	42.5	20.0	132.0	137.0	98.3	95.2	73.0	107.1	22.5
42	Penconazole	0.2	68.1	133.0	103.0	137.0	52.3	98.7	38.4	1.2	91.0	106.0	84.1	132.0	90.9	100.8	17.0
43	Myclobutanil	1.2	104.0	117.0	69.6	77.6	38.4	81.3	37.8	6.0	90.0	96.2	82.4	113.0	85.0	93.3	11.7
44	Imazethapyr	480.0	17.9	48.3	33.1	15.9	15.0	26.0	55.5	2400.0	66.1	83.9	64.9	47.0	38.4	60.1	26.5
45	Paclobutrazol	0.2	48.4	78.2	85.8	83.1	44.6	69.0	30.3	1.0	77.4	92.9	85.0	121.0	77.9	90.8	17.7
46	Fenthion sulfoxide	0.4	50.7	108.0	53.2	74.0	70.8	71.3	32.2	2.0	70.8	97.5	97.1	113.0	67.1	89.1	19.6
47	Triadimenol	4.0	43.1	51.1	46.6	65.8	36.0	48.5	22.9	20.0	70.9	84.4	65.2	98.8	72.6	78.4	15.3
48	Spiroxamine	0.1	78.2	67.3	81.1	73.2	54.4	70.8	14.9	0.4	75.5	72.3	123.0	103.0	71.8	89.1	23.0
49	Tolclofos methyl	120.0	99.2	99.0	79.1	114.0	96.1	97.5	12.8	600.0	80.1	112.0	92.5	116.0	79.7	96.1	16.0
50	Desmedipham	2.0	44.8	92.0	63.8	69.9	50.1	64.1	29.0	10.0	75.0	84.9	72.7	101.0	86.3	84.0	12.0
51	Methidathion	6.0	64.4	91.2	80.9	80.2	84.7	80.3	12.3	30.0	105.0	103.0	119.0	112.0	81.9	104.2	12.0
52	Allethrin	12.0	122.0	112.0	105.0	95.0	89.4	104.7	12.5	60.0	106.0	113.0	87.1	117.0	84.8	101.6	13.1
53	Triallate	40.0	132.0	112.0	114.0	105.0	89.7	110.5	13.9	200.0	106.0	122.0	82.2	163.0	96.1	113.9	24.4
54	Diazinon	0.2	89.7	96.4	75.2	98.1	75.4	87.0	12.8	1.2	92.3	114.0	91.1	133.0	93.8	104.8	15.6
55	Edifenphos	0.8	90.8	89.1	140.0	73.7	77.0	94.1	28.4	4.0	107.0	176.0	163.0	123.0	87.2	131.2	25.5
56	Pretilachlor	0.8	97.5	109.0	94.4	96.0	80.3	95.4	10.7	4.0	90.9	116.0	90.7	121.0	81.4	100.0	15.6
57	Malaoxon	0.4	58.8	64.2	66.3	62.6	62.1	62.8	4.4	2.0	79.4	81.1	89.7	90.2	92.9	86.7	6.2
58	Flusilazole	0.4	69.1	117.0	82.4	100.0	68.1	87.3	24.1	2.0	87.0	96.7	74.3	116.0	92.5	93.3	14.6
59	Methothrin	800.0	174.0	138.0	157.0	114.0	71.0	130.8	30.7	4000.0	86.6	93.1	82.1	162.0	103.0	105.4	27.7

Table 23. Continued

No.	Compound	Level /µg /kg	Oats	Barley	Maize	Rice	Wheat	Ave.	RSD%	Level /µg /kg	Oats	Barley	Maize	Rice	Wheat	Ave.	RSD%
60	Iprovalicarb	0.8	33.5	50.6	33.2	42.4	36.3	42.2	22.9	4.0	38.8	49.9	44.8	52.6	46.5	46.5	10.1
61	Dinitramine	16.0	47.4	56.5	70.0	32.4	15.8	44.4	47.4	80.0	71.0	75.4	73.0	106.0	67.8	78.6	17.7
62	Flutolanil	0.4	30.2	81.0	67.5	110.0	47.1	67.2	45.9	2.0	87.3	89.8	85.5	89.2	53.6	81.1	17.0
63	Famphur	2.4	41.1	62.5	49.8	78.1	62.8	58.9	24.0	12.0	82.8	105.0	82.6	102.0	82.9	91.1	11.2
64	Benalyxyl	0.2	80.1	94.6	78.6	73.0	80.5	81.4	9.8	1.2	98.3	128.0	90.3	98.6	79.7	99.0	16.2
65	Diclobutrazole	0.2	68.7	88.6	61.1	85.1	53.1	71.3	21.4	1.2	80.3	95.6	79.2	113.0	92.2	92.1	13.3
66	Etaconazole	0.4	58.5	92.6	90.1	78.0	72.5	78.3	17.7	2.0	91.1	97.7	75.3	123.0	86.2	94.7	16.8
67	Fenarimol	2.4	50.6	88.3	80.0	99.4	77.6	79.2	22.9	12.0	87.0	101.0	84.7	107.0	79.1	91.8	11.4
68	Phthalic acid,dicyclobexyl ester	2.4	201.0	239.0	182.0	76.0	92.7	158.1	44.7	12.0	89.3	109.0	86.2	161.0	85.0	106.1	27.1
69	Tetramethirn	2.4	147.0	130.0	138.0	97.7	84.6	119.5	22.5	12.0	102.0	145.0	91.5	119.0	96.2	110.7	17.6
70	Dichlofluanid	40.0	N/A	N/A	N/A	N/A	N/A	#DIV/0!	#DIV/0!	200.0	N/A	N/A	N/A	N/A	N/A	#DIV/0!	#DIV/0!
71	Cloquintocet mexyl	0.2	89.6	108.0	93.2	101.0	74.4	93.2	13.6	1.2	88.7	111.0	86.9	122.0	92.5	100.2	13.8
72	Bitertanol	4.0	56.3	91.9	63.3	74.7	66.3	70.5	19.4	20.0	75.4	91.2	61.9	84.3	72.5	77.1	13.1
73	Dimefuron	0.8	26.6	78.7	52.0	57.8	54.4	53.9	34.5	4.0	80.4	98.4	76.4	114.0	64.9	86.8	20.0
74	Chlorprifos methyl	4.0	162.0	79.3	143.0	N/A	N/A	128.1	33.8	20.0	70.2	68.8	107.0	73.5	66.8	77.3	19.5
75	Tepraloxydim	32.0	41.1	80.3	60.3	53.8	39.1	54.9	30.4	160.0	87.3	99.7	79.6	99.9	65.2	86.3	15.2
76	Thiophanate methyl	40.0	N/A	N/A	N/A	N/A	N/A	#DIV/0!	#DIV/0!	200.0	N/A	N/A	N/A	N/A	N/A	#DIV/0!	#DIV/0!
77	Prothiophos	40.0	109.0	138.0	95.4	139.0	81.2	112.5	22.8	200.0	62.7	84.1	65.2	115.0	64.5	78.3	25.5
78	Azinphos ethyl	40.0	97.8	80.3	84.5	57.6	62.1	76.5	21.7	200.0	85.9	97.7	83.4	109.0	84.7	92.1	10.7
79	Tolyfluanid	2000.0	N/A	N/A	N/A	N/A	N/A	#DIV/0!	#DIV/0!	10000.0	N/A	N/A	N/A	N/A	N/A	#DIV/0!	#DIV/0!
80	Profluralin	1200.0	61.6	53.4	58.4	44.4	54.2	54.4	11.9	6000.0	N/A	N/A	N/A	N/A	N/A	#DIV/0!	#DIV/0!
81	Clodinafop propargyl	0.8	82.2	107.0	111.0	113.0	91.4	100.9	13.3	4.0	116.0	158.0	126.0	130.0	86.8	123.4	18.6
82	Hexythiazox	0.4	119.0	168.0	112.0	64.4	74.8	107.6	38.1	2.0	107.0	107.0	81.8	117.0	81.9	98.9	14.6
83	Triflumuron	2.4	53.4	75.4	63.0	74.5	50.9	63.4	18.0	12.0	66.4	73.6	56.8	86.6	71.0	70.9	13.7
84	Isoxaflutole	40.0	56.2	94.7	67.6	83.5	70.3	74.5	20.0	200.0	86.7	95.9	85.7	103.0	81.0	90.5	8.7
85	Anilofos	0.4	110.0	96.8	98.8	80.8	90.8	95.4	11.2	2.0	89.0	151.0	125.0	107.0	75.4	109.5	24.3
86	Thiophanat ethyl	40.0	N/A	N/A	N/A	N/A	N/A	#DIV/0!	#DIV/0!	200.0	N/A	N/A	N/A	N/A	N/A	#DIV/0!	#DIV/0!
87	Heptachlor	40.0	N/A	N/A	N/A	N/A	N/A	#DIV/0!	#DIV/0!	200.0	N/A	N/A	N/A	N/A	N/A	#DIV/0!	#DIV/0!
88	Quizalofop-ethyl	1.2	100.0	113.0	99.7	107.0	100.0	103.9	5.7	6.0	104.0	119.0	100.0	107.0	94.5	104.9	7.8

Table 23. Continued

No.	Compound	Level /µg /kg	Oats	Barley	Maize	Rice	Wheat	Ave.	RSD%	Level /µg /kg	Oats	Barley	Maize	Rice	Wheat	Ave.	RSD%
89	Pyrazophos	0.2	99.2	96.1	103.0	79.3	94.2	94.4	9.6	0.8	111.0	122.0	91.0	133.0	87.7	108.9	16.0
90	Haloxyfop-methyl	0.1	145.0	160.0	124.0	99.0	104.0	126.4	20.7	0.6	91.3	107.0	86.0	118.0	65.5	93.6	19.3
91	Fluazifop butyl	0.4	63.5	65.4	75.2	64.3	55.2	64.7	11.0	2.0	70.9	70.4	54.6	87.8	71.5	71.0	14.8
92	Dialifos	8.0	N/A	N/A	N/A	N/A	N/A	#DIV/0!	#DIV/0!	40.0	N/A	N/A	N/A	N/A	N/A	#DIV/0!	#DIV/0!
93	Diflufenican	0.1	105.0	110.0	95.8	121.0	76.8	101.7	16.4	0.6	87.3	102.0	75.8	109.0	75.3	89.9	15.2
94	Bromfenvinfos	1.6	65.6	98.9	80.7	93.0	91.1	85.9	15.2	8.0	103.0	124.0	92.4	107.0	88.6	103.0	12.1
95	Pyraflufen ethyl	2.4	71.1	69.9	95.9	86.0	72.8	79.1	14.4	12.0	99.0	141.0	128.0	94.6	82.3	109.0	20.1
96	Dioxathion	8000.0	119.0	114.0	128.0	104.0	85.1	110.0	23.3	40000.0	N/A	N/A	N/A	N/A	N/A	#DIV/0!	#DIV/0!
97	Flufenoxuron	1.2	N/A	N/A	N/A	N/A	N/A	#DIV/0!	#DIV/0!	6.0	N/A	N/A	N/A	N/A	N/A	#DIV/0!	#DIV/0!
98	Indoxacarb	2.4	67.3	81.3	77.6	77.8	55.0	73.8	9.7	12.0	86.0	102.0	72.3	110.0	62.1	86.5	20.6
99	Emamectin benzoate	0.2	N/A	N/A	N/A	N/A	N/A	#DIV/0!	#DIV/0!	1.0	N/A	N/A	N/A	N/A	N/A	#DIV/0!	#DIV/0!
B group																	
100	Ethylene thiourea	24.0	N/A	N/A	N/A	N/A	N/A	#DIV/0!	#DIV/0!	120.0	N/A	N/A	N/A	N/A	N/A	#DIV/0!	#DIV/0!
101	Chlormequat	200.0	N/A	N/A	N/A	N/A	N/A	#DIV/0!	#DIV/0!	1000.0	N/A	N/A	N/A	N/A	N/A	#DIV/0!	#DIV/0!
102	Dimefox	8.0	30.2	16.6	43.3	34.8	7.8	31.2	45.7	40.0	39.0	37.8	64.7	73.9	88.5	60.8	36.4
103	Daminozide	1.2	N/A	N/A	N/A	N/A	N/A	#DIV/0!	#DIV/0!	6.0	N/A	N/A	N/A	N/A	N/A	#DIV/0!	#DIV/0!
104	Nicotine	8.0	29.2	81.3	67.0	63.3	70.2	86.5	28.8	40.0	37.1	40.0	73.6	62.5	86.0	59.8	35.4
105	Fenuron	4.0	73.8	75.7	66.3	62.4	60.4	69.6	9.8	20.0	53.7	77.9	87.8	85.2	81.2	77.2	17.7
106	Diphenylamin	0.8	6.1	19.9	2.6	36.9	32.8	16.4	94.1	4.0	79.1	74.3	85.0	87.9	89.1	83.1	7.5
107	Acephate	8.0	35.3	42.1	39.4	47.0	47.8	41.0	12.8	40.0	45.5	48.3	38.6	48.4	66.5	49.5	23.1
108	Molinate	2.0	86.6	72.1	65.7	78.8	79.7	75.8	10.5	10.0	67.1	84.5	82.3	80.1	88.2	80.4	10.0
109	Carbendazim	0.2	48.1	36.3	26.9	62.6	74.6	43.5	44.4	1.0	68.8	79.2	35.0	69.8	75.8	65.7	26.9
110	Desethyl-sebuthylazine	12.0	60.1	53.2	38.5	49.9	49.4	50.4	15.5	60.0	51.4	70.9	44.3	72.7	70.9	62.0	21.3
111	Dinotefuran	160.0	N/A	N/A	N/A	N/A	N/A	#DIV/0!	#DIV/0!	800.0	N/A	N/A	N/A	N/A	N/A	#DIV/0!	#DIV/0!
112	6-chloro-4-hydroxy-3-phenyl-pyridazin	1.2	N/A	N/A	N/A	N/A	N/A	#DIV/0!	#DIV/0!	6.0	N/A	N/A	N/A	N/A	N/A	#DIV/0!	#DIV/0!
113	Propoxur	4.0	91.0	81.3	70.8	73.2	63.8	79.1	13.2	20.0	75.7	99.4	84.3	134.0	104.0	99.5	22.5
114	Isouron	0.4	85.2	68.7	83.7	63.1	61.2	75.2	15.1	2.0	58.5	91.8	87.7	87.6	78.9	80.9	16.5
115	Thiofanox	120.0	78.8	73.5	66.4	71.6	75.8	72.6	6.4	600.0	71.2	85.5	83.9	93.8	89.1	84.7	10.0
116	Bendiocarb	0.8	78.4	68.4	46.3	70.3	54.0	65.9	19.8	4.0	71.3	99.2	83.6	94.3	111.0	91.9	16.5

Table 23. Continued

No.	Compound	Level /μg /kg	Oats	Barley	Maize	Rice	Wheat	Ave.	RSD%	Level /μg /kg	Oats	Barley	Maize	Rice	Wheat	Ave.	RSD%
117	Chlorbufam	80.0	81.2	79.1	61.6	101.0	74.2	80.7	17.7	400.0	82.8	106.0	91.5	99.0	99.3	95.7	9.3
118	Pirmicarb-desmethyl	0.2	67.5	65.1	51.1	43.3	66.9	56.8	19.3	1.0	58.8	72.3	93.4	69.1	77.5	74.2	17.1
119	Terbuthylazine	0.2	71.7	58.0	53.2	77.8	60.7	65.2	15.6	1.0	63.8	85.2	54.7	72.6	71.1	69.5	16.3
120	Propazine	0.4	67.3	56.7	38.8	81.7	72.9	61.1	27.0	2.0	53.2	69.6	46.3	61.7	70.0	60.2	17.2
121	Diuron	0.2	69.7	93.1	84.1	79.1	89.0	81.5	11.2	1.0	84.3	82.3	118.0	116.0	93.9	98.9	17.3
122	Chlormephos	1200.0	70.5	65.2	65.9	57.3	54.4	64.7	10.2	6000.0	59.4	65.7	83.2	92.1	83.4	76.8	17.8
123	3-hydroxy carbofuran	0.8	39.2	45.4	52.1	54.3	47.8	47.8	12.4	4.0	42.7	74.2	74.7	89.6	75.3	71.3	24.2
124	Difenzoquat-methyl sulfate	0.8	43.8	56.4	52.2	54.3	40.2	51.7	13.6	4.0	62.5	64.3	70.1	57.5	96.4	70.2	21.9
125	Clothianidin	24.0	54.0	53.0	54.8	40.0	44.5	50.5	13.1	120.0	39.2	67.0	70.9	80.1	80.4	67.5	25.0
126	Pronamide	1.2	85.0	91.9	71.1	79.0	71.8	81.8	10.8	6.0	65.9	96.3	79.5	106.0	91.6	87.9	17.7
127	Dimethachloro	0.4	79.7	70.4	68.5	78.7	77.8	74.3	7.0	2.0	71.8	81.8	88.9	99.2	93.6	87.1	12.2
128	Methobromuron	1.6	103.0	73.7	59.4	64.1	71.6	75.1	22.7	8.0	90.1	115.0	123.0	98.4	105.0	106.3	12.3
129	Demeton-s	20.0	44.2	67.6	78.6	50.0	40.6	60.1	27.1	100.0	54.8	77.6	91.4	91.7	97.2	82.5	20.7
130	Phorate	32.0	48.3	86.3	55.6	82.6	90.3	68.2	28.2	160.0	57.0	115.0	88.8	92.2	116.0	93.8	25.7
131	Aclonifen	4.0	43.6	65.0	79.0	66.6	86.3	63.6	25.6	20.0	72.4	149.0	95.7	83.5	92.1	98.5	30.0
132	Mephosfolan	0.4	92.0	76.8	81.2	62.8	62.0	78.2	16.3	2.0	61.6	86.3	79.7	82.7	70.3	76.1	13.2
133	Imibenzonazole-des-benzyl	2.0	71.8	66.9	68.0	49.0	57.5	63.9	14.5	10.0	44.2	78.3	69.3	97.0	76.1	73.0	26.1
134	Formothion	8000.0	N/A	N/A	N/A	N/A	N/A	#DIV/0!	#DIV/0!	40000.0	N/A	N/A	N/A	N/A	N/A	#DIV/0!	#DIV/0!
135	Neburon	1.2	113.0	117.0	106.0	112.0	123.0	112.0	5.6	6.0	73.4	127.0	107.0	89.8	97.9	99.0	20.1
136	Mefenoxam	0.2	65.6	65.7	56.9	83.0	44.4	67.8	20.8	1.0	69.1	90.8	87.5	103.0	103.0	90.7	15.4
137	Prothoate	0.8	111.0	76.6	70.4	79.2	95.0	84.3	19.5	4.0	68.7	83.4	93.7	99.2	98.9	88.8	14.5
138	Ethofume sate	320.0	95.6	74.5	80.3	78.7	85.0	82.3	9.8	1600.0	84.8	92.2	95.1	86.4	101.0	91.9	7.2
139	Difenoxuron	0.8	78.2	70.4	60.8	55.9	47.8	66.3	18.0	4.0	56.1	78.5	81.6	80.6	92.7	77.9	17.2
140	Iprobenfos	2.0	78.7	75.5	86.0	80.1	88.4	80.1	6.6	10.0	72.7	101.0	123.0	95.6	89.9	96.4	18.9
141	TEPP	4.0	65.3	64.9	54.9	42.5	44.8	56.9	18.9	20.0	59.8	85.7	84.9	108.0	94.2	86.5	20.3
142	Thiamethoxam	40.0	32.4	44.2	31.4	15.7	41.9	30.9	36.4	200.0	32.4	44.6	36.9	39.2	45.8	39.8	13.9
143	Parathion	240.0	90.2	108.0	90.3	98.1	99.8	96.7	7.7	1200.0	82.5	125.0	104.0	97.6	84.8	98.8	17.4
144	Etrimfos	0.4	110.0	62.8	39.8	102.0	102.0	78.7	38.8	2.0	83.6	94.0	85.6	111.0	93.9	93.6	11.5
145	Cythioate	24.0	66.4	59.4	68.1	49.5	55.0	60.9	12.8	120.0	57.9	72.5	63.7	93.4	67.1	70.9	19.2

Table 23. Continued

No.	Compound	Level /µg /kg	Oats	Barley	Maize	Rice	Wheat	Ave.	RSD%	Level /µg /kg	Oats	Barley	Maize	Rice	Wheat	Ave.	RSD%
146	Phosphamidon	1.2	78.3	70.2	63.2	62.8	52.2	68.6	14.1	6.0	61.0	90.4	93.8	88.0	94.6	85.6	16.3
147	Phenmedipham	2.0	69.4	63.9	61.6	70.8	59.9	66.4	7.2	10.0	62.7	88.6	87.2	90.6	85.0	82.8	13.8
148	Bifenazate	6.0	10.9	19.1	11.9	31.5	32.7	18.4	56.9	30.0	88.1	87.0	66.4	93.9	82.2	83.5	12.5
149	Fenhexamid	0.4	67.5	75.3	59.4	72.9	45.6	68.8	17.5	2.0	64.2	85.2	81.8	90.7	87.0	81.8	12.6
150	Flutriafol	1.2	73.8	55.7	67.5	41.7	53.0	59.7	21.1	6.0	62.7	84.6	84.9	109.0	90.8	86.4	19.2
151	Furalaxyl	0.4	74.0	63.2	67.7	68.0	72.3	63.2	6.2	2.0	64.9	99.1	96.4	87.9	96.2	88.9	15.8
152	Bioallethrin	4.0	38.1	21.9	31.1	25.0	30.1	29.0	21.4	20.0	85.2	80.1	78.3	80.0	88.7	82.5	5.3
153	Cyanofenphos	2.0	78.9	79.2	64.8	115.0	119.0	84.5	28.6	10.0	86.0	121.0	99.4	99.1	95.4	100.2	12.8
154	Pirimiphos methyl	0.2	86.9	93.7	74.5	90.3	85.6	86.4	8.4	1.0	72.2	104.0	96.1	110.0	87.7	94.0	15.7
155	Buprofezin	0.2	76.5	85.7	100.0	93.0	87.6	88.8	9.8	1.0	72.3	116.0	87.2	96.5	83.8	91.2	18.0
156	Disulfoton sulfone	1.2	94.6	74.3	84.0	102.0	67.4	88.7	16.0	6.0	69.9	95.3	95.1	92.0	88.1	88.1	12.0
157	Fenazaquin	0.2	85.9	91.9	87.9	108.0	102.0	93.4	10.2	1.0	68.9	112.0	107.0	91.2	87.9	93.4	18.3
158	Triazophos	0.2	81.9	90.1	80.3	79.4	83.2	82.9	5.1	1.0	71.1	101.0	106.0	90.2	85.8	90.8	15.1
159	DEF	0.4	91.3	91.8	88.8	97.3	110.0	92.3	9.2	2.0	70.4	103.0	94.8	89.4	84.8	88.5	13.8
160	Pyriftalid	0.2	82.6	81.8	73.9	59.0	62.8	74.3	14.5	1.0	83.1	103.0	102.0	86.2	93.5	93.6	9.6
161	Vamidothion sulfone	4.8	63.1	52.4	30.4	54.6	30.5	50.1	29.8	24.0	29.1	68.1	53.7	53.1	48.4	50.5	27.8
162	Metconazole	0.4	89.0	75.9	78.3	75.4	71.2	79.7	8.4	2.0	63.1	87.1	87.6	106.0	88.4	86.4	17.7
163	Pyriproxyfen	0.2	76.3	92.1	76.5	84.0	88.3	82.2	8.6	1.0	73.7	104.0	102.0	129.0	114.0	104.5	19.4
164	Cyazofamid	2.0	96.3	115.0	96.4	93.7	95.7	101.6	8.1	10.0	64.0	95.1	113.0	109.0	85.7	93.4	21.1
165	Cycloxydim	4.0	35.4	46.5	50.4	6.3	8.5	34.6	60.2	20.0	42.0	74.3	76.1	82.3	82.2	71.4	23.6
166	Triphenyl phosphate	0.2	103.0	155.0	136.0	96.8	94.3	122.7	22.0	1.0	79.1	81.3	85.2	91.6	81.9	83.8	5.8
167	Isoxaben	0.2	81.7	62.3	86.7	55.9	55.0	71.7	20.8	1.0	65.0	76.5	92.4	80.5	85.0	79.9	12.8
168	Flurtamone	0.4	70.3	62.1	59.6	43.7	36.8	58.9	23.4	2.0	59.0	80.8	82.9	94.3	83.4	80.1	16.1
169	Trifluralin	200.0	60.3	55.1	40.2	54.8	78.5	52.6	26.2	1000.0	49.2	70.3	66.5	70.8	82.7	67.9	17.8
170	Flamprop methyl	2.0	112.0	107.0	77.0	74.0	90.6	92.5	18.5	10.0	69.4	85.4	90.8	86.9	94.9	85.5	11.4
171	Bioresmethrin	4.0	89.6	78.3	105.0	83.2	87.8	89.0	11.3	20.0	66.1	96.3	91.1	57.5	70.3	76.3	21.9
172	Propiconazole	0.4	95.4	84.3	73.6	58.1	74.3	77.9	17.8	2.0	63.3	90.3	87.4	82.4	91.6	83.0	13.9
173	Fluchloralin	80.0	22.0	39.1	17.2	88.5	28.9	41.7	69.1	400.0	36.1	57.9	56.2	33.7	38.5	44.5	26.1
174	Chlorsulfuron	4.0	36.1	33.1	29.8	9.6	10.6	27.1	47.0	20.0	18.8	48.7	37.5	29.8	26.6	32.3	35.2
175	Clethodim	4.0	50.5	47.4	52.2	8.3	8.3	39.6	57.9	20.0	47.8	70.8	73.2	67.5	76.2	67.1	16.8

Table 23. Continued

No.	Compound	Level /μg /kg	Oats	Barley	Maize	Rice	Wheat	Ave.	RSD%	Level /μg /kg	Oats	Barley	Maize	Rice	Wheat	Ave.	RSD%
176	Flamprop isopropyl	0.4	80.9	76.8	72.9	74.6	78.4	76.3	4.1	2.0	75.8	115.0	112.0	95.9	96.3	99.0	15.8
177	Tetrachlorvinphos	1.2	120.0	92.4	64.5	87.6	114.0	91.1	24.4	6.0	85.8	115.0	104.0	166.0	103.0	114.8	26.6
178	Propargite	8.0	72.6	69.9	91.3	83.8	103.0	79.4	17.2	40.0	91.2	101.0	114.0	86.0	74.5	93.3	16.1
179	Transfluthrin	120.0	N/A	N/A	N/A	N/A	N/A	#DIV/0!	#DIV/0!	600.0	N/A	N/A	N/A	N/A	N/A	#DIV/0!	#DIV/0!
180	Bromuconazole	0.8	61.6	76.1	55.7	69.0	78.2	65.6	14.5	4.0	57.1	88.3	77.4	93.0	99.0	83.0	19.9
181	Picolinafen	0.4	88.6	128.0	71.4	111.0	84.3	99.8	22.7	2.0	64.3	117.0	94.5	104.0	105.0	97.0	20.6
182	Diafenthiuron	1.2	N/A	N/A	N/A	N/A	N/A	#DIV/0!	#DIV/0!	6.0	N/A	N/A	N/A	N/A	N/A	#DIV/0!	#DIV/0!
183	Triasulfuron	1200.0	67.0	65.1	69.1	32.4	25.6	58.4	36.0	6000.0	48.4	79.7	62.0	75.5	71.1	67.3	18.5
184	Fluthiacet methyl	1.2	92.3	87.6	74.3	91.2	89.6	86.4	8.5	6.0	74.5	103.0	104.0	88.0	97.0	93.3	13.2
185	Trifloxystrobin	0.2	73.4	87.3	79.0	74.0	70.1	78.4	8.5	1.0	63.5	103.0	100.0	115.0	99.3	96.2	20.1
186	Chlorimuron ethyl	4.0	58.3	32.3	48.5	38.7	40.0	44.5	22.6	20.0	49.4	90.4	84.0	85.2	61.2	74.0	24.0
187	cloransulam methyl	20.0	N/A	N/A	N/A	N/A	N/A	#DIV/0!	#DIV/0!	100.0	N/A	N/A	N/A	N/A	N/A	#DIV/0!	#DIV/0!
188	Bensultap	200.0	N/A	N/A	N/A	N/A	N/A	#DIV/0!	#DIV/0!	1000.0	N/A	N/A	N/A	N/A	N/A	#DIV/0!	#DIV/0!
189	Hexaflumuron	1.2	N/A	N/A	N/A	N/A	N/A	#DIV/0!	#DIV/0!	6.0	N/A	N/A	N/A	N/A	N/A	#DIV/0!	#DIV/0!
190	Novaluron	0.4	N/A	N/A	N/A	N/A	N/A	#DIV/0!	#DIV/0!	2.0	N/A	N/A	N/A	N/A	N/A	#DIV/0!	#DIV/0!
191	Hydramethylnon	2.0	N/A	N/A	N/A	N/A	N/A	#DIV/0!	#DIV/0!	10.0	N/A	N/A	N/A	N/A	N/A	#DIV/0!	#DIV/0!
192	Tau fluvalinate	8.0	N/A	N/A	N/A	N/A	N/A	#DIV/0!	#DIV/0!	40.0	N/A	N/A	N/A	N/A	N/A	#DIV/0!	#DIV/0!
193	Fluoazuron	0.8	N/A	N/A	N/A	N/A	N/A	#DIV/0!	#DIV/0!	4.0	N/A	N/A	N/A	N/A	N/A	#DIV/0!	#DIV/0!
C group																	
194	Maleic hydrazide	20.0	N/A	N/A	N/A	N/A	N/A	#DIV/0!	#DIV/0!	100.0	N/A	N/A	N/A	N/A	N/A	#DIV/0!	#DIV/0!
195	Propylene thiourea	160.0	N/A	N/A	N/A	N/A	N/A	#DIV/0!	#DIV/0!	800.0	N/A	N/A	N/A	N/A	N/A	#DIV/0!	#DIV/0!
196	Methamidophos	20.0	38.4	19.4	31.6	36.7	44.5	34.1	27.6	100.0	13.5	21.5	28.0	30.7	36.7	26.1	34.1
197	Diethyltoluamide	0.2	103.0	116.0	166.0	106.0	115.0	121.2	21.2	1.0	99.4	85.3	129.0	90.8	109.0	102.7	16.8
198	Simeton	0.8	67.0	65.6	63.8	53.1	54.5	60.8	10.7	4.0	68.4	69.2	67.8	16.6	24.0	49.2	53.9
199	Monuron	8.0	69.9	67.3	74.8	63.7	57.7	66.7	9.7	40.0	73.4	74.7	73.8	61.7	56.6	68.0	12.2
200	Pyrimethanil	0.8	102.0	82.1	86.9	85.6	98.1	90.9	9.5	4.0	79.3	80.5	77.6	80.5	89.0	81.4	5.4
201	Quinoclamine	4.0	77.5	61.6	66.3	73.7	70.0	69.8	8.9	20.0	57.5	49.4	46.4	69.0	69.4	58.3	18.4
202	Propanil	1.6	82.3	75.5	72.0	71.9	99.2	80.2	14.3	8.0	74.8	64.7	70.2	78.0	65.8	70.7	8.1
203	Oxamyl	480.0	N/A	N/A	N/A	N/A	N/A	#DIV/0!	#DIV/0!	2400.0	N/A	N/A	N/A	N/A	N/A	#DIV/0!	#DIV/0!
204	Acetamiprid	1.2	47.6	54.5	52.0	57.0	59.5	54.1	8.5	6.0	41.9	37.7	48.4	49.4	49.7	45.4	11.8

Table 23. Continued

No.	Compound	Level /μg /kg	Oats	Barley	Maize	Rice	Wheat	Ave.	RSD%	Level /μg /kg	Oats	Barley	Maize	Rice	Wheat	Ave.	RSD%
205	Mepanipyrim	0.4	113.0	107.0	84.3	75.4	75.7	91.1	19.5	2.0	74.9	82.6	80.7	75.1	81.3	78.9	4.6
206	Methiocarb	0.8	75.1	59.0	55.9	77.1	71.0	67.6	14.2	4.0	53.0	82.5	70.6	63.9	78.7	69.7	16.9
207	Prometon	0.2	81.2	77.4	26.3	89.9	81.6	71.3	35.9	1.0	77.9	80.5	66.5	64.1	65.3	70.9	10.9
208	Metoxuron	0.8	64.4	62.4	68.3	53.5	55.0	60.7	10.4	4.0	56.4	51.2	63.9	54.0	57.7	56.6	8.4
209	Methfuroxam	8.0	37.4	24.3	66.0	60.5	60.3	49.7	36.1	40.0	51.0	57.2	63.7	45.4	45.9	52.6	14.8
210	Asulam	40.0	37.5	10.1	37.7	12.4	11.3	21.8	66.3	200.0	9.1	21.5	17.8	11.1	13.8	14.7	34.2
211	Monalide	0.4	78.3	88.2	97.9	75.4	86.9	85.3	10.4	2.0	75.3	78.1	78.5	54.3	51.8	67.6	19.8
212	Thiram	400.0	N/A	N/A	N/A	N/A	N/A	#DIV/0!	#DIV/0!	2000.0	N/A	N/A	N/A	N/A	N/A	#DIV/0!	#DIV/0!
213	Fonofos	1.2	113.0	100.0	116.0	83.9	105.0	103.6	12.3	6.0	94.0	109.0	107.0	75.6	67.0	90.5	20.7
214	Etridiazol	160.0	N/A	N/A	N/A	N/A	N/A	#DIV/0!	#DIV/0!	800.0	N/A	N/A	N/A	N/A	N/A	#DIV/0!	#DIV/0!
215	Thiometon	3600.0	63.1	50.8	41.8	71.0	75.1	60.4	24.1	18000.0	N/A	N/A	N/A	N/A	N/A	#DIV/0!	#DIV/0!
216	Dimethametryn	0.2	97.5	84.9	78.4	67.8	84.1	82.5	13.1	1.0	73.0	73.9	70.8	79.7	91.0	77.7	10.5
217	Trichlorphon	12.0	41.2	64.8	44.5	34.6	26.9	42.4	33.5	60.0	60.3	49.7	39.4	58.5	46.3	50.8	17.1
218	Demeton(O+S)	1.6	94.8	50.9	101.0	93.5	70.9	82.2	25.4	8.0	40.9	61.6	82.8	46.5	67.9	59.9	28.1
219	Benoxacor	2.0	101.0	91.6	87.7	94.5	106.0	96.2	7.6	10.0	95.1	75.8	92.4	70.5	78.2	82.4	13.1
220	Bromacil	2.0	89.2	98.6	79.8	67.6	72.2	81.5	15.5	10.0	65.7	63.7	77.7	62.0	53.9	64.6	13.3
221	Phorate sulfoxide	120.0	55.5	41.2	46.3	51.3	33.7	45.6	18.7	600.0	54.7	46.4	52.8	64.1	60.7	55.7	12.4
222	Brompyrazon	6.0	55.4	55.4	43.0	42.5	42.2	47.7	14.7	30.0	37.8	38.3	37.4	47.9	50.7	42.4	15.0
223	Fludioxonil	8160.0	N/A	N/A	N/A	N/A	N/A	#DIV/0!	#DIV/0!	40800.0	N/A	N/A	N/A	N/A	N/A	#DIV/0!	#DIV/0!
224	Oxycarboxin	0.8	71.0	43.0	61.4	57.4	61.1	58.8	17.3	4.0	49.7	37.8	54.0	55.5	57.0	50.8	15.3
225	Mepronil	0.4	88.3	77.3	82.3	85.3	92.7	85.2	6.9	2.0	67.7	62.0	67.5	66.7	69.1	66.6	4.1
226	Fenthion	2.0	73.5	105.0	62.8	80.1	90.2	82.3	19.6	10.0	98.1	129.0	118.0	96.7	103.0	109.0	12.9
227	Metalaxyl	0.4	82.1	70.7	74.5	62.1	68.4	71.6	10.4	2.0	66.3	56.4	60.5	58.5	59.0	60.1	6.2
228	Pendimethalin	1.2	93.1	83.2	84.1	102.0	77.2	87.9	11.0	6.0	78.5	96.2	94.1	73.7	76.7	83.8	12.5
229	Ofurace	1.2	70.6	91.2	100.0	99.8	88.8	90.1	13.3	6.0	55.8	46.9	60.5	60.4	59.5	56.6	10.2
230	Imazamethabenz-methyl	1.2	61.2	58.8	68.5	58.4	68.1	63.0	7.9	6.0	59.7	52.6	65.9	75.1	70.8	64.8	13.8
231	Disulfoton-sulfoxide	0.8	47.1	79.0	70.6	88.8	83.8	73.9	22.2	4.0	59.3	62.7	69.1	58.9	64.9	63.0	6.7
232	Chlorthion	600.0	N/A	N/A	N/A	N/A	N/A	#DIV/0!	#DIV/0!	3000.0	N/A	N/A	N/A	N/A	N/A	#DIV/0!	#DIV/0!
233	Phoxim	8.0	66.4	61.3	50.6	58.3	62.7	59.9	9.9	40.0	82.6	99.0	80.8	114.0	117.0	98.7	17.2
234	Quinalphos	0.8	90.9	78.6	84.6	81.3	74.1	81.9	7.7	4.0	73.1	78.6	80.4	72.3	77.9	76.5	4.7

Table 23. Continued

No.	Compound	Level /µg /kg	Oats	Barley	Maize	Rice	Wheat	Ave.	RSD%	Level /µg /kg	Oats	Barley	Maize	Rice	Wheat	Ave.	RSD%
235	Hexachlorobenzene	1600.0	N/A	N/A	N/A	N/A	N/A	#DIV/0!	#DIV/0!	8000.0	N/A	N/A	N/A	N/A	N/A	#DIV/0!	#DIV/0!
236	Ditalimfos	40.0	86.2	76.3	90.1	83.6	78.6	83.0	6.7	200.0	74.7	82.8	85.3	69.6	71.9	76.9	8.9
237	MCPA-butoxyethyl ester	40.0	89.6	97.8	66.3	105.0	74.4	86.6	18.6	200.0	77.4	81.0	76.3	67.9	73.1	75.1	6.6
238	Chlorthal-dimethyl	2400.0	N/A	N/A	N/A	N/A	N/A	#DIV/0!	#DIV/0!	12000.0	N/A	N/A	N/A	N/A	N/A	#DIV/0!	#DIV/0!
239	Fenoxycarb	0.8	103.0	74.2	74.3	79.0	77.0	81.5	15.0	4.0	85.1	78.9	88.0	76.1	85.9	82.8	6.1
240	Pyrimitate	0.2	93.9	84.5	85.8	95.5	97.4	91.4	6.4	1.0	79.3	91.1	94.2	69.6	75.9	82.0	12.6
241	Imazamox	10.0	N/A	N/A	N/A	N/A	N/A	#DIV/0!	#DIV/0!	50.0	N/A	N/A	N/A	N/A	N/A	#DIV/0!	#DIV/0!
242	Fensulfothin	0.8	76.6	100.0	80.9	67.4	66.5	78.3	17.4	4.0	61.2	65.7	72.9	56.3	53.4	61.9	12.5
243	Fluorochloridone	2.4	81.4	67.5	83.0	88.4	106.0	85.3	16.3	12.0	85.9	96.5	91.5	77.2	84.9	87.2	8.3
244	Bromobutide	1.2	68.3	59.1	68.4	89.7	95.2	76.1	20.3	6.0	N/A	N/A	N/A	N/A	N/A	#DIV/0!	#DIV/0!
245	Imazaquin	40.0	43.0	22.0	35.2	10.9	10.1	24.2	60.3	200.0	18.2	22.8	26.8	22.2	21.4	22.3	13.9
246	Butachlor	4.0	102.0	121.0	88.8	90.3	88.4	98.1	14.2	20.0	96.9	88.7	96.3	61.4	69.3	82.5	19.7
247	Kresoxim-methyl	40.0	80.5	73.9	74.5	74.1	83.4	77.3	5.7	200.0	72.3	72.6	82.9	70.2	73.9	74.4	6.6
248	Dichlofenthion	2.4	170.0	81.5	116.0	84.6	113.0	113.0	31.5	12.0	79.1	118.0	98.5	83.8	88.3	93.5	16.5
249	Triticonazole	0.8	83.1	75.6	77.4	87.9	85.1	81.8	6.3	4.0	66.3	56.9	67.4	72.5	75.6	67.7	10.5
250	Fenamiphos sulfoxide	0.8	66.2	57.1	73.4	56.0	56.9	61.9	12.3	4.0	48.0	40.1	56.9	53.2	50.3	49.7	12.7
251	Thenylchlor	1.6	79.3	71.3	89.3	81.3	80.7	80.4	8.0	8.0	40.0	69.1	65.9	68.6	83.1	65.3	24.0
252	Pyrethrin	80.0	78.0	90.8	77.8	94.3	98.9	88.0	10.9	400.0	77.2	109.0	102.0	66.9	62.2	83.5	25.1
253	Fenoxanil	1.2	86.8	72.9	90.1	80.0	86.3	83.3	11.0	6.0	78.1	64.0	78.5	62.8	63.7	69.4	11.7
254	Epoxiconazole	40.0	80.6	76.8	84.9	78.9	79.1	80.1	3.8	200.0	81.3	62.8	77.4	72.0	70.4	72.8	9.7
255	Fluridone	0.4	75.8	74.7	82.2	74.5	76.3	76.7	4.1	2.0	63.5	57.9	73.7	71.6	64.2	66.2	9.7
256	Chlorphoxim	40.0	116.0	90.9	93.6	73.0	91.5	93.0	16.4	200.0	86.6	107.0	88.2	77.8	74.4	86.8	14.6
257	Fenamiphos sulfone	0.4	73.6	96.6	70.2	43.8	48.4	66.5	32.0	2.0	64.9	39.9	69.0	61.5	69.4	60.9	20.0
258	Oryzalin	20.0	29.7	30.5	27.1	14.5	12.5	22.9	37.9	100.0	22.5	20.6	22.4	18.2	17.2	20.2	12.0
259	Fenpropathrin	40.0	97.9	73.2	73.2	57.6	85.8	77.5	19.5	200.0	74.3	44.9	52.1	47.2	70.4	57.8	23.6
260	Phenothrin	40.0	80.1	118.0	75.1	82.1	72.5	85.6	21.7	200.0	44.9	48.3	45.4	78.5	62.5	55.9	26.0
261	Piperophos	0.2	117.0	116.0	122.0	90.2	77.2	104.5	18.8	1.0	91.9	105.0	87.3	71.9	68.5	84.9	17.6
262	Piperonyl butoxide	0.4	80.2	63.1	86.2	118.0	96.6	88.8	22.9	2.0	82.3	125.0	79.4	73.9	85.5	89.2	22.9
263	Oxyflurofen	8.0	133.0	84.0	105.0	115.0	125.0	112.4	16.9	40.0	79.9	118.0	109.0	71.5	71.7	90.0	24.4
264	Coumaphos	0.4	145.0	96.8	86.7	60.8	45.6	87.0	44.0	2.0	65.0	96.9	81.7	80.9	68.8	78.7	16.0

Table 23. Continued

No.	Compound	Level /μg /kg	Oats	Barley	Maize	Rice	Wheat	Ave.	RSD%	Level /μg /kg	Oats	Barley	Maize	Rice	Wheat	Ave.	RSD%
265	Pyridaben	0.4	86.1	58.4	68.4	95.1	108.0	83.2	24.0	2.0	84.5	87.0	65.2	57.5	61.4	71.1	19.2
266	Benzoylprop-ethyl	4.0	80.8	59.7	58.1	109.0	74.2	76.4	27.0	20.0	68.5	58.7	46.8	57.9	62.4	58.9	13.5
267	Phosalone	8.0	112.0	91.7	82.1	77.5	84.5	89.6	15.1	40.0	80.1	91.5	81.0	75.4	65.8	78.8	11.8
268	Methoxyfenozide	8.0	67.0	100.0	68.6	72.7	72.5	76.2	17.8	40.0	52.0	56.3	63.5	53.8	71.4	59.4	13.5
269	Fluquinconazole	0.2	82.7	68.2	54.3	126.0	177.0	101.6	49.2	1.0	67.4	61.5	85.9	61.4	82.2	71.7	16.2
270	Prochloraz	0.8	111.0	96.8	99.0	73.6	116.0	99.3	16.6	4.0	69.1	79.2	77.8	76.9	77.6	76.1	5.3
271	Naled	800.0	56.1	53.9	47.2	54.1	47.2	51.7	8.1	4000.0	N/A	N/A	N/A	N/A	N/A	#DIV/0!	#DIV/0!
272	Aspon	0.8	154.0	139.0	134.0	91.8	103.0	124.4	20.9	4.0	72.5	80.5	75.5	59.5	60.7	69.7	13.3
273	Thifensulfuron-methyl	2.0	49.7	33.8	41.9	44.5	42.8	44.5	23.1	10.0	34.1	33.5	48.4	51.0	47.9	40.0	11.6
274	Ethoxysulfuron	16.0	60.6	50.9	45.2	33.0	34.9	44.9	25.5	80.0	38.5	40.6	35.4	42.9	43.2	42.1	16.3
275	Dithiopyr	1.2	122.0	58.9	44.0	61.7	90.1	75.3	41.1	6.0	68.5	174.0	51.5	45.1	55.2	78.9	68.3
276	Leptophos	20.0	107.0	151.0	229.0	32.5	7.6	105.4	85.1	100.0	135.0	92.3	69.2	93.0	58.5	89.6	32.8
277	Spirodiclofen	4.0	112.0	99.7	50.8	99.6	110.0	94.4	26.5	20.0	91.1	53.0	90.2	61.8	70.1	73.2	23.2
278	Iodofenphos	12.0	47.3	49.3	51.2	47.2	37.6	46.5	11.3	60.0	83.0	80.2	78.3	82.1	85.5	81.8	3.3
279	Fenpyroximate	0.4	114.0	132.0	132.0	108.0	120.0	121.2	8.9	2.0	93.0	99.9	89.3	59.8	66.1	81.6	21.6
280	Flumiclorac-pentyl	2.0	132.0	84.7	96.8	101.0	113.0	105.5	17.0	10.0	71.4	106.0	104.0	60.4	55.3	79.4	30.3
281	Fluoroglycofen-ethyl	40.0	N/A	N/A	N/A	N/A	N/A	#DIV/0!	#DIV/0!	200.0	70.9	98.0	96.5	51.4	25.4	88.4	35.3
282	Temephos	4.0	136.0	91.3	112.0	90.3	111.0	108.1	17.3	20.0	64.7	94.3	96.4	73.6	66.8	79.2	19.2
283	Butafenacil	2.0	36.0	43.9	35.8	59.8	47.7	44.6	22.2	10.0	39.1	30.5	42.8	28.4	30.1	34.2	18.6
284	Spinosad	0.2	20.9	21.7	22.6	11.3	11.4	17.6	32.5	1.0	15.2	12.6	15.5	10.5	7.5	12.3	27.3
D group																	
285	Dazomet	160.0	N/A	N/A	N/A	N/A	N/A	#DIV/0!	#DIV/0!	800.0	N/A	N/A	N/A	N/A	N/A	#DIV/0!	#DIV/0!
286	Cyromazine	40.0	17.7	24.8	5.0	9.1	9.5	13.2	60.1	200.0	7.5	8.7	2.6	12.9	12.5	8.8	47.5
287	Crimidine	1.2	73.9	86.0	72.3	75.8	115.0	84.6	21.1	6.0	45.6	53.5	75.0	75.5	60.9	62.1	21.2
288	Propamocarb	2.0	29.0	19.4	22.5	24.8	19.8	23.1	17.1	10.0	13.9	12.2	17.5	16.4	16.9	15.4	14.6
289	EPTC	120.0	N/A	N/A	N/A	N/A	N/A	#DIV/0!	#DIV/0!	600.0	N/A	N/A	N/A	N/A	N/A	#DIV/0!	#DIV/0!
290	3,4,5-Trimethacarb	0.2	81.8	85.2	87.6	88.5	113.0	91.2	13.6	1.0	59.3	58.1	79.7	88.8	84.6	74.1	19.5
291	Fenfuram	0.4	61.6	72.0	70.1	78.0	83.5	73.0	11.3	2.0	56.1	52.6	71.6	77.4	85.7	68.7	20.5
292	Carbaryl	2.0	94.1	122.0	114.0	79.0	132.0	108.2	19.8	10.0	86.5	88.1	126.0	109.0	102.0	102.3	15.9
293	Metamitron	0.4	43.8	51.9	52.3	31.9	75.6	51.1	31.3	2.0	35.1	47.1	63.9	75.3	72.0	58.7	29.2

Table 23. Continued

No.	Compound	Level /μg /kg	Oats	Barley	Maize	Rice	Wheat	Ave.	RSD%	Level /μg /kg	Oats	Barley	Maize	Rice	Wheat	Ave.	RSD%
294	Vernolate	40.0	82.5	89.7	80.2	75.4	74.3	80.4	7.7	200.0	71.0	75.0	84.8	50.2	51.2	66.4	22.9
295	Isoproturon	0.4	70.9	75.5	71.5	78.7	73.5	74.0	4.3	2.0	52.4	58.6	73.7	82.7	79.7	69.4	19.2
296	Fenobucarb	0.8	85.3	53.4	77.4	95.7	90.4	80.4	20.6	4.0	66.6	65.6	77.9	69.7	71.2	70.2	6.9
297	Ethirimol	0.8	35.3	33.2	44.3	67.4	83.3	52.7	41.4	4.0	51.2	56.5	59.2	72.2	68.9	61.6	14.2
298	Chlorotoluron	0.4	96.4	84.9	113.0	77.7	105.0	95.4	15.1	2.0	60.6	58.0	84.5	78.4	81.4	72.6	17.0
299	Desmetryn	0.2	75.0	79.3	80.2	84.2	113.0	86.3	17.7	1.0	59.8	58.3	74.5	75.6	69.5	67.5	12.0
300	Aldicarb sulfone	7.2	N/A	N/A	N/A	N/A	N/A	#DIV/0!	#DIV/0!	36.0	N/A	N/A	N/A	N/A	N/A	#DIV/-0!	#DIV/0!
301	Sulfallate	24.0	81.8	97.9	67.9	70.4	66.8	77.0	17.1	120.0	57.4	66.1	73.3	66.4	63.6	65.4	8.8
302	Aziprotryne	0.4	63.3	59.2	64.6	72.9	102.0	72.4	23.9	2.0	76.4	69.4	80.4	83.6	83.3	78.6	7.5
303	Terbumeton	0.2	77.3	85.1	40.5	86.9	69.6	71.9	26.2	1.0	62.1	59.2	35.0	94.5	86.9	67.5	35.2
304	Tebuthiuron	0.2	41.5	59.8	60.4	67.0	70.1	59.8	18.6	1.0	42.5	48.5	67.9	75.7	84.1	63.7	27.8
305	Dimethoate	1.6	73.3	76.8	67.1	72.1	93.8	76.6	13.3	8.0	43.5	49.1	66.2	79.5	70.0	61.7	24.3
306	Trietazine	0.4	83.9	90.3	89.9	107.0	90.7	92.4	9.4	2.0	64.3	60.2	82.9	79.7	75.2	72.5	13.6
307	Fluometuron	6.0	63.1	74.6	77.2	81.8	86.3	76.6	11.5	30.0	60.2	62.0	74.8	82.4	75.6	71.0	13.4
308	Tebutam	0.2	75.6	62.9	90.5	114.0	94.1	87.4	22.2	1.0	66.1	54.2	85.3	79.0	82.8	73.5	17.8
309	Thiofanox-sulfoxide	24.0	49.9	51.6	60.3	44.9	51.5	51.6	10.8	120.0	41.6	56.7	80.0	71.3	61.4	62.2	23.5
310	Carboxin	0.4	63.3	75.3	74.8	80.0	69.0	72.5	8.9	2.0	44.0	41.1	56.9	67.7	62.4	54.4	21.2
311	Fenpiclonil	0.2	93.0	41.0	69.6	51.2	46.9	60.3	35.1	1.0	70.3	63.4	83.1	64.6	63.8	69.0	12.1
312	Carbetamide	1.6	62.7	86.3	67.4	77.9	85.9	76.0	14.1	8.0	41.5	55.4	71.8	56.5	58.8	56.8	19.0
313	Buturon	1.2	62.3	82.3	55.0	55.3	106.0	72.2	30.4	6.0	49.0	54.2	75.8	69.8	67.6	63.3	17.7
314	Dicrotophos	5.6	50.0	61.2	60.4	66.8	78.0	63.3	16.2	28.0	37.9	46.6	73.8	54.7	61.6	54.9	25.1
315	Diphenamid	0.2	76.1	76.2	73.3	75.8	78.7	76.0	2.5	1.0	59.1	59.5	89.3	75.9	80.6	72.9	18.2
316	Methacrifos	40.0	77.1	81.8	98.4	85.8	89.4	85.8	13.0	200.0	67.3	65.7	79.8	56.2	70.1	67.8	12.5
317	Ethoprophos	0.8	85.9	78.4	79.6	76.9	124.0	89.0	22.4	4.0	70.9	68.1	86.9	75.2	70.5	74.3	10.1
318	Demeton-s-methyl	16000.0	N/A	N/A	N/A	N/A	N/A	#DIV/0!	#DIV/0!	80000.0	N/A	N/A	N/A	N/A	N/A	#DIV/0!	#DIV/0!
319	Thionazin	4.0	76.0	80.4	84.6	68.4	98.6	81.6	13.8	20.0	67.0	61.5	80.4	79.1	76.0	72.8	11.3
320	4,4-dichlorobenzophenone	1.2	78.4	141.0	87.4	102.0	76.6	97.1	27.3	6.0	66.5	63.9	82.6	89.5	83.3	77.2	14.6
321	Hexazinone	0.4	57.7	76.2	77.9	69.9	58.5	68.0	14.0	2.0	37.4	49.1	72.3	73.8	67.8	60.1	26.7
322	Thiobencarb	0.4	76.2	58.5	89.5	99.8	101.0	85.0	21.0	2.0	62.8	47.7	88.2	73.3	87.9	72.0	24.0
323	Tri-iso-butyl phosphate	0.2	155.0	95.8	106.0	123.0	80.7	112.1	25.4	1.0	133.0	68.9	130.0	89.8	107.0	105.7	25.7

Table 23. Continued

No.	Compound	Level /μg /kg	Oats	Barley	Maize	Rice	Wheat	Ave.	RSD%	Level /μg /kg	Oats	Barley	Maize	Rice	Wheat	Ave.	RSD%
324	Disulfoton	40.0	N/A	N/A	N/A	N/A	N/A	#DIV/0!	#DIV/0!	200.0	N/A	N/A	N/A	N/A	N/A	#DIV/0!	#DIV/0!
325	Anilazine	200.0	117.0	100.0	116.0	61.5	90.9	97.1	23.4	1000.0	70.8	69.9	87.2	76.9	91.0	79.2	12.1
326	Propetamphos	32.0	86.2	61.3	81.6	109.0	57.9	79.2	26.2	160.0	78.0	59.0	91.9	74.4	84.3	77.5	15.9
327	Dodemorph	0.2	76.2	78.5	82.7	105.0	102.0	88.9	15.3	1.0	60.4	55.8	72.9	85.2	77.5	70.4	17.2
328	Fosthiazate	0.8	67.3	84.4	70.5	103.0	79.1	80.9	17.5	4.0	54.6	58.0	69.9	86.2	90.2	71.8	22.4
329	Terbufos	72.0	105.0	123.0	133.0	78.7	117.0	111.3	18.7	360.0	54.1	74.6	97.4	54.7	65.0	69.2	25.9
330	Isoprothiolane	0.4	84.2	61.1	78.2	119.0	98.4	88.2	24.7	2.0	62.3	52.1	88.3	84.3	87.2	74.8	22.1
331	Crufomate	32.0	77.9	84.6	87.3	94.2	83.8	85.6	6.9	160.0	66.4	63.1	84.9	89.2	81.8	77.1	15.1
332	Cyproconazole	0.3	94.8	100.0	113.0	96.6	75.9	96.1	13.9	1.6	56.7	50.7	82.3	75.6	78.7	68.8	20.6
333	Coumatetralyl	40.0	36.7	45.1	55.6	31.5	31.0	40.0	26.0	200.0	31.0	40.1	64.0	55.8	55.3	49.2	27.1
334	Triadimefon	2.0	62.0	76.0	80.2	54.8	72.9	77.2	15.5	10.0	57.4	58.9	73.7	79.5	80.2	69.9	15.8
335	Simeconazole	0.4	50.2	49.1	33.4	67.3	39.2	47.8	27.0	2.0	47.7	40.6	48.3	59.8	62.5	51.8	17.6
336	Triamiphos	0.8	71.8	84.8	83.8	78.0	82.9	80.3	6.7	4.0	47.2	55.0	74.1	73.8	76.6	65.3	20.4
337	Butralin	0.4	69.7	86.1	79.7	62.0	71.3	73.8	12.7	2.0	68.4	66.0	67.5	83.9	93.1	75.8	16.0
338	Imazalil	1.2	56.5	74.5	80.9	75.0	91.4	75.7	16.8	6.0	49.7	51.7	73.7	83.6	83.9	68.5	24.5
339	Tridemorph	4.0	20.7	18.4	16.0	15.5	15.2	17.2	13.7	20.0	18.7	16.8	16.9	16.2	15.5	16.8	7.1
340	Norflurazon	0.8	69.0	79.7	86.6	78.8	70.6	76.9	9.4	4.0	52.9	54.2	74.7	82.3	96.3	72.1	25.8
341	DE-PCB 52	24.0	155.0	113.0	200.0	132.0	152.0	150.4	21.6	120.0	139.0	80.8	131.0	92.4	102.0	113.0	22.9
342	Nuarimol	2.0	80.2	99.8	83.5	91.1	85.0	87.9	8.8	10.0	59.6	57.1	77.8	76.9	70.7	68.4	14.1
343	Imiprothrin	4.0	N/A	N/A	N/A	N/A	N/A	#DIV/0!	#DIV/0!	20.0	N/A	N/A	N/A	N/A	N/A	#DIV/0!	#DIV/0!
344	Tebupirimfos	40.0	82.3	91.1	88.5	85.3	91.3	87.7	4.4	200.0	75.1	77.5	87.5	78.1	80.7	79.8	6.0
345	Sulfotep	0.8	70.2	87.5	82.6	57.4	91.0	77.7	17.8	4.0	64.1	75.8	73.7	70.3	79.0	72.6	7.9
346	Sulprofos	1.6	67.8	125.0	93.7	89.0	123.0	99.7	24.3	8.0	38.1	94.7	95.8	82.6	79.3	78.1	30.1
347	Clomeprop	1.6	106.0	118.0	103.0	94.8	95.8	103.5	9.1	8.0	73.9	68.4	78.5	70.8	85.5	75.4	9.0
348	Benodanil	0.4	81.4	87.8	70.5	78.2	74.7	78.5	8.4	2.0	65.2	65.9	87.4	81.1	85.2	77.0	13.9
349	Sethoxydim	1.2	74.7	97.0	85.6	41.9	62.5	72.3	29.4	6.0	59.1	46.6	66.7	72.7	49.9	59.0	18.6
350	Pencycuron	0.2	114.0	137.0	124.0	99.2	83.6	111.6	18.7	1.0	57.2	69.0	87.7	79.3	74.7	73.6	15.5
351	Mecarbam	4.0	85.9	87.6	98.3	85.4	56.2	82.7	19.0	20.0	75.8	74.9	93.6	89.2	90.7	84.8	10.4
352	Dacthal	1200.0	99.5	106.0	88.9	103.0	99.8	99.4	6.5	6000.0	N/A	N/A	N/A	N/A	N/A	#DIV/0!	#DIV/0!
353	Fenbuconazole	0.2	101.0	89.6	117.0	60.2	136.0	100.8	28.4	1.0	54.9	56.4	75.1	83.1	76.1	69.1	18.4

Table 23. Continued

No.	Compound	Level /µg /kg	Oats	Barley	Maize	Rice	Wheat	Ave.	RSD%	Level /µg /kg	Oats	Barley	Maize	Rice	Wheat	Ave.	RSD%
354	4,4'-dibromobenzophenone	0.8	N/A	N/A	N/A	N/A	N/A	#DIV/0!	#DIV/0!	4.0	N/A	N/A	N/A	N/A	N/A	#DIV/0!	#DIV/0!
355	Pyridaphenthion	0.8	74.9	72.8	78.0	96.6	91.0	82.7	12.7	4.0	62.9	54.8	81.0	89.5	85.1	74.7	20.1
356	Isofenphos	80.0	104.0	93.0	94.5	91.3	84.3	93.4	7.6	400.0	57.1	68.0	91.7	85.2	74.0	75.2	18.2
357	Triflumizole	1.2	N/A	N/A	N/A	N/A	N/A	#DIV/0!	#DIV/0!	6.0	N/A	N/A	N/A	N/A	N/A	#DIV/0!	#DIV/0!
358	Chlorpyrifos	1.2	60.9	43.3	77.9	90.5	93.5	73.2	28.8	6.0	70.2	78.0	102.0	89.1	87.3	85.3	14.1
359	Fentin-chloride	40.0	N/A	N/A	N/A	N/A	N/A	#DIV/0!	#DIV/0!	200.0	N/A	N/A	N/A	N/A	N/A	#DIV/0!	#DIV/0!
360	Tebufenozide	8.0	68.8	61.3	52.5	49.0	64.7	59.3	14.0	40.0	54.4	91.4	127.0	88.2	83.0	88.8	29.2
361	Chlorfenvinphos	1.2	89.3	119.0	142.0	78.1	81.3	101.9	27.1	6.0	70.4	72.2	101.0	84.7	83.4	82.3	14.9
362	Pyraclofos	1.2	94.7	80.9	82.3	83.0	112.0	90.6	14.6	6.0	65.7	60.7	75.6	81.0	77.6	72.1	11.9
363	Flufenacet	0.4	66.5	67.4	72.1	77.0	66.3	69.9	6.6	2.0	53.2	55.7	75.9	78.6	71.2	66.9	17.5
364	Mefenpyr-diethyl	1.5	94.2	95.9	102.0	91.4	88.3	94.4	5.5	7.5	66.9	66.6	88.9	80.5	87.9	78.2	14.0
365	Profenefos	0.8	75.0	95.6	82.3	82.9	73.2	81.8	10.8	4.0	63.1	65.4	75.4	81.4	77.6	72.6	10.9
366	Ethion	1.6	102.0	103.0	83.9	89.9	75.5	90.9	13.0	8.0	69.1	77.3	97.0	88.5	81.0	82.6	12.9
367	Bromophos-ethyl	10.0	73.3	105.0	73.3	80.2	66.0	79.6	19.0	50.0	59.4	92.1	94.2	107.0	75.6	85.7	21.5
368	Kadethrin	0.8	76.2	86.2	107.0	82.0	79.5	86.2	14.2	4.0	57.5	66.3	92.8	82.4	85.3	76.9	18.9
369	Bensulide	10.0	94.4	181.0	222.0	95.1	105.0	139.5	42.0	50.0	50.5	84.1	80.0	84.6	83.6	76.6	19.2
370	Azoxystrobin	0.2	72.4	66.3	65.7	76.6	94.3	75.1	15.5	1.0	55.2	41.0	64.2	72.8	66.9	60.0	20.6
371	Difenoconazole	0.2	73.1	70.6	80.5	115.0	111.0	90.0	23.7	1.0	68.3	56.5	82.7	75.2	73.6	71.3	13.7
372	Propoxycarbazone-sodium	8.0	N/A	N/A	N/A	N/A	N/A	#DIV/0!	#DIV/0!	40.0	N/A	N/A	N/A	N/A	N/A	#DIV/0!	#DIV/0!
373	Fipronil	8.0	N/A	N/A	N/A	N/A	N/A	#DIV/0!	#DIV/0!	40.0	N/A	N/A	N/A	N/A	N/A	#DIV/0!	#DIV/0!
374	Lactofen	1600.0	49.6	70.7	72.8	72.5	74.6	68.0	15.3	8000.0	N/A	N/A	N/A	N/A	N/A	#DIV/0!	#DIV/0!
E group																	
375	Mepiquat chloride	1.2	13.2	3.9	4.4	19.9	26.1	13.5	71.6	6.0	2.9	0.9	2.7	3.6	3.6	2.7	41.1
376	Allidochlor	40.0	82.7	96.4	62.4	87.7	79.0	81.6	15.4	200.0	62.9	87.1	86.1	96.2	90.6	84.6	15.1
377	Isocarbamid	0.8	87.2	80.3	55.9	80.4	79.4	76.6	15.7	4.0	52.4	88.9	84.5	108.0	105.0	87.8	25.3
378	2,6-dichlorobenzamide	12.0	101.0	82.3	60.2	23.7	28.7	59.2	56.5	60.0	49.9	76.9	74.3	90.5	93.3	77.0	22.4
379	Tricyclazole	0.4	54.3	44.8	39.8	41.3	43.4	44.7	12.7	2.0	38.2	44.7	46.7	53.0	48.0	46.1	11.7
380	Thiabendazole	1.2	81.0	68.5	29.0	54.7	84.0	63.4	35.4	6.0	71.7	67.1	72.7	106.0	108.0	85.1	23.6
381	Pebulate	40.0	90.9	99.0	73.9	95.3	88.9	89.6	10.7	200.0	65.1	69.1	92.6	100.0	100.0	85.4	19.9
382	Dimethirimol	2.0	96.1	36.8	81.6	51.0	42.1	61.5	42.2	10.0	79.1	26.7	69.5	89.8	78.8	68.8	35.8

Table 23. Continued

No.	Compound	Level /µg /kg	Oats	Barley	Maize	Rice	Wheat	Ave.	RSD%	Level /µg /kg	Oats	Barley	Maize	Rice	Wheat	Ave.	RSD%
383	Atratone	0.2	109.0	73.9	86.7	32.3	80.7	86.5	15.5	1.0	57.8	57.4	92.0	108.0	113.0	85.6	31.2
384	Metribuzin	1.2	89.3	113.0	84.1	130.0	78.9	99.1	21.9	6.0	49.1	86.8	65.7	87.1	117.0	81.1	31.5
385	Dmst	2.4	65.0	70.0	49.3	78.7	79.8	68.6	18.1	12.0	75.7	76.6	83.0	96.5	94.7	85.3	11.5
386	Atrazine	0.1	81.2	79.0	104.0	78.7	89.5	86.5	12.4	0.6	90.0	82.0	72.2	87.6	101.0	86.6	12.2
387	Cycloate	2.0	85.4	85.5	67.0	78.0	96.0	82.4	13.0	10.0	65.5	62.0	99.4	97.6	104.0	85.7	23.6
388	Chloridazon	4.0	86.3	102.0	60.5	47.2	53.9	70.0	33.2	20.0	45.4	65.0	64.5	86.8	90.4	70.4	26.2
389	Aldoxycarb	40.0	N/A	N/A	N/A	N/A	N/A	#DIV/0!	#DIV/0!	200.0	55.9	70.5	66.1	172.0	98.5	92.6	50.9
390	Dioxacarb	6.0	N/A	N/A	N/A	N/A	N/A	#DIV/0!	#DIV/0!	30.0	64.9	97.7	70.1	68.7	85.2	77.3	17.8
391	Ethiofencarb	2.0	N/A	N/A	N/A	N/A	N/A	#DIV/0!	#DIV/0!	10.0	N/A	N/A	N/A	N/A	N/A	#DIV/0!	#DIV/0!
392	Cyprazine	0.1	97.0	109.0	81.1	75.5	91.8	90.9	14.5	0.6	67.5	90.1	94.9	83.0	90.8	85.3	12.7
393	Ametryn	0.4	130.0	131.0	108.0	96.8	92.1	111.6	16.3	2.0	69.2	76.5	86.2	94.3	84.8	82.2	11.7
394	Sebutylazine	0.2	117.0	69.0	89.7	81.8	64.3	84.4	24.7	1.0	57.2	60.1	81.2	93.3	85.8	75.5	21.2
395	Dibutyl succinate	16.0	114.0	135.0	93.0	83.6	88.0	102.7	20.9	80.0	72.6	99.9	88.7	96.6	92.8	90.1	11.8
396	Lenacil	8.0	94.4	107.0	79.3	69.5	84.7	87.0	16.5	40.0	61.6	80.6	80.4	98.3	99.7	84.1	18.6
397	Cartap	40.0	110.0	112.0	87.0	87.7	91.4	97.6	12.6	200.0	65.9	79.8	95.1	102.0	97.4	88.0	16.9
398	Terbutryn	0.2	97.5	91.3	89.2	80.6	92.9	90.3	6.9	0.8	69.5	68.2	90.4	87.9	90.9	81.4	14.1
399	Triazoxide	28.0	84.4	94.9	78.2	86.8	89.6	86.8	7.1	140.0	61.1	80.5	78.9	82.9	81.2	76.9	11.6
400	Linuron	4.0	114.0	126.0	111.0	75.1	105.0	106.4	17.5	20.0	78.9	90.1	88.2	103.0	95.3	91.1	9.8
401	Heptanophos	4.0	96.8	114.0	79.6	75.0	76.4	88.6	18.7	20.0	69.8	55.6	88.5	98.0	102.0	82.8	23.7
402	Prosulfocarb	0.2	89.4	124.0	73.0	67.8	111.0	93.0	26.0	1.0	77.2	98.7	101.0	108.0	108.0	98.6	12.8
403	Furmecyclox	40.0	40.2	71.8	86.9	53.4	52.5	61.0	30.1	200.0	49.9	72.0	64.0	61.0	63.5	72.1	14.3
404	Prohydrojasmon	8.0	139.0	119.0	122.0	75.6	84.7	108.1	24.8	40.0	64.0	53.6	92.5	103.0	96.8	82.0	26.6
405	Propyzamide	0.8	94.3	94.3	71.7	67.2	96.7	84.8	16.7	4.0	63.5	76.1	80.7	105.0	90.8	83.2	18.8
406	Phosfolan	4.0	87.4	62.3	77.0	66.0	63.1	71.2	15.2	20.0	N/A	N/A	N/A	N/A	N/A	#DIV/0!	#DIV/0!
407	Dipropetryn	0.1	83.0	80.2	76.5	77.8	97.8	83.1	10.4	0.6	55.8	56.4	73.6	88.4	93.7	73.6	23.9
408	Methiocarb sulfoxid	2.4	78.6	107.0	167.0	147.0	122.0	124.3	27.7	12.0	112.0	91.1	106.0	87.7	90.1	97.4	11.2
409	Imazapyr	12.0	N/A	N/A	N/A	N/A	N/A	#DIV/0!	#DIV/0!	60.0	N/A	N/A	N/A	N/A	N/A	#DIV/0!	#DIV/0!
410	Methyl-parathion	40.0	115.0	109.0	79.2	94.2	164.0	112.3	28.6	200.0	69.4	82.8	83.4	106.0	114.0	91.1	20.1
411	Tri-n-butyl phosphate	0.2	76.8	93.6	118.0	80.9	87.7	91.4	17.7	1.0	79.2	147.0	89.7	99.8	98.4	102.8	25.3
412	Diethofencarb	1.6	119.0	128.0	90.4	75.4	85.9	99.9	22.5	8.0	68.1	84.1	86.3	102.0	96.1	87.3	14.9

Table 23. Continued

No.	Compound	Level /µg /kg	Oats	Barley	Maize	Rice	Wheat	Ave.	RSD%	Level /µg /kg	Oats	Barley	Maize	Rice	Wheat	Ave.	RSD%
413	Alachlor	8.0	92.5	116.0	85.5	89.1	85.8	93.8	13.6	40.0	70.7	90.4	85.6	99.2	100.0	89.2	13.4
414	Cadusafos	0.8	107.0	162.0	125.0	78.8	79.5	110.5	31.5	4.0	90.0	109.0	122.0	105.0	91.9	103.6	12.7
415	Metazachlor	0.4	90.6	104.0	81.4	83.1	108.0	93.4	12.9	2.0	59.0	75.0	86.0	89.7	98.8	81.7	18.7
416	Dinoseb acetate	20.0	N/A	N/A	N/A	N/A	N/A	#DIV/0!	#DIV/0!	100.0	N/A	N/A	N/A	N/A	N/A	#DIV/0!	#DIV/0!
417	Metolachlor	0.2	40.6	57.7	89.1	86.9	122.0	79.3	39.6	1.0	65.8	87.3	63.9	108.0	117.0	88.4	27.2
418	Uniconazole	0.4	92.9	90.4	83.6	71.8	106.0	88.9	14.1	2.0	63.9	89.9	85.8	90.2	91.6	84.3	13.8
419	Phorate sulfone	1.6	98.1	103.0	77.3	74.9	79.4	86.5	15.0	8.0	61.1	78.4	80.4	103.0	93.2	83.2	19.1
420	Pyrifenox	0.4	91.1	116.0	109.0	77.9	94.0	97.6	15.5	2.0	80.0	88.9	64.4	106.0	100.0	87.9	18.8
421	Mefenacet	0.4	137.0	143.0	91.7	88.7	76.3	107.3	28.4	2.0	60.5	74.8	71.6	101.0	117.0	85.0	27.4
422	Azaconazole	0.2	105.0	103.0	70.2	76.3	69.9	84.9	20.8	1.0	71.1	61.2	90.1	100.0	90.8	82.6	19.3
423	Prallethrin	1.6	72.8	79.5	54.7	78.3	76.8	72.4	14.1	8.0	46.8	65.5	73.5	98.9	115.0	79.9	33.9
424	Diofenolan	760.0	89.2	74.5	81.1	96.8	96.1	87.5	11.0	3800.0	N/A	N/A	N/A	N/A	N/A	#DIV/0!	#DIV/0!
425	Phosmet-oxon	16.0	N/A	N/A	N/A	N/A	N/A	#DIV/0!	#DIV/0!	80.0	N/A	N/A	N/A	N/A	N/A	#DIV/0!	#DIV/0!
426	Fenamiphos	0.2	100.0	114.0	73.6	46.7	54.9	77.8	37.0	0.8	68.1	80.6	87.1	78.2	74.5	77.7	9.1
427	Fenpropimorph	0.2	75.6	67.0	55.0	72.9	42.2	62.5	22.2	0.8	60.4	62.3	77.5	65.1	68.3	66.7	10.1
428	Alpha-hch	8.0	99.3	97.6	79.4	81.4	100.0	91.5	11.2	40.0	84.0	88.0	95.8	101.0	86.8	91.1	7.7
429	Tebuconazole	0.2	131.0	144.0	103.0	78.8	97.4	110.8	23.8	1.0	82.1	99.1	97.3	87.2	88.2	90.8	7.9
430	Isopropalin	4.0	99.3	108.0	97.0	73.8	114.0	98.4	15.6	20.0	66.9	118.0	101.0	88.0	77.3	90.2	22.2
431	Isazofos	0.4	113.0	139.0	97.4	98.0	63.5	102.2	26.8	2.0	69.2	72.5	94.7	104.0	100.0	88.1	18.3
432	Hexaconazole	0.2	92.4	114.0	72.4	80.6	82.9	88.5	18.0	1.2	60.6	90.7	74.1	115.0	97.0	87.5	24.0
433	Bupirimate	0.4	84.7	96.4	72.5	90.7	103.0	89.5	13.0	2.0	65.6	87.3	91.5	89.8	90.5	84.9	12.9
434	Phosmet	1.2	104.0	123.0	67.3	66.6	98.8	91.9	26.7	6.0	57.3	82.9	96.3	118.0	120.0	94.9	27.5
435	Azinphos-methyl	40.0	102.0	104.0	70.2	78.5	83.8	87.7	16.9	200.0	80.3	72.4	118.0	91.9	120.0	96.5	22.5
436	Phenthoate	4.0	141.0	186.0	110.0	90.4	103.0	126.1	30.4	20.0	62.0	89.7	117.0	111.0	94.9	94.9	22.7
437	EPN	2.0	110.0	99.9	84.1	89.5	113.0	99.3	12.6	10.0	65.6	64.8	84.5	89.5	118.0	84.5	25.7
438	Azamethiphos	0.4	111.0	123.0	86.1	66.9	78.6	93.1	25.0	2.0	70.3	66.5	92.3	92.7	89.0	82.2	15.5
439	Flumetsulam	0.4	32.9	34.4	30.9	11.9	15.5	25.1	42.1	2.0	21.3	31.5	36.0	28.2	28.4	29.1	18.5
440	Diniconazole	0.4	101.0	111.0	83.9	75.8	73.1	89.0	18.5	2.0	68.7	92.2	102.0	97.0	94.7	90.9	14.2
441	Tralkoxydim	2.0	139.0	140.0	102.0	78.7	70.7	106.1	30.7	10.0	73.6	104.0	98.7	124.0	112.0	102.5	18.3
442	Malathion	1.2	68.0	77.0	80.0	58.1	50.2	66.7	18.8	6.0	N/A	N/A	N/A	N/A	N/A	#DIV/0!	#DIV/0!

Table 23. Continued

No.	Compound	Level /μg /kg	Oats	Barley	Maize	Rice	Wheat	Ave.	RSD%	Level /μg /kg	Oats	Barley	Maize	Rice	Wheat	Ave.	RSD%
443	Athidathion	4.0	127.0	90.1	125.0	83.7	87.4	102.6	20.9	20.0	N/A	N/A	N/A	N/A	N/A	#DIV/0!	#DIV/0!
444	Pyributicarb	0.4	103.0	115.0	91.7	70.5	82.6	92.6	18.7	2.0	71.2	120.0	96.4	103.0	104.0	98.9	18.0
445	Pirimiphos-ethyl	0.1	99.4	110.0	86.7	85.2	94.3	95.1	10.6	0.6	69.9	110.0	83.1	113.0	113.0	97.8	20.5
446	Benfluralin	160.0	34.5	85.0	26.8	21.5	39.6	41.5	61.0	800.0	59.2	60.5	73.8	65.3	66.2	65.0	8.9
447	Alanycarb	2.4	N/A	N/A	N/A	N/A	N/A	#DIV/0!	#DIV/0!	12.0	N/A	N/A	N/A	N/A	N/A	#DIV/0!	#DIV/0!
448	Nitralin	4.0	110.0	102.0	73.0	59.5	97.7	83.4	24.1	20.0	49.9	76.7	92.7	120.0	86.9	85.2	29.8
449	Thiodicarb	4.0	49.0	55.0	63.7	51.2	42.3	52.2	15.1	20.0	61.0	55.7	35.7	67.8	75.1	59.1	25.3
450	Florasulam	1.6	51.0	68.0	49.7	21.0	23.1	42.6	47.2	8.0	N/A	N/A	N/A	N/A	N/A	#DIV/0!	#DIV/0!
451	Chlorthiophos	2.0	87.3	128.0	75.7	131.0	92.8	103.0	24.3	10.0	73.9	89.6	140.0	81.4	107.0	98.4	26.8
452	Picoxystrobin	40.0	94.7	108.0	73.6	92.9	80.6	90.0	14.8	200.0	68.9	95.7	98.5	120.0	126.0	101.8	22.2
453	Tetraconazole	0.4	52.1	43.6	48.4	54.9	38.3	47.5	14.0	2.0	38.6	43.5	48.9	55.0	53.1	47.8	14.2
454	Pyraclostrobin	0.8	53.1	63.2	70.1	80.8	67.2	66.9	15.1	4.0	54.6	89.0	54.9	62.5	116.0	75.4	35.4
455	Dimethomorph	0.4	77.8	127.0	79.1	77.7	78.6	88.0	24.7	2.0	58.4	84.3	76.4	87.7	90.1	79.4	16.1
456	Etofenprox	40.0	70.2	81.1	79.8	72.1	73.8	75.4	6.4	200.0	74.3	97.1	97.5	120.0	137.0	105.2	22.8
457	Thiazopyr	1.2	74.0	86.0	69.4	59.4	86.3	75.0	15.3	6.0	60.5	67.0	65.7	70.5	75.8	67.9	8.4
458	Bensulfuron-methyl	40.0	99.7	113.0	82.0	76.0	66.7	87.5	21.3	200.0	65.0	90.0	90.2	98.3	97.2	88.1	15.3
459	Benfuracarb	40.0	N/A	N/A	N/A	N/A	N/A	#DIV/0!	#DIV/0!	200.0	N/A	N/A	N/A	N/A	N/A	#DIV/0!	#DIV/0!
460	Cinosulfuron	0.8	39.7	76.7	59.0	51.6	42.7	53.9	27.5	4.0	71.0	75.4	47.8	73.7	75.1	68.6	17.1
461	Pyrazosulfuron-ethyl	0.4	92.3	108.0	76.2	55.2	53.5	77.0	30.6	2.0	56.8	78.6	76.4	88.1	99.8	79.9	19.9
462	Flubenzimine	40.0	N/A	N/A	N/A	N/A	N/A	#DIV/0!	#DIV/0!	200.0	N/A	N/A	N/A	N/A	N/A	#DIV/0!	#DIV/0!
463	Metosulam	2.0	N/A	N/A	N/A	N/A	N/A	#DIV/0!	#DIV/0!	10.0	N/A	N/A	N/A	N/A	N/A	#DIV/0!	#DIV/0!
464	Iodosulfuron-methyl	16.0	N/A	N/A	N/A	N/A	N/A	#DIV/0!	#DIV/0!	80.0	N/A	N/A	N/A	N/A	N/A	#DIV/0!	#DIV/0!
465	Chlorfluazuron	4.0	N/A	N/A	N/A	N/A	N/A	#DIV/0!	#DIV/0!	20.0	N/A	N/A	N/A	N/A	N/A	#DIV/0!	#DIV/0!

2. For LC-MS-MS Pesticides

The different grain samples free from pesticide residues were used for fortification precision tests. The samples were held for 30min after adding pesticides so that the pesticides were thoroughly absorbed before proceeding to extraction, cleanup and determination according to the said LC-MS-MS method. The data of recoveries and precision at two fortification levels for 5 grains are given in Table 23. Some of the pesticides had unstable recoveries at low levels, with RSD% larger than 25% but larger than 60% at high levels and RSD% less than 25%, which were found to be involved with 31 pesticides, accounting 8% of the total. : Pyrazosulfuron-ethyl, Cinosulfuron, Benfluralin, Fenamiphos , Methyl parathion , Furmecyclox, Dioxacarb, Thiabendazole , 2,6-dichlorobenzamide , DE-PCB 52 , Butralin, Fenpiclonil, Chlorimuron ethyl, Triasulfuron, Clethodim, Flurtamone, Cycloxydim, Bifenazate, Imibenzonazole-des-benzyl, Clothianidin, 3-hydroxy carbofuran, Diphenylamin, Prothiophos, Chlorprifos methyl, Dinitramine, Triadimenol, Myclobutanil, Terbucarb, Fenitrothion, Ethidimuron, Imidacloprid. Some pesticides had recoveries between 40%~60% at both high and low levels, with RSD% less than 30%, which was found to be involved with 9 pesticides, accounting for 3% of the total: Tetraconazole, Thiodicarb, Coumatetralyl, Ethoxysulfuron, Ethoxyquin, Thifensulfuron-methyl, Vamidothion sulfone, Acephate, Iprovalicarb. Some pesticides was tested at only low level fortification because of the limited quantity of these pesticides, and the recoveries and RSD% turned out to be satisfactory, which was found to be involved with 10 pesticides, accounting for 3%: Athidathion, phosfolan, Tricyclazole, Lactofen, Dacthal, Naled, Bromobutide, Thiometon, Dioxathion, Profluralin. There were 50 pesticides of such kind in total, accounting for 14% of the total, and such pesticides and related chemicals were retained in this method. Limit of detection for these 50 pesticides and related chemicals was raised to high level fortification. Finally, 372 pesticides and chemicals were found to be capable of being analyzed by this method.

REFERENCES

[1] Luke M A, Froberg J E, Masumoto H T (1975) *J. AOAC* 58: 1020-1026.

[2] Luke M A, Froberg J E, Doose G M, Masumoto H T (1981) *J. AOAC* 64,(5) : 1187-1195.

[3] Liao W, Toe T, Cusick W G (1991) J. AOAC 74, (3): 554-565.

[4] Lehotay S J, Mastovska K, Schenck F J (2003) Abstracts of Papers, 225th ACS National Meeting.

[5] Mastovska K, Lehotay S J, Stajnbaher D, Schenck F J (2003) *J.AOAC Int.* 86, (2) : 412-431.

[6] Lehotay S J, Mastovska K (2004) Abstracts of Papers, 228th ACS National Meeting.

[7] Lehotay S J (2007) *J. AOAC Int.* 90: 485-520.

[8] Fillon J, Hindle R, Lacroix M, Selwyn J (1995) *J.AOAC Int.* 78, (5) : 1252-1266.

[9] Fillon J, Sauve F, Selwyn J (2000) *J.AOAC Int.* 83, (3) : 698-713.

[10] Ueno E, Oshima H, Saito I, Matsumoto H, Yoshimura Y, Nakazawa H (2004) *J. AOAC Int.* 87, (4) : 1003-1015.

[11] Ueno E, Oshima H, Saito I, Matsumoto H, (2003) *J. AOAC Int.* 86, (6) : 1241-1251.

[12] Obana H, Kikuchi K, Okihashi M, Hori S (1997) *Analyst* 122: 217.

[13] Obana H, Kazuhiko K, Okihashi M, Kakimoto S, Hori S (1999) *Analyst* 124:1159-1165.

[14] Obana H, Akutsu K, Okihashi M, Hori S (2001) *Analyst* 126:1529-1534.
[15] Stan H J (2000) *J. Chromatogr. A* 892: 347-377.
[16] Stan H J, Linkerhagner M (1996) *J. Chromatogr. A* 750: 369-390.
[17] European Commission's Proficiency Tests on Pesticide Residues in Fruit and Vegetables 1996/1997, Final Report August 1997, National Food Administration, Uppsala,1997.
[18] Method IA-Multi-residue, National Program of Residues, M.A.P.A.(Spanish Ministry of Agriculture, Fisheries and Food), Spain, 1996.
[19] Pang, G.F., Fan, C.L., Liu, Y.M.,et al (2006) *J. AOAC Int.* 89, (3) : 740-771.
[20] Pang, G.F., Liu,Y.M.,Fan, C. L., et al (2006) *Anal. Bioanal. Chem.* 384: 1366-1408.
[21] Pang, G.F., Chao, Y. Z.,Zhang, J. J., et al (2006) *J. Chromatogr. A* 1125: 1-30.
[22] Pang, G. F., Fan, C.L. Liu, Y.M.,et al (2006) *Food Additives and Contaminats* 23, (8) : 777-810.
[23] Richter B E, Jones B A, Ezzell J L, Porter N L, Avdalovic N, Pohl C (1996) *Anal. Chem.* 68: 1033-1039.
[24] SW-846 Test Methods for Evaluating Solid Waste, Method 3545 （USEPA）.
[25] L00.00-34 Analysis of Foodstuffs, Modular Multiple Analytical Method for the Determination of Pesticide Residues in Foodstuffs, 1999.
[26] Mastovska K, Lehotay S J J (2004) *J. Chromatogr. A* 1040: 259-272.
[27] Canadian Food Inspection Agency, Laboratory Services, (2004) PMR-005-V1.1 Determination of Pesticides in Difficult Matrix Fruit and Vegetable (With Solid Phasc Extraction Clean-Up and GC/MSD and HPLC Fluorescence Detection).

In: Progress in Pesticides Research
Editor: Christopher M. Kanzantzakis

ISBN 978-1-60456-201-9

Chapter 2

PRIMARY AND SECONDARY TARGETS OF ACTION-RESPONSE TO ANTICHOLINESTERASE PESTICIDE EXPOSURE IN FISH: TRENDS IN UNDERLYING MOLECULAR MECHANISMS

Ana Ferrari*[*]*, Ana María Pechen de D'Angelo and Andrés Venturino

LIBIQUIMA, Unidad Ejecutora Neuquén, CONICET-Universidad Nacional del Comahue. Neuquén, Argentina

ABSTRACT

Organophosphorus (OP) and carbamate (CB) pesticides are known to inhibit the enzyme acetylcholinesterase (AChE) in vertebrates and invertebrates, leading to the accumulation of the neurotransmitter acetylcholine and the subsequent over-stimulation of cholinergic receptors. In fish, the effects of these compounds on AChE and different esterases have been extensively studied. In most cases, OP are more effective and persistent AChE inhibitors than CB, due to both differences in reactivation rates and the aging process. Brain AChE inhibition has been proposed as the major determinant of OP and CB acute toxicity in mammals and other organisms. However, many fish species are able to survive high percentages of brain AChE inhibition indicating that there are other determinant targets of anticholinesterase pesticides acute toxicity such as muscular ChE. The protective action of carboxylesterases seems to be also variable among fish in relation to the one observed in other organisms.

In addition to esterases, secondary targets are currently studied in order to elucidate molecular mechanisms modulating the action of OP and CB and the cellular response to exposure. In fish, different OP and CB have been reported to induce oxidative stress and affect glutathione, glutathione-related enzymes, and other antioxidant enzymes such as

[*] Mailing address: Ana Ferrari. LIBIQUIMA, Universidad Nacional del Comahue. Buenos Aires 1400. 8300 Neuquen. Argentina.; e-mail: aferrari@uncoma.edu.ar

catalase and superoxide dismutase. It is known that reactive oxygen species modulate cell signaling and transcription factor activity, affecting gene expression. These effects related to pesticide exposure have been mainly reported in mammalian cell lines, whereas research on fish is scarce. Some OP and CB induce glutathione-S transferase enzymes in fish, and this induction may be related to the nuclear factor Nrf-2 binding to the antioxidant response element (ARE). In addition, some CB are weak agonists of the aryl hydrocarbon receptor (AhR), activating the xenobiotic response element (XRE) and inducing cytochrome P450 1A (CYP1A). The differential modulation of gene expression in OP and CB exposures may be crucial to cell fate towards apoptosis or proliferation and sublethal/ chronic toxicity. Thus, the study of these molecular targets in fish is highly relevant for the elucidation of molecular mechanisms of toxicity and the searching of earlier and more sensitive biomarkers of effect and response.

1. INTRODUCTION

Organophosphorus (OP) and carbamates (CB) are the most widely used insecticides in the world and there is still a trend towards the use of both families of insecticides in developing countries. Both compounds are extremely toxic and a threat to the environment, humans and wildlife health. OP and CB are anticholinesterase agents that bind to the active site of the enzyme acetylcholinesterase (AChE) (Matsumura, 1985). AChE inhibition has been widely used as a biomarker of OP and CB exposure in aquatic organisms including fish (Edwards and Fisher, 1991; Gruber and Munn, 1998; Abdel-Halim et al., 2006).

Brain cholinesterase (ChE) is considered the main target of OP and CB insecticides, and the decrease in its activity is generally associated with lethality. However, the correlation between its inhibition and mortality has not been fully established in fish (Fulton and Key, 2001). A more relevant role of muscular ChE inhibition as a determinant of OP and CB acute toxicity has been recently reported (Ferrari et al., 2007a). In addition, other esterases act as suicide detoxifying enzymes of OP and CB. These enzymes, such as the carboxylesterases (CarbE), may sequester these pesticides and then protect ChEs from inactivation (Maxwell, 1992). CarbE are also considered specific biomarkers in OP and CB exposures (Denton et al., 2003; Wogram et al., 2001; Küster and Altenburger, 2006).

Although OP and CB insecticides are primarily recognized by their anticholinesterase action, they have been reported to affect secondary targets including detoxifying and antioxidant responses in various species (Hai et al., 1997; Delescluse et al., 2001; Anguiano et al., 2001; Venturino et al., 2001; Ferrari et al., 2007b). The sensitivity to oxidants is modulated by diverse antioxidant enzymes and compounds such as reduced glutathione (GSH) (Livingstone, 2001, Winston and Di Giulio, 2001). The detoxifying and antioxidant systems protect aquatic species from endogenous and exogenous aggression and impairment of their function could conduce to cell damage. In addition, alteration of redox status may lead to a modification in cell signaling and gene expression (Kamata and Hirata, 1999; Biswas et al., 2006). As the OP and CB are able to induce oxidative stress and affect biotransformation enzymes, they could affect cellular signaling as well as transcription factors and hence cellular homeostasis. In addition, direct effects of those compounds on cell receptors and signaling pathways have been reported (Jett and Lein, 2006).

The underlying mechanisms of OP and CB toxicity have to be investigated in order to get a better understanding of sublethal effects that become relevant in chronic toxicity. Besides, the application of new techniques associated with ecotoxicogenomics will contribute to both the identification of OP and CB molecular targets in fish and the finding of novel and sensitive biomarkers.

This chapter discusses the relationship between fish cholinesterase and mortality and their usefulness as a biomarker of OP and CB exposure. It also describes the antioxidant and detoxifying systems as alternative targets of those compounds in fish. Finally, it highlights some of the current trends of research in the underlying molecular mechanisms of OP and CB toxicity in fish and their relevance in acute and chronic toxicity.

2. Esterases as Primary Target of Action of Organophosphorus and Carbamate Insecticides

Acethylcholinesterase (AChE) enzyme (EC 3.1.1.7) is present in the central and peripheral nervous systems in both vertebrate and invertebrate animals (Matsumura, 1985). This enzyme belongs to the group of B-esterases that can be inhibited by OP and CB pesticides (Jokanović, 2001). AChE effectively catalyzes the cleavage of the neurotransmitter acetylcholine into choline and acetate (Fukuto, 1990). OP and CB inhibit this enzyme by phosphorylating or carbamylating the serine hydroxyl group in its active site. The inhibition of AChE caused by OP insecticides is highly persistent in fish as well as in other species due to the very slow rate of spontaneous hydrolytic reactivation, and because of the aging process which leads to the irreversible inactivation of the enzyme. On the other hand, the carbamylated enzyme shows generally a fast recovery after CB exposure (Costa, 1997).

AChE inhibition/inactivation by OP and CB leads to acethylcholine accumulation with subsequent overstimulation of cholinergic receptors on postsynaptic cells (Pope et al., 2005). In general, the principal symptoms of anticholinesterase poisoning reflect a cholinergic crisis that includes increased secretion, muscle fasciculation, cardiovascular effects and miosis, that could finally lead to tetany, coma and death (Costa, 1997). In fish, the reported OP and CB acute toxic symptoms are muscle paralysis, hyperactivity and loss of equilibrium (Zinkl et al., 1991). Moreover, recent studies have reported several behavioral alterations that are related with AChE inhibition (Beauvais et al., 2000; 2001; Jarrard et al., 2004; Venkateswara Rao et al., 2005).

In addition to AChE, vertebrate ChE comprise butyrylcholinesterase enzyme (BuChE; EC 3.1.1.8), also called pseudocholinesterase or serum esterase (Jokanović, 2001). The physiological function of BuChE is not yet completely known. The brain from different fish species contains mainly AChE, while BuChE is present in other tissues like serum and liver (Habig and DiGiulio, 1991). While most freshwater fish species have absent or low BuChE levels in muscle (Magnotti et al. 1994), some marine fish have significant BuChE activity in this tissue (Sturm, et al. 1999a). It is remarkable that BuChE is more sensitive than AChE to anticholinesterase compounds in some fish species (Wogram et al., 2001, Fulton and Key, 2001).

Carboxylesterases (CarbE; EC 3.1.1.1) are other B-esterases that hydrolyze carboxylic esters and have a role in the detoxification of exogenous and endogenous substances

(Jokanović, 2001; Barron et al., 1999). They are relevant in the detoxification of many OP and CB containing carboxylic esters (Fukuto, 1990; Jokanović, 2001). In addition, CarbE act as alternative targets of action of OP and CB, sequestering those insecticides and thus protecting AChE (Maxwell, 1992; Jokanović, 2001). Consequently, these enzymes are considered to be involved in the development of tolerance to some OP compounds (Dettbarn et al., 1999; Barata et al., 2004, Chambers and Carr, 1995).

2.1. Relationship between Fish ChE Inhibition and Mortality

The sensitivity of brain ChE to inhibition by OP and CB has been proposed as the main determinant of the *in vivo* acute toxicity (Matsumura, 1985). However, the relationship between ChE inhibition and mortality is not clear or at least comparable among different fish species. Some studies pointed out that even a low ChE inhibition was lethal for some fish species (Weiss, 1958) or proposed that in many fish species OP and CB concentrations close to the lethal concentration fifty (LC_{50}) caused a brain ChE inhibition between 70-90% (Coppage, 1972; Zinkl et al., 1991). On the contrary, other species were able to survive with high percentages (90-95%) of brain enzyme inhibition (Chamber and Carr, 1995; Fulton and Key, 2001; Ferrari et al., 2004a, b). Fish species showed a high variability between LC_{50} and EC_{50} (brain ChE inhibition) ratios from different OP exposure (Table 1). On the other hand, most CB insecticides show lower ratios than OP compounds, with the remarkable exception of rainbow trout exposed to carbaryl where the ratio is about 300. Interestingly, both CB and OP LC_{50}/EC_{50} ratios were low in muscle.

The information regarding the relationship between ChE inhibition and lethality is completed by scientific literature reporting significant ChE inhibition at insecticide concentrations that cause either low or no mortality at all (Table 2). Brain ChE inhibition between 70 and 90% did not cause mortality in different fish species. These data differ from an earlier report by Coppage et al. (1972).

Therefore, the relationship between mortality and ChE inhibition differs among compounds and fish species and no rules can explain interspecies variability. Some species of fish apparently do not need high levels of active brain ChE to sustain life (Boone and Chambers, 1996). The relative lack of susceptibility to OP and CB in some fish is unlikely to be related to a reduced uptake or greater detoxification rate because, indeed, ChE activity became significantly inhibited (Ferrari et al, 2004a). Recent studies performed with some fish species and anticholinesterase compounds indicate that muscular ChE inhibition is more closely related to mortality than brain ChE (Ferrari et al., 2007a; Bretaud et al., 2000). This hypothesis is supported by a closer LC_{50}/EC_{50} (muscular ChE inhibition) ratio (Table 1).

In that sense, the main accepted cause of death in fish is asphyxiation due to the combination of factors including gill muscle paralysis and cholinergic-induced bradycardia (Zinkl et al., 1991). In relation to that, some authors have remarked the importance of muscular ChE in the control of muscular function (Payne et al., 1996). Indeed, muscular ChE represents the largest pool of ChE activity in the body (Perkins and Schlenk, 2000). In coincidence, there are information regarding OP and CB concentrations that significantly inhibit brain, but not muscular ChE, without causing fish mortality (Perkins and Schlenk, 2000; Bretaud et al., 2000; Ferrari et al, 2007). Nevertheless, some fish are able to survive high levels of muscular ChE inhibition (Table 2). Whereas low concentrations of OP and CB may have none or little effect on muscular ChE, higher concentrations can effectively inhibit it (Bretaud et al., 2000; Ferrari et al, 2007a). This inhibition is associated to high mortality in some fish, but causes no lethal effects in others. Fulton and Key (2001) have pointed out that fish muscular ChE sensitivity may be either similar or lower than the one determined in brain. Conversely, other authors (Whitehead et al., 2005) have reported muscular ChE sensitivity to be slightly higher than brain ChE in OP exposed fish. In conclusion, the relationship between muscular as well as brain ChE inhibition and mortality cannot be clearly established in fish.

Table 1. Relationship between anticholinesterase LC_{50} and EC_{50} (ChE inhibition) in fish

Fish Compound	96h-LC_{50}*	96h-EC_{50} for ChE inhibition*	Ratio LC_{50}/EC_{50}	References
Carassius. auratus				
Parathion	6.5	B:< 0.05	>130	Ferrari et al., 2004a
Azinphos methyl	7.2	B: 0.05	144	
Carbaryl	13.9	B: 2.62	5.3	
Channa striata				
Diazinon	0.79	B: 0.06[b]	13.2	Van Cong et al., 2006
Dicentrarchus labrax				
Dichlorvos	3.5	B: 0.31	11.3	Varò et al., 2003
		M: 1.33	2.6	
Fundulus heteroclitus				
Azinphos methyl	0.065-0.084	B: < 0.001[b]	74.5	Van Dolah et al., 1997
Gambussia yucatana				
Carbofuran	0.77[a]	M: 0.22[b]	3.5	Rendón-von Osten et al., 2005
Chlorpyrifos	0.011	M: 0.011	1	
Oncorhynchus mykiss				
Azinphos methyl	0.007	B: 0.0004	17.5	Ferrari et al., 2004b
		M: 0.001	6.7	Ferrari et al., 2007a
Carbaryl	5.40	B: 0.019	284	Ferrari et al., 2004b
		M: 0.27	20	Ferrari et al., 2007a
Scianops ocellatus				
Azinphos methyl	0.006-0.007	B: >0.005[b]	1.5	Van Dolah et al., 1997

* mg/L.

[a] 24h-LC50.

[b] 24h-EC50 for ChE inhibition.

B: brain ChE, M: muscular ChE.

Table 2. Levels of fish ChE inhibition by organophosphorus and carbamate at either sublethal concentrations of those that cause low mortality

Fish Compound	Concentratio n tested* (LC_{50})*	Mortality (%)	ChE inhibition (%)	References
Lepomis macrochirus (larvae)				
Chlorpyrifos (48h) (Prolarvae)	0.1 (3.16)	0	About 90	Phillips et al., 2002a
Chlorpyrifos (48h) (Postlarvae I)	0.01 (0.024)	0	About 85	
Lepomis macrochirus				
Malathion (24h)	0.48 (NR)	0	60	Richmond and Dutta, 1992
Carassius. auratus				
Carbofuran (48h)	0.05 (NR)	0	B: 19-28 M: NS	Bretaud et al., 2000
	0.50 (NR)	0	B: 85-87 M: 86-92	
Gambusia affinis				
Parathion (48h)	0.15 (0.35)	0	B: 89 M: 96	Boone and Chambers, 1996
Methyl parathion (48h)	8.0 (13.48)	0	B:89 M: 90	(Chambers and Yarbrough, 1974)
Chlorpyrifos (48h)	0.1 (0.15)	0	B: 77 M:82	
Anguilla anguilla				
Diazinon (96 h)	0.042 (0.085)	0	B: 79 P: 99	Cerón et al., 1996. (Sancho et al., 1992)
Fenitrothion	0.04 (0.2)	0	B: 64	Sancho et al., 1997. (Ferrando et al., 1991)
Oncorhynchus kisutch				
Chlorpyrifos	0.0025 (NR)	0	B: 64 M:67	Sandahl et al., 2005.
Oncorhynchus tshawysyscha				
Chlorpyrifos (96h)	0.0073 (NR)	20	B: 85 M: 92	Eder et al., 2004.
Icatlurus punctatus				
Aldicarb (48h)	1 (9.75)	0	B:85-90	Perkins and Schlenk, 2000.
Catostomus occidentalis				
Diazinon (96 h)	0.010 (0.052)	0	B: about 75 M: about 80	Whitehead et al., 2005.

* mg/L.

B: brain ChE, M: muscular ChE, P: plasmatic ChE. NR: LC_{50} value not reported.

2.2. Cholinesterase Inhibition and Recovery

The inhibition and recovery profiles of fish esterase activities following anticholinesterase exposure show some interesting aspects. Brain ChE is in general quickly inhibited in exposed fish. Brain AChE activity of the European eel (*Anguilla anguilla*) was significantly reduced in only 2 h of sublethal exposure to fenitrothion, even though the maximum decline required 96 h (Sancho et al., 1997). A significant inhibition of mosquitofish (*Gambussia affinis*) brain and muscular ChE occurred between 4-12 h of sublethal exposure to the OP chlorpyrifos, parathion and methyl parathion (Boone and Chambers, 1996). An early (2-6 h) significant ChE inhibition has also been reported in goldfish exposed to the CB carbofuran (Bretaud et al., 2000) and also in the cutthroat trout exposed to carbaryl (Labenia et al., 2007). Significant and high levels of ChE inhibition were observed after the first 24 h of exposure to OP and CB insecticides in goldfish, rainbow trout (Ferrari et al., 2004 a, b) and sunfish (*Lepomis gibbosus*) (Benke et al., 1974).

The recovery process is dependant on both the concentration and the type of compound tested (Sancho et al., 1997). Recovery of ChE activity after insecticide inhibition may occur through spontaneous reactivation and/or new synthesis of enzyme. The first process is much more relevant after ChE inhibition by CB, but is practically insignificant for the phosphorylated enzyme (Costa, 1997). Thus, the recovery of ChE activity after OP inhibition occurs mainly through *de novo* enzyme synthesis.

In that sense, recovery of ChE activity following exposure to OP pesticides has been found to be particularly prolonged in fish (Table 3). No significant recovery of mosquitofish brain and muscular ChEs previously inhibited by OP was reached after 96 h in clean media (Boone and Chambers, 1996). Brain AChE activity in tilapia (Zinkl et al., 1991) and European eel (Sancho et al, 1997) was still inhibited after 7 days of recovery from OP exposure. Juvenile goldfish brain ChE needed more than 35 days of depuration in clear water to recover a substantial level of activity after sublethal exposure to OP (Ferrari et al. 2004a). Similarly, brain ChE of rainbow trout previously exposed to a sublethal concentration of azinphos methyl was only partially recovered after 21 days in uncontaminated media (Ferrari et al., 2003), while no significant recovery was observed in the muscular enzyme (Ferrari et al., 2007a). This difference between brain and muscular ChE responses might be related to a different rate of de novo enzyme synthesis in both tissues.

On the other hand, a fast recovery of fish ChE is normally seen after an exposure to CB. Kuhr and Dorough (1976) reported a half-life of 30–40 min for the reactivation of the carbamylated enzyme, resulting in a complete recovery of the enzyme in several hours. Juvenile goldfish previously exposed to sublethal carbaryl levels during 96 h showed a rapid recovery of the enzyme activity in 8 h. However, extra time was needed for a total recovery of the enzyme activity, suggesting that both reactivation and new synthesis of enzyme participate in the whole process (Ferrari et al., 2004a). Almost 4 days were needed for the recovery of ChE activity in carbaryl treated juvenile rainbow trout and goldfish (Ferrari et al., 2004 a, b; Ferrari et al., 2007a). Other authors have reported a brief period for brain ChE recovery (12-48 h) in rainbow trout and brook trout treated with CB (Haines, 1981; Szeto et al., 1985; Zinkl et al., 1987a).

Table 3. Fish ChE inhibition and recovery after organophosphorus sublethal exposure

Specie	Pesticide Concentration[a]	Time of exposure (hours)	% of ChE inhibition after insecticide exposure	Time of recovery (days)	% of remaining ChE inhibition after the recovery period	Reference
Anguilla	Fenithrotion	96		132		Sancho et al., 1997
anguilla	0.002		B: 44		29	
(European eel)	0.040		B: 64		43	
Carassius	Parathion	96	B: 95	2	82*	Ferrari et al.,
auratus	0. 1			7	77*	2004a
(goldfish)				35	37*	
	Azinphos methyl	96	B: 94	3	82	Ferrari et al.,
	0.1			7	77 *	2004a
				35	16*	
Gambussia	Chlorpyrifos	48	B: 77	4	62	Boone and
affinis	0.1		M: 82		81*	Chambers, 1996
(mosquitofish)						
	Parathion	48	B: 89	4	77*	Boone and
	0.15		M: 96		90*	Chambers, 1996
	Methyl parathion	48	B: 88	4	75*	Boone and
	8		M: 90		93	Chambers, 1996
Lepomis	Methyl parathion	24	B: 85	14	B: 50%	Benke and
gibbosus				28	No total	Murphy, 1974
(sunfish*)*					recovery	
Oncorhynchus	Acephate	24	B: 38	1	43	Zinkl et al., 1987b
mykiss	400		P: 67		74	
(rainbow trout)				15	Partial recovery	
	Methamidophos	24	B: about	12	Total recovery	
	25		75			Zinkl et al., 1987b
Oncorhynchus	Azinphos methyl	96	B: 70	1	56*	Ferrari et al.,
mykiss	0.001					2004b
(rainbow trout)				21	38*	Ferrari et al., 2003
		96	M: 56	1	53	Ferrari et al.,
				21	49	2007a
Oncorhynchus	Chlorpyrifos	96	B: 46	20	23	Eder et al., 2007
tshawysyscha	0.0037			60	Total recovery	
(Chinook salmon)						
Poecilia	Chlorpyrifos	336		4	Significant	Van der Well and
reticulate	(sublethal)				recovery	Welling, 1989
(guppy)				28	60*	

[a] mg/L.

* Indicates significant differences of ChE activity compared to that at the end of exposure time.

B: brain ChE, M: muscular ChE

2.3. Fish Esterases as Biomarkers of Organophosphorus and Carbamate Exposure

Different fish ChE have been proved to be good biomarkers of anticholinesterase exposure in field studies (Gruber and Munn, 1998; Sturm et al., 1999b; Kirby et al., 2000; Napierska and Podolska, 2005). An inhibition of ChE greater than 20% is considered to be indicative of OP or CB exposure (Grue et al., 1991; Oruç and Usta, 2007). The enzyme ChE has been considered a specific biomarker for CB and OP (Edwards and Fisher, 1991). However, its specificity has been recently questioned (Guilhermino et al., 1998) because some studies revealed ChE inhibition by other compounds different from anticholinesterase pesticides. A significant inhibition of ChE has been reported by exposure to metals, surfactants, detergents or different herbicides (Guilhermino et al., 1998; De la Torre et al., 2000, Bretaud et al., 2000, Ferrari et al., 2002, Hassanein, 2002). However, ChE inhibition levels reached after CB and OP sublethal exposure are usually higher than those recorded from other environmental contaminants. Moreover, fish brain ChE has shown to be a very sensitive biomarker, acting as an early indicator of OP and CB exposure even at very low and sublethal concentrations (Ferrari et al., 2004 a, b) as shown in Table 1.

Besides, the prolonged fish ChE inhibition caused by different OP at sublethal concentrations (Table 3) reinforces its usefulness as biomarker of exposure/effect. Actually, an incoming short pulse of OP contamination into a stream may cause a decrease in fish ChE activity that could be stable for a long period of time. Thus, the pesticide impact on exposed fish can be evidenced through ChE inhibition even at prolonged periods after the disappearance of the compound from water. Furthermore, the repeated application of anticholinesterase insecticides in the field, mostly in the spring season, may sustain ChE inhibition and hence affect fish survival.

In order to use ChE as biomarker, several studies have evaluated the effect of different conditions on the normal or basal enzyme activity in diverse fish species (Rattner and Fairbrother, 1991; Beauvais et al., 2002; Phillips et al., 2002b). Usually, fish ChE activity showed to be stable respect to temperature variations, not modified by anesthetics or storage at low temperatures (Richmonds and Dutta, 1992; Beauvais et al., 2002). On the contrary, basal ChE activity could vary with fish length (Flammarion et al., 2002).

Biochemical characterization and identification of cholinesterase enzymes present in different tissues of fish species have been performed in several works (Cunha Bastos et al., 1998; Garcia et al., 2000; Rendón-von Osten et al., 2005; Whitehead et al., 2005; Tortelli et al., 2006). It seems evident from literature that brain of most fish species contains mainly AChE, as have been already mentioned (Cerón et al., 1996; Sturm et al., 1999a; Varò et al., 2003 and 2007; Ferrari et al., 2004a), while BuChE isoforms are predominant in liver and plasma (Habig and Di Giulio, 1991) and generally uncommon in muscle tissue (Magnotti et al., 1994; Garcia et al., 2000) except for marine and estuarine fish species (Varò et al., 2003, Sturm et al., 1999a). The different isoforms of ChE may vary in their sensitivity to OP and CB. Sturm et al. (1999a) have reported a higher sensitivity of marine fish muscle BuChE to the OP paraoxon and dichlorvos than the one observed for AChE. Therefore, differences of ChE isoenzymes sensitivity should be considered in the selection of the most useful biomarker. In addition to ChE, other fish esterases such as CarbE are inhibited by OP and CB (Boone and Chambers, 1997; Denton et al., 2003; Küster and Altenburger, 2006; Ferrari et al., 2007b) and thus being appropriate biomarkers of exposure. Fish CarbE are measured mainly in fish liver, and sometimes have proved to be more sensitive to OP exposure than ChE (Wheelock et al., 2005). However, as these enzymes do not react with positively charged esters their sensitivity to the CB is usually low (Jokanović, 2001).

It is desirable that biochemical biomarkers could be able to predict deleterious effects at population or higher ecological levels of organization (van der Oost et al., 2003). Biomarkers would be more valuable if they could be linked to adverse effects at individual, population or ecosystems levels (Eason and O`Halloran, 2002). The little correlation observed in laboratory tests between fish ChE inhibition and mortality seems to be against this concept. However, the situation may be quite different under field conditions. Indeed, several studies have associated enzyme inhibition with behavioral alterations that are relevant for fish survival, such as the swimming speed, swimming distance, optomotor function, feeding behavior and olfactory physiology (Dutta et al., 1992; Beauvais et al., 2000 and 2001; Roex et al., 2003; Jarrard et al., 2004; Venkateswara Rao et al., 2005; Sismeiro-Vivas et al., 2007). These behavioral responses are relevant for food search, predator avoidance, mate search and migration. Therefore; those alterations of normal behavioral patterns due to anticholinesterase exposure may imply severe risks to the success of fish populations (Scott et al., 2004). Thus, fish ChE inhibition should be considered not only a good biomarker of exposure/effect of OP and CB, i.e. sensitive and early warning signal, but also a biomarker with high ecological relevance.

3. "Secondary" Targets of Organophosphorus and Carbamate in Fish

Most OP and CB are able to interact with other non-cholinesterase targets, also referred to as secondary targets (Pope, 1999; Jett and Lein, 2006). These alternative targets are able to modulate acute and chronic toxicity of anticholinesterase insecticide; therefore, their study become of fundamental importance. Furthermore, the molecular pathway from OP and CB exposure to death has not been clearly established (Peña-Lopis et al., 2002). In fish, the lack of correlation between ChE inhibition and mortality reinforces the hypothesis that non-cholinergic targets of OP and CB must be included in the analysis. Some of them involve alteration of redox status and antioxidant system leading to oxidative damage (Hai et al., 1997; Durmaz et al., 2006; Matos et al., 2007), modification of hematological parameters and metabolic enzymes (Jyothi and Narayan, 1999; Venkateswara Rao, 2006a), histological alteration (Fanta et al., 2003), induction of detoxifying enzymes (Ferrari et al, 2007b, Venkateswara Rao, 2006b), heat shock protein and cytokine induction (Eder et al., 2004) and inhibition of nucleotide hydrolysis (Senger et al., 2005). Additional effects of OP and CB have been reported in other vertebrates such as modification of the activity of serine hydrolases (Casida and Quistad, 2005), protein kinases (Bagchi et al., 1997; Renglin et al., 1999; Niijima et al., 2000; Bomser et al., 2002; Souza et al., 2004), muscarinic receptors (Ward and Mundi, 1996; Olivier et al., 2001; Bomser and Casida, 2001), transcription factors (Schuh et al., 2002; Venturino et al., 2003a) and induction of apoptosis (Carlson et al., 2000).

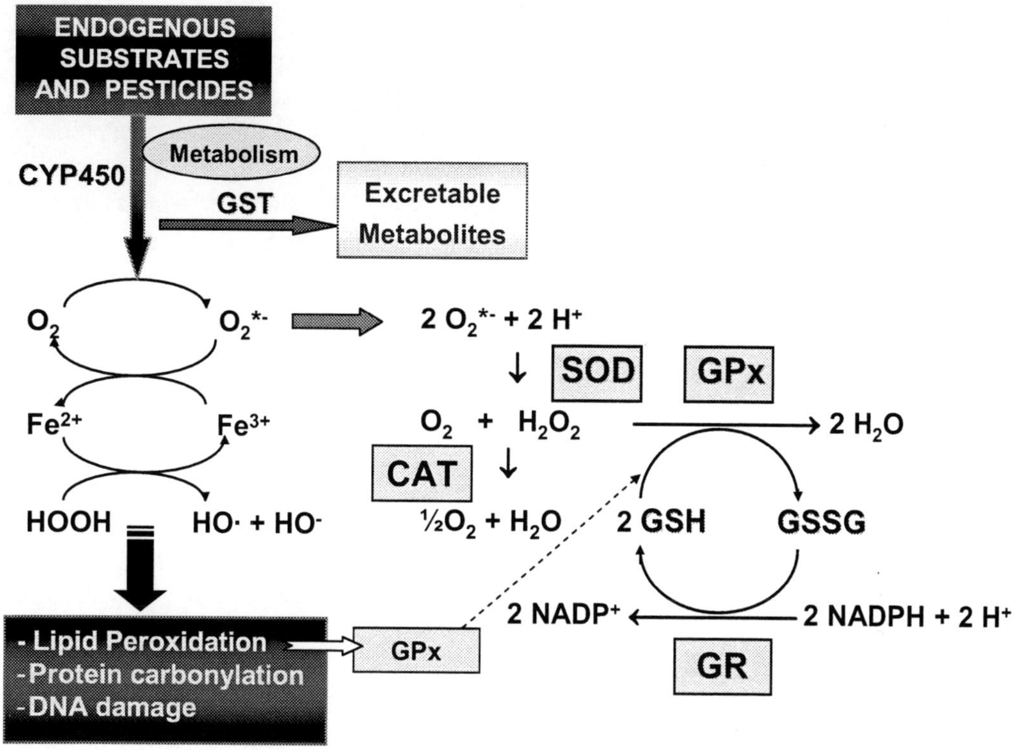

Figure 1. Antioxidant system against reactive oxygen species generated by pesticide or endogenous metabolism.
CAT: catalase, GPx: glutathione peroxidase, GR: glutathione reductase, GSH: reduced glutathione; GSSG: oxidized glutathione, GST: glutathione S-transferase, $O_2{*}^{-}$: superoxide anion, CYPP450: cytochrome P450 system.

3.1. Antioxidant System and Redox Status

Reactive oxygen species (ROS) are commonly formed during aerobic metabolism. Even though, oxygen free radicals are important in the physiological control of cell function (Halliwell and Gutteridge, 1999), an excess of ROS can be toxic and harmful for most living organisms. Cellular sensitivity to oxidants is modulated by diverse antioxidant enzymes (Oruç and Üner, 2000; Stephensen et al., 2002). In addition, low molecular weight antioxidants such as vitamin E, ascorbic acid, reduced glutathione (GSH) and other non protein thiols participate in redox status control (Winston and Di Giulio, 1991).

Among antioxidant enzymes, superoxide dismutase (E.C. 1.15.1.1, SOD) catalyses the reduction of superoxide radicals to hydrogen peroxide. The ubiquitous enzyme catalase (E.C. 1.1.1.6, CAT), decomposes hydrogen peroxide into water and molecular oxygen (van der Oost et al., 2003). Both CAT and SOD provide the first defense against oxygen toxicity (Pandey et al., 2003). Glutathione peroxidase (E.C. 1.11.1.9, GPx) enzymes reduce organic peroxides and hydrogen peroxide to alcohols and water using GSH as cofactor (Regoli et al., 2002). The phospholipid–hydroperoxide specific GPx is particularly relevant, together with phospholipase A2, to restore cell membrane integrity after oxidative damage (Saran et al., 1998). Oxidized GSH (GSSG) is converted to its reduced form (GSH) by the enzyme glutathione reductase (E.C. 1.6.4.2, GR) (Regoli et al., 2002). In addition, GSH acts as a co-substrate of GST enzymes, which catalyze the conjugation of this tripeptide with a variety of electrophilic compounds. Besides to their function in conjugating xenobiotics with GSH as Phase II detoxifying enzymes, GST play an antioxidant role by conjugating cleavage products of lipid peroxides (Stephensen et al., 2002). Other enzymes modulated by xenobiotics that affect redox status are those related to GSH synthesis, mainly γ-glutamylcysteine synthetase (γGCS) (Nishimoto et al., 1995; Stephensen et al., 2002). GSH content is critical for cellular thiols homeostasis and protection against oxidative damage induced by free radicals (Regoli and Winston, 1999). This tri-peptide is found in high concentrations in most of the cells and participates in diverse metabolic reactions (Della Morte et al., 1994; Sies et al., 1999. Figure 1 illustrates the role of antioxidant enzymes in ROS detoxification.

Oxidative stress is generated when cellular antioxidant capacity is overwhelmed by ROS generated by toxins, solar radiation or pesticides and other xenobiotics (Miller and Brzezinska-Slebodzinska, 1993). Therefore, the "redox equilibrium", determined by the balance between the oxidative potential of ROS versus the levels of antioxidant defenses, is shifted into oxidative direction (Saran et al., 1998; Peña-Lopis et al., 2003). Under oxidative stress, ROS cause lipid peroxidation, DNA damage, and additionally they could activate molecular signaling driving to either apoptosis or pre-neoplasic lesions (Panemangalore and Bebe, 2000). Finally, if critical organs are affected, a decrease in growth and eventually death could be observed (Janssens et al., 2002).

The effects of different anticholinesterase insecticides on antioxidant enzymes, GSH levels and lipoperoxidation in fish are summarized in Table 4. Endogenous GSH levels were significantly decreased in different fish species exposed to both OP and CB. It has been proposed that a reduction of 20-30 % in GSH levels can deteriorate antioxidant response, leading to oxidative damage and finally to cellular death (Jokanović, 2001). Peña-Lopis et al. (2003) remark the essential role of GSH metabolism in the tolerance of the European eel (*Anguilla anguilla*) to dichlorvos. GSH decrease induced by OP and some CB has been associated with its role as a direct ROS scavenger or through its participation in the reaction catalyzed by GPx (Almar et al., 1998). ROS are reported to increase during pesticide and other contaminants metabolism (Livingstone, 2001; Bagchi et al., 1995). In addition, the drop of GSH levels caused by several OP and CB could be associated with their own biotransformation performed by GST enzymes (Jokanović, 2001). On the other hand, the GSH increase observed in some fish exposed to anticholinesterase insecticides (Table 4) has been considered an adaptive response to oxidative stress induced by those compounds. Mulcahy et al. (1997) indicate a probable effect of oxidative stress on the enzymes involved in GSH synthesis.

CAT activity is also modified in fish by OP and CB exposure. Control levels of CAT specific activity has been reported to be higher in liver than in other fish tissues, representing a 34-99% of total body CAT activity (Janssens et al., 2000; Ferrari et al, 2007b). However, CAT activity has been measured in kidney, muscle, gill as well as other fish tissues. Many studies reported an increase of this enzyme due to OP or CB exposure (Table 4). Other contaminants lead to CAT activity induction in fish (Strmac and Braunbeck, 2000; Cazenave et al., 2006). Therefore, CAT induction could be considered an adaptive response to xenobiotic contamination (Alves et al., 2002). On the other hand, CAT activity could be reduced in some fish species by exposure to anticholinesterase insecticides (Table 4). These results may be related to an enzyme inhibition in response to ROS increase. Kono and Fridovich (1982) reported CAT inhibition caused by the superoxide radical and the hydrogen peroxide, which promote the oxidation of the iron atom present in the heme group of the enzyme. Thus, CAT inhibition due to OP and CB exposure may be explained by the alteration of redox status that conduce to ROS increase and consequently to a decrease in CAT activity (Ferrari et al., 2007b). A dual behavior has been reported in CAT activity under pesticide exposure in aquatic species (Ferrari et al, 2007b; Venturino 2003b, Dorval et al., 2003). CAT induction and inhibition were described at low and high pesticide concentrations, respectively. Furthermore, CAT inhibition reported in those studies is concomitant with the highest GSH decrease generated by pesticide exposure.

Similarly to CAT, SOD activity significantly increases in fish due to OP exposure (Table 4). Usually both enzymes are simultaneously induced in response to pollutants (Saran et al., 1998; Dimitrova et al., 1994). Induction of SOD may reflect a response to an increase in superoxide anion radical production (Oruç and Üner, 2000). On the contrary, SOD activity was significantly decreased in the Nile tilapia (*Oreochromis niloticus*) after both seven and four days of carbaryl and azinphos methyl exposure, respectively (Table 4). Even though the induction of SOD may be associated to superoxide radical increase, SOD activity could also be decreased by this radical or by hydrogen peroxide through sulphydryl oxidation of the cystein groups present in the enzyme (Dimitrova et al., 1994).

Some authors have reported a lack of significant effect of OP and CB exposure on fish GPx activity (Hai et al., 1997; Oruç and Üner, 2000). However, results are controversial since in other reports the enzyme was either induced or decreased by OP exposure (Table 4). In general, GPx induction is a defensive response towards increased concentrations of lipid hydroperoxides (Banerjee et al., 1999). On the other hand, the decrease in GPx activity may be mediated by the increase of superoxide radical as a consequence of pesticide exposure (Blum and Fridovich, 1985). A reduction in GPx activity suggests a deficient capacity to neutralize peroxide impact (Monteiro et al., 2006).

GPx activity is dependant on GR activity which recycles GSH from GSSG (Banerjee et al. 1999). Fish GR was inhibited by exposure to several OP and CB (Table 4). This could lead to a decrease in GSH content versus the oxidized form GSSG. As mentioned before in this chapter, the decrease of GSH/GSSG ratio is correlated with a reduction in pesticide tolerance in fish (Peña-Lopis et al., 2003).

Some studies have reported a significant increase of lipoperoxides (LPO) in different fish tissues measured as malondialdehide (MDA) also called tiobarbituric acid reactive species (TBARS), or determined by xylenol orange assay (FOX: ferrous oxidation-xylenol orange) (Table 4). The increase in LPO and protein carbonyls in fish tissues due to pesticide exposure is indicative of redox imbalance and also a good biomarker of effect (Almroth et al., 2005; Parvez and Raisuddin, 2005). Moreover, LPO increase is considered one of the most important injuries generated by oxidative stress and a highly relevant toxic effect of some OP in fish (Monteiro et al, 2006). Conversely, no significant increase of LPO may indicate an efficient antioxidant or detoxifying ability to deal with oxidative stress generated by insecticide exposure. The antioxidant capacity could vary from one tissue or organ to other. In consequence, the same compound is able to increase LPO in some fish tissues/organ (i.e. gill and muscle), but not in others (i.e. liver) (Monteiro et al., 2006).

Table 4. Effects of anticholinesterase insecticides on fish antioxidant system

Specie	Pesticide Time of exposure	Enzyme/antioxidant (Tissue)	Main Effect	Reference
Anguilla anguilla	Dichlorvos 1.5 mg/L = LC_{85} 4 days	GST (L) GST (M) GR (L) GSH (L, M)	↓ ↑ ↓ ↓	Peña-Lopis et al., 2003
Brycon cephalus	Methyl parathion 2 mg/L, 1/3 LC_{50}) 4 days	CAT (G, L, M) SOD (G, L, M) GPx (G, M) GST (G, L, M) GSH (G, L, M) LPO FOX (G, M)	↑ ↑ ↓ ↑ ↓ ↑	Monteiro et al., 2006
Cyprinus carpio	Azinphos methyl 0.23 mg/L 4 days	CAT (K) SOD (G) GPx (K) GST (K) LPO MDA	↑ ↑ ↑ ↑ NS	Oruç et al., 2004
	Diazinon 0.0036, 0.018, 0.036 µg/L (sublethal) 5,15,30 days	CAT (M) SOD (G, K, M) GPx (G, M) LPO MDA (G, M)	↑ ↑ ↑ ↑	Oruç and Utsa, 2007
	Dichlorvos 1-5 mg/L 1 day	CAT (L, B) SOD (L, B) GPx (B) GSH (G,K,L,M) (B) LPO FOX (G,K,L,H,M, B)	↑ ↑ ↑ ↓ ↑ ↓	Hai et al., 1997
Gambusia affinis	Monocrotophos LC_{50} 20.49mg/L 4 days	CAT (V) SOD (V) GR (V) LPO (V)	↑ ↑ ↑ ↑	Kavitha and Venkateswara Rao, 2007
Icatalurus nebulosus	Dichlorvos 1-5 mg/L 1 day	CAT (L) SOD (L, M) GPx GST (B, H, M) GSH (G, L, M) LPO FOX (B)	↑ ↑ NS ↑ ↑ ↑	Hai et al., 1997
Oncorhynchus mykiss	Azinphos methyl 2.5, 5.0 µg/L(sublethal) 1, 2, 4 days	CAT (L, K) GST (L, K) GSH (L, K)	↓ NS ↓	Ferrari et al., 2007b

	Carbaryl 1, 3 mg/L (sublethal) 1, 2, 4 days	CAT (L) 1 day 4 day GST (L, K) GSH (L, K) 4 day	↑ ↓ ↑ ↓	Ferrari et al., 2007b
Oreochromis mossambicus	RPR- II 0.017 mg/L (sublethal) 3, 7, 15, 30 days	GST (L) GSH (B, G, L)	↑ ↓	Venkateswara Rao, 2006a
	RPR- V 0.017 mg/L (sublethal) 3, 7, 15, 30 days	GST (L) GSH (B, G , L)	↑ ↓	Venkateswara Rao, 2006b
Oreochromis niloticus	Diazinon 0.1, 1, 2 mg/L (sublethal) 1, 7, 15, 30 days	CAT (G, day 30) CAT(AT) SOD (G, K, M, AT) GPx (K) GPx (AT) LPO MDA (M, K)	↓ ↑ ↑ ↑ ↓ ↑	Durmaz et al., 2006
	Azinphos methyl 0.03 mg/L (sublethal) 4 days	CAT SOD (L) GPx (L) GST GR (L) MDA	↓ ↑ NS NS ↑ NS	Oruç and Üner, 2000
	Azinphos methyl 0.23 mg/L (sublethal) 4 days	SOD (G) GPx (K) GPx (B) GST (K) LPO MDA	↑ ↑ ↓ ↑ NS	Oruç et al., 2004
	Carbaryl 0.25 and 0.50 mg/L (sublethal)	CAT (L) SOD (L) GR (L) GST (L) GSH/GSSG (L)	↓ ↓ ↓ ↓ ↓	Matos et al., 2007

AT: alimentary track, B: brain, G: gill, K: kidney, L: liver, M: muscle, V: viscera.
↑ Increase, ↓ decrease, NS: no significant changes.

The results described above indicate that the antioxidant responses to OP and CB are highly variable and dependant on fish tissue, species, the tested compound and their concentration and time of exposure.

3.2. Detoxifying Enzymes

Lipofilic xenobiotics undergo many biotransformation reactions after their absorption in the organism. Several OP and CB insecticides are metabolized by Phase I enzymes and the resulting electrophilic metabolites can be then conjugated by Phase II enzymes. Phase I enzymes catalyze reactions of oxidation, reduction and hydrolysis (Jokanović, 2001). Among these, the quantitatively most important are the oxidative reactions performed by mixed function- oxidases (MFO), also known as cytochrome P450 (CYP) monooxigenases, and flavin monooxigenases (George et al., 1994; Stegeman and Hahn, 1994). Phase II conjugation reactions are catalyzed by glucuronosyltransferases, sulfotransferases and GST. In addition, some OP and CB can be directly metabolized by the conjugative Phase II enzymes (George, 1994). Some of the most important enzymes involved in anticholinesterase insecticides metabolism are GST and CYP.

OP and CB modified the activities of GST enzymes in fish (Table 4). The dimethylated OP such as azinphos methyl, malathion and methyl parathion as well as other OP are usually metabolized by these detoxifying enzymes (Ecobicon, 1992; Jokanović, 2001). Tolerance to OP by GST detoxification has been reported in amphibians (Anguiano et al., 2001; Venturino et al., 2003a). Even though some variability in GST activities among fish species have been observed, they are comparable to those determined in mammals (George, 1994). Fish exposure to some OP can either increase GST activity or fail to induce or significantly decrease its activity (Table 4). Similarly, the CB carbaryl was also reported to both induce and decrease GST activity in different fish species.

GST induction by different xenobiotics in fish is usually moderate and sometimes it could not be distinguished clearly (George, 1994). Furthermore, a high variability is found among individuals within a particular fish species. Like anticholinesterase insecticides, other compounds have been found to either induce (Sandvik et al., 1997; Egaas et al., 1999) or fail to increase fish GST (Jensen et al., 1991; Lindström-Seppä et al., 1996; Fenet et al., 1998). Moreover, Wiegand et al. (2001) have reported an increase of GST activity in zebrafish (*Danio rerio*) at low concentration of the herbicide atrazine and a significant decrease of its enzyme activity at higher concentrations of this compound. Therefore, GST activity is dependant on inter-species variation, the xenobiotic, and their concentration as well as exposure time (Oruç and Üner, 2000).

It is well known that GST can be induced by electrophilic and/or pro-oxidant compounds through the activation of the antioxidant response element (ARE) (George, 1994; Nguyen et al., 2003). Carbaryl was able to transiently induce GST activity in rainbow trout (Table 4), even tough it is not primarily metabolized by this enzyme (Ecobicon, 1992). This fact could be related to its ability to increase the apo-protein levels of the Phase I detoxifying enzyme cytochrome P450-1A (CYP1A) (Ferrari et al., 2007b). Carbaryl was reported to bind to the aryl hydrocarbon receptor (AHR) and, in consequence, to activate xenobiotic response element (XRE), also called dioxin response element (DRE), inducing some Phase I and phase II enzymes under its control (Denison et al., 1998; Delescluse et al., 2001). Additionally, the CB carbofuran was reported to be an AHR ligand (Ghosh et al., 2000). Nguyen et al. (2003) have stated that some AHR ligands are able to induce not only CYP1A but also Phase II enzymes that are under ARE control, such as GST. The authors named this kind of compounds as "bifunctional inducers". Prooxidant metabolites generated by the reactions catalyzed by Phase I enzymes are the responsible for GST as well as other Phase II enzymes induction. Many studies reported a simultaneous induction of GST and CYP1A in fish (George, 1994; Egaas et al., 1999). Besides, a XRE was found together to the ARE in some regulatory regions of the genes encoding some classes of mammalian GST (George, 1994).

CYP1A induction by carbaryl exposure could be the cause of antioxidant alterations generated by this compound in some fish species (Table 4). It has been proposed that CYP enzymes induction due to pesticide exposure, along with GSH depletion, could be related to ROS increase in cells (Panemangalore and Bebe, 2000; Radice et al., 2001; Sharma et al., 2005; Sivapiriya et al., 2006). Some authors have also reported an increase of ROS production due to other xenobiotic metabolism performed by CYP (Stegeman and Hahn, 1994; Bagchi et al., 1995; Livingstone, 2001). Furthermore, some studies have reported a link between the induction of MFO and the generation of oxidative stress in fish (Palace et al., 1996; Radice et al., 2001)

Conversely, most OP are not CYP1A inducers. Azinphos methyl failed to induce CYP1A levels in rainbow trout (Ferrari et al., 2007b). Furthermore, Wheelock et al. (2005) have reported a decrease in CYP1A levels in Chinook salmon (*Oncorhynchus tshawytscha*) exposed to chlorpyrifos. On the other hand, Flammarion et al. (1998) found no changes in cyprinid fish gudgeon (*Gobio gobio*) protein levels of CYP1A by methidathion exposure. Several works have studied the mammalian CYP involved in OP metabolism, suggesting that CYP1A would be less important than other isoenzymes (Atterberry, 1997; Fabrizi et al., 1999; Mutch and Williams, 2006). On the other hand, some authors have reported CYP induction by OP exposure in rodents (Sharma et al., 2005; Sivapiriya et al., 2006). A great diversity of CYP in fish has been reported (Stegeman and Hahn, 1994). However, there are few data of which isoenzymes are involved in OP metabolism or induce by these insecticides in fish. Nevertheless, OP metabolism by CYP in fish, even in absence of any significant CYP induction, would increase ROS leading to oxidative stress.

3.3. Antioxidant and Detoxifying Systems as Biomarkers in Field Studies

OP and CB insecticides effectively affect antioxidant and detoxifying systems in fish. Thus, some of the enzymes and compounds of those systems might be used as biomarkers in order to evaluate the impact of those pesticides in aquatic environments.

Fish detoxifying enzymes are in general very sensitive biomarkers of xenobiotics exposure and effect (van der Oost et al., 2003). Among these, CYP1A has been extensively studied in different fish species and it has been established as a biomarker for many xenobiotics (Fisk et al., 1997; Soimasuo et al., 2001; Gravato and Santos, 2002; Corsi et al., 2003). Despite the fact that some CB are able to induce hepatic CYP1A levels in fish, many OP failed to induce its activity. Thus, it use as biomarker of OP exposure is not functional.

Antioxidant enzyme activities have been modified in fish by the exposure to diverse pollutants present in the field (Bainy et al., 1996; Ventura et al., 2002; Ferreira et al., 2005). Changes in the levels of GSH, ascorbate, vitamin E and β-carotene have also been proposed as biomarkers of pro-oxidant challenge in aquatic environments (Livingstone, 2001). GSH and acid soluble thiols are useful biomarkers that have been evaluated in both laboratory and field studies (Lindström-Seppä et al., 1996; Almar et al., 1998). Besides, Dorval et al. (2005) have found alterations in antioxidant enzymes and LPO in fish collected in agricultural areas.

It is important to emphasize that the activity of antioxidant enzymes may be either induced or decreased under chemical stress. The trend of the response is dependant on the stress duration and intensity as well as the susceptibility of the exposed organisms (Oruç and Utsa, 2007). In conclusion, the responses of those enzymes observed in fish exposed to anticholinesterase insecticides are variable and not as persistent as ChE inhibition. Conversely, GSH decrease seems to be a common feature after OP and CB exposure (Table 4) and thus, resulting in a more useful biomarker. In general, antioxidant enzymes would be induced in mild or earlier phases of exposure, while oxidative stress is still under control. A subsequent ROS increase, coupled with a GSH decrease, may directly inhibit some antioxidant enzyme activities.

Since the enzymes and compounds of the antioxidant and detoxifying systems are considered as non-specific biomarkers, their use to evaluate the exposure to a particular xenobiotic is not so valuable. However, oxidative stress biomarkers have the advantage of being not only indicators of exposure but also indicators of effect and toxicity (Ferreira et al., 2005). In conclusion, their application as biomarkers of OP and CB exposure/effect should be analized along with the determination of esterase activities. The integrate use of ChE and antioxidant enzymes as biomarkers of OP and CB exposure/effects in aquatic organisms have been recently proposed by some authors (Lionetto et al., 2003; Venturino et al., 2005).

4. Trends in Underlying Molecular Mechanisms

Molecular biomarkers related to receptor binding, signal transduction and genetic response have gained increasing relevance in assessing the effects of xenobiotics on vertebrates (Venturino et al, 2003a; Eder et al., 2004). However, the knowledge of molecular mechanisms of anticholinesterases in fish is scarce and has been better studied in mammalians cells and models.

AChE inhibition by OP could be associated to changes in cellular signaling and gene expression. Acetylcholine accumulation by OP exposure conduces to a muscarinic receptor-mediated upregulation of the transcription factor c-Fos and the expression of AChE gene (Kaufer et al., 1998). On the other hand, several of the non-cholinergic targets of action of OP and CB include alteration of neurotransmitter receptors affecting cell signaling pathways (Pope et al., 2005). For example, Bagchi et al. (1997) reported a significant increase in protein kinase C (PKC) activity produced by two OP, fenthion and chlorpyrifos. This enzyme participates in cell signaling pathways mediating vital cellular process. Besides, in mammalian cells chlorpyrifos is able to activate extracellular signal –regulated kinases (ERK) (Bomser and Casida, 2000; Bomser et al., 2002), alter the Ca^{2+}/cAMP response element binding protein (Schu et al., 2002) and affect the expression and binding activity of transcription factors involved in cell replication and differentiation (Crumpton et al., 2000). In addition, sublethal concentrations of azinphos methyl and carbaryl were reported to modify protein phosphorylation pattern in rainbow trout brain and liver (Ferrari, 2004; Ferrari et al., 2006).

It is now recognized that cell redox status controls many signal pathways and gene expression and their imbalance could lead either to apoptosis or proliferation (Kamata and Hirata, 1999; Biswas et al., 2006). Increase of ROS, such as hydrogen peroxide, lead to the activation of protein tyrosine kinases followed by the stimulation of downstream signaling systems including mitogen activated kinases (MAPK) and phospholipase C-gamma (PLCγ) and those modifications would lead to the activation of several transcription factors (Saran et al., 1998; Kamata and Hirata, 1999). Some transcription factors such as nuclear factor-kB (NF-kB) and activator protein-1 (AP-1) are known to be activated by ROS and inactivated by some reductants. Domenicotti et al. (2000) reported the activation of both novel PKC isoforms and DNA binding capacity of AP-1 as a result of GSH depletion in rat hepatocytes leading to apoptosis. Conversely, oxidative stress could also decrease the activation of redox-sensitive transcription factors which have cystein residues in their binding domains (Sen and Packer, 1996). In those cases, DNA binding is only possible when the residues are in a reduced form (Saran et al., 1998). Thus, there is an evident cross talk between cellular redox state and different cellular signaling pathways, which are subjected *in vivo* to a dual redox regulation (Kamata and Hirata, 1999).

OP and CB insecticides have been reported to affect gene expression inducing some phase I and phase II detoxifying enzymes. As has been mentioned above, carbaryl is a weak AhR ligand and thus is able to bind to XRE in DNA and to induce CYP1A and other genes under its control (Denison et al., 1998; Delescluse et al., 2001). Inactive AhR is located in the cytosol forming a heterodimer with heat shock protein 90 (Hsp90). After binding to an endogenous or exogenous ligand, AhR is able to dissociate from HSP90, translocate to the nucleus and dimerize with the aryl hydrocarbon nuclear translocator (Arnt) factor. This active AhR/Arnt acts as a transcription factor which eventually binds to the XRE/DRE, regulating the expression of a battery of genes (Dalton et al., 2002, Nguyen et al., 2003). Carbaryl and carbofuran have been reported to induce liver CYP1A in fish (Gosh et al. 2000, Ferrari et al., 2007b). However, the activation and translocation to nucleus of AhR due to those CB insecticides exposure have not been clearly established. Besides, other genes under AhR control are expected to be increased in fish due to carbaryl or carbofuran exposure, in addition to CYP1A. Cytosolic SOD gene contains a functional AhR element necessary for its upregulation (Dalton et al., 2002).

The AhR activation by dioxins, considered their model ligand, have been reported to induce oxidative stress, GSH depletion and lipid peroxidation, but also to activate the arachidonic acid cascade, the production of cytokines and grown factors, the ERK pathway, and to increase the expression of transcription factors such as NF-κB (Matsumura, 2003). Thus, those effects in cellular stress associated to dioxins should be also studied in fish exposed to aryl-carbamates to discriminate whether their ability to bind AhR contributes significantly to their toxicological manifestations.

Electrophilic compounds such as OP and also thiol-reactive compounds and phenolic antioxidants are able to induce GST and other phase II enzymes which are under ARE control (Jaiswal et al., 2004). In general, genes under ARE regulation codify proteins that participate in redox control and defend cells against oxidative damage (Nguyen et al., 2003). The nuclear factor E2-related protein (Nrf2), the main effector of ARE-regulated genes, is located in the cytosol as a heterodimer with a repressor protein named Keap1. Under oxidative stress and GSH depletion generated by electrophiles, Keap1 is oxidized, liberating Nrf2 which is translocated into the nucleus (Kobayashi and Yamamoto, 2006). This relocalization requires Nrf2 activation through its phosphorylation. Some protein kinases such as PKC, phosphatidyl inositol 3 kinase (PI3K) and the MAPK called Jun kinase (JNK) seem to be involved in this step (Nguyen et al., 2003; Biswas et al., 2006). It has been suggested that Nrf2 is highly relevant in the cellular protection against environmental stressors and governs a large number of cytoprotective genes regulated via the ARE, including γGCS, CAT, GPx, SOD, thioredoxin, thioredoxin reductase, CarbE (Kobayashi and Shamamoto, 2006). It has been recently reported that Nrf2 is required for the induction of GST in zebrafish (Kobayashi et al., 2002). Several studies reported a GST induction by anticholinesterase insecticides in fish. However, the effect of those compounds on Nrf2 activation, translocation and binding to the ARE has still to be studied.

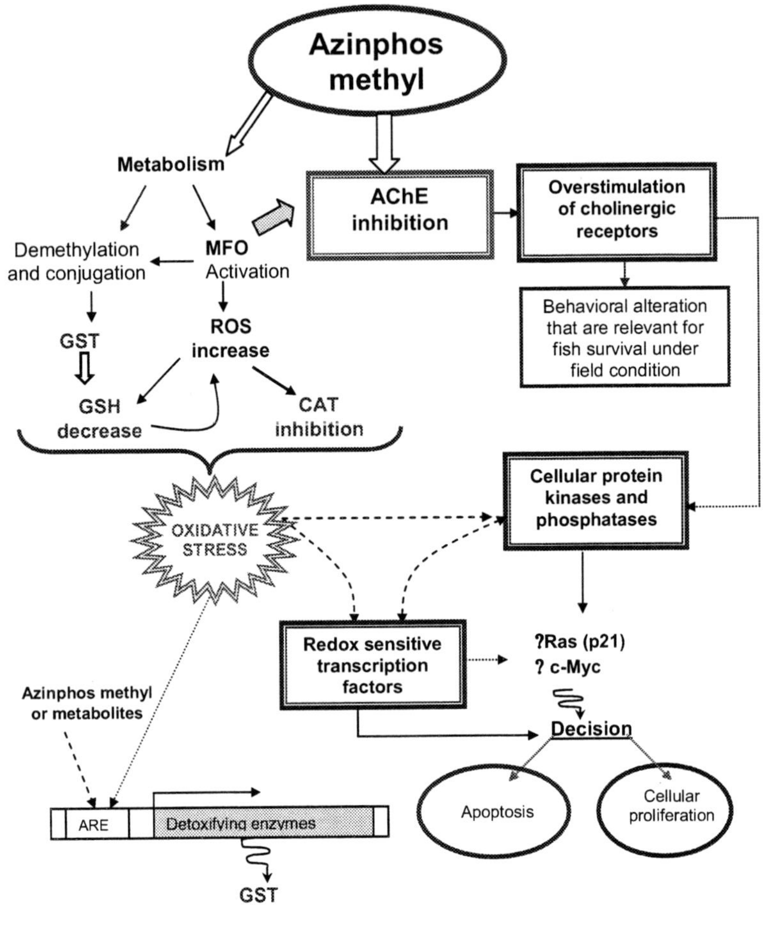

Figure 3. Summary of azinphos methyl observed effects and proposed mechanisms of action in juvenile rainbow trout *O. mykiss.* Sublethal concentrations of *t*he OP azinphos methyl caused a sustained AChE inhibition. In other organisms, some transcription factors (c-Fos) could be regulated trough overstimulation of cholinergic receptors followed by protein phosphorylation. In addition, azinphos methyl metabolism induces GSH decrease and ROS increase associated to a more pronounced GSH depletion and CAT inhibition and finally to oxidative stress. AS a result, several transcription factors and cell signaling pathwaysv are affected. Finally, oxidative stress triggered responses could lead either to apoptosis or cellular proliferation. AChE: acetylcholinesterase, ARE: antioxidant response element, CAT: catalase, c-Myc: transcription factor, GSH: reduced glutathione, GST: glutathione S-transferase, MFO: mixed function oxidases, Ras: monomeric small G protein, ROS: reactive oxygen species. Ferrari, 2004.

In summary, either a direct effect of OP and CB on cell signaling pathways or an indirect effect through oxidative stress induction would lead to alteration in gene expression and cell homeostasis, and in turn to apoptosis or proliferation. Some of the observed effects and proposed mechanisms of action of the OP azinphos methyl and the CB carbaryl on juvenile rainbow trout are summarized in figures 2 and 3.

Finally, the development of genomics through microarray studies and proteomics should provide new and more powerful biomonitoring techniques in aquatic toxicology (Miracle and Ankley, 2005; Ju et al., 2007; Denslow et al., 2007). These are "open" techniques making no *a priori* assumptions about potential biomarkers, and evaluating thousands of molecules simultaneously. Omics have been applied to pesticide effects, mainly in mammalian models. Unexpected responses may be detected from these assays. In any case, further confirmation by more precise techniques such as Real Time-PCR is needed. Recent studies have analyzed the neurotoxicity of some OP by characterizing transcriptional responses in rat brains through microarray technology (Dam et al., 2002; Slotkin and Seidler, 2007). The application of this technology is not as advanced in aquatic toxicology as it is for mammalian models probably because of the lack of sequence information for non-model species (Denslow et al., 2007). However, several studies have been performed applying toxicogenomics and proteomics to assess the effects of contaminants different from OP and CB in fish (Bradley et al., 2002; Williams et al., 2003; Mortensen and Arukwe, 2007; Finne et al., 2007). Moreover, cDNA and olignonucleotide microarrays are now available for some fish used as a sentinel species for pollutant exposure evaluation (Williams et al., 2003; Krasnov et al., 2005; Aleström et al., 2005; Mortensen and Arukwe, 2007; Martyniuka et al., 2007).

The application of these new techniques in fish ecotoxicology would allow the identification of novel biomarkers for specific xenobiotics including anticholinesterase insecticides. In addition, toxicogenomic analysis would help to better understand the molecular mechanisms of action of these compounds in fish.

5. Conclusion

ChE activity has proved to be a sensitive and specific biomarker of OP and CB exposure in fish. Besides, many sublethal behavioral alterations that are relevant for fish survival in field are well correlated with ChE activity, showing that its inhibition could be considered a biomarker of ecological significance. However, the lack of correlation between ChE inhibition and mortality observed in most fish species point out the importance of studying non-cholinergic effects of OP and CB in those vertebrates to better understand the toxic responses elicited by those compounds. Antioxidant and detoxifying systems are reported to be modified by anticholinesterase insecticides. Therefore, antioxidant enzymes and compounds could be used as biomarkers of their exposure or effect. However, as they are not specific, their use together with esterase activities is recommended. Regardless, OP and CB insecticides are able to induce oxidative stress in fish and therefore they could affect cell signaling and gene expression leading to either apoptosis or proliferation. Transcription factor expression and activity as well as protein kinase activities are known to be modified by anticholinesterases in other vertebrates, but the underlying molecular targets of OP and CB in fish are not sufficiently studied. The complexity of the responses and the coexistence of multiple potential molecular targets indicate the necessity of toxicogenomic and proteomic approaches. The use of new technologies such as microarray analysis to evaluate altered gene expression due to xenobiotic exposure is now emerging and has been applied in some fish species. Toxicogenomics and proteomics will, in the future, help to understand the molecular mechanisms of action of OP and CB compounds in fish and allow the identification of potential novel biomarkers to assess adverse effects.

ACKNOWLEDGEMENTS

Grants PI0980 Universidad Nacional del Comahue and PICT Redes 2002-203 Secyt, Argentina.

The authors thank Dr Cristina Monica Montagna for her critical reading of the manuscript.

6. REFERENCES

Abbas, R; Schultz, IR; Doddapaneni, S; Hayton, WL. Toxicokinetics of parathion and paraoxon in rainbow trout after intravascular administration and water exposure. *Toxicology and Appied Pharmacology* 1996 136, 194-199.

Abdel-Halim, KY; Salama, AK; El-khateeb, EN; Bakry, NM. Organophosphorus pollutants (OPP) in aquatic environment at Damietta Governorate, Egypt: Implications for monitoring and biomarker responses. *Chemosphere* 2006 63, 1491–1498.

Aleström, P; Holter, JL; Nourizadeh-Lillabadi, R. Zebrafish in functional genomics and aquatic biomedicine. *Trends in Biotechnology*, 2006 24, 15-21.

Almar, M; Otero, L; Santos, C; Gonzales Gallego, J. Liver glutathione content and glutathione-dependent enzymes of two species of freshwater fish as bioindicators of chemical pollution. *Journal of Environmental Science and Health B*, 1998 33, 769-783.

Almroth, BC; Sturve, J; Berglund, A; Förlin L. Oxidative damage in eelpout (*Zoarces viviparus*), measured as protein carbonyls and TBARS, as biomarkers. *Aquatic Toxicology* 2005 73, 171–180.

Alves, SRC; Severino, PC; Ibbotson, DP; Da Silva, AZ; Lopes, FRAS; Sáenz, L A; Bainy, ACD. Effects of furadan in the brown mussel *Perna perna* and in the mangrove oyster *Crassostrea rhizophorae*. *Marine Environmental Research* 2002 54, 241–245.

Anguiano, OL; Caballero, A; Pechen de D'Angelo, A.M. The role of glutathion conjugation in the regulation of early toad embryo's tolerance to pesticides, *Comparative Biochemistry and Physiology C* 2001 128, 35-43.

Atterberry, TT; Burnett, WT; Chambers JE. Age-Related Differences in Parathion and chlorpyrifos toxicity in male rats: target and nontarget esterase sensitivity and cytochrome P450-mediated metabolism. *Toxicology and Applied Pharmacology*, 1997 147, 411–418.

Bagchi, D; Bagchi, M; Hassoun, EA; Stohs, SJ. In vitro and in vivo generation of reactive oxygen species, DNA damage and lactate dehydrogenase leakage by selected pesticides. *Toxicology* 1995 104 1-3, 129-140.

Bagchi, D; Bagchi, M; Tang, L; Stohs, SJ. Comparative *in vitro* and *in vivo* protein kinase C activation by selected pesticides and transition metal salts. *Toxicology letters* 1997 91, 31-37.

Bainy, ACD; Saito, E; Carvalho, PSM; Junqueira, VBC. Oxidative stress in gill, erythrocytes, liver and kidney of Nile tilapia (*Oreochromis niloticus*) from polluted site. *Aquatic Toxicology*, 1996 34, 151-162.

Banerjee, BD; Seth, V; Bhattacharya, A; Pasha, ST; Chakraborty, AK. Biochemical effects of some pesticides on lipid peroxidation and free-radical scavengers. *Toxicology Letters*, 1999 107, 33–47.

Barata, C; Solayan, A; Porte, C. Role of B-esterases in assessing toxicity of organophosphorus (chlorpyrifos, malathion) and carbamate (carbofuran) pesticides to Daphnia magna. *Aquatic Toxicology* 2004 66, 125-139.

Barron, MG; Charron, KA; Stott, W; Duvall, S. Tissue carboxylesterase activity of rainbow trout. *Environmental Toxicology and Chemistry* 1999 18, 2506-2511.

Beauvais, L; Cole, K; Atchinson, G; Coffey, M. Factors affecting brain cholinesterase activity in Bluegill (*Lepomis macrochirus*). *Water, air, and soil pollution* 2002 135, 249-264.

Beauvais, L; Jones SB; Parris, JT; Brewer, SK; Little EE. Cholinergic and behavioral neurotoxicity of carbaryl and cadmiun to larval rainbow trout (*Oncorhynchus mykiss*). *Ecotoxicology and Environmental Safety* 2001 49, 84-90.

Beauvais, L; Jones, SB, Brewer, SK; Little EE. Physiological measures of neurotoxicity of diazinon and malathion to larval rainbow trout (*Oncorhynchus mykiss*) and their correlation with behavioral measures. *Environmental Toxicology and Chemistry* 2000 19 7, 1875-1880.

Benke, GM; Cheever KL; Mirer FE; Murphy SD. Comparative toxicity, anticholinesterase action and metabolism of methyl parathion and parathion in sunfish and mice. *Toxicology and Applied Pharmacology*, 1974 28, 97-109.

Benke, GM; Murphy SD. Anticholinesterase action of methyl parathion, parathion and azinphosmethyl in mice and fish: Onset and recovery of inhibition. *Bulletin of Environmental Contamination and Toxicology* 1974 12, 117-122.

Biswas, S; Chida, A S; Rahman, I. Redox modifications of protein–thiols: Emerging roles in cell signaling. *Biochemical Pharmacology* 2006 71 5, 551-564.

Blum, J; Fridovich, I. Inactivation of glutathione peroxidase by superoxide radical. *Archives of Biochemistry and Biophysics* 1985 240 2, 500–508.

Bomser, J; Casida, JE. Activation of extracellular signal-regulated kinases (ERK 44/42) by chlorpyrifos oxon in Chinese hamster ovary cells. *Journal of Biochemical and Molecular Toxicology* 2000 14, 346–353.

Bomser, J; Casida, JE. Diethylphosphorylation of rat cardiac m2 muscarinic receptor by chlorpyrifos oxon in vitro. *Toxicology Letters* 2001 119, 21–26.

Bomser, J; Quistad, GB; Casida, JE. Chlorpyrifos oxon potentiates diacylglycerol-Induced extracellular signal-regulated kinase (ERK 44/42) activation, possibly by diacylglycerol lipase inhibition. *Toxicology and Applied Pharmacology,* 2002 178, 29–36.

Boone, JS; Chambers JE. Time course of inhibition of cholinesterase and aliesterase activities, and nonprotein sulphydryl levels following exposure to organophosphorus insecticides in mosquitofish (*Gambusia affinis*). *Fundamental and Applied Toxicology* 1996 29, 202-207.

Bradley, BP; Shrader, EA; Kimmel, DG; Meiller, JC. Protein expression signatures: an application of proteomics. *Marine Environmental Research*, 2002 54, 373–377.

Bretaud, S; Toutant, JP; Saglio, P. Effects of carbofuran, diuron, and nicosulfuron on acetylcholinesterase activity in goldfish (*Carassius auratus*). *Ecotoxicology and Environmental Safety* 2000 47, 117-124.

Carlson, K; Jortner, BS; Ehrich M. Organophosphorus compound-induced apoptosis in SH-SY5Y human neuroblastoma cells. *Toxicology and Applied Pharmacology*, 2000 168 2, 102-113.

Casida, JE; Quistad GB. Serine hydrolase targets of organophosphorus toxicants *Chemico-Biological Interactions* 2005 157–158, 277–283.

Cazenave, J; Bistoni, MA; Pesce, SF; Wunderlin, DA. Differential detoxification and antioxidant response in diverse organs of *Corydoras paleatus* experimentally exposed to microcystin-RR. *Aquatic Toxicology* 2006 76, 1–12.

Cerón, JJ; Ferrando, MD; Sancho, E; Gutierrez-Panizo, C; Andreu-Moliner, E. Effects of diazinon in different tissues of European eel (*Anguilla anguilla*). *Ecotoxicology and Environmental Safety* 1996 35, 222-225.

Chambers, JE; Carr RL. Biochemical mechanisms contributing to species differences in insecticidal toxicity. *Toxicology* 1995 105, 291-304.

Chambers, JE; Yarbrough, JD. Parathion and methyl parathion toxicity to insecticide-resistant and susceptible populations of mosquitofish (*Gambusia affinis*). *Bulletin of Environmental Contamination and Toxicology*, 1974 11, 315–320.

Coppage, DL. Organophosphate pesticides: Specific level of brain AChE inhibition related to death in sheephead minnows. *Transactions of the American Fisheries Society* 1972 101, 534-537.

Corsi, I; Mariottini, M; Sensini, C; Lancini, L; Focardi S. Cytochrome P450, acetylcholinesterase and gonadal histology for evaluating contaminant exposure levels in fishes from a highly eutrophic brackish ecosystem: the Orbetello Lagoon, Italy. *Marine Pollution Bulletin*, 2003 46, 203–212.

Costa, LG. Basic Toxicology of Pesticides. *Occupational Medicine: State of the Art Reviews* 1997 12 2, 251-267.

Crumpton, TL; Seidler, FJ; Slotkin, TA. Developmental neurotoxicity of chlorpyrifos in vivo and in vitro: Effects on nuclear transcription factor involved in cell replication and differentiation. *Brain Research*, 2000 857, 87-98.

Cunha Bastos, VLF; Rossini, A; Ribeiro Pinto, LF; de Lima, LM; Ceccarelli, PS; Coelho, MGP; Cunha Bastos, J. Different sensitivities to paraoxon of brain and serum cholinesterases from pacu, and indigenous Brazilian fish. *Bulletin of Environmental Contamination and Toxicology*, 1998 60, 1-8.

Dalton, TP; Puga, A; Shertzer, HG. Induction of cellular oxidative stress by aryl hydrocarbon receptor activation. *Chemico-biological interactions*, 2002 141, 77-95.

Dam, K; Seidler, FJ; Slotkin, TA. Transcriptional biomarkers distinguish between vulnerable periods for developmental neurotoxicity of chlorpyrifos: Implications for toxicogenomics. *Brain Research Bulletin*, 2003 59 4, 261–265.

De la Torre, FR; Ferrari, L; Salibián A. Freshwater pollution biomarker: response of brain acetylcholinesterase activity in two fish species. *Comparative Biochemistry and Physiology C* 2002 131, 271-280.

Delescluse, ChN; Ledirac, RL; Piechocki, MP; Hines, RN; Gidrol, X; Rahmani, R. Induction of cytochrome P450 1A1 gene expression, oxidative stress, and genotoxicity by carbaryl and thiabendazole in transfected human HepG2 and lynphoblastoid cells. *Biochemical Pharmacology* 2001 61, 399-407.

Della Morte, R; Villani, GRD; Di Martino, E; Squillacioti, C; De Marco, L; Vuotto, P; Belisario, MA; Staiano, N. Glutathione depletion induced in rat liver fractions by seven pesticides. *Bollettino della Società Italiana di Biologia Sperimentale* 1994 70 8-9, 185-192.

Denison, MS; Phelan, D; Winter, GM; Ziccardi, MH. Carbaryl, a carbamate insecticide, is a ligand for the hepatic Ah (dioxin) receptor. *Toxicology and Applied Pharmacology* 1998 152, 406-414.

Denslow, ND; Garcia-Reyero, N; Barber, DS. Fish 'n' chips: the use of microarrays for aquatic toxicology. *Molecular BioSystems*, 2007 3, 172-7.

Denton, DL; Wheelock, CE; Murray, SA; Deanovic, LA; Hammock, BD; Hinton DE. Joint acute toxicity of esfenvalerate and diazinon to larval fathead minnows (*Pimephales promelas*). *Environmental Toxicology and Chemistry*, 2003 22 2, 336–341.

Dettbarn, W-D; Yang, ZP; Milatovic D. Different role of carboxylesterase in toxicity and tolerance to paraoxon and DFP. *Chemico-biological interactions* 1999 119/120, 445-454.

Dimitrova, MST; Tishinova, V; Velchava, V. Combined effect of zinc and lead on the hepatic superoxide dismutase-catalase system in carp (*Cyprinus carpio*). *Comparative Biochemistry and Physiology C*, 1994 108, 43–46.

Domenicotti, C; Paola, D; Vitali, A; Nitti, M; D'Abramo, C; Cottalasso, D; Maloberti, G; Biasi, F; Poli, G; Chiarpotto, E; Marinari, UM; Pronzato, MA. Glutathione depletion induced apoptosis of rat hepatocytes through activation of protein kinase C novel isoforms and dependent increase in AP-1 nuclear binding. *Free Radical Biology & Medicine*, 2000 29, 1280-1290.

Dorval, J; Leblond, VS; Deblois Ch; Hontela, A. Oxidative stress and endocrine endpoints in white sucker (*Catostomus commersoni*) from a river impacted by agricultural chemicals. *Environmental Toxicology and Chemistry*, 2005 24 5, 1273–1280.

Dorval, J; Leblond, VS; Hontela, A. Oxidative stress and loss of cortisol secretion in the adrenocortical cells of rainbow trout (*Oncorhynchus mykiss*) exposed to endosulfan, an organochlorine pesticide. *Aquatic Toxicology*, 2003 63, 229-241.

Durmaz, H; Sevgiler, Y; Üner, N. Tissue-specific antioxidative and neurotoxic responses to diazinon in *Oreochromis niloticus. Pesticide Biochemistry and Physiology* 2006 84, 215-226.

Dutta, HM; Marcelino, J; Richmonds, CR. Brain acetylcholinesterase activity and optomotor behavior in bluegills, *Lepomis machrochirus* exposed to different concentration of diazinon. *Archives Internationales de Physiologie, de Biochimie et de Biophysique* 1992 100 5, 331-334.

Eason, Ch; O'Halloran, K. Biomarkers in toxicology versus ecological risk assessment. *Toxicology* 2002 181-182, 517-521.

Ecobicon DJ. Pesticides and neurological diseases. Boca Raton, Florida: CRC Pres, Inc; 1982.

Eder, KJ; Kohler, HR; Werner, I. Pesticide and pathogen: heat shock protein expression and acetylcholinesterase inhibition in juvenile Chinook salmon in response to multiple stressors. Environmental Toxicology and Chemistry, 2007 26 6, 1233-1242.

Eder, KJ; Leutenegger, ChM; Wilson, BW; Werner, I. Molecular and cellular biomarker responses to pesticide exposure in juvenile Chinook salmon (*Oncorhynchus tshawytscha*). *Marine Environmental Research*, 2004 58,809–813.

Edwards, CA; Fisher SW. The use of cholinesterase measurement in assessing the impact of pesticides on terrestrial and aquatic invertebrates. In: Mineau P, editor. *Cholinesterase Inhibiting Insecticides: Their Impact on Wildlife and the Environment. Volume 2, Chemicals in Agriculture*. New York, USA: Elsevier; 1991; 255-275.

Egaas, E; Sandvik, M; Fjeld, E; Källqvist, T; Goksøyr, A; Svensen, A. Some effects of the fungicide propiconazole on cytochrome P450 and glutathione S-transferase in brown trout (*Salmo trutta*). *Comparative Biochemistry and Physiology C*, 1999 122, 337-344.

Fabrizi, L; Gemma, S; Testai, E; Vittozzi, L. Identification of the cytochrome P450 isoenzymes involved in the metabolism of diazinon in the rat liver, J. *Journal of biochemical and molecular toxicology* 1999 13, 53-61.

Fanta, E; Rios, SAF; Rom, S; Casagrande Vianna, AC; Freiberger, S. Histopathology of the fish *Corydoras paleatus* contaminated with sublethal levels of organophosphorus in water and food. *Ecotoxicology and environmental safety* 2003 54, 119–130.

Fenet, H; Casellas, C; Bontoux, J. Laboratory and field-caging studies on hepatic enzymatic activities in the European eel and rainbow trout. *Ecotoxicology and Environmental Safety*, 1998 40, 137-143.

Ferrando, MD; Sancho, E; Andreu, E. Comparative acute toxicities of selected pesticides to *Anguilla anguilla*. *Journal of Environmental Science and Health B*, 1991 26, 491–498.

Ferrari, A. Evaluación de las respuestas moleculares producidas por la exposición de peces a plaguicidas organofosforados y carbamatos. PhD Thesis, Universidad Nacional del Comahue, Argentina, 2004.

Ferrari, A; Anguiano, L; Soleño, J; Venturino, A; Pechen de D'Angelo, AM. Diferente sensibilidad de especies acuáticas a plaguicidas metil azinfos, paratión y carbaril. In: Herkovits J, editor. *First Book of Latino American Society of Environmental Toxicology and Chemistry: Contributions for a sustainable development*. Buenos Aires, Argentina: SETAC-LA member of SETAC PRESS; 2003; 178-181.

Ferrari, A; Anguiano, OL; Soleño, J; Venturino, A; Pechen de D'Angelo, AM. Different susceptibility of two aquatic vertebrates (*Oncorhynchus mykiss* and *Bufo arenarum*) to azinphos methyl and carbaryl. *Comparative biochemistry and physiology C* 2004b 139, 239-243.

Ferrari, A; Bubach, D; Ribeiro Guevara, S; Arribére, MA; Venturino, A; Pechen de D'Angelo, AM. Distribución de paratión y mercurio y biomarcadores de efecto en la exposición de *Cyprinus carpio* a concentraciones subletales. *Acta Toxicológica Argentina*, 2002 10 1, 5-10.

Ferrari, A; Venturino, A; Pechen de D'Angelo, AM. Effects of carbaryl and azinphos methyl on juvenile rainbow trout (*Oncorhynchus mykiss*) detoxifying enzymes. *Pesticide Biochemistry and Physiology*, 2007b 88, 134–142.

Ferrari, A; Venturino, A; Pechen de D'Angelo, AM. Muscular and brain cholinesterase sensitivities to azinphos methyl and carbaryl in the juvenile rainbow trout *Oncorhynchus mykiss*. *Comparative Biochemistry and Physiology C*, 2007a 146, 308–313.

Ferrari, A; Venturino, A; Pechen de D'Angelo, AM. Time course of brain cholinesterase inhibition and recovery following acute and subacute azinphosmethyl, parathion and carbaryl exposure in the goldfish (*Carassius auratus*). *Ecotoxicology and Environmental Safety*, 2004a 57, 3, 420-425.

Ferrari, A; Venturino, A; Pechen de D'Angelo, AM. Efectos del insecticida carbarilo sobre biomarcadores en juveniles de trucha arcoiris (*Oncorhynchus mykiss*). In: Herkovits J, editor. *Second Book of Latino American Society of Environmental Toxicology and Chemistry:* Salud Ambiental y Humana: Una Visión Holística. Buenos Aires, Argentina: SETAC-LA member of SETAC PRESS; 2006; 118-121.

Ferreira, M; Moradas-Ferreira, P; Reis-Henriques, MA. Oxidative stress biomarkers in two resident species, mullet (*Mugil cephalus*) and flounder (*Platichthys flesus*), from a polluted site in River Douro Estuary, Portugal. *Aquatic Toxicology*, 2005 71, 39-48.

Finne, EF; Cooper, GA; Koop, BF; Hylland, K; Tollefsen, KE. Toxicogenomic responses in rainbow trout (*Oncorhynchus mykiss*) hepatocytes exposed to model chemicals and a synthetic mixture. *Aquatic Toxicology*, 2007 81 3, 293-303.

Fisk, AT; Yarechewski, AL; Metner, DA; Evans, RE; Lockhart, WL; Muir DCG. Accumulation, depuration and hepatic mixed-function oxidase enzyme induction in juvenile rainbow trout and lake whitefish exposed to dietary 2,3,7,8-tetrachlorodibenzo - p-dioxin. *Aquatic Toxicology*, 1997 37, 201-220.

Flammarion P; Bernard, M; Urios, S; Morfin, P; Garric J. Effect of methidathion on the cytochrome P-450 1A in the cyprinid fish gudgeon (*Gobio gobio*). *Aquatic Toxicology*, 1998 42, 93–102.

Flammarion, P; Noury, P; Garric J. The measurement of cholinesterase activities as a biomarker in chub (*Leuciscus cephalus*): the fish length should not be ignored. *Environmental Pollution* 2002 120, 325–330.

Fukuto, RT. Mechanism of action of organophosphorus and carbamate insecticides. *Environmental Health Perspectives*, 1990 87, 245-254.

Fulton, MH; Key, PB. Acetylcholinesterase inhibition in estuarine fish and invertebrates as indicator of organophosphorus insecticide exposure and effects. *Environmental Toxicology and Chemistry*, 2001 20, 37-45.

Garcia, LM; Castro, B; Ribeiro, R; Guilhermino, L. Characterization of cholinesterase from guppy (*Poecilia reticulata*) muscle and its in vitro inhibition by environmental contaminants. *Biomarkers*, 2000 5, 274–284.

George, AG. Enzymology and molecular Biology of Phase II Xenobiotic –Conjugating Enzymes in Fish. In: Malins, DC, Ostrander GK, editors. *Aquatic Toxicology: Molecular, biochemical, and cellular perspectives*. Boca Raton, Florida: CRC Press; 1994; 37-85.

Ghosh, MC; Ghosh, R; Ray AK. Induction of CYP1A by carbofuran in primary culture of fish hepatocytes. *Journal of Biochemical and Molecular Toxicology*, 2000 14 4, 204-209.

Gravato, C; Santos, MA. Juvenile Sea Bass liver P450, EROD induction, and erythrocytic genotoxic responses to PAH and PAH-like compounds. *Ecotoxicology and Environmental Safety*, 2002 51, 115-127.

Gruber, SJ; Munn, MD. Organophosphate and carbamate insecticides in agricultural waters and cholinesterase (ChE) inhibition in common carp (*Cyprinus carpio*). *Archives of Environmental Contamination and Toxicology*, 1998 35, 391-396.

Grue, CE; Hart, ADM; Mineau P. Biological consequences of depressed brain cholinesterase in wildlife. In: Mineau P, editor. *Cholinesterase Inhibiting Insecticides: Their Impact on Wildlife and the Environment. Volume 2, Chemicals in Agriculture*. New York, USA: Elsevier; 1991; 151-210.

Guilhermino, L; Barros, P; Silva, MC; Soares, AMVM. Should the use of inhibition of cholinesterases as a specific biomarker for organophosphates and carbamate pesticides be questioned? *Biomarkers*, 1998 3, 157–163.

Habig, C; Di Giulio, RT. Biochemical characteristics of cholinesterases in aquatic organisms. In: Mineau P, editor. *Cholinesterase Inhibiting Insecticides: Their Impact on Wildlife and the Environment. Volume 2, Chemicals in Agriculture*. New York, USA: Elsevier; 1991; 19-33.

Hai, DQ; Varga, SI; Matkovics, B. Organophosphate effects on antioxidant system of carp (*Cyprinus carpio*) and catfish (*Ictalurus nebulosus*). *Comparative Biochemistry and Physiology C*, 1997 117, 83-88.

Haines, TA. Effect of an aerial application of carbaryl on brook trout (*Salvelinus fontinalis*). *Bulletin of Environmental Contamination and Toxicology*, 1981 27 4, 534-542.

Halliwell, B; Gutteridge JMC. Free radicals in biology and medicine. 3rd edition. Oxford: Oxford University Press; 1989.

Hassanein, HMA. Toxicological effects of the herbicide oxyfluorfen on acetylcholinesterase in two fish species: *Oreochromis niloticus* and *Gambusia affinis*. *Environmental Science and Health A*, 2002 37 4, 521 –527.

Jaiswal, AK. Nrf2 signaling in coordinated activation of antioxidant gene expression. *Free Radical Biology & Medicine*, 2004 36, 1199-1207.

Janssens, B; Childress, JJ; Baguet, F; Rees, J-F. Reduced enzymatic antioxidant defense in deep-sea fish. *The Journal of experimental biology*, 2000 203, 3717-3725.

Janssens, BJ; Le Gall, R; Rees, JF. Peroxide-triggered erythrocytes haemolysis as a model for the study of oxidative damage in marine fishes. *Journal of Fish Biology*, 2002 61, 71-84.

Jarrard, HE; Delaney, KR; Kennedy CJ. Impacts of carbamate pesticides on olfactory neurophysiology and cholinesterase activity in coho salmon (*Oncorhynchus kisutch*). *Aquatic Toxicology*, 2004 69, 133–148.

Jensen, EG; Skaare, JU; Egas, E; Goksøyr, A. Response of metabolizing enzymes in rainbow trout (*Oncorhynchus mykiss*) to endosulfan by enzyme activities and immunological methods. *Aquatic Toxicology*, 1991 21, 81-92.

Jett DA. Lein PJ. Noncholinesterase mechanisms of central and peripheral neurotoxicity: muscarinic receptors and other targets. In: Gupta RC, editor. Toxicology of Organophosphate & Carbamate Compounds. Academic Press Elsevier Inc; 2006; 233-245.

Jokanović M. Biotransformation of organophosphorus compounds. *Toxicology*, 2001 166, 139-160.

Ju, Z, Wells, MC; Walter, RB. DNA microarray technology in toxicogenomics of aquatic models: methods and applications. *Comparative Biochemistry and Physiology C*, 2007 1451, 5-14.

Jyothi, B; Narayan G. Certain Pesticide-induced carbohydrate metabolic disorders in the serum of freshwater fish *Clarias batrachus* (Linn.). *Food and Chemical Toxicology: an International journal published for the British Industrial Biological Research Association*, 1999 37: 417-421.

Kamata, H; Hirata, H. Redox regulation of cellular signaling. *Cellular Signaling*, 1999 11 1, 1-14.

Kaufer, D; Friedman, A; Seidman, S; Soreq, H. Acute stress facilitates long-lasting changes in cholinergic gene expression. *Nature*, 1998 393, 373-377.

Kavitha, P; Venkateswara Rao J. Oxidative stress and locomotor behaviour response as biomarkers for assessing recovery status of mosquito fish, *Gambusia affinis* after lethal effect of an organophosphate pesticide, monocrotophos. *Pesticide Biochemistry and Physiology*, 2007 87, 182–188

Kirby, MF; Morris, S; Hurst, M; Kirby, SJ; Neall, P; Tylor, T; Fagg, A. The use of cholinesterase activity in flounder (*Platichthys flesus*) muscle tissue as a biomarker of neurotoxic contamination in UK estuaries. *Marine Pollution Bulletin*, 2000 40: 780-791.

Kobayashi, M; Itoh, K; Suzuki, T; Osanai, H; Nishikawa, K; Katoh, Y; Takagi, Y; Yamamoto, M. Identification of the interactive interface and phylogenic conservation of the Nrf2-Keap1 system. *Genes Cells*, 2002 7,807–20.

Kobayashi, M; Yamamoto, M. Nrf2–Keap1 regulation of cellular defense mechanisms against electrophiles and reactive oxygen species. *Advances in Enzyme Regulation*, 2006 46 1, 113-140.

Kono, Y; Fridovich I. Superoxide radical inhibits catalase. *Journal of Biological Chemistry*, 1982 257, 5751-5754.

Krasnov, A; Koskinen, H; Rexroad, C; Afanasyev, S; Mölsä H; Oikari A. Transcriptome responses to carbon tetrachloride and pyrene in the kidney and liver of juvenile rainbow trout (*Oncorhynchus mykiss*). Aquatic Toxicology, 2005 74 , 70-81.

Kuhr, RJ; Dorough, HW. Carbamate pesticides: Chemistry, Biochemistry and Toxicology. Cleveland, OH, USA: CRC Press; 1976.

Küster E, Altenburger R. Comparison of cholin- and carboxylesterase enzyme inhibition and visible effects in the zebra fish embryo bioassay under short-term paraoxon-methyl exposure. *Biomarkers*. 2006 11, 341-54.

Labenia, JS; Baldwin, DH; French, BL; Davis, JW; Scholz, NL. Behavioral impairment and increased predation mortality in cutthroat trout exposed to carbaryl. *Marine Ecology Progress Series*, 2007 329, 1–11.

Lindström-Seppä, P; Roy, S; Huuskonen, S; Tossavainen, K; Ritola, O; Marin, E. Biotransformation and glutathione homeostasis in rainbow trout exposed to chemical and physical stress. *Marine Environmental Research*, 1996 42: 323-327.

Lionetto, MG; Caricato, R; Giordano, ME; Pascariello, MF; Marinosci, L; Schettino, T. Integrated use of biomarkers (acetylcholinesterase and antioxidant enzymes activities) in *Mytilus galloprovincialis* and *Mullus barbatus* in an Italian coastal marine sea. *Marine Pollution Bulletin*, 2003 46, 324-330.

Livingstone, DR. Contaminant-stimulated reactive oxygen species production and oxidative damage in aquatic organisms. *Marine Pollution Bulletin*, 2001 42, 656-666.

Magnotti Jr, RA; Zaino, JP; McConnell, RS. Pesticide-sensitive fish muscle cholinesterases. *Comparative Biochemistry and Physiology C*, 1994 108 2, 187-194.

Martyniuka, CJ; Gerriea, ER; Popeskua, JT; Ekkera, M; Trudeau VL. Microarray analysis in the zebrafish (*Danio rerio*) liver and telencephalon after exposure to low concentration of 17alpha-ethinylestradiol *Aquatic Toxicology*, 2007 84, 38-49.

Matos, P; Fontainhas-Fernandes, A; Peixoto, F; Carrola, J: Rocha, E. Biochemical and histological hepatic changes of Nile tilapia *Oreochromis niloticus* exposed to carbaryl. *Pesticide Biochemistry and Physiology*, 2007 89, 73-80.

Matsumura F. On the significance of the role of cellular stress response reactions in the toxic actions of dioxin. *Biochemical Pharmacology*, 2003 66 4, 527-540.

Matsumura, F. Toxicology of insecticides. 2nd edition. New York, USA: Plenum Press, 1985.

Maxwell, DM. The specificity of carboxylesterase protection against the toxicity of organophosphorus compounds. *Toxicology and Applied Pharmacology* 1992 114, 306-312.

Miller, JK; Brzezinska-Slebodzinska E. Oxidative stress, antioxidants, and animal function. *Journal of Dairy Science*, 1993 76, 2812-2823.

Miracle, AL; Ankley GT. Ecotoxicogenomics: linkages between exposure and effects in assessing risks of aquatic contaminants to fish. *Reproductive Toxicology* 2005 19, 321–326

Monteiro, DA; Alves de Almeida, J; Tadeu Rantin, F; Kalinin, A. Oxidative stress biomarkers in the freshwater characid fish, *Brycon cephalus*, exposed to organophosphorus insecticide Folisuper 600 (methyl parathion). *Comparative Biochemistry and Physiology C*, 2006 143, 141–149.

Mortensen, AS; Arukwe, A. Targeted salmon gene array (SalArray): a toxicogenomic tool for gene expression profiling of interactions between estrogen and aryl hydrocarbon receptor signalling pathways. *Chemical Research in Toxicology*, 2007 20 3, 474-88.

Mulcahy, RT; Wartman, MA; Bailey, HH; Gipp, JJ. Constitutive and ß-naphthoflavone-induced expression of the human g-Glutamylcysteine synthetase heavy subunit gene is regulated by a distal antioxidant response element/TRE sequence. *Journal of Biological Chemistry*, 1997 272 11, 7445–7454.

Mutch, E; Williams, FM. Diazinon, chlorpyrifos and parathion are metabolized by multiple cytochromes P450 in human liver, *Toxicology*, 2006 224, 22–32.

Napierska, D; Podolska, M. Biomarkers of contaminant exposure: results of a field study with flounder (*Platichthys flesus*) from the southern Baltic Sea. *Marine Pollution Bulletin*, 2005 50, 758–767.

Nguyen, T; Sheratt, PJ; Pickett, CB. Regulatory mechanisms controlling gene expression mediated by the antioxidant response element. *Annual Review of Pharmacology and Toxicology*, 2003 43, 233-260.

Niijima, H; Nagao, M; Nakajima, M; Takatori, T; Iwasa, M; Maeno, Y; Koyama H; Isobe, I. The effects of sarin-like and soman-like organophosphorus agents on MAPK and JUNK in rat brains. *Forensic Science International*, 2000 112, 171-178.

Nishimoto, M; Eberhart, BTL; Sanborn, HR; Krone, C; Varanasi, U; Stein JE. Effects of a complex mixture of chemical contaminants on hepatic glutathione, l-cysteine and γ-glutamylcysteine synthetase in English sole (*Pleuronectes vetulus*). *Environmental Toxicology and Chemistry*, 1995 14, 461–469.

Olivier, K; Liu, J; Pope, C. Inhibition of forskolin-stimulated cAMP formation in vitro by paraoxon and chlorpyrifos oxon in cortical slices from neonatal, juvenile, and adults rats. *Journal of Biochemical and Molecular Toxicology*, 2001 15, 263-268.

Oruç, EÖ; Üner, N. Combined effects of 2,4-D and azinphosmethyl on antioxidant enzymes and lipid peroxidation in liver of *Oreochromis niloticus*, *Comparative Biochemistry and Physiology C*, 2000 127, 291-296.

Oruç, EÖ; Üner, N. Tissue specific oxidative stress responses in fish exposed to 2,4-D and azinphosmethyl, *Comparative Biochemistry and Physiology C*, 2004 137, 43-51.

Oruç, EÖ; Utsa D. Evaluation of oxidative stress responses and neurotoxicity potential of diazinon in different tissues of *Cyprinus carpio*. *Environmental Toxicology and Pharmacology*, 2007 23 1, 48-55.

Palace, VP; Dick, TA; Brown, SB; Baron, CL; Klaverkamp, JF. Oxidative stress in Lake Sturgeon (*Acipenser fulvescens*) orally exposed to 2,3,7,8-tetrachlorodibenzofuran. *Aquatic Toxicology*, 1996 35, 79–92.

Pandey, S; Parvez, S; Sayeed, I; Haque, R; Bin-Hafeez, B; Raisuddin, S. Biomarkers of oxidative stress: a comparative study of river Yamuna fish *Wallagu attu* (Bl. & Schn.). *The Science of Total Environment*, 2003 309 1-3, 105-115.

Panemangalore, M; Bebe FN. Dermal exposure to pesticides modifies antioxidant enzymes in tissues of rats. *Journal of Environmental Science and Health B*, 2000 35, 399-416.

Parvez, S; Raisuddin, S. Protein carbonyls: novel biomarkers of exposure to oxidative stress-inducing pesticides in freshwater fish *Channa punctata* (Bloch). *Environmental Toxicology and Pharmacology*, 2005 20 1, 112-117.

Payne, JF; Mathieu, A; Melvin, W; Fancey, LL. Acetylcholinesterase, an old biomarker with a new future? Field trials in association with two urban rivers and a paper mill in Newfoundland. *Marine Pollution Bulletin*, 1996 32 2, 225-231.

Peña-Llopis, S; Ferrando, MD; Peña, JB. Fish tolerance to organophosphate-induced oxidative stress is dependent on the glutathione metabolism and enhanced by N-acetylcysteine. *Aquatic Toxicology*, 2003 65 4, 337-360.

Peña-Lopis, S; Ferrando, MD; Peña JB. Impaired glutathione redox status is associated with decreased survival in two organophosphate-poisoned marine bivalves. *Chemosphere*, 2002 47, 485-497.

Perkins, EJ; Schlenk D. In Vivo Acetylcholinesterase inhibition, metabolism, and toxicokinetics of aldicarb in channel catfish: Role of biotransformation in acute toxicity. *Toxicological Sciences: an official journal of the Society of Toxicology*, 2000 53, 308-315.

Phillips, TA; Summerfelt, RC; Atchinson, GJ. Environmental, biological and methodological factors affecting cholinesterase activity walleye (*Stizostedion vitreum*). *Archives of Environmental Contamination and Toxicology*, 2002b 43, 75-80.

Phillips, TA; Wu, J; Summerfelt, RC; Atchinson, GJ. Acute toxicity and cholinesterase inhibition in larval and early juvenile walleye exposed to chlorpyrifos. *Environmental Toxicology and Chemistry*, 2002a 21, 1469-1474.

Pope, C; Karanth, S; Liu J. Pharmacology and toxicology of cholinesterase inhibitors: uses and misuses of a common mechanism of action. *Environmental Toxicology and Pharmacology*, 2005 19, 433–446.

Pope, CN. Organophosphorus pesticides: Do they have all the same mechanism of toxicity? *Journal of Toxicology and Environmental Health B*, 1999 2, 161-181.

Radice, S; Ferraris, M; Marabini, L; Grande, S; Chiesara, E. Effect of iprodione, a dicarboximide fungicide, on primary cultured rainbow trout (*Oncorhynchus mykiss*) hepatocytes. *Aquatic Toxicology*, 2001 54, 51–58.

Rattner, BA; Fairbrother, A. Biological variability and the influence of stress on cholinesterase activity. In: Mineau P, editor. *Cholinesterase Inhibiting Insecticides: Their Impact on Wildlife and the Environment. Volume 2, Chemicals in Agriculture*. New York, USA: Elsevier; 1991; 89-105.

Regoli, F; Pellegrini, D; Winston, GW; Gorbi, S; Giuliani, S; Virno-Lamberti, C; Bompadre, S. Application of biomarkers for assessing the biological impact of dredged materials in the Mediterranean: the relationship between antioxidant responses and susceptibility to oxidative stress in the red mullet (*Mullus barbatus*). *Marine Pollution Bulletin*, 2002 44, 912–922

Regoli, F; Winston, GW. Quantification of total oxidant scavenging capacity (TOSC) of antioxidants for peroxynitrite, peroxyl radicals and hydroxyl radicals. *Toxicology and Applied Pharmacology*, 1999 156, 96-105.

Rendón-von Osten, J; Ortiz-Arana, A; Guilhermino, L; Soares, AMVM. In vivo evaluation of three biomarkers in the mosquitofish (*Gambusia yucatana*) exposed to pesticides. *Chemosphere* 2005 58, 627–639.

Renglin, A; Harmala-Brasken, AS; Eriksson, JE; Onfelt, A. Mitotic aberrations induced by carbaryl reflect tyrosine kinase inhibition with coincident up-regulation of serine/threonine protein phosphatase activity: implications for coordination of karyokinesis and cytokinesis. *Mutagenesis*, 1999 14 3, 327-33.

Richmonds, CR; Dutta, HM. Effects of pH on inhibition and spontaneous reactivation of acetylcholinesterase treated with esters of phosphorus acids and carbamic acids. *The Biochemical Journal*, 1992 105, 171-179.

Roex, EWM; Keijzers, R; Van Gestel, CAM. Acetylcholinesterase inhibition and increased food consumption rate in the zebrafish, *Danio rerio*, after chronic exposure to parathion. *Aquatic Toxicology*, 2003.64: 451–460.

Sancho, E; Ferrando, MD; Andreu, E. Response and recovery of brain acetylcholinesterase activity in the European eel, *Anguilla anguilla*, exposed to fenitrothion. *Ecotoxicology and Environmental Safety*, 1997 38, 205-209.

Sancho, E; Ferrando, MD; Andreu, E; Gamon, M. Acute toxicity, uptake and clearance of diazinon by the European eel (*Anguilla anguilla*, L.). *Journal of Environmental Science and Health B*, 1992 27, 209–211.

Sandahl, JF; Baldwin, DH; Jenkins, JJ; Scholz, NL. Comparative thresholds for acetylcholinesterase inhibition and behavioral impairment in Coho salmon exposed to chlorpyrifos. *Environmental Toxicology and Chemistry*, 2005 24 1, 136–145.

Sandvik, M; Beyer, J; Goksoyr, A; Hylland, K; Egaas, E; Skaare, U. Interaction of benzo[a]pyrene, 2,3,3',4,4',5-hexachlorobiphenyl (PCB-156) and cadmium on biomarker responses in flounder (*Platichthys flesus* L.). *Biomarkers*, 1997 2, 153-160.

Saran, M; Michel; Ch; Bors W. Radical functions in vivo: A critical review of current concepts and hypotheses. *Zeitschrift für Naturforschung. C, Journal of biosciences*, 1998 53, 210-227

Schuh, RA; Lein, PJ; Beckles, RA; Jett, DA. Non cholinergic mechanisms of chlorpyrifos neurotoxicity: altered phosphorylation of Ca2+/cAMP response element binding protein in cultured neurons. *Toxicology and Applied Pharmacology*, 2002 182, 176-185.

Scott, GR; Sloman, KA. The effects of environmental pollutants on complex fish behaviour: integrating behavioural and physiological indicators of toxicity. *Aquatic Toxicology*, 2004 68, 369–392.

Sen, ChK; Packer, L. Antioxidant redox regulation of gene transcription. *The FASEB journal: official publication of the Federation of American Societies for Experimental Biology*, 1996 10, 709-720.

Senger, MR; Rico, EP; de Bem Arizi, M; Rosemberg, DB; Dias, RD; Bogo, MR; Bonan, CD. Carbofuran and malathion inhibit nucleotide hydrolysis in zebrafish (*Danio rerio*) brain membranes. *Toxicology*, 2005 212 2-3, 107-15.

Sharma, Y; Bashir, S; Irshad, M; Gupta, DS; Dogra, TD. Effects of acute dimethoate administration on antioxidant status of liver and brain of experimental rats. *Toxicology*, 2005 206, 49–57.

Sies H. Glutathione and its role in cellular functions. *Free Radical Biology & Medicine*, 1999 27 9/10, 916–921.

Sismeiro-Vivas, J; Abrantes, N; Pereira, JL; Castro, BB; Gonçalves, F. Short-term effects of Quirlan (chlorfenvinphos) on the behavior and acetylcholinesterase activity of *Gambusia holbrooki*. *Environmental Toxicology*, 2007 22, 194-202.

Sivapiriya, V; Jayanthisakthisekaran, Venkatraman S. Effects of dimethoate (O,O-dimethyl S-methyl carbamoyl methyl phosphorodithioate) and ethanol in antioxidant status of liver and kidney of experimental mice. *Pesticide Biochemistry and Physiology*, 2006, 85 2, 115-121.

Slotkin, TA; Seidler, FJ. Comparative developmental neurotoxicity of organophosphates in vivo: transcriptional responses of pathways for brain cell development, cell signaling, cytotoxicity and neurotransmitter systems. *Brain Research Bulletin*, 2007 72, 232-274.

Soimasuo, MR; Werner, I; Villalobos, A; Hinton, DE. Cytochrome P450 1A- and stress protein-induction in early life stages of medaka (*Oryzias latipes*) exposed to trichloroethylene (TCE) soot and different fractions. *Biomarkers*, 2001 6, 166-143.

Souza, MS; Magnarelli de Potas, G; Pechen de D'Angelo, AM. Organophosphorus and organochlorine pesticides affect human placental phosphoinositides metabolism and PI-4 kinase activity. *Journal of Biochemical and Molecular Toxicology*, 2004 18 1, 30-6.

Stegeman, JJ; Hahn, ME. Biochemistry and molecular biology of monooxygenases: current perspectives on forms, functions and regulation of cytochrome P450 in aquatic species. In: Malins, DC, Ostrander GK, editors. *Aquatic Toxicology: Molecular, biochemical, and cellular perspectives*. Boca Raton, Florida: CRC Press;1994; 87-206.

Stephensen, E; Sturve, J; Förling, L. Effects of redox cycling compounds on glutathione content and activity of glutathione-related enzymes in rainbow trout liver. *Comparative Biochemistry and Physiology C*, 2002 133, 435-442.

Strmac, M; Braunbeck, T. Isolated hepatocytes of rainbow trout (*Oncorhynchus mykiss*) as a tool to discriminate between differently contaminated small river systems. *Toxicology in Vitro*, 2000 14, 361-377.

Sturm, A; Assis, HCS; Hansen, P-D. Cholinesterase of marine teleost fish: enzymological and potential use in the monitoring of neurotoxic contamination. *Marine Environmental Research*, 1999a 47, 389–398.

Sturm, A; Wogram, J; Hansen, P-D; Liess, M. Potential use of cholinesterase in monitoring low levels of organophosphates in small streams: Natural variability in three-spined stickleback (*Gasterosteus aculeatus*) and relation to pollution. *Environmental Toxicology and Chemistry*, 1999b 18, 194-200.

Szeto, SY; Sundaram, KM; Feng J. Inhibition of brain AChE in brook trout by aminocarb and its toxic metabolites. *Journal of Environmental Science and Health B*, 1985 20 5, 559-75.

Tortelli, V; Colares, EP; Robaldo, RB; Nery, LEM; Pinho, GLL, Bianchini, A; Monserrat, J.M. Importance of cholinesterase kinetic parameters in environmental monitoring using estuarine fish. *Chemosphere*, 2006 65 4, 560-566.

Van Cong, N; Phuong, NT; Bayley, M. Sensitivity of brain cholinesterase activity to diazinon (Basudin 50ec) and fenobucarb (Bassa 50ec) insecticides in the air-breathing fish *Channa striata* (Bloch, 1793). *Environmental Toxicology and Chemistry*, 2006 25 5, 1418–1425.

Van der Oost, R; Beyer, J; Vermeulen, PE. Fish Bioaccumulation and biomarkers in environmental risk assessment: a review. *Environmental Toxicology and Pharmacology*, 2003 13, 57-149.

Van der Wel, H; Welling, W. Inhibition of acetylcholinesterase in guppies (*Poecilia reticulata*) by chlorpyrifos at sublethal concentrations: methodological aspects. *Ecotoxicology and Environmental Safety*, 1989 17 2, 205-215.

Van Dolah, RF; Maier, PP; Fulton, MH; Scott, GI. Comparison of azinphosmethyl toxicity to juvenile red drum (*Sciaenops ocellatus*) and the mummichog (*Fundulus heteroclitus*). *Environmental Toxicology and Chemistry*, 1997 16, 1488–1493.

Varò, I; Navarro, JC; Amat, F; Guilhermino, L. Effects of dichlorvos on cholinesterase activity of European sea bass (*Dicentrarchus labrax*). *Pesticide Biochemistry and Physiology*, 2003 75, 61-72.

Varò, I; Navarro, JC; Nunes, B; Guilhermino, L. Effects of dichlorvos aquaculture treatments on selected biomarkers of gilthead sea bream (*Sparus aurata* L.) fingerlings. *Aquaculture* 2007 266, 87–96.

Venkateswara Rao J. Toxic effects of novel organophosphorus insecticide (RPR-V) on certain biochemical parameters of euryhaline fish, *Oreochromis mossambicus*. *Pesticide Biochemistry and Physiology*, 2006b 86, 78-84.

Venkateswara Rao, J. Sublethal effects of an organophosphorus insecticide (RPR-II) on biochemical parameters of tilapia, *Oreochromis mossambicus*. *Comparative Biochemistry and Physiology C*, 2006a 143, 492–498.

Venkateswara Rao, J., Begum G, Sridhar V, Reddy NC. Sublethal effects of monocrotophos on locomotor behavior and gill architecture of the mosquito fish, *Gambusia affinis*. *Journal of Environmental Science and Health B*, 2005 40, 813–825.

Ventura, EC; Gaelzer, LR; Zanette, J; Marques, MRF; Bainy, ACD. Biochemical indicators of contaminant exposure in spotted pigfish (*Orthopristis ruber*) caught at three bays of Rio de Janeiro coast. *Marine Environmental Research*, 2002 54, 775-779.

Venturino, A; Anguiano, L; Pechen de D'Angelo, AM. Glutathione reductase and catalase in toad embryogenesis: pesticide and polyamine interaction. Proceedings of XXXIX Annual Meeting of Argentine Society of Biochemical and Molecular Biology Investigation, San Carlos de Bariloche, Argentina. *Biocell*, 2003b 27 (Suppl. I), 123.

Venturino, A; Anguiano, OL; Gauna, L; Cocca, C; Bergoc, RM; Pechén de D'Angelo, AM. Thiols and polyamines in the potentiation of malathion toxicity in larval stages of the toad *Bufo arenarum*. *Comparative Biochemistry and Physiology C*, 2001 130, 191-198.

Venturino, A; Pechen de D'Angelo, AM. Biochemical targets of xenobiotics: Biomarkers in amphibian ecotoxicology. *Applied Herpetology*, 2005 2, 335-353.

Venturino, A; Rosenbaum, E; Caballero de Castro, A; Anguiano, OL; Gauna, LE; Fonovich de Schroeder, T; Pechen de D'Angelo, AM. Biomarkers of effect in toads and frogs. *Biomarkers*, 2003a 8, 167-186.

Ward, TR; Mundi, WR. Organophosphorus compounds preferentially affect second messenger systems coupled to M2/M4 receptors in rat frontal cortex. *Brain Research Bulletin*, 1996 39 1, 49-55.

Weiss, CM. The determination of cholinesterase in the brain tissues of three species of fresh water fish and its inactivation in vivo. *Ecology*, 1958 30, 194-199.

Wheelock, CE; Eder, KJ; Werner, I; Huanga, H; Jones, PD; Brammell, BF; Elskus, AA; Hammock, BD. Individual variability in esterase activity and CYP1A levels in Chinook salmon (*Oncorhynchus tshawytscha*) exposed to esfenvalerate and chlorpyrifos, *Aquatic Toxicology*, 2005 74,172–192.

Whitehead, A; Anderson, SL; Ramirez, A; Wilson, BW. Cholinesterases in aquatic biomonitoring: assay optimization and species-specific characterization for a California native fish. *Ecotoxicology*, 2005 14 6, 597-606.

Wiegand, C; Krause, E; Steinberg, Ch; Plugmacher, S. Toxicokinetics of atrazine in embryos of the zebrafish (*Danio rerio*). *Ecotoxicology and Environmental Safety*, 2001 49, 199-205.

Williams, TD; Gensberg, K; Minchin, SD; Chipman, JK. A DNA expression array to detect toxic stress response in European flounder (*Platichthys flesus*). *Aquatic Toxicology*, 2003 65 2, 141-57.

Winston, GW; Di Giulio, RT. Prooxidant and antioxidant mechanisms in aquatic organisms. *Aquatic Toxicology*, 1991 19,137-161.

Wogram, J; Sturm, A; Segner, H; Liess, M. Effects of parathion on acetylcholinesterase, butyrylcholinesterase, and carboxylesterase in three-spined stickleback (*Gasterosteus aculeatus*) following short-term exposure. Environmental *Toxicology and Chemistry*, 2001 20, 1528-31.

Zinkl, JG; Lockhard, WL; Kenny, SA; Ward, FJ. The effects of cholinesterase inhibiting insecticides on fish. In: Mineau P, editor. *Cholinesterase Inhibiting Insecticides: Their Impact on Wildlife and the Environment. Volume 2, Chemicals in Agriculture*. New York, USA: Elsevier; 1991; 233-254.

Zinkl, JG; She, PJ; Nakamoto, RJ; Callman, J. Effects on cholinesterases of rainbow trout exposed to acephate and methamidophos. *Bulletin of Environmental Contamination and Toxicology*, 1987b 38 1, 22-28.

Zinkl, JG; Shea, PJ; Nakamato RJ; Callman, J. Brain cholinesterase activity of rainbow trout poisoned by carbaryl. *Bulletin of Environmental Contamination and Toxicology*, 1987a 38, 29-35.

In: Progress in Pesticides Research
Editor: Christopher M. Kanzantzakis

ISBN 978-1-60456-201-9

Chapter 3

NEW TRENDS TOWARDS ENVIRONMENTALLY FRIENDLY MICROEXTRACTION TECHNIQUES OF PESTICIDE RESIDUES

R. Romero-González, A. Garrido Frenich, J. L. Martínez Vidal
Research Group "Analytical Chemistry of Contaminants",
Department of Analytical Chemistry, Almeria University, E-04071 Almeria, Spain

ABSTRACT

Analytical process is comprised of several steps such as sampling, sample preparation, separation and quantification. Despite of much attention has been focused on developing advanced determination techniques such as liquid or gas chromatography coupled to mass spectrometric detectors, which provides a rapid and powerful determination, it is well known that sample preparation is still one of the most critical steps in the determination of trace pollutants such as pesticide residue analysis in different matrices. Actually, sample preparation step still influences the reliable and accurate analysis of pesticides in food and environmental samples. Bearing in mind that traditional extraction techniques based on liquid-liquid or solid-phase extraction are time consuming, labor intensive and require the use of large amounts of solvents, which are expensive and generate considerable waste, a lot of progress has been made toward the development of faster, safer and more environmentally friendly techniques for sample extraction and extraction clean up prior to instrumental analysis. Although different alternatives such as QuEChERS (Quick, Easy, Cheap, Effective, Rugged and Safe) method based on acetonitrile extraction and dispersive solid phase extraction have been proposed, recent trends in sample preparation are based on miniaturization, and simplification of sample preparation methods, automation, high throughput performance, on-line coupling with analytical instruments such as liquid, gas chromatographs and capillary electrophoresis, and reduction in solvent volume and time, providing simple, inexpensive, and virtually solvent-free extraction techniques.

Thus, techniques such as solid-phase and liquid-phase microextraction (single-drop microextraction and hollow-fiber supported liquid membrane), and stir bar sorptive extraction have been developed in the last few years. These new microextraction techniques, which usually integrates sampling, extraction and concentration into a single step, are defined as non-exhaustive sample preparation techniques, and they only use microliters of extracting solvent or are solvent-free techniques.

This chapter includes a discussion of the different microextraction techniques, presenting their advantages and shortcomings, extraction devices available and commercial equipment. The main parameters for method optimization and practical considerations such as differences between operating in the kinetic extraction regime or under equilibrium conditions will be discussed as well as quantification procedures used for a reliable determination. Finally the potential of these techniques to be used in monitoring purposes and their application in pesticide residue analysis in environmental and food samples will be given, highlighting some common characteristics such as low detection limits and high selectivity.

1. INTRODUCTION

Pesticides constitute a very important group of chemical compounds which have been widely used throughout the world to protect plants against a range of pests and/or to maintain hygienic conditions in production lines or during storage. They include a variety of compounds, mainly insecticides, herbicides and fungicides, with very different physicochemical characteristics, and large differences in polarity, volatility and persistence. In long term, the use for agriculture of pesticides can cause the contamination of the environment. Besides, these chemicals can remain in the plant in which they were spread, so they can be introduced in the food chain.

The frequent presence of their residues in the environment and fruits and vegetables as well as their high toxicity to the human health has caused a great social and scientific concern in the last years [1]. In this sense, national and international organizations, such as European Union (EU) and US Environmental Protection Agency (EPA) have fixed maximum residues limits (MRLs) in foodstuffs [2,3] as well as they have established directives that remark the necessity to asses the environmental impact of these type of products in order to ensure public health [4,5].

Due to lower detection levels required by regulatory bodies and the complex nature of the matrices in which pesticides are present, the development of sensitive and multiclass analytical methods for the analysis of pesticide residues in food and environmental samples has received increasing attention in the last years [6,7].

In general, the developed analytical procedures have several steps: sampling, sample preparation, separation and quantification, and data analysis [8]. Related to the separation and quantification step, highly efficient instruments such as gas chromatography (GC) [9], liquid chromatography (LC) [10] or capillary electrophoresis (CE) [11] coupled to mass spectrometry (MS) have been used, depending on the thermal stability, polarity and volatility of pesticides.

However, sample preparation is still the bottleneck of such modern procedures because these methodologies cannot handle the sample matrices directly without some preliminary sample preparation, since contaminants are too diluted and/or the matrix is rather complex [12]. Thus, sample preparation is the main source of errors and the most time consuming step, affecting the reliability and accuracy of analysis [13] and over 75 % of analysis time is spent on the sampling and sample preparation steps such as extraction, concentration, clean-up, fractionation and isolation of analytes. In fact, traditional sample preparation techniques, which are mainly based on liquid-liquid extraction (LLE) are time-consuming, labor intensive and multi-step operations. Each step can introduce errors and losses, specially when analyzing volatile compounds. Besides, the use of large amounts of toxic solvents, which are expensive, is an additional problem, generating considerable waste and creates health hazards to the laboratory personnel. In this sense, solid phase extraction (SPE) was developed as a powerful alternative, due to its simplicity and because it needs less solvent than LLE. However, it is still time-consuming, multi-step process and often requires a final concentration step, which may result in a loss of volatile compounds. Besides, other problems such as clogging and evaporation, which is more time consuming in SPE than in LLE because protic solvents are mainly used (methanol), are still present [14].

This has resulted in the development of alternative procedures, trying to fulfill some desired properties of the sample preparation methodologies for pesticide residue analysis, such as precision, accuracy, safe, low costs, easy to use, low time consuming and compatible with a range of analytical instruments [15]. In this sense, alternative procedures, such as the method developed by Anastassiades et al. [16], termed as QuEChERS (quick, easy, cheap, effective, rugged and safe method) and matrix solid phase dispersion (MSPD) [17] try to overcome some practical limitations of traditional methodologies.

Notwithstanding, a great effort has been devoted in the last few years to miniaturize the current procedures, which has led to the development of faster and more powerful and more versatile extraction techniques. These include solid phase microextraction (SPME) [18], stir bar sorptive microextration (SBSE) [19], single-drop microextraction (SDME) [20] and hollow-fiber supported liquid membrane (HF-SLM), also termed HF liquid phase microextraction, (HF-LPME) [21], which are easy and fast to apply, using only few microliters of extracting solvent or are solvent free techniques, so exposure to toxic organic solvent is only minimal. These new microextraction techniques usually integrate sampling, extraction and concentration into a single step, minimizing the source of errors of traditional methodologies. These sample preparation techniques allow reduction of the sample preparation time, solvent purchase and disposal costs, obtaining low detection limits [22], improving laboratory productivity and allowing for the monitoring of systems were analyte concentration rapidly vary over time.

This chapter describes some of the most interesting sample preparation procedures based mainly in microextraction techniques, showing their advantages and shortcomings and presenting their applications in pesticide residue analysis in environmental and food samples.

2. Sample Preparation Techniques

2.1. Conventional Techniques

Despite the advances in separation and quantification techniques, many samples preparation practices are based on traditional technologies such as LLE and Soxhlet extraction. Thus, pesticides have been extracted from several matrices such as water [23,24], fruits and vegetables [25], honey [26] and olive oil [27] using LLE, whereas Soxhlet extraction has been mainly used for environmental samples such as plants [28], soils and sludge [24,29], although other matrices can be analyzed by this technique [30]. Although these extraction techniques are among the most popular procedures in routine sample preparation because their simplicity, robustness and minimal operator training efficiency, they present several disadvantages. They are time consuming, require large volume of solvents, which are hazardous and expensive, enable to extract polar pesticides and require the evaporation of large volumes of solvents to analyze trace level of pesticides.

SPE was developed in the 1970s and has emerged as a powerful alternative to LLE, because its simplicity, higher recovery, more effective concentration, less organic solvent usage, no foaming or emulsion problems, shorter sample preparation time and easier incorporation into an automated process [15]. SPE generally uses an adsorbent material to extract the analytes and basically, the sample is passed through a cartridge (a packed column filled with the sorbent) or a disk, where the pesticides are adsorbed, and then they are eluted with an appropriate organic solvent. The most important factor in SPE is the selection of the sorbent. Although there are several types of sorbents for SPE, including normal-phase, reversed phase, ionic and other special sorbents, C18 bonded silica and styrene/divinyl benzene co-polymers are the most frequently used [31]. It can be used in off-line and on-line modes, obtaining more accurate results with the last one because there is no sample handling between preconcentration and analytical steps. Besides, it can handle small samples (50-100 μL) and consumes small amount of solvent [32].

This technique has been widely used for water samples [33-37] and liquid foods such as fruit juices [38,39], wine [40] or milk [41]. However, for solid samples such as soils, vegetables and fruits, it is necessary an homogenization, centrifugation and/or filtration steps previous SPE [42-45]. Besides, for complex matrices containing interfering substances such as salts, humic acids and proteins, low recovery is obtained [46].

The main drawbacks of SPE procedure are: high blank values, considerable performance variation between the products offered by different manufacturers, lot to lot variation, treatment of large sample volumes is not possible in some cases, clogging if the sample has not been filtrated, and SPE cartridges are normally constructed with plastics, which can adsorb the analyte [47]. To overcome these problems alternative extraction techniques has been developed in the last 20 years.

2.2. Matrix Solid-phase Dispersion (MSPD)

Despite of the advantages of SPE, there are still serious limitations when analysing semi-solid and solid matrices. In this sense, Barker et at. introduced in 1989, MSPD, an extraction and clean-up technique which can be used for solid samples [48]. This technique combines several related analytical techniques, performing sample disruption while simultaneously generating a chromatographic material that possesses a unique character for target analyte isolation, avoiding general drawbacks of SPE for the extraction of analytes from solid and semi-solid samples [49]. This technique offers several advantages such as the analytical procedure is simplified and shortened, elimination of emulsion formation, reduction of solvent consumption and the extraction efficiency of the analytes is enhanced as the entire sample is exposed to the extractant.

This technique involves the dispersion of the sample in a solid sorbent, generally Florisil or C18, which are transferred to a column. The sample is then compressed to form a column packing by using a modified syringe plunger, avoiding channels in the column and not overly compress or compact the material. After that, the analytes are purified and eluted with a relative small volume of solvent, and finally, the extract can be analysed directly, although sometimes further purification is necessary [50], showing in Figure 1 an schematic representation of the whole procedure.

Proper selection of sorbent and eluent affects the specific retention of impurities on the sorbent surface and selective elution of the target compounds. In this sense, the sorbent has to show several properties:

- Working as abrasive compound breaking the physical structure of the sample
- Adsorbing the compounds from the matrix
- Working as a solid support for filling the column
- Allowing the fractionation of the sample

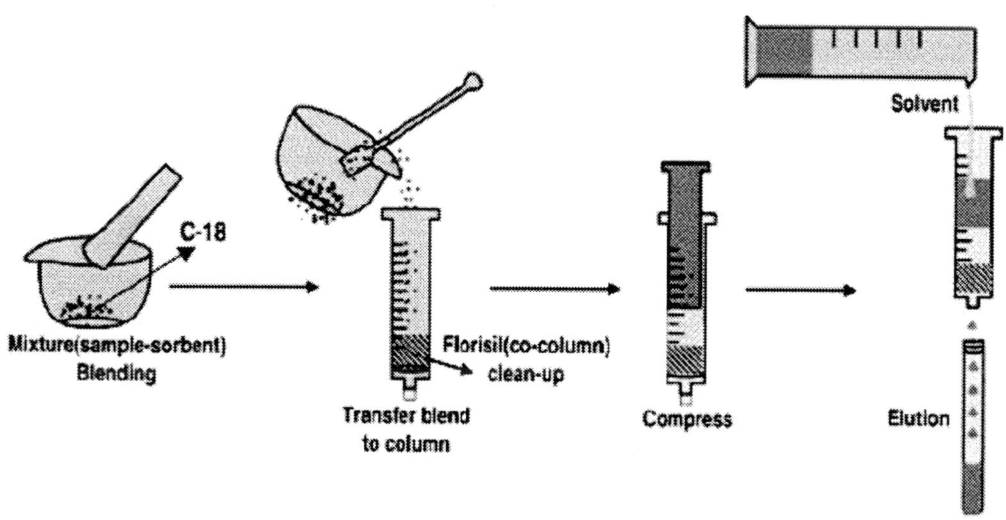

Figure 1. Schematic representation of a typical MSPD extraction procedure. Reprinted from [58] copyright 2005, with permission from Elsevier.

Pesticides are generally eluted with a small amount of solvent (5-10 mL) and the most common solvents are n-hexane, dichloromethane and ethyl acetate. The extraction solvent selection depends on the polarity range of the selected pesticides. In this sense, for the extraction of the most polar and unpolar substances, several extraction solvents with different polarities can be used [49].

MSPD has been applied for the simultaneous extraction and clean-up of pesticides in matrices such as fruits [51], juices [52], vegetables [53] and other foods [17,54,55], indicating in Table 1 some selected MSPD-based methods.

For most of the applications, quantitative recoveries were found, despite the different chemical structures and polarities of the analytes. Furthermore, if a selective and sensitive technique is used for the detection of pesticides, detection limits lower than ppbs can be obtained. For instance, herbicides can be detected in fruit juices at concentrations of 0.1 μg/L [56], or in the range 1-10 μg/kg in fruits and vegetables [57].

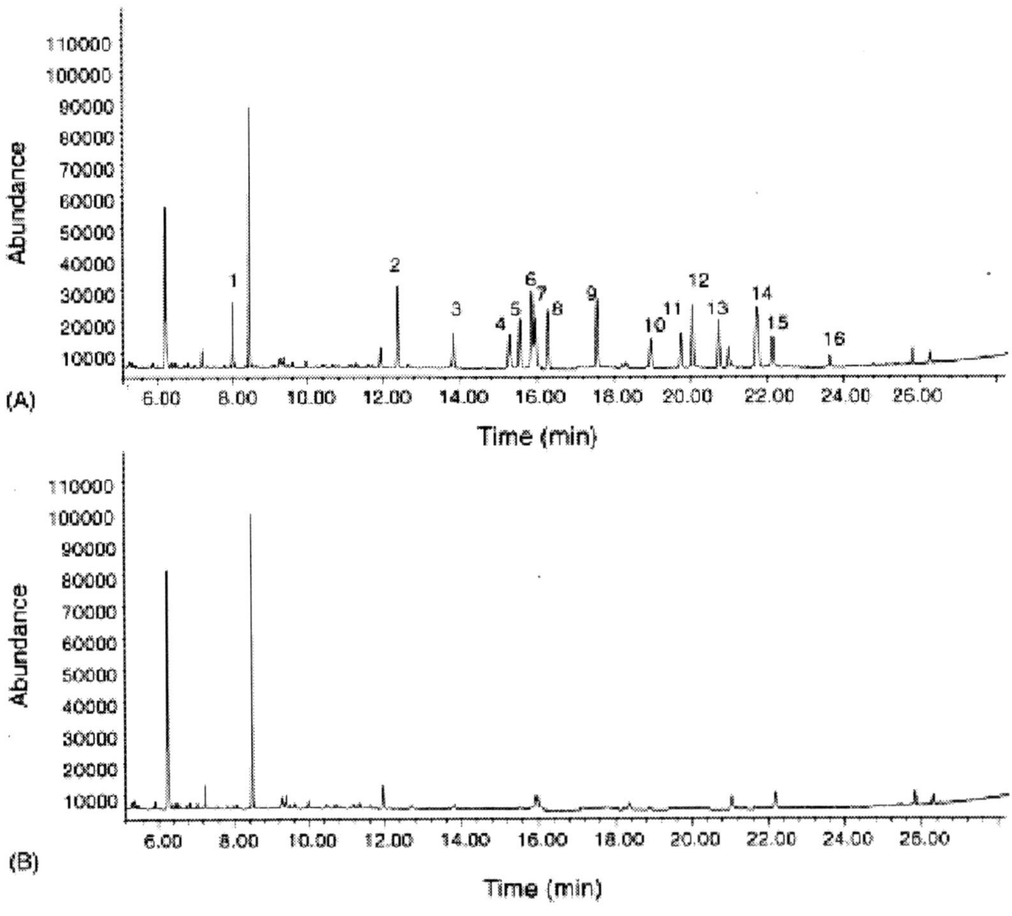

Figure 2. GC–MS-SIM chromatograms of (A) a grape juice sample fortified at 0.01 μg/L and (B) a blank grape juice sample. Peak identifications: 1: EPTC; 2: Propachlor, trifluaralin; 3: Simazine, atrazine; 4: Terbumeton, lindane (Internal Standard); 5: Terbuthylazine; 6: Triallate; 7: Metribuzin; 8: Alachlor, prometryn; 9: Terbutryn; 10: Metolachlor, cyanazine; 11: Pendimethalin. Reprinted from [56], copyright 2004 with permission from Elsevier.

For complex matrices, it is necessary an on-line purification step, by packing a selected sorbent, such as florisil, in the bottom of the extraction column to obtain cleaner extracts [58]. For the analysis of fatty samples more attention is required, and traces of fat should be removed before analytical determination, using sorbents such as alumina, strong ion-exchangers and silica modified with sulphuric acid [59], obtaining relative standard deviations (RSDs) lower than 15 % for the whole analysis of fatty samples.

In order to obtain faster and more efficient extraction of pesticides from several matrices, MSPD with heated and/or pressurized extractants has recently been proposed and has been successfully used for developing LC-mass spectrometry (MS) methods capable of detecting several contaminants in various matrices [17].

Table 1. Selected MSPD methods for pesticides analysis[1]

Analytes	Matrix	Sorbent	Elution	Clean up	Detection	Ref.
Carbamates	Fruits & vegetables	C8	CH_2Cl_2/Acetonitrile (6:4)		LC-MS	[57]
Fungicides	Fruits & vegetables	C8	Ethyl acetate		GC-NPD-ECD	[60]
Dithiocarbamates	Fruits & vegetables	Carbograph	CH_2Cl_2/MeOH (8:2)		LC-MS	[61]
Fungicides	Oranges	C18	CH_2Cl_2/MeOH (1:1)		LC-MS/MS	[62]
OPPs	Juices	Florisil	Ethyl acetate		GC-NPD	[52]
Herbicides	Juices	Florisil	Ethyl acetate		GC-MS	[56]
Multiclass	Apple juice	Diatomaceous earth	n-hexane/CH_2Cl_2 (1:1)		GC-MS	[63]
OCPs & pyrethroids	Tea	Florisil	n-hexane/CH_2Cl_2 (1:1)		GC-ECD	[64]
Multiclass	Honey	Florisil	n-hexane/ethyl acetate (9:1)		GC-ECD	[65]
OCPs	Fish	C18	n-hexane	Florisil	GC-MS	[66]
OCPs	Fish feed & marine species	C18	n-hexane		GC-MS/MS	[67]
OCPs	Milk	C18	n-hexane	Alumina	GC-ECD	[68]
OCPs	Chicken eggs	Florisil	CH_2Cl_2/n-hexane (1:1)	H_2SO_4	GC-ECD	[54]
OCPs	Animal fat	Alumina	n-heptane		LC-DAD	[69]
Multiclass	Animal liver	C18	Ethyl acetate	Florisil	GC-MS/MS	[70]
Multiclass	Olives and olive oil	Aminopropyl	Acetonitrile	Florisil	GC-MS & LC-MS/MS	[58]

[1] Abbreviations: DAD: Diode Array Detection; ECD: Electron Capture Detection; NPD: Nitrogen Phosphorus Detection; OCPs: Organochlorine pesticides; OPPs: Organophosphorus pesticides.

Finally, MSPD miniaturization can be achieved by reducing the amount of sample, the sorbent used for dispersion and the time required per analysis, so this technique can be considered as a valuable alternative to the more classical sample preparation methods because it allows a significant reduction in both the sample size and solvent consumption needed for multiresidue analysis, simplifying the determination of pesticides in solids, semisolids and highly viscous samples. In Figure 2, representative chromatograms of a fortified juice sample with herbicides after MSPD extraction, and a blank sample are shown. It can be observed that the analytes are well-resolved and no interferences affect the determination of the target compounds.

2.3. QUECHERS

As it was indicated before, liquid extraction has been the main method used for the isolation of pesticide residues from environmental and food samples, and several alternatives have been developed, although two extraction procedures have been harmonized at the international level in the past: acetone followed by partitioning with a mixture of dichloromethane and light petroleum (Luke method) [71] and extraction with ethyl acetate at the presence of sodium sulphate [72]. However, both methods have been modified considering health and environmental factors, selecting alternative solvents. In this sense, acetonitrile, acetone and ethyl acetate are three extraction solvents most commonly used in multiresidue methods for the determination of pesticide residues [73], presenting several advantages and disadvantages. Ethyl acetate has the advantage of partial immiscibility with water, so the addition of other nonpolar solvents is unnecessary. However it has the disadvantage that some of the most polar pesticides do not readily partition into ethyl acetate phase. Besides, a large amount of nonpolar co-extractives, such as lipids and epicuticles, are extracted, and these compounds must be removed before the determination step. On the other hand, one advantage of acetonitrile versus acetone is that residual water (after partitioning step) can be removed better by drying agents such as $MgSO_4$. Besides, this solvent it is compatible with LC as well as it can also be served as a medium for GC injection.

Considering these facts, in 2003, Anastassiades et al. [16] introduced a simple, fast and inexpensive method for the determination of pesticide residues in fruits and vegetables. The method was given an acronymic name, QuEChERS (pronounced *catchers*), reflecting its major advantages (Quick, Easy, Cheap, Effective, Rugged and Safe) and it can be used for a wide range of pesticides and food types. This procedure is based on liquid-liquid partitioning with acetonitrile followed by a clean-up step with dispersive SPE, as it is indicated in Figure 3. Basically, this methodology involves initial single-phase extraction of 10 g of samples with 10 mL of acetonitrile, followed by liquid-liquid partitioning formed by addition of 4 g anhydrous $MgSO_4$ and 1 g of NaCl. Removal of residual water and clean-up are performed simultaneously by using a rapid procedure called dispersive solid-phase extraction (dispersive-SPE), in which 150 mg anhydrous $MgSO_4$ and 25 mg of primary secondary amine (PSA) sorbent are mixed with 1 mL of acetonitrile extract. The dispersive-SPE with PSA effectively removes many polar matrix components, such as organic acids, certain polar pigments and sugars. In fact, this step reduces significantly the matrix induced enhancement [74].

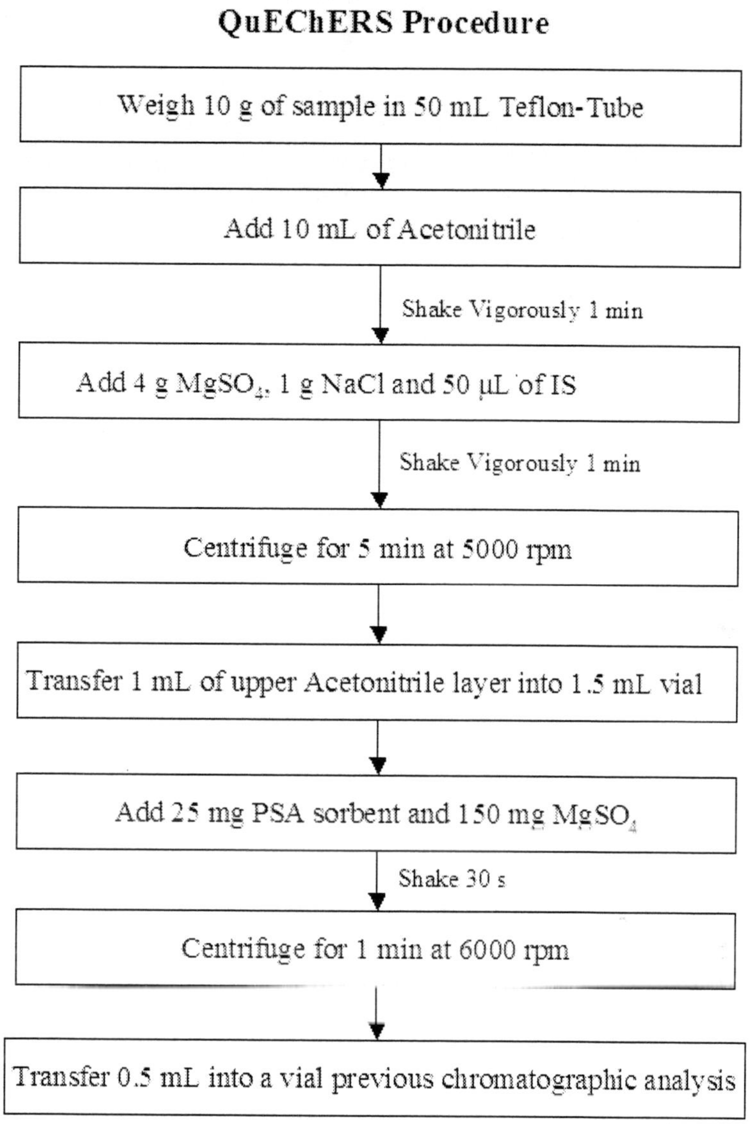

Figure 3. Outline of the protocol in the QuEChERS method.

However, when this procedure is applied, several problems can be found. First, lower recoveries were obtained for basic insecticides such as pymetrozine in orange matrices. Second, degradation of the base sensitive pesticides, captan, folpet, dichlofluanid and chlorothalonil was observed in nonacidic matrices such as lettuce, since PSA clean up step can increase the pH of the extract by 2-3 units. On the other hand, pesticides with acidic groups (phenoxyalcanoinic acids) interact with aminosorbents such as PSA, so lower recoveries can be obtained for these pesticides due to the clean-up step and PSA does not remove chlorophyll and sterols from extracts of leafy vegetables. Finally the main disadvantage of the QuEChERS method compared to other common methods is that the 1 g/mL final extract concentration is lower than the 2-5 g/mL concentrated extract of most traditional methods, so to achieve quantification limits lower than 10 µg/kg in modern GC-

MS, large volume injection is typically used or the final extract can be concentrated and solvent exchanged to toluene [75].

In order to overcome these problems, several modifications have been proposed. Thus, 1 % of acetic acid in acetonitrile in combination with the same amount of sodium acetate in the buffered QuECHERS approach has been indicated to extract basic herbicides, as well as to avoid the degradation of the base sensitive pesticides, stabilizing recovery values of problematic pesticides in different matrices [76]. Graphitized Carbon Black (GBC) can be used as an alternative to PSA as sorbent during "dispersive SPE", and the combination of PSA and GBC is very useful for samples with high content of carotenoids and chlorophyll [76], although losses of certain compounds of low polarity and planar structure (hexachlorobenzene, thiabendazole and chlorothalonil) may also occur.

Despite these problems, this methodology offers several advantages over traditional methods of pesticide residue analysis [77]. High recoveries are obtained for a wide polarity and volatility range of pesticides, such as methamidophos, omethoate, imazalil and pyrethroids; high sample throughput of about 10-20 preweighed samples in about 30-40 min is possible; no chlorinated solvents are used and the only devices needed are chopper, balance and centrifuge. In this sense, a technician can prepare a batch of 6-12 previously chopped samples in about 30 minutes with $1 worth of materials per sample, giving at least fourfold lower material costs and fourfold greater sample throughput per analyst than traditional methods [16]. That is why several applications have been published in the last few years, including other analytes such as pharmaceuticals in blood [78]. This method was originally applied for the determination of pesticides in fruits and vegetables such as tomato, zucchini and apple by GC-MS [16], and in less than two years, this methodology has been developed to analyze more than 220 pesticides by GC-MS/MS and LC-MS/MS in several matrices, lettuce, strawberry, orange, barley [73,77,79-83], increasing its applicability when buffered QuEChERS has been applied [76,84]. Furthermore, this procedure has been slightly modified for the analysis of pesticides in fatty matrices, such as milk, eggs, avocado [85] and olives [86]. In this sense, C18 can be used as sorbent in the "dispersive SPE" instead of PSA, or MSPD clean up can be added after extraction process [85], observing that lower recoveries for the most lipophilic analyte were obtained as fat content of the sample increased. Finally, it has been applied for the determination of pesticides in baby food [87], reaching the requirements established by EU legislation in terms of quantification limits for baby food (< 10 μg/kg), recovery and reproducibility [88].

Considering these aspects, QuEChERS is a powerful multiclass, multiresidue approach to pesticide analysis of foods, which can be applied in routine laboratories, and it has been evaluating as a substitute for the existing official procedures [82].

3. Microextraction Techniques

Despite the advances in sample preparation techniques, some of them have been described above, there is still a great concern over the elimination, or at least, significant reduction of organic solvents, as well as liquid and solid waste generated [89]. There is a current trend in the development of solventless techniques of sample preparation, based on gas extraction, miniaturized solid-phase extraction and membrane based techniques. These reduce considerably or eliminate the consumption of toxic solvents, so there is a great interest in these methodologies, and in this chapter, four sample preparation techniques, which are still developing, will be presented with respect to their usability and effectiveness for the analysis of pesticides in environmental and food matrices: solid phase microextraction (SPME), stir bar sorptive extraction (SBSE), single drop microextraction (SDME) and hollow fiber supported liquid membrane (HF-SLM).

3.1. Solid Phase Microextraction (SPME)

Solid phase microextraction (SPME) is a sample preparation technique, developed by Arthur and Pawliszyn [47] which uses a fused-silica rod with an outer diameter of typically 150 μm which is coated with a polymeric film (5 to 100 μm thick) as the extraction medium. This technique is fast, universal, sensitive and solventless and can be coupled to GC, LC or CE. SPME eliminates the problems associated with SPE, presenting several advantages [90]:

- Solvents are completely eliminated
- Extraction time is reduced to a few minutes in several applications
- It provides good results over a wide range of analyte concentrations
- It does not require complete removal of the analyte from the matrix
- It combines sampling and preconcentration in one single step
- It can be easily automated
- Minimal sample handling

On the other hand, there are several drawbacks such as relatively poor reproducibility, lot-to-lot variations, lack of selectivity, sensibility against organic solvents and cost [18].

SPME has two steps. In the first one, the coated fiber is exposed to the sample or its headspace and the target analytes partition from the sample matrix to the coating. Secondly, the analytes are desorbed from the fiber into a suitable separation and detection system. The effectiveness of analyte preconcentration by SPME depends on many parameters which must be optimised during the development of an analytical procedure: extraction mode, fiber type, sample volume, temperature, extraction time and salting-out among others.

3.1.1. Extraction Mode

Analytes can be extracted from the sample to the fiber in two different modes, headspace (HS-SPME) and direct immersion (DI-SPME). In HS-SPME, the fiber is exposed in the vapor phase above a liquid or solid sample, whereas in DI-SPME, the fiber is directly immersed in liquid samples. HS-SPME sampling is generally used for volatile compounds, and has the advantage of faster extraction times and selectivity is often improved. This mode prolongs the life of the fiber because it is not in direct contact with the sample since the fiber is inserted into the headspace, above the liquid or solid sample. For the extraction of less volatile compounds, DI-SPME is used although the fiber deteriorates quicker, decreasing the lifetime of the fiber and increasing the cost of the technique [22].

Once the analytes are extracted in the fiber using either DI or HS-SPME, the fiber is withdrawn into the needle, and it is removed from the septum and is then inserted directly into the injection port of the GC system or the desorption chamber of the SPME/LC interface. The desorption of analytes from the fiber coating is performed by heating the fiber in the hot injection port of the GC or by loading solvent into the desorption chamber of the SPME/LC interface, and it must be as short as possible. Then the analytes are transferred directly to the separation column for analysis.

Despite of these configurations, another SPME technique, known as in-tube SPME has also been developed for combination with LC using an open tubular fused-silica capillary column as an SPME device instead of SPME fiber [91]. Basically, the concepts of fiber and in-tube SPME methodologies are very similar although there is a significant difference. The, extraction of analytes is performed on the outer surface of the fiber for SPME, whereas for in-tube SPME, the extraction is carried out in the inner surface of the capillary column. Besides, in-tube SPME does not need a special SPME/LC interface for desorption of analytes, so automation is much easier.

3.1.2. Fiber Type

The sensitivity of this technique depends mainly on the value of the partition coefficient of the analyte partitioned between the sample and the fiber. Thus, the type of fiber affects the amount and selectivity of sorbed species, following the general rule "like dissolves like". The most commonly used sorbent is the apolar polydimethylsiloxane (PDMS). This material is frequently selected because it is very inert, reducing the risk of losses of unstable and/or polar analytes by irreversible adsorption or by catalytic reaction, it is thermally stable, the extraction time is considerably less than that required with a solid polymer and it is commercially available in a variety of thickness.

In order to improve the capacity of SPME fibers, alternative coatings are commercially available in different coating combinations, blends or copolymers, film thickness, and fiber assemblies. In particular, other coatings such divinylbenzene (DVB), polyacrylate (PA), Carboxen (CAR) and Carbowax (CW) [92], have been used, although PDMS [93] and PA [94] have been mainly applied for pesticides extraction.

Besides, the efficiency of preconcentration depends not only on the type of fiber used but also on the thickness. Thick layers extracts large amounts of analytes than thin ones, but more extraction time is required to reach the equilibrium. Thick fibers are effective for the extraction of volatile compounds, whereas fibers with a thin film are used for the isolation and preconcentration of analytes with high boiling points.

3.1.3. Extraction Time

Analyte molecules need time for their mass transfer from the sample to the fiber before reaching a thermodynamic equilibrium between sample and fiber, obtaining the maximum sensitivity at the equilibrium point. Thus, longer time favors the occupation of active sites of the fiber by a larger number of analyte molecules, but when all sites are occupied, prolonged time does not affect the preconcentration efficiency, and it can usually cause desorption [95]. Sometimes, reaching the equilibrium is not necessary for accurate and precise determination by SPME, because of the linear relationship between the amount of analyte desorbed by SPME fiber and its initial concentration in the sample matrix in non-equilibrium conditions [96]. Besides equilibrium time depends on the type and thickness of the fiber. In general, equilibrium time is usually less than 1 hour in a 100 μm PDMS coating for compounds with distribution constant less than 10000, whereas thinner PDMS coatings should be considered for compounds with higher constants, since the equilibrium time is shorter [92].

3.1.4. Other Parameters

In order to increase extraction efficiency, other parameters are also important. First, the addition of salts improves the extraction efficiency, since the ionic strength of the samples increases and this decreases the solubility of analytes which are more easily extracted. Although this approach is frequently used for the determination of pesticides [97], this effect is not general and depends on the polarity of the analyte, the concentration of salt and the sample matrix, so it should always be evaluated. Temperature and sample agitation are other parameters which can increase the extraction efficiency of analytes when SPME is used. Thus, an increase of the extraction temperature accelerate the sorption of the compounds on the fiber, decreasing the extraction time. The same effect was observed with agitation. However, higher temperatures can provoke the desorption of analytes and high agitation forms gas bubbles on the fiber surface which affects the mass transfer rates.

3.1.5. Applications

Since SPME was first use as preconcentration procedure for pesticide analysis in 1994 [98,99] a lot of applications in environmental samples such as air, water, seawater, wastewater and soils, as well as food samples have been carried out. In fact, more than 270 articles related to the determination of pesticides in different samples have been published so far, indicating in Table 2, some examples, although more data and examples can be found in several reviews published in the last few years [14,31,100].

It can be observed that a great variety of pesticides (polar and non polar) can be extracted from several matrices using this technique. Organochlorine, organophosphorus and pyrethroids, which are non polar compounds, can be extracted using PDMS or PDMS-DVB fiber, whereas other type of fibers such as Carbowax must be used for the extraction of polar pesticides such as carbamates or phenylureas.

In general, direct immersion is the most frequently used SPME sampling mode in aqueous sample, although good results can be also obtained when HS-SPME is used [101]. For solid samples, a extraction step is necessary previous SPME. For instance, fruit samples are usually extracted with high speed blending using water [93]. Furthermore, in combination with GC-MS or LC-MS/MS, SPME is able to extract and to detect pesticides residues in an easy and automated approach, obtaining detection limits lower than 0.1 μg/L when water is analyzed [102] and lower than 10 μg/kg in solid samples [11]. For instance, Figure 4 presents chromatograms obtained from a honey sample, and it can be observed that the lack of interfering peaks provided unequivocal identification.

3.2. Stir Bar Sorptive Extraction (SBSE)

The sensitivity of SPME is sufficient for most applications, but sometimes it is limited by the small amount of coating material on the needle, which results in low extraction efficiency. To improve the extraction efficiency and the amount of extracted analytes, the amount of extraction phase must be increased. For this purpose, a new miniaturized extraction technique, stir bar sorptive extraction (SBSE) has been introduced by Baltussen et al. in 1999 [121]. In this technique, a magnetic stir bar coated with 55 or 219 μL of PDMS (corresponding to magnets 10 and 40 mm long, respectively), was used. The short bed packed with PDMS was prepared by removing the Teflon coating, reducing the outer diameter of the magnet and coating the magnet with a glass tube and covering it with PDMS. PDMS-coated stir bars are commercially as "Twister" by Gerstel (Müllheim, Germany). Considering the surface area of the stir bar is greater than that of the SPME fiber and the volume of the adsorbent is at least 100 larger, there is a higher phase ratio than SPME, and hence, a higher extraction efficiency, as it can be observed in Figure 5, which results in lower detection limits. Thus, sampling of large volumes can be performed in a relatively short time using stir bar.

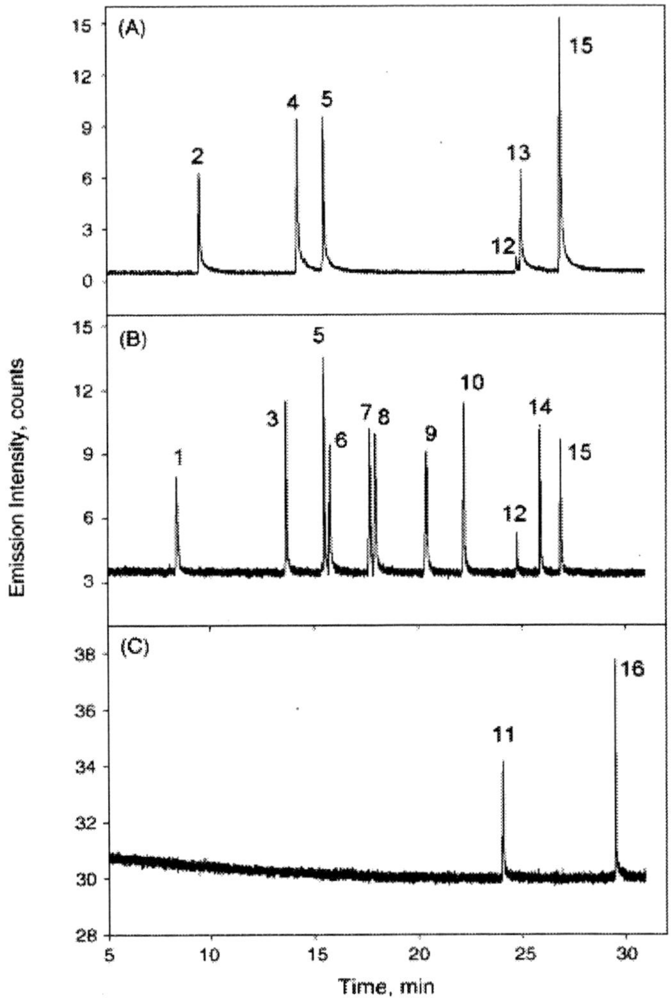

Figure 4. SPME-GC-AED chromatograms obtained from a honey sample previously fortified with a standard mixture of pesticides. Peak identification: 1: Chlordimefom; 2: Dimethoate; 3: Aldrin; 4: Parathion-ethyl; 5: Captan; 6: Chlorfenvinphos; 7: Dieldrin; 8: p,p'-DDE; 9: p,p'-DDD; 10: p,p'-DDT; 11: Bromopropylate; 12: Tetradifon; 13: Azinphos-methyl; 14: λ-cyhalothrin; 15: Coumaphos; 16: Deltamethrin. Reprinted from [119] copyright 2006, with permission from Elsevier.

Table 2. SPME methods for the analysis of pesticides[1]

Analyte	Sample	Type of Fiber	Extraction	Detection	Ref.
OCs & Ops	Water	100 µm PDMS	HS-SPME	GC-MS/MS	[101]
OCs, OPs, Pyrethroids, Triazine herbicides	Water	60 µm PDMS-DVB	DI-SPME	GC-MS/MS	[102]
Phenylurea herbicides	Water	60 µm PDMS-DVB	DI-SPME	LC-UV	[103]
Multiclass	Water	85 µm PA	HS-SPME	GC-MS/MS	[104]
Multiclass	Water	65 µm PDMS-DVB	DI-SPME	GC-MS/MS	[105]
Triazines & Phenylurea	Rainwater	85 µm PA	DI-SPME	GC-MS/MS	[106]
Carbamates & Phenylurea	Water and wine	PPY	In-tube SPME	LC-MS	[107]
Ops	Water	Polysiloxane	In tube SPME	LC-UV	[108]
Atrazine & OCs	Soils	100 µm PDMS	DI-SPME	GC-ECD/MS	[109]
OCs	Sediments	100 µm PDMS	HS-SPME	GC-ECD	[110]
Acidic pesticides	Fruits	50 µm CW/TPR	DI-SPME	CE-MS	[11]
Ops	Fruits & fruit juices	65 µm PDMS-DVB	DI-SPME	GC-FPD	[93]
Fungicides	Green groceries	85 µm PA	HS-SPME	GC-MS	[111]
Ops	Cherry & Strawberry	100 µm PDMS	HS-SPME	GC-MS	[112]
Ops	Fruits	100 µm PDMS	HS-SPME	GC-NPD	[113]
Multiclass	Fruit juices	60 µm PDMS-DVB	DI-SPME	CE-UV	[114]
OPs	Fruit, vegetables & juices	80 µm vinyl crown ether	HS-SPME	GC-FPD	[115]
Carbamates & Phenylurea	Fruit juices	50 µm CW/TPR	DI-SPME	LC-MS/MS	[116]
OCs	Fish	100 µm PDMS	DI-SPME	GC-ECD	[117]
OPs	Honey	100 µm PDMS	DI-SPME	LC-MS	[118]
OCs, OPs & pyrethrins	Honey	100 µm PDMS	DI-SPME	GC-AED	[119]
Multiclass	Milk	65 µm PDMS-DVB	DI-SPME	GC-MS/MS	[120]

[1] Abbreviations: AED: Atomic Emission Detection; CW/TPR: Carbowax Template Resin; ECD: Electron Capture Detection; FPD: Flame photometric detection; OCs: Organochlorine pesticides; OPs: Organophosphorus pesticides; PA: Polyacrylate; PPY: Polypyrrole.

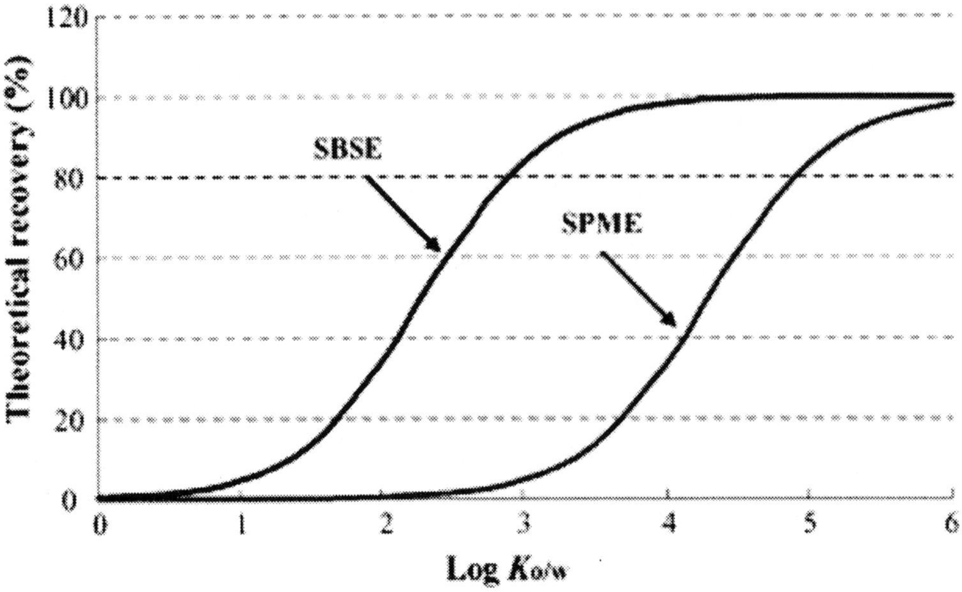

Figure 5. Theoretical recoveries for solutes as a function of the octanol-water partitioning coefficient $K_{o/w}$ for SBSE (10 mL sample, 50 μL PDMS phase) and SPME (10 mL sample, 0.5 μL PDMS phase). Reprinted from [19] copyright 2006, with permission from Elsevier.

For strongly retained compounds, sampling is performed in the breakthrough mode, whereas for weakly retained analytes, or in those cases where maximum sensitivity is desired, sampling is continued until all analytes are in equilibrium with the sorbent. Generally, sampling times can be kept within 30 min, obtaining adequate sensitivity.

After the extraction, the stir bar is removed from the sample, often manually, and transferred to the injection port of a GC for thermal desorption (TD) [122] or the analytes can be desorbed by a liquid, liquid desorption (LD), for LC analysis [123]. In the former, the stir bar was rinsed lightly with distilled water to remove adsorbed proteins or other sample components, preventing the formation of non-volatile materials during TD step. Recently, it was observed that TD provides better sensitivity than LD [124]. This does not cause solute loss because the sorbed solute is present in the PDMS phase [19] Then, the PDMS-coated stir bar is placed inside an empty glass tube, and the solute is thermally desorbed. Desorption temperature depends on the volatility of the analytes, and it is typically between 150 and 300 °C. Recently, a desorption unit enables fully automated analysis of 98 or 196 PDMS coated stir bars has been developed [125]. For LD, 100-200 μL of an organic solvent, such as methanol or acetonitrile, is used for desorption. Desorption time is approximately 5-10 min and the analytes are usually extracted at room temperature or by ultrasonication. However, the main drawback is more manipulation, dilution and loss of analyte, because only a fraction of the extract is injected into the LC system.

Although this technique is relatively new, SBSE has been widely used [19], showing in Table 3 some applications of pesticides determination in several matrices. In general, SBSE is considered to be superior to SPME in terms of sensitivity and accuracy for determinations at trace level in difficult matrices, for example organophosphorus pesticides in honey [118]. Thus, the use of larger amounts of PDMS on the stir bar, resulted in extraction efficiencies up to 100 times higher. For liquid samples, 10-20 mL are usually used for the extraction of target compounds [126], obtaining good sensitivities as it can be observed in Figure 6. For more complex samples, such as biological fluids [127] or solid samples [125], the technique can only be applied after a pre-treatment step. Thus, in terms of sensitivity, it can be noted that in aqueous samples, LODs can be lower than ng/L level, whereas for solid samples, it is possible to detect concentration closed to 1 ng/kg.

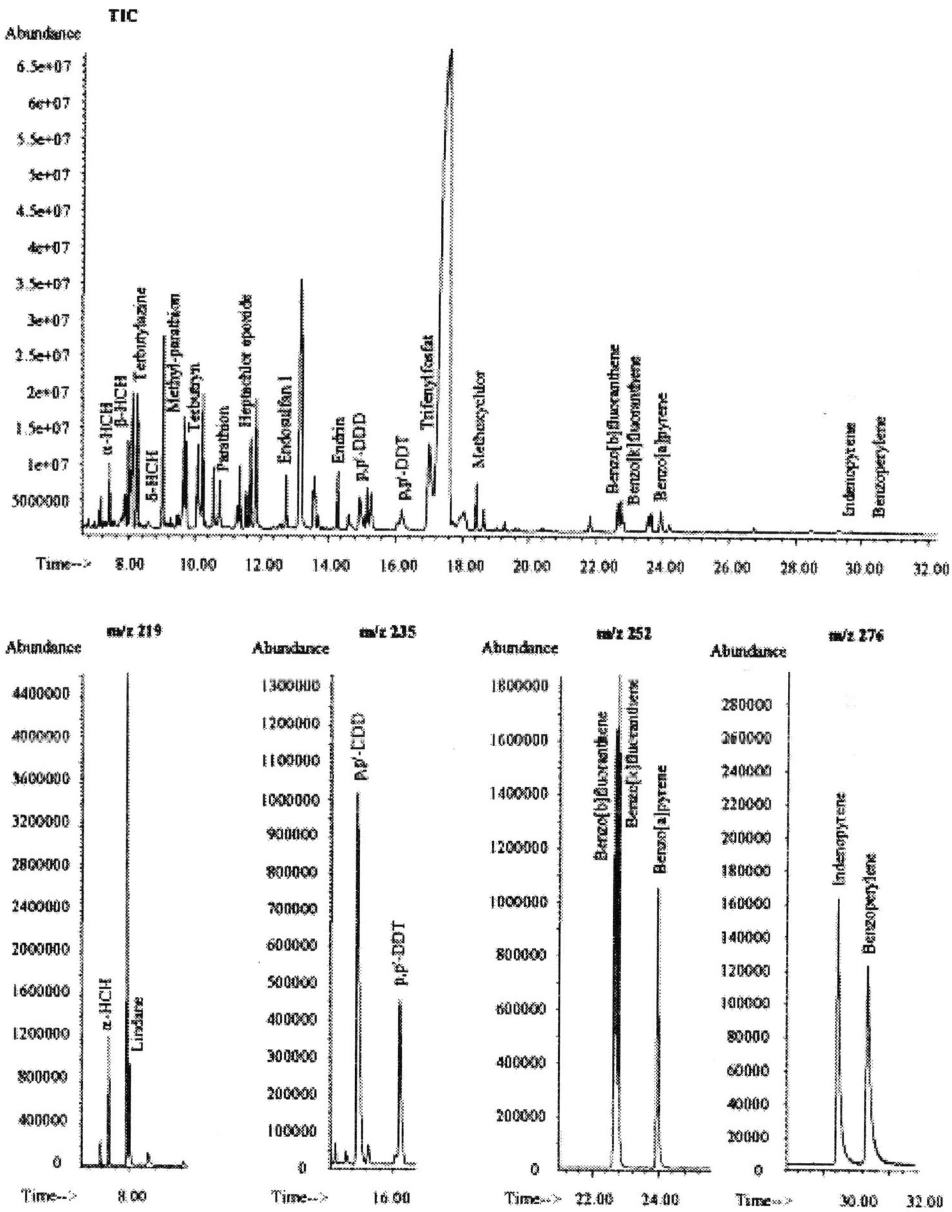

Figure 6. GC-MS chromatogram of a tap water sample fortified with 35 pesticides at 50 ng/L using SBSE. Total ion chromatogram (TIC) is shown at the top and some selected fragment ion chromatograms below. Reprinted from [126] copyright 2003, with permission from Elsevier.

Table 3. Determination of pesticides by SBSE[1]

Analyte	Sample	Desorption	DL	Instrument	*Ref*
Pyrethroids	Water	TD	0.02-1.4 ng/L	GC-MS	[124]
OCPs	Water	TD	0.05-1 ng/L	GC-MS	[128]
Pyrethroids	Water	LD	1.0-2.5 ng/L	GC-MS	[129]
Multiclass	River water	TD	0.2- 2 ng/L	GC-MS	[130]
Multiclass	Water	TD	0.1-10.7 ng/L	GC-MS	[131]
OCPs	Soils	TD	0.002-4.7 μg/kg	GC-MS	[132]
OCPs & OPPs	Wine	TD	< 10 μg/L	GC-MS	[133]
OCPs	Wine	TD	0.3-17.3 ng/L	GC-MS/MS	[134]
OPPs	Honey	LD	0.01-0.1 mg/kg	LC-MS	[118]
Multiclass	Brewed green tea	TD	0.58-100 ng/L	GC-MS	[135]
Multiclass	Oranges	LD	1-50 μg/kg	LC-MS	[136]
Multiclass	Vegetables	TD	0.63-26 μg/kg	GC-MD	[137]
OPPs	Vegetables	TD	1.2-98 ng/kg	GC-TSD	[138]

[1] Abbreviations: DL: Detection limit; OCPs: Organochlorine pesticides; OPPs: Organophosphorus compounds; TSD: Thermionic Specified Detector.

3.3. Single Drop Microextraction (SDME)

The popularity of SPME has spurred the development of similar technologies such as single drop microextraction (SDME), which has been introduced in 1996 by Jeannot and Cantwell [139], and involves the miniaturization of the traditional liquid/liquid extraction by greatly reducing the solvent to aqueous phase ratio. This technique has also been referred to as solvent microextraction, liquid-phase microextraction and liquid-liquid microextraction [140]. This technique is based on the extraction of analytes in headspace or aqueous sample solution into a microdrop of a water immiscible organic acceptor solvent, commonly a few microliters, which is suspended on the tip of GC syringe needle or at the end of a PTFE rod. The analyte diffuse into this droplet in a similar manner as into the SPME fiber, and after the sampling time, the microdrop is retracted into the syringe and transferred into the GC system for further analysis. Unlike fibers, drops can be renewed for each extraction, so it is possible to overcome the limited availability of fiber coatings and carry over effect, and a wide variety of solvents and trapping agents can be used.

The main advantages of this technique are the speed and it does not require any complicated equipment (only a microsyrange), reducing the cost of analysis. However, the main drawback of the technique is that precision of measurements is mainly influenced by the skill and experience of the analyst as well as the instability of the drop when the sample is stirred or sonicated.

Basic principles, theoretical aspects and formats of this technique have been recently reviewed [141]. In this sense, different modes of SDME have been developed, such as direct inmmersion (static mode), headspace and continuous flow microextraction, observing that sensitivity of headspace was significantly higher than that of direct immersion mode [142]. Based on these developments, different approaches have been used to improve selectivity, introduce a degree of automation and make it compatible with more analytical techniques [20].

Basically, the effectiveness of extraction depends on various factors such as:

3.3.1. Solvent Type

Depending on the polarity of the analytes and the nature of the sample, solvent must be selected, using the same criteria than in LLE. In general the extraction solvent has to meet the following requirements: the solvent must have low water solubility, extract analytes well and have good drop stability when stirring and low level of toxicity [141]. Several solvents such as hexane, toluene and undecane can be used. If headspace mode is used, solvents must not be too volatile, because the drop can evaporate within a few minutes of its exposure.

3.3.2. Solvent Drop Volume

In the first paper of SDME [139], the droplet size was 8 μL. However, this volume was too large to inject directly into a GC system, so an aliquot was taken, which limited sensitivity. In the next published papers [143,144], the droplet size was reduced to 1-2 μL by using the tip of a GC syringe needle as the drop holder. In general, an increase in drop volume yields increased extraction efficiency, but drops larger than 2 μL are characterized by lesser stability and lower reproducibility.

3.3.3. Extraction Time

The longer extraction time, the higher extraction is. However, mass transfer rate is reduced when the system is closer to equilibrium conditions. In order to increase the repeatability of the analysis, a time in which equilibrium is reached between the organic solvent and the sample is convenient, but it is not always necessary, since if the time is well fixed, repetitive results can be obtained. However, an increase in exposure time can provoke that the solvent drop can lose touch with the needle due to gravitational force or can absorb water, reducing the repeatability of the system.

3.3.4. Salt Addition

Addition of salt to the sample may have significant effects on the extraction as it increases the ionic strength of the sample. Salting out effect has been widely used in traditional liquid–liquid extraction, because addition of a certain amount of salt can enhance the partition of the analytes into the organic phase. However, for SDME, the addition of salt was found to have different effect depending on the analyte. Generally, the addition of salt to the sample provokes an expected decrease in extraction efficiency for the majority of the analytes, which is more pronounced for the less polar analytes. This can be explained by the change of the physical properties of the extraction film, reducing the diffusion rates of the analytes into the drop [89,145]. This means the diffusion of analytes towards the organic becomes more difficult when salt concentration is increased, limiting extraction.

3.3.5. Stirring

Agitation can enhance extraction efficiency. However, when the microdrop is directly exposed to the aqueous phase there is an upper limit as higher stirring rates, because, air bubbles are formed, losing organic solvent by dissolution in the aqueous phase or by dislodgement, so the sample should be stirred at low speed.

3.3.6. Other Factors

Other factors such as temperature, sample pH, can influenced the extraction efficiency of the analytes. Generally, extraction efficiency can be increased by increasing temperature, but high temperatures can provoke drop damage and decrease the reproducibility of the SDME procedure [146]. Related to sample pH same considerations as in conventional LLE should be taken.

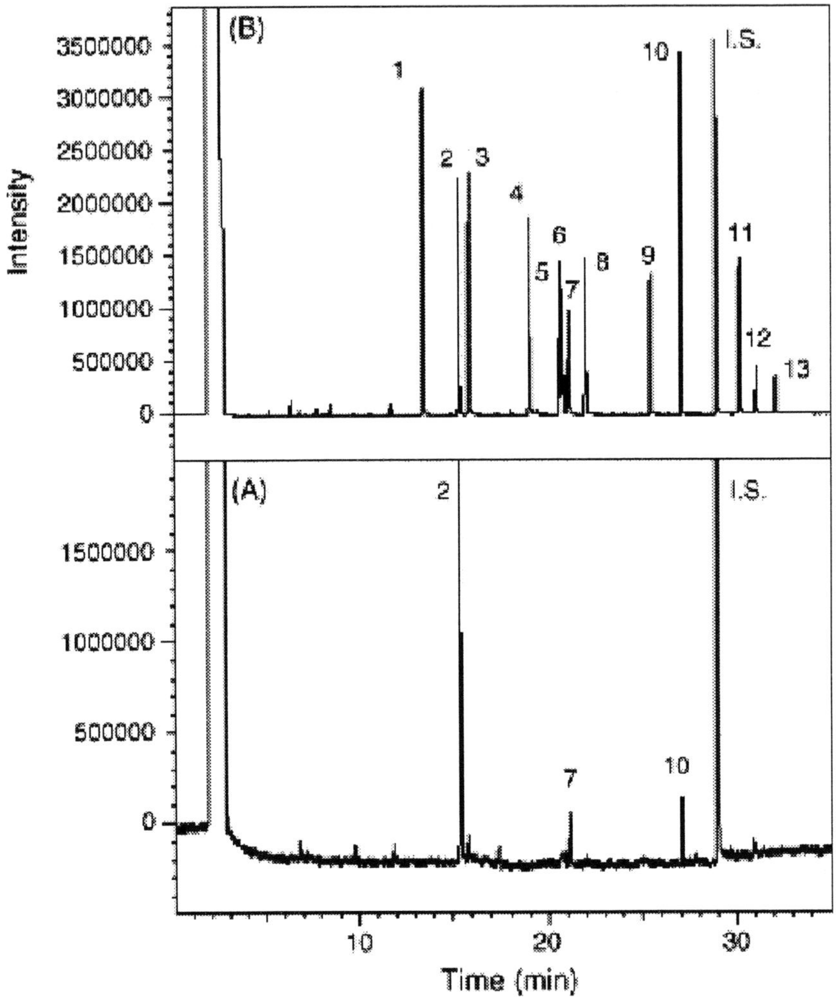

Figure 7. Chromatograms of (A) farm water and (B) spiked farm water at concentration level of 1 μg/L obtained by using SDME combined GC-FPD. Peak identification: (1) phorate, (2) diazinon, (3) disolfotan, (4) methyl parathion, (5) fenithrothion, (6) chlorpyrifos, (7) malathion, (8) fenthion, (9) profenphose, (10) ethion, (11) phosolane, (12) azinphose methyl, (13) co-ral, (I.S.) triphenyl phosphate. Reprinted from [145], copyright 2006, with permission from Elsevier.

Table 4. Applications of SDME technique used in pesticide analysis[1]

Analyte	Matrix	Solvent	Drop Volume (μL)	EF	DL (μg/L)	Detection	Ref
OPPs	Water	Carbon Tetrachloride	0.9	540-830	0.001-0.005	GC-FPD	[145]
Carbamates & OPPs	Water	Iso-octane	1.0	1800	0.2-5.0	GC-NPD	[146]
OCPs	Water	Hexane	2.0	1.6-26.5	0.25	GC-ECD	[150]
OCPs	Water	Hexane	3.0	20-95	0.005-0.2	GC-ECD	[151]
α & β-endosulfan	Water	Isooctane	1.5		0.01	GC-ECD	[152]
Antifouling agents	Water	Hexane	1.5	10.7-32.6	0.00025-0.003	GC-ECD	[153]
OPPs	Water	Toluene	1.5		0.010-0.073	GC-MS	[154]
Fungicides	Water	Toluene	1.6	> 56	0.006-0.010	GC-ECD	[155]
OPPs	Water and juice	Toluene	1.5	23-109	0.21-0.56	GC-FPD	[156]
OPPs	Juices	Toluene	1.6		0.98-2.20	GC-FPD	[149]

[1] Abbreviations: ECD: Electron capture detection; EF: Enrichment factor; FPD: Flame photometric detection; DL: Detection Limit; NPD: Nitrogen phosphorus detection; OPPs: Organophosphorus pesticides.

3.3.7. Applications

Since the introduction of this technique, it has experienced a rapid development, which led to enhanced analyte enrichment, easier automation and a wider range of application (Table 4). It can be observed that SDME can be used for rapid detection of pesticides, with enrichment factors higher than 100. Although the experimental results suggest that the nature of the samples has little effect on the enrichment process for aqueous sample [147], previous filtration of the sample should be used for more complex matrices [148]. Besides, SDME is more suited for volatile or semivolatile analytes which are usually hydrophobic compounds, so GC is the technique of choice to determine these compounds. Figure 7 shows an example of the determination of OPPs, detecting diazinon, malathion and ethion in a farm water sample. For complex matrices such as juices, it was observed that the addition of water enhanced the relative recoveries of pesticides by lowering the concentration of matrix interference [149].

3.4. Hollow Fiber Supported Liquid Membrane (HFSLM)

Although SDME proved to be a simple, inexpensive, fast and virtually solvent-free sample pre-treatment techniques, one of the main problems of this technique is drop stability, so novel microextraction techniques that involve the protection of the organic solvent have been introduced [157]. Among them, the use of a membrane or hollow fiber provides mechanical stability and protection to the organic solvent. Basically, the analytes are extracted from aqueous samples (donor) through a thin layer of organic solvent immobilized within the pores of a hollow fiber and further into an acceptor solution inside the lumen of the hollow fiber. Then the acceptor solution can be directly subjected to a final analysis by GC, LC or CE without any further effort. Analyte molecules are transported from the donor to the acceptor phase by diffusion through the membrane.

One of the main advantages of this technique over SPME and SDME is that the acceptor solution in hollow fiber was effectively protected within the fiber which served as filter, while SDME and SPME suffered from the presence of sample particles [21]. Besides, HFSLM can also show some selectivity because of the pores in its wall. In this sense, large molecules, which can be soluble in extracting solvent, may not be extracted.

This system has more solvent area per unit volume than does a spherical drop. Thus the contact area between sample solution containing target compounds and the extracting solution is substantially greater than would be the case if the solvent were spherical.

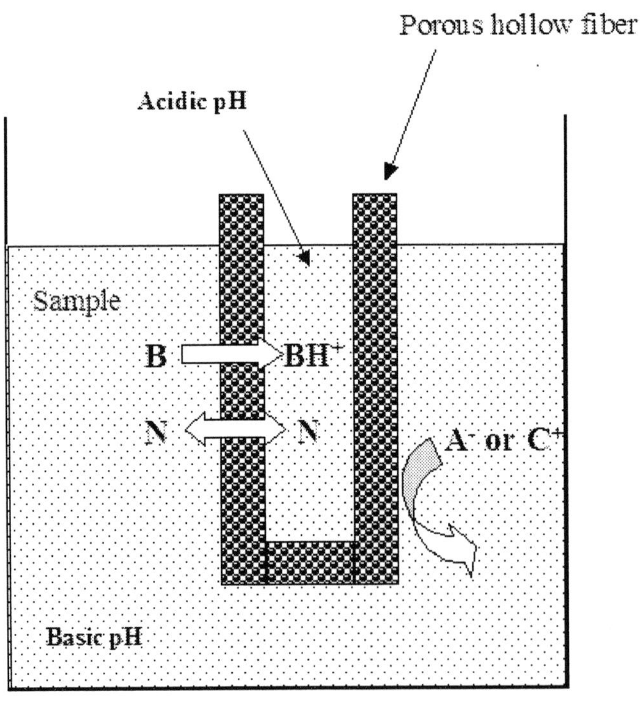

Figure 8. Schematic description of HFSLM principle for the extraction of basic compounds. Abbreviations: B: Basic compound; N: Neutral compound; A and C: Anionic and cationic compounds.

For the extraction, the analyte in the donor phase must be uncharged or undissociated, so it can be extracted into the organic phase (Figure 8). After diffusion to the acceptor phase, the analyte must be "trapped" by ionization in the acceptor phase to avoid back extraction into the membrane. Basically, this is similar to liquid-liquid extraction into an organic solvent, followed by a back-extraction into a second aqueous phase. Neutral compounds, may be extracted but not enriched as the concentration in the acceptor will never exceed in the donor, whereas ionic compounds are not extracted.

The chemistry and principles of this technique are similar to the used for extraction with supported liquid membranes (SLM) [158,159], describing some of the most important aspects below.

3.4.1. Extraction Mode

Depending on the acceptor phase, two sampling modes can be used:

a) Two Phase Sampling Mode

In this mode, the analyte is extracted from a liquid sample through the organic solvent immobilized in the pores of the hollow fiber into the same organic solvent (acceptor phase) present inside the hollow fiber. This mode is applied to most analytes with a solubility in an organic solvent immiscible with water higher than in aqueous phase, such as non polar pesticides (organochlorines, pyrethroids, etc…). Considering the acceptor solution is organic, this mode is compatible with GC.

b) Three Phase Sampling Mode

When this mode is used, the analytes, which are in one aqueous phase, are extracted into another via an organic phase. This is more suitable for polar or ionizable compounds with hydrophobic properties as their neutral forms, that are highly soluble in the organic membrane and their ionic forms are highly soluble in the aqueous acceptor, which is compatible with LC analysis. In order to obtain higher enrichment factors and to avoid back extraction from the acceptor to the donor, once the analytes are extracted in the acceptor phase are converted by reactions, such as protonation or complexation to species that will have very small affinity for the organic phase.

In this mode, analytes are transferred selectively between the two aqueous phases by adjustment of the pH. For example, for the extraction of a basic compound (see Figure 8), at high pH the compound will be uncharged and will be extracted into the organic solvent immobilized in the fiber. On the other hand, the acceptor phase will be acidic, so the basic compound will be protonated and thereby prevented from re-entering the membrane.

It is possible to predict how the hydrophobicity of a compound affects its dissolution into the membrane liquid during elution using the logarithm of octanol-water partition coefficient (log K_{ow}) [160]. Then, it was found that log K_{ow} of 2-4 gives an optimum region with the highest extraction efficiency. Compounds with higher log K_{ow} should be extracted into a two phases system, whereby the acceptor solution is the same as the organic solvent immobilized as a membrane liquid. For low values of log K_{ow} the extraction can be improved by substituting the liquid membrane with a more polar organic solvent or by adding suitable extractants, which can be cationic or anionic exchange carriers [161]. Briefly, the very hydrophilic compound is complexed with the carrier at the sample/membrane interface; then it is partitioned into the organic phase and the complex diffuses at the membrane/acceptor interface to release the compound at the membrane/acceptor interface. Finally the compound is complexed in the acceptor phase to avoid back extraction. This mode could be used for very ionisable compounds such as birypilidium herbicides [162].

3.4.2. Set-up

Basically, two methodologies are used for HFSLM (Figure 9), based on the use of hollow fiber membranes either in U-shaped (Figure 9a) or a rod-like (Figure 9b) configuration. The first set up was introduced by Pedersen-Bjergaard and Rasmussen [163], and each end of the fiber is connected to a medical steel needle leaving the fibre in a U-shape in the sample. One needle served to guide a microsyringe into the lumen of the fiber to deliver the acceptor solution and the other needle is used as an exit tube for the acceptor solution. The main problem with this set-up is the transfer of the acceptor, which is difficult to automate. In order to overcome this problem a rod-like configuration can be used [164]. In this configuration, a tip of the needle of a microsyringe is used to support the fiber. Only one end of the hollow fiber is used for injection/collection of the acceptor solution and the other one is suspended in the sample solution.

HFSLM may be carried out in a dynamic mode, in which a microsyringe is filled with few μL of an organic solvent [165], and a small piece of HF is soaked in the same organic solvent to fill the pores. The piece of HF is attached to the needle of the microsyringe and it is placed in the sample. During the extraction, small volumes of the aqueous sample are pulled in and pushed out of the HF using the syringe plunger, including many cycles. This mode improves extraction speed but adds experimental parameters that must be optimized and controlled [166].

In order to automated the whole process, Gerstel (Mülheim, Germany) has developed an alternative configuration [167] (Figure 9c). The extraction cell consists of a conventional 20 mL headspace vial and the membrane bag is 4 cm long, with a pore size of 0.2 μm and a thickness of 30 μm. It is attached to a metal funnel and fixed with a PTFE ring. The membrane bag is placed into the vial which is then closed with a metallic crimp cap.

The membrane bag is filled with organic solvent and transferred into an agitator. After the optimized agitation time, the organic phase is withdrawn with a syringe from the membrane bag and transferred to a 2 mL autosample vial. However, the main problem is the volume immobilized in the membrane, 800 μL, instead of 10-20 μL with conventional set-ups, so large volume injection is used to reduce detection limits [168].

3.4.3. Type of Fiber

Polypropylene is commonly used because it is highly compatible with a broad range of organic solvents. Generally, HFSLM procedures are based on polypropylene fiber, except for a few cases that use porous polyvinylidene difluoride [169]. Polypropylene provides strong immobilization of organic solvents in the pores of the fiber and excellent mechanical stability. The most common type of fibers has an inner diameter of 600 μm and are compatible with the volume of the acceptor solution required for microextraction, providing good mechanical stability. Recently, thinner fibers, with 280 μm of inner diameter have been used [161,170], reducing extraction time, because large inner diameter results in a low ratio of acceptor volume to surface area of fiber.

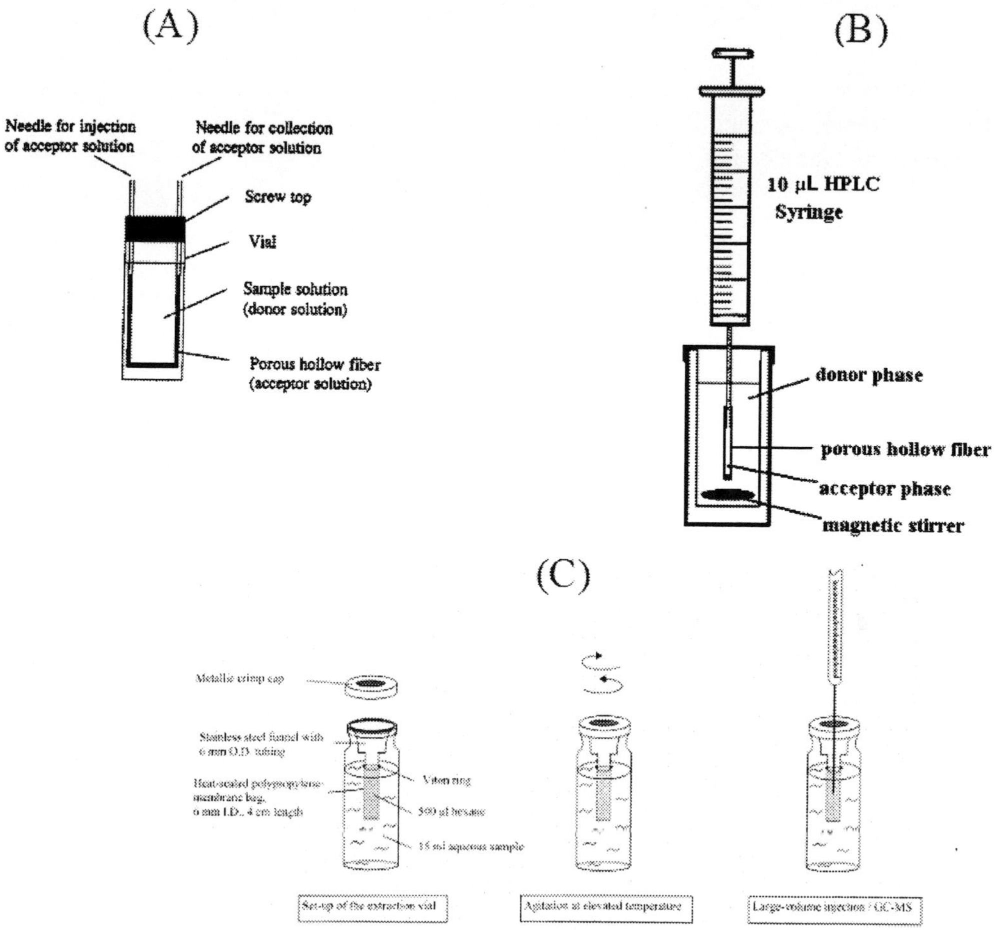

Figure 9. Different HFSLM configurations: (A) U-shape configuration, reprinted with permission from [163], copyright 1999, American Chemical Society, (B) rod configuration, reprinted from [164], copyright 2001, with permission from Elsevier, and (C) membrane assisted extraction, reprinted from [167] copyright 2002, with permission from Elsevier.

3.4.4. Organic Solvent

The choice of an organic solvent is an essential consideration in the extraction process. The organic solvent fills the pores of the membrane and it is held in the pores by capillary forces so it should have to satisfy several criteria:

- It should be immiscible with water to prevent leakage
- It should be strongly immobilized in the pores of the hollow fiber to prevent leakage
- It should provide an appropriate extraction selectivity to obtain high extraction recoveries
- It should be of low volatility to prevent solvent loss
- The solubility of analyte in the solvent should be higher than that in the donor phase and be lower than that in the acceptor phase. This is in order to promote the analyte migration from the donor phase through the pores of the hollow fiber and finally into the acceptor phase.

Based on these factors and basic principle of extraction of "like dissolves like", toluene, cyclohexane, 1-octanol, and di n-hexylehter are mainly used.

Besides, it was found that polar compounds requires the addition of such compounds such as hydrogen bonding reagent (tri-octylphosphine oxide, TOPO) to improve the extraction of these compounds [171].

3.4.5. pH

As it was indicated previously, the adjustment of pH can enhance extraction, because dissociation equilibria are affected. Specially in three phase mode, the adjustment of the donor and acceptor pH is critical, and pHs should be selected to assure that the analyte is not ionized in the donor phase and it must be ionized in the acceptor phase in order to avoid back extraction. In general, pH should differ from pK_a values of the analytes by at least 2-3 units to be sure that the compound is ionized or not [21].

3.4.6. Agitation Speed

One important parameter in HFSLM is the agitation speed during extraction. Agitation decreases the thickness of the unstirred boundary layer adjacent to the hollow fiber membrane, and this is known to be an efficient way to accelerate the extraction. However, if the agitation speed is to high, the extraction efficiency can decrease, probably due to the formation of air bubbles in the hollow fiber and the loss of the organic solvent, which can form microemulsions in the sample and in the acceptor phase [172]. Besides, agitation can be done either by stirring or shaking conditions, observing that extraction is usually more efficient when the sample is shaken [173]. This can be explained considering that during shaking, the fiber itself moves and mixes the acceptor solution in the lumen of the fiber and facilitates the transfer of the analytes into the acceptor solution at a faster rate [171].

3.4.7. Ionic Strength

Extraction of analytes can be enhanced or attenuate by addition of salts [165], depending on the nature of analytes. Basically, increasing ionic strength of the aqueous phase, the water molecules solvate the electrolyte ions and therefore the water solubility decreases. This effect is stronger for analytes with a small octanol-water partitioning coefficient such as parathion methyl, fenitrothion, malathion and fenamiphos. For extraction of triazines, a strong salting out effect has also noticed [174]. However, this parameter should always be studied during optimization step.

3.4.8. Applications

Typical examples of HFSLM as the sample preparation method for pesticide analysis are summarized in Table 5, although more applications have been reported in several reviews [21,175,176]. It can be observed that pesticides have been mainly determined in environmental samples, and the techniques has been applied basically in water (seawater, rainwater, etc…). In this sense, two phase system has been used, considering that most of the analyzed pesticides were non polar. For instance, Figure 10 shows a comparison between LLE and HFSLM extraction of chlorinated compounds in wastewater with a high amount of sulfur, which caused a high back ground even when the SIM mode was used. However it can be observed that this interference is reduced when HFSLM was used instead of LLE.

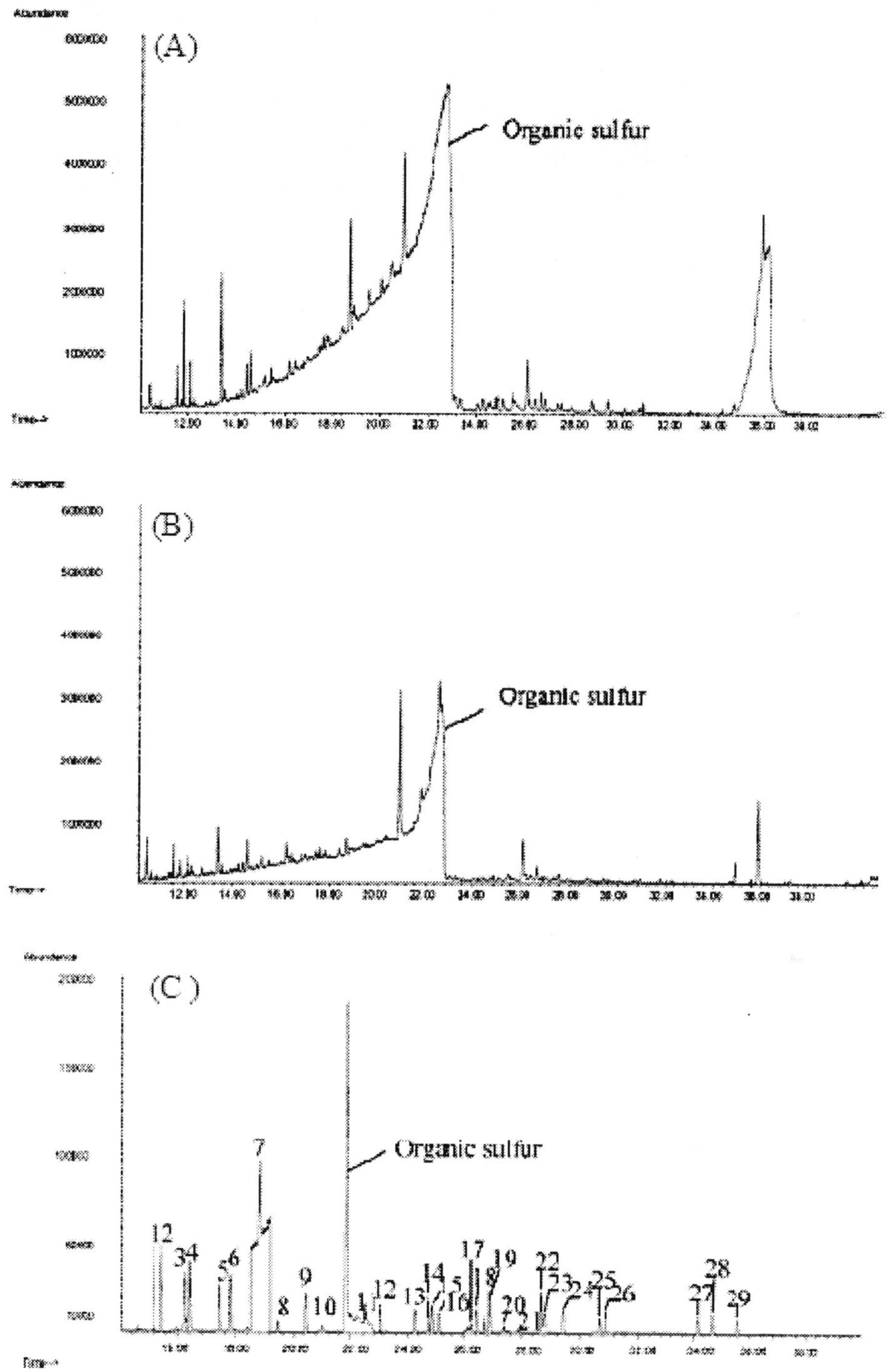

Figure 10. Comparison between (A) LLE extract of spiked wastewater with organochlorines pesticides recorded in scan mode, (B) MASE extract of spiked wastewater with salt and acid recorded in scan mode and (C) MASE extract of spiked wastewater with salt and acid, recorded in SIM mode. Reprinted with permission from [168], copyright 2004 American Chemical Society.

In relation to food samples, few applications have been developed. In this sense, some work have been carried out to determine pesticides in juices, milk and vegetables, as it can be observed in Table 5.

On the other hand, most studies have been focused on liquid solutions (water and juices) and only few applications have been performed on solid samples. In this sense, the general methodology, prior HFSLM extraction, implies a conventional extraction with organic solvent [177] or microwave assisted extraction [178]. However, some works have been described for the determination of pesticides in solid samples such as vegetables or aqueous soil samples without any treatment of the sample prior to the microextraction step [171,179].

Table 5. HFSLM applications on pesticide analysis[1]

Analyte	Matrix	Sampling mode	Organic solvent	Technique	DL	Ref
Dinitrophenols	Water	Three-phase	Undecane	LC-UV	6-8 ng/L	[173]
OCPs	Seawater	Two-phase	Toluene	GC-MS	13-59 ng/L	[180]
OCPs	Rainwater	Two-phase	Toluene	GC-MS/MS	13-59 ng/L	[181]
OCPs	Water	Two-phase	Toluene	GC-MS/MS	1 μg/L	[182]
Carbamates	Water	Two-phase	n-octanol	GC-MS/MS	0.2-0.8 μg/L	[183]
OPPs	Lake water	Two-phase	Cyclohexane	GC-MS/MS	6-200 ng/L	[184]
Vinclozolin	Water and Sediments	Two-phase	Toluene	GC-ECD	1 ng/L & 0.5 μg/kg	[177]
Triazines	Slurry	Two phase	Toluene	GC-MS	0.04-0.18 μg/L	[165]
OCPs	Marine sediments	Two-phase	Toluene	GC-MS	0.1-0.7 μg/kg	[178]
OCPs	Soils	Two-phase	Toluene	GC-MS	50-100 μg/kg	[179]
Triazines, OCPs & OPPs	Orange Juice	Two-phase	Cyclohexane	GC-MS	2-50 ng/L	[168]
OPPs	Wine and Juices	Two-phase	Cyclohexane	GC-MS	1-23 ng/L	[185]
OCPs	Sugarcaine Juice	Two-phase	Cyclohexane	GC-MS	0.004-56 μg/L	[186]
Multiclass	Vegetables	Three-phase	Dihexyl ether + TOPO	LC-MS	0.06-2.70 μg/kg	[171]
OCPs	Green tea leaves	Two-phase	n-octanol	GC-ECD	11-28 μg/kg	[187]
Phenoxy herbicides	Bovine milk	Three-phase	n-octanol	LC-UV	0.5 μg/L	[188]

[1] Abbreviations: ECD: Electron Capture Detection; OCPs: Organochlorine pesticides; OPPs: Organophosphorus pesticides.

Finally it can be concluded that HFSLM is virtually solvent free, simple, fast and reproducible sample cleanup procedure which can be adopted for analysis of pesticides. The simplicity and least requirement of laboratory wares (large volume of solvents, glass wares, concentration devices, etc. required in conventional sample preparation techniques) qualifies this technique most suited not only for off-site but on-site analysis as well. Besides, due to the simplicity and the low cost of the sampling device, the hollow fiber can be discarded after each extraction to avoid carryover effects.

One drawback of this technique is a lack of precision, which may be attributed to the completely manual operation from fiber preparation and conditioning to the handling of small extract volumes so further work is required in order to automate this technique.

3.5. Kinetic or Equilibrium Sampling

Mass transfer is a time-dependent process and its rate is reduced when the system reaches equilbrium conditions. If the concentration of the analyte in the extraction system (fiber, twister, drop of organic solvent or hollow-fiber) is studied versus the extraction time, as shown in Figure 11, it is possible to obtain three different regions of the curve: kinetic, intermediate and equilibrium. For method optimization, it is important to establish the extraction time profiles of target compounds to obtain the time after which equilibrium is obtained. Although longer extraction times result in increased extraction efficiency, sometimes it is not practical to apply extended extraction times [173]. However, it should be noted that when working in the kinetic regime, precise timing is required for good precision [147].

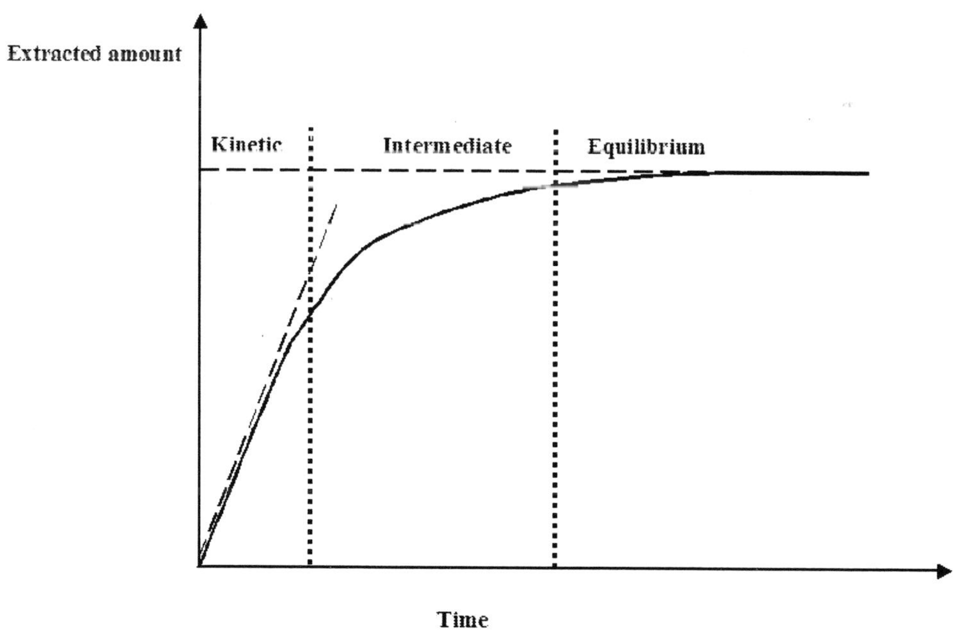

Figure 11. Extraction time profile during the application of microextraction techniques.

These microextraction techniques can be operated and optimized in different ways, in the kinetic uptake regime or under equilibrium conditions [90]. Besides, they can be tuned to extract as much analyte as possible from a sample or alternatively they can only extract a portion of the analyte from the sample. Generally, these techniques are non-exhaustive extraction techniques, considering that in the most of the cases, a significant amount of the analytes still remain in the sample. In this sense, it is important that "depletive" should not be confused with "kinetic" and "non-exhaustive" with "equilibrium sampling". Thus, there are four possibilities to carry out this technique: kinetic-depletive; kinetic-nondepletive; equilibrium-depletive and equilibrium-nondepletive, being the last two options the most used.

These techniques have the potential to be used as an equilibrium sampling devices for measurement of the free concentration of analytes. The basic assumption of this approach is that, because only a very small amount is extracted, the extraction does not influence the existing equilibrium between the bound and free form of a chemical, so the sampler measures the freely dissolved concentration. This methodology has been mainly used with SPME [189] although recently, when HFSLM as used, Equilibrium Sampling Through Membranes (ESTM) has been proposed as a new and powerful approach to measure free or available fraction of contaminants in environmental samples [161,170].

3.6. Quantification

As well as conventional extraction techniques, when microextraction techniques are used, different calibration modes for compensation of matrix interference can be used. For this purpose, traditional calibration methods, such as external standard, internal standard or standard addition methods can be used. The selection of the method mainly depends on the complexity of the sample. For instance, external or internal standard calibration can be used for tap water, considering matrix effects are not significant. However, for complex matrices, such as environmental waters or food samples, in which the matrix effect can contribute to enhance or reduce the analytical signal, matrix matched calibration at three or more levels can be studied. Besides, internal standard addition of deuterated or ^{13}C labeled target solutes can be used. However, the main problem to used this approach is that labeled standards are not commercially available for all the pesticides, so it can be difficult to use in routine analysis.

At the same time that microextraction techniques have been developed, new calibration models has been proposed for non exhaustive techniques. In this sense, typical quantification based on equilibrium-based approach, where extractions are completed when equilibrium is achieved between the sorbed analyte in the sorbent and analyte dissolved in the sample, has been described in the literature [90]. The extracted amount of analyte depends on its concentration in the sample, the amount of sorbent and the sorbent/matrix distribution constant, which can be estimated from extractions performed on standard solutions of the analyte, from physicochemical properties, etc… [190]. An alternative approach, based on kinetic calibration model, was recently developed for the quantification when microextraction techniques are used [191-193], and it is useful to correct matrix effect. This methodology is based on the desorption of standards which are preloaded in the extraction phase to calibrate the extraction of the analytes, and it has been successfully used to correct the matrix effects in the carbaryl analysis in wine analyses by HFSLM [193]. This approach can be very important for the calibration of in-site, because direct spiking of standards into the matrix is typically not possible in these cases.

4. Conclusion and Future Trends

Although solvent extraction is still the most used technique for pesticides extraction, new extraction methods based on reduction of organic solvent or miniaturization, have emerged as viable sample preparation, obtaining acceptable analytical data and diminishing the negative effects of analytical chemistry in the environment.

The combination of miniaturized extraction techniques and chromatographic techniques coupled to mass spectrometry detectors provides good future applications in environmental and food analysis, especially for trace amount of analytes in complex sample matrices.

The main advantages of the miniaturized systems are high speed analysis with high efficiency, environmentally friendly operation due to the minimal or no solvent consumption and highly selective analysis by developing tailored systems designed for particular applications. Besides, they can increase sample throughput, improve the quality of the analytical methods and reduce the cost of analysis.

However each microextraction technique presents some disadvantages. SPME is a suitable technique for GC applications, and although this technique can be coupled to LC systems, is less attractive, so other microextraction techniques such as HFSLM can be used as alternative for these determinations. Other problems related to SPME are carry over, conditioning should always be performed on each new fiber, so poor reproducibility and linearity can be found when complex matrices are analyzed [194]. On the other hand, SDME is an elegant method in which experimental parameters can be controlled to obtain reproducible results, but problems with drop stability can be considered as the main drawback.

When solid or semi-solid samples have to be analyzed, it is required the use of same sample pretreatment in order to minimize sample matrix or to condition the sample to the microextraction step. For instance, solid samples require a previous extraction step, such as solvent extraction, although some attempts have been carried out without prior treatment [171].

As it was observed, the majority of the applications were developed in environmental pesticide analysis, although some applications in food analysis can be found.

Despite of SPME is fully automated, future work should be focused on automation of other microextraction techniques such as HFSLM or SDME, since it requires careful and elaborate manual operation. This will provide valuable tools to measure a broad range of compounds in different matrices. This will simplify, accelerate and miniaturize pretreatment procedures to make these into an on-line and therefore, an integrated part of the total analytical system, which allow the introduction of these techniques into laboratories as routine extraction techniques.

REFERENCES

[1] Planas, C.; Puig, A.; Rivera, J.; Caixach, J. *J. Chromatogr. A 2006*, *1131*, 242-252.

[2] European Council Directives 76/895/EEC, 83/362/EEC, 86/363/EEC and 90/642/EEC, updated on 04/11/2004, European Union, Brussels. Available at http://ec.europa.eu/comm/food/plant/protection/pesticides/index_en.htm.

[3] European Council Directive 86/363/EEC, updated on 31/05/2007, European Union, Available at http://ec.europa.eu/food/plant/protection/pesticides/legislation_en.htm.

[4] European Council Directive 98/83. Official Journal of European Communities, 1998, L330/32.

[5] http://www.epa.gov/pesticides/regulating/laws.htm#regulations

[6] Soler, C.; Picó, Y. *Trends Anal. Chem. 2007*, *26*, 103-115.

[7] Hildebrandt, A.; Lacorte, S.; Barceló, D. *Anal. Bioanal. Chem. 2007*, *387*, 1459-1468.

[8] Kataoka, H. *Anal. Bional. Chem. 2002*, *373*, 31-45.

[9] Arrebola, F. J.; Martínez-Vidal, J. L.; Mateu-Sánchez, M.; Álvarez-Castellón, F. J. *Anal. Chim. Acta 2003*, *484*, 167-180.

[10] Hernández, F.; Sancho, J. V.; Pozo, O. J. *Anal. Bioanal. Chem. 2005*, *382*, 934-946.

[11] Rodríguez, R.; Mañés, J.; Picó, Y. *Anal. Chem. 2003*, *75*, 452-459.

[12] Ahmed, F. E. *Trends Anal. Chem. 2001*, *20*, 649-661.

[13] Smith, R. M. *J. Chromatogr. A 2003*, *1000*, 3-27.

[14] Vas, G.; Vekey, K. *J. Mass Spectrom. 2004*, *39*, 233–254.

[15] Kataoka, H. *Trends Anal. Chem. 2003*, *22*, 232-244.

[16] Anastassiades, M.; Lehotay, S. J.; Stajnbaher, D.; Schenck F. J. *J. AOAC Int. 2003*, *86*, 412-431.

[17] Bogialli, S.; Di Corcia, A. *J. Biochem. Biophys. Methods 2007*, *70*, 163-179.

[18] Dietz, C.; Sanz, J.; Cámara, C. *J. Chromatogr. A 2006*, *1103*, 183-192.

[19] Kawaguchi, M.; Ito, R.; Saito, K.; Nakazawa, H. *J. Pharm. Biomed. Anal. 2006*, *40*, 500-508.

[20] Xu, L.; Basheer, C.; Lee, H. K. *J. Chromatogr. A 2007*, *1152*, 184-192.

[21] Rasmussen, K. E.; Pedersen-Bjergaard, S.; *Trends Anal. Chem. 2004*, *23*, 1-10.

[22] Kataoka, H.; Lord, H. L.; Pawliszyn, J. *J. Chromatogr. A 2000*, *880*, 35-62.

[23] Mahara, B. M.; Borossoy, J.; Torkos, K. *Microchem. J. 1998*, *58*, 31-38.

[24] Fatoki, O. S.; Awofolu, R. O. *J. Chromatogr. A 2003*, *983*, 225-236.

[25] Araoud, M.; Douki, W.; Rhim, A.; Najjar, M. F.; Gazzah, N. *J. Environ. Sci. Health B 2007*, *42*, 179-187.

[26] Blasco, C.; Lino, C. M.; Picó, Y.; Pena, A.; Font, G.; Silveira, M. I. N. *J. Chromatogr. A 2004*, *1049*, 155-160.

[27] Amvrazi, E. G.; Albanis, T. A. *J. Agric. Food Chem. 2006*, *54*, 9642-9651.

[28] Barriada-Perreira, M.; Concha-Graña, E.; González-Castro, M. J.; Muniategui-Lorenzo, S.; López-Mahía, P.; Prada-Rodríguez, D.; Fernández-Fernández, E. *J. Chromatogr. A 2003*, *1008*, 115-122.

[29] Helaleh, M. I. H.; Al-Omair, A.; Ahmed, N.; Gevao, B. *Anal. Bional. Chem. 2005*, *382*, 1127-1134.

[30] Gfrerer, M.; Chen, S.; Lankamyr, E. P.; Quan, X.; Yang, F. *Anal. Bional. Chem. 2004*, *378*, 1861-1867.

[31] Picó, Y.; Fernández, M.; Ruiz, M. J.; Font, G. *J. Biochem. Biophys. Methods 2007*, *70*, 117-131.
[32] Huck, C. W.; Bonn, G. K. *J. Chromatogr. A 2000*, *885*, 51-72.
[33] López-Roldán, P.; López de Alda, M. J.; Barceló, D. *Anal. Bional. Chem. 2004*, *378*, 599-609.
[34] Belmonte-Vega, A.; Garrido Frenich, A.; Martínez Vidal, J. L. *Anal. Chim. Acta 2005*, *538*, 117-127.
[35] Bischoff, G.; Pestemer, W.; Xu, W.; Kofoet, A. *Chromatographia 2006*, *63*, 233-237.
[36] Marín, J. M.; Sancho, J. V.; Pozo, O. J.; López, F. J.; Hernández F. *J. Chromatogr. A 2006*, *1133*, 204-214.
[37] Guardia-Rubio, M. Ruiz-Medina, A, Pascual-Regurea, M. I.; Fernández de Córdova, M. L. *Microchem. J. 2007*, *85*, 257-264.
[38] Tadeo, J. L.; Sánchez-Brunete, C.; Albero, B.; González, L. *Crit. Rev. Anal. Chem. 2004*, *34*, 165-175.
[39] Albero, B.; Sánchez-Brunete, C.; Tadeo, J. L. *Talanta 2005*, *66*, 917-924.
[40] Jiménez, J. J.; Bernal, J. L.; del Nozal, M. J.; Toribio L.; Bernal, J. *J. Sep. Sci. 2007*, *30*, 547-556.
[41] Hernández-Borges, J; Rodríguez-Delgado, M. A.; García-Montelongo, F. J.; Cifuentes, A. *J. Sep. Sci. 2005*, 948-956.
[42] Schenk, F. J.; Donoghue, D. J. *J. Agric. Food Chem. 2000*, *48*, 6412-6415.
[43] Schenk, F. J.; Lehotay, S. J.; Vega, V. *J. Sep. Sci. 2002*, *25*, 883-890.
[44] Stajnbaher, D.; Zupancic-Kralj L. *J. Chromatogr. A 2003*, *1015*, 185-198.
[45] Di Muccio, A.; Fidente, P.; Barbini, D. A.; Dommarco, R.; Seccia, S.; Morrica, P. *J. Chromatogr. A 2007*, *1108*, 1-6.
[46] Amand, A. D.; Girard, L. *Int. J. Environ. Anal. Chem. 2004*, *84*, 739-748.
[47] Arthur, C. L.; Pawliszyn, J. *Anal. Chem. 1990*, *62*, 2145-2148.
[48] Barker, S. A.; Long, A. R.; Short, C. R. *J. Chromatogr. A 1989*, *475*, 353-361.
[49] Barker, S. A. *J. Chromatogr. A 2000*, *880*, 63-68.
[50] Garcinuño, R. M.; Ramos, L.; Fernández-Hernando, P.; Cámara, C. *J. Chromatogr. A 2004*, *1041*, 35-41.
[51] Valenzuela, A. I.; Picó, Y.; Font, G. *J. AOAC Int. 2001*, *84*, 901-909.
[52] Albero, B.; Sánchez-Brunete, C.; Tadeo, J. L. *J. Agric. Food Chem. 2003*, *51*, 6915-6921.
[53] Blasco, C.; Picó, Y.; Font, G. *J. AOAC Int. 2002*, *85*, 704-711.
[54] Barker, S. A. In *Pesticide Protocols*; Martínez Vidal, J. L.; Garrido Frenich, A., Ed.; Humana Press: Totowa, 2006, pp 285-296.
[55] Valsamaki, V. I.; Boti, V. I.; Sakkas, V. A.; Albanis, T. A. *Anal. Chim. Acta 2006*, *573-574*, 195-201.
[56] Albero, B.; Sánchez-Brunete, C.; Donoso, A.; Tadeo, J. L. *J. Chromatogr. A 2004*, *1043*, 127-133.
[57] Fernández, M.; Picó, Y, Mañés, J. *J. Chromatogr. A 2000*, *871*, 43-56.
[58] Ferrer, C.; Gómez, M. J.; García-Reyes, J. F.; Ferrer, I.; Thurman, E. M.; Fernández-Alba, A. R. *J. Chromatogr. A 2005*, *1069*, 183-194.
[59] Ramos, J. J.; González, M. J.; Ramos, L. *J. Sep. Sci. 2004*, *27*, 595-601.
[60] Navarro, M.; Picó, Y.; Marín, R.; Mañés, J. *J. Chromatogr. A 2002*, *968*, 201-209.
[61] Blasco, C.; Font, G.; Picó, Y. *J. Chromatogr. A 2004*, *1028*, 267-276.

[62] Soler, C.; Mañés, J.; Picó, Y. *J. Chromatogr. A 2004*, *1048*, 41-49.

[63] Hu, X.; Yu, J.; Yan, Z.; Ni, L.; Lin, Y.; Wang, P.; Li, J.; Huang, X.; Chu, X.; Zhang, Y. *J. AOAC Int. 2004*, *87*, 972-985.

[64] Hu,Y. Y.; Zheng, P.; He, Y. H.; Sheng, G. P. *J. Chromatogr. A 2005*, *1098*, 188-193.

[65] Albero, B.; Sánchez-Brunete, C.; Tadeo, J. L. *J. AOAC Int. 2001*, *84*, 1165-1171.

[66] Ling, Y. C.; Huang, I. P. *Chromatographia 1995*, *40*, 259-266.

[67] Carro, A. M.; Lorenzo, R. A.; Fernández, F.; Rodil, R.; Cela, R. *J. Chromatogr. A 2005*, *1071*, 93-98.

[68] Yagüe, C.; Bayarri, S.; Lázaro, R.; Conchello, P.; Arino, A.; Herrera, A. *J. AOAC Int. 2001*, *84*, 1561-1568.

[69] Furusawa, N. *Anal. Bioanal. Chem. 2004*, *378*, 2004-2007.

[70] Frenich, A. G.; Bolaños, P. P.; Vidal, J. L. M. *J. Chromatogr. A 2007*, *1153*, 194-202.

[71] Luke, M.; Froberg, J. E.; Masumoto, H. T. *J. AOAC Int. 1975*, *58*, 1020-1026.

[72] Garrido Frenich, A.; Martínez Vidal, J. L.; López-López, T.; Cortés-Aguado, S.; Martínez-Salvador, I. *J. Chromatogr. A 2004*, *1048*, 199-206.

[73] Mastovska, K.; Lehotay, S. J. *J. Chromatogr. A 2004*, *1004*, 259-272.

[74] Anastassiades, M.; Mastovska, K.; Lehotay, S. J. *J. Chromatogr. A 2003*, *1015*, 163-184.

[75] Lehotay, S. J.; In *Pesticide Protocols*; Martínez Vidal, J. L.; Garrido Frenich, A., Ed.; Humana Press: Totowa, 2006, pp 239-261.

[76] Lehotay, S. J.; Mastovska, K.; Lightfield, A. R. *J. AOAC Int. 2005*, 88, 615-629.

[77] Lehotay, S. J.; Kok, A.; Hiemstra, M.; Bodegraven, P. *J. AOAC Int. 2005*, *88*, 595-614.

[78] Plossl, F.; Giera, M.; Bracher, F. *J. Chromatogr. A 2006*, *1135*, 19-26.

[79] Schenk, F. J.; Hobbs, J. E. *Bull. Environ. Contam. Toxicol. 2004*, *73*, 24-30.

[80] Mastovska, K.; Lehotay, S. J.; Anastassiades, M. *Anal. Chem. 2005*, *77*, 8129-8137.

[81] Looser, N.; Kostelac, D.; Scherbaum, E.; Anastassiades, M.; Zipper, H. *J. Verbr. Lebensm. 2006*, *1*, 135-141.

[82] Diez, C.; Traag, W. A.; Zommer, P.; Marinero, P.; Atienza, J. *J. Chromatogr. A 2006*, *1131*, 11-23.

[83] Plaza Bolaños, P.; Fernández Moreno, J. L.; Shtereva, D. D.; Garrido Frenich, A.; Martínez Vidal, J. L. *Rapid Commun. Mass Spectrom. 2007*, *21*, 2282-2294.

[84] Leandro, C. C.; Hancock, P.; Fussell, R. J.; Keely, B. J. *J. Chromatogr. A 2007*, *1144*, 161-169.

[85] Lehotay, S. J.; Mastovska, K.; Yun, S. J. *J. AOAC Int. 2005*, *88*, 630-638.

[86] Cunha, S. C.; Lehotay, S. J.; Mastovska, K.; Fernandes, J. O.; Beatriz, M.; Oliveira, P. P. *J. Sep. Sci. 2007*, *30*, 620-632.

[87] Hercegova, A.; Domotorova, M.; Kruzlicova, D.; Matisova, E. *J. Sep. Sci. 2006*, *29*, 1102-1109.

[88] Commission Directive 1999/39/EC amending Directive 95/5/EC. Official Journal of European Communities, 1999, L124/8.

[89] Wardencki, W.; Curylo, J.; Namiesnik, J. *J. Biochem. Biophys. Methods 2007*, *70*, 275-288.

[90] Pawliszyn, J. *Solid-phase microextraction: theory and practice;* Ed. John Wiley, New York, 1997.

[91] Eisert, R.; Pawliszyn, J. *Anal. Chem. 1997*, *69*, 3140-3147.

[92] Alpendurada, M. F. *J. Chromatogr. A 2000*, *889*, 3-14.

[93] Simplicio, A. L.; Vilas-Boas, L. *J. Chromatogr. A 1999*, *833*, 35-42.
[94] Hwang, B. H.; Lee, M. R. *J. Chromatogr. A 2000*, *898*, 245-256.
[95] Zhang, Z.; Pawliszyn, J. *Anal. Chem. 1999*, *65*, 1843-1852.
[96] Ai, J. *Anal. Chem. 1997*, *69*, 3260-3266.
[97] Fernández, M.; Padrón, C.; Marconi, L.; Ghini, S.; Colombo, R.; Sabattini, A. G.; Girotti, S. *J. Chromatogr. A 2001*, *922*, 257-265.
[98] Eisert, R.; Levsen, K.; Wünsch, G. *J. Chromatogr. A 1994*, *683*, 175–183.
[99] Popp, P.; Kalbitz, K.; Oppermann, G. *J. Chromatogr. A 1994*, *687*, 133–140.
[100] Krutz, L. J.; Senseman, S. A.; Sciumbato, A. S. *J. Chromatogr. A 2003*, *999*, 103-121.
[101] Boyd-Boland, A. A.; Magdic, S.; Pawliszyn, J. B. *Analyst 1996*, *121*, 929-938.
[102] Goncalves, C.; Alpendurada, M. F. *J. Chromatogr. A 2004*, *1026*, 239-250.
[103] Lin, H. H.; Sung, Y. H.; Huang, S. D. *J. Chromatogr. A 2003*, *1012*, 57-66.
[104] Sakamoto, M.; Tsutsumi, T. *J. Chromatogr. A 2004*, *1028*, 63-74.
[105] Arrebola, F. J.; Cortés-Aguado, S.; Sánchez-Morito, N.; Garrido Frenich, A.; Martínez Vidal, J. L. *Anal. Letters 2004*, *37*, 99-117.
[106] Sauret-Szczepanski, N.; Mirabel, P.; Worthman, H. *Environ. Poll. 2006*, *139*, 133-142.
[107] Wu, J.; Tragas, C.; Lord, H.; Pawliszyn, J. *J. Chromatogr. A 2002*, *976*, 357-367.
[108] Chafer-Pericas, C.; Herráez-Hernández, R.; Campíns-Falcó, P. *J. Chromatogr. A 2007*, *1141*, 10-21.
[109] Bouaid, A.; Ramos, L.; González, M. J.; Fernández, P.; Cámara, C. *J. Chromatogr. A 2001*, *939*, 13-21.
[110] Chang, S.; Doong, R. *Chemosphere 62*, *2006*, 1869-1878.
[111] Navalón, A.; Prieto, A.; Araujo, L.; Vílchez, J. L. *J. Chromatogr. A 2002*, *975*, 355-360.
[112] Lambropoulou, D. A.; Albanis, T. A. *J. Chromatogr. A 2003*, *993*, 197-203.
[113] Fytianos, K.; Raikos, N.; Theodoridis, G.; Velinova, Z.; Tsoukali, H. *Chemosphere 2006*, *65*, 2090-2095.
[114] Hernández-Borges, J.; Cifuentes, A.; García-Montelongo, F. J.; Rodríguez-Delgado, M. A. *Electrophoresis 2005*, *26*, 980, 989.
[115] Cai, L.; Gong, S.; Chen, M.; Wu, C. *Anal. Chim. Acta 2006*, *559*, 89-96.
[116] Sagratini, G.; Mañés, J.; Giardina, D.; Damián, P.; Picó, Y. *J. Chromatogr. A 2007*, *1147*, 135-143.
[117] Fidalgo-Used, N.; Centineo, G.; Blanco-González, E.; Sanz-Medel, A. *J. Chromatogr. A 2003*, *1017*, 35-44.
[118] Blasco, C.; Fernández, M.; Picó, Y.; Font, G. *J. Chromatogr. A 2004*, *1030*, 77-85.
[119] Campillo, N.; Peñalver, R.; Aguinaga, N.; Hernández-Córdoba, M. *Anal. Chim. Acta 2006*, *562*, 9-15.
[120] González-Rodríguez, M. J.; Arrebola-Liébanas, F. J.; Garrido Frenich, A.; Martínez Vidal, J. L.; Sánchez-López, F. J. *Anal. Bional. Chem. 2005*, *382*, 164-172.
[121] Baltussen, E.; Sandra, P.; David, F.; Cramers, C. *J. Microcol. Sep. 1999*, *11*, 737-747.
[122] Baltussen, E.; David, F.; Sandra, P.; Cramers, C. *Anal. Chem. 1999*, *71*, 5213-5216.
[123] Popp, P.; Bauer, C.; Wennrich, L. *Anal. Chim. Acta 2001*, *436*, 1-9.
[124] Van Hoeck, E.; David, F.; Sandra, P. *J. Chromatogr. A 2007*, *1157*, 1-9.
[125] Sandra, P.; Tienpont, B.; David, F. *J. Chromatogr. A 2003*, *1000*, 299-309.
[126] León, V. M.; Álvarez, B.; Cobollo, M. A.; Muñoz, S.; Valor, I. *J. Chromatogr. A 2003*, *1999*, 91-101.

[127] Tienpont, B.; David, F.; Desmet, K.; Sandra, P. *Anal. Bional. Chem. 2002*, *373*, 46-55.
[128] Popp, P.; Bauer, C.; Hauser, B.; Keil, P.; Wenrich, L. *J. Sep. Sci. 2003*, *26*, 961-967.
[129] Serodio, P.; Nogueira, J: M. F. *Anal. Bioanal. Chem. 2005*, *382*, 1141-1151.
[130] Nakamura, S.; Daishima, S. *Anal. Bional. Chem. 2005*, *382*, 99-107.
[131] León, V. M.; Llorca-Porcel, J.; Álvarez, B.; Cobollo, M. A.; Muñoz, S.; Valor, I. *Anal. Chim. Acta 2006*, *558*, 261-266.
[132] Rodil, R.; Popp, P. *J. Chromatogr. A 2006*, *1124*, 82-90.
[133] Hayasaka, Y.; MacNamara, K.; Baldock, G. A.; Taylor, R. L.; Pollnitz, A. P. *Anal. Bional. Chem. 2003*, *375*, 948-955.
[134] Chia, K. J.; Huang, S. D. *Rapid Commum. Mass Spectrom. 2006*, *20*, 118-124.
[135] Ochiai, N.; Sasamoto, K.; Kanda, H.; Nakamura, S. *J. Chromatogr. A 2006*, *1130*, 83-90.
[136] Blasco, C.; Font. G.; Picó, Y. *J. Chromatogr. A 2002*, *970*, 201-212.
[137] Ochiai, N.; Sasamoto, K.; Kanda, H.; Yamagami, T.; David, F.; Tiepont, B.; Sandra, P. *J. Sep. Sci. 2005*, *28*, 1083-1092.
[138] Liu, W. M.; Hu, Y.; Zhao, J. H.; Xu, Y.; Guan, Y. F. *J. Chromatogr. A 2005*, *1095*, 1-7.
[139] Jeannot, M. A.; Cantwell, F. F. *Anal. Chem. 1996*, *68*, 2236-2240.
[140] Majors, R. E. *LC-GC Europe 2006*, *5*, 284-292.
[141] Psillakis, E.; Kalogerakis, N. *Trends Anal. Chem. 2002*, *21*, 55-64.
[142] Ye, C. L.; Zhou, Q. X.; Wang, X. M. *Anal. Chim. Acta 2006*, *572*, 165-171.
[143] Jeannot, M. A.; Cantwell, F. F. *Anal. Chem. 1997*, *69*, 235-239.
[144] He, Y.; Lee, H. K. *Anal. Chem. 1997*, *69*, 4634-4640.
[145] Ahmadi, F.; Assadi, Y.; Milani-Hosseini, S. M. R.; Rezaee, M. *J. Chromatogr. A 2006*, *1101*, 307-312.
[146] López-Blanco, C.; Gómez-Alvárez, S.; Rey-Garrote, M.; Cancho-Grande, B.; Simal-Gándara, J. *Anal. Bional. Chem. 2005*, *383*, 557-561.
[147] Psillakis, E.; Kalogerakis, N. *J. Chromatogr. A 2001*, *907*, 211-219.
[148] de Jager, L. S.; Andrews, A. R. J. *J. Chromatogr. A 2001*, *911*, 97-105.
[149] Zhao, E.; Han, L.; Jiang, S.; Wang, Q.; Zhou, Z. *J. Chromatogr. A 2006*, *1114*, 269-273.
[150] de Jager, L. S.; Andrews, A. R. J. *Analyst 2000*, *125*, 1943-1948.
[151] Zhao, L.; Lee, H. K. *J. Chromatogr. A 2001*, *919*, 381-388.
[152] López-Blanco, M. C.; Blanco-Cid, S.; Cancho-Grande, B.; Simal-Gandara, J. *J. Chromatogr. A 2003*, *984*, 245-252.
[153] Lambropoulou, D. A.; Albanis, T. A. *J. Chromatogr. A 2004*, *1049*, 17-23.
[154] Lambropoulou, D. A.; Psillakis, E.; Albanis, T. A.; Kalogerakis, N. *Anal. Chim. Acta 2004*, *516*, 205-211.
[155] Liu, Y.; Zhao, E.; Zhou, Z. *Anal. Lett. 2006*, *39*, 2333-2344.
[156] Xiao, Q.; Hu, B.; Yu, C.; Xia, L.; Jiang, Z. *Talanta 2006*, *69*, 848-855.
[157] Barri, T.; Bergstrom, S.; Hussen, A.; Norberg, J.; Jönsson, J. A. *J. Chromatogr. A 2006*, *1111*, 11-20.
[158] Jönsson, J. A.; Mathiasson, L. *Trends Anal. Chem. 1999*, *18*, 318-325.
[159] Jönsson, J. A.; Mathiasson, L. *J. Sep. Sci. 2001*, *24*, 495-507.
[160] Chimuka, L.; Mathiasson, L.; Jönsson, J. A. *Anal. Chim. Acta 2000*, *416*, 77-86.
[161] Romero, R.; Liu, J.; Mayer, P.; Jönsson, J. A. *Anal. Chem. 2005*, *77*, 7605-7611.

[162] Mulugeta, M.; Megersa, N. *Talanta 2004*, *64*, 101-108.
[163] Pedersen-Bjergaard, S.; Rasmussen, K. E. *Anal. Chem. 1999*, *71*, 2650-2656.
[164] Zhu, L.; Zhu, L.; Lee, H. K. *J. Chromatogr. A 2001*, *924*, 407-414.
[165] Shen, G.; Lee, H. K. *Anal. Chem. 2002*, *74*, 648-654.
[166] Hou, L.; Shen, G.; Lee, H. K. *J. Chromatogr. A 2003*, *985*, 107-116.
[167] Hauser, B.; Popp, P.; Kleine-Benne, E. *J. Chromatogr. A 2002*, *963*, 27-36.
[168] Hauser, B.; Schellin, M.; Popp, P. *Anal. Chem. 2004*, *76*, 6029-6038.
[169] de Jager, L. S.; Andrews, A. R. *Analyst 2001*, *126*, 1298-1303.
[170] Liu, J.; Jönsson, J. A.; Mayer. P. *Anal. Chem. 2005*, *77*, 4800-4809.
[171] Romero-González, R.; Pastor-Montoro, E.; Martínez Vidal, J. L.; Garrido Frenich, A. *Rapid Commun. Mass Spectrom. 2006*, *20*, 2701-2708.
[172] Wen, X.; Tu, C.; Lee, H. K. *Anal. Chem. 2004*, *76*, 228-232.
[173] Berhanu, T.; Liu, J.; Romero, R.; Megersa, N.; Jönsson, J. A. *J. Chromatogr. A 2006*, *1103*, 1-8.
[174] Hauser, B.; Popp, P. *J. Sep. Sci. 2001*, *24*, 551-560.
[175] Psillakis, E.; Kalogerakis, N. *Trends Anal. Chem. 2003*, *22*, 565-574.
[176] Lambropoulou, D.; Albanis, T. A. *J. Biochem. Biophys. Methods 2007*, *70*, 195-228.
[177] Lambropoulou, D. A.; Albanis, T. A. *J. Chromatogr. A 2004*, *1061*, 11-18.
[178] Basheer, C.; Obbard, J. P.; Lee, H. K. *J. Chromatogr. A 2005*, *1068*, 221-228.
[179] Hou, L.; Lee, H. K. *J. Chromatogr. A 2004*, *1038*, 37-42.
[180] Basheer, C.; Lee, H. K.; Obbard, J. P. *J. Chromatogr. A 2002*, *968*, 191-199.
[181] Basheer, C.; Balasubramanian, R.; Lee, H. K. *J. Chromatogr. A 2003*, *1016*, 11-20.
[182] Yan, C. H.; Wu, H. F. *Rapid Commun. Mass Spectrom. 2004*, *18*, 3015-3018.
[183] Zhang, J.; Lee, H. K. *J. Chromatogr. A 2005*, *1117*, 31-37.
[184] Chen, P. S.; Huang, S. D. *Talanta 2006*, *69*, 669-675.
[185] Schellin, M.; Hauser, B.; Popp, P. *J. Chromatogr. A 2004*, *1040*, 251-258.
[186] Gomes-Zuin, V.; Schellin, M.; Montero, L.; Yariwake, J. H.; Augusto, F.; Popp, P. *J. Chromatogr. A 2006*, *1114*, 180-187.
[187] Huang, S. P.; Huang, S. D. *J. Chromatogr. A 2006*, 1135, 6-11.
[188] Zhu, L.; Ee, K. H.; Zhao, L.; Lee, H. K. *J. Chromatogr. A 2002*, *968*, 335-343.
[189] Van der Wal, L.; Jager, T.; Fleuren, R. H. L. J.; Barendregt, A.; Sinnige, T. L.; Van Gestel, C. A. M. *Environm. Sci. Technol. 2004*, *38*, 4842-4848.
[190] Martos, P. A.; Saraullo, A.; Pawliszyn, J. *Anal. Chem. 1997*, *69*, 402-40.
[191] Chen, Y.; Pawliszyn, J. *Anal. Chem. 2004*, *76*, 5807-5815.
[192] Chen. Y.; O'Rilley, J.; Wang, Y.; Pawliszyn, J. *Analyst 2004*, *129*, 702-703.
[193] Ouyang, G.; Pawliszyn, J. *Anal. Chem. 2006*, *78*, 5783-5788.
[194] Negrao, M. R.; Alpendurada, M. F. *J. Chromatogr. A 1998*, *823*, 211-218.

In: Progress in Pesticides Research
Editor: Christopher M. Kanzantzakis

ISBN 978-1-60456-201-9

Chapter 4

ENVIRONMENTAL FRIENDLY METHODS FOR THE EXTRACTION AND DETERMINATION OF ORGANOPHOSPHORUS AND ORGANOCHLORINE PESTICIDES

J.J. Santana Rodríguez*[1]*, Z. Sosa Ferrera and D. Vega Moreno
Department of Chemistry, Faculty of Marine Sciences, University of Las Palmas de Gran Canaria, 35017, Las Palmas de Gran Canaria, Spain

ABSTRACT

Organochlorine pesticides have been used during decades as insecticides and for the control of vector-borne diseases. In 1980s these pesticides were banned in developed countries and were replaced by other kind of compounds like organophosphorus pesticides; however, they are usually more toxic than organochlorine pesticides. Nowadays these compounds are present in the environment; thus the long-term persistence and their metabolites are presented in water, sediments, soils and vegetables.

The methods usually used for the extraction of these compounds from environmental samples are often too time consuming, involve multi-step procedures that are prone to loss of analytes and use high amounts of toxic organic solvents.

Modern strategies in analytical chemistry have been developed towards the simplification of sample's preparation, as well as the minimization of organic solvent used on the extraction process. An alternative to organic solvents is the use of micellar systems as extractants of organic pollutants in liquid and solid environmental samples. Micellar media facilitate the solubilization of different compounds present in different kind of samples. Benefits obtained when micellar media are used, are lesser toxicity and cost than organic solvents.

Cloud-point extraction (CPE) process in liquid samples and microwave assisted micellar extraction (MAME) in solid samples, have been developed for the extraction of

[1] jsantana@dqui.ulpgc.es.

pesticides prior to their determination by high performance liquid chromatography. Moreover, during extraction process, many interfering components are usually co-extracted from solid samples together with target analytes. For increasing sensitivity and selectivity, MAME procedure can be coupled with other extraction and preconcentration techniques, introducing a clean-up step for removing substances that could interfere with the final determination of target analytes in HPLC system. This coupling can be done with solid phase microextraction (SPME), called MAME-SPME, and with solid phase extraction (SPE), MAME-SPE, producing also an intensification of the analytes signals by means of extract clean-up and preconcentration.

In this study an overview of the use of micellar systems well alone, or well combined with the aforementioned extraction/preconcentration techniques, for the determination of organochlorine and organophosphorus pesticides in liquid and solid samples is presented. Fundaments, experimental, variables optimization, quantitative procedures and applications are described and discussed, making emphasis on the advantages of these green methodologies with regard to traditional extraction techniques.

ABBREVIATIONS

ASE: Acelerated Solvent Extraction
CMC: Critical Micellar Concentration
CPE: Cloud Point Extraction
DDT: Dichlorodiphenyltrichloroethane
EPA: Environmental Protection Agency
HPLC: High Pressure Liquid Chromatography
LLE: Liquid-Liquid extraction
LOD: Limit of detection
MAE: Microwave Assisted Extraction
MAME: Microwave Assisted Micellar Extraction
PLE: Pressurized Liquid Extraction
POLE: Polyoxyethylene 10 Lauryl Ether
RSD: Relative Standard Desviation
SFE: Supercritical Fluid Extraction
SPE: Solid Phase Extraction
SPME: Solid Phase Microextraction

INTRODUCTION

Organochlorine and Organophophorus Pesticides

Organochlorine pesticides have been used during decades in agricultural as insecticide and for the control of vector-borne diseases. Although these pesticides have banned in developed countries since 1970s and 1980s, some of them have continued to use DDT, especially for public health applications such as malaria control [1, 2]. Despite not being used by many countries for some years they are present in the environment; in this way, the long-term persistence of DDT and its metabolites in soil have been reported [3-6].

The lipophilic nature, hydrophobicity and low chemical and biological degradation rates of organochlorine pesticides have led their accumulation in biological tissues and subsequent magnification of concentrations in organisms progressing up the food chain with consequences for the humans [7, 8]; for these reasons are listed as US Environmental Protection Agency (EPA) priority pollutants [9].

Organophosphorus pesticides, which may be considered as second generation of synthetic pesticides, after the use of organochlorine pesticides, have become more popular, as they are generally of low persistence, biodegradable and do not tend to accumulate in the food chain. However they are generally more toxic (potent cholinesterase inhibitors) than organochlorine or carbamate compounds [10].

Over recent decades, research into developing new pesticides and analytical methodology has been severely curtailed in favour of alternate means of pest control. Nevertheless, the use of pesticides continues to be our major weapon for controlling pests. Large amounts of pesticide residues remain in agricultural and forest soils after their application [11]. They leach into groundwater and are also found in run-off water. Therefore, the development of new and improved techniques for analysing these agents and their breakdown products is mandatory.

Tradional techniques for analysis of organochlorine and organophosphorus pesticides are based in gas-chromatographic multiresidue procedures [12-14]. However, these methods are often too time consuming, due to the slow chromatographic elution and the intensive pre-analytical clean-up procedures [15-17] followed, in some cases, by a derivatization [17, 18].

They usually involve multi-step procedures, are prone to loss of analytes, and use toxic organic solvents [19, 20]. Consequently, most of the analysis time is spent on the sample preparation steps alone. Furthermore, some pesticides are thermolabile and therefore difficult to volatilize without decomposition [21].

Table 1. Organochlorine and Organophosphorus pesticides studied

Organochlorine Pesticides	Organophosphorus Pesticides
	DIMETHOATE
4,4'-Dichlorodiphenyldichloroethane (4,4'-DDD)	Methidation
Dieldrin	Parathion methyl
4,4'-Dichlorodiphenyltrichloroethane (4,4'-DDT)	Malathion
2,4'-Dichlorodiphenyltrichloroethane (2,4'-DDT)	Ethoprophos
4,4'- Dichlorodiphenyldichloroethylene (4,4'-DDE)	Parathion ethyl
Aldrin	Diazinon
	Chlorpyrifos

New methods therefore have been developed in order to supplement or even replace existing gas-chromatographic methods. These altenative methodologies are based in a high performance liquid chromatography determination combined with a previous micellar extraction: cloud point extraction (CPE) procedure for liquid samples and microwave assisted micellar extraction (MAME) for solid samples. Specifically, the organochlorine and organophosphorus pesticides listed in Table 1 have been studied and determined using these green methods.

Main Conventional Extraction Techniques for Liquid and Solid Samples

The need to determine low concentrations of pollutants in environmental samples, as well as the matrix complexity, means that analytical methods with high sensitivity, selectivity and resolution have to be applied to soil, sediment, water and biological samples. Elaborate sample preparation involving analyte isolation and enrichment is generally necessary before the final analysis, which is usually performed using gas or liquid chromatography.

The most frecuently used methods for the extraction and preconcentration of organochlorine and organophosphorus pesticides in water samples are liquid-liquid extraction (LLE) and solid-phase extraction (SPE) [22,23]. Solvent extraction methods have the disadvantage of poor recoveries, which vary from compound to compound and use a large amount of organic solvent. There is therefore a trend to replace solvent extraction procedure in order to minimised sample manipulation, analyte losses and the use of toxic solvents [24]. So there is an increasing tendency to replace LLE by solid-phase extraction (SPE). SPE is widely accepted as an alternative extraction/cleanup method to LLE in order to determine pollutants in liquid samples [25]. However, SPE have certain disadvantages for water analysis: the cross-sectional area is small, sample processing rates are slow, the particle blockage tolerance and adsorbed matrix components is low and channelling reduces the capacity to retain analytes. Another drawback in SPE, which is the same for LLE, is the considerable amount of time needed and the manual operations involved [26].

Other methodologies have been developed with a view to eliminating or, at least minimising, the use of organic solvents. Solid-phase microextraction (SPME) is a recent technique for the extraction of organic compounds from environmental samples. SPME is based on the partition equilibrium of analytes between a polymeric stationary phase, and the sample matrix. The desorption step is usually carried out by placing the fiber in a hit injector of a gas chromatograph [27] or it can also be performed in an HPLC system by introducing an interface [28]. An inherent disadvantage of SPME is that quantitative work is still rather laborious because severe carry-over may occur between samples. Moreover, a relatively long equilibration time for the absorption of analytes is needed.

An alternative extraction technique for liquid samples is cloud point methodology or cloud point extraction (CPE) [29]. It lets extraction and preconcentration steps based on phase separation using non ionic surfactants as extractants. It offers a convenient alternative to more conventional extraction procedures. From an analytical point of view, one of the most important properties of surfactants is their good capacity to solubilise solutes of different type and nature due to their organised structure and their anphipatic character.

Table 2 gives a description of the aforementioned techniques and lists their advantages and their disadvantages for liquid samples.

Table 2. Extraction techniques for liquid samples [from ref. 54]

	LLE	SPE	SPME	CPE
Brief description	Analyte is partitioned between two immiscible solvents	Analyte retained on a solid adsorbent	Partition equilibrium of analytes between a polymeric stationary phase and the sample matrix	Analyte is portioned between two isotropic phase 'surfactant-rich phase' and 'aqueous phase'
Extraction Time	Up to 24 h	20-30 min	Up to 60 min	20 min
Solvent Usage	Organic Solvent	Organic Solvent	None	Surfactant solution
Cost	Low	Relatively low	Relatively low	Low
Ease of operation	Easy	Relatively easy	Relatively easy	Easy
Disadvantages	Large consumption of solvent and concentration of sample required after extraction	Insufficient retention of very polar compounds and limited selectivity	Choice of fiber and optimisation of procedure required	Optimisation of operating conditions

The extraction of organochlorine and organophosphorus pesticides from solid samples requires the use of organic solvent, which competes in the releasing of the analytes retained owing to the high activity of the matrix. Traditional methods employ large volumes of solvent under aggressive shaking and/or temperature conditions. The most frequently used method for the extraction of organic compound from soils is Soxhlet extraction or the use of an ultrasonic bath. Soxhlet extraction, the most conventional of all methods, is particularly suitable for organic pollutants strongly absorbed in soil matrices but it requires long extraction times (from 24 to 48 hours) and the use of large volumes of frequently toxic organic solvents (300 to 500 ml per sample) [30, 31].

Sonication extraction is faster than Soxhlet extraction (30 to 60 min per sample) and allows the extraction of large amounts of sample, but it still uses about as much solvent as the Soxhlet extraction [32].

The newer extraction techniques such as supercritical fluid extraction (SFE), based in the extraction of target compounds with solvents near supercritical conditions [33]; accelerated solvent extraction or pressured liquid extraction (ASE or PLE), which extracts target compounds using solvents at high pressures and high temperatures [34] and microwave assisted solvent extraction (MAE), based in the extraction by microwave energy and temperature [35,36]. All of them are very attractive because they are faster, use much smaller amounts of solvents and are more environmentally friendly techniques than traditional ones [37].

In the last few years, the number of procedures using extraction of organic compounds from environmental matrices by microwave energy has increased [38, 39]. Microwave assisted extraction (MAE) allows rapid extraction of solutes from solid matrices by employing microwave energy as a heat source, with an extraction efficiency comparable to that of the classical techniques. It exhibits many substantial improvements in analytical sample preparation, as it requires much lower volume of organic solvent (30 ml in MAE versus 300 - 500 ml in Soxhlet extraction), reduces extraction time (20 to 30 min per sample) and lets prepare multiple samples in one step [40-42]. Optimisation of MAE conditions is rather easy, as there are few parameters (nature of solvent, time and power) and it is cheaper than other modern techniques, such as supercritical fluid extraction. The main drawbacks of this system are: extraction solvent has to be able to absorb microwaves, a clean-up procedure may be required before analysis and the vessels need to be cooled to room temperature after extraction and before they can be opened, thus increasing the overall extraction time.

Table 3. Extraction techniques for solid samples [from ref. 54]

	Soxhlet	Sonication	SFE	PLE	MAE
Brief description	Sample is placed in an extraction thimble and leached with hot solvent	Sample is placed in a ultrasonic bath with an organic solvent, with shaking	Sample is placed in a high-pressure cartridge and extracted with a supercritical fluid	Sample is placed in an extraction vessel and pressurized with a solvent heated	Sample is placed immersed in solvent, and heated with microwave energy
Extraction Time	Up to 24- 48 h	30-60 min	30-60 min	15 min	20-30 min
Solvent usage	Organic solvent	Organic solvent	Solvent in its supercritical state	Organic solvent at high temperature and pressure	Organic solvent heated with microwave energy
Cost	Low	Low	High	High	Moderate
Ease of operation	Easy	Easy	Relatively easy	Relatively easy	Relatively easy
Dis-advantages	Large amount of solvent and evaporation of extract are neccesary	Large amount of solvent and filtration are required	Limited sample size, and dependence of matrix and analyte type	Dependence on matrix	Clean-up procedure is required

This methodology has been applied to the extraction of different organic compounds, such as organochlorine insecticides [43], polynuclear aromatic hydrocarbons, [44], arochlors [45], etc. with good results. In all these studies, the extractant has always been an organic solvent.

Table 3. provides a summarised description of these techniques and show their advantages and disadvantages for solid samples.

SURFACTANTS

An alternative to the use of organic solvents would be the use of micellar systems (surfactants) as extractants.

Organic solvents are toxic compounds which are dangerous to researchers who work with them and also for environment. For this reason it is necessary look for new extraction solvents that will be less toxic for environment and work's atmosphere. In this sense, surfactants can be used as extractants in the analysis of organic pollutants like organochlorine and organophosphorus pesticides in different kind of matrices.

Surfactants are known as tensoactives, whose main characteristic is their anphipatic structure; the presence in a same molecule of two or more groups with anthagonic properties respect the solubility of them [46]. A monomer of surfactant has an structure R-X where R is a hydrocarbon chain between 8 and 18 carbon of hydrophobic characteristics and X is a polar group of ionic or non-ionic nature (Figure 1).

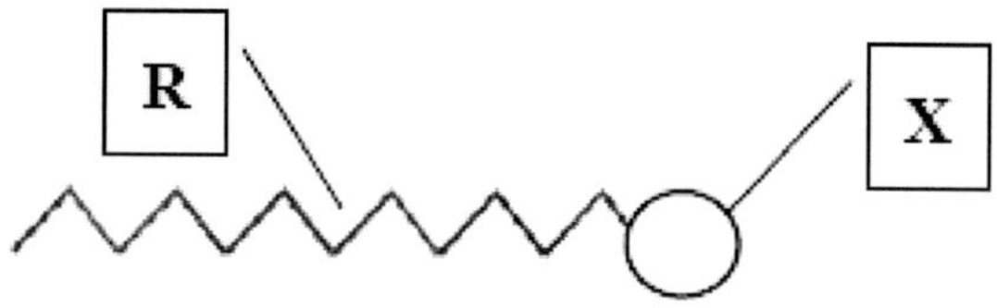

Figure 1. Scheme of surfactant's monomer molecular structure.

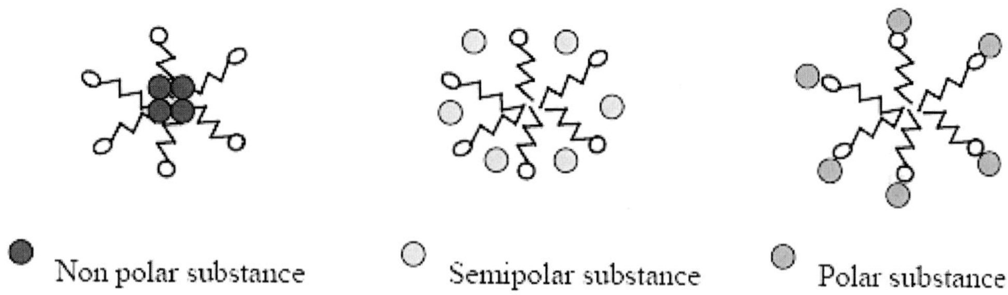

Figure 2. Micellar aggregated depending on substance nature

Depending on the nature of hydrophilic head, surfactants are classified in four groups: Cationic, Anionic, Zwitterionic (positive and negative charge) and Non-ionic [47]:

Due to the anphipatic character of surfactants, their molecules in presence of water are absorbed in interfaces, decreasing interfacial pressure and forming organized structures, called 'micelles' when surfactant concentration are higher than critical micellar concentration (c.m.c.).

Main property of surfactants, related with their capacity to micelle formation, it is analytes solubilization. Micelles, formed in aqueous solution (direct micelle) or in non aqueous solution (reverse micelle) can solubilize polar and non polar molecules. In next scheme it is showed a micellar aggregated depending on their nature, polar, semipolar or non polar (Figure 2).

Solubilization is located in different points depending on the nature of solubilized substance, but also depending on surfactant. In non-ionic surfactants the solubilization is generally in hydrophilic heads [47]. This capacity for solubilization produces that micellar solutions have important applications in fields like chemistry, pharmacological, cosmetics, etc. [48]

From the analytical point of view, this capacity to solubise solutes of different types and nature as well as their compatibility with the mobile phase in HPLC are the most important properties. Moreover, other advantages of surfactants are safety, low cost and no toxicological effects due to the biodegradability [49, 50].

For liquid samples, the use of extraction and preconcentration steps based on phase separation using the cloud point methodology offers a convenient alternative to more conventional extraction systems. Aqueous solutions of some surfactants have been used in cloud point extraction (CPE) of different species prior to their determination by several techniques [51-53]. The small volume of the surfactant-rich phase obtained in CPE process allows to extract and preconcentrate the analytes in one step, prior to gas or liquid chromatographic analysis (Figure 3).

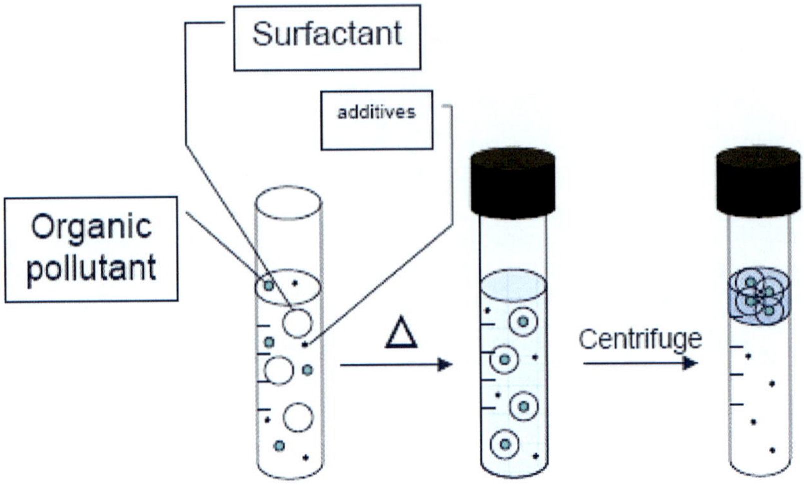

Figure 3. Squeme of the Cloud-Point process .

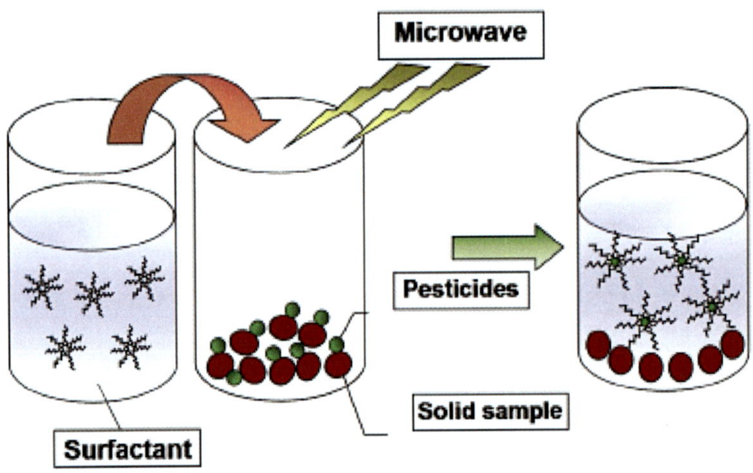

Figure 4. Squeme of Microwave Assisted Micellar Extraction process.

For solid samples, using the combination of the surfactant capacity to solubilise different species and the energy supplied by microwave assisted extraction, hereinafter microwave assisted micellar extraction (MAME), it can be applied to the extraction of different compounds present in solid samples [54], which always offers the benefit of low cost, easy handling, reduced toxic effects and compatibility with aqueous-organic mobile phase in HPLC (Figure 4).

The MAME procedure have been applied to the extraction of several compounds as phenols [55], organochlorine and organophosphorous pesticides [56,57] present in different solid environmental samples such as marine sediments, soils and biological samples.

DETERMINATION OF PESTICIDES BY USING CLOUD POINT EXTRACTION (CPE)

Introduction

When a micellar solution of a non-ionic surfactant is heated, it become turbid over a narrow temperature range, which is referred to as its cloud point temperature [58]. Above the cloud point temperature, such solutions separate into two isotropic phases, one of which will contain a surfactant-rich phase with a small amount of water, "surfactant rich phase", which is separated from the bulk aqueous solution, "aqueous phase", in which the surfactant concentration will be approximately equal to the critical micelle concentration of the non-ionic surfactant present. The phenomenon is reversible and, upon cooling, a single phase is again obtained. Table 4 shows the main characteristics of the non-ionic surfactants most used in CPE.

Table 4. Characteristics of non-ionic surfactants [from ref. 54]

Surfactant	Commercial name	CMC[a]	Cloud point (°C)[b]
Polyoxyethylene 4 lauryl ether	Brij 30 ($C_{12}E_4$)	0.064	2.0
Polyoxyethylene 6 lauryl ether	$C_{12}E_6$	0.068	50.3
Polyoxyethylene 9 lauryl ether	Polidocanol ($C_{12}E_9$)	0.069	86.6
Polyoxyethylene 10 lauryl ether	Pole ($C_{12}E_{10}$)	0.090	77.0
Polyoxyethylene 8 dodecanol	Genapol X80 ($C_{12}E_8$)	0.050	75.0
Polyoxyethylene 7.5 nonylphenyl ether	Ponpe 7.5	0.085	5.0
Polyoxyethylene 9.5 octylphenyl ether	Triton X-100	0.240	64.0
Polyoxyethylene 10 cetyl ether	(Brij 56) $C_{16}E_{10}$	0.001	69.0
Polyoxyethylene 10 stearyl ether	Brij 97 ($C_{18}E_{10}$)	0.002	72.0

a) Critical micellar concentration (CMC) values given in mM/L.

b) Values given are the critical temperatures for micellar solutions in which the surfactants concentration is 1% (w/v).

The solubilisation of organic material in micelles and subsequent cloud point extraction technique thus offers a convenient alternative to conventional liquid-liquid extraction that uses organic solvents. The extraction and preconcentation of organic compounds by CPE prior to their analysis by gas or liquid chromatography has been used for different authors [51-53].

The cloud point temperature depends on the structure of the surfactant and on its concentration. For a homologous series of polyoxyethylene non-ionic surfactant, the cloud point temperature increases with decreasing length of the hydrocarbon chain or increasing length of the ethylene oxide moiety. This temperature can be modified by the presence of salts, alkalis, acids, polymers, urea and other surfactants [58]. The ionic strength can alter the cloud point temperature as well as facilitate the separation of the two phases by altering the density of the bulk aqueous phase.

It has been shown, for example, that for Genapol X-080, the cloud point temperature obtained when NaCl is added in a 5% w/v concentration, is lesser than the temperature obtained in the absence of this salt [59].

In CPE, extraction needs to be carried out under conditions in which the preconcentration factor will be the maximum or the extraction yield will be 100%. The preconcentration factor is defined by the expression $F_c = C_s/C_{in}$, where C_s is the analyte concentration in the surfactant-rich phase after phase separation, and C_{in} is the analyte concentration in the initial solution, before the preconcentration step. This depends on the phase relationship, on the distribution constant of the analyte between the phase and on the surfactant concentration used.

Moreover, the ratio between the volume of aqueous solution to be preconcentrated and the surfactant-rich phase volume (V_w/V_s) increases with the decrease in the surfactant concentration. This shows that the smaller the surfactant concentration, the higher the preconcentration factor; but when the volume of surfactant-rich phase is small, the extraction process become more difficult, and the accuracy and reproducibility probably suffer [29,60]. In this way, a compromise must be reached so that the surfactant concentration will allow a high phase ratio and a manageable volume of surfactant-rich phase.

The extraction process can be also altered by different factors such as equilibration time, pH, concentration and nature of the surfactant and addition of salt [61]. The effect of these factors on the percentage extraction of the analytes studied therefore needs to be established.

Optimization of Variables

Surfactant Concentration

The effect of the surfactant concentration is related with the surfactant capacity for the extraction of target pesticides [62]. It depends on the kind of surfactant used and on the concentration of pesticides in the sample. The range of surfactant concentration can vary between 1-10% (v/v) [56,62].

Table 5. Organophosphorus Pesticides Studied

Identification Number	*Compound*
1	Dimethoate
2	Methidation
3	Parathion methyl
4	Malathion
5	Ethoprophos
6	Parathion ethyl
7	Diazinon
8	Chlorpyrifos

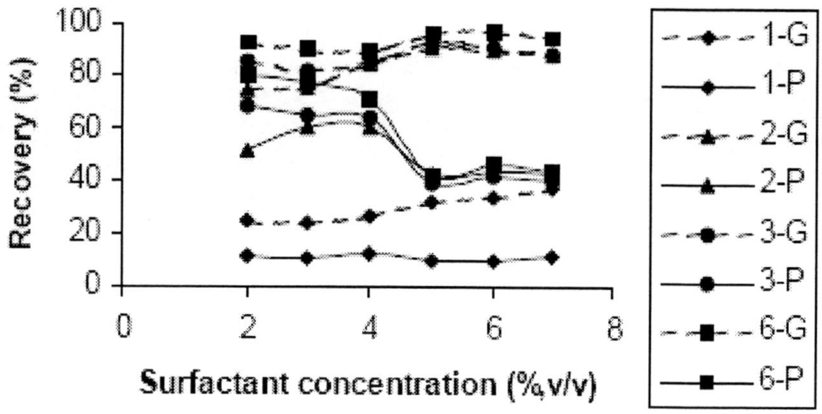

Figure 6. Effect of surfactant concentration on recovery. Dashed line, Genapol X-080 (G); continuos line, POLE (P). CPE conditions: 95°C, 15 min. Identification number in Table 5. [from ref. 56].

As an example, for a mixture of organophosphorus pesticides, it was determined by measuring a series of aqueous solutions containing the mixture of target pesticides (listed in Table 5) and different surfactant concentrations ranging from 2 to 7% (v/v) for each of the surfactants under study, POLE and Genapol (X-080).

Figure 6 shows how the recovery percentages with POLE remain practically constant up to 4% (v/v), from which point onwards they drop sharply, before stabilising again round 5% (v/v). This indicates that pesticides are not extracted as efficiently at high surfactant concentrations. The Genapol X-080 concentration can therefore be considered practically not to influence on the extraction of these analytes in the range studied.

In view of these results and taking into account that the surfactant concentration has a proportionally inverse impact on the preconcentration factor [63], the lowest possible surfactant concentrations must be used, but within a range that allowed the samples to be adequately handled.

Equilibration Time

The recovery percentage depends on the time that the analytes have to interact with the micelles and get into their core. It has been reported that longer equilibration times (more than 30 min) do not have any significant effect on the extraction parameters and that an equilibration time of 20 min is sufficient to obtain a good extraction [59,60]. It has likewise been demonstrated that the analyte preconcentration factor and percent recovery in CPE process increases as the equilibration temperature for phase separation is progressively increased above that of the cloud point temperature of the system [64]. Likewise, an increase in the cloud point temperature leads to a slight decrease in the volume of the surfactant-rich phase [61]. As the temperature, or the equilibration time, increases, the amount of water in a surfactant-rich phase decreases and hence the volume of that phase decreases. This can be interpreted in terms of the fact that as the temperature is increased, the hydrogen bonds are disrupted and dehydratation occurs.

Equilibration time studies normally range between 5 and 25 min, higher extraction times than 20 min generally don't produce better recoveries.

Figure 7 shows the results obtain for the optimization of this variable on the recovery of a mixture of organophosphorus pesticides. For this case, this variable practically does not affect the recovery of the analytes, in the case of POLE and Genapol X-080 surfactants [56].

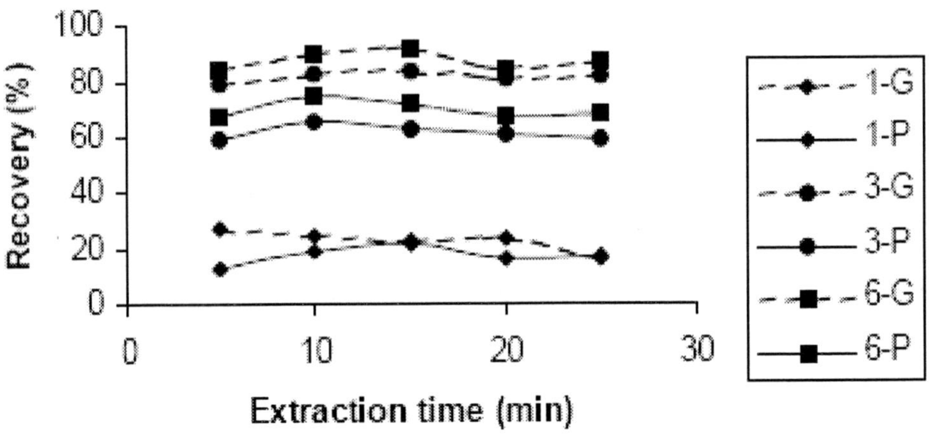

Figure 7. Effect of the extraction time on the recovery when using 2% (v/v) Genapol X-080 and 3% (v/v) POLE with 3 and 4% (w/v) of NaCl, respectively. Dashed line, Genapol X-080 (G); continous line, POLE (P). CPE conditions: 95°C, 15 min. Identification number in Table 5. [from ref. 56].

pH

The ionic form of a neutral molecule normally does not interact with the micellar aggregate as strongly as its neutral form does and a lesser amount of the analyte is therefore extracted [61, 65]. Thus, the pH should be adjusted to ensure that the neutral molecular form of the analytes is present prior to performing the CPE step. The influence of the pH on the recovery percentages is not important for those compounds that do not present ionic forms, such as organochlorine or organophosphorus pesticides. However, in the case of the phenolic derivatives (can appear as degradation products of some pesticides like 2,4-D), when working with pH modified by the addition of 1% acetic acid, the percentage extraction is better for most solutes and is the most noticeable change for more polar compounds [59].

Ionic Strength

The addition of an inert salt can facilitate the phase separation process for some non-ionic surfactant systems since it increases the density of the bulk aqueous phase [58,66]. When the salt concentration is increased, the micelle size and the aggregation number are increased and the critical micellar concentration remains constant [67]. In addition, non-polar analytes may become less soluble in the solution at higher salt concentrations and thus contribute to higher recoveries. The results obtained indicate that the addition of salt produces an increase in the extraction of the more polar solutes while the recoveries of the less polar compounds is not affected [65,68].

In the case of organophosphorus pesticides the maximum extraction for surfactant POLE is obtained with 4% (w/v) NaCl, while with Genapol X-80 practically this parameter does not affect on the recovery. It can be observed in Figure 8.

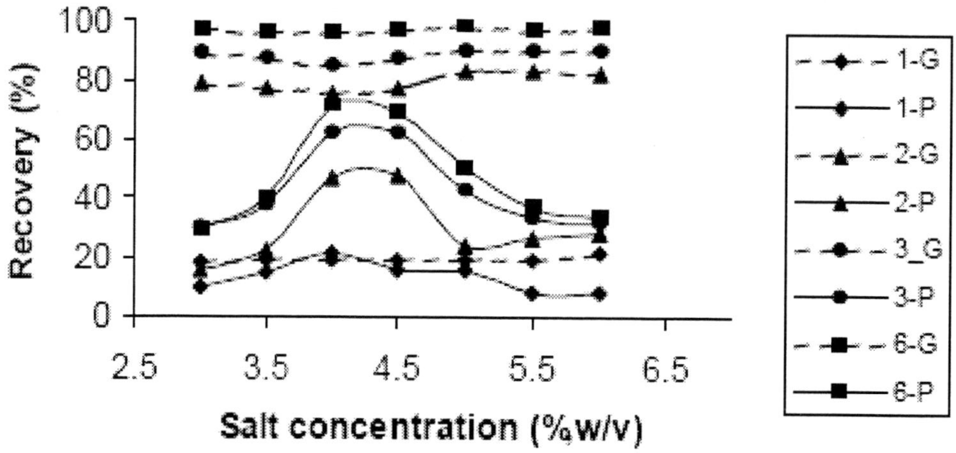

Figure 8. Effect of salt addition on the recovery when using 2% (v/v) Genapol X-080 and 3% (v/v) POLE. Dashed line, Genapol X-080 (G); continous line, POLE (P). CPE conditions: 95°C, 15 min. Identification number in Table 5. [from ref. 56].

Analytical Parameters and Applications

Most analytical applications of CPE methodology for the extraction of organic compounds make use of reversed-phase high performance liquid chromatography (HPLC). The surfactant-rich phase obtained in the extraction process is compatible with the hydro-organic phase usually employed in this chromatographic mode. From this, another important step is the optimization of the chromatographic conditions. There are two important factors to be taken into account in liquid chromatography separations: the analysis time and a good separation of the analytes.

The CPE methodology has been successfully exploited for the preconcentration of pesticides as a step prior to their determination by HPLC using non-ionic surfactants [56]. Figure 9, shows a chromatogram obtained for the mixture of the organophosphrus pesticides in Genapol X-080 solution, after CPE process was applied.

Relative standard deviation values are listed in Table 6, where values $\leqslant$5.5 and 2.6% were obtained for POLE and Genapol X-080 respectively, in the case of organophosphorus pesticides extraction. The limits of detection were also calculated for each analyte using a signal-to-noise ratio (s/n = 3) [69] once the CPE method was fully applied. The results obtained are also listed in Table 6.

In order to corroborate its efficiency, the method was applied to a certified standard containing the pesticides: Parathion-methyl, Parathion-ethyl, Malathion and Diazinon (Pesticide Mix 1, Dr. Ehrenstorfer). The results (Table 7) show the satisfactory recoveries obtained using the optimised CPE procedure.

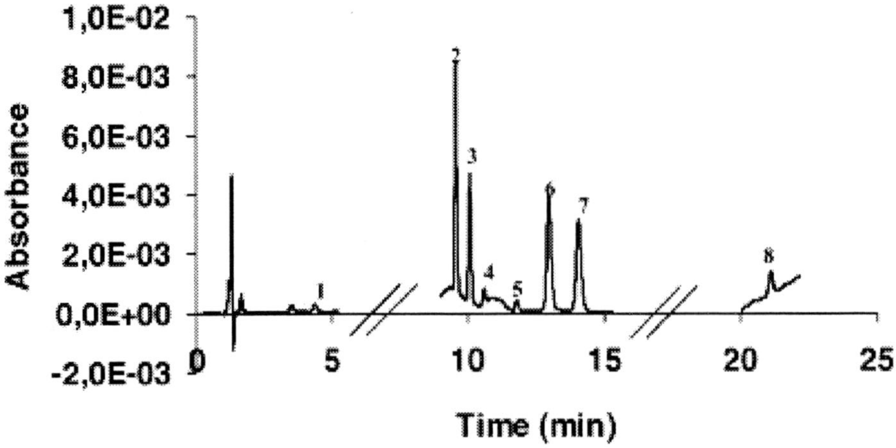

Figure 9. Chromatogram of the organophosphorus pesticides in the presence of 5% (v/v) Genapol X-080. Pesticide concentration: 2mg L^{-1}; Detection wavelength: 247 nm. Identification number in Table 5. [from ref. 56].

Table 6. Analytical parameters for the cloud point extraction method of organophosphorous pesticides [from ref. 56]

	POLE		Genapol X-80	
Compounds	R.S.D[a] (%)	L.O.D.[b](ngml^{-1})	R.S.D[a] (%)	L.O.D.[b](ngml^{-1})
Dimethoate	1.3	1.86	0.6	0.61
Methidathion	2.5	1.65	0.8	1.37
Parathion methyl	3.2	1.86	2.1	0.03
Malation	0.5	28.45	0.7	1.37
Ethoprophos	5.5	0.88	1.4	1.84
Parathion ethyl	1.6	2.03	2.6	0.36
Diazinon	2.8	2.96	1.0	0.47
Chlorpyrifos	2.4	3.54	2.2	2.81

a) Relative standard desviation (n = 6)
b) Limit of detection (s/n = 3)

Table 7. Recoveries obtained on the application of the CPE method using Genapol X-080 to a certified standard pesticide mixture [from ref. 56]

Compound	Recovery (%)[a]	
	Added amount (10 ng·ml^{-1})	Added amount (100 ng·ml^{-1})
Parathion methyl	81.7 ± 2.3	80.5 ± 1.1
Malation	75.0 ± 0.0	66.0 ± 0.7
Parathion ethyl	80.4 ± 1.6	84.5 ± 1.1

a) Mean ± S.D. (n = 3)

The proposed method can be applied to real samples for the determination of pesticide's concentration in them. As example, it was applied to various spiked aqueous samples from different locations in Gran Canaria island (Canary Islands, Spain). Samples of depurated waste water and groundwater were employed in this study. After checking the absence of pesticides in the samples, they were spiked with the analytes under study (6–30 ng.ml^{-1}) and the established method with Genapol X-080 was applied to them. The results obtained for all the samples analysed are set out in Table 8. Recoveries between 70 and 100% were obtained for the majority of cases. Only Dimethoate and Ethoprophos were extracted with recoveries <50%.

Table 8. Recoveries obtained on the application of the CPE method using Genapol X-080 to pesticides added to real aqueous samples [from ref. 56]

Compound	Added amount (ng ml^{-1})	RECOVERY (%)[A]	
		Depurated waste water	Groundwater
Dimethoate	10	27 ± 0.5	29 ± 1.5
Methidathion	10	86 ± 2.2	85 ± 3.1
Parathion methyl	6	92 ± 2.0	83 ± 1.2
Malation	10	87 ± 1.1	81 ± 0.6
Ethoprophos	30	43 ± 1.8	47 ± 2.3
Parathion ethyl	6	105 ± 1.2	100 ± 1.6
Diazinon	6	89 ± 1.0	84 ± 0.0
Chlorpyrifos	30	99 ± 3.6	99 ± 5.0

a) Mean ± S.D. (n = 3)

DETERMINATION OF PESTICIDES BY USING MICROWAVE ASSISTED MICELLAR EXTRACTION (MAME)

Introduction

In recent years, the use of microwave technology for sample preparation has been widely accepted. The reasons for the large dissemination of microwave-assisted extraction (MAE) relate to its clear advantages over more traditional technology:

- a shorter digestion time
- a supposed better recovery of volatile elements and compounds
- lower contamination levels
- minimal volumes of reagent being required
- more reproducible procedures
- better working environment

Ganzler et al. [70] were the first researchers ever to use a microwave oven to extract organic compounds from contamined soil. Since then microwave digestion methods have been developed for different types of matrices to extract organic contaminants, such as PAHs [23, 71], PCBs [72,73], pesticides [73,74], phenols [75, 76], etc. using an organic solvent as extractant.

It has been reported that the combination of MAE with micellar media as extractants (MAME) allows the extraction of different organic compounds from solid samples [77-80]. The extraction process starts with the loading of the sample into the extraction vessel, followed by surfactant addition and closing of the vessel. Microwave radiation is applied and an extraction step is initiated in order to heat the extractant to set values. When the extraction is concluded the samples are allowed to cool down to a temperature that is reasonable to handle. Prior to the analysis, a clean-up step might be needed. In most applications, simple filtration of the extract using a membrane syringe filter is enough.

Optimization of MAME Process

There are different parameters that can influence the extraction process when MAME is used. The most commonly studied parameters are: nature and volume of surfactant, extraction temperature or extraction power and duration of extraction. Factorial, central composite and orthogonal array can be used to find optimal conditions.

Nature and Volume of Surfactant

A correct choice of kind of surfactant is fundamental for obtaining an optimal extraction process. When selecting the extractant, consideration should be given to its interaction with the matrix, as well as the solubility of the analyte. Another important aspect is the compatibility of the extraction surfactant with the analytical method used for the final analysis step. The non-ionic surfactant solutions should be effective in desorbing and extracting organic compounds from solid matrices due to their power of solubilisation. In addition, the surfactant phase is compatible with the hydro-organic mobile phases usually employed in HPLC.

In some cases, extractant solvent volume may be an important parameter for efficient extraction. The surfactant volume must be sufficient to ensure that the entire sample is immersed, especially when having a matrix that will swell during the extraction process. Research has led to the conclusion that the proportion of sample in the extraction solution should not exceed 30-34% (w/v) sediment/solvent ratio. In conventional extraction techniques, a higher volume of solvent will generally increase the recovery, but in the case of MAME, a higher surfactant volume does not influence the extraction efficiency. Therefore, if the sample is full immersed in the extractant, an increase of the surfactant volume does not seem to have any influence on the extraction percentages [77,81]. Normally at extraction volumes lower than 5 ml irreproducible data are obtained (due to the insufficient covering of the extractant) and at volumes higher than 20 ml some evaporation losses take place (due to the high temperatures reached). This last effect maybe due to the high capacity of the aqueous surfactant solutions in absorbing the microwave radiation and transforming it into heat [82].

As example, for MAME extraction of organophosphorus pesticides using POLE and Genapol X-080 surfactants, the optimum extraction volume is around 10 ml [78] and for organochlorine pesticides using POLE, 8 ml [79], a very similar value in both cases.

Surfactant Concentration

Surfactant concentration is one of the most important parameters that needs to be optimized in MAME process. Surfactant concentration have to secure that the amount of micelles is enough for a efficient interaction with analytes and a correct solubilization, but it should be not too high due to possible viscosity problems.

Optimum surfactant concentration can be determined using sequential optimization steps or applying factorial designs.

For example, to study the effect of surfactant concentration on the recovery percentage of organophosphorus pesticides, several samples containing different POLE and Genapol X-080 concentrations, 1, 3, 5 and 7% (v/v), were analyzed using sequential mode. The pesticides from enriched samples were extracted with both surfactant solutions in the microwave system at 300 W during 5 min [78]. The recoveries obtained for some of the compounds studied can be seen in Figure 10.

For the extraction of organochlorine pesticides, the results of the experiment design applied and the correlation between variables with the recovery and each other show that the surfactant concentration is the variable which mostly affect to the recovery and is correlated with the microwave power. The results obtained for the case of 4,4'-DDT using a mixture of surfactants POLE and Polyoxyethylene 10 Stearyl Ether (Stearyl mixture) are shown in Figure 11 [79].

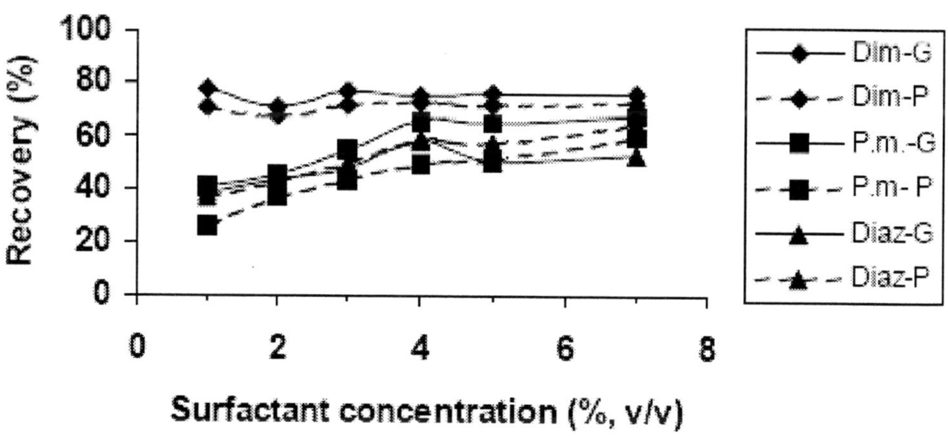

Figure 10. Effect of surfactant concentration on the recovery of the pesticides under study. Continous line (—): Genapol X-080 (G); dashed line (- - -): POLE (P). Dim: Dimethoate, P.m.:parathion-methyl and Diaz:diazinon. [from ref. 78].

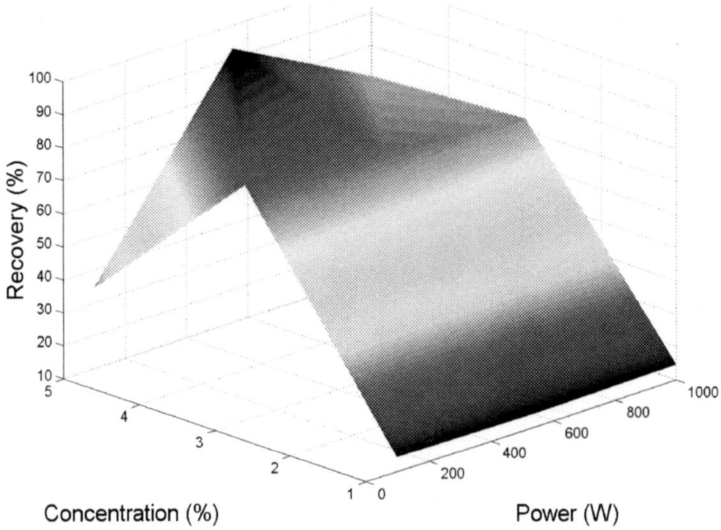

Figure 11. Response surface for the effect of surfactant concentration and power on the extraction of 4,4'-DDT using Stearyl mixture. [from ref. 79].

Extraction Temperature and Extraction Power

The most researched MAME parameter is the extraction temperature or extraction power, as an increase in the power involves a corresponding increase in heat. When MAME is conducted in closed vessels, the temperature may rise well above the boiling point of the solvent at atmospheric pressure, thus enhancing both extraction speed and efficiency. These elevated temperatures result in improved extraction efficiencies, since desertion of analytes from active sites in the matrix will increase. However, the effect of the temperature depend on the type of analyte as well as the type of matrix.

As in other extraction techniques, time is another parameter whose influence needs to be taken in account. Extraction times in MAME are very short compared to conventional techniques where extraction procedures require as much as 24 hours. 30 minutes are often sufficient to obtain good results in MAME. With thermolabile compounds, long extraction times may result in degradation, obtaining lower recovery percentages when increase extraction time [80].

A central composite design can be used in order to study the effect on the recovery, it allows the direct evaluation of the considered variables [83]. The values normally vary between 200 to 800 W during 2 to 14 min [78]. The optimization of both variables can be done using a 3^2 factorial design, where the recovery is the dependent variable. The experiment involves 10 run by duplicate and other variables involved in the extraction process were kept constant in their optimized values.

Figure 12a shows the response surface for 4,4'-DDT using Cetyl mixture (a mixture of POLE and Polyoxyethylene 10 Cetyl ether) as extractant, where we can observe that from 750 W for microwave power and during 14 minutes the best recoveries are obtained; for Stearyl mixture (Figure 12b) the optimum conditions are 775 W and 8 minutes .

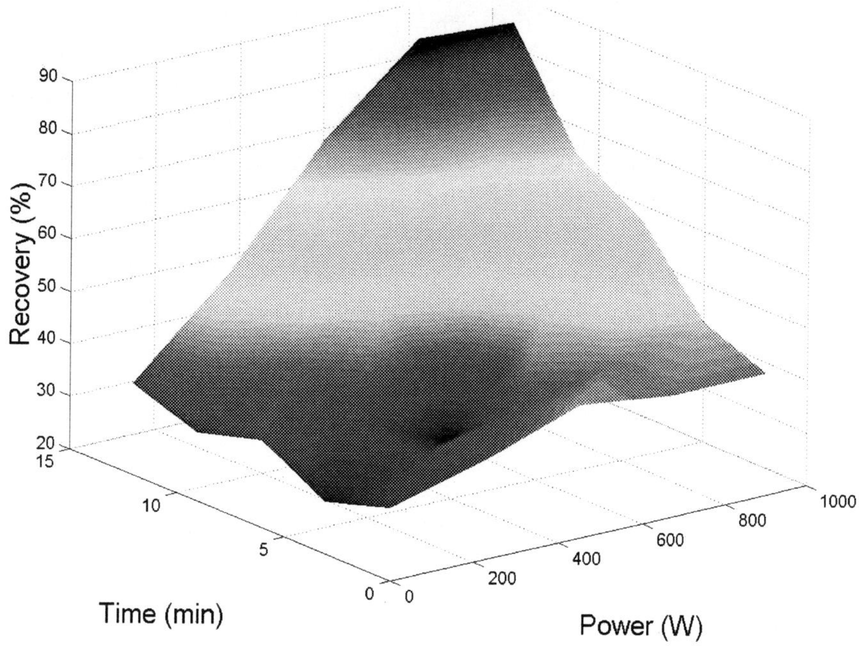

a)

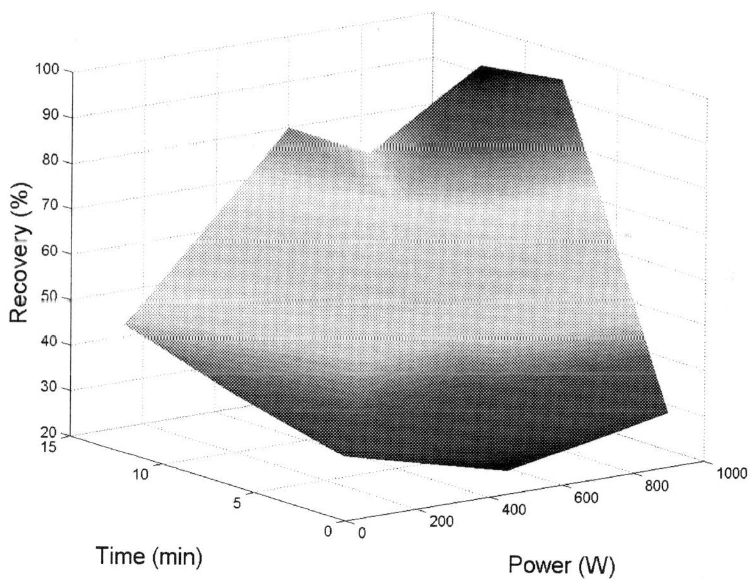

b)

Figure 12. Response surface for the effect of power and time on the extraction of 4,4'-DDT using Cetyl mixture (a-left) and Stearyl mixture (b-right) [from ref. 79].

Matrix Characteristic

The nature of the matrix in which the analytes of interest are bound can have a profound effect on the recoveries of the compounds. The different interactions that the organic pollutants present with the organic and inorganic matter of the matrix can play different roles in their solubility or diffusivity into the micellar medium.

The pH, conductivity and organic content of the sample can vary the recoveries obtained. Moreover, the texture of a soil is extremely important in the sorption process. When the particles are small they present a high superficial area thus increasing the adsorption, although this adsorption process depends on the nature of the analytes. In this sense, a high variability can be observed in the recoveries obtained with the different granulometry of the soil samples. In general, it can be said that the proposed method is applicable in the sense that it enables the extraction of the pesticides under study even if the recovery efficiency depends on the characteristics of the soil as well as the surfactant to be used.

Appling the method to different soil samples containing a mixture of organochlorine pesticides can be observed that recoveries vary depending on the soil characteristics (Table 9) and the surfactant used as extractant [79]. The results obtained are shown in Tables 10a and 10b.

Table 9. Soil characteristics (from ref. [79])

Soil	Agricultural Uses	Particle size (%)				pH	Organic matter (%)	Conductivity (μS/cm)
		0.3 mm	0.2 mm	0.15mm	≤0.1 mm			
Tafira	Garden	44.4	17.2	14.6	23.8	8.3	4.8	292
San Roque	Bananas	52.1	20.0	13.2	14.6	7.86	12.54	3266
Sta Brigida	Pine Forest	56.5	17.8	12.8	12.8	5.9	3.9	483
Valleseco I	Potatoes	36.33	16.43	15.19	32.06	4.84	4.4	202
Valleseco II	Potatoes	25.16	13.17	14.38	47.29	3.91	6.21	100

Table 10a. Recoveries obtained after MAME procedure for organochlorine pesticides in five different soils using Cetyl mixture*. [from ref. 79]

Soil	4,4'-DDD	Dieldrin	4,4'-DDT	2,4'-DDT	4,4'-DDE	Aldrin
Tafira	99.1	66.2	67.7	58.6	65.0	63.0
San Roque	69.4	44.7	78.0	59.4	67.1	52.6
Sta Brigida	96.8	80.7	84.0	75.1	73.1	63.8
Valleseco I	78.3	73.3	76.5	66.3	65.3	64.4
Valleseco II	65.2	72.3	61.7	59.4	56.1	60.2

* Mean of three determinations

Table 10b. Recoveries obtained after MAME procedure for organochlorine pesticides in five different soils using Stearyl mixture*. [from ref. 79]

Soil	4,4'-DDD	Dieldrin	4,4'-DDT	2,4'-DDT	4,4'-DDE	Aldrin
Tafira	103.6	78.6	101.8	62.8	71.8	79.4
San Roque	101.3	46.1	94.4	86.2	76.0	80.0
Sta Brigida	84.7	80.8	78.8	62.4	66.0	65.6
Valleseco I	55.2	56.3	100.7	83.5	44.8	57.5
Valleseco II	54.3	50.2	98.3	68.4	49.2	72.8

* Mean of three determinations

The soils with high organic matter have good recoveries because the surfactant can also extract humic substances which are linked to the pesticides [84]. These recoveries decreases when the temperature is too high (observed in the most extremely conditions in the experimental design).

As we commented before, texture of a soil is quite important in the sorption process [85]. Clay is by far the most adsorbent of the three main soil textures (clay, silt and sand) due to its small particle size, high surface area, and high surface charge [86]. This has a relevant effect on the recovery; the higher clay percent a soil has, the lower the recovery is so; in this case, the Valleseco II soil has a higher amount of particles with smaller size and therefore, lower recoveries are obtained.

The effect of ageing the samples on the recovery of the analytes is a known phenomenon [87]. Decreasing recoveries resulting from aging of matrices is frecuent [88,89]. The analytes present in recent soil samples are more easily extracted than those that have had a longer contant time. This can be explained according to whether the analytes are incorporated by adsorption (short periods) [90] or by sequestration (long periods). The former phenomenon occurs at the early stages of sorption, where H-bonding and Van der Waals forces prevail. On the other hand, sequestration involves sorption at remote microsites within the soil matrix [91]. As the contact time increases between the analyte and the matrix the general recuperation of all the analytes under study was reduced which could be explained by the greater interaction between them [92]. This effect could also be caused by a process of degradation because as is well known the degradability of this family of compounds is far greater than that of organochlorine pesticides. Among the different pathways of organophosphates decomposition (hydrolysis, photolytic oxidation, microbial transformations, etc.) [93,94], hydrolysis is the most common degradation process which can occur at several reactive centres in a given organophosphorus pesticide molecule [95], but also can occur different degradation mechanisms including the presence of humic acids, metal oxides and the soil structure [94-97].

In the case of the extraction of organochlorine pesticides, MAME procedure was applied to four soils for different time periods after conditioning. Figure 13a shows the recoveries obtained for the different pesticides in the case of Valleseco I soil using Cetyl mixture with the time aging and Figure 13b shows the results obtained in the case of San Roque soil using Stearyl mixture with the same aging time.

It can be seen that the recoveries decrease with the aging time for all compounds. Only in the case of dieldrin using Stearyl mixture, the recovery remains practically constant with the time. Moreover, none of the metabolites of DDT increases with the time then we can assert that degrading processes were not happen in these aging time. These results indicate that the developed method leads determine the presence of pesticides in real samples after a contact time and show the viability of the method as alternative to the conventional ones.

Analytical Parameters and Validation

Separation and detection of pesticides have been developed using MAME procedure coupled with HPLC. The calibration curves can be obtained by injecting standard solutions containing a known concentration of the pesticide mixture and the surfactant chosen into the chromatographic system.

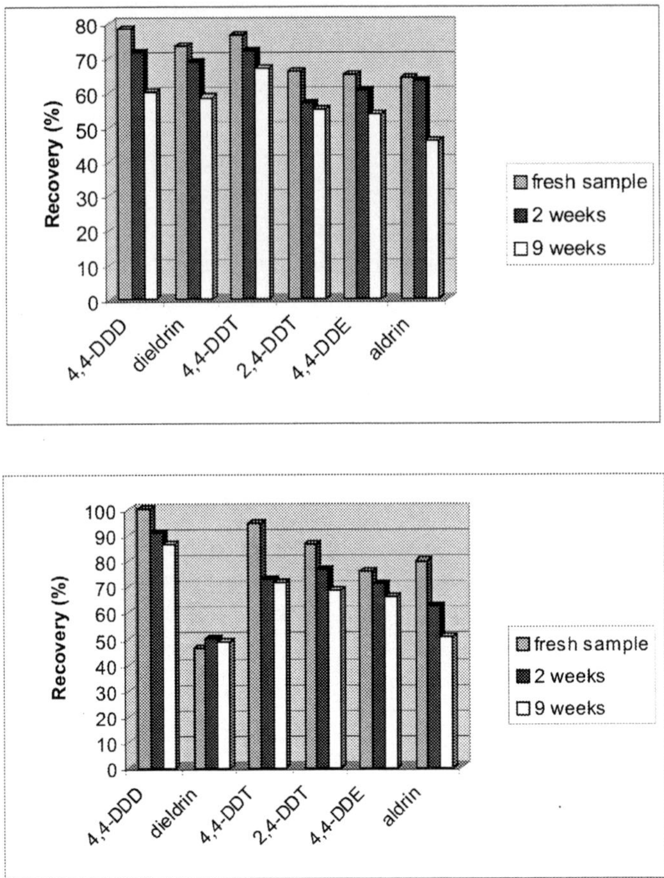

Figure 13. Recoveries obtained after MAME procedure for organochlorine pesticides for three different aging times for Valleseco I soil using Cetyl mixture (a-left) and San Roque soil using Stearyl mixture (b-right) [from ref. 79].

This method provides optimum detection limits and reproducibility in soil matrices. For organophosphorus pesticides, on the application of this process a linear relationship is obtained in the interval 100–2500 ng ml^{-1}, when POLE or Genapol X-80 are used as extractants, with high correlation coefficients (0.999) for all pesticides.

Relative standard deviation lower than 2.1 and 2.6% for Genapol-X080 and POLE repectively and LODs lower than 0.2 in both cases were obtained (Table 11). Figure 14 shows the chromatogram of the pesticides mixture extracted from a soil sample under the optimized conditions for POLE.

Table 11. Analytical parameters for the determination of organophosphorus pesticides under study using MAME procedure [from ref. 78]

	POLE		*Genapol X-080*	
Compound	RSD[a] (%)	LOD[b] ($ng{\cdot}ml^{-1}$)	RSD[a] (%)	LOD[b] ($ng{\cdot}ml^{-1}$)
Dimethoate	0.4	0.2	1.8	2.4
Methidathion	2.2	4.5	2.1	3.0
Parathion/methyl	1.1	2.1	1.2	4.8
Malathion	2.2	14.3	1.1	11.3
Ethoprophos	2.6	95.0	1.0	0.8
Parathion-ethyl	1.6	0.7	1.6	1.8
Diazinon	1.2	2.5	0.8	0.9
Chlorpyrifos	0.3	6.7	1.1	0.2

a) Relative standard deviation (n = 6)
b) Limit of detection

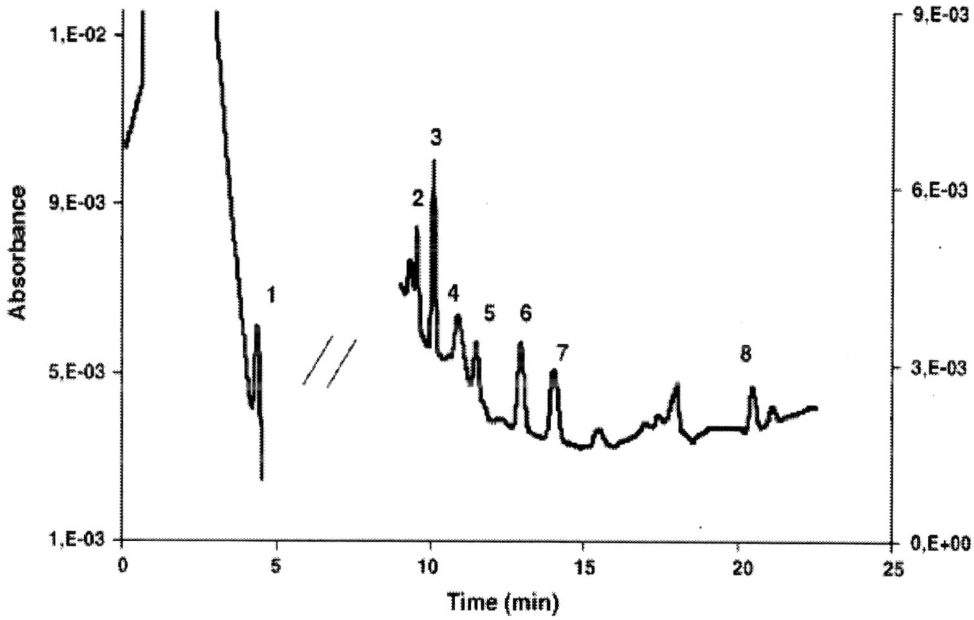

Figure 14. Chromatogram of the organophosphorous pesticide mixture extracted from soil sample n° 1 with POLE at the optimized conditions. Identification number in Table 5 [from ref. 78].

In the case of organochlorine pesticides, a good success was obtained appling MAME-HPLC method with POLE-Cetyl and POLE-Stearyl mixtures as extractants, with RSDs values between 5.2-8.4% for POLE-Cetyl and 5.2-9.7 for POLE-Stearyl and LODs of 86.4-806.4 $ng{\cdot}ml^{-1}$ and 150.8-734.0 $ng{\cdot}ml^{-1}$, respectively [79]. In Figure 15 we can observe the chromatogram obtained for the mixture of organochlorine pesticides extracted of a spiked soil using Steary mixture as extractant.

Recoveries obtained with spiked compounds may not be representative of those found with native compounds. Spiked analytes are generally lightly coated on the surface of the matrix whereas native ones can be strongly adsorbed inside the porous matrix. This can be explained by the diffusional and the kinetic limitations of the sorption process, and the several interactions which may have been simultaneously established between native analyte and the matrix [98]. This is the reason why it is necessary to validate the extraction procedure with certified reference matrices.

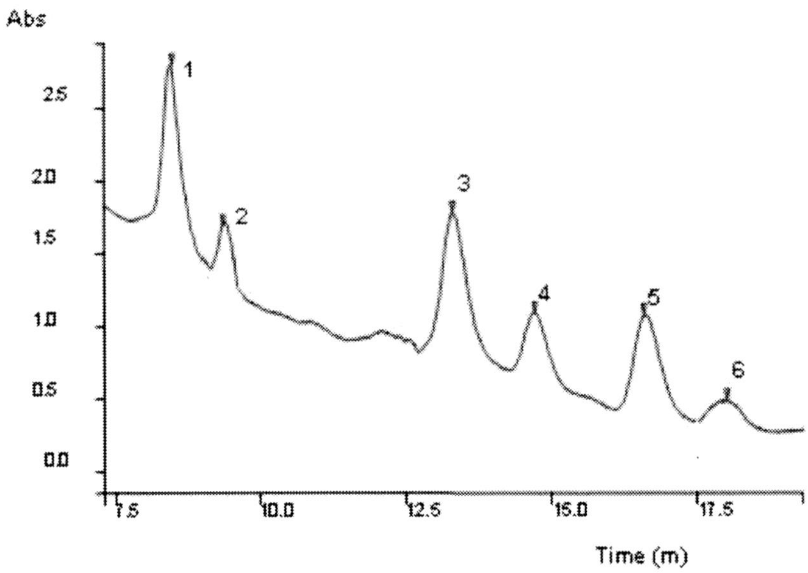

Figure 15. Chromatogram of an extract of a spiked Tafira soil after MAME procedure. (1- 4,4'-DDD; 2- Dieldrin; 3- 4,4'-DDT; 4- 2,4'-DDT; 5- 4,4'-DDE; 6- Aldrin) [from ref. 79].

Table 12. Application of MAME procedure and Soxhlet extraction to a soil sample containing a certified mixture of organophosphorous pesticides, (Pesticide Mix 1, EPA Method 914)* [from ref. 78]

	MAME		*Soxhlet*	
Compound	Added (μg g^{-1})	Found (μg g^{-1})	Added (μg g^{-1})	Found (μg g^{-1})
Parathion/methyl	1.50	1.39 ± 0.02	2.00	2.04 ± 0.02
Malathion	1.50	1.27 ± 0.04	2.00	1.89 ± 0.11
Parathion-ethyl	1.50	1.49 ± 0.12	2.00	1.85 ± 0.02
Diazinon	1.50	1.46 ± 0.02	2.00	1.64 ± 0.05

*) Mean of three determinations

Due to the lack of a certified material available containing target organophosphorus pesticides, the developed extraction method was applied in a soil sample enriched with a certified mixture, Pesticide Mix 1, EPA Method 914, and the results obtained are shown in Table 12.These results were compared with those found using the traditional Soxhlet extraction procedure as proposed by the EPA in the 3540 Method [82], finding an important similarity of results in both methods.

For organochlorine pesticides, a contaminated certified reference soil from an agricultural region of the Western United States (CRM804-050) was used. Table 13 reports the pesticides present in the certified soil, the corresponding certificated values, the confidence intervals and the concentration of each analyte that was obtained using the MAME procedure with Stearyl mixture. The concentrations obtained, with the proposed method, falling within the certified range for all compounds analyzed, indicate that the proposed extraction procedure is suitable for this kind of matrix.

Table 13. Extraction of organochlorine pesticides compounds using MAME with Stearyl mixture in a certified reference soil[a]. [from ref. 79]

Compound	Reference Value[b]	RSD (%)	Confidence Interval	Concentration found
4,4'-DDD	1.531	0.476	1.294 – 1.767	1.733 ± 0.009
Dieldrin	1.863	0.655	1.539 – 2.186	2.242 ± 0.084
4,4'-DDT	1.060	0.275	0.926 – 1.195	1.079 ± 0.015
4,4'-DDE	1.520	0.410	1.325 – 1.715	1.502 ± 0.022

aAll values are expresed in µg·g-1

bThe pesticides values in the sample were certified by USEPA SW846 (3rd edition). Extraction Methods 3540A/3541 (soxhlet), 3550 (sonication) and Analysis Method 8081

Determination of Pesticides by Using Coupling MAME-SPME and MAME-SPE

Coupling MAME with Solid Phase Microextraction (MAME-SPME)

Solid phase microextraction (SPME) is a new variation of extraction techniques which has been used mainly for the analysis of pollutants in environmental water samples [99], but this methodology can be coupled with other techniques, as the microwave extraction, for the analysis of soil samples [100]. SPME allows the simultaneous extraction and preconcentration of analytes from a sample. The principle of SPME is equilibration of the analytes between the sample matrix and the organic polymeric phase usually coating a fused-silica fibre. The amount of the analyte absorbed by the fibre is proportional to the initial concentration. Figure 16 shows the scheme of SPME process.

Normally the application of SPME has been performed in combination with gas chromatography, however, it is possible coupled with HPLC [101]. The advantage using HPLC is that compounds with low volatility or thermal labile compounds can be analysed. An organic solvent (static desorption) or the mobile phase (dynamic mode) is used to disorb the analytes from the SPME fiber.

One application of solid phase microextraction (SPME) procedure can be a clean-up and preconcentration steps applied to extracts obtained with other extraction systems such as microwave assisted micellar extraction (MAME). Using HPLC-UV as determination system, the coupling MAME-SPME has been used for the analysis of five organochlorine pesticides (4,4'-DDD, 4,4'-DDT, 2,4'-DDT, 4,4'-DDE, Dieldrin and Aldrin) in agricultural soil samples

[57,102]. A non ionic surfactant, POLE, was used, and the different variables for the optimization of MAME and SPME procedures were studied. This method was successfully applied to the determination of these pesticides in several kinds of agricultural soil samples, obtaining lower detection limits than determining MAME extract directly.

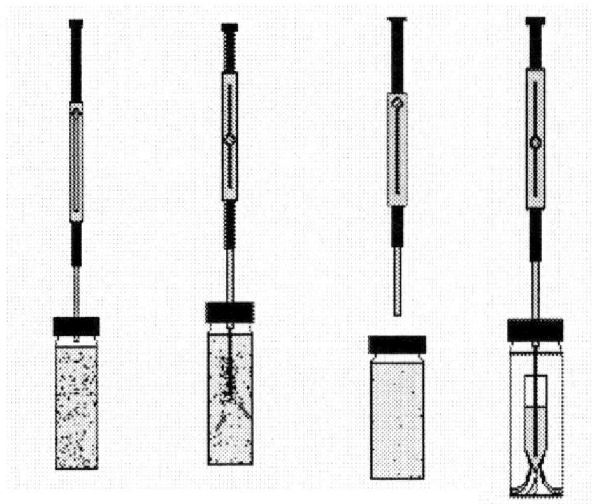

Figure 16. Scheme of the solid phase microextraction (SPME) process.

Microwave Assisted Micellar Extraction

Once the soil sample is transferred to the vessel, the optimized volume of surfactant is added and the solid sample is subjected to the MAME process [15, 78]. The vessels are closed for avoiding the loss of extract during the extraction process at microwave power and time optimized; after this, the vessels are allowed to cool first before being opened.

Optimization of SPME Procedure

For MAME-SPME extraction, a fraction of MAME extract is diluted with water and introduced in a vial for the SPME extraction under the optimized conditions [103]. The desorption of the compounds is done in a glass vial with a small volume of desorption solvent (100 μl of methanol) during several minutes.

The efficiency of analytes extraction by SPME can vary widely depending upon the matrix nature, the time period of absorption, temperature, salt addition to the sample and pH [104,105]. Figure 17 shows different variables optimized for the analytes 4,4'-DDT, 4,4'-DDD and 4,4'-DDE as example [57]. One of the most important steps is the determination of the time needed until it reaches equilibrium between the sample matrix and the coating of the fiber. Most of organochlorine pesticides reach the equilibrium after about 40 min; with this time, it is obtained a good reproducibility and an acceptable analysis time (Figure 17a).

Desorption is done normally using static mode by immersion of the fiber into the vial during enough time for producing the efficienct analytes desorption. This time normally is studied between 1 and 15 min, longer times don't give better recoveries. In the studied case, it was proved that a time of 4 min was enough for obtaining a desorption of target analytes (Figure 17b).

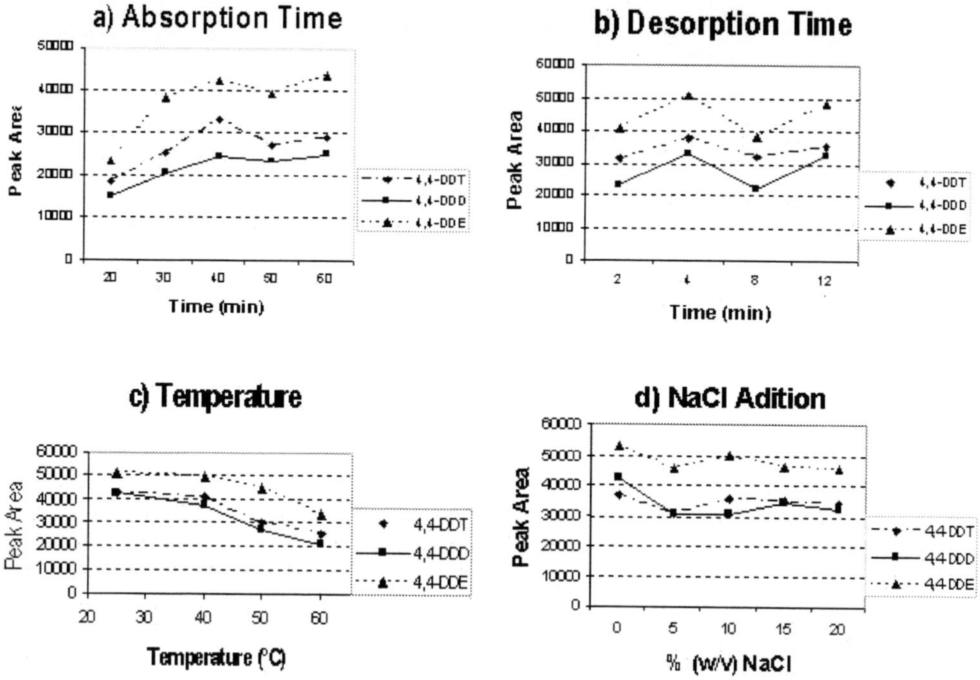

Figure 17. Optimized variables for SPME extraction using POLE coupling with PDMS/DVB fiber. [from ref. 57].

Extraction temperature can play an important role on the absorption of analytes, because it influences the mass transfer rates and the partition coefficients of the analytes. A high temperature could increase the diffusion rate of the analytes, however, absorption is generally an exothermic process and the amount of analytes absorbed could decrease with a temperature increase [106]. For that, it is necessary to search the optimum extraction temperature. In the case of organochlorine pesticides, for temperatures lower to 40 °C extraction efficiency is constant, but to higher temperatures, it decrease due to the exothermic process (Figure 17c).

Another extraction parameter which has a well-establish effect in the extraction procedure is the ionic strength of the sample by addition of salt. An increase of ionic strength can improve extraction efficency due to the reduction of the solubility of least polar analytes. However, an increase of salt concentration can generate damage on the fiber by salt deposition. Figure 17d shows the effect of ionic strength on the peak area of target analytes, between 0% and 20% (w/v) NaCl concentrations. It can be seen that for these compounds salt addition doesn't improve the analyte absorption.

Analytical Parameters and Validation

Liquid chromatography with UV detection can be used in the analysis of organochlorine pesticides after their MAME-SPME extraction. Limits of detection reached using this process are in the range 56-96 ng·g^{-1} for all compounds studied and relative standard desviations are lower than 9% in all cases (Table 14).

Table 14. Analytical parameters for the determination of organochlorine pesticides using MAME-SPME procedure [from ref. 57]

Compound	RSD (%)*	LOD (ng· g^{-1})
4,4'-DDD	7.4	68
Dieldrin	5.5	96
4,4'DDT	7.3	72
2,4'DDT	8.4	84
4,4'-DDE	7.9	56

* n=6

When this process MAME-SPME is applied to the determination of a mixture of pesticides 4,4'-DDD, 4,4'-DDT, 4,4'-DDE and Dieldrin present in a certified soil sample (CRM804-050), recoveries obtained for all compounds analyzed, falling within the certificated range.

Applications

Soil characteristics could also have a great influence on the recovery. However, if we applied the optimized method MAME-SPME-HPLC to the extraction and determination of organochlorine pesticides mixture in different types of agricultural soils, with different characterisitics, extraction efficiency is satisfactory for most compounds in study, independently of the characteristics of the matrix (Figure 18).

This method has been also applied to different mud samples with satisfactories results [107]. Therefore, the versatility and less complex sample's preparation make MAME-SPME a suitable process for solid samples.

Coupling MAME with Solid Phase Extraction (MAME-SPE)

During extraction step, many interfering components are co-extracted from solid samples together with target analytes which could interfere on the final quantification. Removal of interfering substances can be made by SPME [100] as we studied before, and by SPE [108] process, producing also an intensification of the analytes signal by means of extract clean-up and preconcentration.

With the aim to obtain cleaner extracts and lower detection limits, it can be applied the coupling between MAME procedure and solid phase extraction (SPE). This coupling has a simply handling and allows simultaneous analysis. MAME-SPE procedure has been applied to the extraction and determination of organochlorine pesticides from seaweed [109] and mollusc samples [110] with very satisfactories results.

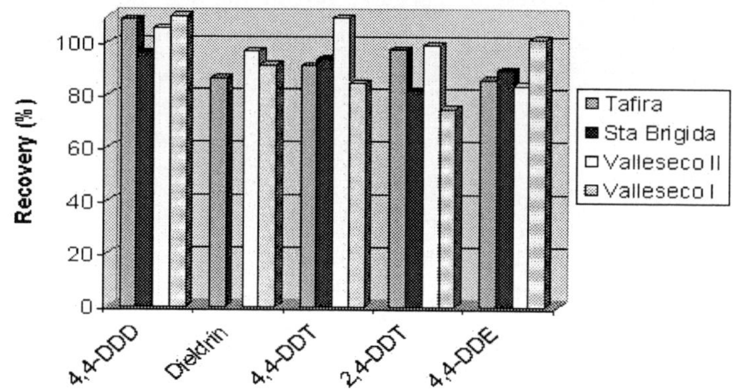

Figure18. Recoveries obtained after MAME-SPME procedure for organochlorine pesticides for four different soils using POLE as extractant [from ref. 57].

MAME-SPE extraction can be divided in two steps: microwave assisted micellar extraction (MAME) to obtain MAME extract under the optimum conditions (optimized before) and the application of solid phase extraction (SPE) clean up procedure to this extract.

Solid phase extraction (SPE) is widely used as preparation process based on the selective retention of interest's compounds in a solid sorbent and then eluted with an organic solvent. But also can be coupled with other extraction process as clean-up step (see Figure 19).

In SPE procedure, before the extraction it is necessary conditioning the cartridge. After absorption, a wash step is done for reducing surfactant remains with ultra-pure water. The retained analytes are eluted with methanol. Finally, 50 μl of extract were injected into chromatographic system.

There are different parameters that may influence in the clean-up and preconcentration step such as type of sorbent, sample volume, composition and volume of washing solution and type and volume of desorption solution.

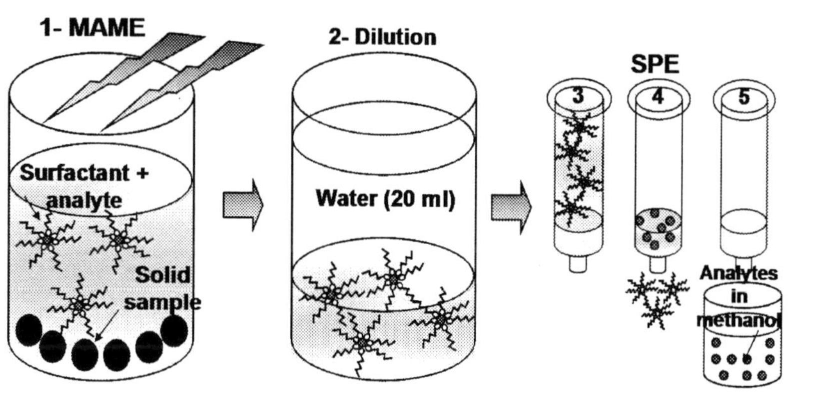

Figure 19. Scheme of the coupling MAME-SPE (1- MAME extract, 2- Extract's Dilution, 3- SPE Absorption, 4- SPE Clean step, 5- SPE Desorption). [from ref. 110].

Optimization of MAME-SPE Procedure

It is necessary to optimize and compare different SPE cartridges for obtaining the best extraction efficiencies [111]. Several papers reported on the evaluation of a number of stationary phases that can be used based on compound's characteristics. For example, in the case of organochlorine pesticides, results obtained (as peak area) for different SPE cartridges show that the best recoveries were obtained with C_{18} and Envirelut-Pesticides cartridges. However, the last one was selected because it gives a better baseline in the chromatogram.

For MAME-SPE procedure it is necessary to study the amount and concentration of MAME extract which are going to be use in SPE step [112]. Sample must be diluted with water to facility the pass of the sample through the solid phase and minimize the competition for analytes between surfactant and cartridge sorbent.

Figure 20 shows a comparison between relative peak areas obtained using 5 ml of MAME extract (5%, v/v of surfactant) without and with dilution with different volumes of ultra-pure water. It can be observed that, results were quite similar for each analyte using different final volume samples. However, from a volume of 25 ml the ratio signal-noise obtained in the chromatogram was better and this volume was selected as the optimum for further experiments.

Analytical Parameters

After MAME-SPE extraction, analytes can be determined by HPLC. This procedure has been applied to the determination of organochlorine pesticides in seaweeds [109] and molluscs [110]. In both cases, average recoveries were in the range between 78.8-107.7% for seaweeds and 68.2-100.4% for molluscs.

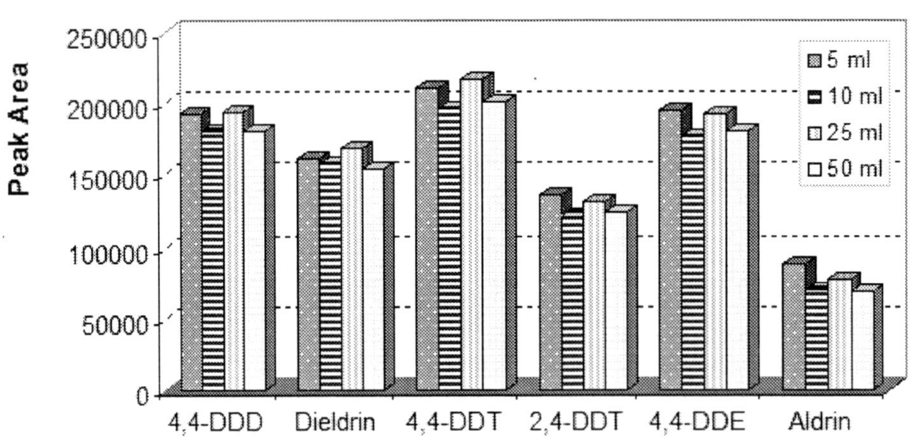

Figure 20. Study of absorption volume, represented as final volume, used in MAME-SPE process. [from ref. 109].

In order to increase preconcentration factor in MAME-SPE procedure, methanol on the desorption volume can be reduced until a value which would allow high sensitivity but with an acceptable reproducibility of the method [109,110].

Precision was determined by reproducibility studies, expressed by relative standard desviation (RSD). Reproducibility was measured by analysis of six spiked seaweed samples under optimum conditions. RSD values were below 10% and limits of detection of all studied compounds were calculated [69], obtaining values ranged from 2 to 38 ng·g^{-1} for seaweeds and from 6 to 38 ng·g^{-1} for molluscs. Results obtained for seaweeds are reported in Table 15.

Characteristic chromatograms for MAME-SPE of spiked (4 µg·g^{-1} for 4,4'-DDD, 4,4'-DDT, 2,4'-DDT and 4,4'-DDE and 8 µg·g^{-1} for Aldrin and Dieldrin) (a) and non-spiked (b) seaweed samples are shown in Figure 21.

Table 15. Analytical parameters in the determination of organochlorine pesticides using MAME-SPE [from ref. 109]

Compound	Recovery (%)	RSD[a] (%)	LOD[b] (ng· g^{-1})
4,4'-DDD	92.9	3.5	12
Dieldrin	100.5	3.3	22
4,4'DDT	92.6	2.8	2
2,4'DDT	91.2	3.6	6
4,4'-DDE	107.7	5.3	2
Aldrin	78.8	4.6	38

a) Relative Standard Desviation (n=6)

b) Limit of Detection

c) Concentration of spiked seaweed: 4 µg/g for 4,4'-DDD; 4,4'-DDT; 2,4'-DDT and 4,4'-DDE, and 8 µg/g for Aldrin and Dieldrin

Applications and Validation

When this method is applied to several seaweed samples or different kind of molluscs, recoveries obtained in all cases were equal or higher than 73.9 and 68.3%, respectively. It is demonstrated that the proposed method MAME-SPE-HPLC can be an alternative to the determination of organochlorine pesticides in complex solid samples [109,110].

If we compared traditional Soxhlet extraction [113] with MAME-SPE process developed, results obtained by both methods are comparable (Table 16). A clean-up of Soxhlet extracted was not done because it was only used to validate our method, and with direct injection was enough for obtaining recovery's results but with bad baseline and higher detection limits than in MAME-SPE. Moreover, MAME-SPE offers as advantages that is a faster process and it needs much less organic solvent.

General Conclusions and Future Trends

New analytical methods are continually being developed for sample preparation. Although LLE continues to be widely used, CPE methodology is a good alternative extraction technique for liquid samples and offers a series of highly interesting advantages from an analytical point of view, such as the possibility of extracting and preconcentrating the analytes of different polarities in only one step.

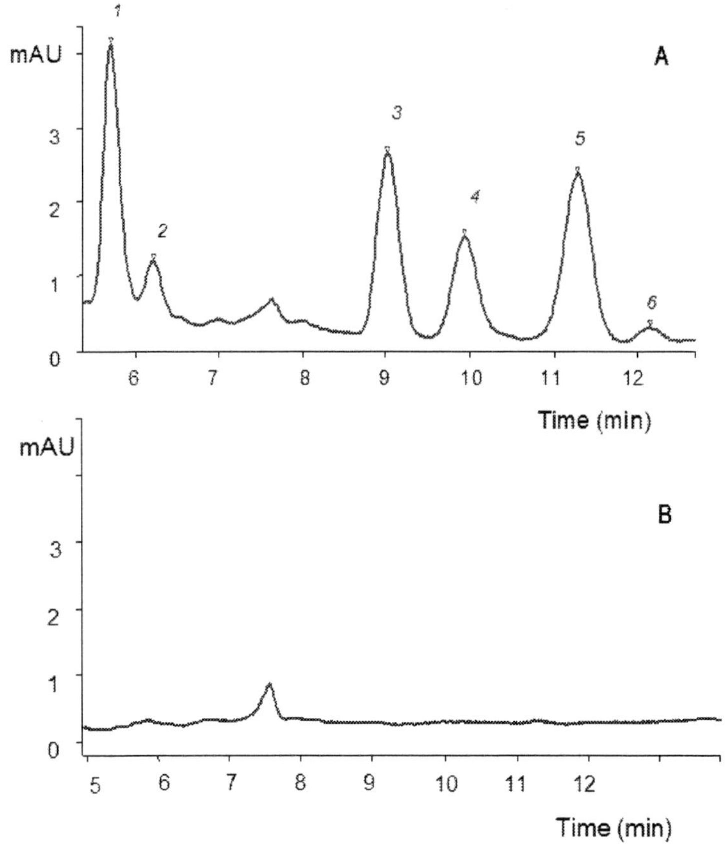

Figure 21. Chromatogram of an extract of a spiked (a) and non-spiked (b) Ulva seaweed sample after MAME-SPE procedure. λ=238 nm. (*1*- 4,4'-DDD; *2*- Dieldrin; *3*- 4,4'-DDT; *4*- 2,4'-DDT; *5*- 4,4'-DDE; *6*- Aldrin). [from ref. 109].

Table 16. Application of MAME-SPE procedure and Soxhlet extraction to a seaweed sample containing a certified mixture of pesticides[a,b] [from ref. 109]

Compound	MAME-SPE ($\mu g \cdot g^{-1}$)	Soxhlet ($\mu g \cdot g^{-1}$)
4,4'-DDD	3.72 ± 0.08	4.12 ± 0.24
Dieldrin	4.04 ± 0.14	3.94 ± 0.30
4,4'DDT	3.82 ± 0.10	4.04 ± 0.18
2,4'DDT	3.52 ± 0.18	3.44 ± 0.22
4,4'-DDE	3.78 ± 0.14	3.76 ± 0.15

a) Concentration of spiked seaweed: 4 μg/g for all compounds

b) Mean of two determinations

The pre-concentration factor can be optimised by modifying the type and the concentration of surfactant as well as the experimental conditions under which extraction and phase separation are carried out. Surfactants are lesser toxic and cheaper than the extractants used in LLE. The most commonly used surfactants are commercially available and, since it is not necessary to evaporate off the solvents, no analyte is lost in the process. The experimental operations involved in CPE methodology are very simple and the surfactant-rich phase is compatible with the mobile phases used in HPLC.

MAME procedure could replace the conventional Soxhlet extraction, offering several advantages over traditional extraction techniques applied to solid samples. The major benefits are:

- a reduction in extraction time (typically 10–30 min)
- a reduction in solvent consumption (around 25–50 ml, reduced by 5–10-fold compared to traditional extractions); and,
- the opportunity to perform multiple extractions.

MAME procedure is easy to use and is cheaper than other modern techniques. Furthermore, the use of micellar medium reduces exposure to solvent vapours and wastes. When considering economical and practical aspects, the use of micellar systems for the extraction of organic compounds in solid matrices offers strong competition to other extraction techniques.

This competitiveness may be improved in the future in favour of two-directional micellar extraction. It may also be applied to a great number of types of solid real samples (plants, marine organisms, etc.), using other types of surfactants, as well as a mixture of them as extractants. In addition, combining the use of surfactants with other extraction techniques and with highly sensitive chromatographic techniques, such as LC–MS and GC–MS or GC-ECD, may be of great interest for the extraction and the determination of organochlorine and organophosphorus pesticides that, in many occasions, are found in the environment in concentrations too small to be detected using conventional techniques.

Finally, the combination of microwave assisted micellar extraction (MAME) with SPME or SPE as a clean-up and preconcentration step of MAME extract, provided sensitive and selective methods for the extraction of organochlorine and organophosphorus pesticides residues in environmental samples. Specifically, the results obtained for MAME-SPE-HPLC indicate that this technique is more sensitive for the extraction and determination of this kind of compounds where the LODs are from 10 to 100 times lower than those obtained using MAME-SPME-HPLC.

REFERENCES

[1] Kusvuran, E. and Erbatur, O. (2004) Degradation of aldrin in adsorbed system using advanced oxidation processes: comparison of the treatment methods. *Journal of Hazardous Materials,* 106B, 115-125.

[2] Stockholm Convention on Pops: A Global Public Health Treaty. (2001) Sweden.

[3] Shivaramaiah, H.M., Odeh, I.O.A., Kennedy, I.V. and Skerritt, J.H. (2002) Mapping the distribution of DDT residues as DDE in the soils of the irrigated regions of Northern New South Wales, Australia using ELISA and GIS. *Journal of Agricultural and Food Chemistry*, 50, 5360-5367.

[4] Chen, W., Zhang, L., Xu, L., Wang, X., Hong, L. and Hong, H. (2002) Residue levels of HCHs, DDTs and PCBs in shellfish from coastal areas of east Xiamen Island and Minjiang Estuary, China. *Marine Pollution Bulletin*, 45, 385-390.

[5] Nawab, A., Aleem, A. and Malik, A. (2003) Determination of organochlorine pesticides in agricultural soil with special reference to gamma-HCH degratation by Pseudomonas strains. *Biosesource Technology*, 88, 41-46.

[6] Megharaj, M., Kantachote, D., Singleton, I. and Naidu, R. (2000) Effects of long-term contamination of DDT on soil microflora with special reference to soil algae and algal transformation of DDT. *Environmental Pollution*, 109, 35-42.

[7] Mwevura, H., Othman, O. and Mhehe, G.L. (2002) Organochlorine pesticide residues in sediments and biota from the coastal area of Dar es Salaam city, Tanzania. *Marine Pollution Bulletin*, 45, 262-267.

[8] Maroni, M., Colosio, C., Ferioli, A. and Fait, A. (2000) Organochlorine pesticides. *Toxicology*, 143, 61-75.

[9] U.S. Environmental Protection Agency. Toxic Substance Control Act, US EPA. (1979) Washington DC.

[10] Anson Moye, H. (Ed.) Analysis of Pesticide Residues (1981) vol 58. John Wiley and Sons, New York/Chichester/Brisbane/Toronto. EEUU.

[11] Woodwell, G.M., Craig, P.P. and Johnson, H.A. (1971) DDT in the Biosphere: Where Does It Go?. *Science*, 174, 1101-1107.

[12] Hernández, F., Pitarch, E., Beltran, J. and López, F.J. (2002) Headspace solid-phase microextraction in combination with gas chromatography and tandem mass spectrometry for the determination of organochlorine and organophosphorus pesticides in whole human blood. *Journal of Chromatography* B, 769, 65-77.

[13] Zulin, Z., Huasheng, H., Xinhong, W., Jianqing, L., Weiqui, C. and Li, X. (2002) Determination and load of organophosphorus and organochlorine pesticides at water from Jiulong River Estuary, China. *Marine Pollution Bulletin*, 45, 397-402.

[14] Viñas, P., Campillo, N., López-García, I., Aguinaga, N. and Hernández-Córdoba, M. (2002) Determination of pesticides in waters by capillary gas chromatography with atomic emission detection. *Journal of Chromatography* A, 978, 249-256.

[15] Torres, C.M. and Mañes, Y.P.J. (1996) Determination of pesticide residues in fruit and vegetables. Journal of Chromatography A, 754, 301-331.

[16] Tadeo, J.L., Sánchez-Brunete, C., Pérez, R.A. and Fernández, M.D. (2000) Analysis of herbicide residues in cereals, fruits and vegetables. *Journal of Chromatography* A, 882, 175-191.

[17] Carabias-Martínez, R., Rodríguez-Gonzalo, E., Paniagua-Marcos, P.H. and Hernández-Méndez, J. (2000) Analysis of pesticide residues in matrices with high lipid contents by membrane separation coupled on-line to a high-performance liquid chromatography system. *Journal of Chromatography* A, 869, 427-439.

[18] Caballo-López, A. and Luque de Castro, M.D. (2003) Continuous ultrasound-assisted extraction coupled to on line filtration–solid-phase extraction–column liquid chromatography–post column derivatisation–fluorescence detection for the

determination of N-methylcarbamates in soil and food. *Journal of Chromatography* A, 998, 51-59.

[19] Rastrelli, L., Totaro, K. and De Simone, F. (2002) Determination of organophosphorus pesticide residues in Cilento (Campania, Italy) virgin olive oil by capillary gas chromatography. *Food Chemistry*, 79, 303-305.

[20] Russo, M.V., Campanella, L. and Avino, P. (2002) Determination of organophosphorus pesticide residues in human tissues by capillary gas chromatography-negative chemical ionization mass spectrometry analysis. Journal of Chromatography B, 780, 431-441.

[21] Jeong, M.L. and Chesney, D.J. (1999) Determination of *N*-methylcarbamate pesticides from spiked matrices using supercritical fluid extraction. *Analytica Chimica Acta,* 389, 53-57.

[22] Hennion, M.C., Cau-Dit-Coumes, C. and Pichon, V. (1998) Trace analysis of polar organic pollutants in aqueous samples: Tools for the rapid prediction and optimisation of the solid-phase extraction parameters. *Journal of Chromatography* A, 823, 147-191.

[23] Hennion, M.C. (1999) Solid-phase extraction: method development, sorbents, and coupling with liquid chromatography. *Journal of Chromatography* A, 856, 3-54.

[24] Chiron, S., Fernández Alba, A. and Barceló, D. (1993) Comparison of on-line solid-phase disk extraction to liquid-liquid extraction for monitoring selected pesticides in environmental waters. *Environmental Science Technology*, 27, 2352-2359.

[25] Ahmed, F.E. (2001) Analyses of pesticides and their metabolites in foods and drinks. *Trends in Analytical Chemistry*, 20, 649-661.

[26] [26] Rodriguez, I., Llompart, M.P. and Cela, R. (2000) Solid-phase extraction of phenols. *Journal of Chromatography* A, 885, 291-304.

[27] Lores, M., Llopart, M., González-García, R., González-Barreiro, C. and Cela, R. (2002) On-fibre photodegradation studies of polychlorinated biphenyls using SPME-GC-MS-MS: a new approach. *Chemosphere*, 47, 607-615.

[28] Eisert, R. and Pawliszyn, J. (1997) Automated In-Tube Solid-Phase Microextraction Coupled to High-Performance Liquid Chromatography. *Analytical Chemistry*, 69, 3140-3147.

[29] Moreno Cordero, B., Pérez Pavón, J.L., García Pinto, C. and Fernández Laespada, E. (1993) Cloud point methodology: A new approach for preconcentration and separation in hydrodynamic systems of analysis. *Talanta*, 40, 1703-1710.

[30] Heemke, O.P., Norbert, T. and Wenclawiak, B.W. (1997) Comparison of ASE and SFE with Soxhlet, Sonication, and Methanolic Saponification Extractions for the Determination of Organic Micropollutants in Marine Particulate Matter. *Analytical Chemistry*, 69, 2171-2180.

[31] Snyder, J., Grob, R., McNally, M. and Gostdyk, T. (1992) Comparison of supercritical fluid extraction with classical sonication and Soxhlet extractions for selected pesticides. *Analytical Chemistry*, 64, 1940-1946.

[32] Fernández, P., Grifoll, M., Solanas, A.M., Bayona, J. and Albaigés, J. (1992) Bioassay-directed chemical analysis of genotoxic components in coastal sediments. *Environmental Science Technology*, 26, 817-829.

[33] Punín Crespo, M. and Lage Yusty, M. (2006) Comparison of supercritical fluid extraction and Soxhlet extraction for the determination of aliphatic hydrocarbons in seaweed samples. *Ecotoxicology and Environmental Safety*, 64, 400-405.

[34] Li, K.; Landrialult, M.; Fingas, M. and Llompart, M. (2003) Accelerated solvent extraction (ASE) of environmental organic compounds in soils using a modified supercritical fluid extractor. *Journal of Hazardous Materials*, 102, 93-104.

[35] Zuloaga, O., Etxebarria, N., Fernández, L.A. and Madariaga, J.M. (2000) Fresenius *Journal Analytical Chemistry*, 367, 733-737.

[36] Sanghi, R. and Kannamkumarath, S. (2004) Comparison of extraction methods by soxhlet, sonicator, and microwave in the screening of pesticides residues from solid matrices. *Journal of Analytical Chemistry*, 59, 1032-1036.

[37] Sporring, S.; Bowadt, S.; Svensmark, B. and Björklund, E. (2005) Comprehensive comparison of classic Soxhlet extraction with Soxtec extraction, ultrasonication extraction, supercritical fluid extraction, microwave assisted extraction and accelerated solvent extraction for the determination of polychlorinated biphenyls in soil. *Journal of Chromatography* A, 1090, 1-9.

[38] Pastor, A., Vazquez, E., Ciscar, R. and de la Guardia, M. (1997) Efficiency of the microwave-assisted extraction of hydrocarbons and pesticides from sediments. *Analytica Chimica Acta*, 344, 241-249.

[39] Shu, Y.Y., Tey, S.Y. and Wu, D.K.S. (2003) Analysis of polycyclic aromatic hydrocarbons in airborne particles using open-vessel focused microwave-assisted extraction. *Analytica Chimica Acta*, 495, 99-108.

[40] López-Ávila, V. and Young, R. (1994) Microwave-assisted extraction of organic compounds from standard reference soils and sediments. *Analytical Chemistry*, 66, 1097-1106.

[41] Esteve-Turrillas, F.A., Aman, C.S., Pastor, A. and de la Guardia, M. (2004) Microwave-assisted extraction of pyrethroid insecticides from soil. *Analytica Chimica Acta*, 522, 73-78.

[42] Barriada-Pereira, M., Concha-Graña, E., González-Castro, M.J., Muniategui-Lorenzo, S., López-Mahía, P., Prada-Rodríguez, D. and Fernández-Fernández, E. (2003) Microwave-assisted extraction versus Soxhlet extraction in the analysis of 21 organochlorine pesticides in plants. *Journal of Chromatography* A, 1008, 115-122.

[43] Onuska, F.I. and Terry, K.A. (1993) Extraction of pesticides from sediments using a microwave technique. *Chromatographia*, 36, 191-194.

[44] Letellier, M., Budzinski, H. and Garrigues, P. (1999) Focused microwave-assisted extraction of polycyclic aromatic hydrocarbons. *LC-GC International*, 12, 222-225.

[45] López-Avila, V., Young, R., Benedicto, J., Ho, P., Kim, R. and Beckert, W.F. (1995) Extraction of organic pollutants from soild samples using microwave energy. *Analytical Chemistry*, 67, 2096.

[46] Rosen, M.J. (1986) Surfactants and Interfacial phenomena, John Wiley and sons. USA.

[47] Pelizzetti, E. and Pramauro, E. (1985) Analytical applications of organized molecular assemblies. *Analytica Chimica Acta*, 169, 1-29.

[48] Uhrich, K.E. (1999) *Chemical Review*, 99, 3181-3198.

[49] Goto, T., Ito, Y., Oka, H., Saito, I., Matsumoto, H. and Nakazawa, H. (2003) Simple and rapid determination of N-methylcarbamate pesticides in citrus fruits by electrospray ionization tandem mass spectrometry. *Analytica Chimica Acta*, 487, 201-209.

[50] Namera, A., Yashiki, M., Liu, J., Okajima, K., Hara, K., Imamura, T. and Kojima, T. (2000) Simple and simultaneous analysis of fenfluramine, amphetamine and methamphetamine in whole blood by gas chromatography–mass spectrometry after

headspace–solid phase microextraction and derivatization. *Forensic Science International,* 109, 215-223.

[51] Paleologos, E. K., Giokas, D. L. and Karayannis, M. I. (2005) Micelle-mediated separation and cloud point extraction. *Trends in Analytical Chemistry*, 24, 426-436.

[52] Faria, A. M., Dardengo, R. P. Lima, C. F., Neves, A. and Queiroz, M. E. (2007) Determination of disulfoton in surface water samples by cloud-point extraction and gas chromatography. *International Journal of Environmental Analytical Chemistry*, 87, 249-258.

[53] Shen, J. and Shao, X. (2006) Determination of tobacco alkaloids by gas chromatography-mass spectrometry using cloud point extraction as a preconcentration step. *Analytica Chimica Acta*, 561, 83-87.

[54] [54] Sosa Ferrera, Z., Padrón Sanz, C., Mahugo Santana, C. and Santana Rodríguez, J.J. (2004) The use of micellar systems in the extraction and pre-concentration of organic pollutants in environmental samples. *Trends in Analytical Chemistry*, 23 (7), 469-479.

[55] Mahugo Santana, C., Sosa Ferrera, Z. and Santana Rodríguez, J.J. (2004) Use of polyoxyethlylene-6-lauryl ether and microwave-assisted extraction for the determination of chlorophenols in marine sediments. *Analytica Chimica Acta*, 524, 133-139.

[56] Padrón Sanz, C., Halko, R., Sosa Ferrera, Z. and Santana Rodríguez, J.J. (2004) Micellar extraction of organophosphorus pesticides and their determination by liquid chromatography. *Analytica Chimica Acta*, 524, 265-270.

[57] Vega Moreno, D., Sosa Ferrera, Z. and Santana Rodríguez, J.J. (2006) Microwave assisted micellar extraction coupled with solid phase microextraction for the determination of organochlorine pesticides in soil simples. *Analytica Chimica Acta*, 571, 51-57.

[58] Hinze, W.L. and Pramauro, E. (1993) A critical review of surfactant-mediated phase separations (cloud point extractions): theory and applications. *Critical Reviews in Analytical Chemistry*, 24(2), 133-177.

[59] [59] Mahugo Santana, C., Sosa Ferrera, Z. and Santana Rodríguez, J.J., (2002) Use of non-ionic surfactant solutions for the extraction and preconcentration of phenolic compounds in water prior to their HPLC-UV detection. *The Analyst*, 127, 1031-1037.

[60] Eiguren Fernández, A., Sosa Ferrera, Z. and Santana Rodríguez, J.J., (1997) *Química Analítica*, 16(2), 283.

[61] Frankewich, R.P. and Hinze, W. (1994) Evaluation and Optimization of the Factors Affecting Nonionic Surfactant-Mediated Phase Separations. *Analytical Chemistry*, 66, 944-954.

[62] Carabias-Martínez, R., Rodríguez-Gonzalo, E., Domínguez-Álvarez, J., García Pinto, C., Hernández-Méndez, J. (2003) Prediction of the behaviour of organic pollutants using cloud point extraction. *Journal of Chromatography* A, 1005 (1-2), 23-34

[63] Carabias, R., Rodríguez, E., Moreno, B., Pérez, J.L., García, C. and Fernández, E. (2000) Surfactant cloud point extraction and preconcentration of organic compounds prior to chromatography and capillary electrophoresis. *Journal of Chromatography* A, 902, 251-265.

[64] Silva, M.F., Fernández, L., Olsina, R.A. and Stacchiola, D. (1997) Cloud point extraction, preconcentration and spectrophotometric determination of erbium(III)-2-

(3,5-dichloro-2-pyridylazo)-5-dimethylaminophenol. *Analytica Chimica Acta*, 342, 229-238.

[65] Eiguren Fernández, A., Sosa Ferrera, Z. and Santana Rodríguez, J.J. (1998) Determination of polychlorinated biphenyls by liquid chromatography following cloud-point extraction. *Analytica Chimica Acta*, 358, 145-155.

[66] Faria, A.M., Dardengo, R.P., Lima, C.F., Neves, A.A., Queiroz, M.E.L.R. (2007) Determination of disulfoton in surface water samples by cloud-point extraction and gas chromatography. *International Journal of Environmental Analytical Chemistry* 87 (4), 249-258

[67] Froschl, B., Stangl, G. and Niessner, R. (1997) Combination of micellar extraction and GC-ECD for the determination of polychlorinated biphenyls (PCBs) in water. *Fresenius Journal Analytical Chemistry*, 357, 743-746.

[68] Eiguren Fernández, A., Sosa Ferrera, Z. and Santana Rodríguez, J.J. (1999) Application of cloud-point methodology to the determination of polychlorinated dibenzofurans in seawater by high-performance liquid chromatography. *Analyst*, 24, 487-491.

[69] Lindsay, S. (1992) High Performance Liquid Chromatography, John Wiley and Sons: New York, pp. 337.

[70] Ganzler, K., Salgo, A. and Lebensm-Unter, Z. (1987) *Forsch*, 184, 274

[71] Pineiro Iglesias, M., López Mahía, P., Vázquez Blanco, E., Muniategui Lorenzo, S., Prada Rodríguez, D. and Fernández Fernández, E. (2000) Microwave assisted extraction of polycyclic aromatic hydrocarbons from atmospheric particulate samples. *Fresenius Journal Analytical Chemistry*, 367, 29-34.

[72] López-Avila, V., Benedicto, J., Charan, C. and Young, R. (1995) Determination of PCBs in soils/sediments by microwave-assisted extraction and GC/ECD or ELISA. *Environmental Science Technology*, 29, 2709-2712.

[73] Hummert, K., Vetter, W. and Luckas, B. (1996) Fast and effective sample preparation for determination of organochlorine compounds in fatty tissue of marine mammals using microwave extraction. *Chromatographia*, 42, 300-304.

[74] Pastor, A., Vázquez, E., Ciscar, R. and de la Guardia, M. (1997) Efficiency of the microwave-assisted extraction of hydrocarbons and pesticides from sediments. *Analytica Chimica Acta*, 344, 241-249.

[75] Egizabal, A., Zuloaga, O., Etxebarria, N., Fernández, L.A. and Madariaga, J.M. (1998) Comparison of microwave-assisted extraction and Soxhlet extraction for phenols in soil samples using experimental designs. *The Analyst*, 123, 1679-1684.

[76] Llompart, M.P., Lorenzo, R.A., Cela, R., Li, K., Belanger, J.M.R. and Pare, J.R.J. (1997) Evaluation of supercritical fluid extraction, microwave-assisted extraction and sonication in the determination of some phenolic compounds from various soil matrices. *Journal of Chromatography* A, 774, 243-251.

[77] Padrón Sanz, C., Sosa Ferrera, Z. and Santana Rodríguez, J.J. (2004) Extraction of PCDDs from marine sediments using polyoxyethylene 10 lauryl ether and oligoethylene glycol monoalkyl ether surfactants. *Analytical Letters*, 37, 1385-1399.

[78] Padron Sanz, C., Halko, R., Sosa Ferrera, Z. and Santana Rodríguez, J.J. (2005) Combination of microwave assisted micellar extraction and liquid chromatography for the determination of organophosphorous pesticides in soil simples. *Journal of Chromatography* A, 1078, 13-21

[79] Vega Moreno, D., Sosa Ferrera, Z. and Santana Rodríguez, J.J. (2006) Use of polyoxyethylene surfactants for the extraction of organochlorine pesticides from agricultural soils. *Journal of Chromatography* A, 1104, 11-17.

[80] Padrón Sanz, C., Eiguren Fernández, A., Sosa Ferrera, Z. and Santana Rodríguez, J.J. (2002) Determination of Organochlorinated Compounds in Marine Organisms by Microwave-Assisted Extraction with Molecular Organized Systems and Liquid Chromatography with Fluorescence Detection. *Journal of AOAC Internacional*, 85(1), 44-49.

[81] Bianco Prevot, A., Gulmini, M., Zelano, V. and Pramauro, E. (2001) Microwave-Assisted Extraction of Polycyclic Aromatic Hydrocarbons from Marine Sediments Using Nonionic Surfactant Solutions. *Analytical Chemistry*, 73, 3790-3795.

[82] U.S. Environmental Protection Agency (1996), Method 3540 C, Washington, DC.

[83] Llompart, M.P., Lorenzo, R.A. and Par´e, J.R. (1997) Optimization of a Microwave-assisted Extraction Method for Phenol and Methylphenol Isomers in Soil Samples Using a Central Composite Design. *Analyst*, 122, 133-137.

[84] Huang, W., Peng, P., Yu, Z. and Fu, J. (2003) Effects of organic matter heterogeneity on sorption and desorption of organic contaminants by soils and sediments. *Applied Geochemistry*, 18, 955-972.

[85] Thiele-Bruhn, S., Seibicke, T., Schuelten, H.R. and Leinweber, P. (2004) Sorption of Sulfonamide Pharmaceutical Antibiotics on Whole Soils and Particle-Size Fractions. *Journal of Environmental Quality*, 33, 1331-1342.

[86] Aquino, A. J., Tunega, D., Haberhauer, G., Gerzabek, M. and Lischka, H. (2003) Adsorption of Organic Substances on Broken Clay Surfaces: A Quantum Chemical Study. *Journal of Computational Chemistry*, 24(15), 1853-1863.

[87] Shu, Y.Y., Lai, T.L., Lin, H.S., Yang, T.C. and Chang, C.P. (2003) Study of factors affecting on the extraction efficiency of polycyclic aromatic hydrocarbons from soils using open-vessel focused microwave-assisted extraction. *Chemosphere*, 52, 1667-1676.

[88] Kiflom, G., Wandiga, S.O., Ng'ang'a, P.K. and Kamau, G.N. (1999) Variation of plant p,p'-DDT uptake with age and soil type and dependence of dissipation on temperature. *Environmental Internacional*, 25(4), 479-487.

[89] Hawthorne, S.B., Björklund, E., Bowadt, S. and Mathiasson, L. (1999) Determining PCB sorption/desorption behavior on sediments using selective supercritical fluid extraction. 3. Sorption from water. *Environmental Science Technology*, 33, 3152-3159.

[90] Dec, J. and Bollag, J. (1997) Determination of covalent and noncovalent binding interactions between xenobiotic chemicals and soil. *Soil Science*, 162, 858-874.

[91] Kopinke, F., Pörschmann, J. and Stottmeister, U. (1995) Sorption of organic pollutants on anthropogenic humic matter. *Environmental Science Technology*, 29, 941-950.

[92] Stevenson, F.J. (1982) Humus Chemistry, John Wiley and Sons Inc., USA.

[93] Racke, K.D. in: Chambers, J.E. and Levi, P.E. (Eds.) (1992) Organophosphates: Chemistry, Fate, and Effects. Academic Press, San Diego, pp. 47–73.

[94] Pehkonen, S.O. and Zhang, Q. (2002) The degradation of organophosphorus pesticides in natural waters: A critical review. Critical Review Environmental *Science Technology*, 32, 17-72.

[95] Dannenberg, A. and Pehkonnen, S.O. (1998) Investigation of the Heterogeneously Catalyzed Hydrolysis of Organophosphorus Pesticidas. *Journal of Agricultural and Food Chemistry*, 46, 325-334.

[96] Dowling, K.C. and Lemley, A.T. (1995) Organophosphate insecticide degradation by non-amended and cupric ion-amended Fenton's reagent in aqueous solution. *Journal Environmental Science Health* B, 30 (5), 585-604.

[97] Noblet, J.A., Smith, L.A. and Suffet, I.H. (1996) Influence of natural dissolved organic matter, temperature, and mixing on the abiotic hydrolysis of triazine and organophosphate pesticidas. *Journal of Agricultural and Food Chemistry*, 44, 3685-3693.

[98] Dupeyron, S., Dudermel, P.M. and Counturier, D. (1997) Focused microwave assisted extraction (FMAE) of polynuclear aromatic hydrocarbons from contaminated soil: Role of acetone and water content impact on microwave efficiency. *Analusis*, 25, 286-292.

[99] Peñalver, A., Pocurull, E., Borrull, F. and Marcé, R.M. (1999) Trends in solid-phase microextraction for determining organic pollutants in environmental samples. *Trends in Analytical Chemistry*, 18(8), 557-567.

[100] Ho, W. and Hsieh, S. (2001) Solid phase microextraction associated with microwave assisted extraction of organochlorine pesticides in medicinal plants. *Analytica Chimica Acta*, 428, 111-120.

[101] Aulakh, J. and Malik, A. (2005) A review on solid phase microextraction-high performance liquid chromatography (SPME-HPLC) analysis of pesticides. *Critical Reviews in Analytical Chemistry*, 35, 71-85.

[102] Vega Moreno, D., Sosa Ferrera, Z. and Santana Rodríguez, J.J. (2006) Sample extraction method combining micellar extraction-SPME and HPLC for the determination of organochlorine pesticides in agricultural soils. *Journal of Agricultural and Food Chemistry*, 54, 7747-7752.

[103] Beltrán, J., López, F. and Hernández, F. (2000) Solid-phase microextraction in pesticides residue analysis. Journal of Chromatography A, 885, 389-404.

[104] Krutz, L., Senseman, S. and Sciumbato, A. (2003) Solid-phase microextraction for herbicide determination in environmental samples. *Journal of Chromatography* A, 999, 103-121.

[105] Hernández, F., Beltrán, J., López, F. and Gaspar, (2000) Use of solid-phase microextraction for the quantitative determiantion of herbicides in soil and water samples. *Journal Analytical Chemistry*, 72, 2313-2322.

[106] Arthur, C.L., Killiam, L.M., Buchholz, K.D. and Pawliszyn, J. (1992) Automation and optimization of solid-phase microextraction. *Journal Analytical Chemistry*, 64, 1960-1966.

[107] Vega Moreno D, Sosa Ferrera Z, Santana Rodríguez JJ (2008) Application of microwave assisted micellar extraction combined with solid phase microextraction and HPLC-UV for the determination of organochlorine pesticides in diferents mud samples. *International Journal on Environmental Analytical Chemistry*, in press.

[108] Mutavdzic, D., Horvat, A., Babic, S. and Kastelan-Macan, M. (2005) SPE- Microwave-assisted extraction coupled system for the extraction of pesticides from water samples. *Journal of Separation Science*, 28, 1485-1492.

[109] Vega Moreno, D., Sosa Ferrera, Z. and Santana Rodríguez, J.J. (2007) SPME and SPE comparative study for coupling with microwave-assisted micellar extraction in the

analysis of organochlorine pesticides residues in seaweed samples. *Microchemical Journal*, 87, 139-146.

[110] Vega Moreno D, Sosa Ferrera Z, Santana Rodríguez JJ (2008) Analysis of organochlorine pesticides in mollusc samples by HPLC after microwave assisted micellar extraction coupled with solid phase extraction. *Journal of Liquid Chromatography and Related Technologies*, in press.

[111] Deger, A., Gremm, T. and Frimmel, F. (2000) Problems and solutions in pesticides analysis using solid-phase extraction (SPE) and gas chromatography ion-trap mass spectrometry detection (GC-MS). Acta Hydrochim. *Hydrobiol.*, 28, 292-299.

[112] Papadakis, E. and Papadopoulou-Mourkidou, E. (2006) LC-UV determination of atrazine and its prinicipal conversion products in soil after combined microwave-assisted and solid-phase extraction. *International Journal on Environmental Analytical Chemistry*, 86, 573-582.

[113] AOAC Official Method 970.52, AOAC Official Methods of Analysis, Pesticide and Industrial.

In: Progress in Pesticides Research
Editor: Christopher M. Kanzantzakis

ISBN 978-1-60456-201-9

Chapter 5

RESIDUE LEVEL OF ORGANOCHLORINE PESTICIDES IN FISH FROM ITALIAN FISHERY AND AQUACULTURE

Orban Elena* and Masci Maurizio
National Institute for Food and Nutrition Research (INRAN), via Ardeatina 546, 00178 Rome, Italy

1. ABSTRACT

The risk analysis represents the basis of any food safety policy. The three components of the risk analysis are the following: risk assessment, risk management (through regulations and food inspections) and risk communication. As regards the first point, in the Fish Quality Unit of the Italian National Institute for Food and Nutrition Research (INRAN) works on fishery safety are carried out. Since 2001 our Laboratory has been involved in checking for the presence of 34 selected Organochlorine Pesticides (OC Pesticides) in fish samples collected either from the wild environment (Mediterranean Sea, Lakes, Rivers) or from Italian Sea Farms.

This type of trace analysis requires high-level equipment, well-planned sample preparation techniques and regular Quality Controls to ensure that accurate results are produced. A powerful instrumental procedure developed in our Laboratory and routinely used is described.

Pollutants selected are included in the current Italian regulatory status and are also cited in the Stockholm Conference as "POP's" (Persistent Organic Pollutants). These Chlorine compounds, largely used in the past, are still present in the Food Chain.

The aquatic environment is almost always the final destination of all waste products; therefore the fishery is particularly exposed to this kind of contamination. Most of the Pesticides we analyse have been banned for use since the 1980s; nevertheless they are not degradable compounds and now represent a big problem. Some of them are well-known substances such as DDT, Dieldrin, Endrin, Chlordane, Aldrin, Lindane, and Heptachlor. Their toxicity was confirmed by a great number of publications and it acts with "short

* orban@inran.it, masci@inran.it

term effects" (acute effects derived from exposition at high concentrations) and "long term effects" (chronicle action, included carcinogenesis).

In our studies we observed different fish contamination levels (from low to medium) depending on the sample origin, size, species, and age.

Fish from Italian Sea Farms seem to be of poor contamination, mainly due to the small age of these products. We found that a relationship between the contamination of farmed fish and the contamination of fish feedstuffs probably exists.

Wild fish collected in three lakes of Central Italy were of good safety, but sometimes a medium contamination was observed especially when the fat content was high. This was the case, for example, of Eels (*Anguilla anguilla*)

Finally an evaluation of OC Pesticide content in Tunas deriving from the new practice of "Bluefin Tuna Sea Ranching" is reported.

2. INTRODUCTION

In the last two centuries the number of people in the world increased significantly. This fact maked necessary a better use of soil and advanced technologies in order to improve food crops.

Today in large areas of many countries only one type of vegetable (maize, wheat, rice, and others) can be seen. No other species seems to be allowed to live there.

It should be hard to believe that such a situation is a spontaneous event: a natural environment generally shows hundreds of species per km^2.

We also notice that many health diseases and epidemics are now absent in some areas where they were largely widespread in the past (Malaria, for example).

A big effort was made by scientists in the last century to improve people's lives. Chemistry supported this work with a lot of substances showing their effectiveness against pests, epidemics and unwanted vegetables, but along with their efficiency they later showed their negative aspects. The compounds we are speaking about were based on Chlorine (Organochlorine Pesticides, also named OC Pesticides) and are now banned for use: they were replaced by other chemicals (Organophosphorous Pesticides and others).

Amounts of OC Pesticides produced and used in the past are impressive. Some examples are cited below:

- global production of Aldrin and Dieldrin was estimated to be 13000 tonnes/year in 1972 [1]
- worldwide production of DDT in 1974 was 60 000 tonnes [2]
- worldwide production of pure HCB was estimated to be 10 000 tonnes/year for the years 1978-1981 [3]

By considering all OC pesticides the mean total production per year could be estimated at 100 000 tonnes, or more, during a lapse of time of at least thirty years, from the 1940s to the 1970s. To Pesticides other OC Pollutants such as Polychlorobiphenyls, used in large extent, have to be added. Before they were banned we can reasonably estimate that the total amount of Organochlorines released in the environment was of several millions of tonnes.

Main disadvantages deriving from the use of Organochlorines were their persistency (not degradable chemicals) and adverse effects on humans, animals and the environment.

OC Pesticides, together with other Chlorine-based compounds such as Dioxins and PCBs (Polychlorobiphenyls), were considered in "Stockholm Convention on Persistent Organic pollutants", POP's, in 2001, where a controlled disposal of Organochlorines still available and a global ban for use were proposed.

The 12 "Stockholm Convention" POP's are:

Aldrin, Chlordane, DDT, Dieldrin, Endrin, Heptachlor, Mirex, Toxaphene, and HCB (OC Pesticides)

PCBs

Dioxins

Furans

For each pesticide its isomers, metabolites and degradation products have to be taken into account and determined. The analysis of Chlordane, for example, requires the analysis of at least five different compounds: cis-Chlordane, trans-Chlordane, Oxychlordane, cis-Nonachlor and trans-Nonachlor. We must consider that the original chemical employed was subsequently subjected to enzymatic or environmental degradation; moreover when it is synthesized a pesticide may result in a mixture of isomers.

Many reports about toxic effects of POP's caused OC Pesticides to be banned in most Western countries during the 1970s and 1980s. During the 2001 Stockholm Convention many countries, in which Malaria is still now a rising problem, requested exemption from the ban for OC Pesticide use, mainly DDT.

The Convention entered in force in May, 2004.

2.1. Toxicological Aspects

An overview of OC Pesticide impact on humans seems to be needed now.

Aldrin and Dieldrin were produced since 1950 up to the 1970s and were used for the control of many soil pests and in the treatment of seed. Many insects were controlled by these chemicals: termites, grasshoppers, wood borers, beetles, textile pests, tse-tse fly and other vectors of debilitating tropical diseases.

Aldrin and Dieldrin are very toxic for humans: the lowest quantity judged as a deadly dose is 10 mg/kg body weight, but only a few accidental poisonings resulted in a fatal event and survivors of these acute intoxications recovered completely without irreversible effects or residual pathologies. Diet does not seem an important source of acute contamination for the population; occupational exposure is the main hazard for industrial workers. Regarding chronic exposure and possible carcinogenetic effects, there should be no correlation with Aldrin and Dieldrin as shown by an epidemiological mortality study carried out at a manufacturing plant in the USA on 870 workers exposed to aldrin, dieldrin, and endrin. With almost 25 000 man-years of observation, no specific cancer risk associated with employment at this plant could be identified [1].

DDT is the most known OC Pesticide used worldwide in agriculture and in the prevention of malaria, typhus and other vector-born diseases. During 1959 in the USA more than 35,000 tonnes of DDT were used and it is known that DDT has been manufactured in many parts of the world including the developing countries. Subsequently the use of DDT decreased until it was banned. The term DDT refers to 1,1'-(2,2,2-trichloroethylidene)-bis(4-chlorobenzene), but it is also applied to commercial products consisting predominantly of 4,4'-DDT together with some 2,4'-DDT and smaller amounts of other compounds. A typical example of technical DDT has the following composition: 4,4'-DDT, 77.1%; 2,4'-DDT, 14.9%; 4,4'-DDD, 0.3%; 2,4'-DDD, 0.1%; 4,4'-DDE, 4.0%; 2,4'-DDE, 0.1%; and unidentified compounds, 3.5%. The main degradation product of 4,4'-DDT is 4,4'-DDE. Effects of 4,4'-DDT on human beings are well documented, with an observed little acute toxicity, if we except such cases of high poisonings. Doses as high as 285 mg/kg did not cause death, however one dose of 10 mg/kg can result in illness in some people. In laboratory animals DDT can cause death, increases liver weight and is carcinogenic, mainly causing liver tumours. DDT poisoning in humans results in dizziness, headache, tremor, confusion and fatigue. A high DDE concentration in breast milk has shown an association with a shortened duration of lactation. Other toxicological evidences show possible disruptions in semen quality, influence on menstruation, and on gestational length, with preterm birth [4]. Although a great number of studies were carried out in order to evaluate possible carcinogenic effects of DDT on humans, none of them led to conclusive results, neither positive nor negative. DDT is a possible carcinogen as it is classified by IARC, International Agency for Research on Cancer (Group 2B) [5] and further studies are needed. A weak correlation between DDT exposure and pancreatic cancer risk was repeatedly reported [4], but it was not always clearly shown. The same is referred for DDT-liver cancer correlation [4] and for lung cancer [5]: neither these works showed fully convincing results. Some aspects have to be ascertained (the role of 2,4'-DDT, for example, the most estrogenic isomer). At now we may say that DDT is only a suspected carcinogenic agent.

Chlordane has been used for more than 35 years as a broad-spectrum contact insecticide, mainly on non-agricultural crops and on animals. In the USA its use is now restricted to underground termite control. It was first prepared in the 1940s, described as an insecticide in 1945, and produced commercially in the USA in 1947. Chlordane is not produced in Europe nor has it ever been manufactured in Japan. In Japan, the only permitted use of the compound is for the control of termites. Both the amounts of chlordane produced and used have decreased considerably in recent years.

The acute lowest dose of Chlordane leading to a fatal outcome is estimated to be 25 - 50 mg/kg body weight. Effects of acute ingestions are mainly at gastrointestinal and neurological level with tremors, ataxia (incoordinaton) and convulsions. Survivors of poisonings recovered completely. No sufficient data are available to judge the potential carcinogenicity of Chlordane for man: in certain works only small excesses of bladder, skin or lung cancer were observed, but these were not statistically significant. Workers in a Chlordane-manufacturing plant were studied in 1981 and a slight excess of stomach cancer was observed, but the number of workers studied was small and further follow-up of the cohort was recommended by the authors [6]. IARC Group: 2B [5].

Endrin was used in Agriculture as a foliar insecticide against insects, mainly Lepidoptera. Crops protected included cotton, rice, sugar, maize.

Endrin is a very toxic compound. Several episodes of fatal and non-fatal poisoning have occurred, mostly from accidental contamination of food and also from intentional (suicidal) ingestion. The lethal oral dose is estimated to be 10 mg/kg body weight. In non-fatal cases, recovery is rapid and complete within a few days.

In some cases of poisoning, dizziness, weakness of the legs, abdominal discomfort and nausea, were observed. Severe poisoning is manifested by epileptiform movements and convulsions. The oral dose that causes a single convulsion is estimated to be 0.25 mg/kg body weight, and that which induces repeated convulsions, 1.0 mg/kg body weight. No sufficient data are available to judge the carcinogenicity of Endrin: The International Agency for Research on Cancer (IARC) concluded in 1974 and 1987 that there was inadequate evidence for the carcinogenicity of endrin in experimental animals and that the evidence from studies in humans was inadequate. Endrin was therefore classified in Group 3: not classifiable as to its carcinogenicity to humans [7].

HCB (Hexachlorobenzene) had Agriculture and Industry applications. The main use in Agriculture was in preventing growth of fungi in seeds. In industry it was employed in the manufacture of pyrotechnics, tracer bullets and as a fluxing agent in the manufacture of aluminium. Other uses involved wood-preservation, porosity-control in the manufacture of graphite anodes, and peptization in the production of nitroso and styrene rubber for tyres. After HCB production ceased in most countries, it was still generated inadvertently as a by-product and/or impurity in several chemical processes. HCB is a persistent chemical that bioaccumulates owing to its lipid solubility and resistance to breakdown. An accidental poisoning that occurred in Turkey in the 1950s revealed effects of the acute exposure to HCB. Disturbances in metabolism, dermatological lesions, hyperpigmentation, hypertrichosis, enlarged liver, enlargement of the thyroid gland and lymph nodes, and (in roughly half the cases) osteoporosis or arthritis was observed, mainly in children. There is also limited evidence that the same symptoms occur in humans with relatively high exposure to HCB in the workplace or in the general environment. Breast-fed infants of mothers exposed to HCB in the "Turkey incident" developed a disorder called pembe yara (pink sore), and most died within a year. Even though the few available epidemiological studies of cancer are limited by small size, poorly characterized exposures to HCB and exposure to numerous other agents (and are therefore insufficient to assess the carcinogenicity of HCB to humans) animal studies have shown that HCB causes cancer and affects a wide range of organ systems including the liver, lungs, kidneys, thyroid, reproductive tissues and nervous and immune systems. For these reasons Total Daily Intakes (TDI) of 0.16 μg/kg body weight/day for neoplastic effects and 0.17 μg/kg body weight/day of HCB for non-cancer effects are the values suggested by the WHO Task Group on Environmental Health Criteria [3].

Heptachlor was isolated in 1946 from technical chlordane in both the USA and the Federal Republic of Germany. It has been used as an insecticide, mainly in the control of termites and soil insects. In 1970, the world use of heptachlor was as follows: Africa 5%, Asia 15%, Canada and the USA 5%, Europe 60%, and South America 15%.

There is no information about acute poisonings caused by Heptachlor and so symptoms are quite unknown. Studies on occupational exposure were carried out, but no conclusive results were achieved. For example small excesses of bladder cancer in some operators and of skin and lung cancer in other operators were observed, but these were not statistically significant. In 1982, an IARC Working Group concluded that the above studies were inadequate to evaluate the possible carcinogenicity of heptachlor for human beings. Three recommendations of the WHO Task Group on Environmental Health Criteria can be cited:

a) more information on human exposure to Heptachlor from various sources is required

b) further research is required in order to better assess the significance for man of the carcinogenic findings in mice
c) continuing epidemiological studies should be made on workers who, in the past, have been exposed to heptachlor [8].

γ-HCH (LINDANE) and other HCHs were mostly used as insecticides.

HCHs are a series of isomers with the same chemical formula and different spatial arrangements of Chlorine atoms. At least four isomers constitute technical HCH, widely used: α-HCH, β-HCH, δ-HCH and the most toxic isomer γ-HCH (Lindane). In case of poisoning it's difficult to evaluate the toxicity of each isomer in a separate way, especially as regards α-, β- and δ-HCH, in fact Lindane is much more toxic of them even at trace level. Due to this fact it's easier to evaluate the Toxicity of Lindane (γ-HCH) than the toxicity of other isomers (impurities of α-HCH, β-HCH, and δ-HCH in formulations mainly composed by Lindane are not toxicological relevant with respect Lindane itself).

Several cases of fatal poisoning and of non-fatal illness caused by Lindane have been reported, which were either accidental, intentional (suicide), or due to gross neglect of safety precautions or improper uses of medical products containing Lindane. Symptoms included nausea, restlessness, headache, vomiting, tremor, ataxia, and convulsions and/or changes in the electroencephalographic pattern. These effects were reversible after discontinuation of exposure or symptomatic treatment. There is no evidence for the relationship suggested in some publications between exposure to Lindane and the occurrence of blood dyscrasias. A few acute and short-term studies in humans indicate that a dose of approximately 1.0 mg/kg body weight does not induce poisoning; however, a dose of 15-17 mg/kg body weight resulted in severe toxic symptoms. Approximately 10% of a dermally applied dose is absorbed, although more passes through damaged skin. There is no evidence for possible carcinogenetic effects of Lindane, but Long-term carcinogenicity tests according to present-day standards should be conducted [9].

We must emphasize that all pesticides here mentioned clearly showed carcinogenic effects and reproductive toxicology on experimental animals, mainly mice (with the exception of Endrin, for which data are not exhaustive). However is of big relevance the fact that these pathologies are peculiar to rodents and, therefore, interpretation of their significance for humans is difficult.

2.2. Diffusion, Transport and Biomagnification

Human exposure to OC Pesticides occurs both from environmental and food-chain uptake.

High lipid solubility, low water solubility and resistance to breakdown lead to the retention of Organochlorine pollutants and their metabolites in fatty tissues; subsequently they tend to migrate worldwide, therefore toxic effects can occur in organisms remote in time and geographical area from the point of exposure. In Figure 1 a scheme of Organochlorine transport and diffusion is shown: some important sources of pollution are waste products, agriculture, and human activity on the sea. As a result of evaporation, deposition, water contamination and atmospheric transport the presence of these pollutants, at trace level, is global and uniform in both hemispheres.

Marine Ecosystems are particularly exposed to pollution being at the lowest altitude in the terrestrial Environment and so acting as an "end-point" of any type of pollutant created on Dry Lands (even if the vapour form of POP's can be transported by atmospheric diffusion also in quite-unbelievable sites [10]). A similar reasoning is valid for rivers and lakes that are at lower level than fields and hills around them.

The concentration in water of OC Pollutants is generally very low, but the organic matter tends to capture and concentrate them. For this reason from a low concentration in water a high concentration in fishes can be reached, especially in predatory fishes.

This phenomenon is called "Biomagnification" and can be explained by taking into account that in humans and animals food is mostly consumed in performing biochemical reactions needed to Life (energy and other). Pollutants are not needed to Life (Xenobiotics) and tend to accumulate in organisms without any specific function. During some years an adult predator eats many other animal species together with their fat and possible contaminants. In the same period the weight of predator can stay quite constant, this means that all victims' fat was consumed by predator in performing biochemical reactions while pollutants accumulated and so increased Organochlorine contamination of predator's fat. This fact explains why a predator always increases his contamination even in eating a species with a lower contamination than itself.

Moreover that predator is often the prey of another species. That's why in raising the Food Chain we observe raising levels of POP's (Biomagnification).

More the energy required for metabolic processes higher the food requirement. Homoeotherm predators (Mammals) need larger quantities of food than Poikilotherm predators (Fish) and therefore higher pollutant uptakes can be expected in marine mammals and birds. Such reasoning is confirmed by a number of studies reporting the largest Biomagnification Factors for OC pesticides usually observed between fish and marine mammals or sea birds [11][12][13][14]. The Biomagnification Factor is the concentration fat-basis of a contaminant in predator divided by the concentration fat-basis of the same contaminant in prey:

Biomagnification Factor = Pesticide Concentration fat basis in predator/ Pesticide Concentration fat basis in prey

A parameter that may be useful in predicting OC Pesticide Biomagnification Factors is K_{OW}, the octanol-water partition coefficient. Octanol is an organic solvent that is used in replacing natural organic matter in laboratory research studies. In a mixture of Water and Octanol a chemical will reach an equilibrium with a definite concentration in water and a definite concentration in Octanol. K_{OW} is expressed as:

K_{OW} = Pesticide Conc. in Octanol / Pesticide Conc. in Water

Greater the K_{OW} higher the Biomagnification Factor. In this way it is possible to predict which chemicals will accumulate more easily at the top of the Trophic Chain by means of controlled experimental measures [15].

2.3. Fish Products

Currently the dietary habits of industrialised countries are characterised by elevated intakes of high-calorie food, particularly those rich in simple sugars and saturated fatty acids, responsible of the incidence of obesity, cardiovascular diseases as well as of metabolic disorders like diabetes, hypercholesterolemia, hypertriglyceridemia. Therefore the dietary guidelines of international health societies suggest to lower the total calorie and saturated fatty acid intake, increase the consumption of fibre-rich items, like fruits and vegetables, use alcohol moderately and increase the assumption of fish, rich of n-3 fatty acids, in place of red meat.

The health benefits associated with a diet rich in seafood products have been documented in several epidemiological and clinical studies in the recent years. In fact, due to the peculiar composition of their lipids, seafood products are the only significant sources in the human diet of polyunsaturated fatty acids, in particular those of the n-3 family (eicosapentaenoic acid, 20:5; docosahexaenoic acid, 22:6), precursors of prostaglandins and tromboxanes with platelet anti-aggregating activity and vasodilator effects. Their consumption has been positively correlated with the low incidence of atherosclerosis and cardiovascular diseases in some populations.

In addition, seafood products are a source of proteins of high biological value, fat-soluble and water-soluble vitamins and minerals, and therefore represent a valid alternative to other protein-rich food of animal origin (mammalian or avian meat, eggs, and dairy products). The tender, delicate texture and high digestibility of its flesh, both related to the reduced presence of connective tissue, make fish a food suitable for children, athletes and elderly people. All these factors may have contributed to the increment of fish consumption registered worldwide in the recent years.

3. Experimental

Experimental section is of big relevance and requires a detailed explanation.

We must consider that the substances we search for are generally at a very low level: 10^{-6} - 10^{-4} g per kg of edible portion or less, that's to say 0,0000001% - 0,00001%. In this situation is very difficult both to purify and detect these pollutants. A complex preparation step is required to eliminate interfering compounds present in very large quantities and capable to completely hinder instrumental detection of Pesticides. Losses can occur if work is not properly done or the method is not suitable (False Negatives). On the other hand we must remember that OC Pesticides residues at trace levels may be present in soil, air, dust and other and so they may accidentally added to the sample during the analysis (False Positives) if attention is not paid that a totally free-contamination procedure is applied [16].

The same is valid for instrumental analysis when incorrect settings are applied and the poor sensitivity (or peak overlapping) generates a False Negative or when only one retention time is used to affirm that a Pesticide is present in the sample (False Positive). In particular we observed that no accurate and repeatable responses are obtained if a perfect and regular instrumental maintenance is not carried out.

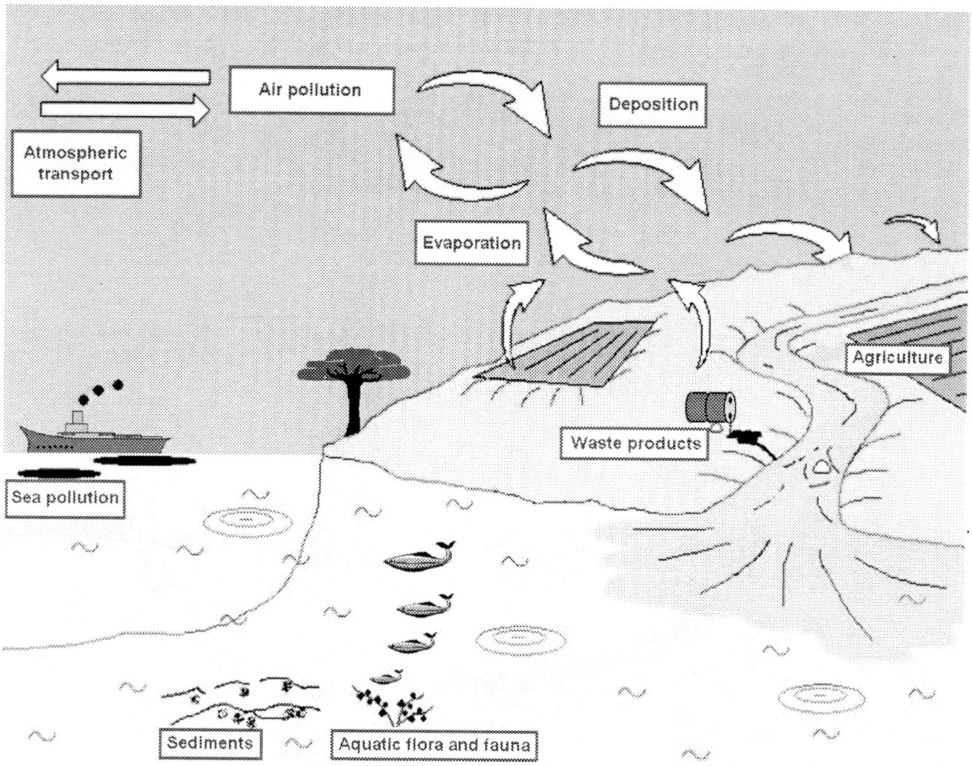

Figure 1. Environmental transport and diffusion of POP's (Persistent Organic pollutants).

3.1. Materials

Aquaculture

Four different Sea Farms located in central and south Italy were considered in this study. The Aquaculture product most requested by the Italian market is the species *Sparus aurata* (Sea Bream) and to a certain extent *Dicentrarchus labrax* (Sea Bass).

For each Sea Farm multiple samplings were carried out. In each sampling three individuals of Sea Bream were selected, pooled and prepared for the analysis.

In this study also feedingstuffs used in farming Fish were analyzed. The sample preparation was modified a lot, especially as regards the fat extraction step, in order to develop a suitable method for Organochlorine analysis in fishfeed samples, quite different from fish fillets. After fat extraction efficiency was repeatedly demonstrated, recovery tests on real samples were carried out to ensure that accurate data are obtained.

Fish from Lakes

Ecosystems such as lakes could be of high concern as regards the presence of persistent pollutants, being limited the volume and the recirculation rate of the entire water body. If any evenience occurs that leads to the immission of contaminants in the basin it is very likely that the contamination remains quite stationary for many years with deposition and resolubilization processes that can occur in a cyclic way.

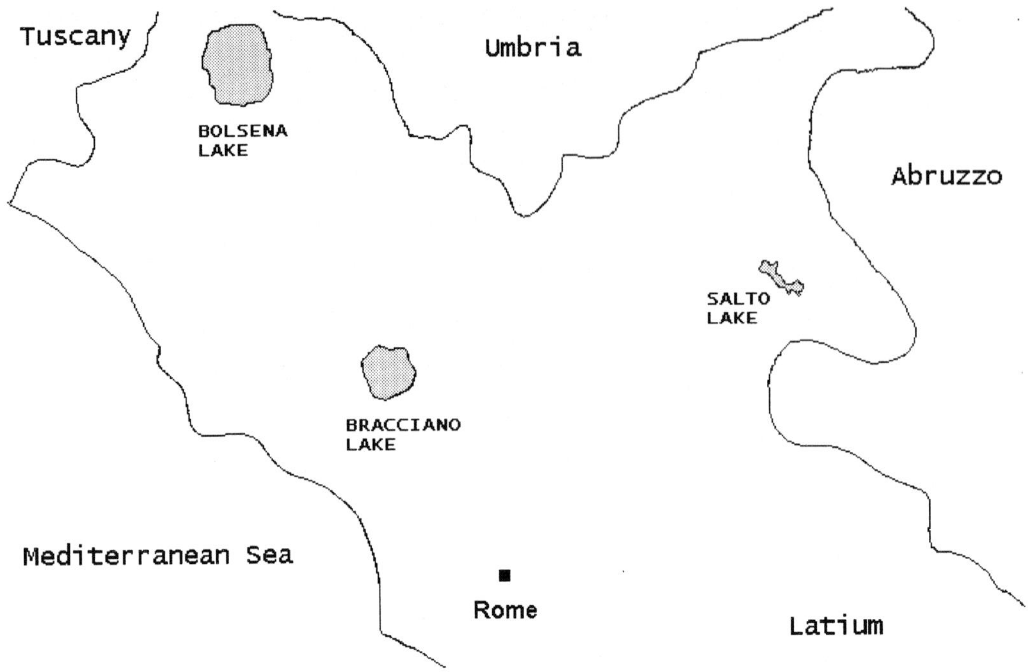

Figure 2. Map of Central Italy.

Fishes from three Italian lakes were analyzed in order to ascertain the possible presence of Organochlorine Pesticides. Lakes considered are situated in Central Italy, in the Latium Region (see Figure 2) and are of some relevance as regards Fishery; they are Bolsena lake, Bracciano lake and Salto lake.

Bolsena and Bracciano lakes are of volcanic origin with surfaces of 114 and 58 Km^2 respectively, deeps of 150 and 164 meters, and located 305 and 164 meters above sea level. They are nice places also known as touristic sites. Bracciano lake is an important drinking water reservoir.

Salto lake is an artificial basin mainly used in producing hydroelectric energy. It measures 8 km^2 in surface and 90 meters in deep and is located 535 meters above sea level.

Species collected for analyses were the most known and marketed among the ones living in these lakes. They were Eel (*Anguilla anguilla*), European perch (*Perca fluviatilis*) and European whitefish (*Coregonus lavaretus*). No samples of Eel were collected in the Salto lake.

Bluefin Tuna Sea Ranching

In recent years a new activity in the field of Fish-Commerce became of great relevance: the "Bluefin Tuna Sea Ranching".

Among Tunas Bluefin Tuna (*Thunnus thynnus*) is the species of largest size being 600 kgs or more the weight of adult individuals. It's a migratory fish, which can easily swim the entire Atlantic Ocean covering a mean distance per day of about 180 kms. *Thunnus thynnus* is a widespread species that can be found either in the Pacific or in the Atlantic Ocean. The North Atlantic population has two main reproductive areas, the Gulf of Mexico and the Mediterranean Sea. Historically the Mediterranean Bluefin Tunas were captured by means of big stationary fishing nets in which fishes could only enter and there were trapped and caught. An increasing demand by Japanese consumers for Mediterranean *Thunnus thynnus* made this type of fishing inadequate to the requirements of the Market that employees this fish in making typical Japanese products named "Sushi" and "Sashimi".

A new approach ("Bluefin Tuna Sea Ranching") was developed that allows big quantities of Bluefin Tuna be available for sale. Tunas are massively captured by using big fishing boats and modern search devices, and then are transported in special cages not far away from the coasts (with the exception of big individuals that are readily filleted and deep-freezed). In cages Bluefin Tunas are fed (more exactly "fattened") with large quantities of bait such as mackerels, sardines, anchovies and other, until they reach the wanted size of 200-300 kgs: it generally needs a period of 4-6 months.

Only peripherals portions of each Tuna were available for analyses. Anyhow we believe that the collected parts can give a good indication about OC Pesticide contents in this type of products.

We executed the sampling in November 2004, when the selected Tunas were fed in cage for four months and were therefore ready to be marketed.

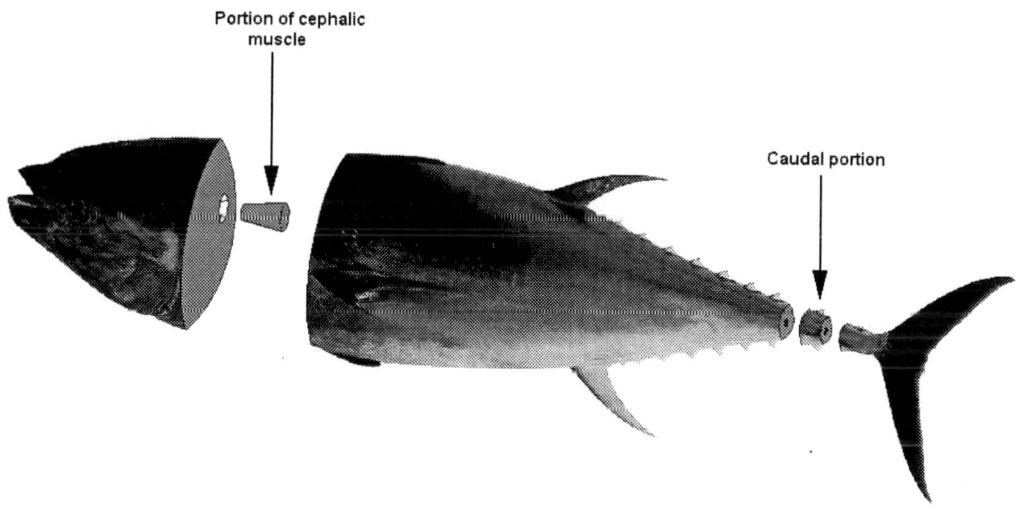

Figure 3. Portions of Bluefin Tuna collected for analyses.

Four *Thunnus thynnus* caught from a cage located 3,5 kms away from coasts of south Italy were chosen. From them samplings of a caudal section and a portion of cephalic muscle situated between the eye and the rear edge of the operculum were carried out (see Figure 3).

Analysis of the eight samples collected in this way was achieved in our Laboratory, as usual. As reference samples caudal portions of two individuals of *Thunnus thynnus* not in cage-fed were delivered to us (wild individuals with big size, as reported above).

Table 1. Organochlorine Pesticides* selected for analysis in the Fish Quality Unit of INRAN (some isomers are not reported)

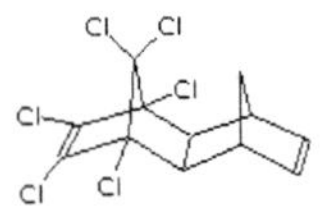 Aldrin	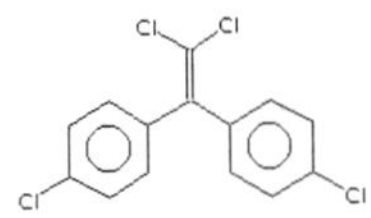 4,4' - DDE	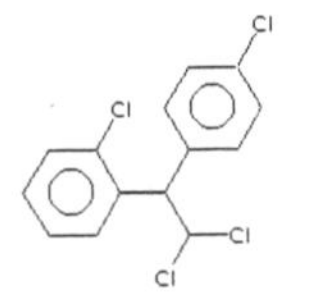 2,4' - DDD	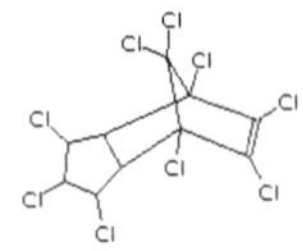 Nonachlor
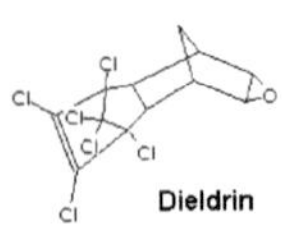 Dieldrin	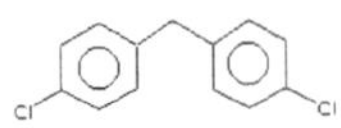 4,4' - DDM	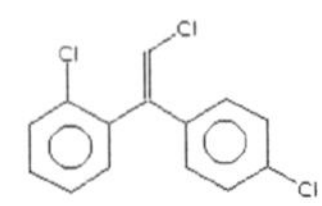 2,4' - DDD olefin	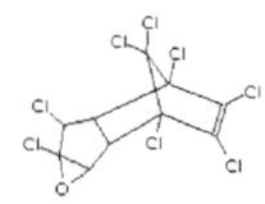 Oxychlordane
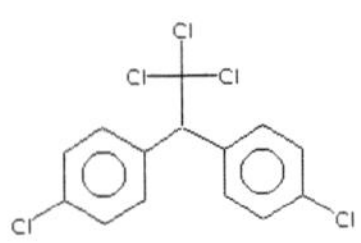 4,4' - DDT	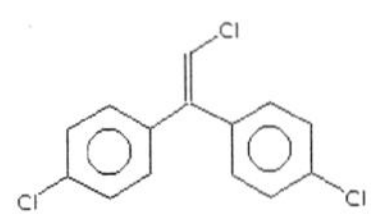 4,4' - DDMU	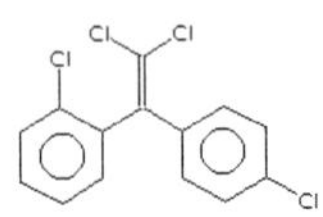 2,4' - DDE	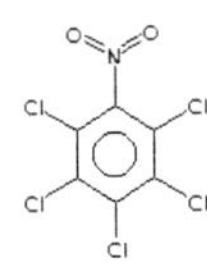 Quintozen (PCNB)
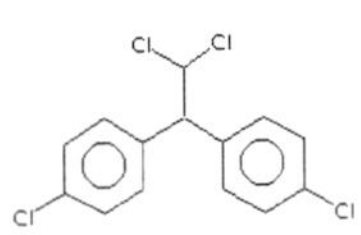 4,4' - DDD	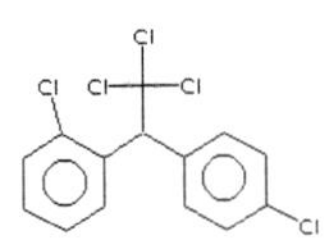 2,4' - DDT	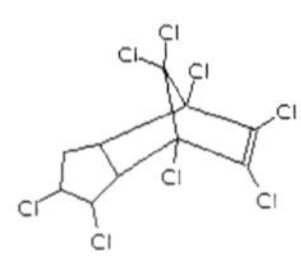 Chlordane	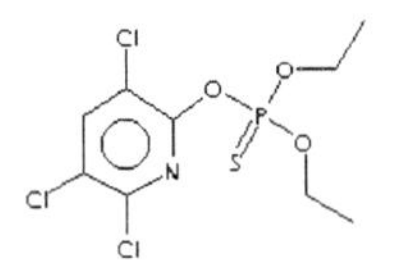 Chlorpyriphos
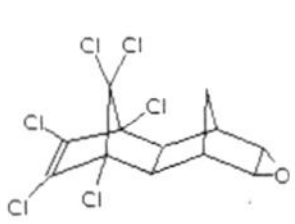 Endrin	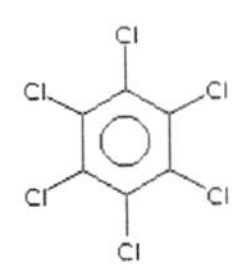 HCB	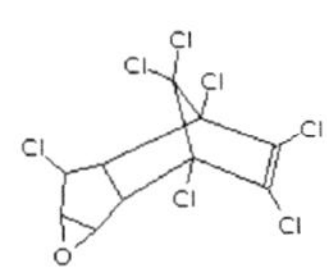 Heptachlor epoxide	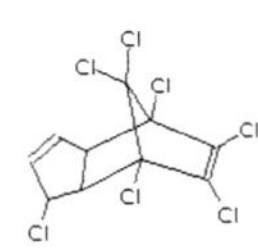 Heptachlor
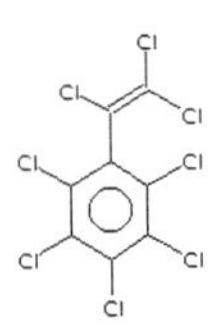 Octachlorostyrene	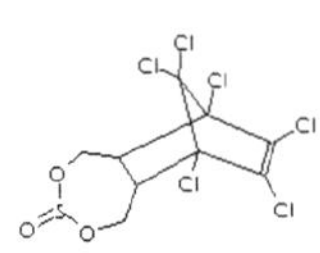 alpha - Endosulfan	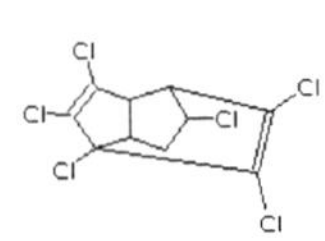 alpha - Chlordene	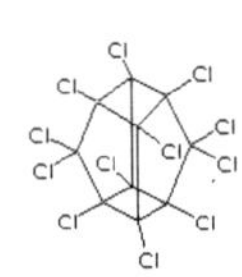 Mirex

* NIST 98 MS Database © United States Government.

As regards chemicals Acetonitrile, Acetone, Isooctane, n-Hexane, Petroleum ether 40-60°C, Toluene, methyl Alcohol, Dichloromethane, Ethyl acetate, Sodium sulphate and Florisil 60-100 mesh of Pesticide-Grade purity were purchased from Carlo Erba® (Milan, Italy).

The Supelclean LC-18 solid phase was from Supelco® (Bellefonte, PA, USA).

Pure standards of OC Pesticides (Table 1) were purchased from different Producers, mainly in solution form, along with their certificates of analysis. Certified solutions, as single standard or mixtures, were from Dr. Ehrenstorfer® (Augsburg, Germany), AccuStandard® Inc. (New Haven, CT, USA), Supelco® (Bellefonte, PA, USA) and Riedel-de Haen/Fluka/ SIGMA ALDRICH® (Switzerland).

3.2. Sample Preparation

Fish samples were collected directly by us or by well-trained personnel, stored on ice and immediately transported to our Laboratory. Here they were subjected to deep-freezing (-30°C) until the day of analysis.

Three individuals were mostly pooled for the determination of residual Organochlorine Pesticides. After the viscera were separated from the muscle (by accurately avoiding any type of cross-contamination, especially from the liver that generally tends to accumulate Xenobiotics) the edible portion was prepared as following.

Samples were filleted by means of contamination-free tools then homogenised in a Waring® blender 38BL40 (Waring®, New Hartford, CT - USA).

Owing to their lipophilic nature the pollutants were extracted from the samples together with the fat. For sample purification and clean up the method of Di Muccio et al. [17], with minor modifications [18], was applied.

Extraction of Pesticides together with the Fat was achieved by using Acetone: Petroleum Ether 1:1 for three times and with an Ultra Turrax homogenizer (Model T25B, IKA®, Staufen, Germany) as dispersing tool.

A first purification was obtained by using Diatomaceous Earth (Extrelut NT3 and NT1, Merck® KgaA, Darmstadt, Germany) and a C18 phase (Supelclean LC-18, Supelco®, Bellefonte, PA, USA).

In this way most of the Fat was trapped and separated from the Pesticides that were recovered in Acetonitrile.

The final purification was achieved by employing Florisil 60-100 mesh (Carlo Erba®, Milan, Italy) placed in a glass column with such a diameter that a Liquid Chromatographic separation with a flow of mobile phase equal to 0.5 mL/min at atmospheric pressure took place. Most of OC Pesticides were eluted by means of n-Hexane (30 mL) but some of them needed 30 mL of a n-Hexane/Toluene/Ethyl acetate mixture (79.6/19.9/0.5 v/v/v) to be recovered. The method here described allows good sample purification and a preliminary separation of pesticides (clean up).

After the sample was brought to a small volume and the internal standard was added the instrumental analysis had to be performed.

3.3. Instrumental Analysis

In analyzing pollutants at trace levels in a complex matrix like a fish a great attention was paid by us in setting and routinely applying powerful instrumental methods.

We developed a procedure that extensively uses all possibilities of the Mass Spectrometer equipped with the Ion Trap as analyzer device.

Even if this method can be a lot time-consuming in some complex cases, in this way reliable results are always obtained. In any case, and in order to save time, we always took advantage of all hardware and software automation available.

To our knowledge no one has still applied the entire instrumental procedure here presented when OC Pesticides in fish samples are determined.

Instrumental analyses were performed with a Varian® 3800 Gas Chromatograph (Varian®, Walnut Creek, CA, USA) equipped with the Mass Spectrometer Saturn 2000, an ECD Detector, two injectors and two columns having different polarities (both 30 m, 0.25 mm i.d., 0.25 μm film thickness). The first column (CP-SIL 24CB from CHROMPACK®, Middelburg, the Netherlands) was connected to ECD detector and the second (CP-SIL 8CB, from CHROMPACK®, Middelburg, the Netherlands) was connected to the Mass Spectrometer.

Instrumental GC conditions were as follows: helium as carrier gas supplied through Electronic Flow Control (EFC) in constant flow mode. Sample introduction was carried out in splitless mode by using the Varian® CP-8400 autosampler. Both injectors were maintained to a constant temperature of 250°C, while the ECD detector was settled to a temperature of 300°C. Injection volume was 1 μL or 2.5 μL in the "ECD-column" and 2 μL or 2.5 μL in the "Mass-column".

Instrumental MS parameters were adjusted and calibrated each time according to the type of investigation needed (GC/MS, GC/MS/SIS, GC/MS/MS, other). In any case Ion Trap and Transfer Line temperatures were always maintained constant at 210 and 250°C respectively, and Electronic Ionization, EI, was always used.

Qualitative and quantitative analyses were carried out by using a number of standard reference solutions. Each contaminant, with respect the pure standard, had to match a lot of confirmation criteria for a positive identification: ECD retention time, MS retention time, mass spectrum. When necessary MS/MS determinations were performed.

Each sample was injected in both columns. Then, after the injections of some standard solutions were achieved, both detectors (ECD and MS) were searched for contaminant responses in the sample.

3.3.1. Identification and Confirmation

When ECD indicated that a contaminant may was present in the sample (Figure 4), GC/MS analysis followed the procedure here described (example: Hexachlorobenzene). Full scan and extracted ion chromatograms (Figure 5) at the exact retention time of Hexachlorobenzene, were first evaluated. Then comparison between the mass spectrum of the sample peak (Figure 6) and the spectrum of Hexachlorobenzene standard injected by us (Figure 7) was made. Spectrum present in the NIST© database was also taken into account. We often noted (especially at very low concentrations and with intense matrix interferences) that the full scan GC/MS chromatogram does not show the peak of substances at trace levels; however the same substances exhibit their peaks when the chromatogram is reviewed in extraction ion mode. This is just the case of the sample in Figure 5.

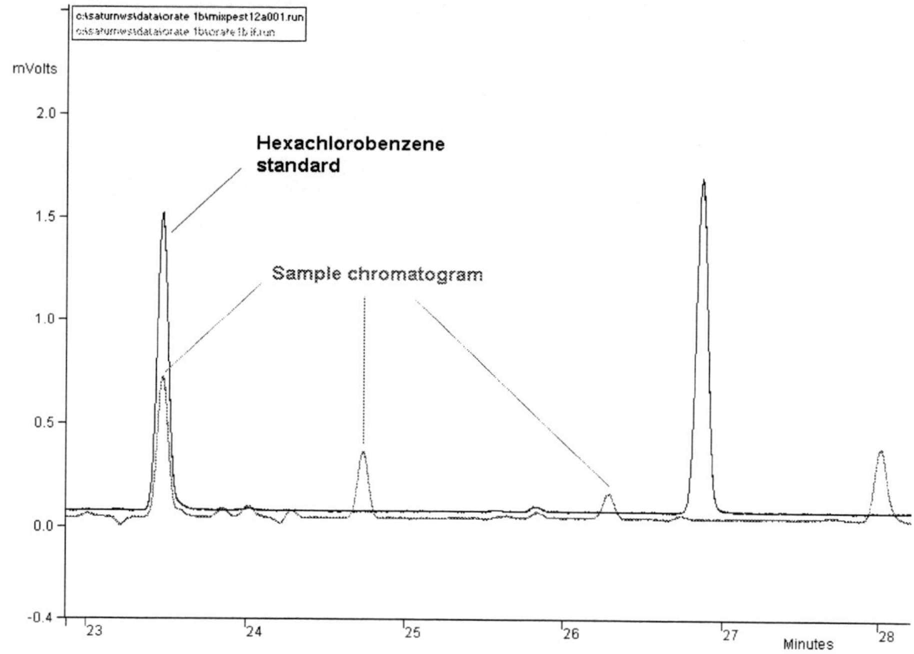

Figure 4. ECD Chromatograms of a pure standard solution (blue) and of a real sample of Sea Bream (red).

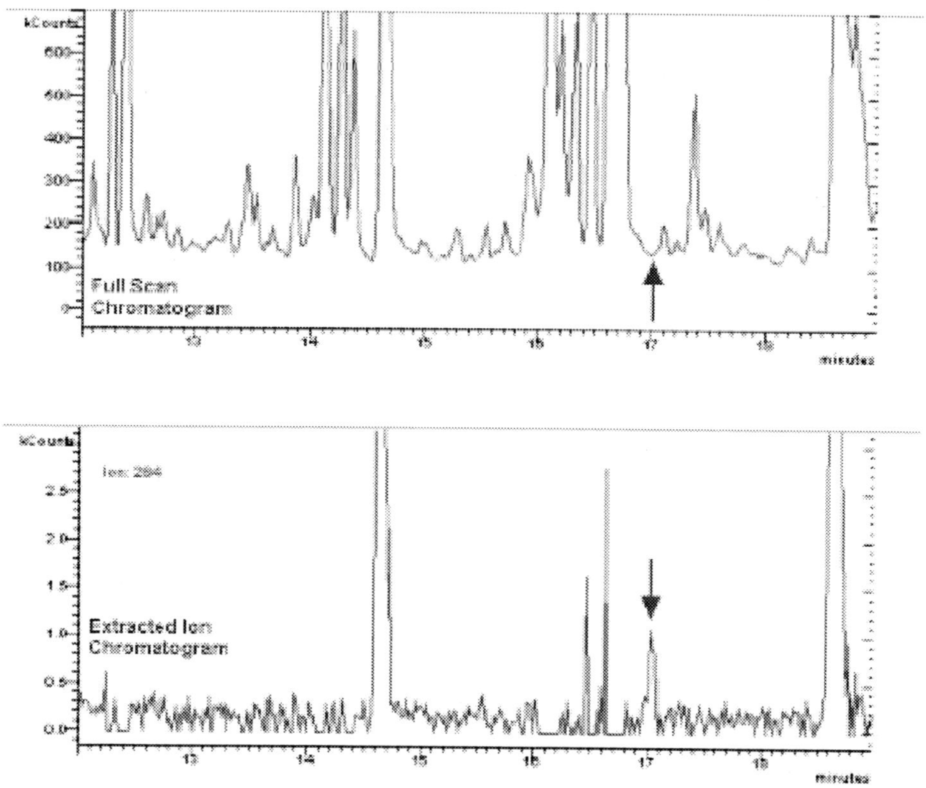

Figure 5. Full scan (upper) and Extr. ion (lower) MS Chromatograms of a real sample of Sea Bream.

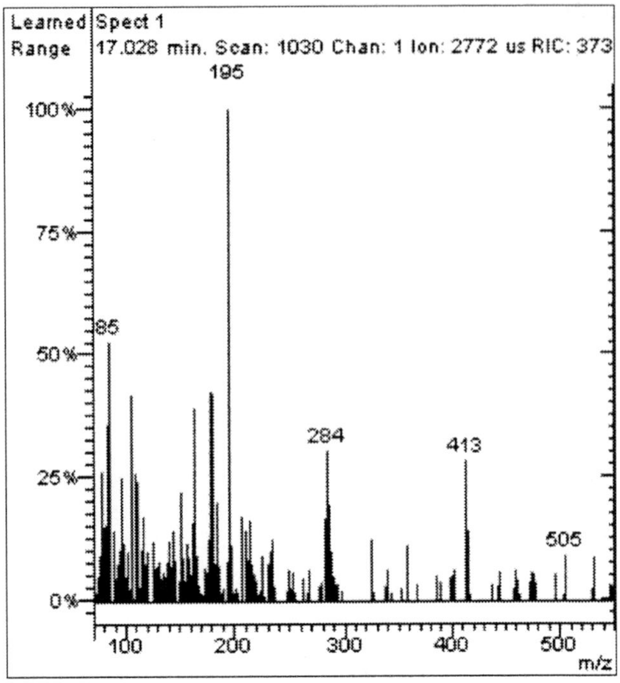

Figure 6. Mass spectrum at the retention time indicated by the arrow in Figure 5.

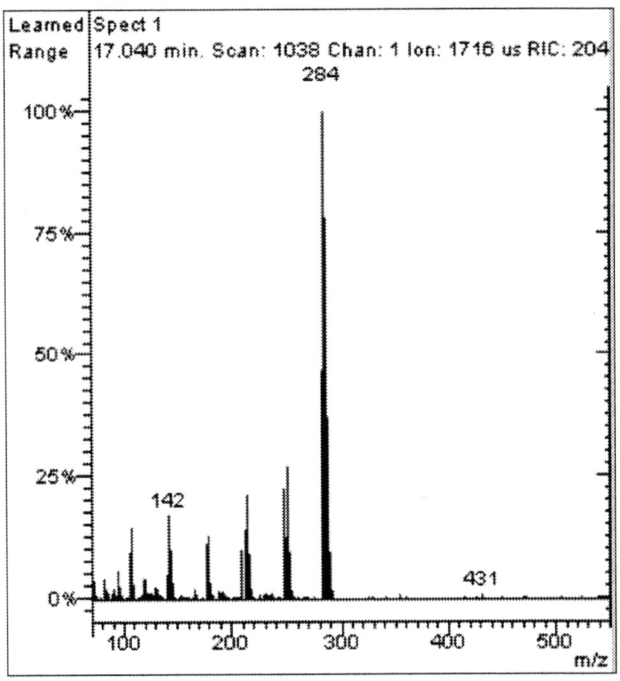

Figure 7. Mass spectrum of Hexachlorobenzene standard injected.

Order	Sim.	RSim.	Name (InLib probability = -1844)
1	409	462	4a.alpha.,4b.beta.-Gibbane-1,1,10.beta.-tricarbo›
2	393	460	Acetylcodein
3	392	463	Dinaphtho[1,2-b:1',2'-d]thiophene
4	391	490	Morphinan-6-ol, 7,8-didehydro-4,5-epoxy-3-methc
5	390	415	L-Lysine, N6-acetyl-N2-[N-[N-[N-(N2-acetyl-N,N,N2-
6	389	497	Aspidofractinine-3-methanol, 17-methoxy-, (2.alpl
7	387	450	Cobalt,[(2-pyridyl)-(1-diethylboryl-1,5-cyclooctadie
8	387	449	Thebacon
9	378	419	Ruthenium, bis[(1,2,3-.eta.)-2-methyl-2-propenyl]t
10	376	442	Naltrexone
11	373	443	Dinaphtho[2,3-b:2',3'-d]thiophene, 12,13-dihydro-
12	370	437	Benzo[c]thiophene, 1,3-diphenyl-
13	362	380	Borane, [3,8-diethyl-1,4,6,9-tetrakis(trimethylsilyl)-
14	361	414	Naltrexone
15	356	376	3,10b-Ethanonaphtho[1,2-c]pyran, 1,3,4,6,6a,7,8,
16	354	424	Morphinan-6-ol, 4,5-epoxy-3-methoxy-17-methyl-,
17	354	479	Condyfolan, 1-acetyl-11-methoxy-
18	346	610	Benzo[b]selenophene-2-carboxylic acid, 3-metho:
19	345	375	Morphin, 6-acetyl-3-O-trimethylsilyl-
20	342	398	Molybdenum, (.sigma.-allyl)(.eta.-3-butenyl)(tolue
21	341	554	Benzene, hexachloro-

Figure 8. Response of the NIST© database for the Mass spectrum showed in Figure 6.

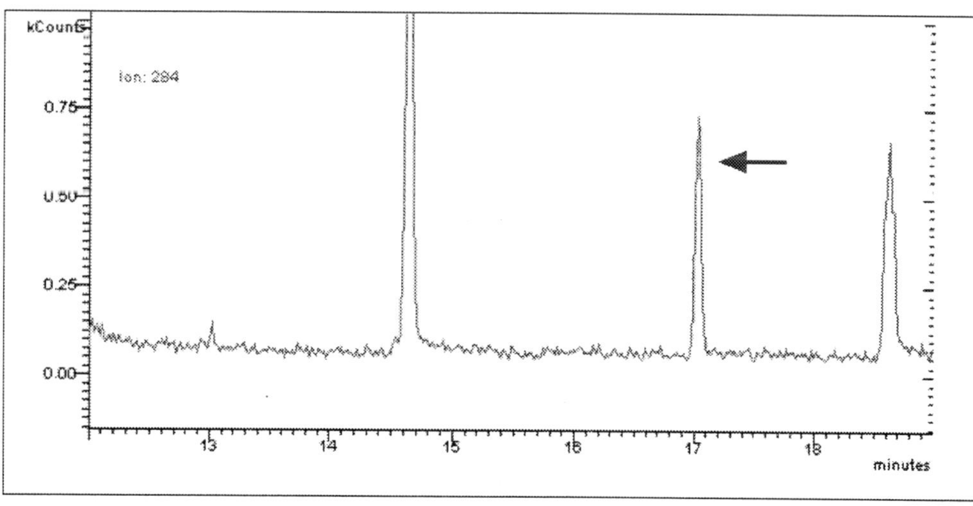

Figure 9. GC/MS-SIS analysis of a real sample of Sea Bream. Ion selected: 284.

We observe that the NIST© database (Figure 8) gave an affirmative response (note the intense 284 ion in spectrum of Figure 6) for the presence of Hexachlorobenzene in the sample, even if it was not indicated as the most probable compound. This is due to the low concentration and to the ionic interference of the matrix.

If this first evaluation was not deemed satisfactory (in any case we must remember the affirmative response of the ECD detector) the GC/MS analysis continued with runs in SIS mode (Selected Ion Storage: 284) and in GC/MS/MS mode (ion 284). In SIS mode (the storage of ions of interest and the discharge of the others) we may see the intense peak of ion 284 at the exact retention time (Figure 9). In GC/MS/MS mode (refragmentation of ions of interest) it is evident the peak at the exact retention time in the Chromatogram of the 284 ion fragments (Figure 10). Finally the MS/MS spectrum (Figure 11) of ion 284 of the sample peak of Figure 10 was compared with a MS/MS spectrum of the ion 284 relative to a pure standard of Hexachlorobenzene (Figure 12): they are practically identical thus confirming that the compound present in the real sample is really Hexachlorobenzene. For a complete investigation we wanted to perform a GC/MS/MS analysis of another peak in the same sample of Figure 10. This peak (Figure 13) is from a substance that certainly has the 284 ion fragment (that's the condition to be seen in the chromatogram of Figure 10). MS/MS spectrum of this substance (the fragments of another ion 284) is shown in Figure 14: the differences from the standard spectrum in Figure 12 are evident (because different are the ions 284).

This confirms the high specificity of the MS/MS technique even in trace analysis of pesticides.

3.3.2. Detection and Resolution of Coelutions and Interferences

GC/MS/MS technique is very powerful also in revealing complicated coelutions and interferences. In Figs. 15, 16 and 17 it is shown the GC/MS/MS analysis of Lindane (γ-HCH) in a real sample. Figure 15 reports two MS/MS chromatograms of the ion 182 (an important ion of Lindane) the upper relative to a fish real sample and the lower relative to a standard solution of Lindane. The retention times seem to indicate the presence of Lindane in the real sample and the same was for the ECD detector. Figure 16 is simply a magnification of Figure 15. In Figure 17 the MS/MS spectra at different retention times are reported. We may observe that the apex spectrum of the Lindane standard peak (Figure 17-2) is very similar to the spectrum at the left side of the real sample peak (Figure 17-1), but the apex spectrum of the real sample peak (Figure 17-4) is totally different from them.

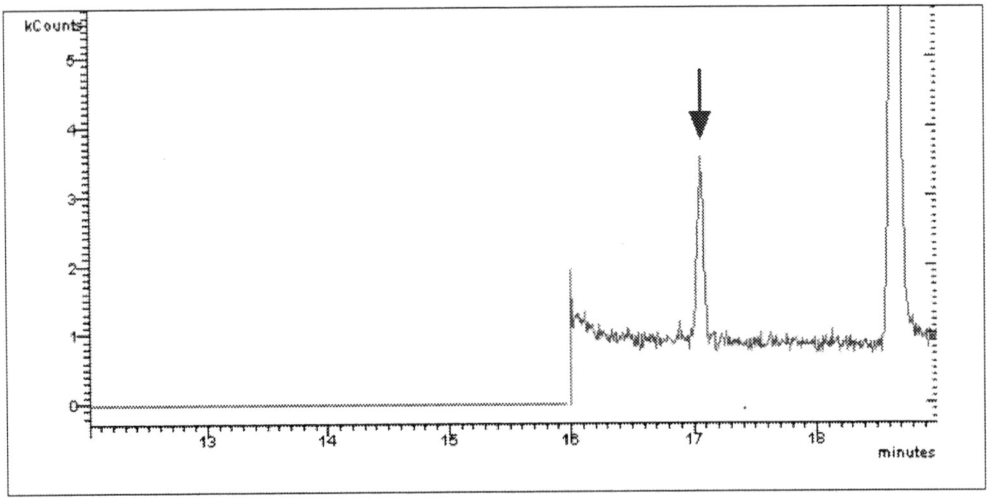

Figure 10. GC/MS/MS analysis of a real sample of Sea Bream. Ion selected: 284.

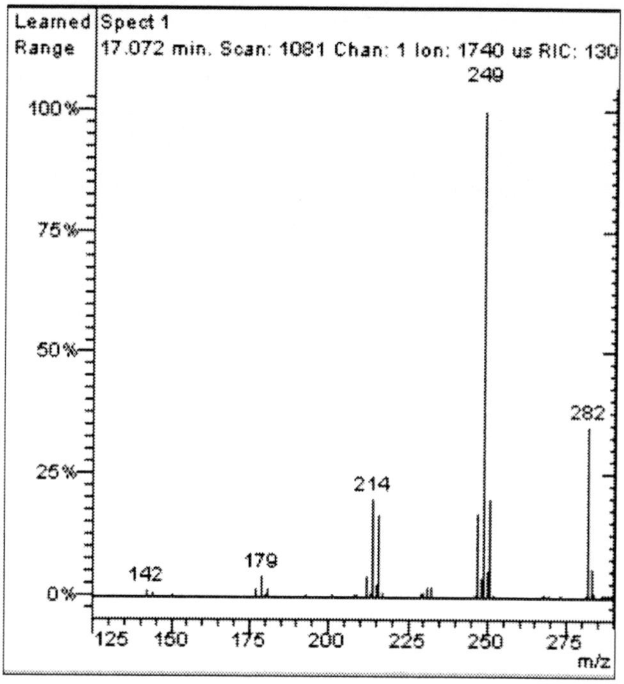

Figure 11. Mass spectrum at the retention time indicated by the arrow in Figure 10.

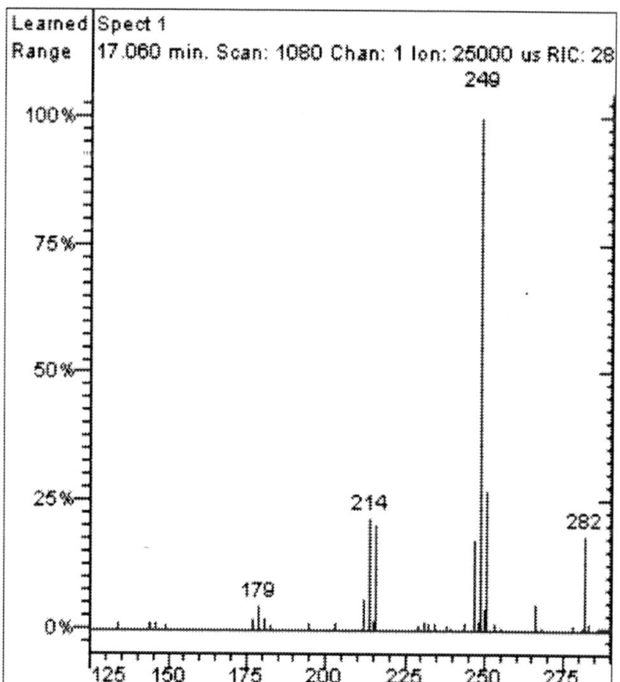

Figure 12. GC/MS/MS spectrum of the ion 284 relative to a pure standard of Hexachlorobenzene.

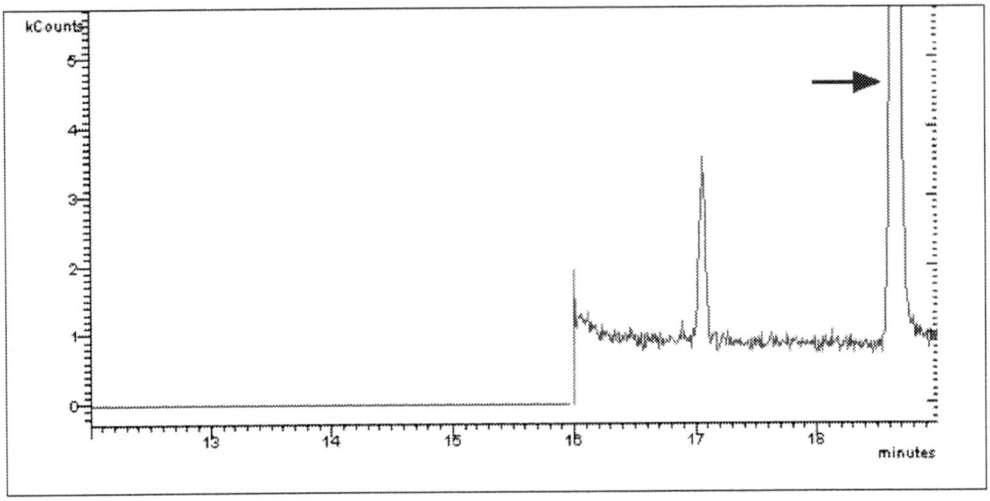

Figure 13. Chromatogram of Figure 10. The arrow points to a different peak than Hexachlorobenzene; this peak has an intense 284 ion in its Mass Spectrum.

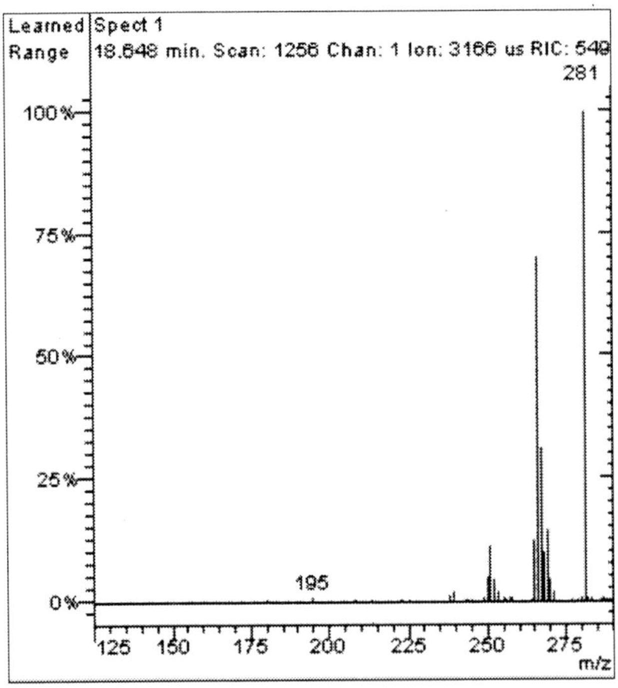

Figure 14. GC/MS/MS spectrum of the ion 284 relative to the peak indicated by the arrow in Figure 13.

The spectrum of the Lindane standard peak is the same along the entire peak (Figure 17-3). We may conclude that the sample peak is the sum of Lindane and another unexpected compound. Moreover this interfering compound has the 182 ion in its Mass Spectrum!

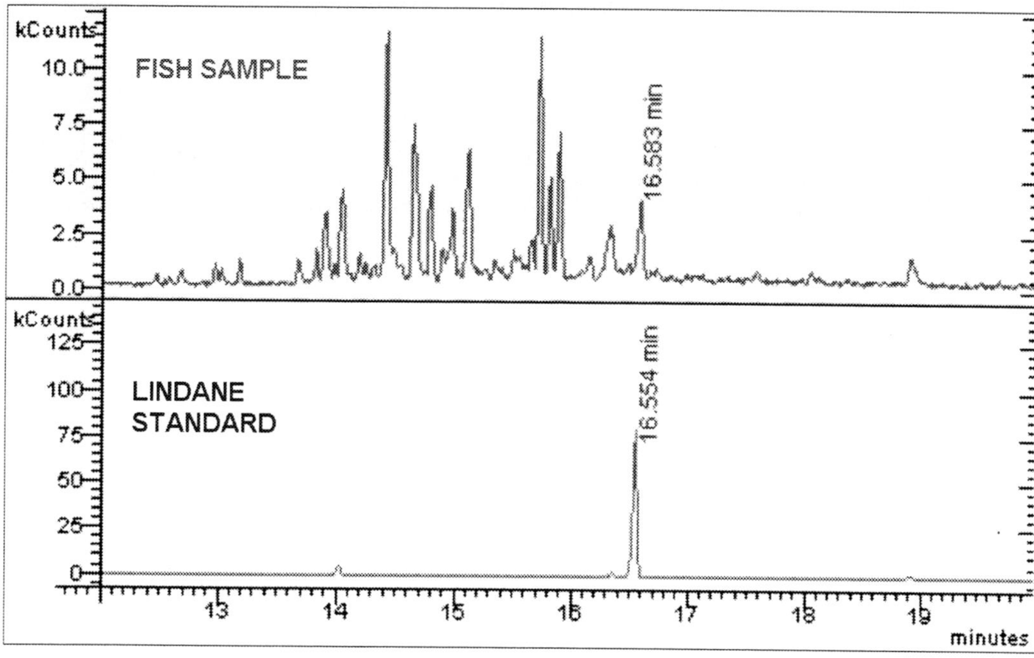

Figure 15. GC/MS/MS analysis of a real sample of Fish (upper) and of a standard solution of Lindane (lower). Ion selected: 182.

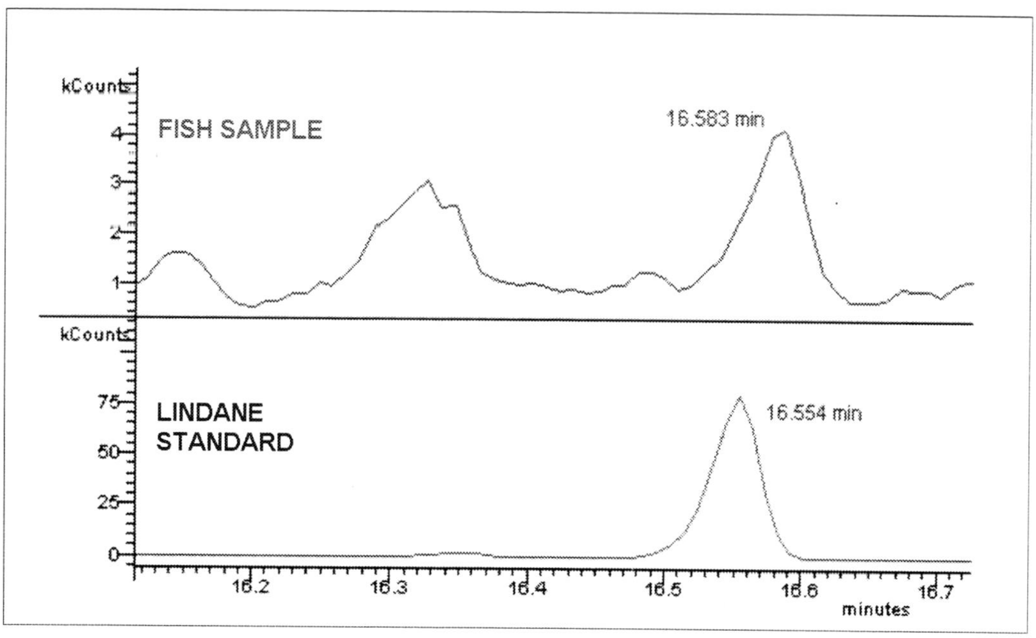

Figure 16. Magnification of Figure 15.

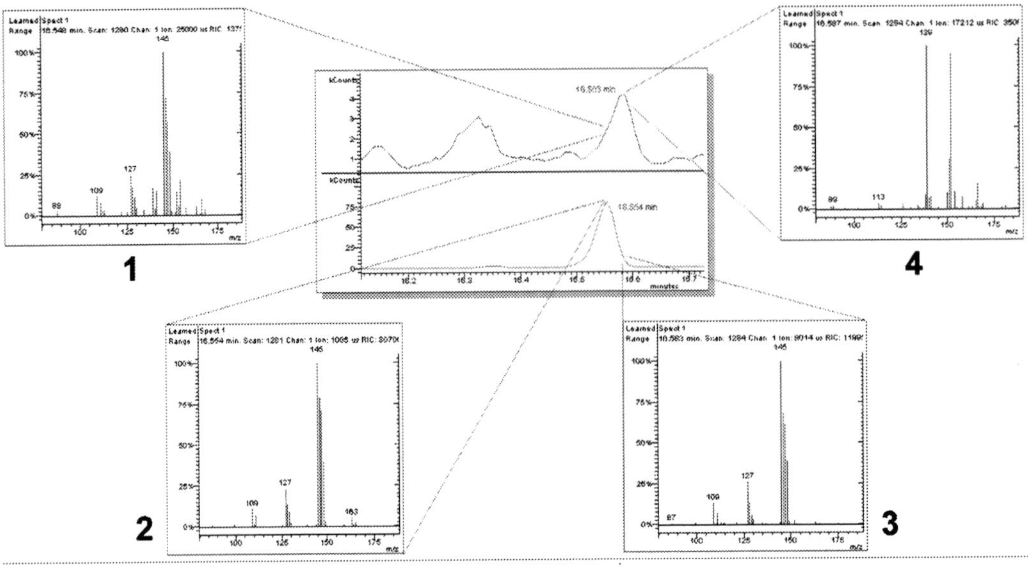

Figure 17. Mass spectra of peaks showed in Figure 16 at different retention times along the peak itself.

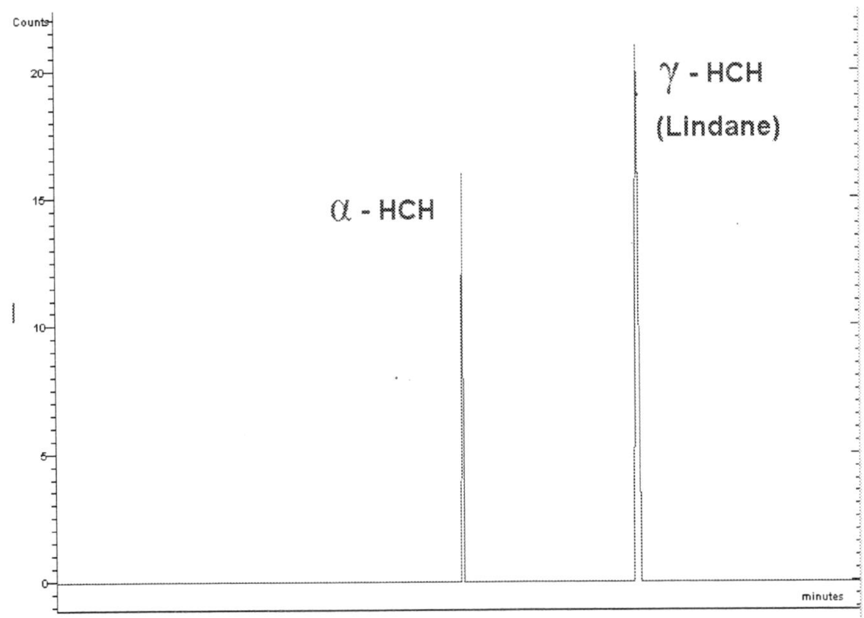

Figure 18. GC/MS/MS analysis of a real sample of Fish. Ion selected: 182. Ions held in the Ion Trap: 145, 146, 147 and 148.

When this very particular situation is observed settings of Mass Spectrometer can be useful adjusted to isolate only the substance of interest. We notice in Figure 17-4 that when the ion 182 of the interfering compound is *re-fragmented* (undergoes to a second collision to achieve MS/MS experiment) three main groups of ions are generated, 139, 152, 166. Among these there is not the ion group 145-146-147-148. On the contrary, when the ion 182 of Lindane is fragmented the very intense group of ions 145,146,147,148 is generated (Figures 17-1, 17-2, 17-3). So, sample was reinjected by adjusting the Mass Spectrometer Saturn 2000 Varian® as follows. After the first EI ionization was carried out ion 182 was selected and re-fragmented. The Ion Trap was set to eject all ionic fragments generated, except the four ions 145, 146, 147 and 148: only Lindane, if present in the sample, can generate a chromatographic peak (at that retention time).

In this way Mass specificity and sensitivity became very high: we obtained a peak of Lindane, with a signal to noise ratio of 5, by injecting 2.5 μL of a standard solution containing Lindane at 290 ng/L. This is due to the fact that the noise is eliminated, even in a very complex matrix (Figure18). In this experiment it was necessary to set the Electron Multiplier to a value of 1600 Volts by means of solvent injections in order to obtain a perfect baseline prior to the real sample injection.

We must cite here that the Ion Trap allows MS^n determinations, probably of higher specificity (even if the absolute signal intensity could become too low with these type of trace analyses). However we didn't need of them until now.

It's a recommended practice to achieve the maximum chromatographic performances (mainly resolution) before using all detectors' possibilities.

When the working chromatographic conditions were used a little number of coelutions were observed, some occurring in the GC/MS column, others in the GC/ECD column. This was evident by preliminary injections of pure standards.

Only few coelutions occur in both columns. To solve these cases we developed a second chromatographic method holding the same capillary columns. Figure 19 shows the development of this second method at the ECD column: it is reported the resolution between 4,4'-DDT and PCB 138 (Polychlorobiphenyl 138, a non-pesticidic Pollutant). In executing all analyses sample and standards were always reinjected by using these new chromatographic conditions. The same was done for the cis-Nonachlor/PCB 188 coelution.

3.4. Quality controls and GLP (Good Laboratory Practice)

A number of guidelines are reported to be followed by Laboratories to ensure that their work is carried out in compliance with GLP (Good Laboratory Practice) established for the analysis of residue levels of Organochlorine contaminants. Among these the use of Certified Standards and Certified Reference Materials are indicated as "Intralaboratory tests" while participations in "Ring Test Analyses" are indicated as "Interlaboratory tests".

In the Fish Quality Unit of the Italian National Institute for Food and Nutrition Research (INRAN) we currently use Certified Standards and Certified Reference Materials. As regards Interlaboratory tests we participated in the *CSL-MAFF proficiency testing FAPAS© series 05, round 32, OC Pesticide Residues and PCBs'*, an international Ring Test about food analysis performance assessment. As regards the qualitative analysis, all contaminants present in the test sample were correctly identified. As regards the quantitative analysis a Z-score between -2 and +2 is generally considered of good quality. Z-scores obtained in our Laboratory, for Pesticides present in the test sample, were as follows: α-HCH= -1.0 and Dieldrin = -1.2 [19].

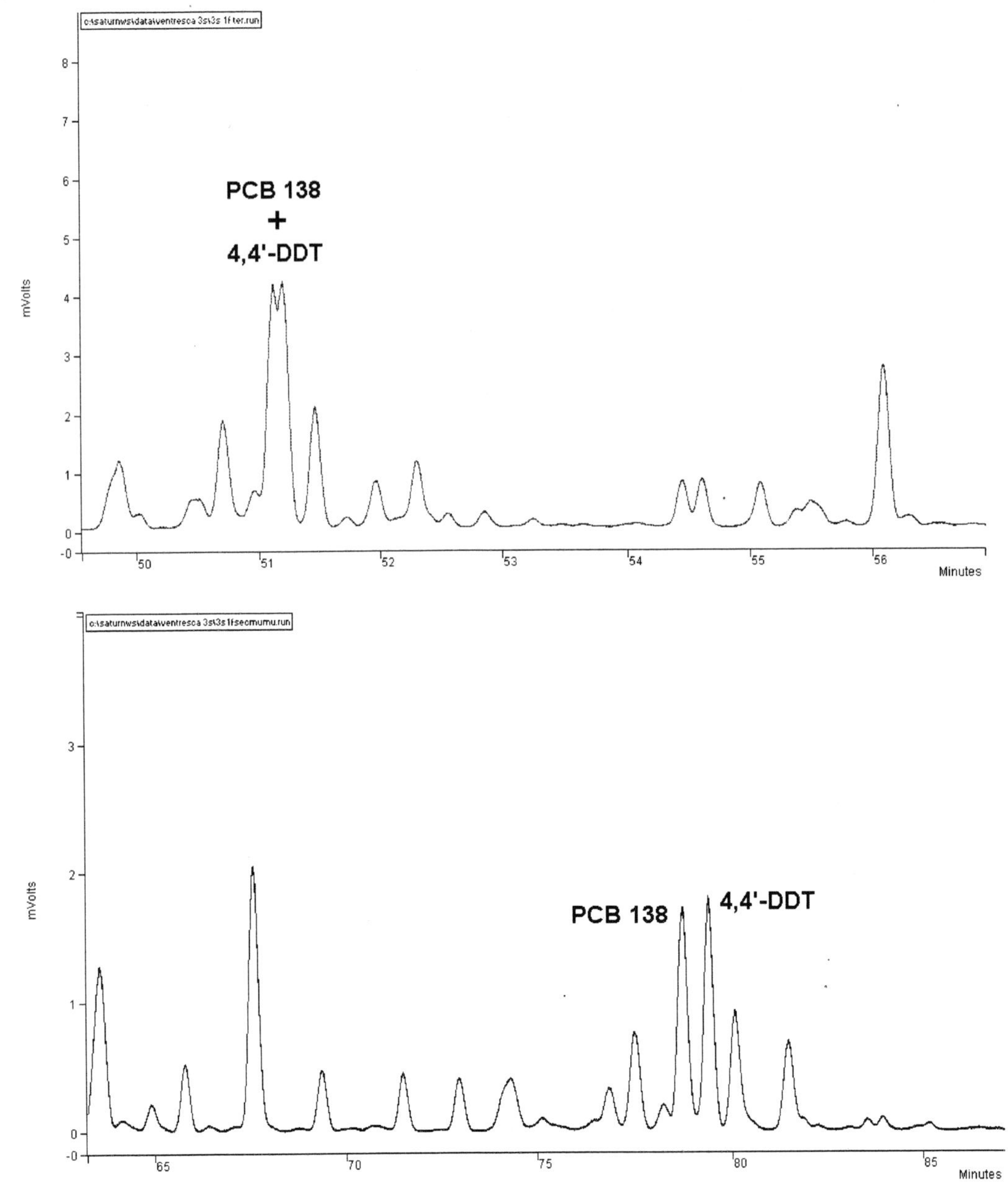

Figure 19. Development of a second Chromatographic method in order to resolve the coelution between 4,4'-DDT and PCB 138. Detector: ECD.

We regularly conduct other intralaboratory tests, such as analyses of blank materials, checks for detector response (linear range) or recovery analyses of spiked samples.

Recovery percentages were repeatedly determined on test fat materials and showed a mean value of 89.7%. The maximum observed value was 102.2% for Lindane with a Standard deviation (SD) of 3.5% while the minimum was 79.0% (SD=4.7%) for Aldrin. Recoveries similar to these ones were observed when the *FAPAS*© *proficiency testing* [19] was carried out.

When two consecutive analyses were carried out on the same homogenized sample results showed a mean Relative Standard Deviation (RSD) of 12.0% with maximum values of about 20% (Dieldrin, Heptachlor exo epoxide, Octachlorostyrene) and minimum values of 5-10%.

As regards the Limit Of Detection (LOD), that means the minimum amount of OC Pesticide detectable, two different situations must be considered: the instrumental limit and the minimum analysable amount of Pesticide in the sample due to the entire method employed.

As regards instrumental limits we observed a greater sensitivity of ECD detector with respect MS detector, but the last is of greater identification capabilities.

ECD showed a mean LOD of 0.05 micrograms of Pesticide, per litre of solution. For example by injecting 2.5 μL of a Lindane solution at 0.07 μg/L a peak with a signal to noise ratio equal to 4 was obtained, that corresponds to a quantity of 0.175 picograms injected. This is ECD instrumental LOD for Lindane.

The maximum MS sensitivity is obtained when MS/MS determinations are carried out and by holding in the Ion Trap only the ions of interest (see 3.3.2 section). In these conditions a mean LOD of 0.20 micrograms of Pesticide, per litre of solution, was shown by the Mass Spectrometer. For example by injecting 2.5 μL of a Lindane solution at 0.29 μg/L a peak with a signal to noise ratio equal to 5 was obtained, that corresponds to a quantity of 0.725 picograms injected. This is the MS instrumental LOD for Lindane.

In sample preparing an amount of 15 grams of homogenized sample was always weighed for analysis. At the end of the sample preparation all pollutants present in those 15 grams were brought into a vial having a volume of 1 mL, which means a weight of about 1 gram (more exactly less than 1 gram). In this way a concentration factor of about 15 times was achieved. It is clear that a vial of 1 mL that shows a concentration of Lindane equal to 0.07 μg/L (i.e. instrumental ECD LOD) contains 0.00007 micrograms of Lindane. These 0.00007 micrograms come from 15 grams of homogenized sample: this means that the level of Lindane in the sample is 0.005 μg/Kg=ng/g edible portion. This is the method LOD for Lindane when ECD is used as detector. In general the mean ECD Method LOD was about 0.004 μg/Kg=ng/g edible portion.

In the same way the mean MS Method LOD was about 0.014 μg/Kg=ng/g edible portion.

4. Results

4.1. Fish from Aquaculture

Table 2 summarizes results of this monitoring programme on Sea Bream samples collected from September 2001 to September 2002. Maximum and minimum values showed refer to pools with the highest and lowest levels among all pools analyzed.

Table 2. Organochlorine Pesticide contents in Sea Bream samples (*Sparus aurata*) from five Italian Sea Farms

Pesticide	ng/g edible portion (min. / max.)	sum (ng/g edible portion) of isomers and metabolites (min. / max.)	Italian tolerance levels (ng/g edible portion). December 2004 [20] [27]. (Tolerances depend on the fat content)	U.S. FDA action levels (ng/g edible portion) 2007 [21]	ng/g fat basis (min. / max.)
Aldrin	n. d. / 0. 01	n. d. / 0. 99	5 - 40	300	n. d. / 17. 03
Dieldrin	n. d. / 0. 98				
4,4'-DDT	n. d. / 1. 38	4. 94 / 9. 39	50 - 500	5000	84. 97 / 168. 45
4,4'-DDD	1. 21 / 2. 67				
4,4'-DDE	2. 20 / 4. 42				
2,4'-DDT	0. 06 / 0. 09				
2,4'-DDD	n. d. / 0. 06				
2,4'-DDE	0. 01 / 0. 03				
2,2'-DDE	n. d. / n. d.				
4,4'-DDM	n. d. / n. d.				
2,4'-DDD olefin	n. d. / n. d.				
4,4'-DDMU	0. 84 / 1. 14				
cis-Chlordane	n. d. / 0. 40	0. 50 / 1. 32	5 - 40	300	6. 31 / 38. 81
trans-Chlordane	n. d. / 0. 11				
Oxychlordane	n. d. / 0. 20				
cis-Nonachlor	0. 17 / 0. 37				
trans-Nonachlor	0. 28 / 0. 66				
Endrin	n. d. / 0. 17		1		n. d. / 2. 92
α-HCH	n. d. / 0. 14	n. d. / 1. 37	10 - 80		n. d. / 13. 61
β-HCH	0. 15 / 1. 23				
δ-HCH	n. d. / n. d.				
γ-HCH (Lindane)	n. d. / 0. 25		25 - 100		n. d. / 2. 76
HCB	0. 28 / 0. 59		10 - 80		3. 78 / 14. 41
Heptachlor	n. d. / n. d.	0. 11 / 0. 21	5 - 40	300	1. 51 / 4. 12
Heptachlor endo epox	n. d. / n. d.				
Heptachlor exo epox.	0. 11 / 0. 21				
Octachlorostyrene	n. d. / 0. 02				n. d. / 0. 34
α-Endosulfan	n. d. / 0. 65				3. 28 / 6. 46
α-Chlordene	n. d. / 0. 02				n. d. / 0. 34
γ-Chlordene	n. d. / n. d.				n. d. / n. d.
Mirex	n. d. / 0. 03			100	n. d. / 0. 40
Quintozen	n. d. / n. d.				n. d. / n. d.
methyl-Chlorpyriphos	n. d. / 0. 05				n. d. / 0. 86
Chlorpyriphos	n. d. / 0. 36				n. d. / 4. 86

Size of Sea Bream sampled ranged from 265 to 540 grams.

n. d. = not detected.

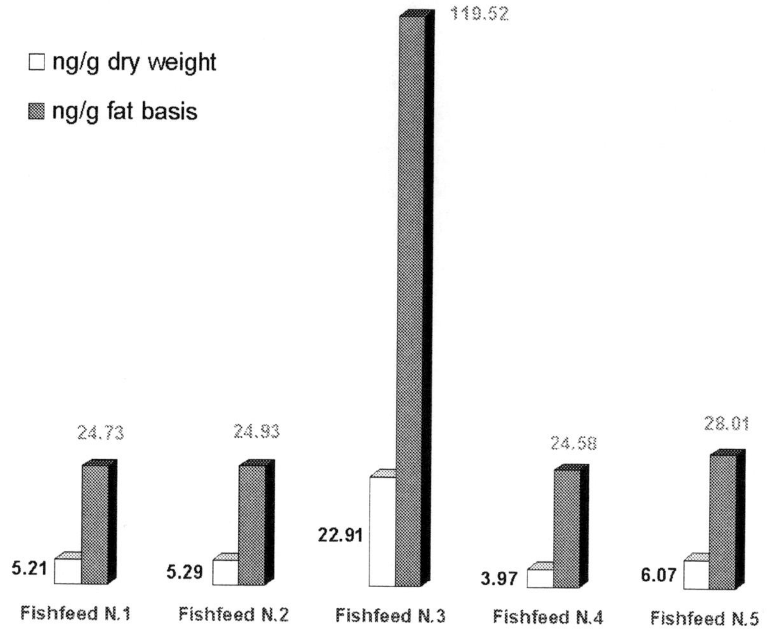

Figure 20. DDT level in five fishfeeds analyzed in the Fish Quality Unit of INRAN.

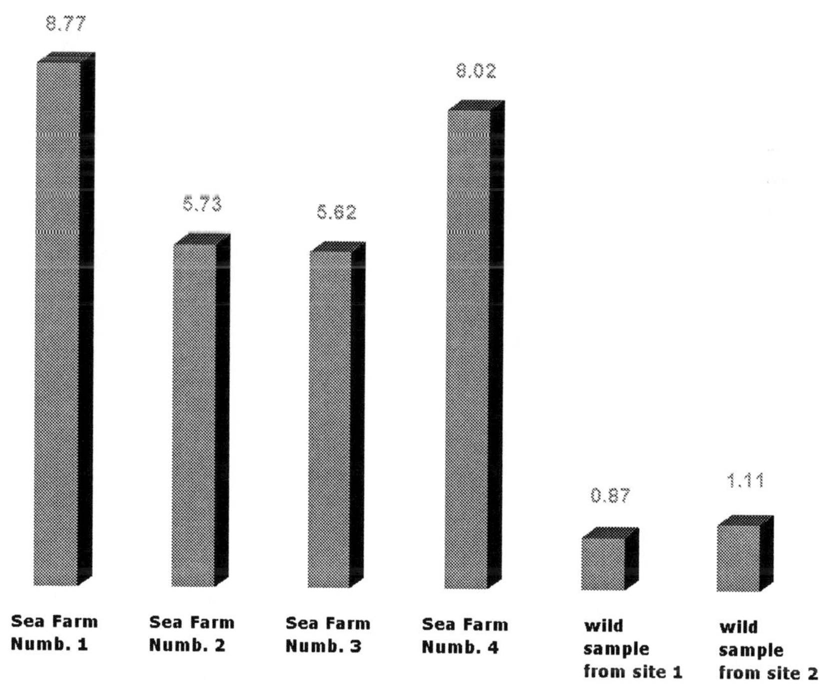

Figure 21. Mean DDT level (ng/g edible portion) in pools of Sea Bream collected both from Sea Farms and from natural marine environment (site 1 and site 2). Each pool was created by grouping three of four (site 1) individuals.

As regards Food Safety assessment Italian tolerance levels and U.S. FDA Action levels for OC pesticides in Fish are listed [20][21]: both Italian and U.S. FDA values are based on edible portion concentrations, but they are quite different being the Italian Tolerance level "fat dependent" (more the fat content in the sample higher the Tolerance level). Along with edible portion-based also fat basis results are always shown in this work. It must be noted that edible portion results are the main parameter when Food safety evaluations have to be carried out (fishes with a low fat content that is highly contaminated are probably more safe than fishes with a high fat content that is of medium contamination) but fat basis results are of a big relevance when environmental pollution estimations are done (the fat contamination is strongly dependent from the environmental pollution).

As we can see pollutant concentrations are low in all samples analyzed, well below Action or Tolerance levels listed. In many cases OC Pesticides were not detected at all (i.e. were probably at a level below the Limit of Detection of the method -see Experimental section-). As regards Italian Tolerance levels it must be cited that fat contents of Sea Bream samples collected were mostly in the range 5-10% and so the Tolerance reference values are mostly the following (ng/g edible portion): Aldrin+Dieldrin=10, Heptachlor=10, HCB=20, Chlordane=10, α-HCH+β-HCH+δ-HCH=20, Lindane (γ-HCH)=50, DDT=100, Endrin=1.

One type of fish feedstuff was mainly employed in growing up Sea Breams of the Table 2: it showed a level of DDT of 22.91 ng/g dry weight (equal to 119.52 ng/g fat basis). This value is well below the DDT Tolerance level established by European Community for feedingstuffs [22]. Nevertheless this concentration is quite higher than the concentrations usually observed by us in other fishfeed samples (Figure 20). We suspect that this amount of DDT could be able of increasing DDT content in fishes fed with that particular feedingstuff, even in the brief period of farming (less than one year). A comparison should be made with Sea Bream samples fed with low content-DDT fishfeeds.

For now we can only compare DDT levels in Sea Breams fed with the feedingstuff N.3 of Figure 20 with DDT levels of wild Sea Bream samples collected in non-polluted areas (Figure 21): it is very likely that the higher level of DDT in farmed Sea Breams was due to the fishfeed employed.

The contamination level of fish feedingstuffs was already reported [23] to be in relationship with Aquaculture product contamination.

4.2. Fish from Lakes

Data are shown in Table 3. Only the five groups of most concentrated OC Pesticides are reported, the others were observed at very low level and are not reported: we preferred to compare pollutions between the three lakes and the three species by reporting the results in a unique table than listing all pollutants analyzed in more than one table. Fat contents ranged from 13 to 32% for Eel, 2-4% for Whitefish and 0.5-0.6% for Perch.

All samples analyzed were below the current Italian tolerance level even when the total DDT content in a Eel sample were at 108 ng/g edible portion: due to the high fat content in this particular case the tolerance level is of 150 ng/g edible portion. Nevertheless such a value cannot be judged of good safety, especially if the sum of total Organochlorine pollutants in the sample are considered (this is a lack in most current action or tolerance levels).

Table 3. Organochlorine Pesticide contents in fish samples from Bolsena, Bracciano and Salto Italian lakes (mean values)

		Aldrin + Dieldrin	Total DDT	Total Chlordane	HCB	Heptachlor
ng/g edible portion	Eel from Bolsena lake	3. 20	61. 66	3. 20	2. 09	0. 83
	Eel from Bracciano lake	2. 27	48. 84	6. 52	2. 23	0. 75
ng/g fat basis	Eel from Bolsena lake	13. 78	287. 31	14. 61	8. 87	3. 55
	Eel from Bracciano lake	11. 43	242. 82	31. 48	10. 78	3. 77
ng/g edible portion	Whitefish from Bolsena lake	0. 24	5. 93	0. 27	0. 21	0. 07
	Whitefish from Bracciano lake	0. 37	5. 70	0. 73	0. 32	0. 11
	Whitefish from Salto lake	0. 19	2. 63	0. 31	0. 22	0. 05
ng/g fat basis	Whitefish from Bolsena lake	10. 44	258. 04	11. 75	9. 14	3. 05
	Whitefish from Bracciano lake	10. 57	162. 88	20. 86	9. 14	3. 14
	Whitefish from Salto lake	9. 76	135. 12	15. 93	11. 30	2. 57
ng/g edible portion	Perch from Bolsena lake	0. 04	2. 32	0. 07	0. 08	0. 01
	Perch from Bracciano lake	0. 03	0. 38	0. 03	0. 04	0. 01
	Perch from Salto lake	0. 02	0. 65	0. 05	0. 34	0. 01
ng/g fat basis	Perch from Bolsena lake	7. 81	453. 13	13. 67	15. 63	1. 95
	Perch from Bracciano lake	4. 68	59. 31	4. 68	6. 24	1. 56
	Perch from Salto lake	3. 35	108. 86	8. 37	56. 94	1. 67

Mean sample sizes:

Eels = 171 g (Bolsena lake), 166 g (Bracciano lake).

Whitefish = 235 g (Bolsena lake), 250 g (Bracciano lake), 158 g (Salto lake).

Perch = 84 g (Bolsena lake), 92 g (Bracciano lake), 83 g (Salto lake).

As regards the species we can see that Whitefish and Perch are of more safety with respect Eels (see edible portion concentrations) and this fact can entirely be attributed to the difference in fat contents.

When the Pesticide concentration on fat basis is considered a certain "level-alignment" among the species of the same lake can be noted. If the Bolsena lake is considered, for example, we note that the mean edible portion DDT content is 61.66 ng/g in Eels and 5.93 ng/g in Whitefishes, but in these samples the mean DDT content is quite the same if concentrations are expressed on fat basis: 287.31 ng/g in Eels and 258.04 ng/g in Whitefishes. This could be explained by the fact that the two species lived in the same ecosystem probably for a similar lapse of time. Other examples of this type can be observed in Table 3. Such an observation confirms what above supposed (see 4.1 section): when concentrations are expressed on edible portion Food Safety can be estimated, but for environmental evaluations levels on fat basis have to be considered (one gram of fat from the Species A "captures" an amount of contaminants not too different from that captured by one gram of fat from the Species B if the two species live in the same lake for the same period and are in the similar position in the trophic chain).

By examining the DDT column in Table 3 slight higher levels in samples from Bolsena lake, with respect Bracciano and Salto lake, can be noted. This is valid both for fat basis and edible portion concentrations and for all the three species considered. Figure 22 summarizes what we have just said. DDT seem to be more concentrated in the Bolsena lake ecosystem if the species *Anguilla anguilla*, *Coregonus lavaretus* and *Perca fluviatilis* can be considered as reliable Bioindicators of environmental pollution in their habitat. On the other hand such a result could be a reasonable occurrence if the main economic resource of this area is taken into account: historically in the North district of Latium region extensive agriculture was largely carried out. Nevertheless the possible past use of DDT do not represent now a topic of concern as regards safety of species caught in Bolsena lake, if we except Eels individuals of big size and fat content. In this last case a little attention should be useful.

4.3. Bluefin Tuna Sea Ranching

Results for "in cage-fed" Tunas (Bluefin Tuna Sea Ranching) are reported in Table 4.

Fat contents ranged from 11 to 35 % being <20% for some samples and >20% for the others. According to this two different sets of Italian tolerance levels exist: one set valid for samples with the lipid percentage <20% and another set valid for samples with the lipid percentage >20% (see 4.1 section). We must cite that OC Pesticides analysed do not exceed current Italian tolerance levels, if we except Chlordane. Three samples were a little above the Italian Tolerance level of 20 ng/g edible portion established for Chlordane, these samples were the caudal (22 .71 ng/g) and cephalic portion (30. 95 ng/g) of Tuna number 4 and the cephalic portion (21. 34 ng/g) of Tuna number 1. Tuna number 1 and Tuna number 4 were the individuals of biggest size among the four chosen.

As regards the other OC Pesticides we can observe a medium-to-low ranging level of contamination. Comparison seems to be appropriate with other works that analysed OC Pesticides in liver [24] and muscle [25] of wild Bluefin Tunas. In these studies, and as regards HCB and HCH, similar levels to the ones reported in Table 4 can be seen, while as regards DDT the concentrations measured by us in "Sea Ranching Activity" seem generally lower. Chlordane is at the same high levels [24] than the levels observed by us and this fact probably means that Tunas are particularly exposed to this pollutant. Other evidences, not cited here, suggested us that the last assertion is a very likely occurrence; as regards this important point further research should be carried out.

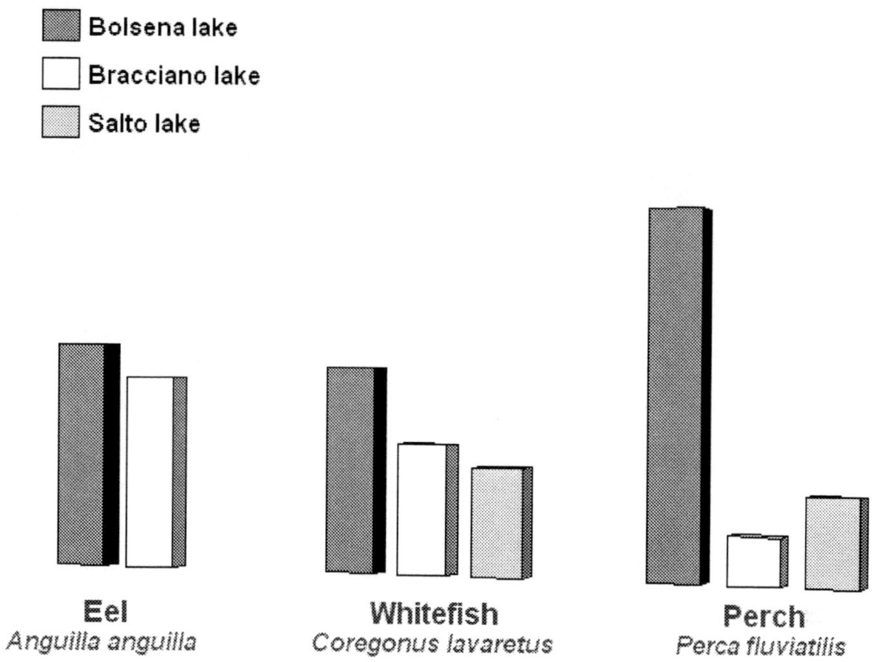

Figure 22. Mean DDT concentration (ng/g fat basis) in Fish samples from three lakes of Central Italy.

Analyses of the two samples from "not in cage-fed" Tunas were also carried out. Only slight differences with the Tunas in cage-fed were observed mainly when concentrations fat basis are considered. Wild individuals seem to show higher fat basis levels of Organochlorine pollutants, even if the number of measures is too low to get definite conclusions. On the other hand the lower fat basis Organochlorines content in reared Tunas with respect the wild ones is a reasonable hypothesis: individuals in cages are allowed to perform only limited movements and the fat eaten, less contaminated of the fat already present, is mostly accumulated and not consumed. A process of dilution of contaminants in the fat is the final result, even if the total absolute content of pollutants in the entire Tuna, in micrograms, increases.

When the sampling steps were carried out it was possible to distinguish the sex of Tunas collected. In Table 5 are reported the concentrations fat basis of total DDT and total Chlordane the two Pesticides that were observed to be at the highest levels in Bluefin Tunas. All samples analysed, along with sex, size and portion collected (caudal or cephalic) are also listed in such a way that makes it possible to compare Male and Female individuals of the same typology (size, portion sampled, wild or rearing).

It seems to exist a strong relationship between DDT and Chlordane levels and sex, being higher (or very higher) the contamination of Female individuals with respect Male individuals. A similar relationship is valid for the other Organochlorines. We don't know how this fact can be explained; it's likely that behaviour or metabolic reasons there are, however more analyses are needed being low the number of measures carried out in this work.

Table 4. Organochlorine Pesticide contents in Bluefin Tunas *(Thunnus thynnus)* from "Bluefin Tuna Sea Ranching"

Pesticide	ng/g edible portion (min. / max.)	sum (ng/g edible portion) of isomers and metabolites (min. / max.)	Italian tolerance levels (ng/g edible portion). December 2004 [20] [27]. (Tolerance levels depend on the fat content)	U.S. FDA action levels (ng/g edible portion) 2007 [21].	ng/g fat basis (min. / max.)
Aldrin	n. d. / n. d.	2. 23 / 6. 46	5 - 40	300	10. 82 / 22. 97
Dieldrin	2. 23 / 6. 46				
4,4'-DDT	1. 89 / 8. 43	26. 81 / 94. 55	50 - 500	5000	122. 35 / 336. 14
4,4'-DDD	2. 99 / 13. 79				
4,4'-DDE	17. 40 / 64. 05				
2,4'-DDT	n. d. / 2. 15				
2,4'-DDD	1. 25 / 3. 14				
2,4'-DDE	0. 08 / 0. 34				
2,2'-DDE	n. d. / n. d.				
4,4'-DDM	n. d. / n. d.				
2,4'-DDD olefin	n. d. / 0. 62				
4,4'-DDMU	0. 19 / 5. 45				
cis-Chlordane	0. 76 / 4. 10	8. 51 / 30. 95	5 - 40	300	40. 85 / 110. 03
trans-Chlordane	0. 20 / 2. 50				
Oxychlordane	1. 05 / 3. 68				
cis-Nonachlor	3. 60 / 11. 66				
trans-Nonachlor	1. 44 / 11. 88				
Endrin	n. d. / n. d.		1		n. d. / n. d.
α-HCH	n. d. / 0. 20	n. d. / 0. 20	10 - 80		n. d. / 0. 57
β-HCH	n. d. / n. d.				
δ-HCH	n. d. / n. d.				
γ-HCH (Lindane)	0. 03 / 0. 54		25 - 100		0. 15 / 1. 53
HCB	0. 80 / 2. 01		10 - 80		3. 65 / 8. 50
Heptachlor	n. d. / n. d.	0. 27 / 1. 45	5 - 40	300	1. 31 / 5. 15
Heptachlor endo epox	n. d. / n. d.				
Heptachlor exo epox.	0. 27 / 1. 45				
Octachlorostyrene	n. d. / 0. 08				n. d. / 0. 53
α-Endosulfan	n. d. / n. d.				n. d. / n. d.
α-Chlordene	n. d. / n. d.				n. d. / n. d.
γ-Chlordene	n. d. / n. d.				n. d. / n. d.
Mirex	0. 32 / 1. 78			100	2. 27 / 6. 94
Quintozen	n. d. / n. d.				n. d. / n. d.
methyl-Chlorpyriphos	n. d. / n. d.				n. d. / n. d.
Chlorpyriphos	n. d. / 1. 29				n. d. / 4. 38

Size of Bluefin Tunas ranged from 190 to 300 kilograms.

n. d. = not detected.

Table 5. DDT and Chlordane contents* in Bluefin Tunas (*Thunnus thynnus*). Comparison between Male and Female individuals

Sample	Total DDT (ng/g fat basis)	Total Chlordane (ng/g fat basis)
Tuna N.1, 300 kg, MALE, cephalic muscle (in cage-fed)	180. 57	60. 37
Tuna N.4, 280 kg, FEMALE, cephalic muscle (in cage-fed)	336. 14	110. 03
Tuna N.1, 300 kg, MALE, caudal portion (in cage-fed)	122. 35	40. 85
Tuna N.4, 280 kg, FEMALE, caudal portion (in cage-fed)	228. 55	77. 03
Tuna N.14, 190 kg, MALE, cephalic muscle (in cage-fed)	203. 12	53. 90
Tuna N.12, 190 kg, FEMALE, cephalic muscle (in cage-fed)	243. 16	64. 57
Tuna N.14, 190 kg, MALE, caudal portion (in cage-fed)	226. 45	63. 54
Tuna N.12, 190 kg, FEMALE, caudal portion (in cage-fed)	237. 60	73. 23
Tuna N.6, 226 kg, MALE, caudal portion (not in cage-fed)	147. 58	32. 59
Tuna N.2, 309 kg, FEMALE, caudal portion (not in cage-fed)	394. 08	72. 22

* Sum of all isomers and metabolites.

5. Conclusion

5.1. Fish from Aquaculture

Analyses of Sea Bream samples collected from some Italian Sea Farms during one year from September 2001 to September 2002, showed a good safety. The product size requested by the market led to this result. Nevertheless great attention has to be always paid to the main parameter of concern: fish feedingstuffs. In this monitoring programme we observed a higher level of pollutants in fish from marine Aquaculture than the level observed in wild fish from non-polluted areas (in any case all samples analysed showed pesticide concentrations well below the Tolerance levels). Only the limited size (and the short culturing time needed) prevented the contamination from rising too much in these farmed fish.

5.2. Lakes

The three fish species monitored from Bolsena, Bracciano and Salto Italian lakes exhibited a good safety as regards OC Pesticides. Due to the complexity and extension of the lakes, the data presented here cannot be viewed as definitive results; nevertheless they represent a good approach to evaluate Fishery Safety. Some attention has to be paid to Eels (*Anguilla anguilla*) of big size, age and fat content.

From the data obtained a hypothesis about the pollution of the Bolsena lake is reported.

5.3. Bluefin Tuna Sea Ranching

On the basis of the limited number of analyses executed here and taking into account that parts of Tunas collected are peripherals (out of market) we may conclude that products deriving from "Bluefin Tuna Sea Ranching" activity are medium contaminated and at the same level of wild tuna caught in the free environment. Such levels of contamination are similar to the levels observed in other species of the same size and at the same position in the Trophic Chain [26]. Nevertheless just these levels have to be monitored in view of the large consumption of these types of products capable of leading to high human Organochlorine uptakes during a period of several years. A great attention has to be paid to the level of Chlordane and its metabolites: Tunas seem to be particularly exposed to this xenobiotic and further research can be useful.

In this work an interesting correlation was observed between OC Pesticide concentrations and the sex of individuals analysed: more measures are needed.

6. References

[1] World Health Organization, United Nations Environment Programme, International Labour Organisation: *ICPS International Programme on Chemical Safety*, Environmental Health criteria 91 Aldrin and Dieldrin (1989).

[2] World Health Organization, United Nations Environment Programme, International Labour Organisation: *ICPS International Programme on Chemical Safety*, Environmental Health criteria 9 DDT and its derivatives (1979).

[3] World Health Organization, United Nations Environment Programme, International Labour Organisation: *ICPS International Programme on Chemical Safety*, Environmental Health criteria 195 Hexachlorobenzene, (1997).

[4] Rogan W.J. and Chen A., *The Lancet* 2005, 366, 763-773.

[5] IARC working group on the evaluation of the carcinogenic risk of chemicals to humans, "Occupational exposures in insecticide application, and some pesticides. IARC monographs on the evaluation of carcinogenic risks to humans", volume 53, *International Agency for Research on Cancer,* World Health Organization, Lyon, France (1991), pp. 179–250. http://monographs.iarc.fr/ENG/Monographs/vol53 /volume53.pdf

[6] World Health Organization, United Nations Environment Programme, International Labour Organisation: *ICPS International Programme on Chemical Safety*, Environmental Health criteria 34 Chlordane, (1984).

[7] World Health Organization, United Nations Environment Programme, International Labour Organisation: *ICPS International Programme on Chemical Safety*, Environmental Health criteria 130 Endrin, (1992).

[8] World Health Organization, United Nations Environment Programme, International Labour Organisation: *ICPS International Programme on Chemical Safety*, Environmental Health criteria 38 Heptachlor, (1984).

[9] World Health Organization, United Nations Environment Programme, International Labour Organisation: *ICPS International Programme on Chemical Safety*, Environmental Health criteria 124 Lindane, (1991).

[10] Schmida P., Kohlera M., Gujera E., Zennegga M. and Lanfranchi M., *Chemosphere* 2007, 67, S16-S21.

[11] Muir D.C.G., Wagemann R., Hargrave B.T., Thomas D.J., Peakall D.B. and Norstrom, R.J., *Sci Total Environ.* 1992, 122, 75–134.

[12] Muir D., Braune B., DeMarch B., Norstrom R., Wagemann R., Lockhart L., Hargrave B., Bright D., Addison R., Payne J. and Reimer K., *Sci. Tot. Environ.* 1999, 230, 83-144.

[13] Borgå K., Gabrielsen G.W., Skaare J.U., *Environ. Poll.* 2001, 113, 187-198.

[14] Fisk A.T., Hobson K.A., Norstrom R.J., *Environ. Sci. Techn.* 2001, 35, 732-738.

[15] James Pontolillo and Robert P. Eganhouse, *U.S. Department of the Interior, U.S. Geological Survey, U.S. Geological Survey Water-Resources Investigations* Report 01-4201. Reston, Virginia 2001 http://pubs.water.usgs.gov/wri01-4201/ .

[16] Wells D.E., Chiffoleau J.-F., Klungsøyr J., *Mar. Poll. Bullettin* 1997, 35,109-124.

[17] Di Muccio A., Stefanelli P., Funari E., Attard Barbini D., Generali T., Pelosi P., Girolimetti S., Amendola G., Vanni F., Di Muccio S., *Food Additives and Contaminants* 2002, 19, 1148-1161.

[18] Orban E. and Masci M., “Pollutant levels in Fish and Fishfeed from Aquaculture”, in *Aquaculture Europe 2002 Conference*, Trieste October 16-19 2002, Book of Abstracts 2002, 400-401.

[19] *FAPAS© Report*, series 5, Round 32, June-July 2003, Sand Hutton, York (UK).

[20] *Gazzetta Ufficiale della Repubblica Italiana*, Number 207, September the 5th 2000 (Supplemento Ordinario n.144). Istituto poligrafico e Zecca dello Stato, Roma (confirmed in the year 2004 [27]) .

[21] U. S. Food and Drug Administration, *National Shellfish Sanitation Program - Guide for the Control of Molluscan Shellfish.Food commodity: all fish;* 2007. http://www.cfsan.fda.gov/~ear/nss4-42d.html (see also the Compliance Policy Guide: http://www.fda.gov/ora/compliance_ref/cpg/cpgfod/cpg575-100.html#DDT).

[22] *Official Journal of the European Communities,* Directive 2002/32/EC of the European Parliament and of the Council of 7 May 2002 on undesirable substances in animal feed, L 140/10-21, 30.5.2002.

[23] Serrano R., Simal-Julian A., Pitarch E., Hernandez F., Varó I., Navarro J.C., *Environ. Sci. Techn.* 2003, 37, 3375-3381.

[24] Ueno D., Iwata H., Tanabe S., Ikeda K., Koyama J., Yamada H., *Mar. Poll. Bulletin* 2002, 45, 254-261.

[25] Corsolini S., Focardi S., Kannan K., Tanabe S., Borrell A., Tatsukawa R., *Mar. Environ. Res.* 1995, 40, 33-53.

[26] Stefanelli P., Ausili A., Di Muccio A., Fossi C., Di Muccio S., Rossi S. and Colasanti A., *Mar. Poll. Bulletin* 2004, 49, 938-950.

[27] *Gazzetta Ufficiale della Repubblica Italiana*, Number 292, December the 14th 2004 (Supplemento Ordinario n.179). Istituto poligrafico e Zecca dello Stato, Roma.

In: Progress in Pesticides Research
Editor: Christopher M. Kanzantzakis

ISBN 978-1-60456-201-9

Chapter 6

VALIDATION OF NEUROBEHAVIORAL STUDIES FOR EVALUATING THE PERINATAL EFFECTS OF SINGLE AND MIXTURE EXPOSURE TO PESTICIDES

Vera Lucia S. S. de Castro[1,*] ***and Paulo Afonso Silva***[2,†]

[1]Embrapa Environment, Laboratory of Ecotoxicology and Biosafety, Rodovia SP 340, km 127.5, 13820-000, Jaguariúna, SP, Brazil

[2]Instituto Militar de Engenharia, Praça General Tibúrcio, 80 – Urca, 22290-270 - Rio de Janeiro , RJ, Brazil

ABSTRACT

Laboratory studies have clearly revealed neurotoxic damage after exposure to specific pesticides and in some studies at concentrations equivalent to the environmental exposure. There are numerous opportunities during gestation where pesticides can alter the purpose of a cell, tissue, organ, or system function in the CNS. It is not uncommon for many classes of pesticides, such as insecticides, herbicides, and fungicides, to be used concomitantly. Additional work with environmentally-based mixtures is needed to test the hypothesis of dose-additivity or synergistic effects. Also, the maternal exposure to pesticides during the reproductive life may lead to animal developmental damage. Consequently, the evaluation of the possible effects due to this exposure is important to be evaluated during different developmental periods, especially those considered critical by the influence on nervous system. However, the quality and quantity of the data about the risk posed to humans by individual pesticides vary considerably and few are know about mixture pesticides effects. Consequently, sometimes it is difficult to quantify and compare neurodevelopmental impairment. The potential of pesticides mixtures to affect human toxicity or to pose a reproductive hazard to female rats exposed during critical periods of development needs more research attention. Facing this, the present study provides an overview of the theoretical discussion on potential neurodevelopmental

* E-mail address: castro@cnpma.embrapa.br

effects of perinatal exposure to single or various chemical substances, considering different experimental situations and their limitations. Knowledge of data quality aids in the correct interpretation of the potential effects of a chemical pollutant exposure and opens perspectives of its use in public health. In this way, some criterions selected to the data evaluation and to guarantee the quality control are discussed, especially the dose-response effect and the variability obtained in a laboratorial test. Based on the findings from the data, a methodology was written with an overall decision rule for the acceptable behavior presented by the experimental units. It is also important to observe the possible influence of others factors interfering on results may cause experimental bias and establish a laboratorial historical control for test procedures. The criticism on the evaluation of them needs to be assessed in light of the end use of neurobehavioral results for improving the process in risk analysis to simple and mixture exposure.

INTRODUCTION

When two or more chemicals are applied simultaneously to a living system or unit, the combined effect may modify the individual toxic effect since they may have common cellular targets or metabolic pathways. Contaminated media at hazardous waste sites may contain hundreds of organic and inorganic compounds. Pesticide exposure can come from a variety of sources, including diet, drinking water, and both indoor and outdoor residential use. Despite the fact that almost all human chemicals exposure are mixtures, and that these mixed exposures occur in the context of numerous other risk modifiers, the current understanding of human health risks is based almost entirely on the evaluation of chemicals studied in isolation. Under realistic environmental conditions, concomitant or sequential exposure to pesticides mixtures dictates the necessity of exposure assessment, hazard identification, and risk assessment of them (Cory-Slechta, 2005). For example, exposure to the mixture of metals in water can carry for harmful effects in reproductive parameters of mice males (Jadhav et al., 2007). The acute and chronic health effects of individual chemicals may not accurately estimate the toxicological significances of mixtures. Methods are needed to investigate the interactions of complex mixtures (Donnelly et al., 2004).

Mixtures toxicology is proving to be different from single-chemical toxicology in several fundamental but barely recognized ways (Robinson and MacDonell, 2004). A complex chemical mixtures may consist of thousands of (often unidentified) components, each often at very low doses, but together constituting significant exposure levels and the exposure is nearly always via multiple routes, pathways. Other stressors such as noise, heat, infection, etc., may play a significant role in the overall environmental health response. The interactions are potentially many and varied: pharmacokinetic and pharmacodynamic interactions may occur at the same site, or at different sites via complex physiological processes (including defense mechanisms). Consequently, cumulative effects of different exposures/stresses over time need to be considered (altering the "baseline" susceptibility of the individual).

Attempts to deal with the problem of chemical mixtures have largely been restricted to classes of chemicals that are structurally related. While a within-class focus may be a logical starting point for evaluating the toxicity of chemical mixtures, it encompasses only a small

† E-mail: pauloafonsolopes@uol.com.br

fraction of the aggregate or overall exposure problem. The breadth of classes and types of chemicals to which humans are normally exposed are not structurally-related chemicals (Cory-Slechta, 2005).

Development and validation of mechanism-based models and predictive tools are essential for improving current risk assessment processes for mixtures. With mixtures, there is an added degree of complexity that requires explicit integration of multiple biochemical and physiological systems in order to fully understand many interaction mechanisms for even simple mixtures. As such mechanisms of biological processes can be embodied in inter-connected mathematical models; the scientists will develop a continually improving scientific infrastructure, which will allow one to generate better hypotheses based on plausible interaction mechanisms. Methods need to be developed to understand and integrate experimental data from the molecular to the whole organism level for understanding multiple data from mixed exposures. Understanding and prediction of precursors to adverse health effects will inevitably lead to identification of useful biomarkers of effect, and to earlier and more effective intervention strategies (Robinson and MacDonell, 2004). Therefore, it is recommendable for the modeler and toxicologist to include as much information as possible to limit uncertainties in the model. However, the models up to date have their limitations, as they cannot integrate every relevant biological mechanism, and mechanisms of activated toxicants might be quite different (Groten et al., 2004).

In managing risks to the environment and human health in modern economies, scientific evidence is a key knowledge input for decision-making (Allio et al., 2006). Many design elements impact data quality and therefore must be managed carefully for optimum outcome. These elements need to be managed in a consistent manner for optimum results (e.g., to reduce variability and facilitate interpretation of the results) (Crofton et al., 2004; Slikker Jr et al., 2005).

The most frequently studied situation in mixture toxicology is primarily to determine situations where the effects of combinations of chemicals are additive. Situations that differ from additive effects need a great deal of effort in creating statistical methods for assessing differences such as synergism or potentiation of the chemicals effects given individually. Better understanding of the patterns of exposure, the underlying variability within the human population, and the links between the animal toxicology data and human health effects will improve the evaluation of the risks to human health posed by pesticides mixtures (Mumtaz et al. 2004).

Some efforts were made to evaluate a complex mixture of pesticides. In this sense, a dietary study in rats with a combination of different groups of pesticides commonly found as residues in Denmark was made. The pesticides were alphacypermethrin, bromopropylate, carbendazim, chlorpyrifos and mancozeb. Some of these pesticides share common target organs in oral toxicity studies in rats including liver, kidney, testis, brain and nervous system. It was concluded that inhibition of acetylcholinesterase activity in plasma and brain by chlorpyrifos was not enhanced by coadministration of the other four pesticides. Effects were seen in liver, thyroid, thymus and blood in the combination groups. However, identification of the pesticide(s) responsible for these changes would require further studies of the individually pesticides as well as various combinations of the pesticides (Jacobsen et al., 2004).

Other study in relation to complex mixture aimed to analyze the alterations in the cortical and peripheral electrophysiological activity of rats acutely treated with combinations of insecticides. Young adult male Wistar rats were treated with of insecticides dimethoate, propoxur, cypermethrin and amitraz, given alone or in triple or quadruple combinations. All treatments changed the cortical activity spectrum. The results indicate that simultaneous exposure by various pesticide agents deserves further investigation in, among others, neurotoxicological points of view (Lengyel et al., 2007).

Although standard reference texts, regulatory guidance documents, and journal articles contains useful theoretical concepts, clear definitions of most terminology, and well developed protocols for study design and statistical analysis, no general theoretical basis for the mechanisms and interactions of mixture toxicity could be discerned. There is also a poor understanding of the relationship between exposure-based and internal received dose metrics. Similarly, responses/adverse effects should be linked so that some understanding of how various effects are expressed at various levels of biological organization is also specified. Although the relative simplicity which might be achieved by focusing only an a single level of biological organization, such as the whole organism, has considerable attraction, the reality of both available data and toxicological concerns across the spectrum of the levels of biological organization dictates that multiple levels must be considered (McCarty and Borgert, 2006).

Prior workshops, symposia, and roundtable discussions have discuss a variety of aspects of the developmental and reproductive toxicological studies of a single chemical exposure, including some appropriate endpoints. Therefore, the objective of this chapter is focus on the data quality and validation aspects of methods for evaluate perinatal neurobehavior toxicity due to single or mixture exposure to pesticides. Beyond that, there is a growing concern about the quality of the test results emitted by the laboratories which is promoting the implantation of quality systems and laboratory accreditation programs by the official organs.

In this way, due to the fact that the tests for reproductive and neurobehavioral toxicity parameters of evaluation sometimes are not specific for a determined toxicant, some of the criteria generally selected for the laboratory data evaluation and quality control are the dose-response effect and the observed variability of the response, as the studies of Castro et al. (2007) and Presibella et al. (2005). The tests planning and the data interpretation obtained in the accomplished behavioral evaluations should be done carefully, considering aspects related to experimental drawing and to the method used, the tests validation, to the control of intercurrences, the variability of the data, among others.

Maternal Exposure and Perinatal Effects

The female agricultural work in developmental countries produces around of 60 and 80% of the food. Beyond that, 70% of the infantile work is related to agricultural sector (Dinham and Sapna Malik, 2003). In this context, there is the possibility that children's neurobehavioral performance in preschool age, as sons of rural workers, can be affected by the exposure to the insecticides in small doses (Rohlman et al., 2005).

Poisoning of newborns has been associated with maternal exposure to pesticides at the place of residence or occupation. Multiple pesticides may be present at the same time in breast milk. The quantity of pesticide that is passed to the infant via breast milk is influenced by many variables (Weiss et al., 2004). In general, in the familiar agriculture at small agricultural communities, all take part in the planting process, combat to pests and diseases and harvest. In this way, the pregnant women are also exposed to pesticides during the gestational period (Castro et al., 1999; Araújo et al., 2007).

Age related effects on susceptibility appear to depend on the chemical of concern, the effect that is observed, the dose that is received, its duration and the period of development during which exposure occurred (Scheuplein et al. 2002). It is also noteworthy that infants may be more susceptible to neurobehavioral effects, which may not be apparent until later in life (Moser et al. 2001, Scheuplein 2002, Costa et al. 2004, Ladics et al., 2005). The consequences of the exposure during are target of preoccupation because the young organisms are more susceptible to the pesticides effects due to differences in physiology of different phases of life (Cohen-Hubal et al., 2000). For example, developing animals are more susceptible to the acute toxicity of the organophosphorus (OP) chlorpyrifos in experimental tests that may result in neurobehavioral abnormalities (Richardson and Chambers 2005).

The identification of environmental agents that have adverse effects on reproductive health and animal development is particularly challenging. Prolonged exposure to environmental contaminants at apparently nontoxic doses might represent a major risk factor for the health of children (Ricceri et al., 2006). Protecting them from the developmental hazards of environmental agents requires a strategy capable of monitoring the patterns of toxic effects around weaning age, as well as in adulthood.

The central nervous system appears to be especially susceptible to toxic insults during development and there is evidence that functional changes can be induced at a lower exposure level than those resulting in toxicity in adults. Some developmental neurotoxicants are structural teratogens as well, but behavioral dysfunctions may be more serious than structural defects under certain circumstances (Hass, 2006).

Damage to a particular structure in the circuit or a connecting pathway in the developing nervous system, more susceptible than the adult one, may produce structural or functional changes that could result in behavioral changes (Costa et al., 2004). Furthermore, the vulnerability of the developing brain to toxic insults is dependent on exposure issues and on the stage of development of the potential target organ or system. For an accurate evaluation of the effects on development induced by xenobiotic exposure, the developing organism must be exposed to the product at times when the target organ under investigation is most likely to be affected, since also the maturity of enzyme systems can influence the effects of this product (Clewell et al., 2004). The evaluation of neurobehavioral performance can be a sensitive biomarker to assess the neurodevelopmental consequences of environmental exposures, as reported in relation to pesticide exposure (Dam et al., 2000).

Neurobehavioral evaluations are widely used to examine the potential neurotoxicity of pesticides and other chemicals (Ehman and Moser, 2006). There is a variety of methodologies that can be utilized to assess these processes, which depend on factors such as cross-species generality and parallels to human behavior (Hass, 2006).

In the literature, numerous agrochemicals are described as able to change the endocrine system, since the differentiation and development of the reproductive system are dependent of the hormones action. So, chemicals with potential to affect the endocrine system can interfere in the hormones production or action compromising the sexual identity, fertility or behavior (Castro et al., 2007; Castro, 2000; Vinggaard et al., 2005). Endocrine disruptors are agents and chemical substances that promote alterations in the endocrine system. Many of them are persistent in the environment, can be found at the rivers sediment and are easily transported at the long distances by the atmosphere (Waissmann, 2002).

Endocrine active compounds are a significant public health concern since these compounds interfere with normal function of pathways responsible for both reproduction and development. Several modes of action could contribute to the same outcome, including aromatase inhibition, anti-estrogenicity, testosterone biosynthesis disruption, and anti-androgens that alter upregulation of aromatase in the target regions within the brain. More complex biological responses will generally represent combinations of several physiological processes integrated through multiple biological pathways (Andersen and Dennison, 2004).

The developing nervous system is extremely sensitive to the organizational actions of gonadal steroids, actions that are manifested as permanent changes in both functional and structural characteristics of the brain. It is known that any manipulation of the hormonal milieu during the perinatal critical period has significant consequences on brain organization for those regions controlling reproductive behaviors and also on the modulation of behaviors not directly linked to reproduction, such as cognition, memory, and stress response (Calamandrei et al., 2006).

Neurobehavioral outcomes are important endpoints for assessing potential human health risks of environmental contaminants since endogenous estrogens, synthesized from androgens by the enzyme aromatase, are involved in organizing both the brain and sexually differentiated behavior during development and activating sexual behavior in adult rodents (Laessig et al., 2007). Behavior represents an integrated response of the nervous system that can reveal functional changes important to the overall fitness and survival of the organism exposed to sigle or mixtures pesticides.

Deltamethrin, methiocarb, prochloraz, simazine, and tribenuron-methyl are all commonly dissimilarly acting pesticides used for agricultural and horticultural purposes. Their mixture effects were analyzed for antiandrogenic effects in vitro and in vivo. The pesticides were found to act additively in vitro. In vivo, the organ weight changes indicated that the pesticides had an accumulating effect that was not observed for the individual pesticides (Birkhøj et al., 2004). Also, it was observed that the potential of organophosphorous insecticides used in combination with pyrethroid insecticides have enhanced human toxicity needs more research attention (Perry et al., 2007).

Data Quality and Critical Evaluation of Experimental Data

In order to accomplish a good experimental planning, some points before the beginning of the study should be observed, as: a) the problem should be defined carefully based on previous evidences (association cause-effect consistent); b) the study goal should be clear; c) the data collection should be rigorously planned with emphasis in criteria's choice of the experimental and statistical methods aiming at reproducibility of the obtained data; d) the possibility of adequate human and physical infrastructure and; e) the adequacy of the data interpretation in view of the experimental delineation.

The principal purpose of a study is an important consideration for test selection and study design; namely, the term validation is used in many ways and includes animal-to-human predictive validity and construct validity. Validation can be local (specific for that laboratory) or global (generalizable and reproducible across laboratories). The validation process should include data to show that the equipment works as it should (hardware and software), that the equipment tests the behavioral domain of interest, and that the results are robust, reproducible, and relevant to the behavior (or effect) of interest. The validation should include evaluation of appropriate installation as well as defining the range of use of equipment. Also, it should be accomplished the facilities environmental controls (Slikker Jr. et al., 2005).

The discussion of experimental subject variables that can confound treatment effect includes species, strain, and developmental age. Differences in developmental age in breeding studies with postnatal components were seen as a common source of variability that could confound interpretation of effects, e.g., developmental delays.

One of the oldest principles in biomedical research is to guarantee comparability and reproducibility of results within and between laboratories. This is especially true for testing for regulatory purposes. Among others, experimental conditions are believed to affect experimental results (Verwer et al., 2007, Slikker Jr. et al., 2005, Festing and Altman, 2002).

In view of that, housing variables including diet, single vs. group housing, wire vs. plastic cages, and whether animals of the opposite sex were housed in the same room. Depending on the endpoints of the study, the type of housing condition must be taken into consideration since they could have great implications for the interpretation and validity of results from toxicological assays and the number of animals needed to detect significant effects of toxic compounds (Verwer et al., 2007). To ensure the scientific validity of experimental data, scientists must be aware of the complex nature of the environment in which their animals are maintained (Weed and Raber, 2005).

Diet was identified as an important factor to be examined, as illustrated by the potential effects of soy (high estrogen content) in diets or in food rewards/reinforcers/treats as confounders in behavioral/hormonal studies. Group housing of same sex subjects seems to be the preferred choice, as it was thought to be a more humane and natural environment (Slikker Jr. et al., 2005, Festing and Altman, 2002). Counterbalancing the order of testing among test groups is a critical control point to reduce the bias on results. The variability introduced by the experimenter may arise from two sources: the conduct of procedures involving the animal (e.g., injection, oral dosing and surgical intervention) and lack of precision with measurements (Slikker Jr. et al., 2005).

Thus, regarding the tests with animals it owes pay attention at a) the number of animals used b) the biological meaning of the observed alterations, c) doses used are generally established between those that do not cause effects and those that produce the studied toxic effect (dose-answer); carrying in consideration the exposure real probability in the environment, and d) animal age or developmental critical period that affects the experimental planning. Since several factors can affect the experimental result (Aldridge et al., 2003), it is important during the tests accomplishment to standardized the correct animal manipulation and a data control that allows the identification of the animal used.

Predicted effects following exposure to various mixtures of chemicals are initially evaluated for the non-interaction of toxicants. The dose levels combinations may be selected based on selected effect as reproductive toxicology studies previously conducted in laboratory with each pesticide separately (Presibella et al., 2005). It also may be selected based on the presence of pesticides in biological samples as urine (Perry et al., 2007), food, water and the environment. Beyond that, the estimated dose to be used during mixture pesticides tests may be calculated based on mathematical models as those using DL_{50} and toxic-unit concept calculated by a dose-response curve for each chemical. The toxic-unit was then used to classify each mixture response as additive, greater than additive, or less than additive (Mahar and Watzin, 2005).

Other important aspect is to define comparability factors before initiate the evaluation, verifying if the profiles used in the study are the same. Thus, the results obtained with the same experimental delineation can be compared. If a scientist wishes to explore measurements or assessments of several types of behavior (or any other information), it is important to realize that the information types could be correlated (Howard, 2002). Sometimes responses to the same essential questions are sought in several independent experiments or trials from different investigators.

It is important that the observed effects occur due to exposure to the substance test and not mother toxicity. In mammals, the normal performance of the maternal functions for the nestling health is indispensable. The development of major regulatory systems underlying behavior and physiology in the neonatal rat is primarily determined by the mother, who serves as the primary source of nutrition, grooming and warmth required for immediate survival (Huot et al., 2004), thereby playing a crucial role in the development of the architecture of the brain after birth. The quality of mother–infant relationships and the juvenile's subsequent social milieu are considered to be crucially formative for adjustment and social competence in adulthood. In most mammals, the mother–infant relationship is terminated at weaning (Ferdman et al., 2007). It is also important to note that the neurobehavioral genetics and some almost undetectable environmental differences may have large behavioral consequences (Smith and Corrow, 2005).

In the same direction, given the close integration of HPA function and behavior and particularly behavioral responses to stress; prenatal glucocorticoid manipulation leads to modification of behavior, brain and organ morphology, as well as altered regulation of other endocrine systems. Exposure of the fetal brain to excess glucocorticoid can have life-long effects on neuroendocrine function and behavior (Owen et al., 2005). The development of the HPA-axis, limbic system, and the prefrontal cortex are likely to be affected by antenatal maternal stress and anxiety (van der Bergh et al., 2005).

The experimental delineation have to consider the impact of maternal toxicity on hazard identification and/or risk assessment in reproductive toxicology; since maternal toxicity, particularly with the use of a maximum tolerated dose, has an impact on the various studied endpoints (Ladics et al., 2005). The traditionally signals observed of maternal toxicity in experimental studies are decrease in the weight gain (the females weight generally is observed during the whole test period), gestational period, water and food consumption decrease, clinical signals and mortality. For that, there are some methods described to evaluate the maternal behavior (Champagne et al., 2003, Slamberová et al., 2005).

The influence of litter must be taken into account in the allocation of test animals as well as the statistical analyses in developmental neurotoxicity studies (Holson et al., 2007). In respect of the litter, it is consider as the experimental unit. It can be used in case of time or cost limitations, one animal for sex per litter randomly chosen (in the risk evaluation of a substance) or to consider the average response of two or three pups as the unit of the experimental answer (Buelke-Sam et al., 1985; Holson and Pearce, 1992; Haseman et al., 2001); due to the response variability among pups of the same litter (Ulbrich, 2001). Litter must remain a factor in analysis throughout the study, not just in young animals.

The method validation is designed to determine the operational characteristics of a test, that is, its reliability and relevance, in addition to its strengths and limitations. For a method validation, it is important to verify, using test subjects, that the results are reproducible across time. The performing laboratory should maintain a historical control database to track any changes in the data over time in the animals and/or in the equipment. The value of historical data depends on its quality and its reliability. Many factors (e.g., strain, origin, associated microflora, housing, husbandry, and methods of measurand each outcome) can influence individual results so that in nearly all studies, contemporary controls are almost essential, and historical data, particularly from another laboratory should be treated with considerable caution.

Positive control data are instrumental in evaluating laboratory proficiency in detecting chemically-induced changes in measured endpoints of developmental neurotoxicity testing that involve functional and neuropathological assessments in offspring during and following maternal exposure. Positive control data are valuable in a weight-of-evidence approach to help determine the biological significance of results and provide confidence in negative results from these studies. Comparison of historical controls levels and effects of positive control, both within and between laboratories over time, will assist in interpretation of results, and will serve to further increase the level of confidence in the proficiency of the testing laboratory (Krofton et al., 2007).

To compare results obtained at different occasions and/or laboratories, methods of meta-analysis may be appropriate in some cases (Trksak et al., 2007). Formal methods of "meta-analysis" have been developed that attempt to combine the results of different experiments taking account of sample sizes and apparent quality of the data. Meta-analysis provides a tool to statistically aggregate data from existing experiments, so that the results can be summarized across a range of conditions and an increased pool of experimental data can be subjected to statistical analysis (Moher et al., 1999). However, when similar experiments are performed repeatedly in the same laboratory with the standardization of the same strain and sex and using a standard protocol that includes contemporary controls; there will often be scope for using historical data.

The sensibility of a test refers to the ability of this test in detect alterations in the answer that is being evaluated, preferably in doses below of that produces evident signals of toxicity by the agent in study. The behavioral biological responses can present great variability. In fact, some parameters present great variation among individuals.

Discussion of how to define a desirable/achievable level of variability in control responses in common behavioral studies proved to be too specific to address without having specified a set of tests or having provided data for evaluation (Slikker Jr. et al., 2005). Because of their size and complexity, traditional validation studies can consume valuable resources and use large numbers of research animals. This is particularly true when prenatal exposures to chemicals are used to validate multiple behavioral tests in the context of developmental neurotoxicity studies (Marable and Maurissen, 2004).

VARIABILITY OBTAINED IN THE EXPERIMENTAL DATA

In case of experimental data with high variability of response, the difference among treatments cannot be easily detected. In order to work with a considered high variability it should used an experimental delineation in which the variation between individuals is minimized in order to evidence the statistical meaning of the interest effects for the study (Haseman et al., 2001).

The approaches developed for chemical analyses in relation to uncertainty estimation and method validation are not directly applicable to biological essays. Therefore, if rigorous/statistically valid calculation of uncertainty is not possible, the uncertainty components are to be identified, and to be reasonably estimated.

Uncertainty in variables of an exposure model comes from many sources including instrument error and population variability; and can be grouped as systematic error (i.e., bias) and random error. The systematic error refers to the exactness that a determination is made. The exactness lack promotes the called bias or tendencies. It can cite as example problems with the calibration, equipment or reagents, etc. Then, the random error refers to the precision of a method, evaluated by the repeatable analysis performance of a sample. He provokes inconsistent changes in the system, causing dispersion around the respective average value. The examples are sampling mistake, pipetting, mistaken annotations, etc. However in neurobehavioral methods, it is difficult to establish a consistent uncertainty estimative due to the large variability of the population. It should be developed a control plan to assure that the variability keeps inside limits allowed for each methodology to establish the control of important experimental parameters.

The factors related to data variability were seen as the same variables that broadly influence behavioral testing in general, other than the obvious parameters of the test procedure (Slikker Jr. et al., 2005). A question was raised about defining a desirable/ achievable level of variability in control responses in behavioral studies.

A process can be considered reliable if the registered data is situated inside the control limits and the variability of the results is stable. In order to accomplish the conformity evaluation of the control data it should be know the analysis process and establish the criteria to verify whether the observed alterations are caused by inherent variations stable through time of the system. Checking on a determined day of test if all controls remain inside of a selected criterion of variability, the possibility of detecting anything other than the normal animal behavior variability would be negate.

The point is establish what are the principal confounds to be considered in behavioral studies and how are they to be dealt with and what are the principal variables and how they can be evaluated and controlled. Following, are presented some considerations about of the theme.

The standardized maintenance of the controls organism test and the experimental conditions aid the retrospective comparison of the method data through different occasions; facilitating the analysis of the observed response variability in the groups control accomplished at the same laboratory (Festing and Altman, 2002). Due to several possible interpretations, it is important to supply specific instructions for the controls analysis, explaining how to interpret the results and what to do with them. For that, it is necessary to define rules of data acceptance and rejection by the clear establishing of these limits. Thus, a laboratory can build historical control graphs to measure and to analyze the variability in the procedures and methods used.

Biomarkers are used in numerous areas of research. The biomarkers use presupposes the determination of what the measurand is and your numeric referential for its comparison. The biomarkers owe preferentially reflect an increasing answer as the exposure dose increases, considering the frequency normal distribution (Gaussian) in a population, where the answer is small arriving to a maximum and declining again.

In general, the superior and inferior limits are established as 3 deviations-standard around of the average, with 99.73 % of chance of all measured are inside this band according to the normal distribution. Results with values outside the established limits will be considered as outside the standard after data careful evaluation.

However, in a complex biological system, sometimes it is not possible to obtain monotonic or biphasic curve in response to a toxicant. The absence of such a response may result from reactions of a complex biological system to a toxicant, nonmonotonic (biphasic) dose-effect relationships (Conolly and Lutz, 2004), since not all brain regions develop in the same time course. Indeed, there are gradients of maturation across modalities and hierarchical levels of processing within a given pathway and neuronal circuits are shaped by experience during critical periods of early postnatal life (Costa et al., 2004, Hensch 2004). The dose-response curves of different endpoints may be best described with nonlinear models or have different shapes or regions of activity (e.g. increasing and decreasing functions, different slopes or plateaus, etc.) (Coffey et al., 2007). Therefore, it is not always possible to express direct correlations between the observed effects and the exposure (Alonso et al., 2004) since multiple types of outcomes are sometimes measured on each animal in toxicology dose-response experiments.

EXAMPLE OF QUANTIFYING A NEUROBEHAVIORAL TEST: OPEN-FIELD

There are several parameters used to characterize motor performance. Of these, the open field seems to be able to provide a good measure of the approach response toward novelty (exploration) and describe influences of drug exposure (Prut and Belzung, 2003, Ehman and Moser, 2006), even if exploratory activity and emotional reactivity can interfere with one another (Lehmann et al., 2000, Zimmermann et al., 2001). A number of potential confounders could have produced or masked differences among the experimental groups. In addition, motor and/or sensory changes could either mask or exaggerate cognitive dysfunctions (Ehman and Moser, 2006).

One can observe that so many factors can affect the response in this test. For example, in a crowding cage that is considered to be a model of social stress, it appears that the effect of stocking density differs depending on the developmental stage of the animal: Juvenile rats increased anxiety following limiting space, whereas adult rats increased activity following increase in social tension (Arakawa, 2005). Also, litter is a significant source of variation in results obtained from all behavioral measures either on physical endpoints or during the pre-weaning period by potential maternal behavior components (Buelke-Sam et al., 1985; Holson and Pearce, 1992).

As to the face and construct validity of the open-field model, one may propose that it is fulfilled into the criteria of predictive, face and construct validity to be a relevant model of human behavior. In fact, face validity implies that the anxiety response (the phenomenological aspect) observed in the animal is identical to the one observed in humans. In the open field, the observed behavior is avoidance of threatening places, which can also be observed in humans. In rodents, forced confrontation with novelty is stressful. Stress induces anxiety-like behaviors, as it does in humans. So, the model may also fit construct validity (similar etiology) and may be a rodent model of normal anxiety (Prut and Belzung, 2003).

The open-field test is also interesting because locomotor´s activity can serve as a tool to evaluate the effects of stressors that alter the functioning of the hypothalamic-pituitary-gonadal (HPG) axis (McCormick et al., 2005), that is one of the targets of chemical endocrine disruptors. In addition, the medial hypothalamus could play an important role in the modulation of defense responses measured by rat exploratory behavior in open-field and elevated plus-maze tests (Jardim and Guimarães, 2004).

The results of the open-field arena locomotion test can reveal changes related to fungicide fenarimol exposure during the perinatal and lactation periods, with a consequent decrease in locomotion, mainly during lactation (Castro et al., 2007). Fenarimol acts as a genotoxin, and possesses estrogenic properties and acts both as an estrogen agonist and as an androgen antagonist (Castro et al., 2005; 2007). These findings could indicate long-term neurotoxicity (Viberg et al., 2004), although they do not shed light on the question of whether the lower number of units recorded is related to a motor effect or to other factors such as alterations to the physiological systems that control anxiety, exploratory activity and emotional reactivity (Zimmermann et al., 2001).

However, some important analytical key steps need to be controlled in order to avoid many sources of error and bias. It is important to control full range of laboratory quality control measures, to limit variability and demonstrate that methods are in control. Keeping these points in view, it is important to determine the normal fluctuation of a tested behavior, as the general activity in an open-field arena. The normal fluctuation can be considered as the variability in behavior due to animal biology and not due to some non-random cause as a stressful situation or environment. So, this fluctuation can be established from the laboratory historical data in each behavioral test.

To determine the normal fluctuation of a control group in open-field behavior, one can use the coefficient of variation (CV). However, the CV is good only for comparing variability of samples, independently of the time of the observations. As we are studying effects of pesticides during several days, although the CV detects changes in the average, it cannot show the real modifications; for example, a mean equal to 100 and standard deviation (SD) 10 give a CV of 10%. If the mean changes to 1000, and the SD to 100, both mean and variability have changed but the CV remains the same.

For this reason, we propose a new approach for interpreting the results of behavioral experiment, as provided by the present example of open-field data. We know that animal behavioral variability depends on several causes, and we need to simplify the problem. So, after the exposure to the same test condition, if the behavior of a group of rats is different from another group of rats not exposed to chemicals or other physical and environmental parameters in the same experimental conditions, we need to investigate if there is an assignable cause or it is only a natural variation. The aim is to propose a limit for this decision.

In this experiment, rearing, locomotion and immobilized time frequencies were observed during the gestational period in two phases: Phases 1, rats exposed during days 1 to 6 of pregnancy, and Phase 2 exposed during days 6 to 15 of pregnancy.

After understanding the experiment, we begin search a statistical model for modeling the data, and the first one was to observe the individual behavior of the rats. For example, the rearing frequency, data of 44 rats whose mothers were manipulated during the gestational period are showed in the Figure 1, where D21, D30, D60 and D90 are the days of observation.

However, how we deal with rearing frequency that surpasses other observations for the same group? For this reason, as we have studied, initially, the parameter rearing, observing the rats' behavior, and constructing the following behavior models:

(a) individual by using a control-chart for individual measurements,
(b) litter average, using a control-chart for grouped data, and
(c) litter averages, using an analysis of variance (ANOVA).

We have studied individual behaviors for days 21, 30, 60 and 90, Phases 1 and 2, and Phase 1 for the litter averages and ANOVA, as following described.

a. Individual as an Experimental Unit

For individual observations, the sample size is 1, and for measuring the parameter we have used the moving range of two successive observations.

The moving range is defined as

$$MR_i = |x_i - x_{i-1}|$$

which is the absolute value of the first difference of the data (the difference between two consecutive data points).

The control-chart lines for individual measurements are calculated using the following expressions:

$$UCL = \bar{x} + 3\frac{\overline{MR}}{1.128}$$

$$\textit{Center}\ \text{Line} = \bar{x}$$

$$LCL = \bar{x} - 3\frac{\overline{MR}}{1.128}$$

UCL stands for upper control limit, LCL for lower control limit, $\bar{x}$ is the average of the rearing frequency measurements of all rats, and $\overline{MR}$ is the average of all moving ranges of two rats observed in a random manner.

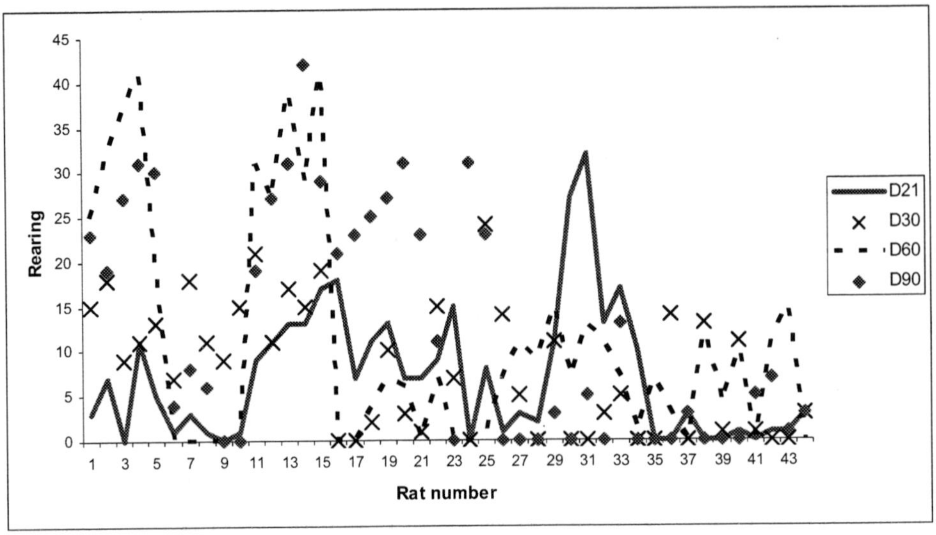

Figure 1. Rearing of rats in an open-field arena at different postnatal days (individual observations).

Using the data for rearing, D21, Phases 1 and 2, we have: $\bar{x}$=7.16 and $\overline{MR}$ = 4.65. Then,

UCL = 16.3211, rounded to 17.
Center Line = 5.505495
LCL = 0 (if the calculation is negative, LCL is set to zero)

Figure 2 illustrates this control-chart for individual measurement for rearing, Day 21, Phases 1, exposed during days 1 to 6 of pregnancy, and Phase 2 exposed during days 6 to 15 of pregnancy, whose data are on the Annex.

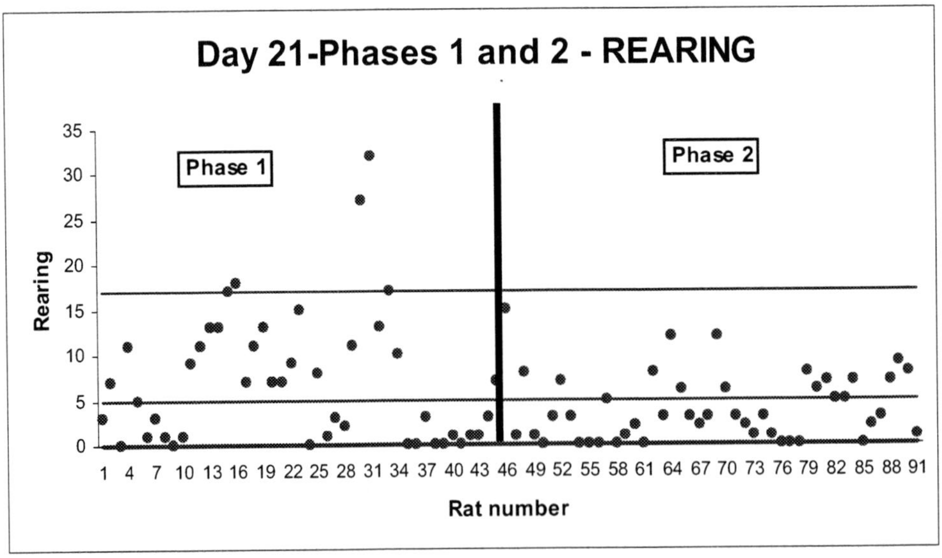

Figure 2. Control-chart for individual measurement of rearing frequency for Day 21, Phases 1 and 2.

It's reasonable to consider that the limits of decision are 0 and 17, that is, we accept a maximum of 17 and a minimum of 0 for a normal behavior. Using these limits, the rats with a non-normal behavior are three, quantifying 3.3% (3/91 x 100%0) of all rats. If we define that a maximum of 5% of the rats' out-of-control is acceptable, then we do not need to redo the experiment.

b. Grouped Data by Litter as an Experimental Unit

Control limits for grouped data is similar as we get samples of size n, when use n=4 in this example.

The control-chart lines are calculated using the following expressions:

$$\text{UCL} = \bar{\bar{x}} + 3\frac{\bar{s}}{\sqrt{n}}$$

$$\textit{Center}\ \text{Line} = \bar{x}$$

$$\text{LCL} = \bar{\bar{x}} - 3\frac{\bar{s}}{\sqrt{n}}$$

UCL stands for upper control limit, LCL for lower control limit, $\bar{\bar{x}}$ is the average of all litter averages, and $\bar{s}$ is mean of all standard deviations of the rearing measurements of all rats observed in a random manner.

Using the data for rearing, D21, Phases 1 and 2, for litter whose mothers were manipulated during the gestational period, we have: $\bar{\bar{x}} = 7.30$ and $\bar{s} = 3{,}65$ Then,

$$\text{UCL} = 7.30 + 3\frac{3.65}{\sqrt{4}} = 7.30 + 5.48 = 12.78, \text{ rounded to 13.}$$

Center Line = 5.505495

$$\text{LCL} = 7.30 - 3\frac{3.65}{\sqrt{4}} = 7.30 - 5.48 = 1.82, \text{ rounded to 1.}$$

Figure 3 illustrates this control-chart for 4 pups from the same litter, Day 21.

It's reasonable to consider that the limits of decision are 1 and 13, that is, we accept a maximum of 13 and a minimum of 1 for an average normal behavior. Using these limits, and considering a sample of nine litters (litter 8 was excluded, as it had only 3 observations), there is only one without a normal behavior. So, the percentage of litters with a non-uniform behavior is 11% (1/9 x 100%) of all of them.

If we define that a maximum of 5% for litters' out-of-control is acceptable, then we need to redo the experiment.

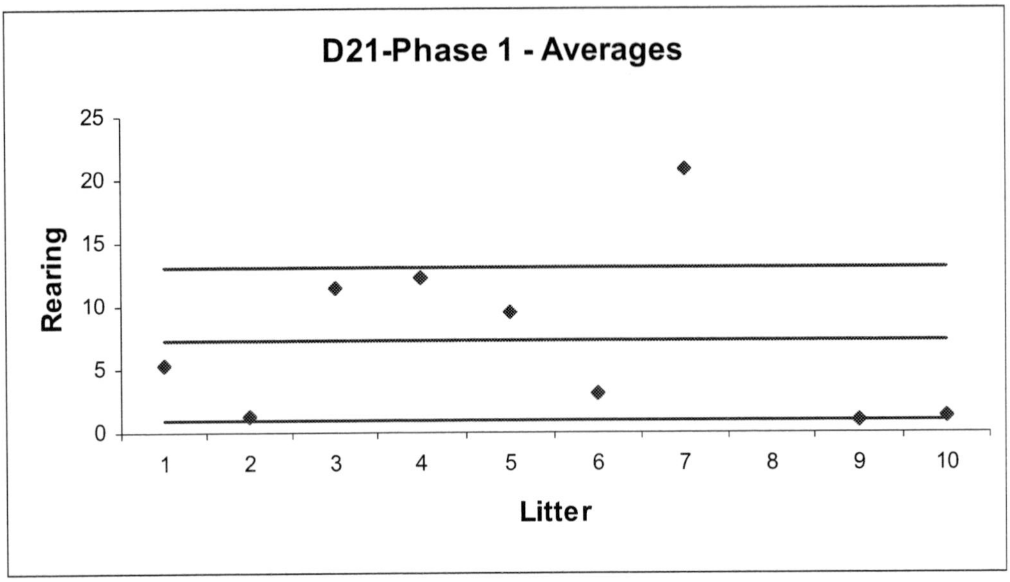

Figure 3. Control-chart for litter averages rearing frequency for D21.

c. Analysis of Variance (ANOVA) for the Litter Averages

The results for this line of thought is on Table 1.

If the null hypothesis is rejected, significance level of 5%, the decision rule is the following: we do not to redo the experiment.

As we see in this example, the null hypothesis that all the litter averages are equal is rejected. For this reason, at least one of the averages is different from the others, and we need to redo the experiment.

A summary for the rearing frequency data, whose mothers were manipulated during the gestational period on D21, is the following:

Individual behavior, Phases 1 and 2	do not repeat
Litter averages, Phase 1	repeat
ANOVA, Phase 1	repeat

However, for not losing all the data, we have applied the Tukey´s test to identify which litter has an average different from the others. For this example, the results were: L1 different from L6, L7, L8, L9, and L10; L2 different from L7, L8, L9, and L10; L3 different from L8, L9, and L10, and L4 not different from L5 to L10. For this reason, we can make a partition in two groups: (L1, L2, L3) and (L4, L5, L6, L7, L8, L9 and L10). For this reason, the candidates for a new experiment are L1, L2 and L3, because their group is smaller than the other one.

Table 1. ANOVA for rearing, Day 21, Phases 1 and 2

Source	SS	df	MS	F	P-value	Critical F
Between groups	1725,119697	9	191,6799663	9,833202518	3,18795E-07	2,169562317
Within groups	662,7666667	34	19,49313725			
Total	2387,886364	43				

Based on these findings from the data of the samples for rearing, we can write the following methodology with an overall decision rule for the acceptable behavior presented by the rats and, in consequence, if we redo or not an experiment:

Step 1. Observe the rats regarding the characteristics being studied, and register the results of them.

Step 2. For each characteristic, calculate the control limits for the individual observations and litter average, according the following expressions:

individual rats

$$UCL = \bar{x} + 3\frac{\overline{MR}}{1.128}$$

$$Center\text{ Line} = \bar{x}$$

$$LCL = \bar{x} - 3\frac{\overline{MR}}{1.128}$$

litter

$$\text{UCL} = \bar{\bar{x}} + 3\frac{\bar{s}}{\sqrt{n}}$$

$$Center\text{ Line} = \bar{x}$$

$$\text{LCL} = \bar{\bar{x}} - 3\frac{\bar{s}}{\sqrt{n}}$$

Step 3. Verify how many rats for individual control or by litter are out of the limits, and calculate the respective percentage.

Step 4. If the percentage is greater than 5 (five) % in both situations, the experiment must be performed again.

Step 5. If the percentage is less than 5 (five) % in one or both situations, perform an analysis of variance (ANOVA) regarding the litter averages.

Step 6. If the ANOVA null hypothesis is rejected, perform the Tukey´s test, and identify the different sets of behavior. Redo all the experiment for the smaller set.

Step 7. Continue the study in such a way that the assignable cause for the difference in the litters can be found

This chapter emphasizes that it is necessary to have a clear decision rule, based on the data, because statistics is a real experimental science. For this reason, if the model diverges from the reality, the model is wrong, not the nature. It can be concluded that on neurobehavioral tests, the diminution of the variability in the animal performance response to a stimulus is a big challenge to the scientists. In a near future, criteria related to the accepted behavioral needs improve when more data are available. The approach to be used to estimate the behavior change obtained during mixture pesticides tests and its potential applicability in additive and synergistic responses needs careful evaluation.

PERSPECTIVES OF NEUROBEHAVIORAL PERINATAL STUDIES

It can be conclude that the utilization of an experimental protocol containing indices related to reproduction and animal development can identify initial damages due to the environmental pollutants exposure. The protection of possible damages due to pesticides at the development of the young organisms, like the newborn and children; requires an integrated strategy able to monitor the standard use of these products and the integration in the potential effects evaluation improvement.

The near future requires the ability to see how individual compounds affect signaling networks and how mixtures of compounds affect a common physiological endpoint by either similar or diverse modes of action in the body (Andersen and Dennison, 2004). In this sense, the study of subtle effects through new batteries of experimental tests promoting biomarkers more specific should be a complement of the traditional evaluation approach for pesticides in young animals. Sharing this information is important for the improvement of the perinatal evaluation, in order to the implementation of a regulatory approach that minimizes the possible risks to what the young organisms.

In order to reduce the uncertainty of the estimates regarding the exposure to these products in order to protect adequately the children, it can be suggest to address these aspects in various areas of research as: a) kinetic differences of an agrochemical in young and adult in order to subsidize information regarding the exposure of these products in children; b) appropriated investigation to sustain the safety use of the compounds regarding the habits of the infantile population, for example, the diet (Pennycook et al., 2004); c) improve the pre and postnatal developmental parameters evaluation; d) promote more specific biomarkers study; and e) promote advances in ecogenetics through the interaction gene-environment-health study (Hubal et al., 2000).

ANNEX: DATA FOR REARING, DAY 21, PHASE 1 AND 2

Control Group			Rat	D21	D30	D60	D90
Phase	Dose	Litter					
1	1	1	1	3	15	25	23
		1	2	7	18	32	19
		1	3	0	9	37	27
		1	4	11	11	41	31
		1	5	5	13	19	30
		2	1	1	7	0	4
		2	2	3	18	0	8
		2	3	1	11	0	6
		2	4	0	9	0	0
		2	5	1	15	0	0
		3	1	9	21	31	19
		3	2	11	11	27	27
		3	3	13	17	39	31

		3	4	13	15	29	42
		3	5	17	19	41	29
		4	1	18	0	0	21
		4	2	7	0	0	23
		4	3	11	2	4	25
		4	4	13	10	7	27
		4	5	no data	no data	no data	no data
		5	1	7	3	6	31
		5	2	7	1	0	23
		5	3	9	15	7	11
		5	4	15	7	0	0
		5	5	no data	no data	no data	no data
		6	1	0	0	0	31
		6	2	8	24	1	23
		6	3	1	14	7	0
		6	4	3	5	11	0
		6	5	2	0	9	0
		7	1	11	11	15	3
		7	2	27	0	7	0
		7	3	32	0	13	5
		7	4	13	3	11	0
		7	5	17	5	7	13
		8	1	10	0	1	0
		8	2	0	0	7	0
		8	3	0	14	3	0
		8	4	no data	no data	no data	no data
		8	5	no data	no data	no data	no data
		9	1	3	0	0	3
		9	2	0	13	12	0
		9	3	0	1	4	0
		9	4	1	11	10	0
		9	5	no data	no data	no data	no data
		10	1	0	1	0	5
		10	2	1	0	11	7
		10	3	1	0	15	1
		10	4	3	3	0	3
		10	5	no data	no data	no data	no data
2	1	1	1	7	12	27	5
		1	2	15	8	13	15
		1	3	1	9	19	27
		1	4	8	3	32	21
		1	5	no data	no data	no data	no data
		2	1	1	3	15	0
		2	2	0	2	18	0
		2	3	3	2	9	0
		2	4	7	9	32	0

Table. Continued

Control Group			Rat	D21	D30	D60	D90
Phase	Dose	Litter					
		2	5	no data	no data	no data	no data
		3	1	3	0	13	0
		3	2	0	0	7	0
		3	3	0	0	8	0
		3	4	0	0	9	0
		3	5	5	0	1	0
		4	1	0	1	3	1
		4	2	1	3	5	0
		4	3	2	7	9	0
		4	4	0	9	1	4
		4	5	8	1	0	4
		5	1	3	5	1	11
		5	2	12	9	3	1
		5	3	6	3	1	7
		5	4	3	8	5	1
		5	5	2	0	0	0
		6	1	3	5	31	18
		6	2	12	17	23	29
		6	3	6	0	0	15
		6	4	3	11	11	21
		6	5	2	10	9	17
		7	1	1	1	0	20
		7	2	3	3	8	11
		7	3	1	5	0	13
		7	4	0	4	0	9
		7	5	0	0	0	23
		8	1	0	11	27	13
		8	2	8	19	33	15
		8	3	6	7	0	21
		8	4	7	19	9	17
		8	5	no data	no data	no data	no data
		9	1	5	13	11	17
		9	2	5	17	10	1
		9	3	7	9	3	15
		9	4	0	8	1	19
		9	5	2	6	5	7
		10	1	3	0	0	11
		10	2	7	4	0	13
		10	3	9	0	0	17
		10	4	8	0	0	9
		10	5	1	4	1	3

REFERENCES

Aldridge, J., Gibbons, J., Flaherty, M., Kreider, M., Romano, J., Levin, E. (2003) Heterogeneity of Toxicant Response, Sources of Human Variability. *Toxicological Sciences, 76*, 3–20.

Allio, L., Ballantine, B., Meads, R. (2006) Enhancing the role of science in the decision-making of the European Union. *Regulatory Toxicology and Pharmacology, 44,* 4–13.

Alonso, A. & Geys, H., Molenberghs, G., Kenward, M., Vangeneugden, T. (2004) Validation on surrogate markers in multiple randomized clinical trials with repeated measurements: canonical correlation approach. *Biometrics, 60*, 845-853.

Andersen, M. & Dennison, J. (2004) Mechanistic approaches for mixture risk assessments—present capabilities with simple mixtures and future directions. *Environmental Toxicology and Pharmacology, 16,* 1–11.

Arakawa, H. (2005) Age dependent effects of space limitation and social tension on open-field behavior in male rats. *Physiology & Behavior, 84,* 429–436.

Araújo, A, Lima, J, Moreira, J, Jacob, S, Soares, M, Monteiro, M, Amaral, A, Kubota, A, Meyer, A, Cosenza, A, Neves, C, Markowitz, S. (2007) Exposição múltipla a agrotóxicos e efeitos à saúde: estudo transversal em amostra de 102 trabalhadores rurais, Nova Friburgo, RJ. *Ciência & Saúde Coletiva, 12(1),* 115-130.

Birkhøj, M., Nellemann, C., Jarfelt, K., Jacobsen, H., Andersen, H., Dalgaard, M., Vinggaard, A. (2004) The combined antiandrogenic effects of five commonly used pesticides. *Toxicology and Applied Pharmacology, 201,* 10– 20.

Buelke-Sam, J., Kimmel, C., Adams, J., Nelson, C., Vorhees, C., Wright, D., ST Omer, V., Korol, B., Butcher, R., Geyer, M., Holson, J., Kutscher, C., Wayner, M. (1985) Collaborative behavioral teratology study. *Neurobehavioral Toxicology and Teratology, 7*, 691-624,.

Calamandrei G, Maranghi F, Venerosi A, Alleva E, Mantovani E. (2006) Efficient testing strategies for evaluation of xenobiotics with neuroendocrine activity. *Reproductive Toxicology, 22*, 164–174.

Castro, V.; Chiorato, S.; Pinto, N. (2000) Relevance of developmental testing of exposure to methamidophos during gestation to its toxicology evaluation. *Toxicology Letters, 118(1-2)*, 93 – 102.

Castro, V.; Mello, M.A.; Diniz, C.; Morita, L.; Zucchi, T.; Poli, P. (2007) Neurodevelopmental effects of perinatal fenarimol exposure on rats. *Reproductive Toxicology, 23*, 98-105.

Castro, V.; Mello, M.A.; Poli, P.; Zucchi, T. (2005) Prenatal and perinatal fenarimol-induced genotoxicity in leukocytes of in vivo treated rats. *Mutation Research, 583(1),* 95-104.

Castro, V.; Silveira, M.; Perez, M. (1999) Application of clinical indicators of exposition in the evaluation of family agriculture health: the Sumaré case – Brazil. *The International Journal of Sustainable Development and World Ecology, 6 (3),* 172 – 184.

Champagne, F., Francis, D., Mar, A., Meaney, M. (2003) Variations in maternal care in the rat as a mediating influence for the effects of environment on development. *Physiology and Behavior, 79*, 359-371.

Clewell H. Gentry P, Covington T, Sarangapani R, Teeguarden J. (2004) Evaluation of the potential impact of age- and gender-specific pharmacokinetic differences on tissue dosimetry. *Toxicology Science*, *79*, 381-393.

Coffey, T., Gennings, C., Moser, V. (2007) The simultaneous analysis of discrete and continuous outcomes in a dose-response study: using desirability functions. *Regulatory Toxicology and Pharmacology, In Press, Corrected Proof, Available online 12 January 2007.*

Cohen Hubal EA, Sheldon LS, Burke JM, McCurdy TR, Berry MR, Rigas ML, et al. (2000) Children's exposure assessment: a review of factors influencing children's exposure, and the data available to characterize and assess that exposure. *Environmental Health Perspectives, 108*, 475–86.

Conolly, R. & Lutz, W. (2004) Nonmonotonic dose-response relationships, mechanistic basis, kinetic modeling, and implications for risk assessment. *Toxicological Sciences,* 77, 151-157.

Cory-Slechta, D. (2005) Studying Toxicants as Single Chemicals: Does this Strategy Adequately Identify Neurotoxic Risk? *NeuroToxicology, 26*, 491–510.

Costa, L., Aschner, M., Vitalone, A., Syversen, T., Soldin, O. (2004) Developmental neuropathology of environmental agents. *Annual Review of Pharmacology and Toxicology, 44*, 87-110.

Crofton K.M., Makris S.L.,. Sette W.F, Mendez E., Raffaele K.C. (2004) A qualitative retrospective analysis of positive control data in developmental neurotoxicity studies. *Neurotoxicology and Teratology, 26*, 345–352.

Crofton, K.M., Foss, J.A., Hass, U., Jensen, K., Levin, E.D., Parker, S.P. (2007) Undertaking positive control studies as part of developmental neurotoxicity testing. *Neurotoxicology and Teratology, doi: 10.1016/j.ntt.2007.06.002*

Dam K, Seidler F, Slotkin T. (2000) Chlorpyrifos exposure during a critical neonatal period elicits gender-selective deficits in the development of coordination skills and locomotor activity, *Developmental Brain Research, 121*, 179-187.

Das, R. (2002) Role of ancillary variables in the design, analysis, and interpretation of animal experiments. *ILAR Journal, 43 (4)*, 214-222.

Dinham, B. & Sapna Malik, L. (2003) Pesticides and Human Rights. *International Journal of Occupational and Environmental Health, 9*, 40–52.

Donnelly, K., Lingenfelter, R., Cizmas, L., Falahatpisheh M.H., Qian, Y, Tang, Y., Garcia, S., Ramos, K., Tiffany-Castiglioni, E., Mumtaz, M. (2004) Toxicity assessment of complex mixtures remains a goal. Environmental *Toxicology and Pharmacology, 18*, 135–141.

Ehman K.D. & Moser V.C. (2006) Evaluation of cognitive function in weanling rats: A review of methods suitable for chemical screening. *Neurotoxicology and Teratology, 28*, 144 – 161.

Festing, M. & Altman, D. (2002) Guidelines for the design and statistical analysis of experiments using laboratory animals. *ILAR Journal, 43 (4)*, 244-258.

Ferdman, N., Murmu, R.P., Bock J., Braun, K., Leshem, M. (2007) Weaning age, social isolation, and gender, interact to determine adult explorative and social behavior, and dendritic and spine morphology in prefrontal cortex of rats. *Behavioural Brain Research, 180*, 174–182.

Groten, J., Heijne, W., Stierum, R., Freidig, A., Feron, V. (2004) Toxicology of chemical mixtures: a challenging quest along empirical sciences. *Environmental Toxicology and Pharmacology, 18*, 185–192.

Haseman, J., Bailer, A., Kodell, R., Morris, R., Portier, K. (2001) Statistical issues in the analysis of low-dose endocrine disruptor data. *Toxicological Sciences, 61*, 201-210.

Hass U. (2006) The need for developmental neurotoxicity studies in risk assessment for developmental toxicity. *Reproductive Toxicology. 22*, 148–156.

Hensch, T. (2004) Critical period regulation. *Annual Review Neuroscience*, 27, 549–79.

Holson, R. & Pearce, B. (1992) Principles and pitfalls in the analysis of prenatal treatment effects in multiparous species. *Neurotoxicology and Teratology, 14*, 221-228.

Holson, R., Freshwater, L., Maurissen, J., Moser, V., Phang, W. (2007) Statistical issues and techniques appropriate for developmental neurotoxicity testing. *Neurotoxicology and Teratology, doi: 10.1016/j.ntt.2007.06.001*

Howard B. R. (2002) Control of Variability. *ILAR Journal, 43(4)*, 194-201.

Hubal, E., Sheldon, L., Burke, J., Mc Curdy, T., Berry, M., Rigas, M., Zartarian, V., Freeman, N. (2000) Children's Exposure Assessment, A Review of Factors Influencing Children's Exposure, and the Data Available to Characterize and Assess That Exposure. *Environ Health Perspectives, 108*, 475-486.

Huot R, Gonzalez M, Ladd C, Thrivikraman K, Plotsky P. (2004) Foster litters prevent hypothalamic-pituitaryadrenal axis sensitization mediated by neonatal maternal separation. *Psychoneuroendocrinology, 29,* 279–289.

Jacobsen, H., Østergaard G., Lam H.R., Poulsen M.E, Frandsen H., Ladefoged O., Meyer O. (2004) Repeated dose 28-day oral toxicity study in Wistar rats with a mixture of five pesticides often found as residues in food: alphacypermethrin, bromopropylate, carbendazim, chlorpyrifos and mancozeb. *Food and Chemical Toxicology, 42*, 1269–1277.

Jadhav, S, Sarkar, S, Kataria, M, Tripathi, H. (2007) Subchronic exposure to a mixture of groundwater-contaminating metals through drinking water induces oxidative stress in male rats. *Environmental Toxicology and Pharmacology, 23*, 205–211.

Jardim M. & Guimarães F. (2004) Role of glutamate ionotropic receptors in the dorsomedial hypothalamic nucleus on anxiety and locomotor behavior. *Pharmacolology,Biochemistry and Behaviour*; *79(3),* 541-546.

Ladics, G., Chapin, R., Hastings, K., Holsapple, M., Makris, S., Sheets, L., Woolhiser, M., Burns-Naas, L. (2005) Developmental Toxicology Evaluations—Issues with Including Neurotoxicology and Immunotoxicology Assessments in Reproductive Toxicology Studies. *Toxicological Sciences, 88(1),* 24–29.

Laessig, S., Auger, A., McCarthy, M., Silbergeld, E. (2007) Effects of prenatal chlordecone on sexually differentiated behavior in adult rats. *Neurotoxicology and Teratology, 29*, 255–263.

Lehmann J, Stohr T, Feldon J. (2000) Long-term effects of prenatal stress experiences and postnatal maternal separation on emotionality and attentional processes. *Behaviour Brain Research, 107*,133 - 144.

Lengyel, Z., Szabó, A., Papp, A., Nagymajtényi, L. (2007) Functional neurotoxic effects in rats acutely exposed to insecticide combinations. *Pesticide Biochemistry and Physiology, 87*, 94–102

Marable, B.R. & Maurissen, J.P.J (2004) Validation of an auditory startle response system using chemicals or parametric modulation as positive controls. *Neurotoxicology and Teratology, 26*, 231–237

Mahar, A.M. & Watzin, M.C. (2005) Effects of metal and organophosphate mixtures on Ceriodaphnia dubia survival and reproduction. *Environmental Toxicology and Chemistry, 24(7),* 1579-1586.

McCarty, L.S. & Borgert, C.J. (2006) Review of the toxicity of chemical mixtures: Theory, policy, and regulatory practice. *Regulatory Toxicology and Pharmacology, 45*, 119–143.

McCormick C, Robarts D, Kopeikina K, Kelsey J. (2005) Long-lasting, sex- and age-specific effects of social stressors on corticosterone responses to restraint and on locomotor responses to psychostimulants in rats. *Hormones and Behaviour, 48(1),* 64-74.

Moher, D., Cook, D.J., Eastwood, S., Olkin, I., Rennie, D., Stroup, D.F. (1999) Improving the quality of reports of meta-analyses of randomized controlled trials: the QUOROM statement. *Lancet*, *354,* 1896–1900.

Moser, V., Shafer, T., Ward, T., Meacham, C., Harris, M., Chapin, R. (2001) Neurotoxicological outcomes of perinatal heptachlor exposure in the rat. *Toxicological Sciences, 60*, 315-326.

Mumtaz, M., De Rosa, C., Cibulas, W., Falk, H. (2004) Seeking solutions to chemical mixtures challenges in public health. *Environmental Toxicology and Pharmacology, 18*, 55–63.

Owen, D. Andrews, M.H. Matthews, S.G. (2005) Maternal adversity, glucocorticoids and programming of neuroendocrine function and behaviour. *Neuroscience and Biobehavioral Reviews,* 29, 209–226.

Pennycook, F., Diamand, E., Waterson, A., Howard, CV. (2004) Modeling the dietary pesticide exposures of young children. *International Journal of Occupational and Environmental Health, 10*, 304-309.

Perry, M., Venners, S., Barr, D., Xu, X. (2007) Environmental pyrethroid and organophosphorus insecticide exposures and sperm concentration. *Reproductive Toxicology, 23*, 113–118.

Presibella, K., Kita, D., Carneiro, C., Andrade, A., Dalsenter, P. (2005) Reproductive evaluation of two pesticides combined (deltamethrin and endosulfan) in female rats. *Reproductive Toxicology, 20,* 95–101.

Prut, L. & Belzung C. (2003) The open field as a paradigm to measure the effects of drugs on anxiety-like behaviors: a review. *European Journal of Pharmacology, 463,* 3– 33.

Ricceri, L., Venerosi A, Capone F, Cometa M, Lorenzini P, Fortuna S, Calamandrei G. (2006) Developmental neurotoxicity of organophosphorous pesticides: fetal and neonatal exposure to chlorpyrifos alters sex-specific behaviors at adulthood in mice. *Toxicological Sciences, 93(1),*105–113.

Richardson, J. and Chambers, J. (2005) Effects of repeated oral postnatal exposure to chlorpyrifos on cholinergic neurochemistry in developing rats. *Toxicological Sciences, 84*, 352 - 359.

Robinson, P. & MacDonell, M. (2004) Priorities for mixtures health effects research. *Environmental Toxicology and Pharmacology, 18*, 201–213.

Rohlman, D., Arcury, T., Quandt, S., Lasarev, M., Rothlein, J., Travers, R., Tamulinas, A., Scherer, J., Early, J., Marín, A., Phillips, J., MCCauley, L. (2005) Neurobehavioral

Performance in Preschool Children from Agricultural and Non-Agricultural Communities in Oregon and North Carolina. *NeuroToxicology, 26*, 589–598.

Scheuplein, R., Charnley, G., Dourson, M. (2002) Differential sensitivity of children and adults to chemical toxicity i. biological basis. *Regulatory Toxicology and Pharmacology. 35*, 429–447.

Slamberová, R., Charousová, P., Pometlová, M. (2005) Maternal behavior is impaired by methamphetamine administered during pre-mating, gestation and lactation. *Reproductive Toxicology, 20*, 103–110.

Slikker Jr., W., Acuff, K., Boyes, W., Chelonis, J., Crofton, K., Dearlove, G., Li, A., Moser, V., Newland, C., Rossi, J., Schantz, S., Sette, W., Sheets, L., Stanton, M., Tyl, S., Sobotka, T. (2005) Behavioral test methods workshop. *Neurotoxicology and Teratology, 27*, 417 – 427.

Smith, A.L. & Corrow, D.J. (2005) Modifications to husbandry and housing conditions of laboratory rodents for improved well-being. *ILAR Journal, 46*, 140–147.

Trksak, G., Glatt, S., Mortazavi, F., Jackson, D. (2007), A meta-analysis of animal studies on disruption of spatial navigation by prenatal cocaine exposure. *Neurotoxicology and Teratology, doi: 10.1016/j.ntt.2007.06.003*

Ulbrich, B. (2001) Litter variability in behavioural testing, is testing of one animal per litter enough? *Frontiers in Fetal Health, 3*, 273-274.

Van den Bergha, B.R.H., Mulderb, E.J.H., Mennesa, M., Gloverd, V. (2005) Antenatal maternal anxiety and stress and the neurobehavioural development of the fetus and child: links and possible mechanisms. A review. *Neuroscience and Biobehavioral Reviews, 29*, 237–258.

Verwer, C.M.; van der Ven, L.T.; van den Bos, R.; Hendriksen, C.F. (2007) Effects of housing condition on experimental outcome in a reproduction toxicity study. *Regulatory Toxicology and Pharmacology, 48*, 184–193.

Viberg H, Fredriksson A, Eriksson P. (2004) Investigations of strain and/or gender differences in developmental neurotoxic effects of polybrominated diphenyl ethers in mice. *Toxicological Sciences, 812*, 344-353.

Vinggaard A, Jacobsen H, Metzdorff S, Andersen H, Nellemann C. (2005) Antiandrogenic effects in short-term in vivo studies of the fungicide fenarimol. *Toxicology, 207*, 21–34.

Waissmann, W. (2002) Health surveillance and endocrine disruptors. *Cadernos de Saúde Pública, 18 (2),* 511-517.

Weed, J.L. & Raber, J.M. (2005) Balancing Animal Research with Animal Well-being: Establishment of Goals and Harmonization of Approaches. *ILAR Journal, 46*, 118-128.

Weiss, B., Amler, S., Amler, R. (2004) *Pesticides Pediatrics, 113 (4)*, 1030-1036.

Zimmermann A, Stauffacher M, Langhans W, Würbel H. (2001) Enrichment-dependent differences in novelty exploration in rats can be explained by habituation. *Behaviour Brain Res*earch,*121*,11-20.

In: Progress in Pesticides Research
Editor: Christopher M. Kanzantzakis

ISBN 978-1-60456-201-9

Chapter 7

TRACKING THE DISPERSAL OF DDT RESIDUE-BEARING SEDIMENTS ONTO THE FLOOR OF SANTA MONICA BASIN

C. K. Paull[1,*], W. Ussler III[1], W. Normark[2], R. Keaten[1]

[1]Monterey Bay Aquarium Research Institute
Moss Landing, CA 95039-9644, USA
[2]United States Geological Survey
Menlo Park, CA 94025, USA

ABSTRACT

The concentrations of the pesticide DDT and its residues (DDTr) were measured in sediment samples collected from the seafloor using a remotely operated vehicle in 286 to 875 m water depths offshore Los Angeles, California. The northwestward decrease in DDTr concentrations from the Whites Point outfall off Palos Verdes, California suggests dispersal pathways are linked to a concentrated point source of DDTr-contaminated particles. This outfall is known to have carried DDTr released from a DDT production facility. The concentration of DDTr in these sediment samples show that DDTr-bearing sediment extends farther down slope into the adjacent 900 m deep Santa Monica Basin and northwestward along the continental slope than previously known. DDTr concentrations in near-surface sediment are higher in samples from the northeastern flank of the Santa Monica Basin (200 to 900 m water depths) than on the adjacent continental shelf (water depths <200 m) at equivalent distances from the point source. Thus, previous surveys to track the dispersion of DDTr from the Whites Point outfall have not encompassed a broad enough area to properly outline and define the dispersion pathways and depocenters. Even with the new observations from deep water, the area that has been surveyed to date is still inadequate to fully understand the dispersion of DDTr released from the Whites Point outfall.

* Paull@mbari.org

INTRODUCTION

Fate of DDT in the Environment

DDT (1,1,1-trichloro-2,2-bis(p-chlorophenyl) ethane) is an organochlorine pesticide used for the control of insects, and does not occur naturally in the environment (World Health Organization, 1979). DDT's effectiveness for insect control was realized during World War II, and after 1945 DDT was widely used as an agricultural insecticide (Brooks, 1979; Mischke et al., 1985). Some DDE (1,1-dichloro-2,2-bis(p-chlorophenyl) ethylene) and DDD (1,1-dichloro-2,2-bis(p-chlorophenyl) ethane) were minor contaminates introduced by the manufacturing of DDT. To a limited extent DDD was also used as a pesticide. Environmental problems associated with DDT's persistence in the environment resulted in it being banned in the United States in 1972.

DDT breaks down to DDE and DDD with a half-life estimated to vary between 2 to 15 years depending on local conditions (Phillips et al., 1975; US Environmental Protection Agency, 1989). DDT breakdown products have similar chemical and physical properties to DDT and are also highly persistent in the environment (World Health Organization 1989). The sum of the concentrations of DDT and its breakdown products (DDD and DDE) is referred to here as DDTr. DDTr has an extremely low solubility in water, but is relatively soluble in organic solvents. DDTr adheres to the organic matter common in fine-grained sediment.

There was no intentional use of DDT in the ocean. Thus, the presence of DDTr in marine sediment requires a terrestrial source and identifies sediment that has accumulated since 1945. Thus, DDTr makes an excellent tracer for following transport of fine sediment from the continent into the ocean. Recent studies have used DDTr as a tracer of recent sediment transport on the continental shelf (Wu et al., 1999; Yang et al., 2005; Pandit et al., 2006; Wafo et al., 2005). Less effort has been spent tracing DDTr farther along its potential dispersal path from the continental shelf into the deep-sea (Paull et al., 2002), in part because of the difficulty in obtaining appropriate samples.

Sources of DDTr Near Los Angeles, California

From 1947 until 1983 DDT was produced at the Montrose Chemical Corporation plant in Torrance, California. DDTr from this plant was carried in the wastewaters that emptied into the ocean from an outfall located in 60 m of water ~2.5 km off Whites Point on the Palos Verdes Peninsula (Figure 1). Over 1,700 tons of DDT is estimated to have been discharged from this outfall (http://www.epa.gov/region09/features/pvshelf/) making the Whites Point outfall a significant point source for DDTr. As a consequence, the area of DDTr contamination on the Palos Verdes Shelf became identified by the United States Environmental Protection Agency as a Super Fund site. The area of particular concern involves ~40 km^2 on the continental shelf around the outfall where sediment DDTr concentrations commonly exceed 50 ppm (http://www.epa.gov/region09/features/pvshelf/).

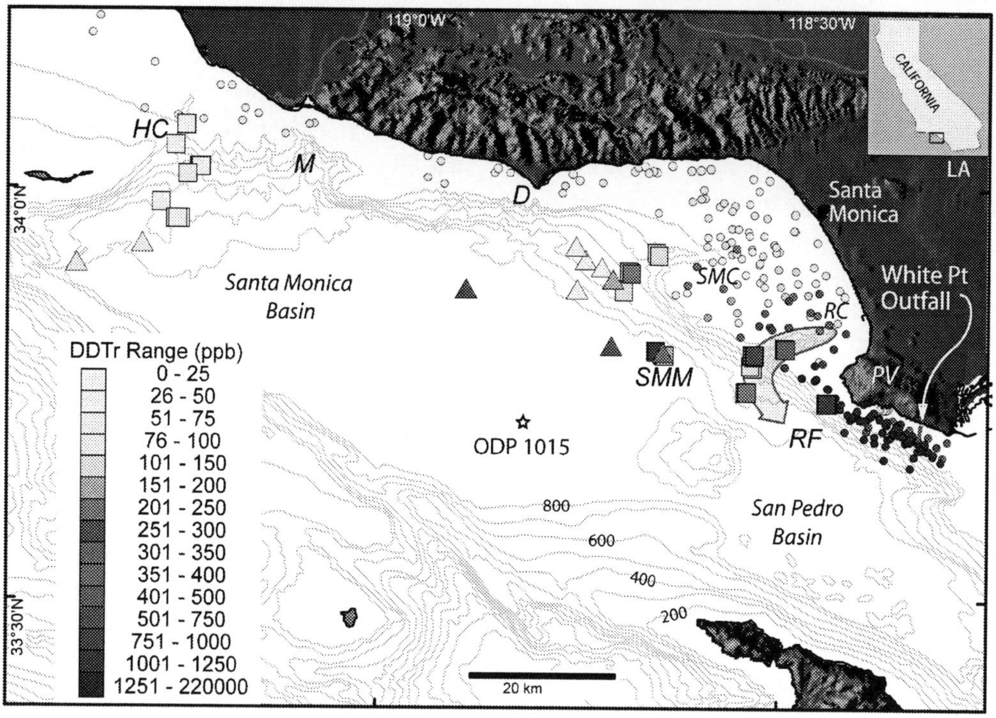

Figure 1. Map showing the location of DDTr concentration measurements of seafloor sediment samples from the continental margin off Los Angeles, California. The filled squares are ROV collected samples, filled triangles are newly reported values from archived piston cores, and filled circles are previously reported values (CH2M Hill, Inc., 2003; Schiff, 2003) with concentration ranges given in color key. The previously published values were reported as tDDT which is equivalent to DDTr. These data are from sediment samples collected from 0 to 15 cmbsf [sources are Appendix 2 EPA report (CH2M Hill, Inc., 2003) and Schiff, 2003]. For clarity, previous samples in water depths <11 m were excluded from the figure. Water depths for the previously reported data were estimated using NOAA charts and the reported position when no water depth was reported. Samples from the harbors were also excluded from the plot. The bathymetry of the Santa Monica and San Pedro Basins (100 m contours), the location of the Whites Point outfall on the Palos Verdes Shelf (red line), and the rivers within the Los Angeles and Oxnard areas are indicated with green lines. Red open arrow indicates route of Redondo Canyon across the lower slope and into San Pedro Basin. HC=Hueneme Canyon; SMC=Santa Monica Canyon; RC=Redondo Canyon; PV=Palos Verdes; LA=Los Angeles; RF=Redondo Fan; D=Dume Canyon; M=Mugu Canyon; SMM=Santa Monica Mound

Studies of DDTr dispersal within the offshore regions of the southern California Borderland are complicated by other possible terrestrial sources of DDTr. These include the Los Angeles sewage outfalls near the head of Santa Monica Bay (Bay et al., 2003) and coastal rivers that drain into Santa Monica Bay. DDT was once extensively used on agricultural lands in this region and may still adhere to soil particles that enter the ocean through the fluvial drainage systems. Other DDTr sources include the Santa Clara River, which currently drains land extensively used for agriculture. Many of the other coastal rivers drain areas that have been converted to urban use, although they may still transport soils with residual DDTr.

Dispersal of DDTr discharged from the Whites Point outfall has been a topic of numerous studies (see summary in Lee and Wiberg, 2002). DDTr is preferentially adsorbed onto organic matter and is transported with fine-grained sediment. While DDTr-bearing fine-grained sediment might initially have been deposited only on the Palos Verdes Shelf near the outfall, repeated re-suspension by surface waves, coastal currents, and internal waves has resulted in the contaminated sediment being transported further away (Wiberg et al., 2002).

Existing physical oceanographic data suggest that DDTr-bearing sediment likely moved to the northwest along the shelf (Nobel et al., 2002). The Southern California Countercurrent flows to the northwest, past the Palos Verdes Peninsula and along the eastern margin of the Santa Monica Bay (Hickey, 1992). Some off-shelf sediment transport will periodically occur away from the canyons (Nobel et al., 2002). DDTr samples collected on the continental shelf in Santa Monica Bay reveal that some DDTr-bearing sediment has crossed Redondo Canyon to the shelf on the northwest side (Sommerfield and Lee, 2003). Unknown amounts of DDTr-bearing sediment might have moved downslope within the submarine canyons.

Submarine Canyons as Sediment Transport Conduits

The flanks of Santa Monica Basin are cut by several large submarine canyons (Gardner et al., 2003) including Redondo, Santa Monica, and Hueneme Canyons (Figure 1). Submarine canyons are believed to be major conduits for sediment transport from the land and continental shelf into the deep-sea and the occurrence of large fans at their bases attest to their long-term significance for sediment transport (Normark et al., 1998; 2006a). Submarine canyons have been very difficult to monitor and little direct evidence exists about how sediments move through them because they have steep walls, are in relatively inaccessible deep water, and are associated with complicated morphologies. Huge amounts of material periodically move through the canyons and onto the adjacent basin floors in episodic energetic turbidity-flow events (Normark et al., 1998; 2006a). Distal channels that emerge from Hueneme and Santa Monica Canyons remain within Santa Monica Basin, whereas, the distal fan-channel that emerges from Redondo Canyon turns toward the southeast at the base of the continental slope and enters San Pedro Basin (Figure 1). The extent to which transport of fine-grained sediment is preferentially focused within the axes of these canyons in the intervening time between turbidity currents is unknown.

Here we report on DDTr concentrations measured within near-seafloor sediment collected from the continental slope off the Palos Verdes Peninsular and within the Santa Monica Basin, offshore southern California. Most of these samples were collected during a remotely operated vehicle (ROV) diving cruise that examined processes within the axes of Redondo, Santa Monica, and Hueneme Canyons as well as a slope-failure scar on the Palos Verdes Slope and gas vents on the floor of Santa Monica Basin. An objective of many of these ROV dives was to refine our understanding of sediment transport in submarine canyons using DDTr as a tracer for recent sediment movement. These DDTr samples were collected as a piggyback effort associated with other objectives that determined the ROV dive sites. Additional samples were obtained from archived piston cores taken in support of the United States Geological Survey (USGS) geologic hazards program.

Table 1. DDTr measured in near seafloor samples from the eastern flank of the Santa Monica Basin

Sample	Core	Interval (cmbsf)	DDD (ppb)	DDE (ppb)	DDT (ppb)	DDTr (ppb)	Lat. (°N)	Long. (°W)	W.D. (m)	Area
T783	PC63	0-2	3.2	24	1.8	29	34.0256	-119.2227	563	HC
T783	PC63	10-12	2.7	14	0	16.7	34.0256	-119.2227	563	HC
T783	PC63	18-20	3.2	17	0	20.2	34.0256	-119.2227	563	HC
T783	PC56	0-2	1.8	17	1.3	20.1	34.0255	-119.2210	579	HC
T783	PC56	8-10	1.1	7.3	0	8.4	34.0255	-119.2210	579	HC
T783	PC50	0-2	3.2	24	1.3	28.5	34.0253	-119.2214	578	HC
T783	PC71/78	0-2	2.9	16	1	19.9	34.0253	-119.2239	558	HC
T786	PC64	0-2	50	730	4.9	784.9	33.7993	-118.5317	542	RC
T786	PC64	10-12	53	740	29	822	33.7993	-118.5317	542	RC
T786	PC64	14-16	50	710	11	771	33.7993	-118.5317	542	RC
T786	PC62	0-2	18	500	0	518	33.7981	-118.5307	576	RC
T786	PC62	10-12	59	700	47	806	33.7981	-118.5307	576	RC
T787	PC71	0-2	42	540	0	582	33.7970	-118.5319	580	RC
T787	PC71	10-12	65	810	20	895	33.7970	-118.5319	580	RC
T787	PC57	0-2	31	400	0	431	33.7992	-118.5349	521	RC
T787	PC69	0-2	2.9	360	0	362.9	33.7974	-118.5324	577	RC
T788	PC70	0-2	2.2	30	0	32.2	33.7829	-118.5340	602	RC
T788	PC62	0-2	10	120	0	130	33.7856	-118.5368	575	RC
T788	PC62	8-10	30	300	7.4	337.4	33.7856	-118.5368	575	RC
T789	PC49	0-2	12	150	1.2	163.2	33.8986	-118.6879	535	SMC
T789	PC49	10-12	26	210	18	254	33.8986	-118.6879	535	SMC
T789	PC46	0-2	11	140	3.1	154.1	33.9003	-118.6899	470	SMC
T789	PC46	10-12	1.2	21	1.7	23.9	33.9003	-118.6899	470	SMC
T789	PC52	0-2	14	140	4.7	158.7	33.8981	-118.6880	545	SMC
T789	PC52	10-12	9.4	190	9.6	209	33.8981	-118.6880	545	SMC
T790	PC69	0-2	7.7	91	1.6	100.3	33.8978	-118.6872	545	SMC
T790	PC69	8-10	13	120	1.8	134.8	33.8978	-118.6872	545	SMC
T790	PC69	10-12	16	211	7.6	234.6	33.8978	-118.6872	545	SMC
T790	PC70	0-2	19	180	4.7	203.7	33.8974	-118.6862	536	SMC
T790	PC70	10-12	23	290	11	324	33.8974	-118.6862	536	SMC
T790	PC66	10-12	6.5	32	14	52.5	33.8957	-118.6843	485	SMC
T791	PC44	0-2	7.2	93	0	100.2	33.9190	-118.6511	464	SMC
T791	PC44	10-12	17	107	4.2	128.2	33.9190	-118.6511	464	SMC
T791	PC68	0-2	7.1	87	0	94.1	33.9196	-118.6516	455	SMC
T791	PC68	10-12	10	90	8.1	108.1	33.9196	-118.6516	455	SMC
T791	PC49	0-2	10	130	0	140	33.9225	-118.6541	384	SMC
T791	PC49	10-12	17	14	22	53	33.9225	-118.6541	384	SMC
T794	PC49	0-2	9.3	99	0	108.3	33.8756	-118.6959	654	SMC
T794	PC44	0-2	4.7	47	1.8	53.5	33.8756	-118.6959	655	SMC
T794	PC44	4-6	8.2	90	1.8	100	33.8756	-118.6959	655	SMC
T794	PC54	10-12	24	330	5.7	359.7	33.8755	-118.6939	572	SMC
T796	PC71	0-2	16	180	16	212	33.8036	-118.6553	820	SMM
T796	PC71	4-6	120	1300	57	1477	33.8036	-118.6553	820	SMM
T796	PC71	10-12	0	4.2	1.6	5.8	33.8036	-118.6553	820	SMM
T796	PC51	0-2	200	700	340	1240	33.8024	-118.6549	825	SMM
T796	PC51	4-6	87	200	57	344	33.8024	-118.6549	825	SMM
T796	PC51	10-12	0	0	0	0	33.8024	-118.6549	825	SMM
T796	PC46	0-4	42	460	3	505	33.8034	-118.6557	822	SMM
T796	PC46	0-4 *	0	35	0	35	33.8034	-118.6557	822	SMM

Table 1. Continued

Sample	Core	Interval (cmbsf)	DDD (ppb)	DDE (ppb)	DDT (ppb)	DDTr (ppb)	Lat. (°N)	Long. (°W)	W.D. (m)	Area
T796	PC55	0-4	42	400	1.6	443.6	33.8030	-118.6553	822	SMM
T796	PC55	0-4 *	4.3	58	0	62.3	33.8030	-118.6553	822	SMM
T797	PC50	0-2	19	160	0	179	33.7537	-118.5407	677	RF
T797	PC64	0-2	15	160	0	175	33.7537	-118.5407	677	RF
T797	PC47	0-2	8.6	140	0	148.6	33.7536	-118.5437	650	RF
T797	PC68	0-2	15	190	0	205	33.7536	-118.5414	680	RF
T797	PC68	2-4	19	200	5.3	224.3	33.7536	-118.5414	680	RF
T797	PC68	10-12	11	77	0	88	33.7536	-118.5414	680	RF
T797	PC41	0-2	23	200	7.3	230.3	33.7546	-118.5423	653	RF
T797	PC41	10-12	20	130	2.7	152.7	33.7546	-118.5423	653	RF
T797	PC61	0-2	24	200	4.3	228.3	33.7536	-118.5418	683	RF
T798	PC73	0-2	71	1000	55	1126	33.7404	-118.4408	314	PV
T798	PC69	0-2	43	630	15	688	33.7403	-118.4400	315	PV
T798	PC55	0-2	56	780	52	888	33.7409	-118.4419	286	PV
T798	PC46	0-2	140	1500	760	2400	33.7402	-118.4391	320	PV
T798	PC53	0-2	11	890	12	913	33.7409	-118.4419	286	PV
T798	PC60	0-2	65	1000	14	1079	33.7404	-118.4408	314	PV
T798	PC58	0-2	27	630	9.6	666.6	33.7403	-118.4400	315	PV
T799	PC48	0-2	2.3	170	1.4	173.7	33.7977	-118.6439	824	SMM
T799	PC68	0-2	12	160	4.8	176.8	33.7984	-118.6451	822	SMM
T799	PC66	0-2	14	160	4.2	178.2	33.7987	-118.6457	818	SMM
T800	PC73	0-2	14	210	1.3	225.3	33.8069	-118.4937	503	SMM
T800	PC46	0-2	19	250	2.6	271.6	33.8072	-118.4940	509	SMM
T801	PC48	0-2	14	310	1.6	325.6	33.8058	-118.4926	486	RC
T801	PC51	0-2	16	410	0	426	33.8055	-118.4921	484	RC
T802	PC62	0-2	4.2	25	0	29.2	33.9620	-119.2532	711	HC
T802	PC59	0-2	4.1	26	0	30.1	33.9618	-119.2521	714	HC
T803	PC64	0-2	4.5	29	0	33.5	33.9619	-119.2506	713	HC
T803	PC62	0-2	2.9	22	1.2	26.1	33.9619	-119.2504	713	HC
T803	PC60	0-2	4.4	37	1.7	43.1	33.9618	-119.2467	704	HC
T804	PC50	0-2	3.2	32	1.7	36.9	33.9820	-119.2713	661	HC
T805	PC57	0-2	1.8	15	0	16.8	34.0761	-119.2397	419	HC
T805	PC41	0-2	0.7	15	0	15.7	34.0763	-119.2399	421	HC
T806	PC50	0-2	3.6	22	1.9	27.5	34.0162	-119.2380	592	HC
T806	PC67	0-2	4	28	1.8	33.8	34.0165	-119.2379	594	HC
T807	PC61	0-2	2.5	17	1.2	20.7	34.0517	-119.2528	476	HC
A2-98-SC	366P1	0-3	67	97	9.9	173.9	33.8870	-118.7097	539	SMS
A2-98-SC	365P1	0-3	1.4	36	0	37.4	33.9107	-118.7441	544	SMS
A2-98-SC	364P1	0-3	75	430	14	519	33.8067	-118.7116	873	SMS
A2-98-SC	363P1	0-3	0.86	5.1	0	5.96	33.8746	-118.7543	711	SMS
A2-98-SC	362P1	0-3	0	1.7	0	1.7	33.9019	-118.7245	541	SMS
A2-98-SC	361P1	0-3	2.1	46	0	48.1	33.9268	-118.7556	570	SMS
A1-03-SC	SMB2-P1	0-3	12	66	2.7	80.7	33.9064	-119.3744	797	SMS
A1-03-SC	SMB3-P1	0-3	56	200	9.7	265.7	33.8764	-118.8908	550	SMS
A1-03-SC	SMB4-P1	0-3	7.3	56	0	63.3	33.9303	-119.2944	769	SMS
A1-03-SC	HEIN2-P1	0-3	20	280	2	302	33.7992	-118.6464	550	SMM

* clam core after clams removed; W.D. – water depth; HC-Hueneme Canyon; RC-Redondo Canyon; SMC-Santa Monica Canyon; SMM-Santa Monica Mound; SMS-Santa Monica Slope; RF-Redondo Fan; PV-Palos Verdes Slope.

Figure 2. Image showing ROV *Tiburon*'s mechanical arm holding a push core tube used to collect samples of the seafloor sediment. One core tube has already been inserted into the sediment. Drawer for holding cores appears in the lower left.

METHODS

Ninety-five fine-grained sediment samples from the Santa Monica Basin (Figure 1) were analyzed for their DDTr concentration (Table 1). Eighty-five samples came from 57 push cores collected by Monterey Bay Aquarium Research Institute's (MBARI) ROV *Tiburon*, and ten samples came from piston cores collected by the USGS.

The samples were collected during *Tiburon* dives T-783 to T-807 in February 2005. *Tiburon* is equipped with a mechanical arm that can collect sediment push cores up to 22 cm in length (Figure 2). ROV-collected push core samples offer the advantages of knowing their position with respect to the local seafloor topography, providing visual observations to help select sites where recent fine-grained sediment is accumulating, and allowing greater confidence that the core tops are not disturbed during core collection. Core sites were selected along each ROV dive transect based on real time video images provided by the ROV. Many of these cores were collected on transects across the axes of Redondo, Santa Monica, and Hueneme Canyons.

While up to 16 push cores were collected on each ROV dive, cores were only selected for DDTr sub-sampling when they were identified as containing accumulations of fine-grained sediment that appeared to be geologically recent. Thus, samples that appeared to be composed of pre-1945 materials were intentionally excluded from DDTr sampling. Cores were extruded and cut into 2-cm-thick intervals. Samples of 20-40 g wet weight were placed into pesticide-free screw-top jars using specially cleaned spatulas, and refrigerated. Samples were typically taken from successive 2-cm-thick intervals until the first sandy horizon was encountered. No measurements were made on sand-bearing samples. In total, 216 samples were collected, however, only the uppermost sample from most cores and selected deeper intervals were analyzed. The sampling procedure was the same as that used in a previous study of DDTr dispersal in Monterey Canyon (Paull et al., 2002).

An additional 10 samples were taken from the top of archived piston cores in the USGS refrigerated core repository in Menlo Park, California. These samples were selected to augment the core distribution along the lower continental slope (Figure 1).

Of the samples analyzed for this study, sixty-seven samples were from core tops [53 samples from 0-2 cmbsf (centimeters below seafloor); 10 piston core samples from 0-3 cmbsf, and 4 samples from 0-4 cmbsf for cores obtained in the middle of a chemosynthetic biological community, which make it difficult to separate the sediment from the living clams]. An additional twenty-eight samples were taken from 2 to 20 cmbsf in these push cores and most of these came from near 10 cmbsf.

Organochlorine pesticides were extracted from dried bulk sediment and analyzed by gas chromatography using an electron capture detector at the Pacific Agricultural Laboratory, Portland, Oregon, with a detection limit of 0.67 ppb (μg/kg) for DDD and DDE and 1.2 ppb (μg/kg) for DDT. We report the DDT, DDE, and DDD concentrations and DDTr, which is the sum of the DDT, DDE, and DDD concentration, in Table 1.

RESULTS

ROV Observations of Sediment Cover

MBARI's ROV *Tiburon* was used to identify sites of recent sediment accumulation and to collect samples from the Palos Verdes Slope, Redondo Canyon, Redondo Fan, Santa Monica Mound, Santa Monica Slope, Santa Monica Canyon, Hueneme Canyon, and Hueneme Fan. ROV observations from the northern Palos Verdes Slope in a gully that probably formed as a submarine landslide show that the seafloor within this slide scar is mostly sediment-bare; places where visual observations suggest sediment has accumulated on the surface of the slide scar were rare. Drapes of fine sediment were common within the axes and on the lower flanks of the canyons, but extensive areas within the canyons were also sediment-bare suggesting recent erosion. Areas identified as having at least a drape of recent sediment cover were nearly ubiquitous on the fans and the flanks of the Santa Monica Mound (Hein et al. 2006; Paull et al., 2008). ROV observations of the seafloor outside the canyons did not show evidence for contour-parallel bottom currents.

DDTr in Near-seafloor Sediment

DDTr concentrations in the 95 sediment samples from the study area are shown in Figure 1 along with previously published data. DDTr concentrations ranged from 0 ppb to 2,400 ppb with a mean±1σ of 286±381 ppb. These samples came from water depths between 286 m and

873 m with a mean depth of 590±143 m. They ranged from 10 to 98 km from the Whites Point outfall off Palos Verdes, California, all along the continental margin to the northwest of the outfall.

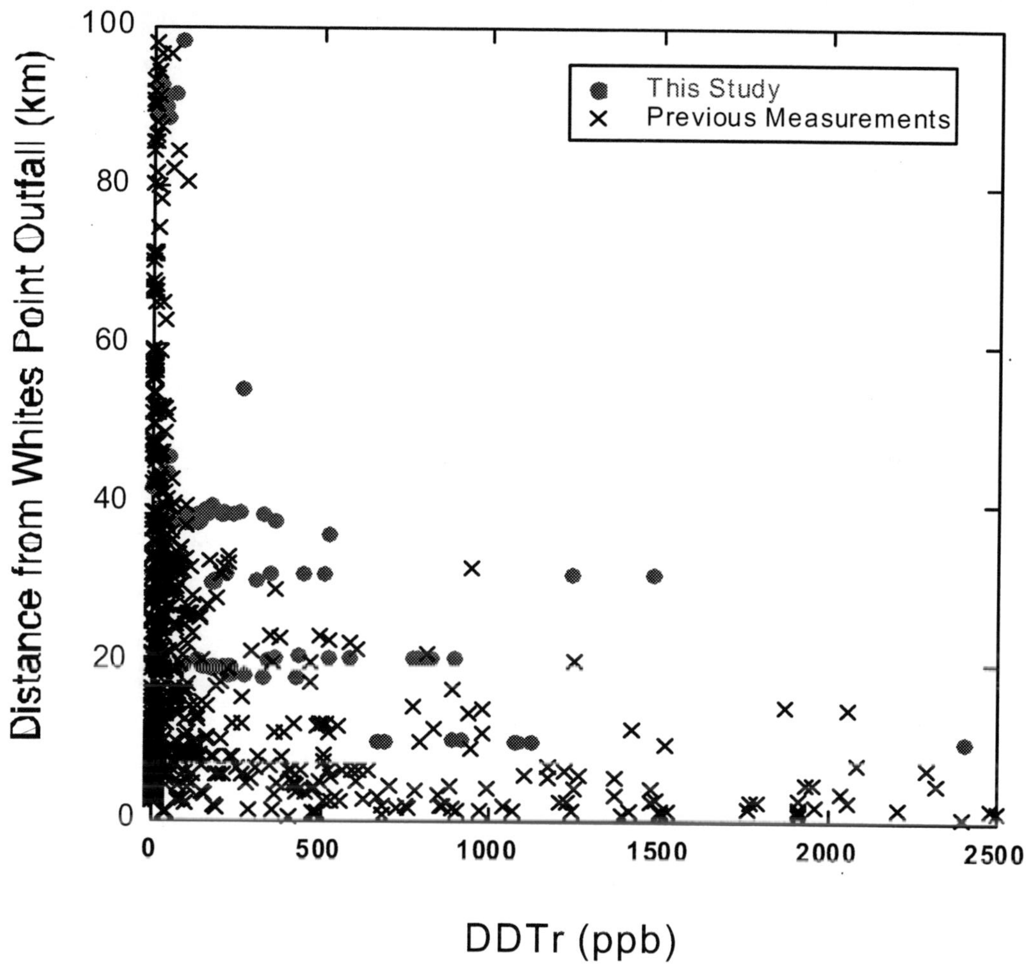

Figure 3. Plot showing the DDTr values with distance from the northernmost of the Whites Point outfalls. The samples marked with X are the 1040 tDDT (=DDTr) values reported in EPA report (CH2M Hill, Inc., 2003; Schiff, 2003). Samples indicated with red circles are reported here. Sample locations are indicated in Figure 1.

Clear geographic patterns exist in DDTr values. Although there is considerable scatter between nearby samples, samples closest to the Palos Verdes Shelf in general have more DDTr (Figures 1 and 3) than those further away. For example, the highest measured DDTr concentration (2,400 ppb) in these samples was from the top of a push core collected from 320 m water depth within a large landslide scar off the Palos Verdes Shelf, 10 km from the outfall and downslope from the US EPA Palos Verdes Super Fund site (Table 1). The mean DDTr concentration for the 7 samples from the Palos Verdes slope is 1,109±595 ppb. Mean values for samples from each of the three canyons decrease systematically toward the northwest— from the highest in Redondo Canyon (516± 27 ppb, n=14), an intermediate value in Santa Monica Canyon (148±86 ppb, n=22), and to the lowest in Hueneme Canyon and Fan (25±9 ppb, n=17).

To assess whether higher DDTr concentrations are found preferentially in sediments along the axes of the canyons, DDTr concentration in the 0-2 cmbsf interval was compared with altitude above the canyon floor. Altitude was calculated as the difference between the depth measured by the ROV pressure sensor at the sampling location and the deepest point along the dive transect. No systematic relationship was observed between the DDTr concentration in sediments recovered from directly within the canyon axis and those from the sides of the canyon.

Individual concentration ratios of DDD, DDE, and DDT to DDTr show the relative importance of each component (Figure 4). The major portion of the DDTr occurs as DDE, which represents 86±14% (±1σ) of the total DDTr. DDD is next most abundant, representing 9±6% of the total DDTr. DDT represents only 4±7% of the total DDTr.

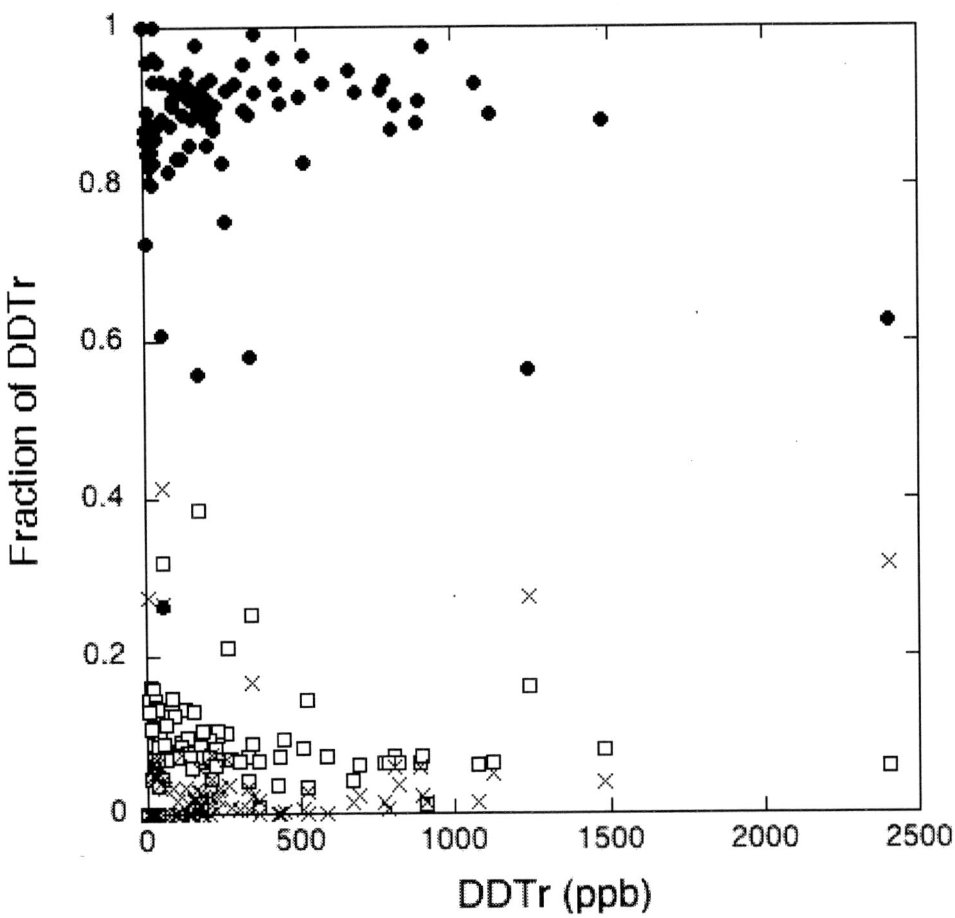

Figure 4. Relative proportions of DDT (crosses), DDD (open squares), and DDE (filled circles) in DDTr, versus the DDTr concentration measured for samples from the Santa Monica Basin.

Sediments from below 4 cmbsf (26 samples in total and 18 samples from the 10-12 cm interval, and 2 samples below this level) contain DDTr in all but one sample. Apparently DDTr persists to ≥10 cmbsf throughout most of this region. The only DDTr-free sample came from 10-12 cmbsf in a core collected near a local topographic high on the floor of the Santa Monica Basin (Hein et al. 2006), which we refer to as the Santa Monica Mound (Paull et al., 2008; Figure 1A). Two samples at shallower depths in this push core (0-2 and 4-6 cmbsf) contained DDTr. Apparently, this core penetrated into pre-1945 sediments by 10 cmbsf. In 14 of the 17 cores with multiple DDTr measurements, the highest DDTr concentrations measured in the cores were not in the core-top sample, but at greater depth (e.g., 10 cmbsf).

Discussion

DDTr in Sediments from the Santa Monica Basin

These new samples augment the database of 1,040 DDTr concentration measurements on near seafloor sediments collected to delineate the fate of the DDTr discharged onto the Palos Verdes continental shelf (Schiff, 2003; Sommerfield and Lee, 2003; CH2M Hill, Inc., 2003; Schiff et al., 2006) supported primarily by Southern California Coastal Water Research Project (SCCWRP; http://www.sccwrp.org/). Previously, only 18 samples were from more than 200 m water depths and 15 of these were directly downslope of the Palos Verdes Shelf (Figure 1). Our new DDTr dataset provides increased coverage within the canyons, along the slope, and within the basin floor to the northwest of the Palos Verdes Peninsula. Apparently, a substantial amount of DDTr-bearing sediment released onto the Palos Verdes Shelf has moved off the shelf and into the basin. The additional data from our study clearly shows that the geographic extent of the DDTr dispersal area associated with the Whites Point outfall still has not been adequately outlined.

Northwest Movement

DDTr concentrations measured in near-surface sediment from water depths ≥256 m collected within the submarine canyons and along the continental slope on the northeastern flank of the Santa Monica Basin are considerably higher than those measured in shelf sediment at equivalent distances from the Whites Point outfall (Figures 1 and 3). The DDTr concentration data suggest either a tongue of higher values extends to the northwest of the Palos Verdes Peninsula along the continental slope on the northeastern flank of the Santa Monica Basin for more than 40 km from the Whites Point outfall (Figure 3) or mixing and/or burial of DDTr-bearing sediments has been more rapid on the continental shelf. The decrease in DDTr concentrations with distance from the Whites Point outfall to the northwest is consistent with dispersal and dilution from a point source (Figures 1 and 3).

Whether DDTr from the Whites Point outfall has reached Hueneme Canyon (~90 km from Whites Point) is unclear. Other point (Raco-Rands and Steinberger, 2000) and non-point sources from the greater Los Angeles region also have contributed DDTr and may be the dominant source for the DDTr measured in Hueneme Canyon and Fan. For comparison, the mean value of DDTr concentration found in Hueneme Canyon and Fan sediment (25±9 ppb, n=17) is not statistically different from the mean for sediment from Monterey Bay collected within the same water depth range (14±20 ppb, n=56). The major source of DDTr in Monterey Bay is believed to be primarily non-point source agricultural use in the adjacent land areas (Paull et al., 2002).

The existence of higher concentrations of DDTr in the near seafloor sediments along the continental slope within the northeastern flank of the Santa Monica Basin, compared to those on the adjacent shelf, suggest that these samples are lithologically different, have experienced less dilution and mixing with respect to the shelf sediments, or lie along a preferential transport pathway.

Lithologic Control

The observed shelf-to-basin variation in DDTr concentrations may be influenced by lithologic controls and differences in sampling techniques. ROV observations of the seafloor (Figure 2) assisted sampling by identifying sites where recent sediment accumulation has occurred and helped avoid sampling areas of recent seafloor erosion. Without the assistance of real-time video observations, cores might have been collected in pre-1945 DDTr-free sediments, or the core tops might have been disturbed.

Although the grain size associated with these new samples was not measured, only samples that appeared to be fine-grained were selected for measurement in this study. Also, no DDTr measurements were made underneath sand layers found in some of the cores. Thus, the samples analyzed reflect accumulations under regular ambient conditions rather than rapid accumulations associated with episodic turbidity currents (Normark et al., 2006b).

Because DDTr preferentially adheres to fine-grained sediment, lower DDTr concentrations will occur in coarser-grained sediment. DDTr concentrations were routinely measured on samples collected on the Santa Monica Shelf as part of the SCCWRP studies (Schiff, 2003; Sommerfield and Lee, 2003; CH2M Hill, Inc., 2003) and many of these sediment samples from the shelf contained silt and coarser material (Alexander and Venherm, 2003). However, many fine-grained sediment samples were also collected from across the shelf during the SCCWRP studies and these samples also have characteristically lower DDTr concentration than those in the basin (Figure 3). Thus, grain-size does not appear to be the primary control on the observed, shelf-to-basin variation in DDTr concentrations.

Canyons as Conduits for Transporting DDTr into the Basins

The geographic distribution of the DDTr-bearing sediment will depend on the mechanisms by which it has moved to the northwest and passed downslope into Santa Monica Basin. Presumably DDTr-bearing sediment moved either downslope within the submarine canyons and was then transported northwestward along the slope by bottom currents, or it was transported in suspension in northwestward moving waters and settled-out over the slope along the northeastern flank of the Santa Monica Basin.

Because Redondo and Hueneme Canyons essentially cut across the entire continental shelf bordering Santa Monica Basin (Gardner et al, 2003), the potential of these canyons (as well as Dume and Mugu Canyons northwest of Santa Monica Canyon) to act as significant sediment traps, which would funnel material downslope, has been suggested (Lee et al., 2002; Normark, 2006a). The available DDTr concentration data along the margin and in the Santa Monica Basin do not show abrupt changes across the canyons (Figures 1 and 3) or patterns that can be ascribed to sample position within the canyons. DDTr appears to blanket the continental slope and basin without being closely controlled by proximity to the axial channels within the canyons.

ROV-collected vibracores (Paull et al., 2001), which sampled sediments up to 1.5 m below seafloor in the axes of Hueneme, Santa Monica, and Redondo Canyons complement the DDTr push core data and show that the facies of the sediments on the canyon floors differ from each other (Normark et al., 2006b; submitted). Vibracores from Hueneme and Redondo Canyons contain multiple silt and sand beds, which is an indication that these canyons have experienced sediment gravity flows and/or turbidity-current events in recent times, which presumably delivered coarse sediments to the basin floor (Sommerfield and Lee, 2003; Normark et al., 2006b). Radiocarbon dating of materials from these cores shows that multiple turbidity-current events have occurred during the last two thousand years and this activity has continued to the present (defined as radiocarbon ages of less than 300 years with some samples too young to determine a calibrated age). In contrast, cores from Santa Monica Canyon do not contain coarse sediment and thus this canyon does not appear to have experienced gravity flows in recent times. Normark et al. (2006b) suggested that the difference in gravity flow activity might be related to the position of the canyon heads. The heads of Redondo and Hueneme Canyons cross into the middle shelf, while the head of Santa Monica Canyon is stranded on the outer shelf, isolated from littoral sand sources. Although the penetration of sand may be restricted to Redondo and Hueneme Canyons, DDTr-bearing fine sediments cover the sides and floors of all three canyons indicating that the movement of DDTr-bearing sediment into the basin is not restricted to gravity flow events.

The levee valley that emerges from the mouth of Hueneme Canyon extends across the floor of the Santa Monica Basin. The smaller fan valley of Santa Monica Canyon terminates relatively close to the mouth of the canyon. The distal fan-channel that emerges from Redondo Canyon turns toward the southeast at the base of the continental slope and enters San Pedro Basin (Figure 1). Thus, if DDTr has been carried downslope in turbidity currents through these canyons, DDTr should be found throughout the floor of these basins at concentrations that presumably reflect their source. DDTr concentration data further to the southwest in Santa Monica Basin and within the San Pedro Basin are needed to evaluate whether these distal channels represent important pathways for DDTr-bearing turbidity currents.

While existing data on Late Quaternary deposits within the Santa Monica and San Pedro Basins show considerable influence of downslope channeling (Normark et al, 1998, 2006a), there were no observations made on the seafloor outside the canyons during the ROV dives that suggest contour currents have substantially modified the seafloor within these basins. Although this study focused on the impact of the submarine canyons on delivery of sediment to the basins, we have found little evidence these canyons have played a major role in transport of DDTr into the basins. The distribution pattern of all the DDTr data (Figure 1) is consistent with widespread dispersal of DDTr within Santa Monica Bay. Unfortunately, the existing data are inadequate to definitively distinguish between focused conduits and diffuse dispersal pathways.

Sediment-accumulation Rates

Data on the thickness of the DDTr-bearing deposits help to outline the effects of dilution and mixing and will be necessary to ultimately establish the inventory of DDTr within the basins. The new data show that more than 10 cm of DDTr-bearing sediment has accumulated on the slope and floor of the Santa Monica Basin and help to constrain sediment-accumulation rates along the landward flank of this basin. Widespread use of DDTr started in 1945 and the point source associated with the Montrose plant may have begun discharging as early as 1947. Thus, the occurrence of DDTr identifies sediment that has accumulated since 1945. Delays associated with the transport of the DDTr-bearing sediment from where the sediment enters the ocean to the sites of deposition will lead to underestimation of sediment-accumulation rates. Conversely, sediment mixing caused by bioturbation will increase the depth of DDTr occurrence and lead to an overestimation of sediment-accumulation rates. If 1945 is used as a datum, recent sedimentation rates for fine-grained sediments on the continental slope based on the presence of DDTr are greater than 1.6 mm per year. This rate is slightly higher than the Holocene hemipelagic sediment-accumulation rate of 0.85 mm per year measured at ODP Site 1015 further offshore on the floor of the Santa Monica Basin (Brian Romans, written communication, Sept. 2007). On the Santa Monica Shelf sediment-accumulation rates are much higher. The thickness of the DDTr-bearing zone is commonly up to 60-cm thick, and the maximum sediment concentrations of DDTr are usually found below the seafloor (Lee et al., 2002; Alexander and Venherm, 2003; Schiff, 2003).

Preservation of DDTr-bearing Sediments as an Event Horizon

The geologic record contains a number of discrete events that occur as distinct marker horizons within stratigraphic sequences. Examples include impact ejecta, carbon-isotope spikes, and volcanic ash layers (Walliser, 1996). Such events are associated with the introduction of materials that are produced or distinctly marked as a consequence of a geologically brief event. These may provide analogs for the dispersal and ultimate fate of the DDTr that was introduced onto the Palos Verdes Shelf over a few decades in the 20th century. Initially, as the event-related materials are dispersed, the highest concentrations of event tracers occur within the uppermost sediment. With time, continued sedimentation and burial, the tracers will be left behind as a distinct geologic layer. This appears to be what is happening on the Santa Monica Shelf and Basin. Measurements of DDTr concentrations within the samples from more than 4 cmbsf within the Santa Monica Basin frequently have slightly greater DDTr levels than the core top sediments (0-2 cmbsf). This suggests that the level of maximum concentration within the DDTr-bearing zone is now in the subsurface and the DDTr-bearing layer is starting to be buried. Sediment-accumulation rates on the shelf are high enough to bury the maximum concentrations within the subsurface. While less dramatic, the DDTr-bearing layer within the basin is being buried, or least has been diluted during the last few decades. This also indicates that DDTr measurements only made on surface sediment (0-2 cmbsf) may now underestimate the amount within the DDTr-bearing sediment zones in slope, canyon, and basin cores.

CONCLUSIONS

Measurements of DDTr concentration in sediment collected from the seafloor within the canyons, slope and basin floor of the Santa Monica Basin help outline the dispersal of DDTr-bearing sediment within this area. A signal of high DDTr concentrations, apparently from the Whites Point sewage outfall off the Palos Verdes Peninsula, can be tracked over 40 km to the northwest along the slope. Surface sediments (0-2 cmbsf) in water depths greater than 256 m characteristically contains higher DDTr concentrations than the sediment in shallower water that is an equivalent distance away from the Whites Point outfall. These data suggest that DDTr have dispersed from Whites Point outfall to the northwest and occur along the slope and landward flank of the Santa Monica Basin.

ACKNOWLEDGEMENTS

The David and Lucile Packard Foundation provided support. We thank Homa Lee and Jingping Xu for helpful reviews.

REFERENCES

Alexander, C.R., and Venherm, C., 2003, Modern sedimentary processes in Santa Monica California continental margin: sediment accumulation, mixing and budget, *Marine Environmental Research*, v. 56, p. 177-204.

Bay, S.M., Zeng, E.Y., Lorenson, T.D., Tran, K., and Alexander, C., 2003, Temporal and spatial distributions of contaminated sediments in Santa Monica Bay, California, *Marine Environmental Research*, v. 56, p. 255-276.

Brooks, G.T., 1979, *Chlorinated Insecticides, Volume I*. Boca Raton, Florida, CRC Press, Inc., p. 1-46.

CH2M Hill, Inc., 2003, Ecological Risk Assessment for the Palos Verdes Shelf, Report Prepared for the US Environmental Protection Agency. Contract No. 68-w-98-225, 340 p.

Gardner, J.V., Dartnell, P., Mayer, L.A., and Hughes Clarke, J.E., 2003, Geomorphology, acoustic backscatter, and processes in Santa Monica Bay from multibeam mapping, *Marine Environmental Research*, v. 56, p. 15-46.

Hein, J.R., Normark, W.R., McIntyre, B.R., Lorenson, T.D., and Powell, C.L. II, 2006, Methanogenic calcite, ^{13}C-depleted bivalve shells, and gas hydrate from a mud volcano offshore southern California, *Geology* v. 34, p. 108-112.

Hickey, B.M., 1992, Circulation over the Santa Monica-San Pedro basin and shelf, *Progress in Oceanography*, v. 30, p. 37-115.

Lee, H.J., and Wiberg, P.L., 2002, Character, fate, and biological effects of contaminated, effluent-affected sediments on the Palos Verdes margin, southern California: an overview, *Continental Shelf Research*, v. 22, p. 835-840.

Lee, H.J., Sherwood, C.R., Drake, D.E., Edwards, B.D., Wong, F., and Hamer, M., 2002, Spatial and temporal distributions of contaminated, effluent-affected sediment on the

Palos Verdes margin, southern Californian, *Continental Shelf Research*, v. 22, p. 859-880.

Mischke, T., Brunetti, K., Acosta, V., Weaver D., 1985, Agricultural sources of DDT residues in California's environment, Environmental Hazards Assessment Program, A Report Prepared in Response to House Resolution No 53 (1984). California Department of Food and Agriculture, (http://www.cdpr.ca.gov/docs/empm/pubs/ehapreps/ddt/sources.htm).

Noble, M.A., Ryan, H.F., and Wiberg, P., 2002, The dynamics of subtideal poleward flow over a narrow continental shelf, Palos Verdes, CA, *Continental Shelf Research*, v. 22, p. 923-944.

Normark, W.R., Piper, D.J.W., and Hiscott, R.N., 1998, Sea level controls on the textural characteristics and depositional architecture of the Hueneme and associated submarine fan systems, Santa Monica Basin, California, *Sedimentology,* v. 45, p. 53-70.

Normark, W.R., Piper, D.J.W., and Sliter, R., 2006a, Sea-level and tectonic control of middle to late Pleistocene turbidite systems in Santa Monica Basin, offshore California, *Sedimentology,* v. 53, p. 967-897.

Normark, W.R., Paull, C.K., Ussler, W., III., Mitts, P., and Romans, B., 2006b, Redondo and Hueneme sand-rich canyon-fan systems, offshore southern California, *AGU Ocean Sciences Meeting*, Honolulu, HI, February 20-24, 2006.

Normark, W.R., Paull, C.K., Caress, D.W., Ussler, W., III, Keaton, R., and Sliter, R., submitted, Fine-scale relief related to late Holocene channel shifting within the floor of upper Redondo fan valley in San Pedro Basin, offshore southern California, *Sedimentology*.

Pandit, G.G., Sahu, S.K., Sharma, S. and Puranik, V.D., 2006, Distribution and fate of persistent Organochorides pesticides in coastal marine environments of Mumbai, *Environment International*, v. 32, p. 240-243.

Paull, C.K., Stratton, S., Conway, M., Brekke, K., Dawe, T.C., Maher, N., and Ussler, W., 2001, Deep sea vibracoring system improves ROV sampling capabilities, *EOS*, v. 82, p. 325-326.

Paull, C.K., Greene, H.G., Ussler, W., III, and Mitts, P.J., 2002, Pesticides as tracers of sediment transport through Monterey Canyon, *Geo-Marine Letters*, v. 22, p. 121-126.

Paull, C.K., Normark, W.R., Ussler, W., III., Caress, D.W., and Keaten, R., 2008, Association among active seafloor deformation, mound formation, and gas hydrate growth and accumulation within the seafloor of the Santa Monica Baisn, offshore California, *Marine Geology*, v. 250, p. 258-275.

Raco-Rands, V.E., and Steinberger, A., 2000, Characteristics of effluents from large municipal wastewater treatment facilities in 1997, *Southern California Coastal Water Resources Project*, p. 28-43.

Phillips, J.H., Haderlie, E.E., and Lee, W.L., 1975, An analysis of the dynamics of DDT in marine sediments. Corvallis, Oregon, US Environmental Protection Agency, Report EPA-660/3-75-013, 58 p.

Schiff, K., 2003, Sediment chemistry on the mainland shelf of the Southern California Bight, *Marine Pollution Bulletin*, v. 40, p. 268-276.

Schiff, K.C., Maruya, K., and Christenson, K., 2006, Southern California Bight 2003, Regional Monitoring: II. Sediment Chemistry, Technical Report 492, *Southern California Coastal Water Research Project*, Westminster, California.

Sommerfield, C.K. and Lee, H.J., 2003, Magnitude and variability of Holocene accumulations in Santa Monica Bay, California, *Marine Environmental Research*, v. 56, p. 151-176.

US Environmental Protection Agency, 1989, Environmental Fate and Effects Division, Pesticide Environmental Fate One Line Summary: DDT (p, p'). Washington, DC, *Properties Reviews of Environmental Contamination and Toxicology*, 137 p.

Wafo, E., Sarrazin, L., Diana, C., Schembri, T., Lagadec, V., and Monod, J., 2005, Polychlorinated biphenyls and DDT residues distributions in sediments of Cortiou (Marseille, France), *Marine Pollution Bulletin*, v. 52, p. 104-107.

Walliser, O.H., 1996, Global Events and Event Stratigraphy, Springer-Verlag, Berlin, 333 p.

Wiberg, P.L., Drake, D.E., Harris, C.K., and Noble, M., 2002, Sediment transport on the Palos Verdes shelf over seasonal to decadal time scales, *Continental Shelf Research*, v. 22, p. 987-1004.

World Health Organization, 1979, Environmental Health Criteria 9, DDT and its Derivatives. Geneva, *World Health Organization*, 194 p.

World Health Organization, 1989, Environmental Health Criteria 83, DDT and its Derivatives—Environmental Effects. Geneva, *World Health Organization*, 98 p.

Wu, Y., Zhang, J., and Zhou, Q., 1999, Persistent organochlorine residues in sediments from Chinese river/estuary systems, *Environmental Pollution*, v. 105, p. 143-150.

Yang, R., Jiang, G., Zhou, Q., Yuan, C., and Shi, J., 2005, Occurrence and distribution of organochlorine pesticides (HCH and DDT) in sediments collected from East China Sea, *Environment International*, v. 31, p. 799-804.

In: Progress in Pesticides Research
Editor: Christopher M. Kanzantzakis

ISBN 978-1-60456-201-9

Chapter 8

RECENT PROGRESS IN METHODS FOR DETERMINATION AND AOM REMOVAL OF ORGANOPHOSPHORUS AND NEONICOTINOID INSECTICIDES

Polonca Trebše*, Mladen Franko, Mojca Bavcon Kralj, Urh Černigoj, Aleš Boškin
Laboratory for Environmental research, University of Nova Gorica, P.O.Box 301, 5000 Nova Gorica, Slovenia

ABSTRACT

This review summarizes the most recent investigations in the field of insecticides and methods for their degradation as well as determination in waters and some foodstuff. Due to their mode and extent of application insecticides pose a serious threat to the environment and humans. This is why the insecticides are extensively survailed groups of chemicals and are subject to several residue monitoring programs. Organophosphorus compounds (OP) were more than fifty years applied in crop, fruit and vegetable production as well as in veterinary more than fifty years. Because of this, they appear so frequently (within first three positions) on the top lists of pesticide residues monitoring in European and American programs. However, during last years there are strong attempts to replace them with neonicotinoids.

The application of several different AOMs in the removal of insecticides is described in this contribution. The photodegradation processes were evaluated not only in terms of kinetics, but also product identification (GC-MS, HPLC-DAD, HPLC-MS) and the degree of mineralisation is compared.

The research in the field of insecticide degradation was, in addition, combined with the development of analytical methods and tools for the determination of trace amounts of pesticide residues in different environmental samples. Very low concentration of OPs (malathion, diazinon, chlorpyrifos, azinphos-methyl) and neonicotinoides (thiamethoxam, imidacloprid, acetamiprid and thiachloprid) were determined by coupling chromatographic techniques and flow injection bioanalytical techniques to a very sensitive detector system based on thermal lens spectrometry (TLS). The achived LODs for both OPs and neonicotinoids are well below 50 ppb, which is approx. 10-times lower than maximal residue limits (MRLs) set in EU and USA legislation. Beside already mentioned classical analytical techniques, the newly developed analytical techniques were also applied in the degradation studies of selected pesticides.

Keywords: neonicotinoids, organophosphates, pesticides, AOM, ozonation, photocatalysis, HPLC-TLS, FIA-AChE-TLS

1. INTRODUCTION

Organophosphorus (OP) compounds were widely used in agricultural practice for more than four decades. Nowadays, they are being slowly replaced by other biocides, mostly from the group of neonicotinoids. Chlorpyrifos and chlorpyrifos-methyl, however, are the only representatives from the OP group, which will be used till the year 2016 (2005/72/EC amending 91/414/EEC). On the other hand, the use of OPs in general, is not diminishing in non-European countries, where they are commonly used, i.e. China (Wu et al., 2007), or USA (Agricultural Chemical Usage 2005 Field Crops Summary, May 2006).

Continuing use of pesticides raises concerns due to increasing pollution of the environment where the main routes of exposure for general public remain the food and drinking water consumption.

There are four major causes of water pollution by pesticides (Chiron and Fernandez-Alba, 1997):

- application of pesticides in agricultural practice (concentration range of few ppb),
- rinse water from containers and spray equipment (10-100 ppm),
- wastewater from agricultural industries (10-100 ppm),
- wastewater from pesticide formulating or manufacturing plants (1-1000 ppm).

The last three point sources of pollution may be efficiently treated in small-scale treatment units including physical entrapment, biodegradation and chemical reactions (Chiron et al., 2000). Unfortunately, entrapment cannot be a final solution, due to low efficiency and high concentration of pollutants in wastewaters and because of the need for destruction of entrapped pollutants. On the other hand, biological treatment techniques are susceptible to toxic compounds that inactivate waste-degrading microorganisms. In some cases, a useful approach is the oxidation pretreatment of pollutants to produce intermediates that are readily biodegradable. Development in the field of chemical water treatment has offered several oxidative degradation procedures based on the generation of highly reactive intermediates that initiate a sequence of reactions resulting in the destruction and removal of organic pollutants. They are generally referred to as advanced oxidation methods (AOMs), which are

* Corresponding author: polonca.trebse@p-ng.si

described later on, and can be sorted into two main categories: photochemical processes and ozonation processes. Photochemical processes can be further divided into direct photolysis, photosensitized oxidation and photocatalysis (Burrows et al., 2002). Recent research in this field has focused on studies of several relevant pesticides, their stability in the environment as well as methods for their removal, with particular emphasis on metabolites, formed during degradation in the environment and during application of AOMs, as will be discussed in more detail later.

Because of continuous input of pesticides into the environment, and due to their adverse effects on wildlife and human health, it is necessary to monitor their presence in the environment, foodstuffs and drinking water sources. There exists a number of classical analytical methods for the determination of pesticides in different environmental and food samples, which are based on separation techniques such as liquid or gas chromatography coupled with mass spectrometry (GC-MS) (Schellin et al., 2004), (LC-MS) (Rodrigues et al. 2007), spectrophotometry (HPLC-DAD) (Nogueira et al., 2003) or other detection principles such as flame ionization (GC-FID) (Bavcon et al., 2003) or electron capture detection (GC-ECD) (Li et al., 2003). Unfortunately these methods are time consuming, labour intensive, expensive and usually not environmental friendly since they are mostly based on extraction procedures with organic solvent. Because of this, the research in this field is lately dedicated to the development of new fast response techniques, which can be used for rapid screening tests, and in addition to their simplicity, offer the advantage of cost efficiency and low consumption of solvents.

The development of new, highly sensitive analytical methods is however still required to offer the possibility of investigating environmental phenomena and processes, which cannot be studied by commercial methods. Such examples are the negative effects of low concentrations of neonicotinoids and their degradation products on certain species like, for example, honey bees (Iwasa, 2000). Various analytical methods have been proposed for the determination of neonicotinoides: gas chromatography-mass spectrometry (GC-MS) (Vilchez et al., 1996) and liquid chromatography with diode array detector and mass spectrometry (HPLC-DAD; HPLC-MS) (Obana et al., 2002; Mandić et al., 2005; Obana et al., 2003; Fidente et al., 2005; Seccia et al., 2005), but analysis of their residuals in traces needs further development.

2. Advanced Oxidation Methods for Degradation of OPS and Neonicotinoids

The use of photocatalysts, such as TiO_2, for the removal of a large variety of organic and inorganic pollutants from waste waters has been the subject of intensive studies for more than 20 years.

In the recent years there were several studies on degradation of OPs employing TiO_2, which included pesticides such as malathion (Muszkat, et al. 1995), monochrotophos (Hua, et al. 1995, Ku and Jung, 1998), dichlorvos, monocrotophos, parathion, and phorate (Mengyeu, et al., 1995), phorate, methamidophos, malathion, diazinon, and EPN (Doong and Chang, 1997), dimethoate and azinphos-methyl (Dominguez et al., 1998), methamidophos (Malato, et al. 2000), and dichlorvos (Naman et al., 2002). In the listed studies authors compared different factors governing the efficiency of photocatalysis in removing the starting compounds, with no attention paid to the analysis of formed by-products.

A better insight into formation of by-products was given by their qualitative determination in photocatalytic studies of diazinon (Mansour, et al. 1999; Kouloumbos, et al. 2003). The authors proposed oxidation of P=S bond and the formation of oxons as primary products, followed by the rupture of P–O bond in the leaving group (Kouloumbos, et al. 2003). Isomerisation took place in irradiated diazinon water/soil suspension, resulting in the formation of isodiazinon (Mansour, et al. 1999). Oxidation and photohydrolysis pathways were proposed also in photocatalysis of ethyl bromophos and dichlofenthion (Konstantinou, et al. 2001), where different products with quinonidal structure and also alkyl phosphorothioate esters were found. Similar products were found in photocatalytic degradation studies of fenthion and parathion (Sakkas, et al. 2002).

Ozone, on the other hand, has been applied efficiently to drinking water treatment and wastewater treatment due to its powerful oxidizing potential. When it is transferred from the gaseous phase into solution, it can react in two different ways: as an initiator of free radicals formed by the decomposition of ozone ($OH^{\cdot}$, $OH_2^{\cdot}$) or as a direct oxidation reagent (Laplanche et al., 1984). In the literature (Laplanche et al., 1984) different reaction schemes are proposed. In the case of phosphorothionates (parathion, diethylchlorophosphate) a cycloaddition of ozone occurs on the P=S double bond before the breakdown of the molecule. In the case of phosphorodithioates (malathion, phosalone, *O,O*-diethylphosphorodithioate), a competitive reaction between cycloaddition on P=S and electrophilic attack on the second sulfur atom (P–S–R) takes place, whereas in the case of phosphoroamidates (dimefox, hexamethylphosphorotriamide), the electrophilic attack on the nitrogen dominates.

Applications of ozone for degradation of OPs in solution were performed on diazinon (Ku et al., 1998), pirimiphos-methyl (Chiron et al., 1998), and on phorate (Ku and Lin, 2002). The degradation efficiency for diazinon, as main focus of the study, was evaluated by the degree of mineralization (formation of sulphate, nitrate, phosphate, and carbonate), and was almost 100% within 1 hour. In the case of phorate, the main focus was on degradation mechanism, which starts with the breakage and subsequent oxidation of the P=S double bond, followed by the isomerisation and final oxidation of various organic intermediates. Under ozone treatment pirimiphos-methyl also degrades to several compounds. Three major pathways including: *N*-oxime formation, *N*-dealkylation, and oxidation of methyl groups, were proposed.

Application of gaseous ozone was reported for degradation of two OP nerve agents (VX - *O*-ethyl *S*-[2-(diisopropylamino) ethyl] methylphosphonothioate and GD - pinacolyl methyl phosphonofluoridate) (Wagner et al., 2000). Under similar experimental conditions VX and GD degraded differently. Ozonation of VX resulted in several oxidation products with O-ethyl S-[2-(diisopropylamino) ethyl] methylphosphonothioate as major product, while GD was completely resistant to ozone and did not degrade or transform.

The efficiency of ozone in degradation of OPs is well understood. However, the stability and the toxicity of their intermediates have not been adequately evaluated. In a recent paper, ozonated water was even employed for effective washing of vegetables in order to remove residual pesticides (Wu et al., 2007a, b). The authors claimed, that, “Ozonation is a safe and promising process for the removal of tested pesticides (diazinon, parathion, methyl-parathion, and cypermethrin) from aqueous solution and vegetable surface under domestic use.” (Wu et al., 2007a).

Several studies demonstrated that oxon analogues form during initial steps of photocatalysis and ozonation, but as degradation proceeds, oxons degrade further. Complete mineralisation is the final goal of AOMs, but studies have suggested the formation of smaller, resistant and toxic degradation products. Such compounds include *O*-, and *S*- trimethyl phosphates, which were generally classified as toxic substances already in early 80-s. An important consideration is also whether additives in commercial formulations will inhibit the reactions of active pesticide ingredients. Additives consist of various inert solids, wetting agents and other substances, which are needed to facilitate handling and application.

Since chlorpyrifos (1) (Table 1) and malathion (3) are still widely used in agriculture and there is rather no information regarding the persistence of their transformation products, recent research aimed to evaluate the photodegradation of malathion and chlorpyrifos, and their degradation products chlorpyrifos-oxon (2A), malaoxon (4A), isomalathion (5A), as well as their commercial product Pyrinex and Radotion during TiO_2 photocatalysis and ozonation (Figure 1, Table 1). The photoprocesses were compared in terms of degradation kinetics (with high performance liquid chromatography – HPLC), product identification (gas chromatography-mass spectrometry – GC-MS), and degree of mineralisation (with ion chromatography - IC). The product formation and the degree of mineralization were followed in order to achieve not only the efficient degradation of the starting compound but also an environmentally safe photodegradation with no toxic degradation end-products.

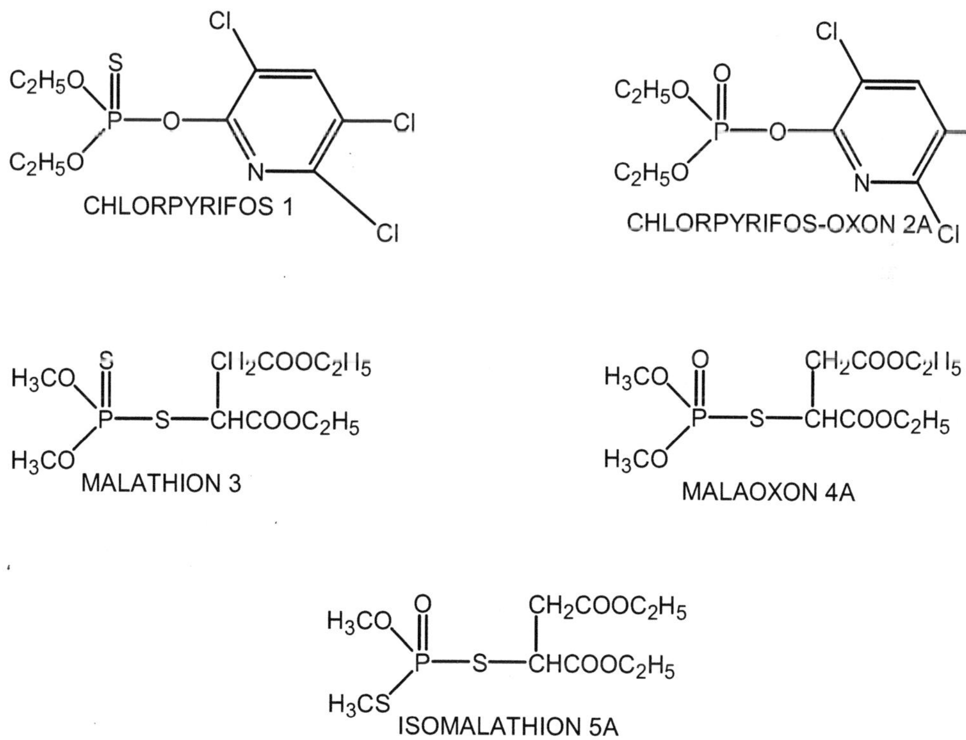

Figure 1. The chemical structures of OPs compounds, studied in AOM degradation experiments.

Table 1. Compounds identified according to mass spectra, listed by numbers, names, characteristic ions and chemical structures

Compound name and index	Characteristic ions (m/z)	Chemical structure	Compound name and index	Characteristic ions (m/z)	Chemical structure
1 chlorpyrifos	316 (34%), 314 (47%), 199 (88%), 197 (91%), 97 (100%)	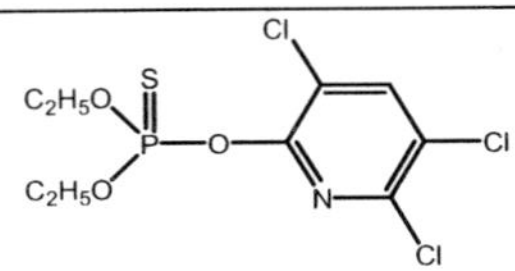	*2A* chlorpyrifos-oxon	300 (37%), 298 (52%), 272 (48%), 270 (73%), 199 (71%), 197 (77%)	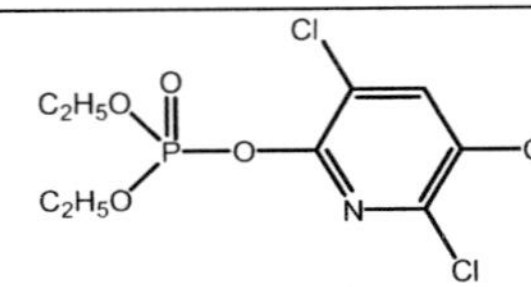
3 malathion	331 [M^+] (16%), 285 (31%), 173 (90%), 127 (100%);	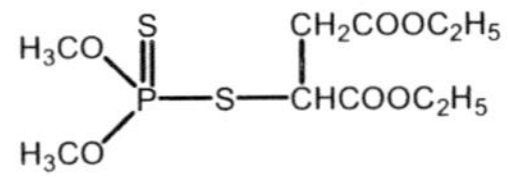	*4A* malaoxon	315 [M^+] (15%), 268 (27%), 173 (30%), 127 (100%), 99 (43%)	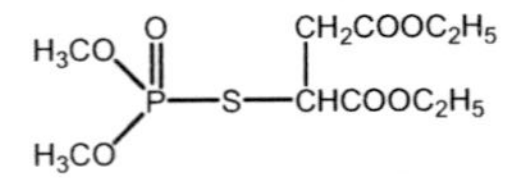
5A isomalathion	331 [M^+] (3%), 283 (42%), 173 (90%), 127 (100%), 99 (37%);	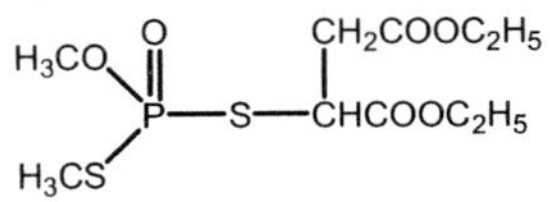	*8A* dimethoxyphosphinyl)oxo - butanedioic acid diethyl ester	315 [M^+] (3%), 269 (10%), 173 (100%), 127 (70%), 99 (27%)	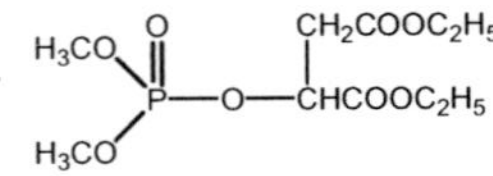
9A O,O,S-trimethyl phosphorothioate [OOS(O]	156 [M^+] (49%), 110 (100%), 79 (63%)	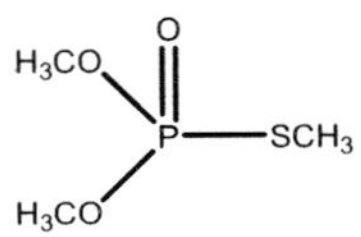	*10A* O,O,S-trimethyl phosphorodithionate [OOS(S)]	172 [M^+] (97%), 125 (78%), 93 (100%), 109 (25%), 79 (32%)	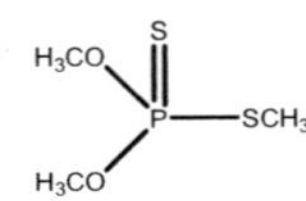
11A O,S,S-trimethyl phosphorodithioate [OSS(O)]	172 [M^+] (60%), 125 (100%), 79 (97%), 47 (90%)	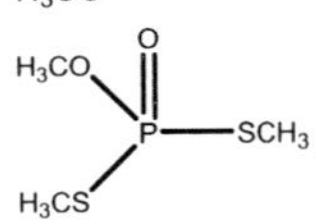	*12* 2-butenedioic acid (Z) diethylester – diethyl maleate	127·(50%), 99 (100%)	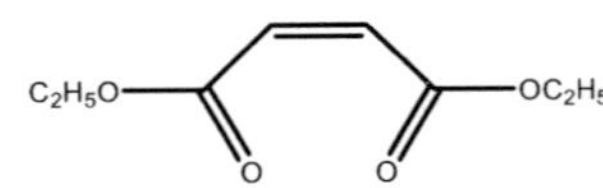
13 2-butenedioic acid (E) diethylester – diethyl fumarate	127 (100%), 99 (56%)	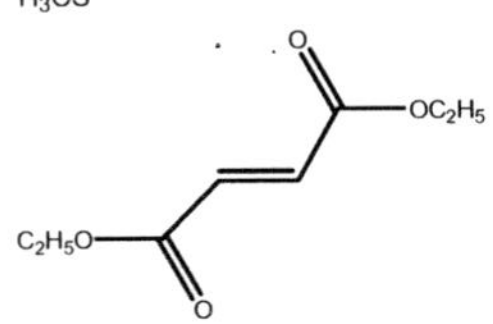	*14* butanedioic acid diethylester	129 (75%), 101 (100%)	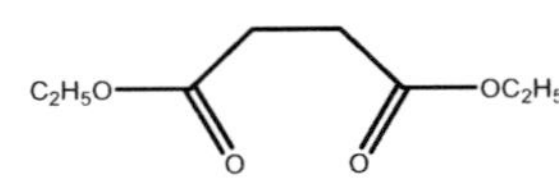

15 sulphanyl butanedioic acid diethyl ester	132 (60%), 104 (60%), 59 (55%), 45 (100%)	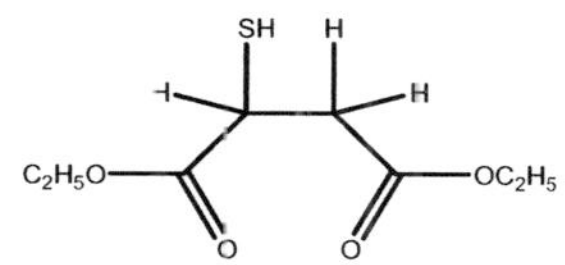	***16*** hydroxy butanedioic acid diethylester	145 (25%), 117 (100%), 89 (51%), 71 (83%)	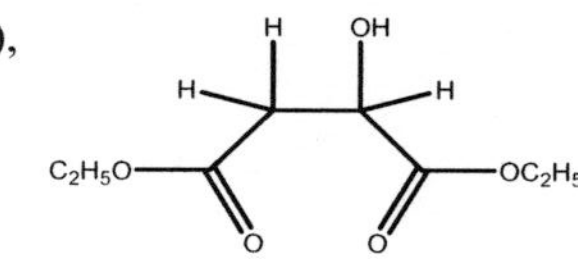
17 2,3-dihydroxy butanedioic acid diethylester	160 (70%), 132 (100%), 104 (35%), 87 (25%)	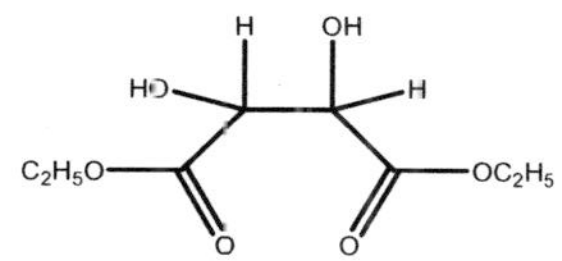	***18*** hydroxy butenedioic acid diethylester	141 (28%), 127 (100%), 99(22%)	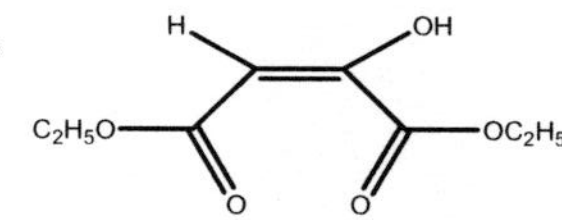
19 cis-2-hydroxy-3-sulphanylbutenedioic acid diethylester	220 (46%), 174 (75%), 146 (100%)	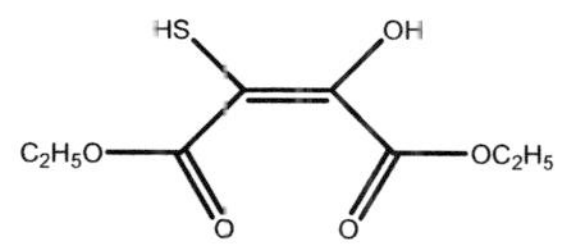	***20*** diethyl methylsulfonyl-succinate	252 [M^+] (48%), 206 (76%), 178 (83%), 132 (100%), 103 (31%)	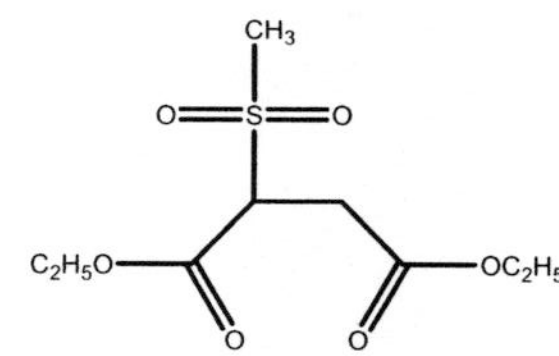
21 3,5,6-trichloro-2-pyridinol	201(26%), 199 (87%), 197 (91%) [M^+], 173 (29%), 171 (91%),169 (100%)	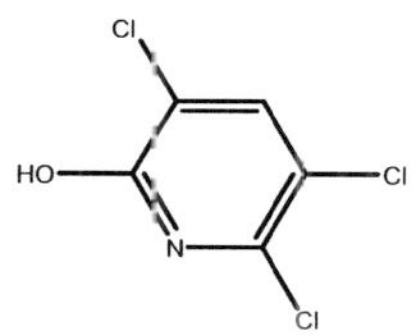			

Notes:

thio-OP	weak AChE inhibiting compound	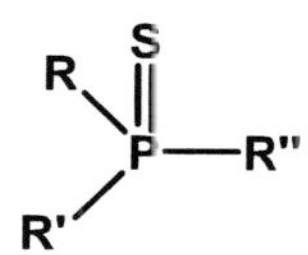	***oxo-OP***	powerful AChE inhibiting compound *marked with index **A***	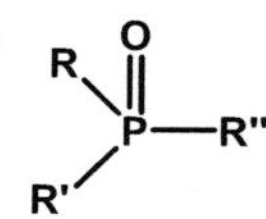

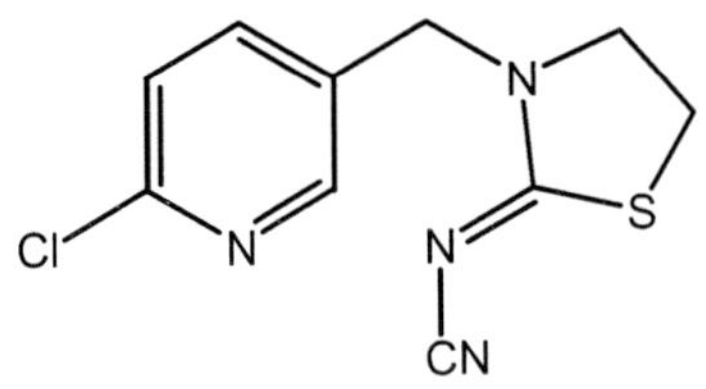

THIACLOPRID 6

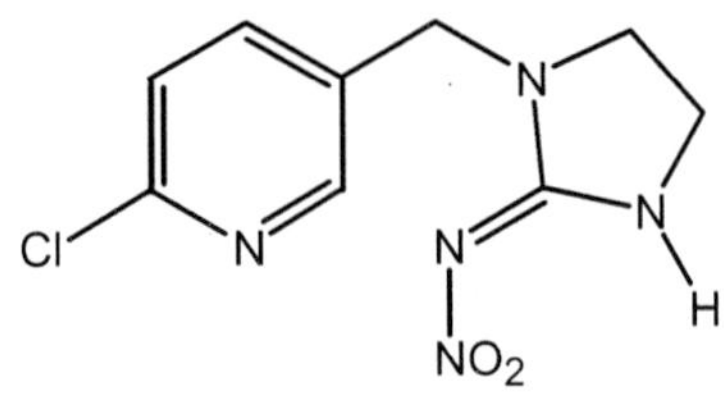

IMIDACLOPRID 7

Figure 2. The chemical structures of neonicotinoid compounds studied in AOM degradation experiments.

Several studies of environmental behavior of imidacloprid and thiacloprid (neonicotinoid compounds) were performed in the past (Roberts et al., 1999; Krohn, 2001) and it was shown that both molecules are resistant to the degradation in water by hydrolysis at neutral or acidic pH values. On the contrary, imidacloprid is susceptible for the attack of hydroxyl anions, resulting in hydrolytic decomposition of the molecule at alkaline pH (Guzsvany, et al., 2006, Černigoj et al., 2007b). Photodegradation of imidacloprid resulted in quantitative transformation of the molecule when the high pressure Hg lamp ($\lambda > 290$ nm) was used as a radiation source (Moza et al., 1998). Additionally, it is well known that O_2/TiO_2/UV-A (photocatalysis) is an effective process for the degradation of imidacloprid (Agüera et al., 1998; Malato et al., 2001).

Comparing the photo- and chemical stabilities of, thiacloprid and imidacloprid, in water (Figure 2), it was concluded that thiacloprid is more stable in aqueous solution than imidacloprid and as such exhibits at least as high threat for the aquatic systems as imidacloprid. It is therefore of great importance to develop effective methods for its removal from aqueous solutions by application of different AOMs. This triggered the studies of photostability of thiacloprid under UVA radiation (Černigoj et al., 2007b). Furthermore, O_3 treatment and three different photochemical advanced oxidation processes, namely O_3/UV (ozonation coupled with UV-A radiation), O_3/TiO_2/UV-A (photocatalytic ozonation) and O_2/TiO_2/UV-A were applied to assess the suitability of these processes to promote the degradation and mineralization of thiacloprid in water at different pH values. The effect of ozone dosage on TiO_2 photocatalysis was evaluated and explained (Černigoj et al., 2007b) as it will be discussed later.

Photocatalysis of Malathion, Radotion, Malaoxon and Isomalathion

Photocatalytic experiments on malathion, Radotion, malaoxon and isomalathion were performed in a recently developed photoreactor, consisting of a DURAN glass tube closed on the lower side with a glass frit and the valve for purging with gas, the spinning basket made of Teflon, on which six glass slides with immobilized catalyst were fastened around the axis by the help of two holders as it is described in original publications (Černigoj et al., 2007a).

Table 2. Calculated half-life times from fitting experimental data to first order decay curves

Compound/photodegradation process/($\tau_{1/2}$) [min]	TiO_2 photocatalysis, λ= 355 nm	photolysis, λ= 355 nm
Malathion	8.3 ± 0.6	2900 ± 230
Radothion	10.8 ± 1.7	420 ± 23
Malaoxon	36.5 ± 5.5	370 ± 15
Isomalathion	26.7 ± 1.5	320 ± 11
Chlorpyrifos	15.0 ± 0.8	/
Chlorpyrifos-oxon	27.7 ± 1.3	/

By monitoring the disappearance of the starting compound by HPLC, first-order degradation curves were derived with half-live times presented in Table 2. Malathion was degraded faster in the form of a pure analytical standard than as malathion in commercial formulation Radotion. In the case of TiO_2 photocatalysis, the rate constants were lower, 0.083 min^{-1} for malathion, and 0.064 min^{-1} for Radotion. The difference between Radotion and malathion was, however, very well pronounced. The 7-times higher degradation rate of malathion in Radotion in comparison with pure analytical standard solution, is the result of the photosensitizing effect of additives present in the commercial product.

The respective half-life times in photocatalytic experiments of malaoxon and isomalathion were considerably longer than in case of malathion. In comparison to photocatalysis of malathion, over 4-times slower degradation was observed in case of malaoxon with the rate constant 0.0190 min^{-1} and over 3-times slower degradation in case of isomalathion with the rate constant 0.026 min^{-1}. Photocatalysis resulted in formation of several transformation products which are listed in Table 3. All the compounds were identified by GC-MS technique. Among the OP degradation products trimethyl phosphate esters *9A* - [OOS(O]; *10A* - [OOS(S)]; *11A* - [OSS(O)]; *8A* – (dimethoxyphosphinyl)oxo - butanedioic acid diethyl ester; *4A* – malaoxon; *3* – malathion; and *5A* – isomalathion were identified. It is known from the literature that they all inhibit the activity of cholinesterases (Talcott et al., 1979). The acute toxicity of *9A-11A* is a result of rapid aging of AChE (Clothier et al. 1981).

In case of malathion photocatalysis (initial concentration 14.7 ppm), [OOS(O] and malaoxon (*9A* and *4A*, respectively) are formed. The maximal concentration of malaoxon was measured between 15th (470 ppb) and 60th (630 ppb) minute of the experiment. [OOS(S)] (*10A)* was present from the beginning and [OOS(O] is probably its oxidation product as also product of malathion oxidation. Its maximal observed amount was at the 15th minute of the experiment. Two AChE inhibiting compounds were detected in non-irradiated samples of Radotion; [OOS(S)] and isomalathion *(5A)*, whereas [OOS(O] *(9A)* and malaoxon *(4A)* were formed during photocatalysis. The maximal amount of *4A* was at the 30th minute (690 ppb) of the experiment. The same is observed in case of *9A*, which expressed maximal concentration between 30th and 45th minutes of experiment.

No trimethyl phosphate esters are formed in photocatalysis of malaoxon and isomalathion. The maximal amounts of *9A* - [OOS(O] in case of malaoxon and *11A* - [OSS(O)] in case of isomalathion were observed in non-irradiated solutions and can be considered as impurities of the standards. No AChE inhibiting compound were formed during the photocatalysis of malaoxon and isomalathion (Bavcon et al., in press - a).

Table 3. Detected transformation products* in irradiated solutions of malathion, Radothion, malaoxon and isomalathion: product analysis by GC-MS based on retention times (t_R) and their MS spectral characteristics (/absent, ~ starting compound)

Compound name and index	Malathion	Radotion	Malaoxon	Isomalathion
3 malathion	~	~	/	/
4A malaoxon	30<max<60	max 30	~	/
5A isomalathion	max 0	max 0	/	~
8A (dimethoxyphosphinyl)oxo - butanedioic acid diethyl ester	/	/	/	/
9A [OOS(O] – O,O,S-trimethylphosphorothioate	max 15	30<max<45	max 0	/
10A [OOS(S)] – O,O,S-trimethylphosphorodithionate	max 0	max 0	/	/
11A [OSS(O)] – O,S,S-trimethylphosphorothioate	/	/	/	0<max<5
12 diethyl maleate	15<max<45	15<max<30	/	0<max<30
13 diethyl fumarate	max 0	max 0	max 0	max 0
14 butanedioic acid diethylester	30<max<45	max 30	/	max 0

16 hydroxy butanedioic acid diethylester	45<max<60	max 60	/	45<max 60
17 2,3-dihydroxy butanedioic acid diethylester	max 15	5<max<15	/	/
18 hydroxy butenedioic acid diethylester	/	/	max 45	/
19 2-hydroxy 3-sulphanyl butenedioic acid diethylester	/	/	/	/
20 diethyl(methyl) sulphonyl succinate	15<max<30	max 0	/	5<max<15

* They are divided in two groups; in AChE inhibiting (compound with index A) and AChE non-inhibiting compounds (compounds without index). The type A compounds express inhibiting potentials toward enzyme AChE, whereas the other non-labelled compounds belong to the group of diethyl esters analogs of different buta(e)ne dioic acids.

Ozonation and Photocatalysis of Chlorpyrifos and Chlorpyrifos-oxon

Several authors have dealt with chlorpyrifos photodegradation (Waila et al., 1988; Barcelo et al., 1993; Penuela and Barcelo, 1997), but no degradation studies were devoted to chlorpyrifos-oxon. In the recent degradation experiments the half-live times were determined as (15.0 ± 0.8) min for chlorpyrifos and (27.7 ± 1.3) min for chlorpyrifos-oxon (Table 2). However the degradation kinetics strongly depends upon experimental conditions (Bavcon et at., submitted - b).

Since chlorpyrifos is a member of phosphorothioates, a cycloaddition of ozone on the double bond P=S before the breakdown of the molecule (Laplanche et al., 1984) was expected in the ozonation experiments. It was noticed, that in pure analytical standard solution of chlorpyrifos the conversion to chlorpyrifos-oxon was rapid. Within 15 minutes a complete conversion into oxon was observed under given experimental conditions (O_3 flow rate: 1.50 ± 0.1) g h^{-1}. When technical product (Pyrinex) was ozonated, the conversion was slower compared to pure analytical standard, and it was achieved in one hour.

The concentration of formed chlorpyrifos-oxon did not drop significantly with additional ozonation process, which indicated, that ozone is rather inefficient for degradation of oxons. This was further confirmed, by ozonation of larger amounts of chlorpyrifos-oxon (analytical standard (11.9 mg L^{-1}) under same experimental conditions. In fact, only 21.5% of the starting compound was degraded in two hours, and the relative enzyme activity increased for 6%. Chlorpyrifos-oxon was further degraded (9.4 %) into trichloropirimidol (TCP). Diethyl phosphate was identified by LC-MS as the second hydrolysis product of chlorpyrifos-oxon (characteristic fragments m/z 153 (M-H) and 125 ($M-C_2H_5$)) (Bavcon et at., submitted - b).

OZONATION AND PHOTOCATALYSIS OF THIACLOPRID

Different AOMs including ozonation, O_3/UV-A and O_3/TiO_2/UV-A processes for degradation of thiacloprid (2 x 10^{-4} M) were investigated in detail, at different pH (3.2 - 11.0) (Černigoj et al., 2007b).

O_3, O_3/UV-A and O_3/TiO_2/UV-A processes with 355 nm broad maximum light source were strongly pH dependent, showing lower efficiencies for degradation of thiacloprid at acidic pH, what is in accordance with the stability of the ozone molecule in aqueous solutions at acidic pH (Hoigne, 1998). All disappearance curves indicated first-order kinetics. The calculated disappearance rate constants are collected in Table 4. The degradation becomes much faster with the increase of pH whenever the ozone is included in the process.

O_3/TiO_2/UV-A was found to be the most efficient process irrespective of pH. Molecular ozone acts as a better scavenger of conduction band electrons in TiO_2 semiconductor compared to molecular oxygen, therefore the recombination of reactive photoproduced holes and electrons in TiO_2 is decreased. This leads to the synergistic effect of ozone and TiO_2 photocatalysis at acidic and neutral pH. But, the synergism is lost at basic pH probably due to faster self-decompositon of ozone under alkaline conditions (the actual concentration of molecular ozone in the solution is lower than the ozone dosage) and nonfavorable adsorption of ozone on titania surface at alkaline pH. The difference between O_3 and O_3/UV-A processes was explained as the consequence of the photodegradation of ozone molecules by UV irradiation.

The formation of sulphate(VI) and chloride anions, monitored by ion-chromatography as a measure of mineralisation, followed the first-order kinetics at all pHs applied. Sulphur from thiacloprid was oxidized quantitatively to sulphate(VI) in all AOMs and no other anions containing sulphur were observed. On the contrary, chlorine atom was initially transformed to chloride in all experiments where degradation was evident. Further oxidation of chloride anion to chlorate(V) was noticed in O_3/TiO_2/UV-A and in O_3/UV-A processes at acidic pH. The oxidation of chloride was more efficient in O_3/TiO_2/UV-A process and the increase of ozone dosage further accelerated the oxidation. No chlorate(V) was observed in O_3 and O_2/TiO_2/UV-A system at similar conditions. No formation of nitrite was noticed during the experiments and the formation of nitrate was a slower process compared to formation of chloride and sulphate(VI). Formation of nitrate was not quantitative in any of AOMs studied. Mineralization in the O_2/TiO_2/UV-A system was slower than in O_3/UV-A and O_3/TiO_2/UV-A systems under neutral or alkaline conditions, even in the cases where the disappearance rates of the pollutant were similar. Evidently, the degradation products are more susceptible to the attack of ozone and ozone-generated oxygen species compared to the parent molecule and the mineralization is more efficient than it can be observed only from HPLC determination of starting compound.

Table 4. The influence of pH on disappearance rate constants of thiacloprid

pH	O_3	O_3/UV-A	O_3/TiO_2/UV-A	O_2/TiO_2/UV-A
3.2	0.002	0.03	0.078	0.033
5.6	0.004	0.04	0.119	0.055
8.1	0.04	0.059	0.126	0.056
11	0.088	0.12	0.152	0.085

The ozone dosage was 0.11 g h^{-1}.

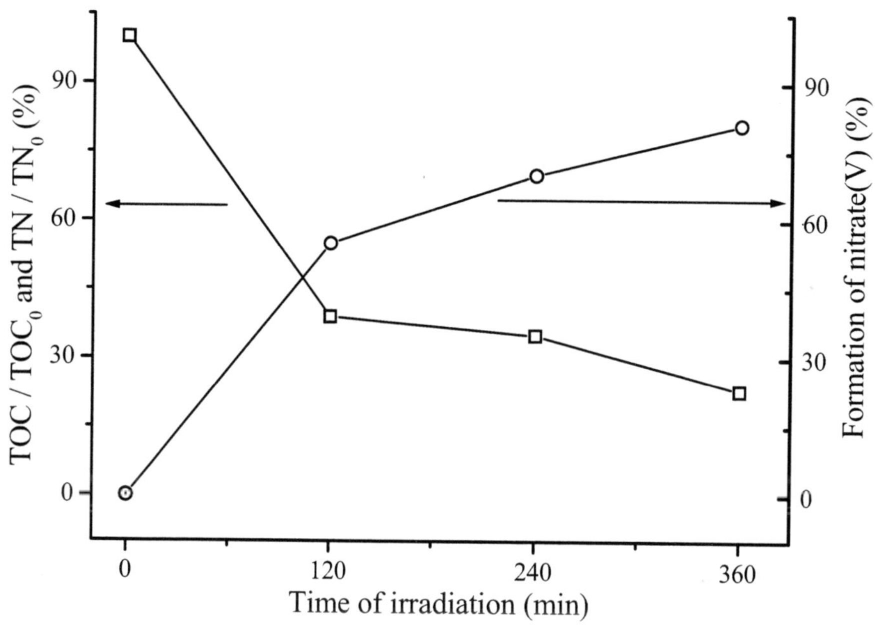

Figure 3. Monitoring of TOC during the O_3/TiO_2/UV treatment at pH = 11.0. Ozone dosage was 1.4 g h^{-1}.

TOC measurements were performed during the O_3/TiO_2/UV-A experiment (ozone dosage of 1.4 g h^{-1} at pH 10.8). TOC value decreased during the degradation process from 23.7 mg L^{-1} (also a calculated value) to 5.6 mg L^{-1} after 6 hours of irradiation (Figure 3). The mineralization was not completed even after 6 hours, since 23 % of the initial organic carbon remained in solution. The possible explanation could be in the formation of some recalcitrant nitrogen containing compounds, which are formed during the O_3/TiO_2/UV-A treatment and slow down the rate of mineralization of the organic matter.

3. Determination of OPS Employing TLS Detection

Most of daily-based analysis of pesticide residues is nowadays carried out by employing highly sophisticated and time-consuming chromatographic techniques. Biosensors and related bioanalytical techniques are however becoming more and more important alternative for approaching the large amount of samples. Most bioassays for organophosphorus pesticides exploit the inhibition of enzymatic activity of acetylcholinesterases, acetylcholine esterase (AChE) and butyrylcholine esterase (BuChE) in particular (Mello and Kubota, 2002). Generally, the amount of detected pesticides is reported as a cumulative concentration of all present cholinesterase inhibiting compounds, expressed in equivalents of pesticide used for calibration (usually paraoxon).

OPs and their metabolites in environmental matrices are generally present in small quantities (ppb-, rarely in ppm- concentration range). Therefore, the extraction of these compounds from solid matrices or their preconcentration from aqueous media is frequently needed. Furthermore, the thio-OPs (P=S), i.e. phosphorothionates and phosphorodithionates, which are the dominant form of OPs in agricultural practice, and consequently in environmental matrices, are much weaker inhibitors or do not inhibit AChE at all. For this reason, oxidation of thio-OPs into their oxo analogues is required before detection. (Kumaran and Tran-Minh 1992; Kumaran and Morita 1995; Lee et al., 2002; Marty et al., 1995; Herzsprung et al., 1990; Barcelo et al., 1995; Schultze et al., 2002, 2003).

H. Schulze and coworkers have developed a method for detection of phosphorothionates in foodstuffs, using N-bromosuccinimide (NBS) as an oxidizing reagent (Schultze et al., 2002, 2003, 2004). However the complete assay could not be performed in less than 2 h (Schultze et al., 2003). Other methods of oxidations were also developed till now, but were mainly intended for synthesis of oxons from thio-phosphates. Modifications of existing oxidation methods were therefore needed to avoid false negative results and to provide adequate means for a rapid routine bioanalytical assay as required for low cost screening of large number of samples.

An AChE bioanalytical system based on flow injection analysis (FIA) (Leon-Gonzalez and Towsend, 1991) and thermal lens spectrometric (TLS) detection (Franko and Tran, 1996), which offers low limits of detection (0.1 ppb paraoxon equivalents or lower) and relatively short time of analysis was developed recently, and is described in detail elsewhere (Pogačnik and Franko, 1999, 2001, 2003). The bioanalytical FIA-AChE-TLS system was applied in practice for determination of oxo OP pesticides in a large set of fruit juice concentrates and blended fruits prior to their processing in fruit juice production. As confirmed by parallel determinations of OPs by GC and GC-MS, no false negative results were observed. This is of high relevance, since a reliable screening test should not give clearance to a sample containing OP compounds (Franko et al., 2005).

To investigate the efficiency of oxidation in matrices such as water and fruit juice several oxidants (NBS, sodium hypochlorite, peroxide, CAN, mCPBA, iodine and Oxone) and pesticides in different molar ratios were tested in order to determine the most appropriate conditions for oxidation of OPs. Since the kinetics of oxidation is favourable in acidic conditions (reaction within 5 min), the pH of oxidizing solution (including the sample) must be adjusted to pH 2.0 (Bavcon et al., 2006).

The comparison of data obtained from GC-MS chromatograms and spectra showed that an efficient conversion into oxons in water samples was achieved using NBS in 1:10 sample to oxidant molar ratio. The only exception in oxidation of investigated OPs was noticed in case of diazinon, which expressed promoted hydrolysis under experimental conditions governed by the low pH of the juice (pH 3.2). For this purpose, sodium hypochlorite, which was adjusted to basic pH 12.0 to avoid induced hydrolysis, was used (Bavcon et al., 2006). In fact, diazinon is susceptible to acidic hydrolysis, whereas the majority of OPs are hydrolyzed in alkaline pH, like most esters (Chambers and Levi, Eds., 1992).

After the oxidation, inhibitions of AChE activity from 10 to 80% were observed for chlorpyrifos in 35 – 350 ppb concentration range, for malathion in 100 – 1000 ppb concentration range, and for azinphos-methyl in the 12.5 – 100 ppb concentration range. The lower limits of given concentration ranges are also representing the LOD values for determination of corresponding pesticides by FIA-AChE-TLS, which are in all cases well below the maximal residue limits for OPs in foodstuffs according to European legislation (Official Journal of the European Union, L 70/1, 16.3.2005).

Novel oxidation methods for thio-OPs based on NBS and sodium hypochlorite can be completed within five minutes. The oxidizing and reducing agents are directly added to water or juice. Since no preconcentration, extraction or other analytical step is needed, the time of analysis is reduced in contrast to previously known oxidation procedures. The AChE bioassay based on the described oxidation procedure enables determination of thio-OPs in the ppb concentration range in less than 15 minutes (Bavcon et al., 2006). This way, the analysis of real samples containing OPs was shortened by 8 times compared to previously existing methods (Schultze et al., 2002, 2003, 2004). The FIA-AChE-TLS was successfully applied to evaluate the toxicity of products formed during the AOMs degradation experiments of various OPs (Bavcon et al. 2007a). The trends of AChE inhibition observed in the toxicity curves were shown to agree very well with the results of degradation experiments followed by GC-MS to reveal the presence of toxic (decreased AChE activity) or non-toxic (preserved AChE activity) degradation products.

Recently, the oxidation of OPs was also attempted by using chloroperoxidase (CPO) enzyme from marine fungi *Caldariomyces fumago*, which is known to transform thio-OP to their oxo analogs (Hernandez et al., 1998). Oxidations performed on malathion, chlorpyrifos and parathion-methyl in diluted apple juice (pH = 2.9) using the CPO enzyme resulted in equimolar conversion of pesticides into oxo analogues (Figure 4). Differently from this, the yield of diazinon oxidation was not comparable to the equimolar concentration of its oxo-analog diazoxon. However, diazinon was not detected after the completion of oxidation reaction indicating either non-specificity of oxidation reaction in terms of oxidation product or rapid acid hydrolysis of diazoxon.

Using the CPO enzyme even less time is needed for oxidation (< 1 min.) and complete sample analysis, compared to oxidation with NBS . The pH of the reaction medium should however be adjusted to about 3 before the oxidation, and back to neutral pH values after the reaction is finished. Most importantly, CPO enzyme can be immobilised on inert porous material and incorporated into the FIA-AChE-TLS detection system. Such experiments, which aim to further shorten the time of analysis and to allow repeated use of enzymes, are currently underway in authors' laboratory. Initial results obtained with FIA-AChE/CPO-TLS detection system have also confirmed quantitative oxidation of malathion, chlorpyrifos and parathion-methyl. At present the LOD values for determination was thus lowered 850 and 270 times for malathion and chlorpyrifos, while (without the oxidation step) methyl-parathion was not detected even at 144 mg/L (Figure 5).

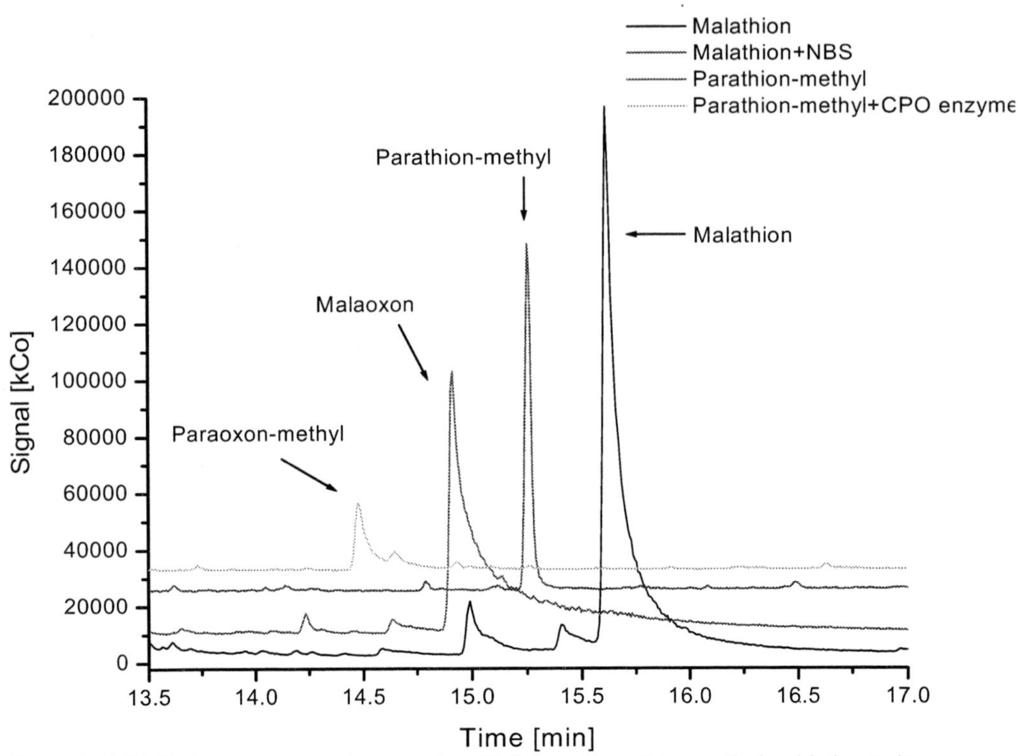

Figure 4. GC-MS chromatograms showing the efficiency of parathion methyl oxidation to its oxo-analog by CPO enzyme and malathion to its oxo analog by NBS.

4. Determination of Neonicotinoids Employing TLS Detection

Highly sensitive combination of thermal lens spectrometry and reversed phase C18 liquid chromatography (HPLC-TLS) has been recently applied (Guszvany et al., 2007 in press) as an alternative to existing HPLC-DAD and LC-MS methods for the determination of neonicotinoids in different samples. In the literature one can find reports on the application of HPLC-TLS method in the visible range for the analysis of different biological and environmental samples (Franko 2001, Luterotti et al. 2002, Kožar-Logar et al. 2003, Passamonti et al. 2005). With the laser source, providing stable continuous wave emission with over 100 mW of single line power in the spectral range of 229–257 nm, determination of compounds absorbing in the UV range, such as neonicotinoids, is also possible.

The HPLC-TLS determinations of imidacloprid, acetamiprid, thiacloprid and thiametoxam were made on a dual beam (pump/probe configuration) thermal lens spectrometer. A frequency doubled Ar–ion laser (Coherent, Sabre MotoFred) operating at 244 nm (100 mW) was used as excitation source (pump beam). For the purpose of comparison, the same samples were analyzed using HPLC-DAD technique.

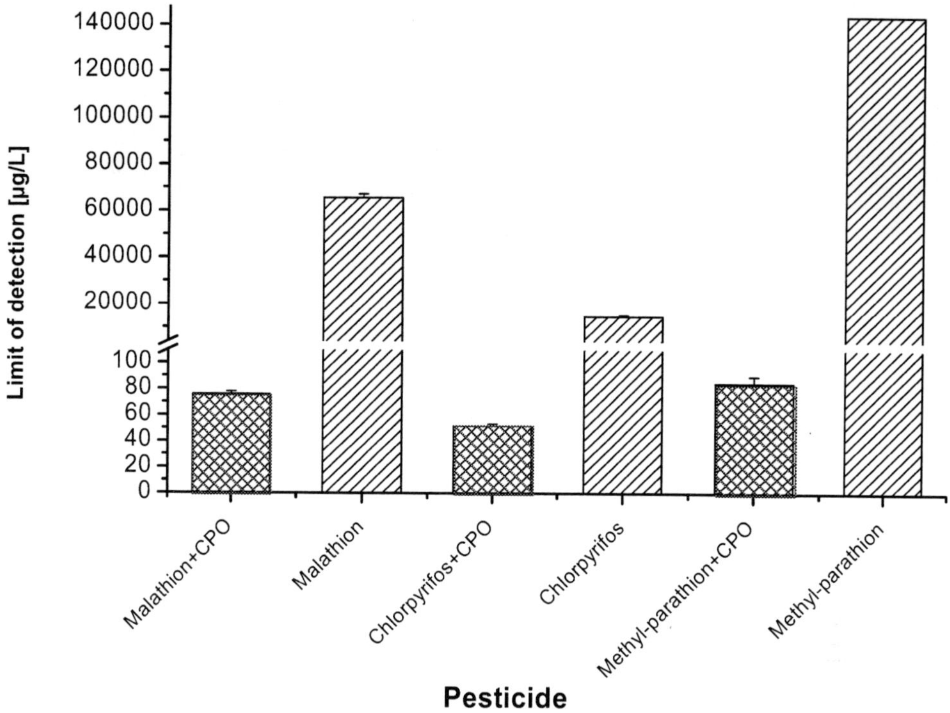

Figure 5. LODs (defined as concentration resulting in 10 % inhibition of AChE) achieved for thio-OP pesticides oxidised with immobilized CPO enzyme (crosswise texture) and without oxidation (sidewise texture) in FIA-AChE/CPO-TLS. No inhibition was observable for methyl-parathion at 144 mg/L concentration.

It was demonstrated, that the HPLC-TLS technique is sensitive enough for the determination of neonicotinoids at the ppb level. In the case of imidacloprid and thiamethoxam the essential analytical parameters obtained with TLS and DAD detectors are in good agreement. The higher sensitivities achieved for determination of acetamiprid and thiacloprid can be explained by the fact that their broad absorption bands with maxima at 245 and 242 nm, respectively, matched well with the excitation laser wavelength of 244 nm. The LOQs for acetamiprid and thiacloprid provided by HPLC-TLS technique, which is known to enable measurement of absorbances as low as 10^{-7}, are almost 10 times lower compared to HPLC-DAD (Table 5).

Table 5. Analytical parameters for the HPLC/TLS and HPLC/DAD determinations

Insecticide	Technique					
	HPLC/TLS			HPLC/DAD		
	LOD (µg/L)	LOQ (µg/L)	RSD (%)	LOD (µg/L)	LOQ (µg/L)	RSD (%)
Thiamethoxam	15	50	0.9	39	130	0.9
Imidacloprid	27	89	1.1	27	89	1.0
Acetamiprid	3.2	10	0.8	26	85	0.8
Thiacloprid	7.5	25	0.7	23	76	0.9

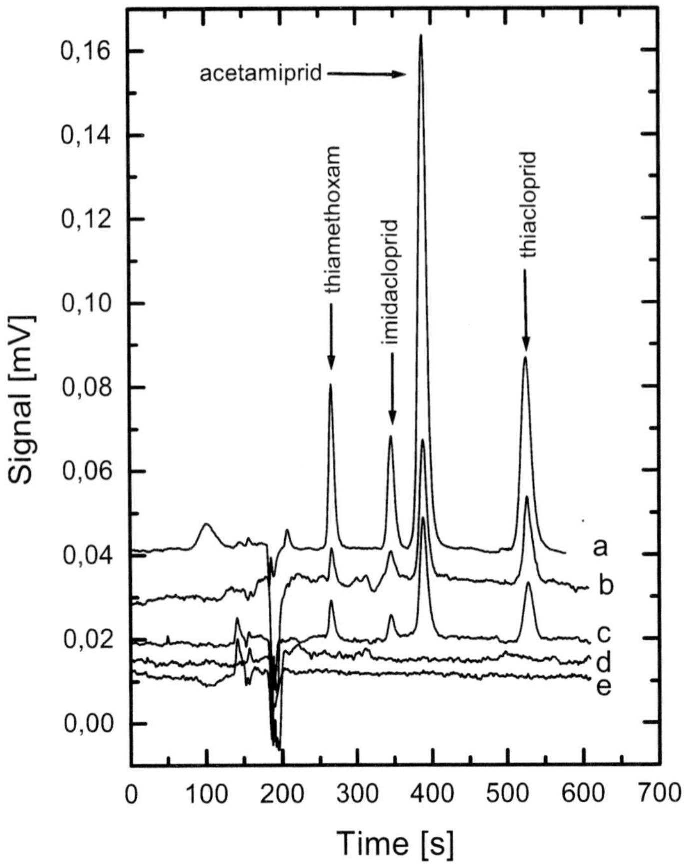

Figure 6. HPLC-TLS chromatograms of neonicotinoid mixed standard (1 mg/L) (a) and of extracts from 100 μg/kg level multiresidue spiked potatoe (b), spiked river water (c), potatoe – blank (d), and river water – blank (e): (1) thiamethoxam; (2) imidacloprid; (3) acetamiprid and (4) thiacloprid. Ecitation laser power: 100 mW, time constant: 1 s; modulation freqency; 80 Hz. Mobile phase: 7:3, v/v water (containing 0.2% phosphoric acid): acetonitrile, flow rate: 1.0 mL/min. Chromatograms are shifted along the Signal axis for clarity.

The applicability of the optimized HPLC-TLS method for the determination of the neonicotinoids' mixture was confirmed on spiked samples of potato and river water (Figure 6). The concentrations of 0.01 mg/kg imidacloprid or thiamethoxam, 0.002 mg/kg acetamiprid and 0.004 mg/kg thiacloprid can be considered as the LOQ limits of the method. These levels are low enough to allow monitoring of the investigated neonicotinoid insecticides in vegetables and surface water samples.

5. Conclusion

Advanced oxidation processes including photodegradation reactions and ozonization are very promising techniques for removal of pollutants such as OP and neonicotinoid insecticides. Recent research has provided important new information on mechanisms and formation of byproduct, which was lacking so far.

It was demonstrated that in the case of OPs the application of AOMs must be considered carefully since more toxic compounds can be formed during degradation processes. The degradation pathways as well as the efficiency of ozonation and photocatalysis differ significantly for individual OP compounds and depend upon variety of parameters, which need to be selected properly in order to enable rapid and possibly complete degradation of OPs. Monitoring the disappearance of parent compounds in AOMs might not give the appropriate information on the efficiency of the processes, since more persistent and/or toxic degradation products can be formed, as demonstrated by several works discussed in this contribution. It is therefore essential to monitor the formation of intermediates and degradation products by methods such as GC-MS and LC-MS at different stages of AOMs. Continuous monitoring of toxic degradation products, which is also needed for optimisation of an AOM into a sufficiently fast and cost efficient process, is however only possible by the application of bioanalytical assays. In this regard important progress has been made in detection of OPs by the development of the FIA bioanalytical assay, based on inhibition of AChE and previous on-line oxidation of thio-OPs by immobilised CPO enzyme.

Since more specific insecticides such as neonicotinoids are lately replacing the OP pesticides, more and more research is being conducted on the stability and degradation of neonicotinoids as well as on their determination. It was demonstrated that laser based detection methods such as TLS can be successfully applied to reduce the LODs for neonicotinoid pesticides in vegetables and natural waters when these compounds are determined by HPLC. Important recent findings indicated that the stability of some neonicotinoids was previously underestimated. From this prospective AOMs were confirmed as efficient methods for degradation of neonicotinoids as well.

6. REFERENCES

Agüera, A; Almansa, E, Malato, S; Maldonado, MI; Fernandez-Alba, AR; 1998. Evaluation of photocatalytic degradation of imidacloprid in industrial water by GC-MS and LC-MS. *Analusis,* 26, 245–251.

Barcelo, D; Durand, G; De Bertrand, N; 1993. Photodegradation of the organophosphorus pesticides chlorpyrifos, fenamiphos and vamidon in water. *Toxicol. environ. chem,* 38, 183-199.

Barcelo, D; Lacorte, S; Marty, JL; 1995. *Validation of an ezymatic biosensor with liquid chromatography for pesticide monitoring.* TrAC, 14, 7.

Bavcon, M; Trebše, P; Zupančič-Kralj, L; 2003. Investigations of the determination and transformations of diazinon and malathion under environmental conditions using gas chromatography coupled with a flame ionization detector. *Chemosphere,* 50, 595-601.

Bavcon, M; Trebše, P; Franko, M; 2006. Oxidation as a pre-step in determination of organophosphorus compounds by the AChe-TLS bioassay. *Acta chim. slov.,* 53, 43-51.

Bavcon Kralj, M; Franko, M; Trebše, P; 2007a. Photodegradation of organophosphorus insecticides - Investigations of products and their toxicity using gas chromatography-mass spectrometry and AChE-thermal lens spectrometric bioassay. *Chemosphere,* 67, 99-107.

Bavcon Kralj, M; Černigoj, U; Franko, M; Trebše, P; 2007b, in press. *Comparison of photocatalysis and photolysis of malathion, isomalathion, malaoxon and commercial malathion – products and toxicity studies.* Water Research.

Bavcon Kralj, M; Franko, M; Trebše, P; 2007c, submitted - b. *Application of Bioanalytical techniques in Evaluation of Advanced oxidation processes for Pesticide Determination.* TrAC.

Burrows, HD; Canle, M; Santaballa, JA; Steenken, S; 2002. Invited Review, Reaction pathways and mechanisms of photodegradation of pesticides. *J. photochem. photobiol., B Biol.,* 67, 71–108.

Chambers, JE; Levi, PE; (Eds.), *Organophosphates, Chemistry, Fate, and Effects.* Academic Press, San Diego, 1992.

Chiron, S; Fernandez-Alba, A; Rodriguez, A; 1997. *Pesticide chemical oxidation processes: an analytical approach.* TrAC, 16, no. 9.

Chiron, S; Rodriguez, A; Fernandez-Alba, A; 1998. Application of gas and liquid chromatography-mass spectrometry to the evaluation of pirimiphos methyl degradation products in industrial water under ozone treatment. *J. Chromatogr.* A, 823, 97-107.

Chiron, S; Fernandez-Alba, A; Rodriguez, A; Garcia-Calvo, E; 2000. Review Paper, Pesticide Chemical Oxidation: *State-Of-The-Art. Water Research,* 34, 366-377.

Černigoj, U; Lavrenčič Štangar, U; Trebše, P; 2007a. Evaluation of a novel Carberry type photoreactor for the degradation of organic pollutants in water. *J. photochem. photobiol., A Chem.,* 188 (2-3), 169-176.

Černigoj, U; Lavrenčič Štangar, U; Trebše, P; 2007b. Degradation of neonicotinoid insecticides by different advanced oxidation processes and studying the effect of ozone on TiO_2 photocatalysis. *Applied Catalysis B: Environmental,* 75, 231-240.

Dominguez, C; Garcia, J; Pedraz, MA; Torres, A; Galan, MA; 1998. Photocatalytic oxidation of organic pollutants in water. *Catal. today,* 40, 85-101.

Doong, RA; Chang, WH; 1997. Photoassisted titanium dioxide mediated degradation of organophosphorus pesticides by hydrogen peroxide. *J. photochem. photobiol.,A Chem.,* 107, 239-244.

Duirk, SE; and Collette TW; 2006. Degradation of chlorpyrifos in aqueous chlorine solutions: pathways, kinetics, and modeling. *Environ.Sci.Technol.,* 40, 546-551.

Fidente, P; Seccia, S; Vanni, F; Morrica, P; 2005. Analysis of neonicotinoid insecticides residues in honey by solid matrix partition clean-up and liquid chromatography-electrospray mass spectrometry. *J Chromatogr A* 1094, 175–178.

Franko, M; Tran, C. D.; 1996. Analytical thermal lens instrumentation, *Rev. Sci. Instrum.* 67, 1-18.

Franko, M; 2001. *Recent applications of thermal lens spectrometry in food analysis and environmental research.* Talanta, 54, 1-13.

Franko, M; Sarakha, M; Čibej, A; Boškin, A; Bavcon, M; 2005. Photodegradation of pesticides and application of bioanalytical methods for their detection. *Pure appl. chem.,* Vol. 77, No. 10, 1727-1736.

Guzsvany, V; Csanadi, J; Gaal, F; 2006. MNR Study of the Influence of pH on the persistence of Some Neonicotinoids in Water. *Acta Chim. Slov.* 53, 52–57.

Guzsvany, V; Madžgalj, A; Trebše, P; Gaal, F; Franko, M; 2007, in press. *Determination of selected neonicotinoid insecticides by liquid chromatography with thermal lens spectorimetric detection.* Environmental Chemistry Letters.

Hernandez, J; Robledo NR; Velasco, L; Quintero, R; Pickard, MA; Vazquez-Duhalt, R; 1998. Chloroperoxidase-mediated oxidation of organophosphorus pesticides. *Pesticide Biochemistry and Physiology,* 61, 87-94.

Herzsprung, P; Weil, L; Quentin, KE; Zombola, I; 1990. Determination of Organophosphorus Compounds and Carbamates by their Inhibition of Cholinesterase, Part 2: Estimation of Detection limits for Insecticide-determination by Concentration, Oxidation and Inhibition Values, V*om Wasser,* 74, 339-35.

Hoigne, J.; 1998 Chemistry of aqueous ozone and transformation of pollutants by ozonation and advanced oxidation processes, in: O. Hutzinger (ed.), *The handbook of environmental chemistry,* Vol. 5, Part C, Springer-Verlag, Berlin, pp. 84–141.

Hua, Z; Manping, Z; Zongfeng, X; Low, GKC; 1995. Titanium dioxide mediated photocatalytic degradation of monocrotophos. *Water Research,* 29, 2681-2688.

Iwasa, T; Motoyama, N; Ambrose, JT; Michael Roe, R; 2004. Mechanism for the different toxicity of neonicotinoid insecticides in the honey bee, Apis mellifera. Crop Protection 23, 371–378.

Li, HP; Li, GC; Len, JF; 2003. Determination of organochlorine pesticides in water using microwave assisted headspace solid-phase microextraction and gas chromatography. *Journal of Chromatography A,* 1012, 129-137.

Konstantinou, IK; Sakellarides, TM; Sakkas, VA; Albanis, TA; 2001. Photocatalytic degradation of selected s-triazine herbicides and organophosphorus insecticides over aqueous TiO_2 suspensions. *Environ. Sci. Technol.*, 35, 398-405.

Kouloumbos, VN; Tsipi, DF; Hiskia, AE; Nikolic, D; van Breemen, RB; 2003. Identification of Photocatalytic Degradation Products of Diazinon in TiO_2 Aqueous Suspensions Using GC/MS/MS and LC/MS with Quadrupole Time-of-Flight Mass Spectrometry. *J Am Mass Spectrom,* 14, 803-817

Kousba, AA; Sultatos, LG; Poet, TS; Timchalk, C; 2004. Comparison of chlorpyrifos-oxon and paraoxon Acetylcholinesterase inhibition dynamics: potential role of a peripheral binding site. *Toxicol. sci.*, 80, 239-248.

Kožar Logar, J;, Malej, A; Franko, M; 2003. Hyphenated high performance liquid chromatography-thermal lens spectrometry technique as a tool for investigations of xanthophyll cycle pigments in different taxonomic groups of marine phytoplankton. *Rev. sci. instrum.,* 74, 776-778.

Krohn, J; 2001. Behaviour of thiaclorpid in the environment. *Planzenschutz-Nachrichten Bayer,* 54, 281–290.

Ku, Y; Chang, J-L; Shen, Y-S; Lin, S-Y; 1998. Decomposition of diazinon in aqueous solution by ozonation. *Water Research* 32, 1957-1963.

Ku, Y; Jung, IL; 1998. Decomposition of Monocrotophos in Aqueous Solution by UV Irradiation in the Presence of Titanium Dioxide. *Chemosphere,* 37, 2589-2597.

Ku, Y; Lin, HS; 2002. Research note, Decomposition of phorate in aqueous solution by photolytic ozonation. *Water Research* 36, 4155-4159.

Kumaran, S; Morita, M; 1995. Application of a cholinesterase biosensor to screen for organophosphorus pesticides extracted from soil. *Talanta,* 42, 649-655.

Kumaran, S; Tran-Minh, C; 1992. Determination of Organophosphorus and Carbamate Insecticides by Flow Injection Analysis. *Anal. Biochem.,* 200, 187-194.

Laplanche, A; Martin, G; Tonnard, F; 1984. Ozonation schemes of organo phosphorus pesticides application in drinking water treatment. *Ozone: sci. eng.*, 6, 207-219.

Lee, HS; Kim, YA; Cho, YA; Lee, YT; 2002. Oxidation of organhophosphorus pesticides for the sensitive detection of cholinesterase-based biosensor. *Chemosphere,* 46, 571-576.

Leon-Gonzalez, MF; Townshend, A; 1991. Determination of organophosphorus and carbamate pesticide standards by liquid chromatography with detection by inhibition of immobilized acetylcholinesterase. *J. Chromatogr., A.* 539, 47-54.

Luterotti, S; Franko, M; Šikovec, M; Bicanic, D; 2002. Ultrasensitive assays of trans- and cis-β-carotenes in vegetable oils by high-performance liquid chromatography-thermal lens detection. *Anal Chim. Acta* 460, 193–200.

Malato, S; Caceres, J; Agüera, A; Mezcua, M; Hernando, D; Vial, J; Fernandez-Alba, AR; 2001. Degradation of imidacloprid in water by photo-fenton and TiO_2 photocatalysis at a solar pilot plant: A comparative study. *Environ. Sci. Technol.,* 35, 4359–4366.

Malato, S; Blanco, J; Richter, C; Maldonado, MI; 2000. Optimization of pre-industrial solar photocatalytic mineralization of commercial pesticides Application to pesticide container recycling. *Applied Catalysis, B Environment,* 25, 31-38.

Mandić, IA; Lazić, DS; Ökrész, NSz; Gaál, FF; 2005. Determination of the insecticide imidacloprid in potato (Solanum tuberosum L) and onion (Allium cepa) by high-performance liquid chromatography with diode-array detection. *Journal of Analytical Chemistry* 60, 1134–1138.

Mansour, M; Feicht, EA; Behechti, A; Schramm, KW; Kettrup, A: 1999. Determination photostability of selected agrochemicals in water and soil. *Chemosphere,* 39, 575-585.

Marty, JL; Garcia, D; Rouillon, R; 1995. Biosensors: potential in pesticide detection. *TrAC,* 14, no. 7, 1995.

Mello, DL; Kubota, LT; 2002. Review of the use of biosensors as analytical tools in the food and drink industries, Analytical, Nutritional and Clinical Methods. *Food chem.,* 77, 237–256.

Mengyeu, Z; Shifu, C; Yaowu, T; 1995. Photocatalytic Degradation of Organophosphorus Pesticides Using Thin Films of TiO_2. *J. Chem. Tech. Biotechnol.,* 64, 339-344.

Muszkat, L; Bir, L; Feigelson, L; 1995. Solar photocatalytic mineralization of pesticides in polluted waters. *J. photochem. photobiol., A Chem.,* 87, 85-88.

Moza, PN; Hustert, K; Feicht, E; Kettrup, A; 1998. Photolysis of imidacloprid in aqueous solution. *Chemosphere,* 36, 497–502.

Naman, SA; Khammas, ZA-A; Hussein, FM; 2002. Photo-oxidative degradation of insecticide dichlorovos by a combined semiconductors and organic sensitizers in aqueous media. *J. photochem. photobiol., A Chem.,* 153, 229-236.

Nogueira, JMF; Sandra, T; Sandra, P; 2003. Considerations on ultra trace analysis of carbamates in water samples. *Journal of Chromatorgraphy A,* 996, 133-140.

Obana, H; Okihashi, M; Akutsu, K; Kitagawa, Y; Hori, S; 2002. Determination of acetamiprid, imidacloprid and nitenpiram residues in vegetables and fruits by high-performance liquid chromatography with diode-array detection. *J. Agric. Food Chem.* 50, 4464–4467.

Obana, H; Okihashi, M; Akutsu, K; Kitagawa, Y; Hori, S; 2003. Determination of neonicotinoid pesticide residues in vegetables and fruits with solid phase extraction and liquid chromatography mass spectrometry. *J. Agric. Food. Chem.* 51, 2501–2505.

Official Journal of the European Union L 70/1, 16.3.2005. Regulation (EC) No. 396/2005 of the European parliament and the Council of 23 Fabruary 2005 on maximum residue levels of pesticides in or on food and feed of plant and animal origin and amending Coucil Directive 91/414/EEC.

Passamonti, S; Terdoslavich, M; Margon, A; Cocolo, A; Medic, N; Micali, F; Decorti, G; Franko, M; 2005. Uptake of bilirubin into HepG2 cells assayed by thermal lens spectroscopy : function of bilitranslocase. *FEBS journal,* 272, 5522-5535.

Penuela, GA; Barcelo, D; 1997. Comparative degradation kinetics of chlorpyrifos in water by photocatalysis with $FeCl_3$, TiO_2 and photolysis using solid-phase disk extraction followed by gas chromatographic techniques. *Toxicol. environ. chem,* 62, 135-147.

Pogačnik, L; Franko, M; 1999. Determination of organophosphate and carbamate pesticides in spiked samples of tap water and fruit juices by a biosensor with photothermal detection. *Biosens. Bioelectr.* 14, 569-578.

Pogačnik, L; Franko, M; 2001. Optimisation of FIA system for detection of organophosphorus and carbamate pesticides based on cholinesterase inhibition. *Talanta* 54, 631-641.

Pogačnik, L; Franko, M; 2003. Detection of organophosphate and carbamate pesticides in vegetable samples by a photothermal biosensor. *Biosens. Bioelectr.* 18, 1-9.

Roberts, TR; Hutson, DH; Jewess, PJ; Lee, PW; Nicholls, PH; Plimmer, JR; 1999. Metabolic pathways of agrochemicals. Part 2: Insecticides and fungicides, *The Royal Society of Chemistry,* Cambridge, 107–120.

Rodrigues, AM; Ferreira, V; Cardoso VV; Ferreira, E; Benoliel, MJ; 2007. Determination of several pesticides in water by solid-phase extraction, liquid chromatography and electrospray tandem mass spectroscopy. *Journal of Chromatography A,* 1150, 267-278.

Sakkas, VA; Lambropoulou, DA; Sakellarides, TM; Albanis, TA; 2002. Application of solid-phase microextraction for monitoring the photocatalytic decomposition of fenthion and parathion in aqueous TiO_2 suspensions. *Anal. chim. acta,* 467, 233-243.

Schellin, M; Hauser, B; Popp, P; 2004 Determination of organophosphorus pesticides using membrane-assisted solvent extraction combined with large volume injection–gas chromatography–mass spectrometric detection. *Journal of Chromatography A,* 1040, 251-258.

Schulze, H; Schmid, RD; Bachmann, TT; 2002. Rapid detection of neurotoxic insecticides in food using disposable acetylcholinesterase-biosensors and simple solvent extraction. *Anal. Bioanal. Chem.,* 372, 268-272.

Schulze, H; Schmid, RD; Bachmann, TT; 2004. Activation of Phosphorothionate Pesticides Based on a Cytochrome P450 BM-3 (CyP102 A1) Mutant for Expanded Neurotoxin Detection in Food Using Acetylcholinesterase Biosensors. *Anal. Chem.,* 76, 1720-1725.

Schulze, H; Vorlova′, S; Villatte, F; Bachmann, TT; Schmid, RD; 2003. Design of acetylcholinesterases for biosensor applications. *Biosens. bioelectron.,* 18, 201-209.

Seccia, S; Fidente, P; Barbini Attard, D; Morrica, P; 2005. Multiresidue determination of neonicotinoid insecticide residues in drinking water by liquid chromatography with electrospray ionisation mass spectrometry. *Anal Chim Acta* 553, 21–26.

Shemer, H; Sharpless, CM; Linden, KG; 2005. Photodegradation of 3,5,6-trichloro-2-pyridinol in aqueous solution. *Water air soil poll.,* 168; 145-155.

Wagner, GW; Bartram, PW; Brickhouse, MD; Connell, TR; Creasy, WR; Henderson, VD; Hovanec, JW; Morrissey, KM; Stuff, JR; Williams, BR; 2000. Reaction of VX and GD with gaseous ozone. *J.Chem.Soc., Perkin Trans.,* 2, 1267-1272.

Walia, S; Dureja, P; Mukerjee, SK; 1988. New photodegradation products of chlorpyrifos and their detection on glass, soil, and leaf surfaces. *Arch.Environ.Conmta.Toxicol.,* 17, 183-188.

Vilchez, JL; El-Khattabi, R; Fernández, J; González-Casado, Navalón, A; 1996 Determination of imidacloprid in water and soil samples by gas chromatography-mass spectrometry. *J Chromatogr A* 746, 289–294.

Wu, J; Luan, T; Lan, C; Lo, TWH; Chan, GYS; 2007a. Removal of residual pesticides on vegetable using ozonated water. *Food control*, 18, 466-472.

Wu, J; Luan, T; Lan, C; Lo, TWH; Chan, GYS; 2007b. Efficacy evaluation of low-concentration of ozonated water in removal of residual diazinon, parathion, methyl-parathion and cypermethrin on vegetable. *J. food eng.,* 79, 803-809.

Wu, J; and Laird, DA; 2003. Abiotic transformation of chlorpyrifos to chlorpyrifos oxon in chlorinated water. *Environ. toxicol. chem.,* 22, No. 2, 261-264.

In: Progress in Pesticides Research
Editor: Christopher M. Kanzantzakis

ISBN 978-1-60456-201-9

Chapter 9

SOLID-PHASE MICROEXTRACTION (SPME) FOR THE ANALYSIS OF PESTICIDE RESIDUES IN FOOD: AN OVERVIEW

Gianni Sagratini, Dario Giardinà and Sauro Vittori
Department of Chemical Sciences, Faculty of Pharmacy, University of Camerino, via S. Agostino 1, 62032 Camerino, Italy

ABSTRACT

The analysis of pesticide residues in food has received great attention in the last years, as testified by the high consideration of the international food safety authorities on the problem and by the large number of papers published in this field. Incorrect use of pesticides can threat the consumer's health, especially if the food is directed to baby population, and for this reason robust and sensible analytical methods are required to detect pesticide traces in food.

Depending on the physical-chemical characteristics of the analytes, instrumental analysis has been normally carried out using gas-chromatography (GC), liquid chromatography (LC) or capillary electrophoresis (CE), each one preceeded by a preliminary extraction process, such as liquid-liquid extraction (LLE), supercritical fluid extraction (SFE), solid-phase extraction (SPE) and matrix solid-phase dispersion (MSPD). A faster extraction technique with respect to those listed above, that doesn't require a long and expensive treatment of sample and that can integrate into a single step sampling, extraction, pre-concentration, and sample introduction, is the solid-phase microextraction (SPME), introduced by Pawliszyn and coworkers in 1990. SPME uses a fused silica fiber externally coated with an appropriate stationary phase such as poly(dimethylsiloxane) (PDMS), poly(acrylate) (PA), PDMS/divinylbenzene (DVB), Carboxen/PDMS. Two basic types of extractions can be performed using SPME: direct immersion (DI) and headspace (HS); the choice depends on sample matrix composition and on volatility of the analytes. Different kind of pesticides, as organophosporus, organochlorine, carbamate and phenylurea, have been monitored in fruit and fruit juices matrices using SPME coupled with GC/MS system or, using a specific interface, with

LC/MS and LC/MS/MS systems. In addition, phenylurea and aniline herbicides were analyzed by SPME/GC/MS in solid vegetable matrices as potatoes, carrots and onion, whereas carbamate and phenylurea were monitored in liquid food samples as wine or drinking water by SPME/LC/MS. The aim of this review is to describe the current trends of SPME applied to analysis of pesticide residues in food, underlining all positive and negative aspects that this technique, coupled with chromatographic systems, posses with respect to other more classic extraction procedures.

1. INTRODUCTION

The analysis of pesticide residues in food has received great attention in the last years, as testified by the high consideration of the international food safety authorities on the problem and by the large number of papers published in this field [1]. According to the list of all active pesticide substances, on the European Union (EU) market [2] more than 1100 pesticides are currently registered. In recent years, the increasing public concern about possible health risk due to pesticide residues in the diet has deeply modified the strategy for the crop protection, with emphasis on food quality and safety, and the widespread social concern about health led to the strict regulation of maximum residue limits (MRLs) of pesticide traces in food commodities. To date, more than 17,000 European Community MRLs have been set for 133 active pesticide substances in various commodities [3, 4]. Incorrect use of pesticides can threat the consumer's health, especially if the food is directed to baby population, and for this reason robust and sensible analytical methods are required to detect pesticide traces in food.

The determination of trace contaminants in complex matrices, such as food, often requires extensive sample extraction and preparation regimens prior to instrumental analysis. The needed amount of sample preparation depends on the kind of matrix and on the properties and level of analytes to be determined [5]. Depending on the physical-chemical characteristics of the analytes, instrumental analysis was normally carried out using gas-chromatography (GC), liquid chromatography (LC) or capillary electrophoresis (CE), each one preceeded by a preliminary extraction process. The typical steps for sample preparation include sampling/homogenization, extraction, clean-up, and concentration, followed by the final analysis. In recent years, many innovations in the processes devoted to food samples preparation for extraction and determination of pesticide residues have been developed [6-14]. This has resulted in the recognition that classical methods can now be replaced with procedures that are faster, less expensive, and possessing the same or higher efficiency than classical methods. Thus, in order to obtain a practical and rapid method for determination of pesticide residues in food complex matrices, several sample preparation procedures have been developed, including liquid-liquid extraction (LLE) [15], supercritical fluid extraction (SFE) [16], solid-phase extraction (SPE) [17], and solid-phase microextraction (SPME) [18]. However, in LLE and SPE disadvantages are the large quantities of utilized solvent, the needed multiple operation steps, the required preconcentration of the extract prior the analysis, and the interfering compounds that, likely, are coextracted. SPME, introduced in 1990 by Arthur and Pawliszyn [19], is an extraction technique using a fused silica fiber externally coated with an appropriate stationary phase such as poly(dimethylsiloxane) (PDMS), poly(acrylate) (PA), PDMS/divinylbenzene (DVB), Carbowax (CW)/DVB,

Carboxen (CAR)/PDMS. With respect to previous it constitutes a convenient alternative extraction method because integrates into a single step sampling, extraction, concentration, and sample introduction, without the use of solvents. Subsequently to the introduction of the SPME, a number of books have been written on the technique [20-22]. SPME is essentially a two step process: firstly, the partitioning of analytes between the sample matrix, which can be a liquid sample or headspace vapour, and the fiber coating, and then the release of the concentrated extract from the fiber into the analytical instrument, usually a GC, where the sample components are termally desorbed. The fiber can also be desorbed into a LC eluent using a static or dynamic mode for which several commercial interfaces are available [23, 24]. However, for LC automation, in-tube SPME devices are generally more suitable [12].

In agricultural and agroindustrial samples the solid-phase microextraction has been mainly applied to the characterization of commodities such as tobacco [25], alcoholic beverages [26, 27], juices [28], herbs and spices [29]. In water samples, SPME has been used for the analysis of pollutants such as polycyclic aromatic hydrocarbons [30], fatty acids [31], and phenols [32, 33].

In the scientific articles published in the last ten years, different kind of pesticides, such as organophosporus, carbamates, and phenylurea, were monitored in fruit and fruit juices matrices, using SPME coupled with GC/MS systems or, using a specific interface, with LC/MS and LC/MS/MS systems [34-37]. Phenylurea and aniline herbicides were analyzed by SPME/GC/MS also in solid vegetable matrices as potatoes, carrots and onions [38], whereas carbamates and phenylurea were monitored in liquid food samples, such as wine or drinking water, by SPME/LC/MS [39]. Many other methods can be developed using SPME to extract and concentrate different kinds of analytes, especially in complex mixtures like food samples.

The goal of this review is to describe the current trends of SPME applied to the analysis of pesticide residues in food, underlining all positive and negative aspects that SPME, coupled with chromatographic systems, possess with respect to other more classic extraction procedures.

2. Solid-phase Microextraction Optimization

As in any other solid-phase extraction (SPE)-based procedure, SPME consists of two separate stages: absorption, that is retention of the analytes on the stationary phase, and desorption. The development of a particular procedure for determination of pesticides using the SPME technique usually requires the optimization of the variable parameters related to both absorption and desorption steps. In this way, most of the studies include a specific section for procedure optimization as reported in Table 1, where the studied variables are listed.

Using the SPME, two main types of extraction are available, *HS- (head space) and DI- (direct immersion) SPME.* Extraction efficiency, with each technique, depends on the properties of the analytes and the sample matrix. In general, it is possible to say that DI-SPME is more sensitive than HS-SPME for analytes present in a liquid sample. However, HS-SPME is suitable for the extraction of volatile compounds in most gaseous, liquid and solid samples. DI-SPME is suitable for the extraction of semi- or low-volatile analytes in liquid samples. The fiber SPME is used with GC system to detect volatile and not thermally labile compounds, while fiber SPME is interfaced to LC system to analyze not volatile compounds. In-tube SPME is a relatively new micro-extraction and preconcentration technique, which can be easily coupled on-line with LC, increasing the automation in the analysis. With the in-tube SPME technique, organic compounds are directly extracted from the sample into the internally coated stationary phase of a capillary column, and then desorbed by introducing a moving stream of mobile phase or static desorption solvent.

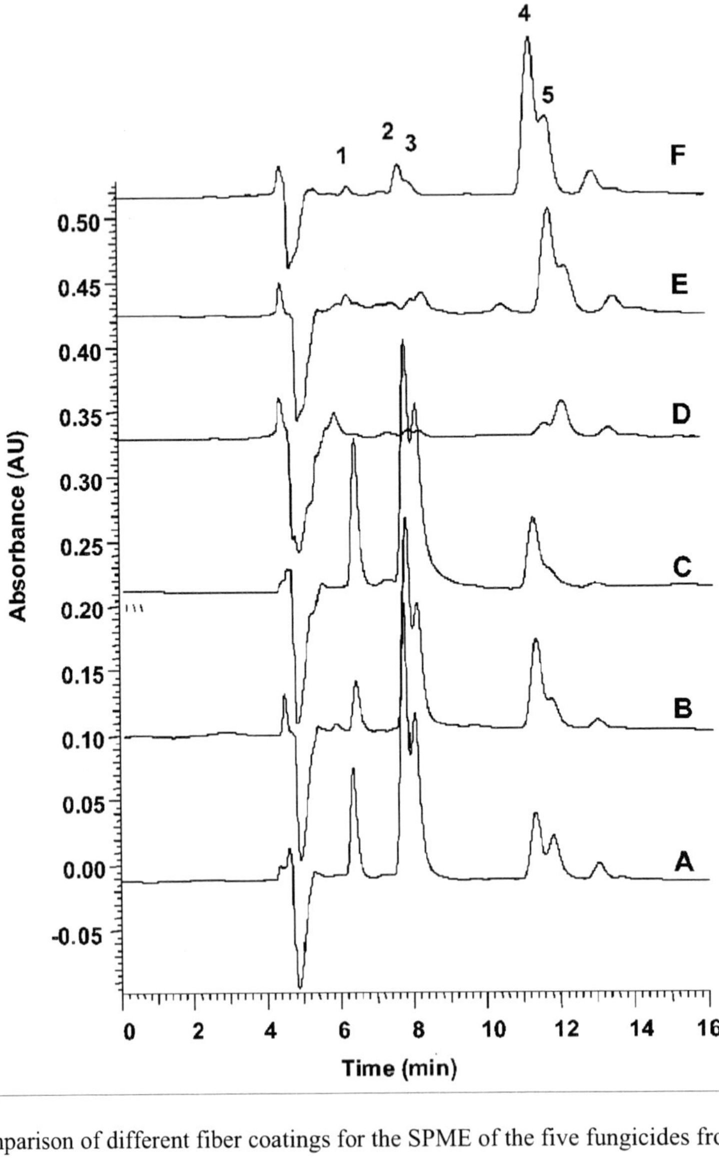

Figure 1. Comparison of different fiber coatings for the SPME of the five fungicides from water: (A) CW/TPR; (B) polyacrylate; (C) PDMS/DVB; (D) PDMS, 7 μm; (E) PDMS, 30 μm; (F) PDMS, 100 μm. The amount of each fungicide spiked into 1 mL of water was 0.25 μg for o-phenylphenol and 0.5 μg for other compounds. Peak identification: 1, flutriafol; 2, o-phenylphenol; 3, dichloran; 4, prochloraz; 5, tolclofos methyl. Figure from Blasco et al. [18].

A critical point of the analysis is the *choice of the fiber*. The affinity of a fiber for an analyte depends on the principle of "like dissolves like", and coating fibers having different properties or thickness are selected in accordance with physical properties of different compounds. The number of commercially available type of fiber coatings is rather limited. There are only few types of stationary phase, among non-polar poly(dimethylsiloxane) (PDMS), polar poly(acrylate) (PA), carboxen/divinylbenzene (CAR/DVB) and carbowax/templated resin (CW/TPR), mixed polarity (PDMS/DVB), (CAR/PDMS), (DVB/CAR/PDMS) [40]. Since the number of SPME applications has become very different, researchers have turned to alternative procedures for manifacturing fiber coatings, to increase their choice of stationary phase. Sol/gel procedures, in-tube extraction, and electrochemical coating deposition are some of these alternative production methods. These developments have been recently reviewed [41]. In Figure 1 it is reported a comparison of extraction capacity referred to different fiber coatings for the analysis of five fungicides in water.

In the SPME, the amount of the analyte extracted by the fiber depends not only on the polarity and thickness of the stationary phase, but also on the extraction time and the concentration of analyte in the sample. Extraction of analyte is also typically improved by stirring, addition of salt to the sample, optimization of pH and temperature.

Extraction time is mainly related to the agitation rate and the partition coefficient of the analyte between the fiber coating and sample matrix. Although SPME has maximum sensitivity at the equilibrium point, full equilibration is not necessary for accurate and precise analysis by SPME, because of the linear relationship between the amount of analyte adsorbed by the fiber and its initial concentration in the sample matrix in non-equilibrium conditions [42, 43]. Lambropoulou et al. [34] for the analysis of organophosphorus in fruit juices established 45 min of extraction time using PDMS fiber in HS-SPME, while Zambonin et al. [36] kept PA fiber in immersion for 30 min to analyze organophosphorus in the same food matrix.

Magnetic stirring is widely used for agitation in both HS- and DI-SPME. Agitation accelerates the transfer of analytes from the sample matrix to the coating fiber. Although the equilibration time progressively decreases with increasing agitation rate, faster agitation tends to be uncontrollable and the rotational speed might cause a change in the equilibrium time and poor measurement precision. On the contrary organophosphorus pesticides were analyzed by Simplicio et al. [35] studying the speed of magneting stirring from 200 to 1250 rpm; the authors concluded that the efficiency of the extraction can be increased by stirring the aqueous solution and that it is important to maintain a constant stirring speed in order to obtain reproducible results.

Another extraction parameter, which effect is well established in other extraction techniques (liquid-liquid partition and SPE) is the salting out effect obtained by *adding ionic salts* to the sample. This effect has been studied in SPME applications mainly by addition of NaCl and, alternatively, divalent salts such as Na_2SO_4 [44]. Most authors agree on the positive effect of the NaCl addition to the sample over extraction efficiency of most compounds, even if some discrepancies have been found in this direct relationship [45-49]. Blasco et al. [18] reported that in fruit juice solution different fungicides (tolclofos methyl and prochloraz) are influenced in opposite way by the salt addition, establishing a compromise that fixes at 30% the addition of NaCl. Additionally, it has been reported that high salt concentrations can led to negative effect on fiber stability using the CW/DVB fiber [50, 51].

Table 1. Performing parameters in SPME method optimization for pesticides analysis in food samples

Parameter	Remarks	Refs.
Extraction step		
Extraction time (min)	15-120	[34]
	1-80	[35]
	30	[36, 58]
	30-120	[37]
	60	[38, 55]
	10-90	[59]
	15-60	[61]
	20-160	[18]
Magnetic stirring (rpm)	960	[34, 55]
	200-1250	[35]
	1000-1400	[59]
	700	[66]
Salt addition (%)	Na_2SO_4 15	[34]
	NaCl 30	[37]
	NaCl 14	[38]
	NaCl 0-30	[58, 66]
	NaCl 5-15	[59]
	NaCl 10-40	[18]
Extraction temperature (°C)	25-85	[34]
	room temperature	[35, 58]
	20-80	[36]
	20-60	[37]
	22 ± 2	[38]
	75	[55]
	20-95	[59, 66]
	25-60	[61]
pH	3.5-5.7	[34]
	4-11	[36, 38]
	4	[58]
	5.5-8	[59]
	4-9	[18]
	4.3-7.5	[66]
Dilution with water (fold)	1-2	[34]
	0-100	[35]
	0-25	[36]
	0-50	[37]
	2	[59]
Extracting sample volume (ml)	5	[34, 36]
	3	[35, 58]
	1-5	[37]
	2	[38]
	1-3	[61]
	1-5	[18]

Desorption step		
GC Injector temperature (°C)	240	[34]
	250	[35, 36, 55]
	300	[38]
	200-280	[59]
	260	[61]
	280	[66]
Desorption time (min)	2	[34, 35]
	5	[36, 38, 61]
	7	[55]
	1-10	[59]
	3-20	[66]

Optimization of *extraction temperature* is generally more important when dealing with headspace SPME [48, 52, 53] than when working by direct immersion of the fiber in the solution. An increase in extraction temperature causes an increase in extraction rate and, simultaneously, a decrease in the distribution constant. Therefore, an adequate temperature, which provides satisfactory sensitivity and extraction rate, should be used. Lambropoulou et al. [34] tested the influence of the temperature in the range 25-85°C for the analysis of organophosphorus with HS-SPME, and choose 75°C as the most efficient temperature. Sagratini et al. [37] evaluated in DI-SPME the effect of the temperature for the extraction of carbamate and phenylurea pesticides in fruit juices, working in the range 20-60°C; the highest recoveries were obtained at 20°C, underlining that at 50 and 60°C significant modifications of the chromatographic peak structure were observed for some pesticides.

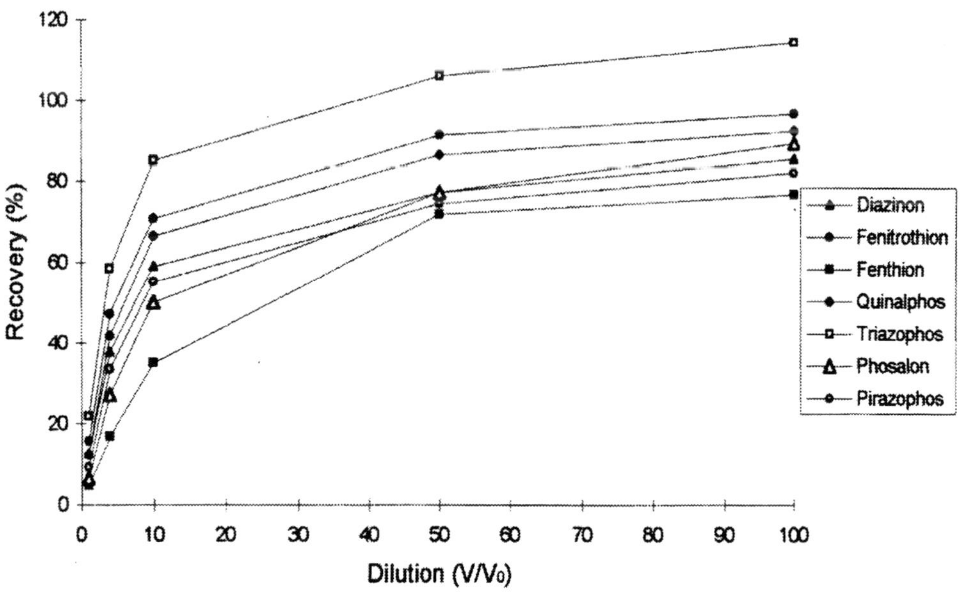

Figure 2. Effect of dilution of juices on recovery (juice fortified at 100 μg/ l, stirred at 1250 rpm and analysed after extraction with the PDMS fiber dipped into the solution for 20 min). Figure from Simplicio et al. [35].

The chemical form of analytes present in the sample mainly depends on the *pH* of the matrix relative to the analyte and influences the extraction efficiency. In general, the sample should be acidified for the extraction of acidic analytes and should be made alkaline for the extraction of basic analytes. Nevertheless, some authors found that the modification of pH values, from acidic to basic, did not change the recovery percentages [34, 36]; it is important to underline that extreme values of pH can damage the fiber coating.

Using SPME for pesticide analysis in food samples, often it is common to *dilute the sample* (normally liquid) by adding water in various ratios from 1:1 to 1:100. In general the effect of dilution is an improvement of analytes recovery up to 70% more [34-36], because of the dilution of the suspended matter that can interfer with the extraction process of the fiber. This effect is shown in Figure 2. Nevertheless, increasing the dilution of the solution, the sensitivity of the method could be compromised, causing negative effect on the value of LOD (limit of detection) and LOQ (limit of quantification). Some authors prefer to obtain lower recoveries diluting the solution 1:1 without decreasing the method sensitivity [18, 37].

The *sample volume* to be extracted with the fiber is another important parameter that has to be considered both in DI and HS. Some authors optimized this parameter in the proposed method for analysis of pesticides in food [18, 37] and they concluded that minimum volume is preferable with respect to large volume to increase the fiber performances.

The *desorption* of the analytes depends on the chromatographic system that is connected with the SPME, in this case GC or LC systems. If the desorption takes places in a GC injection port, some factors can affect the efficiency of the process, like analyte volatility, thickness of the fiber coating, injection depth, injector temperature, and exposure time. Generally, the optimal desorption temperature is approximatively equal to the boiling point of the less volatile analyte. In solvent desorption using SPME-HPLC interface there are two different techniques that can be used to remove molecules from the fibers: dynamic desorption and static desorption. In the first, analytes can be removed by a moving stream of mobile phase, while in the second, mobile phase or another kind of solvent is in contact with the fiber in the desorption chamber for the specific time necessary to remove the analytes from the fiber.

In most cases optimization of extraction and desorption procedure is carried out by a step by step procedure, changing a variable at a time.

3. Application of Solid-Phase Microextraction to Pesticide Analysis in Food

Most of scientific articles that appeared in the last ten years regarding the analysis of pesticide residues in food samples using SPME coupled with a chromatographic system (GC or LC), involved four major classes of pesticides: organophosphorus, orgachlorine, carbamate and phenylurea. Just few works reported analysis of other kind of pesticides, that in this report will go under the generic name of "various". Preferentially, the food matrices that appeared in the mentioned articles are liquid, especially fruit juices or fruit derivatives, where the extraction with SPME can be performed in HS, for volatile compounds, and in DI, for high-boiling or more polar compounds. The summarized applications of pesticide analysis in food samples using SPME are reported in Tables 2-6.

3.1. Organophosphorus

One of the first articles published on the analysis of organophosphorus pesticides with SPME in food samples (fruit juices) was by Simplicio et al. [35]; in this work 7 pesticides were analyzed in DI-SPME, monitored with a GC equipped with a flame photometric detector (phosphorus mode), extracted from the matrix with good recoveries, using preferentially PDMS fiber. It has been remarked that fruit is a very complex matrix in which the suspended matter can interfer with the analysis; the dilution with water can help in this kind of analysis.

The GC-MS system was used in two works [34, 36] to detect organophosporus pesticides extracted with SPME in fruit juices and in wine, chosing PDMS (100 μm) [34] and PA (85 μm) [36] as the best fibers, respectively; a good sensitivity was obtained [34], as testified by the LOD values ranging between 0.025 and 0.050 μg/l.

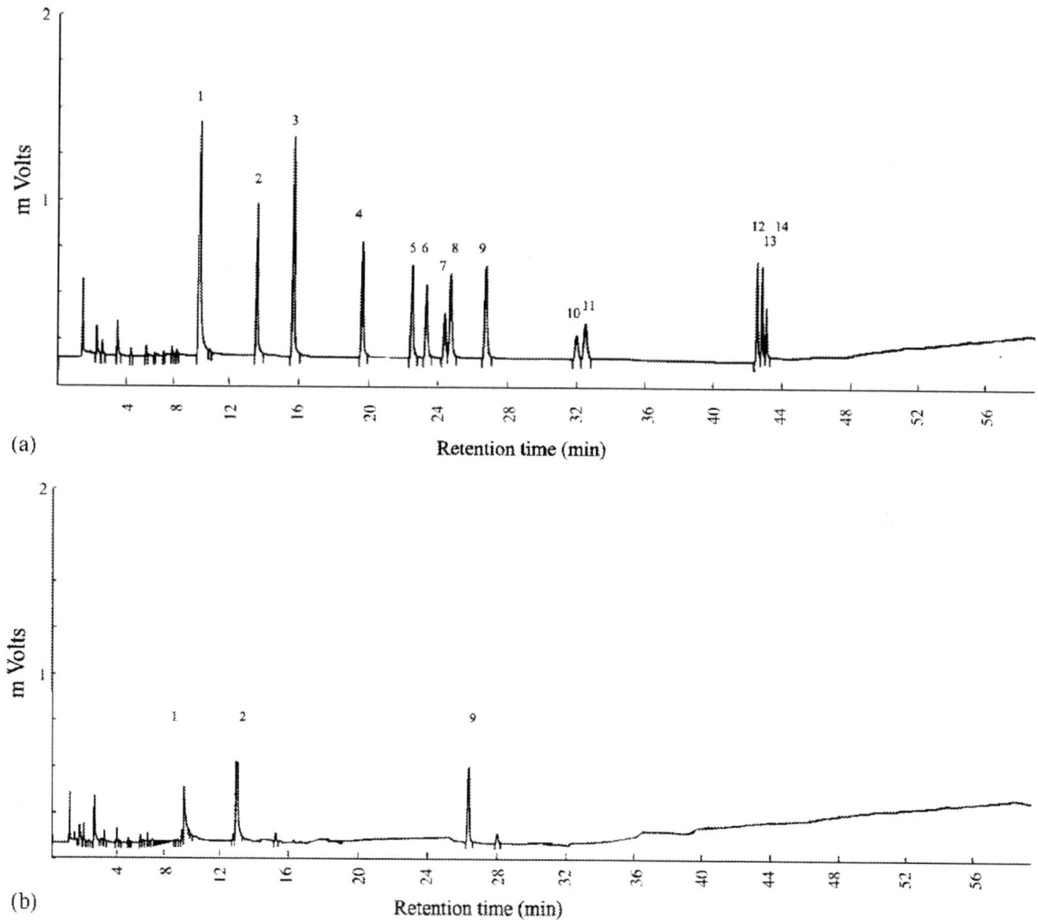

Figure 3. GC/FTD chromatograms obtained after HS-SPME extraction of a 0.050 mg kg^{-1} spiked olive oil sample (a) and a contaminated olive oil sample (b). Identification peaks of organophosphorus pesticides: (1) Omethoate; (2) dimethoate; (3) diazinon; (4) malaoxon; (5) fenitrothion; (6) malathion; (7) fenthion; (8) ethyl parathion; (9) ethyl bromophos (I.S.); (10) methidathion; (11) methyl bromophos; (12) ethion; (13) fenthion sulfoxide; (14) fenthion sulfone. Figure from Tsoutsi et al. [55].

Table 2. SPME applications for organophosphorus pesticide analysis in food

Food matrix	Testing fiber[a]	Application mode[b]	Instrumentation[c]	LOD $\mu g\ l^{-1}$ or $\mu g\ kg^{-1}$	Recovery %	Precision RSD %	Refs
Fruit, Fruit juices	PDMS 100 µm (S), CW/DVB 65 µm, PDMS/DVB 65 □m, PA 85 µm	DI	GC-FPD (phosphorus mode)	0.003-0.014	51-103	1-13	[35]
Fruit juices	PDMS 100 µm (S), CW/DVB 65 µm, PDMS/DVB 65 µm, PA 85 µm	HS	GC-FTD GC-MS	0.025-0.050	28-117	5-19	[34]
Fruit juices, wine	PA 85 µm	DI	GC-MS	2-90	5-100	3-10	[36]
Fruit	PDMS 100 µm, PA 85 µm (S)	HS	GC-NPD	0.07-6	1-3	3-8	[54]
Olive oil	CAR/PDMS 75 and 85 µm, PDMS 7, 30 and 100 µm (S), PDMS/DVB 65 µm	HS	GC-FTD	0.006-0.010	80-106	2-12	[55]
Whole milk	PA 100 µm	-	GC-NPD	0.052-0.060	-	-	[56]
Fish, water, potatoes, guava, coffee	PDMS 100 µm	-	GC-NPD	0.05-8.37	-	-	[57]
Herbal and tea infusion	PDMS 100 µm	DI	GC-AED	0.03-11.9	74-108	5-14	[59]

[a] S = selected fiber.

[b] DI = direct immersion, HS = head space.

[c] FPD = flame photometric detector, FTD = flame thermoionic detector, MS = mass spectrometer, NPD = nitrogen phosphorus detector, AED = atomic emission detector.

PA fiber was chosen as the best fiber, in the work by Fytianos et al. [54], for the analysis of organophosporus in fruit juices by HS-SPME-GC-NPD (nitrogen phosporus detector).

The applicability of SPME in other interesting liquid matrices was also tested. One of this matrix is the olive oil [55], in which organophosphorus insecticides and their metabolites were extracted using PDMS fiber (100 µm) and analyzed in HS-GC-FTD (flame thermoionic detector). In this work it was found that only acidity and total amount of sterols are the main factors influencing the SPME efficiency. In Figure 3 are reported two chromatograms referring to an olive oil sample spiked with organophosphorus pesticides and to a naturally contaminated olive oil sample.

In another work coumaphos and dichlorvos were extracted with SPME from milk samples and PA fiber was chosen to perform the analysis [56]. The authors observed that the residues of pesticides are not removed from the matrix by boiling the milk.

Solid food samples were analyzed with SPME to extract organophosphorus pesticides by Capobiango et al. [57]; in detail, fish, potatoes and coffee were exposed to SPME analysis using PDMS fiber and GC-NPD as chromatographic system.

3.2. Organochlorine

The presence of organochlorine fungicides (nuarimol, triadimenol, triadimefon, folpet, vinclozolin, penconazole) in wine was investigated using SPME-HPLC-DAD by Millàn et al. [58]. In the proposed method, that uses the suitable SPME-HPLC interface, both dynamic and static desorption were evaluated; the selected fiber was the PDMS/DVB and the method showed a sufficient sensitivity, with LOD values ranging from 4 to 27 μg l^{-1}. The presence of ethanol in the solution decreased the efficiency of the extraction.

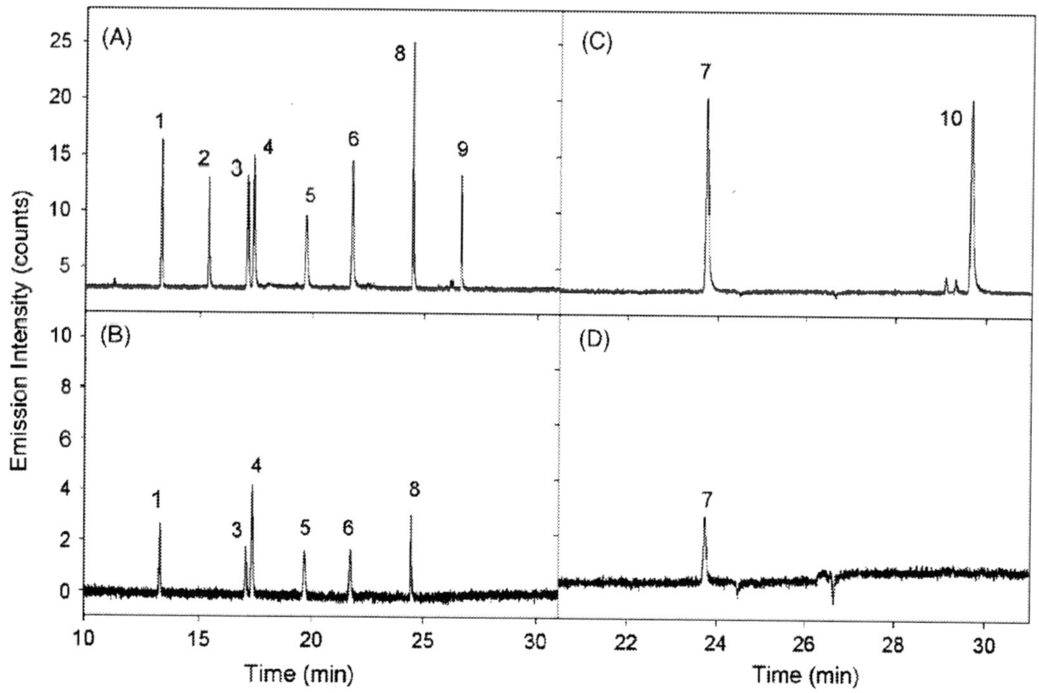

Figure 4. SPME–GC–AED chromatograms obtained from a black tea infusion spiked with the pesticides (A and C) and a non-spiked camomile infusion (B and D). (A and B) Cl 479 nm and (C and D) Br 478 nm. Identification of organochlorine pesticides: (1) Aldrin; (2) chlorfenvinphos; (3) dieldrin; (4) *p,p_*-DDE; (5) *p,p*-DDD; (6) *p,p*-DDT; (7) bromopropylate; (8) tetradifon; (9) coumaphos; (10) deltamethrin. Figure from Campillo et al. [59].

In works by Campillo et al. [59] and by Ning et al. [60] organochlorine pesticides were analyzed in herbal/tea infusion (liquid matrix [59]) and in tea leaves (solid matrix [60]) using SPME coupled to GC. Herbal infusion as camomile and tea are consumed for both therapeutical and pleasure purposes; the proposed methods represent a valid alternative to other classical method to analyze pesticides in this kind of food samples. Figure 4 shows the chromatograms referring to a black tea infusion spiked with a pesticide mixture and to non-spiked camomile infusion analyzed in a SPME-GC-AED apparatus.

SPME coupled with a GC-ECD system was also applied to the analysis of organochlorine pesticides in fish samples [61]. Good recoveries (over 70% in all cases) were achieved from samples spiked at a concentration level of 10 ng g^{-1}. PDMS (100 μm) was the fiber selected for this kind of analysis.

Another liquid matrix that was tested for the applicability of SPME technique was the milk [62]; organochlorine pesticides as HCB (hexachlorobenzene) and DDT (1,1,1-trichloro-2,2-bis(4-chlorophenyl)ethane) were analyzed by SPME-GC-ECD.

An interesting application of SPME was that concerning the analysis of organochlorine compounds in ethanol/water mixtures used as food simulants in experiments of food contact material [63].

Table 3. SPME applications for organochlorine pesticide analysis in food

Food matrix	Testing fiber[a]	Application mode[b]	Instrumentation[c]	LOD μg l^{-1} or μg kg^{-1}	Recovery %	Precision RSD %	Refs
Wine	PDMS/DVB 60 μm	DI	HPLC-DAD	4-27	95-103	2-26	[58]
Herbal and tea infusion	PDMS 100 μm	DI	GC-AED	0.03-11.9	74-108	5-14	[59]
Tea	-	-	GC	0.001-0.05	63-121	10-23	[60]
Fish	PDMS 100 μm	DI	GC-ECD	0.1-0.7	70-104	5-34	[61]
Breast milk	PA 85 μm (S), PDMS 7 and 100 μm	HS	GC-ECD	0.06-3.41	-	-	[62]
Ethanol/water (food simulants)	PDMS 7 and 100 μm (S), PA 85 μm (S), CW/DVB 65 μm	DI	GC-MS	0.02-205.5	78-117	<20	[63]

[a] S = selected fiber.

[b] DI = direct immersion, HS = head space.

[c] DAD = diode array detector, AED = atomic emission detector, ECD = electron capture detector, MS = mass spectrometer.

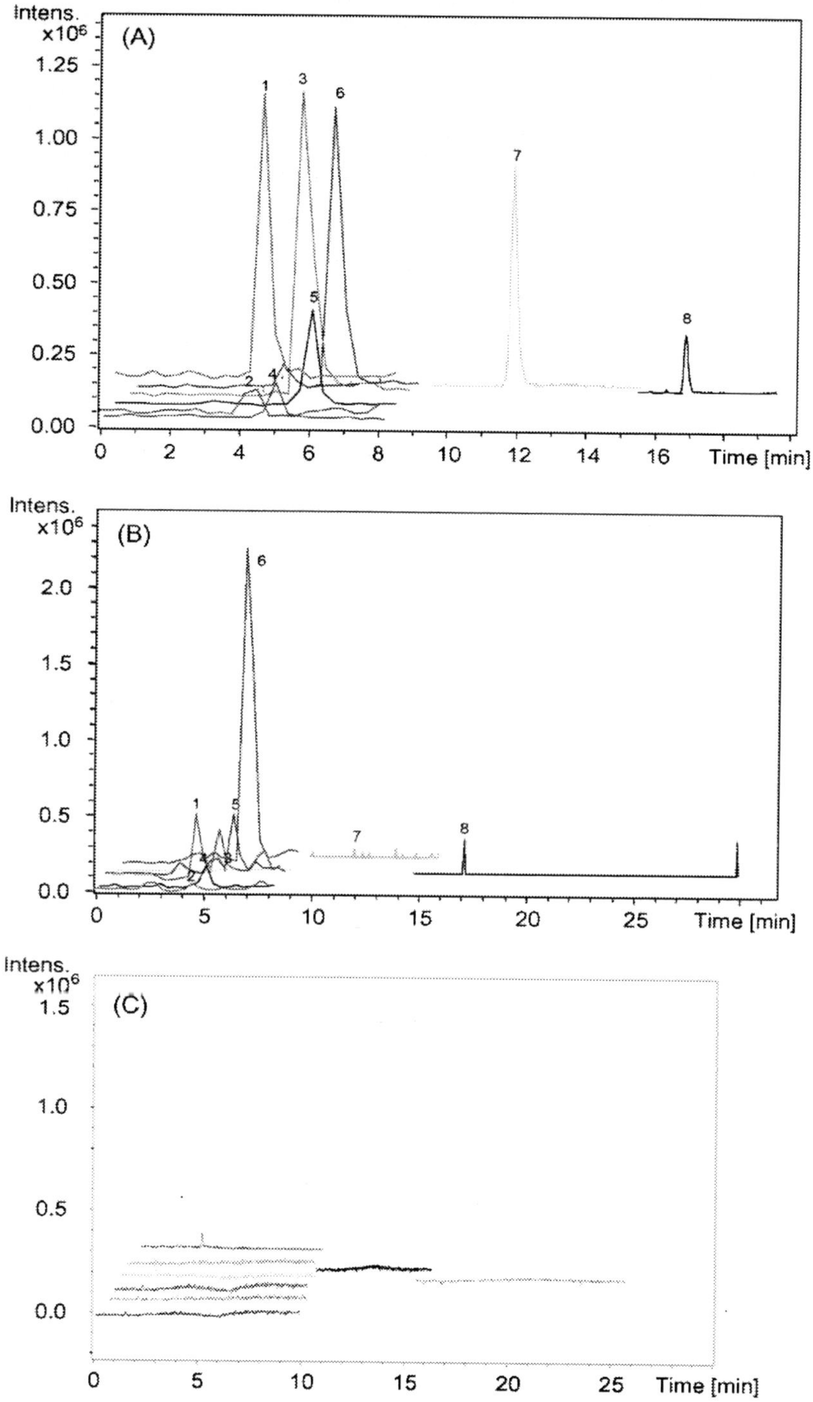

Figure 5. Representative determinative LC-MS-MS chromatograms obtained by ion-trap of (A) 10 ng of standard mixture of tested pesticides, (B) orange juice spiked with 50 ng of pesticides standard mixture and extracted with CW/TPR fiber, (C) orange juice extracted with CW/TPR fiber. Pesticides list as regards to retention time: 1) Carbofuran, 2) Monuron, 3) Pirimicarb, 4) Monolinuron, 5) Diuron, 6) Diethofencarb, 7) Benfuracarb, 8) Carbosulfan. Figure from Sagratini et al. [37].

Table 4. SPME applications for carbamate pesticide analysis in food

Food matrix	Testing fiber[a]	Application mode[b]	Instrumentation[c]	LOD $\mu g\ l^{-1}$ or $\mu g\ kg^{-1}$	Recovery %	Precision RSD %	Refs
Water	Various GC capillaries (in-tube)	DI	HPLC-UV	1-15	97-100	2-5	[64]
Water, wine	PPY- and PMPY-coated capillary, other commercial capillaries (in-tube)	DI	HPLC-UV HPLC-MS	0.38-8.2 0.01-1.2	1-34	<6	[39]
Fruit juices	PA 85 μm, CW/TPR 50 μm (S), PDMS/DVB 60 μm (S)	DI	HPLC-MS HPLC-MS/MS	0.005-0.01 0.005-0.01	1-82	1-17	[37]

[a] S = selected fiber.

[b] DI = direct immersion.

[c] UV = ultra violet spectrophotometer, MS = mass spectrometer.

3.3. Carbamate

In literature there are very few works in which SPME has been applied to the analysis of polar carbamate pesticides in food matrices. This could be due to the polar nature of these analytes and, especially for someone, to their high solubility in water, that is connected to the real difficulty to be extracted by SPME fiber.

Gou et al. [64] used in-tube SPME to analyze six carbamates in water samples performing instrumental analysis with an HPLC-DAD.

In another work by Wu et al. [39], the polar carbamates were extracted from water and wine using the new fibers coated with PPY (polypyrrole) and PMPY (poly-N-methylpyrrole) as stationary phases, and using HPLC-DAD and HPLC-ESI-MS to monitor pesticides. The good performance of the method was mined by the not excellent recoveries, that were obtained also with other fibers.

Sagratini et al. [37] developed a method to analyze carbamates in fruit juices using SPME-HPLC-MS-MS. The best performances were obtained using CW/TPR and PDMS/DVB fibers; the method sensitivity was confirmed by LOQs obtained for all pesticides, which were lower than MRLs (maximum residue limit) set by the legislation. Typical HPLC-MS-MS chromatograms obtained in this work are reported in Figure 5.

3.4. Phenylurea

This kind of pesticides was analyzed in liquid samples, in detail wine [39] and fruit juice [37], using SPME and HPLC-MS apparatus.

In another work, phenylurea herbicides were analyzed in natural waters at concentrations below 1 ng l^{-1} using SPME-GC-MS [65]. The fiber used was a PA (85 μm). The developed method was able to detect the presence of phenylureas and their metabolites that often from surface waters (lakes) go into the drinking water, becoming a worrying risk for health.

Berrada et al. [38] have determined phenylurea herbicides in vegetables, such as carrots, onion, and potatoes, using SPME-GC-NPD method. The authors reported that obtaining a juice from foods with high water content and diluting it makes it possible to perform SPME of such kind of samples.

3.5. Various

In literature there are articles reporting the analysis of various pesticides with SPME in food samples.

Picò et al. [18] described analysis of postharvest fungicides as nitro derivatives or imidazoles in fruits with SPME-HPLC-MS-MS; the research investigated some particolar steps that are decisive in the SPME analysis, like the composition of the solvent mixture that is used to fill the desorption chamber of the interface in the static mode. The authors conclude that different chromatographic profile were obtained changing this composition; the optimum was methanol/water (70/30 v/v). Figure 6 shows an overlay of three HPLC-UV chromatograms obtained filling the desorption chamber with different solvent mixture.

Table 5. SPME applications for phenylurea pesticide analysis in food

Food matrix	Testing fiber[a]	Application mode[b]	Instrumentation[c]	LOD μg l^{-1} or μg kg^{-1}	Recovery %	Precision RSD %	Refs
Carrots, Onion, Potatoes	PA 85 μm	DI	GC-NPD GC-MS	0.8-2.2 0.1-0.7	76-95	3-8	[38]
Water, wine	PPY- and PMPY-coated capillary, other commercial capillaries (in-tube)	DI	HPLC-UV HPLC-MS	0.38-8.2 0.01-1.2	1-34	<6	[39]
Fruit juices	PA 85 μm, CW/TPR 50 μm (S), PDMS/DVB 60 μm (S)	DI	HPLC-MS HPLC-MS/MS	0.005-0.01 0.005-0.01	1-82	1-17	[37]

[a] S = selected fiber.

[b] DI = direct immersion.

[c] NPD = nitrogen phosphorus detector, MS = mass spectrometer, UV = ultra violet spectrophotometer.

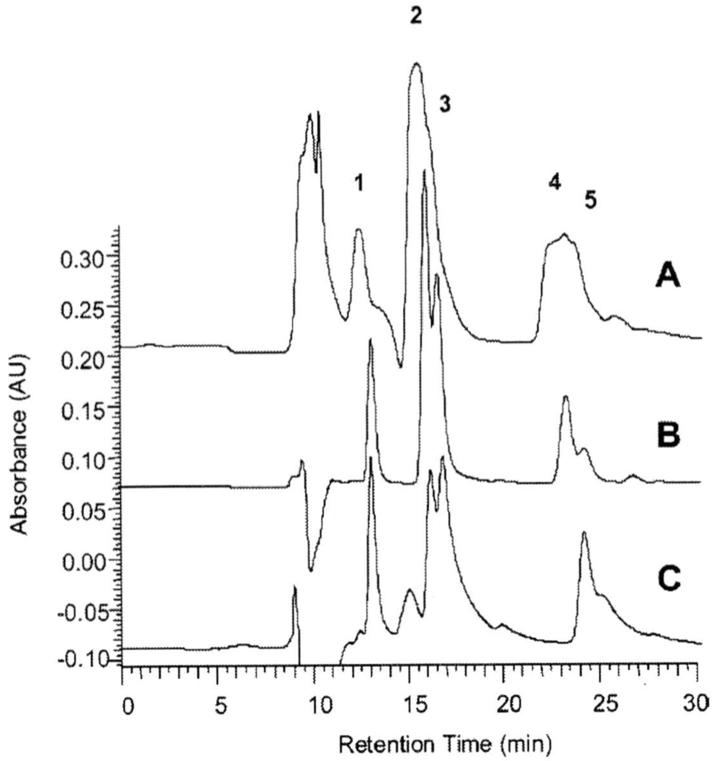

Figure 6. Comparison of chromatograms obtained from different desorption solvent mixtures: (A) 100% methanol, (B) methanol/water (70:30 v/v), and (C) methanol/water (40:60 v/v). Peak identification: 1, flutriafol; 2, o-phenylphenol; 3, dichloran; 4, prochloraz; 5, tolclofos methyl. Figure from Blasco et al. [18].

Table 6. SPME applications for various pesticide analysis in food

Food matrix	Testing fiber[a]	Application mode[b]	Instrumentation[c]	LOD μg l^{-1} or μg kg^{-1}	Recovery %	Precision RSD %	Refs
Fruit	PDMS 7 and 30 and 100 μm, PA 85 μm, PDMS/DVB 60 μm, CW/TPR 50 μm (S)	DI	HPLC-MS/MS	0.0005-0.01 (LOQ)	5-79	3-15	[18]
Honey	PDMS 100 μm (S), PA 85 μm	DI	GC-AED	0.02-10	91 (average)	15 (average)	[66]
Wine	PDMS/DVB 60 μm,	DI	MEKC-UV	17.6-32.3	5-25	1-8	[68]

[a] S = selected fiber.

[b] DI = direct immersion.

[c] MS = mass spectrometer, AED = atomic emission detector, MEKC-UV = micellar electrokinetic chromatography with ultra violet detection.

Campillo et al. [66] extracted various classes of pesticides; among them pyrethrins from honey using SPME equipped with PDMS and PA fibers. The adding of acetone to the sample before extraction process was studied; for some pesticides an increase of recovery was obtained, as reported in other papers [67].

Other kinds of pesticides as anilinopyrimidine and dicarboximide were studied by Molina-Mayo et al. [68] testing SPME; PDMS/DVB was the chosen fiber for the analysis in wine and CE (capillary electrophoresis) was the detection system adopted, in this case, for the first time in wine.

CONCLUSION

From the papers reviewed we can conclude that SPME, from its commercial introduction in 1993, has received increasing attention on its applicability for residue analysis, i.e. pesticides, in food. In general, SPME showed advantages and disadvantages with respect to other extraction techniques, especially in complex matrix (liquid and solid) as food. Principal advantages of this technique can be reasumed in the following easy concepts: SPME is a) simple to use, b) inexpensive, c) fast, d) sensitive, e) low polluting (solvent elimination), f) self-cleaning, g) it has automation possibilities. On the contrary, there are some disadvantages that until now have emerged in various applications: SPME a) has a moderate precision, b) is not quantitative, c) generally furnish a low recovery, d) exposure times of fiber and temperatures need to be carefully controlled, e) few extracting phases are available, f) the instrumentation, especially some kind of fibers, is fragile, and it is common to strip it off, for example, in mechanical part of instrumentation like SPME-HPLC interface.

Moreover, its applicability for pesticide analysis in food has been directed especially toward liquid samples, which in general permit an easier sample preparation using SPME. In the maiority of applications, SPME is in combination with GC system, but the introduction of apposite interfaces permits an easy coupling with LC system too. Since SPME is a not-quantitative technique, selective detectors, such as ECD or NPD, and sensitive detectors, such as MS or MS/MS, are required to detect pesticides in complex matrices as food using SPME.

REFERENCES

[1] Picó, Y; Fernández, M; Ruiz, MJ; Font, G. Current trends in solid-phase extraction techniques for the determination of pesticides in food and environment. *Journal of Biochemical and Biophysical Methods,* 2007 70, 117-131.

[2] Status of active substances under EU review (doc. 3010), http://europa.eu.int/comm/food/plant/protection/evaluation/stat_active_subs_3010_en.xls

[3] Hercegovà, A; Dömötörovà, M; Matisovà, E. Sample preparation methods in the analysis of pesticides residues in baby food with subsequent chromatographic determination. *Journal of Chromatography A*, 2007 1153, 54-73.

[4] Introduction to EC pesticides residues legislation, http://europa.eu.int/comm/food/plant/protection/resources/intro_en.pdf

[5] Ridgway, K; Lalljie, SPD; Smith, RM. Sample preparation techniques for the determination of trace residues and contaminants in foods. *Journal of Chromatography A*, 2007 1153, 36-53.

[6] Kataoka, H. Automated sample preparation using in-tube solid-phase microextraction and its application-a review. *Analytical Bioanalytical Chemistry*, 2002 373, 31-45.

[7] Motohashi, N; Nagashima, H; Parkanyi, C. Supercritical fluid extraction for the analysis of pesticide residues in miscellaneous samples. *Journal of Biochemical and Biophysical Methods*, 2000 43, 313-328.

[8] Picò, Y; Font, G; Moltò, JC; Mañes, J. Solid-phase extraction of quaternary ammonium herbicides. *Journal of Chromatography A*, 2000 885, 251-271.

[9] Kristenson, EM; Ramos, L; Brinkman, UAT. Recent advances in matrix solid-phase dispersion. *Trends in Analytical Chemistry*, 2006 25, 96-111.

[10] Sabik, H; Jeannot, R; Rondeau, B; Multiresidue methods using solid-phase extraction techniques for monitoring priority pesticides, including triazines and degradation products, in ground and surface waters. *Journal of Chromatography A*, 2000 885, 217-236.

[11] Barker, SA. Applications of matrix solid-phase dispersion in food analysis. *Journal of Chromatography A*, 2000 880, 63-68.

[12] Kataoka, H; Lord, HL; Pawliszyn, J. Applications of solid-phase micro-extraction in food analysis. *Journal of Chromatography A*, 2000 880, 35-62.

[13] Beltran, J; Lopez, FJ; Hernandez, F. Solid-phase micro-extraction in pesticide residue analysis. *Journal of Chromatography A*, 2000 885, 389-404.

[14] Aulakh, JS; Malik, AK; Kaur, V; Schmitt-Kopplin, P. A review on solid phase microextraction-high performance liquid chromatography (SPME-HPLC) analysis of pesticides. *Critical Review in Analytical Chemistry*, 2000 35, 71-85.

[15] Pylypiw, HM. Rapid gas chromatographic method for the multiresidue screening of fruits and vegetables for organochlorine and organophosphate pesticides. *Journal of AOAC International*, 1993 76, 1369-1373.

[16] Lehotay, S. Supercritical fluid extraction of pesticides in foods. *Journal of Chromatography A*, 1997 785, 289-312.

[17] Torres, C; Picó, Y; Mañes, J. Comparison of octadecylsilica and graphitized carbon black as materials for solid-phase extraction of fungicide and insecticide residues from fruit and vegetables. *Journal of Chromatography A*, 1997 778, 127-137.

[18] Blasco, C; Font, G; Mañes, J; Picó, Y. Solid-phase microextraction liquid chromatography/tandem mass spectrometry to determine postharvest fungicides in fruits. *Analytical Chemistry*, 2003 75, 3606-3615.

[19] Arthur, C; Pawliszyn, J. Solid phase microextraction with thermal desorption using fused silica optical fibers. *Analytical Chemistry*, 1990 62, 2145-2148.

[20] Pawliszyn, J. Solid phase microextraction: theory and practice. New York: Wiley; 1997.

[21] Wercinski, SAS. Solid phase microextraction: a practical guide. New York: Marcel Dekker; 1999.

[22] Pawliszyn, J. Application of solid phase microextraction (RSC Chromatography Monographs). Cambridge: Royal Society of Chemistry; 1999.

[23] Lord, H; Pawliszyn, J. Microextraction of drugs. *Journal of Chromatography A*, 2000 902, 17-63.

[24] Saito, Y; Nakao, Y; Imaizumi, N; Morishima, Y; Kiso, Y; Jinno, K. Miniaturized solid-phase extraction as a sample preparation technique for the determination of phthalates in water. *Analytical and Bioanaytical Chemistry,* 2002 373, 81-86

[25] Yang, S; Huang, CB; Smetena, I. Optimization of headspace sampling using solid-phase microextraction for volatile components in tobacco. *Journal of Chromatography A,* 2002 942, 33-39.

[26] Ng, Lay-Keow. Analysis of vodkas and white rums by SPME-GC-MS. *Applications of Solid-phase microextaction*, 1999, 393-406.

[27] Lay-Keow, N; Hupe, M; Harnois, J; Moccia, D. Characterisation of commercial vodkas by solid-phase microextraction and gas chromatography/mass spectrometry analysis. *Journal of the Science of Food and Agriculture,* 1996 70, 380-388.

[28] Steffen, A; Pawlizyn, J. Analysis of flavor volatiles using headspace solid-phase microextraction. *Journal of Agricultural and Food Chemistry*, 1993 44, 2187-2193

[29] Miller KG; Poole, CF; Pawloski, X. Classification of the botanical origin of cinnamon by solid-phase microextraction and gas chromatography. *Chromatographia,* 1996 42, 639-646.

[30] Zhang, Z; Yang, MJ; Pawlizyn, J. Headspace solid-phase microextraction. *Analytical Chemistry*, 1993 65, 1843-1852.

[31] Pan, L; Adams, M; Pawliszyn, J. Determination of fatty acids using solid-phase microextraction. *Analytical Chemistry*, 1995 67, 4396-4403.

[32] Eisert, R; Levsen, K. Solid-phase microextraction coupled to gas chromatography: a new method for the analysis of organics in water. *Journal of Chromatography A*, 1996 733, 143-157.

[33] Schafer, B; Engwald, W. Enrichment of nitrophenols from water by means of solid-phase microextraction. *Fresenius Journal of Analytical Chemistry*. 1995 352, 535-536.

[34] Lambropoulou, DA; Albanis , TA. Headspace solid-phase microextraction applied to the analysis of organophosphorus insecticides in strawberry and cherry juices. *Journal of Agricultural and Food Chemistry,* 2002 50, 3359-3365.

[35] Simplicio, AL; Boas, LV. Validation of a solid-phase microextraction method for the determination of organophosphorus pesticides in fruits and fruit juice. *Journal of Chromatography A*, 1999 833, 35-42.

[36] Zambonin, CG; Quinto, M; De Vietro, N; Palmisano, F. Solid-phase microextraction-gas chromatography mass spectrometry: a fast and simple screening method for assessment of organophosphorus pesticides residues in wine and fruit juices. *Food Chemistry,* 2004 86, 269-274.

[37] Sagratini, G; Mañes, J; Giardinà, D; Damiani, P; Picó, Y. Analysis of carbamate and phenylurea pesticide residues in fruit juices by solid-phase microextraction and liquid chromatography-mass spectrometry. *Journal of Chromatography A*, 2007 1147, 135-143.

[38] Berrada, H; Font, G; Moltó, JC. Application of solid-phase microextraction for determining phenylurea herbicides and their homologous anilines from vegetables. *Journal of Chromatography A*, 2004 1042, 9-14.

[39] Wu, J; Tragas, C; Lord, H; Pawliszyn, J. Analysis of polar pesticides in water and wine samples by automated in-tube solid-phase microextraction coupled with high-performance liquid chromatography-mass spectrometry. *Journal of Chromatography A*, 2002 976, 357-367.

[40] Stashenko, EE; Martinez, JR. Sampling volatile compounds from natural products with headspace/solid-phase micro-extraction. *Journal of Biochemical and Biophysical Methods*, 2007 70, 235-242.

[41] Dietz, C; Sanz, J; Càmara, C. Recent developments in solid-phase microextraction coatings and related techniques. *Journal of Chromatography A*, 2006 1103, 183-192.

[42] Ai, J. Solid-phase microextraction for quantitative analysis in nonequilibrium situations. *Analytical Chemistry*, 1997 69, 1230-1236.

[43] Ai, J. Headspace solid-phase microextraction. Dynamics and quantitative analysis before reaching a partition equilibrium. *Analytical Chemistry*, 1997 69, 3260-3266.

[44] Guan, F; Watanabe, K; Ishii, A; Seno, H; Kumazawa, T; Hattori, H; Suzuki, O. Headspace solid-phase microextraction and gas chromatographic determination of dinitroaniline herbicides in human blood, urine and environmental water. *Journal of Chromatography B*, 1998 71, 205-213.

[45] Beltran, J; Lopez, FJ, Cepria, O; Hernandez, F. Solid-phase microextraction for quantitative analysis of organophosphorus pesticides in environmental water samples. *Journal of Chromatography A,* 1998 808, 257-263.

[46] Boyd-Boland, AA; Magdic, S; Pawliszyn, J. Simultaneous determination of 60 pesticides in water using solid phase microextraction and gas chromatography-mass spectrometry *Analyst,* 1996 121, 929-937.

[47] Jinno, K; Muramatsu, T; Saito, Y; Kiso, Y; Magdic, S; Pawliszyn, J. Analysis of pesticides in environmental water samples by solid-phase micro-extraction-high-performance liquid chromatography. *Journal of Chromatography A,* 1996 754, 137-144.

[48] Page, BD; Lacroix, G. Application of solid-phase microextraction to the headspace gas chromatographic analysis of semi-volatile organochlorine contaminants in aqueous matrices. *Journal of Chromatography A,* 1997 757, 173-182.

[49] Eisert, R; Levsen, K; Wünsch, G. Element selective detection of pesticides by gas chromatography-atomic emission detection and solid-phase microextraction. *Journal of Chromatography A,* 1994 683, 175-183.

[50] Moder, M; Popp, P; Eisert, R; Pawliszyn, J. Determination of organophosphorus pesticides in soil by headspace solid-phase microextraction. *Fresenius Journal of Analytical Chemistry,* 1999 363, 680-685.

[51] Hernandez, F; Beltran, J; Lopez, FJ; Gaspar, JV. Use of solid-phase microextraction for the quantitative determination of herbicides in soil and water samples. *Analytical Chemistry,* 2000 72, 2313-2322.

[52] Namera, A; Yashiki, M; Nagasawa, N; Iwasaki, Y; Kojima, T. Rapid analysis of malathion in blood using head space-solid phase microextraction and selected ion monitoring. *Forensic Science International,* 1997 88, 125-131.

[53] Ng, WF; Teo, MJK, Lakso, HA. Determination of organophosphorus pesticides in soil by headspace solid-phase microextraction. *Fresenius Journal of Analytical Chemistry,* 1999 363, 673-679.

[54] Fytianos, K; Raikos, N; Theodoridis, G; Velinova, Z; Tsoukali, H. Solid phase microextraction applied to the analysis of organophosphorus insecticides in fruits. *Chemosphere,* 2006 65, 2090-2095.

[55] Tsoutsi, C; Konstantinou, I; Hela, D; Albanis, T. Screening method for organophosphorus insecticides and their metabolites in olive oil samples based on

headspace solid-phase microextraction coupled with gas chromatography. *Analytica Chimica Acta,* 2006 573-574, 216-222.

[56] Cardeal, ZDL; Paes, CMD. Analysis of organophosphorus pesticides in whole milk by solid phase microextraction gas chromatography method. *Journal of Environmental Science and Health, Part B,* 2006 41, 369-375.

[57] Capobiango, HLV; Cardeal, ZL. A solid-phase microextraction method for the chromatographic determination of organophosphorus pesticides in fish, water, potatoes, guava and coffee. *Journal of the Brazilian Chemical Society,* 2005 16, 907-914.

[58] Millàn, S; Sampedro, MC; Unceta, N; Goicolea, MA; Rodriguez, E; Barrio, RJ. Coupling solid-phase microextraction and high-performance liquid chromatography for direct and sensitive determination of halogenated fungicides in wine. *Journal of Chromatography A,* 2003 995, 135-142.

[59] Campillo, N; Peñalver, R; Hernandez-Cordoba, M. Pesticide analysis in herbal infusions by solid-phase microextraction and gas chromatography with atomic emission detection. *Talanta,* 2007 71, 1417-1423.

[60] Ning, Y; Binbin, Y; Maosheng, Z; Jingbin, Z; Xi, C. Simultaneous determination of residues of organochlorine and pyretroid pesticides in tea by microwave assisted extraction solid phase microextraction-gas chromatography. *Sepu,* 2006 24, 636-640.

[61] Fidalgo-Used, N; Centineo, G; Blanco-Gonzalez, E; Sanz-Medel, A. Solid-phase microextraction as a clean-up and preconcentration procedure for organochlorine pesticides determination in fish tissue by gas chromatography with electron capture detection. *Journal of Chromatography A,* 2003 1017, 35-44.

[62] Rohrig, L; Meisch, HU. Application of solid phase microextraction for the rapid analysis of chlorinated organics in breast milk. *Fresenius' Journal of Analytical Chemistry,* 2000 366, 106-111.

[63] Batlle, R; Sanchez, C; Nerin, C. A systematic approach to optimize solid-phase microextraction. Determination of pesticides in ethanol/water mixtures used as food simulants. *Analytical Chemistry,* 1999 71, 2417-2422.

[64] Gou, Y; Eisert, R; Pawliszyn, J. Automated in-tube solid-phase microextraction-high performance liquid chromatography for carbamate pesticide analysis. *Journal of Chromatography A,* 2000 873, 137-147.

[65] Gerecke, AC; Tixier, C; Bartels, T; Schwarzenbach, RP; Müller, SR. Determination of phenylurea herbicides in natural waters at concentrations below 1 ng l^{-1} using solid-phase extraction, derivatization, and solid-phase microextraction-gas chromatography-mass spectrometry. *Journal of Chromatography A,* 2001 930, 9-19.

[66] Campillo, N; Peñalver, R; Aguinaga, N; Hernandez-Cordoba, M. Solid-phase microextraction and gas chromatography with atomic emission detection for multiresidue determination of pesticides in honey. *Analytica Chimica Acta,* 2006 562, 9-15.

[67] Fernandez, M; Padron, C; Marconi, L; Ghini, S; Colombo, R; Sabatini, AG; Girotti, S. Determination of organophosphorus pesticides in honeybees after solid-phase microextraction. *Journal of Chromatography A,* 2001 922, 257-265.

[68] Molina-Mayo, C; Hernandez-Borges, J; Borges-Miquel, TM; Rodriguez-Delgado, ff.

In: Progress in Pesticides Research
Editor: Christopher M. Kanzantzakis

ISBN 978-1-60456-201-9

Chapter 10

PESTICIDE DEGRADATION BY LACCASES

Susana Rodríguez Couto[1*] ***and José L. Toca Herrera***[2†]
[1]Department of Chemical Engineering, Universitat Rovira i Virgili, Av. Països Catalans 26, 43007 Tarragona, Spain
[2]Biosurfaces Unit, CICBiomaGUNE, Paseo Miramón 182, 20009 San Sebastián, Spain

ABSTRACT

Pesticides are the only chemicals intentionally made to be toxic and introduced directly into the environment. Only 5% of pesticides reach the target organism. The rest runs off into water or dissipates in the soil or air. Such compounds are extremely resistant to biodegradation by native micro-organisms. In addition, the implementation of more and more stringent environmental regulations on hazardous wastes has impelled the search for novel and eco-friendly treatment technologies to complement or substitute the conventional ones. Biodegradation appears a promising technology, particularly the use of oxidative enzymes as biocatalysts, included in a micro-organism or used as a free enzyme.

White-rot fungi produce various types of oxidative enzymes, which are useful for the remediation of environmental contaminants. Laccase [EC: 1.10.3.2] is one of such enzymes. Compared to other oxidative enzymes that need various additives (e.g., hydrogen peroxide for peroxidases), dissolved oxygen is the only requirement for laccase-catalysed reactions. Enzyme-mediated degradation of pesticides has two advantages: (i) the degradation reaction proceeds under relatively mild reaction conditions and (ii) the reaction rate is faster than that by micro-organisms. In addition, they also present some other interesting properties such as low energy requirements, easy process control and operation over a wide range of pH, temperature and ionic strength. Furthermore, enzymes are active in the presence of high concentrations of organic solvents in which hydrophobic molecules are soluble. The purpose of this paper is to review the use of laccases in pesticides degradation.

* Tel.: +34 97755 9617; E-mail: susana.rodriguez@urv.cat
† Tel.: +34 943005313; E-mail: jltocaherrera@cicbiomagune.es

INTRODUCTION

Intensive industrial and agricultural activities during the 20th century have led to considerable pollution of soil and water with toxic compounds, which are harmful both for the environment and the human health. Among such compounds pesticides are of special concern since due to their synthetic origin are highly recalcitrant to biodegradation by native microorganisms and, as a result, their residues persists in the environment at toxic levels causing serious detrimental effects on ecosystems. Table 1 shows the persistence in the environment of common pesticides. The implementation of more and more stringent environmental regulations on hazardous wastes has impelled the necessity of finding innovative and environmentally friendly treatment technologies to complement or substitute the conventional ones. The development of enzymes-based processes seems an appalling alternative. From an environmental point of view, the use of enzymes instead of chemicals or micro-organisms presents the following advantages (Ahuja et al., 2004): the bio-transformation does not generate toxic sub-products, as often occurs with chemical and some microbiological processes, and the enzymes are digested *in situ* by the native micro-organisms after the treatment; the requirement to enhance bio-availability by the introduction of organic co-solvents or surfactants is much more feasible from an enzymatic point of view than using whole cells; the potential to produce enzymes at a higher scale with enhanced stability and/or activity and at a lower cost by using recombinant-DNA technology.

Table 1. Pesticide persistence in the environment (Alikhanidi and Takahashi, 2004)

Low persistence (half-life ≤30 days)	Moderate persistence (30 >half-life ≤100 days)	High persistence (half-life > 100 days)
aldicarb	athrazine	aldrin
azinphos-methyl	carbophenothion	benomyl
barban	diazinon	bentazon
captan	fonofos	bromacil acid
chlorpyrifos	linuron	chlordane
cresol	parathion	cyprodinil
dalapon	permethrin	DDT
2,4-DB acid	phorate	dieldrin
dicamba	picloram	diuron
dichlorvos	pirimicarb	lindane
disulfoton	simazine	methoxychlor
DNOC		paraquat
MCPA		pentachlorophenol
1-naphtol		terbacil
oxamyl		trifluralin*
rotenone		

2,4-DB acid = 4-(2,4-dichlorophenoxy)butyric acid.
DDT = 1,1,1-trichloro-2,2-bis (4-chlorophenyl) ethane.
DNOC = 2-methyl-4,6-dinitrophenol.
MCPA = 4-chloro-2-methylphenoxyacetic acid.

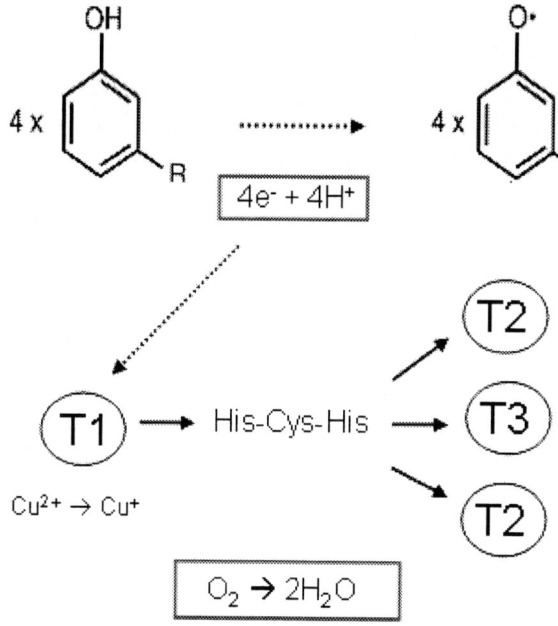

Figure 1. Catalytic cycle of laccase (adapted from Badrian, 2006). Note the laccase four copper ions classified into three types: T1, T2, and T3. The T1 copper is responsible for the blue colour of the enzyme while T2 copper can be detected by EPR. The bi-nuclear T3 copper is diamagnetic and displays a characteristic fluorescence spectrum. The T1 copper centre is involved in the oxidation of the reducing substrate and the generated electrons are then transferred back to the T2 and T3 copper centres, where molecular oxygen is reduced to water.

Among the different existing oxidative enzymes, laccases (*p*-diphenol:dioxygen oxidoreductases; EC 1.10.3.2) are the most interesting ones because they only use dissolved molecular oxygen for their catalytic activity. Additionally, laccases might be produced *via* the use of less costly growth substrates (e.g. wastewater, agricultural and food wastes) that provide low-cost sources of enzymes (Ikehata et al., 2004).

Laccases belong to the group of blue multi-copper oxidases. They catalyse the oxidation of a great variety of aromatic compounds, such as diphenols, arylamines or aminophenols, with concomitant reduction of molecular oxygen to water. Laccases are widespread in nature, so they have been found in fungi, plants, some bacteria (Thurston, 1994) and insects (Barrett, 1991). They have a broad substrate specificity, which can be extended to non-phenolic aromatic compounds with the use of redox mediators. Thus, in the presence of some synthetic and natural mediators, laccases can oxidise compounds such as veratryl and benzyl alcohols, non-phenolic groups of the lignin polymer and lignin model substances (Bollag and Leonowicz, 1984; Xu et al., 1997; Johannes and Majcherczyk, 2000). The full range of laccase substrates is not known; even less known are substrates that can be oxidised in the presence of mediators. Lu and Xia (2004) have recently reviewed the applications of the laccase-mediated systems (LMSs), which comprise pulp bleaching, textile bio-finishing and environmental protection processes. However, despite the LMS has been studied extensively there are still unsolved problems related to mediator recycling, cost and toxicity. More recently, Camarero et al. (2007) have proposed the use of natural mediators to solve the above-mentioned problems. Thus, they used three plant phenols, syringaldehyde, acetosyringone and *p*-coumaric acid, as natural mediators for delignification of paper pulp by laccase. Syringaldehyde and acetosyringone led to very promising results although the delignification increase and the improvement of pulp properties were lower than those obtained with the synthetic mediator *N*-hydroxybenzotriazole (HBT) but the pulp treatment conditions have still to be optimised, as previously done for HBT.

Table 2. Chemical structure of several pesticides

Pesticide	**Chemical structure**
Atrazine	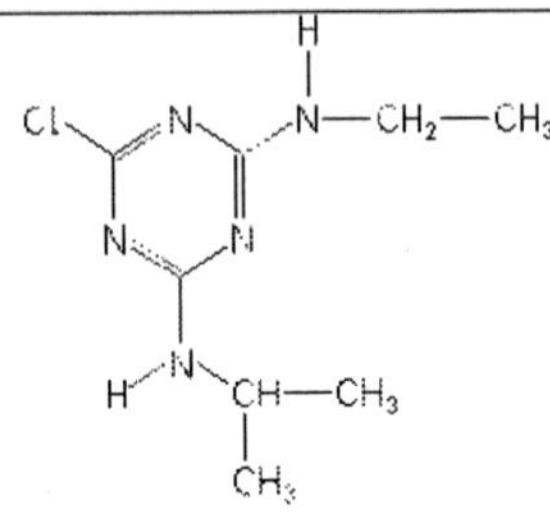
Cyprodinil	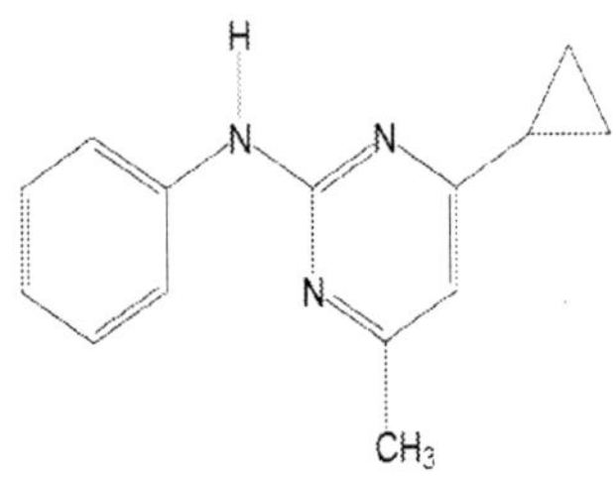
DDT	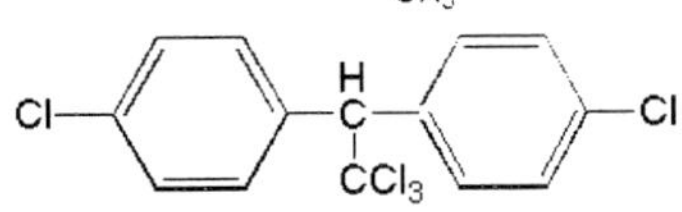
Dieldrin	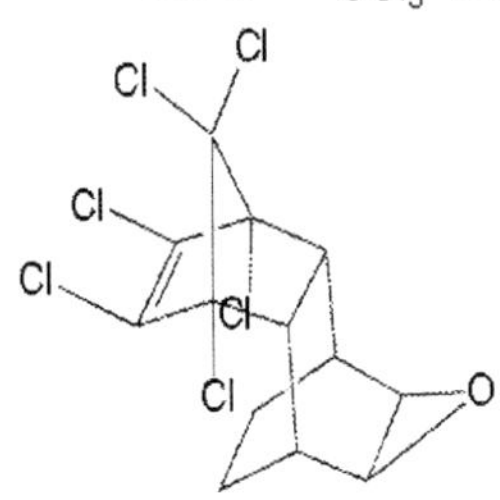
Diketonitrile	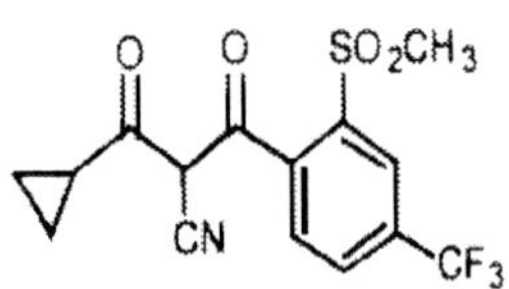
Dymron	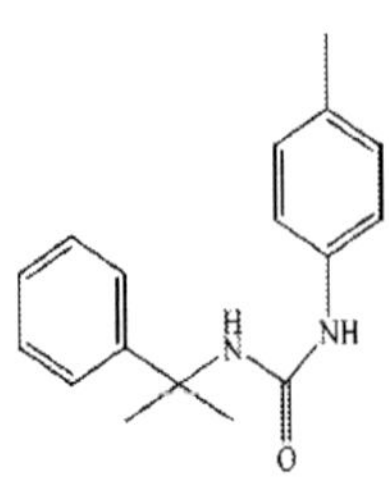

Imazalil

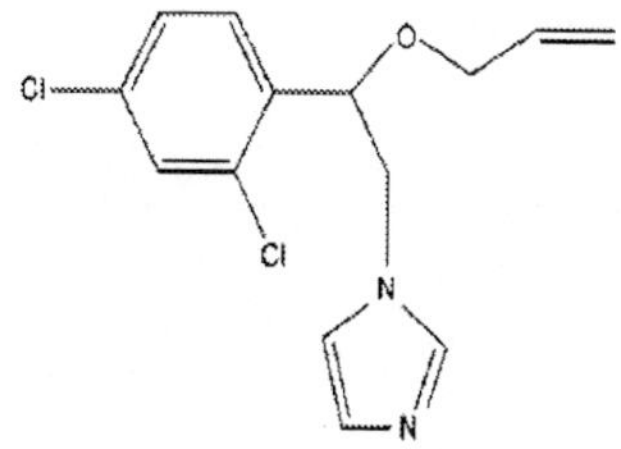

Isoxaflutole

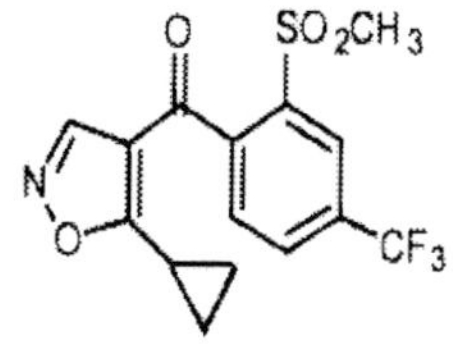

Methoxychlor

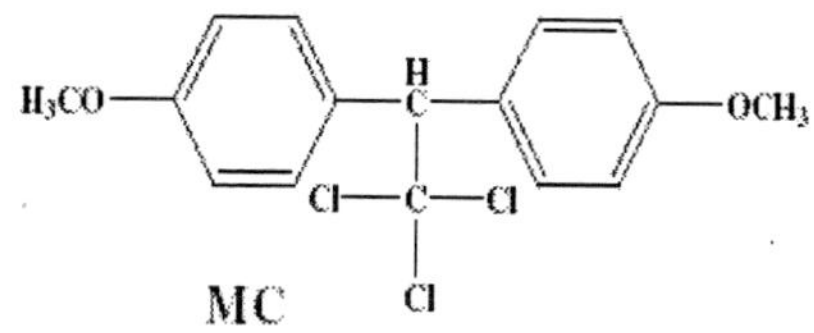

Pentachlorophenol

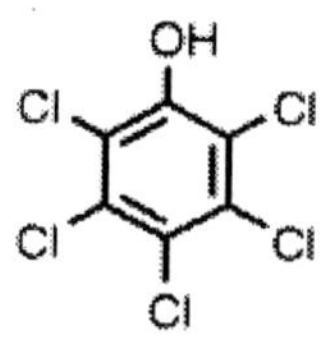

Simazine

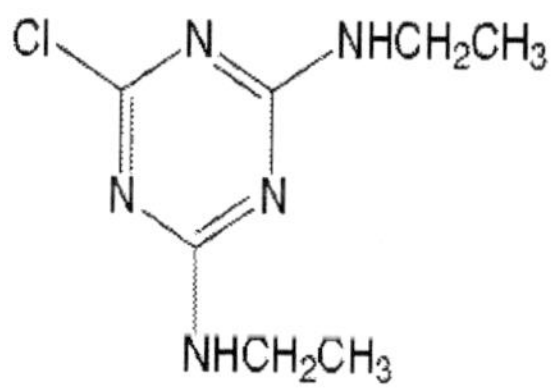

Trifluralin

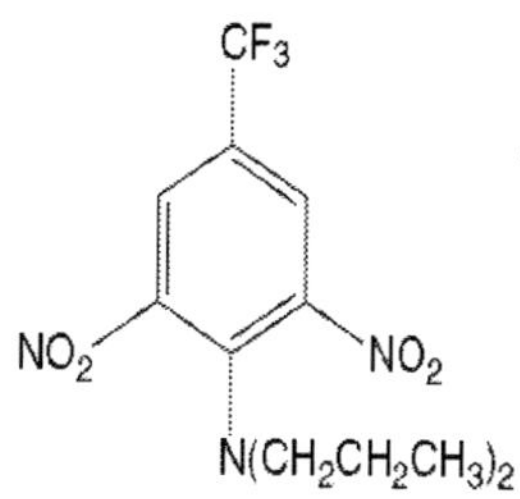

Laccases contain four copper ions classified into three types, referred to as T1, T2, and T3 (Yaropolov et al., 1994; Solomon et al., 1996). The T1 copper is responsible for the blue colour of the enzyme and has a characteristic absorbance around 610 nm. The T2 copper cannot be detected spectrophotometrically; however, it generates a characteristic electron paramagnetic resonance (EPR) signal (Solomon et al., 1996, Leontievsky et al., 1997, Koroljova-Skorobogat'ko, 1998). The bi-nuclear T3 copper is diamagnetic. It displays a spectral absorbance shoulder in the region of 330 nm and also displays a characteristic fluorescence spectrum (Shin and Lee, 2000). The T1 copper centre is involved in the oxidation of the reducing substrate and the generated electrons are then transferred back to the T2 and T3 copper centres, where molecular oxygen is reduced to water (Messerschmidt, 1997) (Figure 1).

So far, the three-dimensional structure of five fungal laccases has been reported: *Coprinus cinereus* (in a copper type-2-depleted form) (Ducros et al., 1998), *Trametes versicolor* (Bertrand et al., 2002; Piontek et al., 2002), *Pycnoporus cinnabarinus* (Antorini et al., 2002), *Melanocarpus albomyces* (Hakulinen et al., 2002) and *Rigidoporus lignosus* (Garavaglia et al., 2004), the latter four enzymes with a full complement of copper ions. Moreover, the three-dimensional structure of the CotA laccase from *Bacillus subtilis* endospores has also been published (Enguita et al., 2003, 2004). Despite the amount of information on laccases neither the precise electron transfer pathway nor the details of dioxygen reduction in laccases are fully understood (Garavaglia et al., 2004).

At present, the main technological applications of laccases are in the textile, dye or printing industries – in processes related to decolouration of dyes (Claus et al., 2002) – and in the pulp and paper industries – for the delignification of woody fibres, particularly during the bleaching process (Bajpai, 1999; Leonowicz et al., 2001). The purpose of this paper is to review the use of laccases in pesticide degradation. Table 2 shows the chemical structure of several pesticides.

Laccases in Pesticide Degradation

Despite the enormous potential of laccases to degrade different recalcitrant pollutants such as polycyclic aromatic hydrocarbons (PAHs) (Pozdnyakova et al., 2006), polychlorinated biphenyls (PCBs) (Keum and Li, 2004), synthetic dyes (Soares et al., 2002; Kandelbauer et al., 2004; Rodríguez Couto et al., 2005; Mechichi et al., 2006; Osma et al., 2007; Rodríguez Couto, 2007), endocrine disrupting chemicals (EDCs) (Auriol et al., 2007; Cabana et al, 2007), etc, there are few studies related to degradation of pesticides by laccases (Table 3).

Bollag et al. (1977) found that the laccase from the fungus *Rhizoctonia praticola* polymerised the phenolic intermediates of various pesticides. Thus, the enzyme catalysed the formation of oligomeric products from halogenated phenolic intermediates of phenoxyalkanoate herbicides and from naphtholic products derived from carbamate insecticides.

Sariaslani et al. (1984) showed that the laccase from *Polyporus anceps* plus the redox mediator chlorpromazine, transformed rotenone to a single, more polar product. Different analyses indicated that rotenone was transformed to 6aβ, 12aβ-rotenolone.

Dawel et al. (1997) described a model for coupling of 2,4,6-trinitrotoluene (TNT) metabolites to a humic organic soil matrix by a commercial laccase from *Pyricularia oryzae* (Sigma Aldrich Chemie GmbH, Deisenhofen, Germany). Thus, they showed the coupling of 2,4-diamino-6-nitrotoluene (2,4-DANT) to a guaiacol dimer. Recently, Nyanhongo et al. (2006) found that during degradation of trinitrotoluene (TNT) by a laccase from *Trametes modesta*, addition of humic monomers prevented the accumulation of all major stable TNT metabolites (aminodinitrotoluenes (AMDNT)) by at least 92%. Although all humic monomers oxidation products were able to co-precipitate the different TNT metabolites, ferulic acid was much more effective in binding TNT metabolites in both *in vitro* and in *T. modesta* cultures and, in addition, its products were less toxic. Apart from 4-hydroxylaminodinitrotoluene (4-HADNT) all the other TNT metabolites were less toxic than the parent TNT compound. It is assumed that the immobilised TNT metabolites underwent covalent binding to humic monomers and are, thus, not easily removed.

Table 3. Pesticide degradation by laccases from different sources

Laccase	Pesticide	Reference
Rhizoctonia praticola	phenoxyalkanoate, carbamates	Bollag (1977)
Polyporus anceps	rotenone	Sariaslani et al. (1984)
Commercial from Pyricularia oryzae *(Sigma Aldrich Chemie GmbH, Deisenhofen, Germany)*	TNT metabolites	Dawel et al. (1997)
P. ostreatus	diisopropyl-Amiton	Amitai et al. (1998)
Polyporus pinsitus	bentazon	Kim et al. (1998)
T. versicolor	phenylurea	Jolivalt et al. (1999)
T. versicolor	phenylurea	Jolivalt et al. (2000)
Phanerochaete chrysosporium *and* T. versicolor	diketonitrile(derivated from isoxaflutole)	Mougin et al. (2000)
Coriolus versicolor	pentachlorophenol	Ullah et al. (2000)
Commercial (Novo Nordisk, Franklinton, NC)	cresols	Vermete (2000)
Trametes villosa	cyprodinil	Kang et al. (2002)
Sub-tropical white-rot fungus DSPM95*	lindane	Tekere et al. (2002)
T. versicolor	methoxychlor	Hirai et al. (2004)
Formes sclerodermeus	Malachite Green	Papinutti and Forchiassin (2004)
9 New Zealand white-rot fungi	pentachlorophenol	Walter et al. (2004)
T. versicolor*	pesticide mixture (simazine, trifluralin and dieldrin)	Fragoeiro (2005)
*T. versicolor**	pesticide mixture (simazine, trifluralin and dieldrin)	Fragoeiro and Magan (2005)
*Pleurotus ostreatus**	lindane	Rigas et al. (2005)
Tobacco plants*	pentachlorophenol	Sonoki et al. (2005)
New Zealand T. versicolor *isolate**	pentachlorophenol	Walter et al. (2005)
Coriolus hirsutus, Coriolopsis fulvocinerea, Cerrena maxima*	atrazine	Gorbatova et al. (2006)
T. versicolor	phenylurea	Jolivalt et al. (2006)
Trametes sp.	dymron	Maruyama et al. (2006)
Trametes modesta	TNT	Nyanhongo et al. (2006)
Commercial from *Trametes sp.* (Daiwa Kasei Co Ltd, Japan)	imazalil	Maruyama et al. (2007)

* Degradation in the presence of the microorganism.

Amitai et al. (1998) showed that a purified laccase from the white rot fungus *Pleurotus ostreatus* plus the redox mediator 2,2'-azinobis(3-ethylbenzthiazoline-6-sulfonate) (ABTS) completely degraded the insecticide diisopropyl-Amiton.

Kim et al. (1998) showed that the herbicide bentazon (3-isopropyl-1*H*-2,1,3-benzothiadiazine-4(3*H*)-one-2,2-dioxide) and several of its less reactive metabolites underwent negligible transformation when incubated with laccase unless reactive humic monomers such as guaiacol or ferulic acid were present. Thus, at pH 3.0 and in the presence of laccase and guaiacol, the concentrations of bentazon and its metabolites 2-amino-N-isopropyl-benzamide, des-isopropyl-bentazon and 8-chloro-bentazon decreased by 27, 57, 20 and 4%, respectively.

Jolivalt, et al. (1999) reported the transformation of of *N′*,*N′*-dimethyl-*N*-(hydroxy-phenyl)ureas (a metabolite of phenylurea herbicide) by purified laccase of *T. versicolor*. The reaction was highly dependent on the pH of the medium. Thus, the chemical was exclusively oxidised into *p*-benzoquinone at pH 3 and into mainly *N′*,*N′*-dimethyl-*N*-[(2,5-cyclo-hexadiene-1-one)-4-ylidene]urea at pH 6. Subsequently (Jolivalt et al., 2000), they reported the immobilisation of crude laccase from *T. versicolor* onto a hydrophilic PVDF microfiltration membrane by covalent linkage and its application for removing a herbicide derivative, *N'*,*N'*-(dimethyl)-*N*-(2-hydroxyphenyl)urea (2-HF), from wastewater. 2-HF was transformed *via* an oxidation reaction catalysed by laccase mostly into an insoluble product, which was simultaneously separated by filtration through the membrane. It was shown that a membrane with a 60cm^2 effective filtration area allowed to transform 200 mL of a 2-HF solution (0.1 g/L) in less than 5 min.

Mougin et al. (2000) showed that a purified laccase from *Trametes versicolor* was able to cleave diketone compounds (derivatives from the herbicide isoxaflutole). Thus, laccase in the presence of ABTS (2 mM) and at pH 3 cleaved the diketone derivatives at a rate of 0.46 nmol/h unit.

Ullah et al. (2000) found that laccase, purified from *Coriolus versicolor*, removed pentachlorophenol (PCP) from solution at pH 5, depending on initial PCP concentration and amount of laccase. Thus, with 100 units of laccase, 100% of 25 μg/mL PCP and 60% of 200 μg/mL PCP were removed, respectively, over 72 h. No free chloride was released in the reaction. Products were mainly polymers, which were easily removed from solution due to their high molecular weight.

Vermette (2000) reported the degradation of cresols from wastewater by a commercial laccase (Novo Nordisk, Franklinton, NC). Under optimal conditions, nearly 90% of *o*-cresol, 80% of *p*-cresol and 70% of *m*-cresol were removed.

Kang et al. (2002) investigated the transformation of cyprodinil (4-cyclopropyl-6-methyl-*N*-phenyl-2-pyrimidinamine), a broadspectrum fungicide, by a laccase from *Trametes villosa* plus a mediator. No transformation of cyprodimil occurred in the absence of the mediator.

Tekere et al. (2002) studied the degradation of the organochlorine pesticide lindane by a sub-tropical white-rot fungus in stationary batch and packed-bed bioreactor systems. Biodegradation of 82 % was achieved in batch cultures at pesticide concentrations of 5 and 10 mg/L. The highest percent degradation achieved in the packed-bed reactor was 81% at lindane concentrations of 1 and 2 mg/L. Also, recently Rigas et al. (2005) studied the degradation of lindane in liquid-agitated cultures of *Pleurotus ostreatus*. The biodegradation was accomplished with the action of extracellular oxidative enzymes produced by the fungus to decompose woody substrates. The maximum biodegradation of lindane, expressed as the extent of biodegradation relative to initial lindane mass and to final biomass was of 25.8 mg/g/g (degraded lindane/initial lindane/biomass) in 12.45 days for an initial lindane concentration of 2.03 mg/L.

Hirai et al. (2004) reported for the first time the degradation of the pesticide methoxychlor (MC) and its metabolite, methoxychlor olefin (MCO), by laccase-HBT treatment. *In vivo*, MC is converted to hydroxy metabolites (e.g., 1,1,1-trichloro-2,2-bis(*p*-hydroxyphenyl)ethane), which have much higher estrogenic activities than MC does (Welch et al., 1969; Bulger et al., 1978). No hydroxylated metabolite of MC was detected in reaction mixtures after laccase-HBT treatment.

Papinutti and Forchiassin (2004) found that *Fomes sclerodermeus* was able to decolourise and detoxify the fungicide dye Malachite Green (MG). In addition, MG treated with supernatants containing high laccase activity in the presence or absence of HBT gave a colourless product , decolourised Malachite Green (DMG), that was not toxic to *P. chrysosporium* and other white-rot fungi tested.

Walter et al. (2004) studied the PCP biodegradation potential of nine New Zealand native white-rot fungi. They showed the capacity of New Zealand fungal strains to degrade and mineralise PCP and the potential of New Zealand native *T. versicolor* isolates for remediation of PCP polluted soil sites stemming from past industry practices. However, the conditions of the study were quite different to those that might be expected in a field remediation situation, both with respect to PCP concentration, and fungal growth conditions. Therefore, scale-up and the effect of pollutant concentration on PCP degradation need to be further clarified to determine engineering requirements and the concentration threshold, where pollutant concentration impedes degradation processes. More recently, Walter et al. (2005) showed that the *T. versicolor* isolate could biodegrade PCP from aged soils in a field situation.

Fragoeiro (2005) showed that the white-rot fungus *T. versicolor* produced high levels of laccase in soil in the presence of a pesticide mixture (dieldrin, trifluralin and simazine). In addition, the pesticides were degraded by 46, 57 and 51%, respectively. These results are very interesting, since there is little information on the degradation of mixtures of pesticides, which in environmental conditions are more common than contamination with single pesticides. Also, Fragoeiro and Mahan (2005) found that *T. versicolor* had the ability to degrade different groups of pesticides (i.e. dieldrin, trifluralin and simazine) due to its capacity for secreting laccase enzyme even under an osmotic stress of -2.8 MPa. Complete degradation of dieldrin and trifluralin was observed, while about 80% of the simazine was degraded.

Sonoki et al. (2005) introduced a cDNA encoding laccase of *C. versicolor* into the genome of tobacco (*Nicotiana tabacum* Sumsun NN) plants. One transgenic plant, designated FL4, produced laccase that was secreted into the rhizosphere. FL4 was able to remove 20 μmol PCP per gram dry weight. Therefore, plants producing a fungal laccase provide useful tools for the degradation and detoxification of environmental pollutants.

Gorbatova et al. (2006) studied the effect of oxidoreductase inducers guaiacol and syringaldazine on the ability of *Coriolus hirsutus*, *Coriolopsis fulvocinerea*, *Cerrena maxima* and co-cultivated *C. hirsutus*/*C. maxima* to degrade atrazine in submerged cultures. All the basidiomycetes reduced atrazine concentration with and without syringaldazine or guaiacol, although atrazine degradation was higher in the induced cultures (77-98% *vs* 68-94% without induction). Of the four cultures, the highest detoxifying potential was observed in *C. fulvocinerea* with and without an inducer.

Jolivalt et al. (2006) investigated the ability of laccases, produced by the white-rot fungus *T. versicolor*, to catalyse *in vitro* the transformation of five hydroxyphenylureas (HFs). HFs are the first main metabolites formed in the environment from pesticide and biocide urea compounds. Their results established that the pH of the reaction had a strong influence on both the kinetics of the reaction and the nature of the transformation products. Thus, they showed that laccases oxidised the HFs to quinones at pH 3. By contrast, at higher pH values, these compounds were further transformed into oligomers by chemical processes. The polymerisation process is of special importance for the environmental fate of HF derivatives. First, when the oxidation coupling reaction occurs at the chlorinated site of the substrate the induced dehalogenation contributes to the overall detoxification effect, since it is generally recognised that toxicity decreases after dechlorination. Second, as the mechanism of covalent bond formation involves the addition of a nucleophile, leading to the formation of heterodimers, the same mechanism is likely to take place between quinones resulting from the two electron oxidation of HFs and humic acids present in soils. Potentially toxic quinones formed from laccase-catalysed oxidation of HF derivatives are, thus, likely to be bound to the organic matter of soils losing their bio-availability and, as a consequence, their ecotoxicity.

Maruyama et al. (2006) reported the effective degradation of the persistent herbicide dymron catalysed by laccase plus ABTS. It was found that about 90% of dymron was degraded within 24 h, while the half-life of dymron is 50 days in soil. In addition, analysis of the degradation products suggested that oxidation of dymron produced compounds with a single aromatic ring. Moreover, the selections of the reaction medium and the mediator allowed the degradation of dymron in mild conditions without negative effects on the ecological system.

Maruyama et al. (2007) reported the degradation of the postharvest fungicide imazalil by a commercial laccase from *T. versicolor* (Daiwa Kasei Co Ltd, Japan) in the presence of different natural redox mediators (4-hydroxybenzoic acid (HBA), ethyl 4-hydroxybenzoate (EHB), methyl 4-hydroxybenzoate (MHB), reduced glutathione (GSH), cysteine (Cys), methionine, 4- hydroxybenzaldehyde (HBal) and oxidised glutathione (GSSG)). Imazalil was degraded most effectively in the presence of HBA while the absence of a mediator resulted in no degradation.

OUTLOOK

Laccases are promising enzymes to degrade pesticides present in soil and water, although the addition of redox mediators is required in most cases for degradation to occur. Problems related to cost and toxicity of synthetic redox mediators have to be solved before their industrial implementation. Further investigations should consider different and less polluting mediators such as the natural mediators produced by laccase in a bio-environment during lignin degradation.

On the other hand, the development of an effective system for laccase immobilisation is worthy since it would enhance its stability and it would allow its reuse, easy separation from the reaction mixture and the prevention of enzyme contamination in products. In addition, immobilisation would protect laccases from deactivating agents usually encountered in polluted sites. We are currently working on new strategies to immobilise laccase (Rodríguez Couto et al., 2007), such as the use of Al_2O_3 pellets as an immobilisation matrix followed by the sequential adsorption of oppositely charged polyelectrolytes (layer by layer technique). We have already shown that the immobilisation of laccase improves its stability properties with time. Several questions are still open, namely, the efficiency of immobilised laccase as a function of temperature and also its reuse. Moreover, all this will considerably reduce the process cost.

ACKNOWLEDGMENTS

SRC and JLTH are *Ramón y Cajal Senior Research Fellows*. The authors thank the Spanish Ministry of Education and Science for promoting the *Ramón y Cajal Programme.*

REFERENCES

Ahuja, S.K., Ferreira, G.M. & Moreira A.R. (2004) Utilization of enzymes for environmental applications. *Crit Rev Biotechnol* 24, 125–154.

Alikhanidi, S. & Takahashi, Y. (2004). Pesticide Persistence in the Environment - Collected Data and Structure-Based Analysis. *J Comput Chem Jpn* 3, 59-70.

Amitai, G., Adani, R., Sod-Moriah, G., Vincze, A., Leader, H., Chefetz, B., Leibovitz-Persky, L., Friesem, D. & Hadar, Y. (1998). Oxidative biodegradation of phosphorothiolates by fungal laccase. *FEBS Letters* 438, 195-200.

Antorini, M., Herpoel-Gimbert, I., Choinowski, T., Sigoillot, J.C., Asther, M., Winterhalter, K. & Piontek, K. (2002). Purification, crystallisation and X-ray diffraction study of fully functional laccases from two ligninolytic fungi. *Biochim Biophys Acta* 1594, 109-114.

Auriol M, Filali-Meknassi Y, Tyagi RD., Adams C D. (2007). Laccase-catalyzed conversion of natural and synthetic hormones from a municipal wastewater. *Water Res.* In Press.

Bajpai, P. (1999). Application of enzymes in the pulp and paper industry. *Biotechnol Prog* 15, 147-157.

Baldrian, P. (2006). Fungal laccases-occurrence and properties. *FEMS Microbiol Rev* 30, 215-242.

Barrett, F.M. In: Binnigton K., Retnakaram A. editors. Physiology of the Insect Epidermis. CISRO, East Melbourne, Australia, 1991; 194-196.

Bertrand, T., Jolivalt, C., Briozzo, P., Caminade, E., Joly, N., Madzak, C. & Mougin, C. (2002). Crystal structure of a four-copper laccase complexed with an arylamine: Insights into substrate recognition and correlation with kinetics. *Biochemistry* 41, 7325-7333.

Bollag, J.M. & Leonowicz A. (1984). Comparative studies of extracellular fungal laccases. *Appl Environ Microb* 48, 849-854.

Bollag, J.M., Sjobald, R.D. & Minard, R.D. (1977). Polymerization of phenolic. *Experientia* 33, 1564-1566.

Bulger, W.H., Muccitelli, R.M. & Kupfer, D. (1978). Studies on the in vivo and in vitro estrogenic activities of methoxychlor and its metabolites. Role of hepatic mono-oxygenase in methoxychlor activation. *Biochem Pharmacol* 27, 2417-2423.

Cabana, H., Jiwan, J.L.H., Rozenberg, R., Elisashvili, V., Penninckx, M., Agathos, S.N. & Jones, J.P. (2007). Elimination of endocrine disrupting chemicals nonylphenol and bisphenol A and personal care product ingredient triclosan using enzyme preparation from the white rot fungus *Coriolopsis polyzona*. *Chemosphere* 67, 770-778.

Camarero, S., Ibarra, D., Martínez, A.T., Romero, J., Gutiérrez, A. & del Río, J.C. (2007). Paper pulp delignification using laccase and natural mediators. *Enzyme Microb Tech* 40, 1264-1271.

Claus, H., Faber, G. & Konig, H. (2002). Redox-mediated decolorization of synthetic dyes by fungal laccasse. *Appl Microbiol Biotechnol* 59, 672-678.

Dawel, G., Kastner, M., Michels, J., Poppitz, W., Günther, W. & Fritsche, W. (1997). Structure of a laccase-mediated product of coupling of 2,4-diamino-6-nitrotoluene to guaiacol, a model for coupling of 2,4,6-trinitrotoluene metabolites to a humic organic soil matrix. *Appl Environ Microb* 63, 2560-2565.

Ducros, V., Brzozowski, A.M., Wilson, K.S., Brown, S.H., Ostergaard, P., Schneider, P., Yaver, D.S., Pedersen, A.H. & Davies, G.J. (1998). Crystal structure of the type-2 Cu depleted laccase from *Coprinus cinereus* at 2.2 angstrom resolution. *Nat Struct Biol* 5, 310-316.

Enguita, F.J., Marcal, D., Martins, L.O., Grenha, R., Henriques, A.O., Lindley, P.F. & Carrondo, M.A. (2004). Substrate and dioxygen binding to the endospore coat laccase from *Bacillus subtilis*. *J Biol Chem* 279, 23472-23476.

Enguita, F.J., Martins, L.O., Henriques, A.O. & Carrondo, M.A. (2003). Crystal structure of a bacterial endospore coat component – a laccase with enhanced thermostability properties. *J Biol Chem* 278, 19416-19425.

Fragoeiro, S. & Magan, N. (2005). Enzymatic activity, osmotic stress and degradation of pesticide mixtures in soil extract liquid broth inoculated with *Phanerochaete chrysosporium* and *Trametes versicolor*. *Environmental Microbiology* 7, 348-355.

Fragoeiro, S.I. (2005). Use of fungi in bioremediation of pesticides. Ph.D. Thesis. Cranfield University, Bedford, UK.

Garavaglia, S., Cambria, M.T., Miglio, M., Ragusa, S., Lacobazzi, V., Palmieri, F., D'Ambrosio, C., Scaloni, A. & Rizzi, M. (2004) The structure of *Rigidoporus lignosus* laccase containing a full complement of copper ions, reveals an asymmetrical arrangement for the T3 copper pair. *J Mol Biol* 342, 1519-1531.

Gorbatova, O.N., Koroleva, O.V., Landesman, E.O., Stepanova, E.V. & Zherdev, A.V. (2006). Increase of the detoxification potential of basidiomycetes by induction of laccase biosynthesis. *Prikladnaia Biokhimiia i Mikrobiologiia* 42, 468-474.

Hakulinen, N., Kiiskinen, L.L., Kruus, K., Saloheimo, M., Paananen, A., Koivula, A. & Rouvinen, J. (2002). Crystal structure of a laccase from *Melanocarpus albomyces* with an intact trinuclear copper site. *Nature Struct Biol* 9, 601-605.

Hirai, H., Nakanishi, S. & Nishida, T. (2004). Oxidative dechlorination of methoxychlor by ligninolytic enzymes from white-rot fungi. *Chemosphere* 55, 641-645.

Ikehata, K., Buchanan, I.D. & Smith, D.W. (2004). Recent developments in the production of extracellular fungal peroxidases and laccases for waste treatment. *J Environ Eng Sci* 3, 1-19.

Johannes, C. & Majcherczyk, A. (2000). Natural mediators in the oxidation of polycyclic aromatic hydrocarbons by laccase mediator systems. *Appl Environ Microb* 66, 524-528.

Jolivalt, C., Brenon, S., Caminade, E., Mougin, C. & Pontie, M. (2000). Immobilization of laccase from Trametes versicolor on a modified PVDF microfiltration membrane: characterization of the grafted support and application in removing a phenylurea pesticide in wastewater. *J Membrane Sci* 180, 103-113.

Jolivalt, C., Neuville, L., Boyer, F.D., Kerhoas, L. & Mougin, C. (2006). Identification and Formation Pathway of Laccase-Mediated Oxidation Products Formed from Hydroxyphenylureas. *J Agric Food Chem* 54, 50465054.

Jolivalt, C., Raynal, A., Caminade, E., Kokel, B., Le Goffic, F. & Mougin, C. (1999). Transformation of *N′*,*N′*-dimethyl-*N*-(hydroxyphenyl) ureas by laccase from the white rot fungus*Trametes versicolor*. *Appl Microbiol Biotechnol 51*, 676-681.

Kandelbauer, A., Erlacher, A., Cavaco-Paulo, A. & Gübitz, G.M. (2004). Laccase-catalyzed Decolorization of the Synthetic Azo-dye Diamond Black PV 200 and of some Structurally Related Derivatives. *Biocatalysis and Biotransformation* 22, 331-339.

Kang, K.H., Dec, J., Park, H. & Bollag, J.M. (2002). Transformation of the fungicide cyprodinil by a laccase of *Trametes villosa* in the presence of phenolic mediators and humic acid. *Water Res* 36, 4907-4915.

Keum, Y.S, & Li, Q.X. (2004). Fungal laccase-catalyzed degradation of hydroxy polychlorinated biphenyls. *Chemosphere* 56, 23-30.

Kim, J.E., Wang, C.J. & Bollag, J.M. (1998). Interaction of reactive and inert chemicals in the presence of oxidoreductases: Reaction of the herbicide bentazon and its metabolites with humic monomers. *Biodegradation* 8, 387-392.

Koroljova-Skorobogat'ko, O.V., Stepanova, E.V., Gavrilova, V.P., Morozova, O.V., Lubimova, N.V., Dzchafarova, A.N., Jaropolov, A.I. & Makower, A. (1998). Purification and characterization of the constitutive form of laccase from the basidiomycete *Coriolus hirsutus* and effect of inducers on laccase synthesis. *Biotechnol Appl Biochem* 28, 47-54.

Leonowicz, A., Cho, N.S., Luterek, J., Wilkolazka, A., Wojtaswasilewska, M., Matuszewska, A., Hofrichter, M., Wesenberg, D, & Rogalski, J. (2001). Fungal laccase: properties and activity on lignin. *J Basic Microbiol* 41, 185-227.

Leontievsky, A.A., Vares, T., Lankinen, P., Shergill, J.K., Pozdnyakova, N.N., Myasoedova N.M., Kalkkinen, N., Golovleva, L.A., Cammack, R., Thurston, C.F. & Hatakka, A. (1997). Blue and yellow laccases of ligninolytic fungi, *FEMS Microbiol Lett* 156, 9-14.

Lu, R. & Xia, L.M. (2004). Advances in research and application of laccase mediator system. *Xianweisu Kexue Yu Jishu* 12, 37-44.

Maruyama, T., Komatsu, C., Michizoe, J., Ichinose, H. & Goto, M. (2006). Laccase-Mediated Oxidative Degradation of the Herbicide Dymron. *Biotechnol Progr* 22, 426-430.

Maruyama, T., Komatsu, C., Michizoe, J., Sakai, S. & Goto, M. (2007). Laccase-mediated degradation and reduction of toxicity of the postharvest fungicide imazalil. *Process Biochem* 42, 459-461.

Mechichi, T., Mhiri, N., & Sayadi, S. (2006). Remazol Brilliant Blue R decolourization by the laccase from *Trametes trogii*. *Chemosphere* 64, 998-1005.

Messerschmidt, A. (Ed.) (1997). Multi-Copper Oxidases. World Scientific, Singapore.

Mougin, C., Boyer, F.D., Caminade, E. & Rama, R. (2000). Cleavage of the diketonitrile derivative of the herbicide isoxaflutole by extracellular fungal oxidases. *J Agric Food Chem* 48, 4529-4534.

Nyanhongo, G.S., Rodríguez Couto, S. & Guebitz, G.M. (2006). Coupling of 2,4,6-trinitrotoluene (TNT) metabolites onto humic monomers by a new laccase from *Trametes modesta*. *Chemosphere* 64, 359-370

Osma, J.F., Toca Herrera, J.L. & Rodríguez Couto, S. (2007). Banana skin: A novel waste for laccase production by *Trametes pubescens* under solid-state conditions. Application to synthetic dye decolouration. *Dyes and Pigments* 75, 32-37.

Papinutti, V. & Forchiassin, F. (2004). Modification of malachite green by *Fomes sclerodermeus* and reduction of toxicity to *Phanerochaete chrysosporium*. *FEMS Microbiology Letters* 231, 205-209.

Piontek, K., Antorini, M. & Choinowski, T. (2002). Crystal structure of a laccase from the fungus *Trametes versicolor* at 1.90- angstrom resolution containing a full complement of coppers. *J Biol Chem* 277, 37663-37669.

Pozdnyakova, N.N., Rodakiewicz-Nowak, J., Turkovskaya, O.V. & Haber, J. (2006). Oxidative degradation of polyaromatic hydrocarbons catalyzed by blue laccase from *Pleurotus ostreatus* D1 in the presence of synthetic mediators. *Enzyme Microb Tech* 39, 1242-1249.

Rigas, F., Marchant, V.D.R., Papadopoulou, K., Avramides, E.J. & Hatzianestis, I. (2005). Biodegradation of lindane by *Pleurotus ostreatus* via central composite design. *Environment International* 31, 191-196

Rodríguez Couto S. (2007). Decolouration of industrial azo dyes by crude laccase from *Trametes hirsuta. Journal of Hazardous Materials. In Press* (available at doi:10.1016/j.jhazmat.2007.06.123)

Rodríguez Couto, S., Osma, J.F., Saravia, V., Gübitz, G. & Toca-Herrera, J.L. (2007). Coating of immobilised laccase for stability enhancement: a novel approach. *Applied Catalysis A General* 329, 156-160.

Rodríguez Couto, S., Sanromán, M.A. & Gübitz G.M. (2005). Influence of redox mediators and metal ions on synthetic acid dye decolourization by crude laccase from *Trametes hirsuta*. *Chemosphere* 58, 417-422.

Sariaslani, F.S., Beale Jr., J.M. & Rosazza, J.P. (1984). Oxidation of rotenone by *Polyporus anceps* laccase. *Journal of natural products* 47, 692-697.

Shin, K.S. & Lee, Y.J. (2000). Purification and characterization of a new member of the laccase family from the white-rot basidiomycete *Coriolus hirsutus*. *Arch Biochem Biophys* 384, 109-115.

Soares, G.M.B., Pessoa Amorim, M.T., Hrdina R., & Costa-Ferreira, M. (2002). Studies on the biotransformation of novel disazo dyes by laccase. *Process Biochem* 37, 581-587.

Solomon, E.I., Sundaram, U.M. & Machonkin, T.E. (1996). Multicopper oxidases and oxygenases. *Chem Rev* 96, 2563-2605.

Sonoki, T., Kajita, S., Ikeda, S., Uesugi, M., Tatsumi, K., Katayama, Y. & Limura, Y. (2005). Transgenic tobacco expressing fungal laccase promotes the detoxification of environmental pollutants. *Appl Microbiol Biotechnol* 67, 138-142.

Tekere, M., Ncube, I., Read, J. & Zvauya, R. (2002). Biodegradation of the organochlorine pesticide, lindane by a sub-tropical white rot fungus in batch and packed bed bioreactor systems. *Environmental Technology* 23, 199-206.

Thurston, C.F. (1994) The structure and function of fungal laccases. *Microbiology* 140, 19-26.

Ullah, M.A., Bedford, C.T. & Evans, C.S. (2000). Reactions of pentachlorophenol with laccase from *Coriolus versicolor*. *Appl Microbiol Biotechnol* 53, 230-234.

Vermette, A.B. (2000). Removal of cresols from synthetic wastewater using laccase. Master thesis. University of Windsor, Ontario, Canada.

Walter, M., Boul, L., Chong, R. & Ford, C. (2004). Growth substrate selection and biodegradation of PCP by New Zealand white-rot fungi. *Journal of Environmental Management* 71, 361-369.

Walter, M., Boyd-Wilsona, K., Boul, L., Ford, C., McFadden, D., Chong, B. & Pinfold J. (2005). Field-scale bioremediation of pentachlorophenol by *Trametes versicolor*. *International Biodeterioration & Biodegradation* 56, 51-57.

Welch, R.M., Levin, W. & Conney, A.H. (1969). Estrogenic action of DDT and its analogs. *Toxicol Appl Pharmacol* 14, 358-367.

Xu, H., Lai, Y.Z., Slomczynsky, D., Nakas, J.P. & Tannenbaum, S.W. (1997). Mediator assisted selective oxidation of lignin model compounds by laccase from *Botrytis cinerea*. *Biotechnol Lett* 19:957-963.

Yaropolov, A.I., Skorobogat'ko, O.V., Vartanov, S.S. & Varfolomeyev, S.D. (1994). Laccase. Properties, catalytic mechanism, and applicability. *Appl Biochem Biotechnol* 49, 257-280.

In: Progress in Pesticides Research
Editor: Christopher M. Kanzantzakis

ISBN 978-1-60456-201-9

Chapter 11

CYTOTOXIC EFFECTS OF PESTICIDES ON MICROALGAE DETERMINED BY FLOW CYTOMETRY

Carmen Rioboo, Raquel Prado, Concepción Herrero and Angeles Cid *

Laboratorio de Microbiología, Facultad de Ciencias. Universidad de A Coruña
Campus da Zapateira s/n. 15071 A Coruña, Spain

ABSTRACT

As much as *ca.* 99.7% of the applied load of pesticides is dispersed in the environment, not reaching the target pests. In this way, pesticides enter into aquatic ecosystems from agricultural runoff or leaching and, as a consequence, have become some of the most frequently ocurring organic pollutants in aquatic ecosystems.

Most phytotoxicological research with herbicides has been conducted on target plants. The sensitivity of algae to many herbicides is very high, and a better understanding of their environmental effects is probably acquired by using test species representing non-target groups. Because of their short generation times, microalgae respond rapidly to environmental changes. Furthermore, microalgal tests are generally sensitive, rapid and low-cost effective. For these reasons, the use of microalgal toxicity tests is increasing, and today these tests are frequently required by authorities for notifications of chemicals and are also increasingly being used to manage chemical discharges.

Cytotoxic effects of aquatic pollutants on microalgae are very heterogeneous, and they are influenced by the environmental conditions and the test species. Growth, photosynthesis, chlorophyll fluorescence and others parameters reflect the toxic effects of pollutants on microalgae; however, other relevant endpoints are less known because experimental difficulties, especially under *in vivo* conditions.

Flow cytometry offers multiparametric analysis of cells on a cell-by-cell basis in near *in vivo* conditions. Applied in the study of the aquatic environments since the 80's,

---

* Corresponding author: Prof. Dr. A. Cid. E-mail: cid@udc.es

this technique has achieved extensive use for the study of microalgae and has been introduced as an alternative to more traditional techniques of analyzing cells in culture and from natural populations.

Several parameters determined by flow cytometry, such as cell volume and granularity, chlorophyll *a* fluorescence, cell viability, cell proliferation, oxidative stress, membrane potential or intracellular calcium level, were employed to assess changes in the physiological status of different microalgae as a consequence of the toxic action of herbicides. The variety of results obtained in the present study reveals that flow cytometry is a useful tool in the toxicity tests with microalgae.

INTRODUCTION

Aquatic environments, including freshwater, estuaries, and coastal marine waters, are often contaminated with numerous organic and inorganic compounds, including pesticides. As much as *ca.* 99.7% of the applied load of pesticides is dispersed in the environment, not reaching the target pests. In this way, pesticides enter into aquatic ecosystems from agricultural runoff or leaching and, as a consequence, have become some of the most frequent organic pollutants in aquatic ecosystems. The increasing occurrence of pollutants in general, and herbicides in particular, has stimulated many studies on their toxicity to aquatic microorganisms, and the need for convenient methods for assayed pollutants toxicity has become evident.

Microalgae as Environmental Monitoring Markers

Most phytotoxicological research with herbicides has been conducted on target plants. The sensitivity of algae to many herbicides is very high, and a better understanding of their environmental effects is probably acquired by using test species representing non-target groups. Because of their short generation times, microalgae respond rapidly to environmental changes. Furthermore, microalgal tests are generally sensitive, rapid and low-cost effective. For these reasons, the use of microalgal toxicity tests is increasing, and today these tests are frequently required by authorities for notifications of chemicals and are also increasingly being used to manage chemical discharges.

Growth, photosynthesis, chlorophyll fluorescence and others parameters reflect the toxic effects of pollutants on microalgae; however, other relevant endpoints are less known because experimental difficulties, especially under *in vivo* conditions. Cytotoxic effects of aquatic pollutants on microalgae are very heterogeneous, and they are influenced by the environmental conditions and the test species.

Flow Cytometry Technique

Flow cytometry (FCM) offers multiparameter analysis of cells on a cell-by-cell basis in near *in vivo* conditions [1]. Since the overview of FCM in biological oceanography published by Legendre and Yentsch in 1989 [2], the aquatic environmental research has changed quite drastically. In a first period, the use of FCM was focused to fundamental ecological questions, mainly abundance and distribution of phytoplankton in the ecosystems. In the 1990s, the flow cytometric studies becoming progressively involved in programs to address some of the major socioecological problems that threaten human societies, such as the release of contaminants in aquatic systems.

The response of microalgae to a toxicant is typically measured using population-based parameters, such as specific growth rate, biomass, cell yield, chlorophyll fluorescence, and primary production. The bulk population based endpoints used in algal toxicity tests did not supply information on the distribution of responses among the individual cells within the population. Flow cytometry is an alternative to the standard algal population-based endpoints, since it allows the rapid and quantitative measurement of responses of individual algal cells to a toxic stress. The development and use of fluorescent probes and dyes allow researches to assess changes in their physiological *status*, however their application is still far from routine.

Microalgae are ideal for flow cytometric analysis as they are unicellular and contain photosynthetic pigments such as chlorophyll *a* which emits fluorescence when excited by blue light [3]. FCM has the advantage in that it can perform multi-parameter analysis on a wide range of cell properties by measurement of algal cell densities, light scatter signals and chlorophyll autofluorescence, after excitation at 488 nm.

The aim of this chapter is demonstrate the usefulness of the flow cytometry technique combined with microalgae to the assessment of potential cytotoxic effects of different herbicides to non-target species. Here we show some experimental results obtained in the study of toxic effects of paraquat on *Chlamydomonas moewusii* and terbutryn on *Chlorella vulgaris*. For this, data results presented in this chapter were obtained using a Coulter EPICS XL4 (Beckman Coulter Inc.) flow cytometer, equipped with an argon ion laser (15 mW) as light source with an emission wavelength of 488 nm (blue light), and the analysis software EXPO 32 ADC 1.1C (Beckman Coulter Inc.). At least 10.000 cells were analyzed for each parameter. Fluorescent emission ranged the visible spectrum (400-700 nm) detected by 4 different photomultipliers (FL1: <550 nm; FL2: 550-600 nm; FL3: 600-645 nm; FL4: >650 nm).

EFFECTS OF PARAQUAT ON *CHLAMYDOMONAS MOEWUSII*

Paraquat (1,1'-dimethyl-4,4'-bipyridilium dichloride) is a non-selective herbicide widely used to prevent the growth of broad leaf weeds and grasses, mainly in railways and roadsides [4]. This quaternary ammonium compound is a cationic species in aqueous solution, and after application, this chemical can be adsorbed into the soil or transported to the aquatic environment by runoff or leaching [5]. Paraquat in cells leads to the generation of free oxygen radicals, leading to lipid peroxidation damaging cell membranes and cell death [6, 7]. Information on the influence of these chemicals on the lowest levels of the food chain is scarce [8, 9, 10].

Chlamydomonas moewusii cultures were carried out during 96 hours as described in Franqueira *et al.* (1999) [7], using a paraquat concentration of 200 nM. Changes in cell volume, internal granularity, chlorophyll *a* fluorescence intensity and the evaluation of cell viability were analyzed by flow cytometry in cells after 96 hours of exposure to paraquat, respect to control cells cultured without paraquat. Aliquots of the different cultures were centrifuged and cells were resuspended in 0.5x PBS buffer (pH 7.2) to obtain cell densities of 2 x 10^5 cells ml^{-1} for their analysis by FCM. Shifts in forward scatter light (<20°) and side scatter light (90°) indicate cell size and internal granularity, respectively. Similarly, chlorophyll *a* fluorescence is detected at FL4 log channel (>650 nm). The fluorescence of cells stained with propidium iodide (PI; Sigma Chemical Co.) was measured to study cell viability. PI is a fluorescent dye that intercalates with double-stranded nucleic acids to produce red fluorescence when excited by blue light. It is unable to pass through intact cell membranes; however, when the cell dies the integrity of the cell membrane fails, and then PI is able to enter and stain the nucleic acids [11]. In this way, PI can be used to discriminate between live non-fluorescent cells and non-viable fluorescent cells; the orange fluorescent emission of this compound was collected in the FL3 channel (605-635 nm). Aliquots of 1 x10^6 cells were stained with PI dissolved in PBS buffer (pH 7.4) to a final concentration of 2.5 μg ml^{-1} during an incubation period of 10 min, in darkness and room temperature. For each cytometric parameter investigated, at least 10^4 cells were analyzed per culture and fluorescence measurements were expressed in a logarithmic scale.

Since an increase in the forward light scatter (FSC) signal can be correlated with an increase in cell volume [12], paraquat concentration assayed caused a slight increase in the cell volume of *C. moewusii*, respect to control cells (Figure 1). Increases of cell volume were reported for different microalgal species exposed to high concentrations of different pollutants [7, 13, 14]. Variations in the side light scatter (SSC) signal indicate variations in the internal granularity of cells [12]. A paraquat concentration of 200 nM provoked an important increase of the internal granularity of microalgal cells (Figure 2).

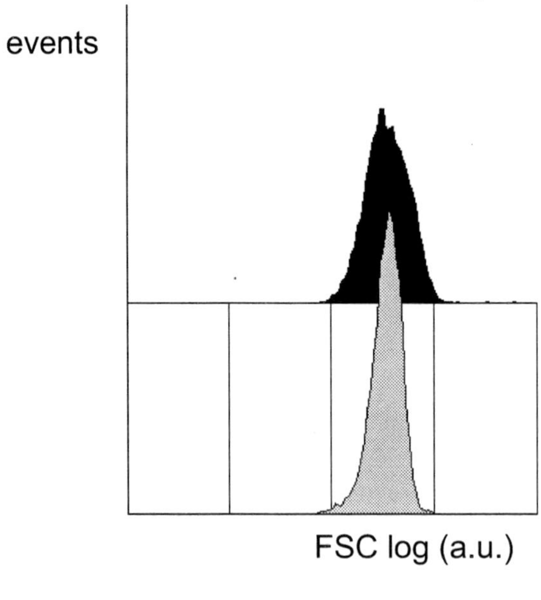

Figure 1. Overlay of FSC signal (a.u., arbitrary units) histograms showing the profiles of *Chlamydomonas moewusii* cells of control cultures (grey histogram) and cells exposed to 200 nM paraquat (black histogram) after 96 h of culture.

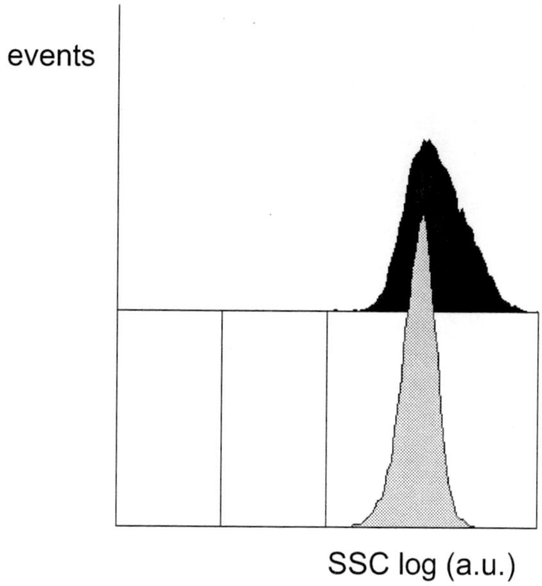

Figure 2. Overlay of SSC signal (a.u., arbitrary units) histograms showing the profiles of *Chlamydomonas moewusii* cells of control cultures (grey histogram) and cells exposed to 200 nM paraquat (black histogram) after 96 h of culture.

In cultures exposed to paraquat, cells formed palmelloid colonies (clusters of non-flagellated cells closed in a common wall), observed by light microscopy. The volume and complexity increases detected could be related to palmelloid colony formation, probably due to the incapacity to finish cell division as well as failures of regulation of cell volume because of the attack of oxidative radicals formed to membranes. Bray *et al.* [9], using transmission electron microscopy techniques, detected that *Chlamydomonas* cultures with 0.075 μM paraquat showed a small proportion of cells swollen and that they were less electron-dense than the normal cells, suggesting an increase in the cell moisture. At 0.15 μM paraquat, the proportion of cells with lower electrondensity had increased considerably and at 0.30 μM, all the cells appeared swollen and were of low electron-density. These authors stated that in addition to the swelling and the low cellular electrondensity observed as a consequence of paraquat addition, other ultrastructural alterations were present that could explain the increase in the granularity that was detected using FCM at 96 h of culture.

Fluorescence of chlorophyll *a* was also affected after 96 hours of paraquat exposure (Figure 3). After 96 hours of culture, control cells showed a single and well defined peak of chlorophyll *a* fluorescence. However, *Chlamydomonas* cultures exposed to 200 nM of paraquat presented a bimodal distribution of chlorophyll *a* fluorescence, showing a differential cell response pattern. In the fluorescence histogram (Figure 3), we can distinguish two subpopulations of cells: one of them with a chlorophyll fluorescence signal close to that observed for control cells, and a second subpopulation showing an important decrease of chlorophyll fluorescence signal. Bulk measurements of fluorescence could not permit this assessment of a different behaviour of cells respect to this parameter, only achieved by a cell-to-cell analysis. The decrease of chlorophyll fluorescence observed for *C. moewusii* cells exposed to paraquat could be related to an important decrease of the chlorophyll cellular content (unpublished data).

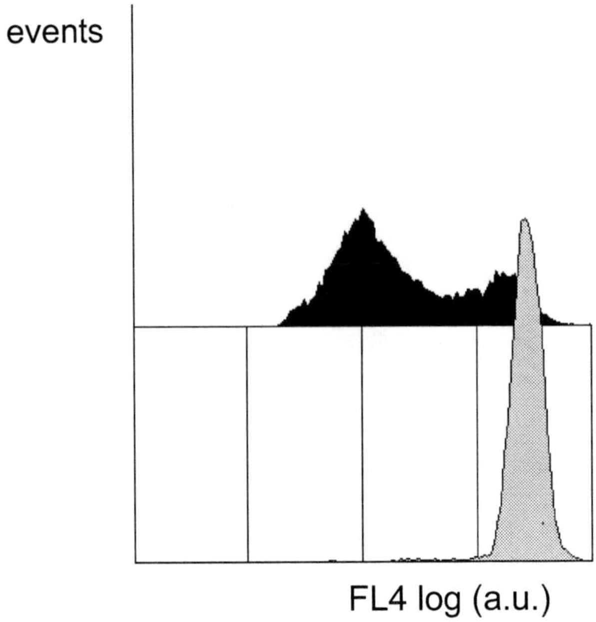

Figure 3. Overlay of chlorophyll *a* fluorescence (a.u., arbitrary units) histograms showing the profiles of *Chlamydomonas moewusii* cells of control cultures (grey histogram) and cells exposed to 200 nM paraquat (black histogram) after 96 h of culture.

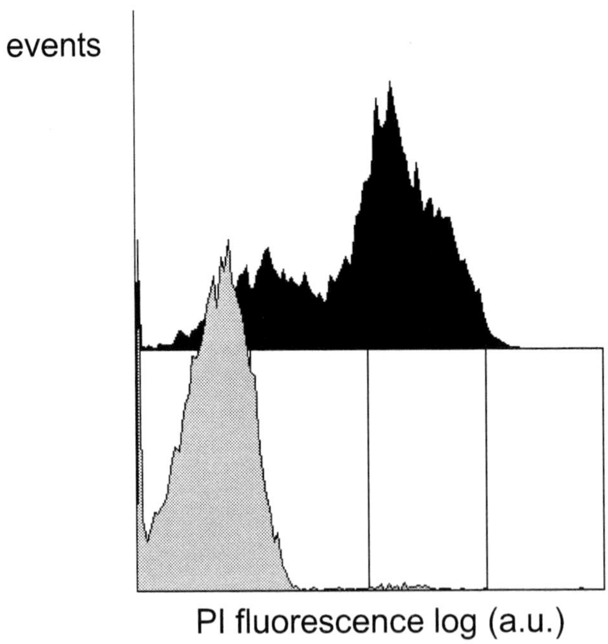

Figure 4. Overlay of propidium iodide (PI) fluorescence (a.u., arbitrary units) histograms showing the profiles of *Chlamydomonas moewusii* cells of control cultures (grey histogram) and cells exposed to 200 nM paraquat (black histogram) after 96 h of culture.

After 96 hours of culture, control cultures showed a 98.75% of viable cells, represented by non-fluorescent cells in the histogram (Figure 4), while the percentage of cell viability was only 26.09% cultures exposed to 200 nM of paraquat. Cellular membranes are selective, dynamic barriers that play an essential role in regulating biochemical and physiological events. The cell viability decrease after 96 h of paraquat exposure (Figure 4), showing a loss of their membrane integrity, like occurs in the degenerative processes driven by exogenous factors [15]. Standard microalgal bioassays are based in bulk population endpoints, as we indicated above, sacrificing information regarding the distribution of responses among individual cells within a population by providing the average response to a toxicant. Here we can distinguish between live of non viable cells using the flow cytometry technique which has the ability to analyze the real heterogeneous microalgal population.

EFFECTS OF A TRIAZINE ON *CHLORELLA VULGARIS*

Triazines are widely used herbicides that are highly toxic and frequently appear in natural watercourses. Terbutryn (N^2-tert-butyl-N^4-isopropyl-1,3,5-triazine-2,4-diamine) is a pre- or post-emergence systemic triazine herbicide, used to control most grasses and many annual broad-leaved weeds in winter cereals, potatoes, legumes, sunflowers, maize, sugar cane and citrus fruit, and as an aquatic herbicide for controlling submerged and free floating weeds and algae in watercourses [16, 17]. This herbicide affects the main energy production process in green cells, the photosynthesis [18], and this mode of action makes them potentially lethal to a wide variety of non-target species of primary producers, including phytoplankton and benthic algae.

The aim of this study is to assess the effect of terbutryn on the freshwater microalga *Chlorella vulgaris* from different flow cytometric approaches. The parameters used to monitor the herbicide effect are: the measurement of the evolution of cell progeny, the analysis of the oxidative stress, the changes in cell membrane potential and the intracellular calcium level.

Microalgal cultures were carried out during 96 hours as described in Rioboo *et al.* (2002) [19], using a terbutryn concentration of 500 nM. Cultures without herbicide were also included as control. Red fluorescence of chlorophyll *a* (FL4; 645-675 nm) was used to exclude non-algal particles, to allow the analysis of *C. vulgaris* growth. Aliquots of the different cultures were centrifuged and cells were resuspended in 0.5x PBS buffer (pH 7.2) to obtain cell densities of 1×10^6 cells ml^{-1} for their analysis by FCM. For each cytometric parameter investigated, at least 10^4 cells were analyzed per culture and fluorescence measurements were expressed in a logarithmic scale.

Evolution of Cell Progeny

During the last ten years, the stable intra-cytoplasmatic dye, 5-,6-carboxyfluorescein diacetate succimidyl ester (CFSE) has been widely used as a quantitative method to measure cell divisions in analysis of inmunological lymphocyte proliferation. However, no data exists about its aplication in vegetable cells as microalgae are. CFSE allowed the identification of cell progeny and tracking and analysis of the division history of individual cells that had undergone multiple rounds of division. When incubated with cells, CFSE crosses the cell membrane and attaches to free amines groups of cytoplasmic cell proteins where it is transfomed in fluorescein by intracellular esterases. Upon subsequent cell division, cytoplasmic fluorescence per cell is equally redistributed among daughter cells at mitosis, providing an ideal tool for monitoring cell proliferation [20, 21].

The rapid and precise determination of proliferation rates and generation times by CSFE staining allow us to develop a model for assessing *in vivo* the effects of pollutants at microalgal cell level, non reached by radiolabeled techniques used until now in this kind of studies.

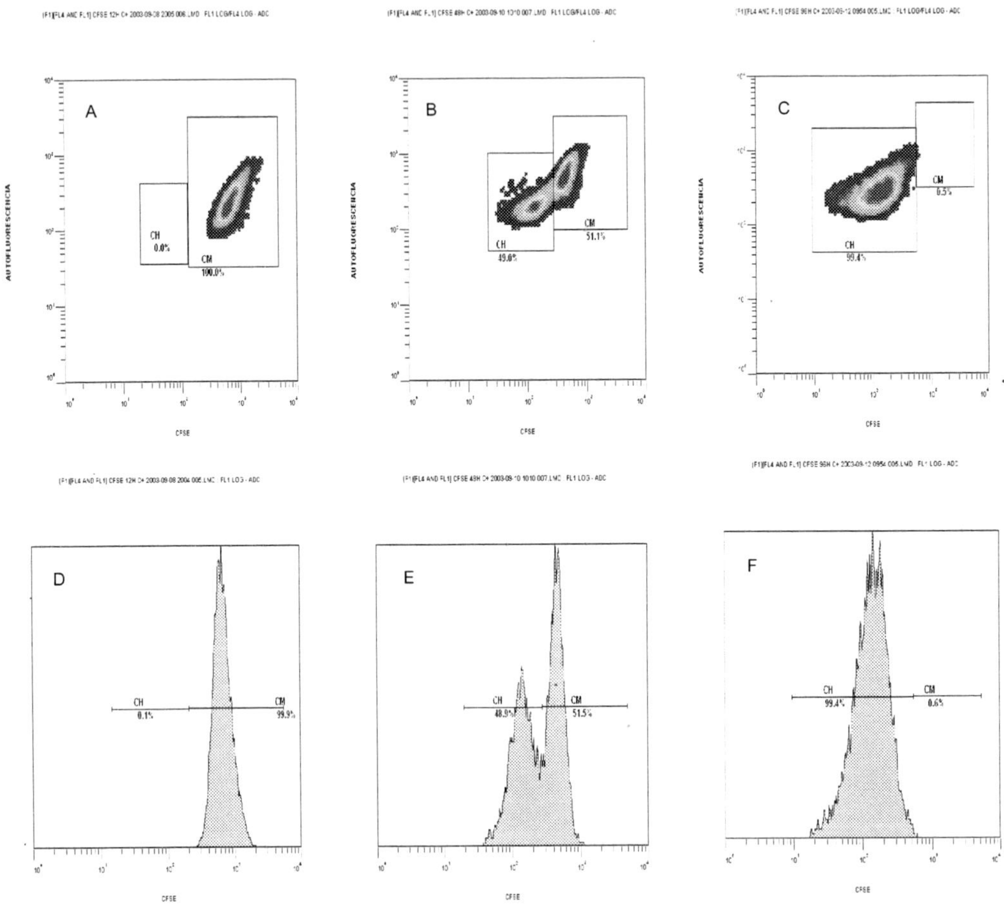

Figure 5. Cytograms of CFSE fluorescence *vs.* autofluorescence of *Chlorella vulgaris* cells cultured in control conditions at different times (12, 48 and 96 h; A, B and C, respectively). The histograms (D, E, F) show the CFSE fluorescence signal (a.u., arbitrary units). CH: daughter cells; CM: mother cells.

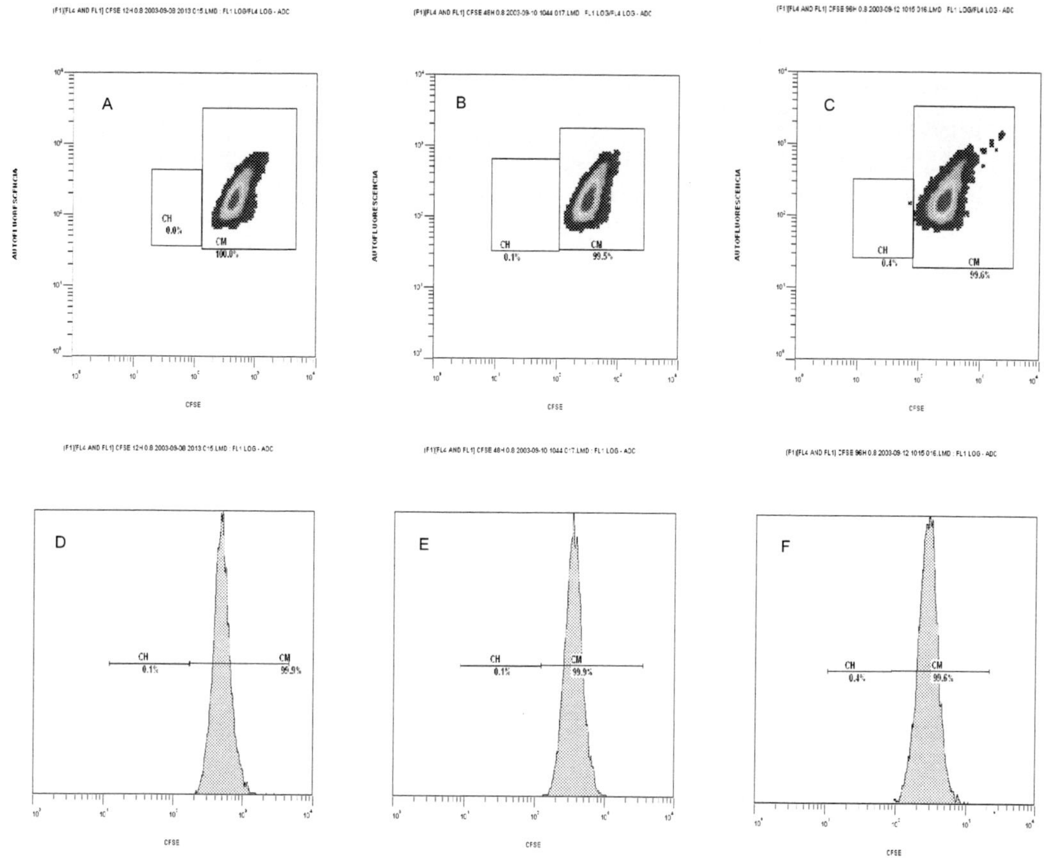

Figure 6. Cytograms of CFSE fluorescence *vs.* autofluorescence of *Chlorella vulgaris* cells exposed to 500 nM terbutryn at different times (12 (A), 48 (B) and 96 (C) hours). The histograms (D, E, F) show the CFSE fluorescence signal (a.u., arbitrary units). CH: daughter cells; CM: mother cells.

Cells were stained with CFSE dissolved in DMSO, at a final concentration of 32 μg CFSE per 150 x 10^6 cells per ml, in darkness at 20°C during 30 min before culture. Aliquots of cultures were analyzed by FCM after 12, 48 and 96 hours, when the green fluorescent emission of this compound was collected in the FL1 channel (<550 nm) (Figure 5 and 6). Figure 5 shows the evolution of cell progeny of a control culture of *Chlorella vulgaris*. After 48 hours of culture (Figure 5B, E) we can distinguish 2 subpopulations corresponding to cells that have not yet divided (CM; >50%) and daughter cells (CH); after 96 hours (Figure 5C, F), only one population appeared, being all of them daughter cells, with a minor level of CFSE fluorescence. However, cells exposed to 500 nM of terbutryn are not able to divide during this period (Figure 6), remaining constant the CFSE fluorescence. The CFSE results on the growth kinetics were consistent with those determined by cell counting [19].

Oxidative Stress

Aerobic organisms produce reactive oxygen species (ROS) in their metabolic process, as oxygen peroxide o superoxide. The level of these ROS increases during several cytotoxic process, leading to the named oxidative stress. The fluorogenic oxidation of hydroethidine (dihydroethidium; HE) to ethidium has been used as a measure of O_2^-. Microalgal suspensions were stained with hydroethidine (HE). HE is a chemically reduced fluorophore able to cross the cell membrane and it is oxidized by superoxide ion in the cytoplasm of cells to a red product (ethidium) [22]. Ethidium binds to the DNA inside cells and has a red fluorescence when is excited with blue light [12].

Aliquots of cells were stained with HE dissolved in DMSO, at a final concentration of 2.5 $\mu g\ ml^{-1}$ in darkness at 20ºC during 30 min. The red fluorescent emission of this compound was collected in the FL3 channel (600-645 nm).

After 96 hours of exposure to 500 nM of terbutryn, an important increase of the oxidative stress level, respect to control cells, can be observed (Figure 7). Membranes could be expected to be highly prone to free radical attack inasmuch as unsaturated fatty acids are major components of most membrane lipid bilayers [14]. The consequences of free radical attack on membranes are numerous and include the induction of lipid peroxidation [23], lysis [24], and fatty acid deesterification [25]. Senescence is an active process initiated by some combination of internal and environmental triggers, and membrane deterioration is an early and fundamental feature of this process.

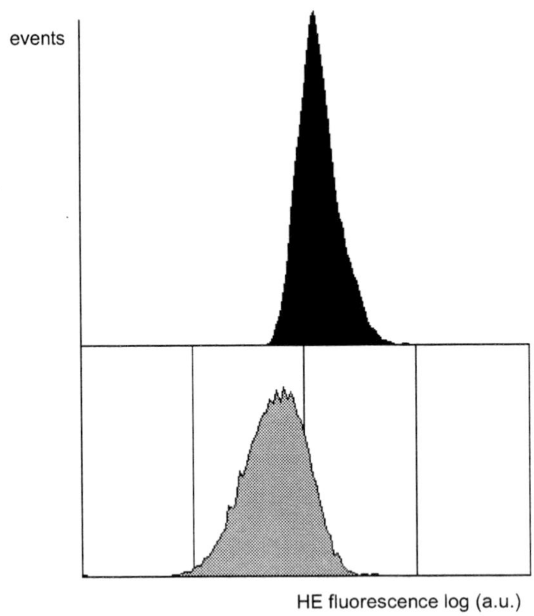

Figure 7. Overlay of hydroethidine (HE) fluorescence (a.u., arbitrary units) histograms showing the profiles of *Chlorella vulgaris* cells of control cultures (grey histogram) and cells exposed to 500 nM terbutryn (black histogram) after 96 h of culture.

Intracellular Calcium Level

Calcium ion plays an important role as a mediator in the transmembrane signal transduction, being its intracellular concentration increase part of the regulation of several cell processes [26].

The development of fluorescent probes that show a spectral response upon binding Ca^{2+} have enabled researchers to investigate changes in intracellular free Ca^{2+} concentrations [27]. These fluorescent indicators are derivatives of Ca^{2+} chelators. Fluo-3 is a Ca^{2+} fluorescent indicator excited with visible light and it is essentially non-fluorescent unless bound to Ca^{2+}. After its union to Ca^{2+}, Fluo-3 increases its fluorescence between 100 and 200 times [28].

Aliquots of cells were stained with Fluo-3 pentammonium salt dissolved in DMSO, at a final concentration of 5 $\mu g\ ml^{-1}$ in presence of 0.1% β-escine, in darkness at 20ºC during 30 min. The green fluorescent emission of this compound was collected in the FL1 channel (505-545 nm).

The intracellular calcium level is closely related with the photosynthetic metabolism of control cultures cells, since the $[Ca^{2+}]_i$ increases in darkness and decrease in the light period (Figure 9A). However, those cells treated with 500 nM of terbutryn showed a drastically different pattern, remaining more or less constant the level of calcium after the first light period (Figure 9B). This terbutryn effect confirms that changes in the intracellular calcium level in this kind of cells are a consequence of the alteration in the photosynthetic process, and an increase in the $[Ca^{2+}]_i$ can be interpreted as an early signal of cell stress. Cessation of photosynthetic electron transport by light-off or by inhibitors (DCMU, DBMIB) causes the Ca^{2+} release from the chloroplast [29].

Changes in Cell Membrane Potential

Microorganisms, and microalgae in particular, are the first organisms affected by pesticides discharges in aquatic environments because they are directly in contact with the medium, separated only by the cytoplasmic membrane and the cell wall. Cellular membranes are selective, dynamic barriers that play an essential role in regulating biochemical and physiological events, so any alteration produced in the environment provokes changes in microorganisms membranes.

Potentiometric optical probes enable researchers to perform membrane potential measurements in cells (or organelles) too small for microelectrodes. The plasma membrane of a cell typically has a transmebrane protential as a consequence of K^+, Na^+ and Cl^- concentration gradients that are maintained by active transport processes. Potentiometric probes offer an indirect method of detecting the translocation of these ions. Increases and decreases in membrane potential -referred to as membrane hyperpolarization and depolarization, respectively- play a central role in many physiological processes, including cell signalling.

$DiBAC_4(3)$ (bis-(1,3-dibutylbarbituric acid trimethine oxonol) is a slow-response probe, which exhibits potential-dependent changes in its transmembrane distribution that is accompanied by a fluorescence change. The dye enters depolarized cells where it binds to intracellular lipids. Increased depolarization results in more influx of the anionic dye and thus an increase in fluorescence. Conversely, hyperpolarization is indicated by a decrease in fluorescence. This dye is excluded from mitochondria because its overall negative charge, simplifying the membrane potential measurement in eukaryotic organisms [30].

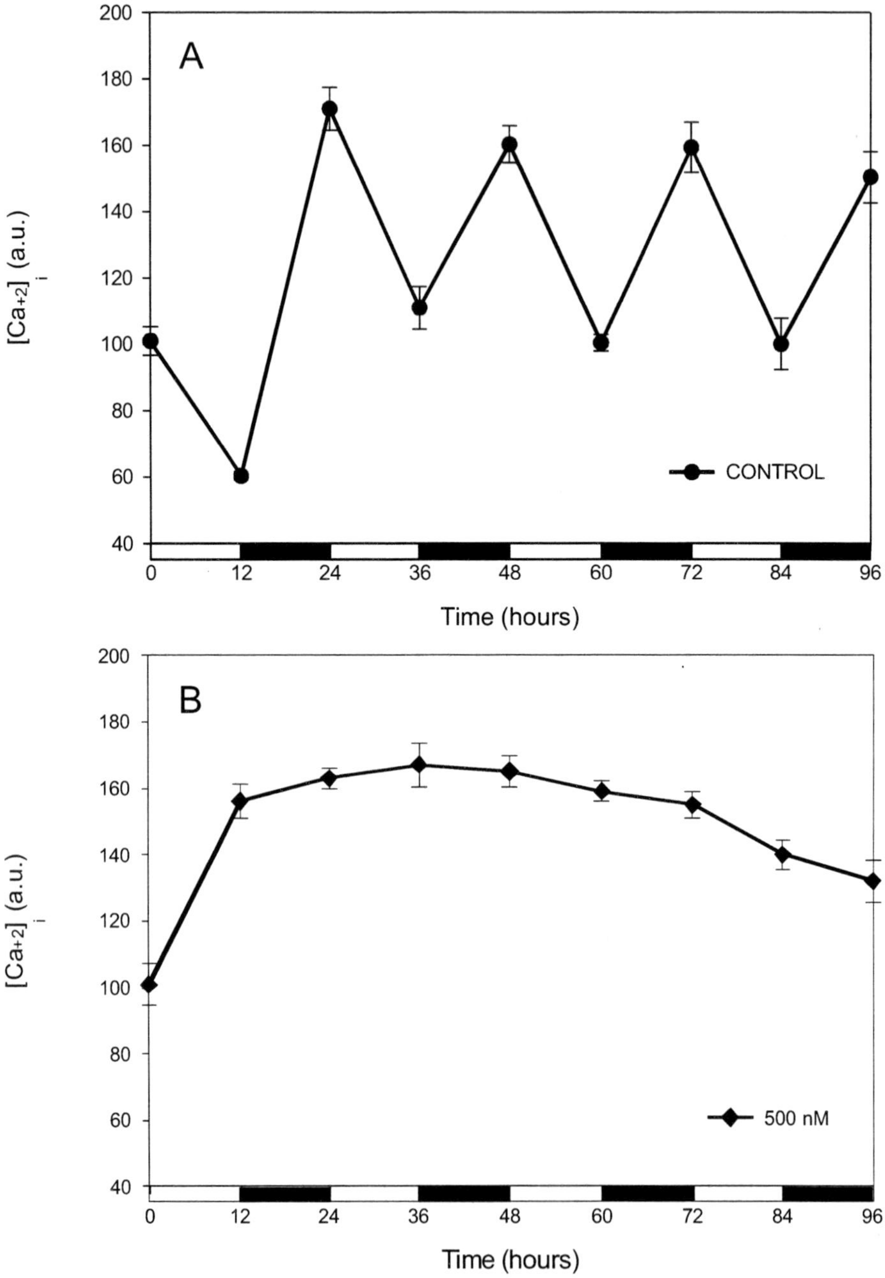

Figure 9. Evolution of the intracellular calcium level inferred from Fluo-3 fluorescence emission of *Chlorella vulgaris* cells of control cultures (A) and cells exposed to 500 nM terbutryn (B) after 96 h of culture. *X* axis represents time of culture, being white the light period and black the dark period.

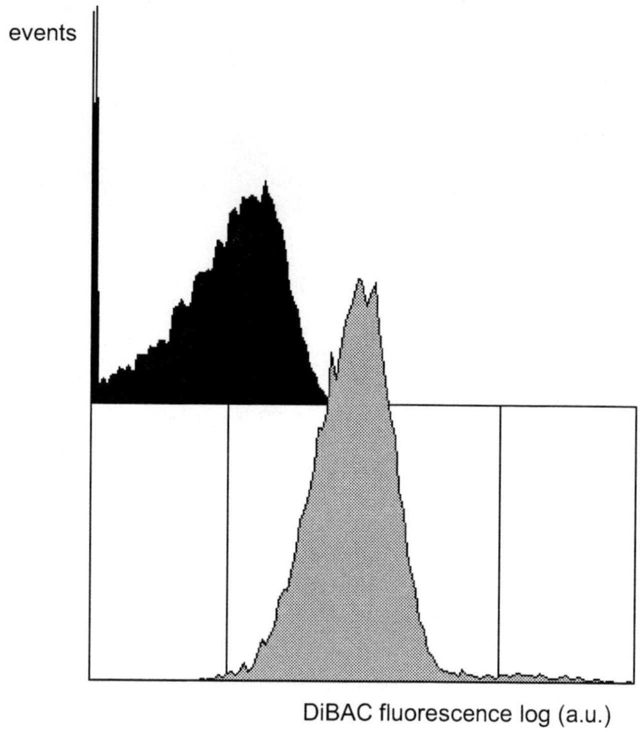

Figure 8. Overlay of DiBAC$_4$(3) fluorescence (a.u., arbitrary units) histograms showing the profiles of *Chlorella vulgaris* cells of control cultures (grey histogram) and cells exposed to 500 nM terbutryn (black histogram) after 96 h of culture.

Aliquots of cells were stained with DiBAC$_4$(3) dissolved in DMSO, at a final concentration of 2.5 μg ml^{-1} in darkness at 20ºC during 30 min. The green fluorescent emission of this compound was collected in the FL1 channel (505-545 nm).

A decrease of DiBAC$_4$(3) fluorescence respect to control cells can be observed in *Chlorella vulgaris* cells exposed to 500 nM terbutryn after 96 hours (Figure 8), indicating the hyperpolarization of the cell membrane of these affected cells. Using eloctrophysiological and microfluorimetric techniques, it has been reported that increases in the intracellular calcium level can be accompanied by a hyperpolarization of cytoplasmic membrane in vegetal cells [29], being in agreement with our results obtained by flow cytometry and showed here.

CONCLUSION

The potential for flow cytometry as a tool in ecotoxicological studies has not been yet fully explored. The variety of results showed in the present study confirms that flow cytometry is very useful in the toxicity tests carried out with microalgae to assess the potential cytotoxicity of herbicides at very low concentrations. Changes in environmental conditions are often most easily and rapidly recognized by the reaction of unicellular organisms like microalgae. Then, taking into account the obtained results, it is expected that FCM could improve the ability of the researchers to address the major environmental challenges that are confronting human societies.

ACKNOWLEDGEMENT

Results showed in this chapter were carried out with the financial support of Xunta de Galicia (PGIDIT04RFO103946PR) and the Spanish Ministerio de Educación y Ciencia (CGL2004-02037BOS). R.P. acknowledges a F.P.U. from the Spanish Ministerio de Educación y Ciencia.

REFERENCES

[1] Carter NP, Meyer EW (1990) Introduction to the principles of flow cytometry, In *Flow Cytometry. A Practical Approach,* Ormerod MG (Ed). Oxford University Press, Oxford. pp: 1-28.

[2] Legendre L, Yentsch CM (1989) Overview of flow cytometry and image analysis in biological oceanography and limnology. *Cytometry* 10:501–510.

[3] Franqueira D, Orosa M, Torres E, Herrero C, Cid A (2000) Potential use of flow cytometry in toxicity studies with microalgae. *Sci Total Environ* 247:119–126.

[4] Kalayanova FP, El Batawi MA (1991) *Human toxicology of pesticides.* CRC Press, Boca Raton, FL.

[5] Ibáñez M, Picó Y, Mañes J (1996) Influence of organic matter and surfactans on solid-phase extraction of diquat, paraquat and difenzoquat from waters. *J Chromatogr A 727:* 245–252.

[6] Tissut M, Ravanel P, Nurit F, Deslandres C, Bourguignon J (1987) Effects of LS82556 on thylakoid activities and photosynthesis: a comparison with paraquat and acifluorfen. *Pestic Biochem Physiol* 29:209–216.

[7] Franqueira D, Cid A, Torres E, Orosa M, Herrero C (1999) A comparison of the relative sensitivity of structural and functional cellular responses in the alga *Chlamydomonas eugametos* exposed to the herbicide paraquat. *Arch Environ Contam Toxicol* 36: 264-169.

[8] Ibrahim EA (1990) The influence of the herbicide paraquat ''gramoxon'' on growth and metabolic activity of three chlorophytes. *Water Air Soil Poll* 51: 89–93.

[9] Bray DF, Bagu JR, Nakamura K (1993) Ultrastructure of *Chlamydomonas reinhardtii* following exposure to paraquat: comparison of wild type and a paraquat-resistant mutant. *Can J Bot* 71:174–182.

[10] Saenz ME, Alberdi JL, Di Marzio WD, Accorinti J, Tortorelli MC (1997) Paraquat toxicity to different green algae. *Bull Environ Contam Toxicol* 58: 922–928.

[11] Ormerod MG (1990) Analysis of DNA. General methods. In *Flow Cytometry. A Practical Approach,* Ormerod MG (Ed). Oxford University Press, Oxford, U.K, pp 69-87.

[12] Shapiro HM (1995) *Practical Flow Cytometry,* 3rd ed. Wiley-Lyss Inc., New York, USA.

[13] Cid A, Herrero C, Torres E, Abalde J (1995) Copper toxicity on the marine microalga *Phaeodactylum tricornutum:* effects on photosynthesis and related parameters. *Aquatic Toxicol* 31: 165–174.

[14] Cid A, Fidalgo P, Herrero C, Abalde J (1996) Toxic action of copper on the membrane system of a marine diatom measured by flow cytometry. *Cytometry* 25: 32–36.

[15] Thompson JE (1988) The molecular basis for membrane deterioration during senescence. In *Senescence and Aging in Plants,* Nooden LD & Leopold AC (Eds). Academic Press, Inc., San Diego, California, pp 51-83.

[16] Cremlyn RJ (1991) *Agrochemicals: Preparation and Mode of Action.* Wiley, New York, USA.

[17] Tomlin C (1994) *The Pesticide Manual. A World Compendium.* The Bath Publisher, Bath, UK.

[18] Védrine C, Leclerc, J-C, Durrieu C, Tran-Minh C (2003) Optical whole-cell biosensor using *Chlorella vulgaris* designed for monitoring herbicides. *Biosens Bioelectron 18:* 457-463.

[19] Rioboo C, González O, Herrero C, Cid A (2002) Physiological response of freshwater microalga (*Chlorella vulgaris*) to triazine and phenylurea herbicides. *Aquat Toxicol 59:* 225-235.

[20] Lyons AB, Hasbold J, Hodgkin PD (2001) Flow cytometric analysis of cell division history using dilution of carboxyfluorescein diacetate succinimidyl ester, a stably integrated fluorescent probe. In *Methods in Cell Biology 63:* Cytometry, 3rd edition, Darzynkiewicz Z, Crissman A, Robinson JP (Eds). pp: 375-398.

[21] Hara T, Ishii T, Fujishiro M, Masuda M, Ito T, Nakajima J, Inoue T, Matsuse T (2004) Glutathione S transferase P1 has protective effects on cell viability against camptothecin. *Cancer Lett 203: 1*99-207.

[22] Benov L, Sztejnberg L, Fridovich I (1998) Critical evaluation of the use of hydroethidine as a measure of superoxide anion radical – a simple assay for superoxide dismutase. *Free Radic Biol Med 25*: 826-831.

[23] Kellog EW, Fridovich I (1975) Superoxide, hydrogen peroxide and singlet oxygen in lipid peroxidation by the xanthine oxidase system. *J Biol Chem 250:* 8812-8817.

[24] Goldstein IM, Weissmann G (1977) Effects of generation of superoxide anion on permeability of liposomes. *Biochim Biophys Res Commun 70:* 452-458.

[25] Niehaus WG 1978 A proposed role of the superoxide anion as a biological nucleophile in the deesterification of phospholipids. *Bioorg Chem 7:* 77-84.

[26] Tsien RY, Pozzan T, Rink TJ (1982) Calcium homeostasis in intact lymphocytes: cytoplasmic free calcium monitored with a new, intracellulary trapped fluorescent indicator. *J Cell Biol 94:* 325-334.

[27] Tepikin AV (Ed) (2001) *Calcium Signalling. A Practical Approach,* 2nd ed. Oxford University Press, 272 pp.

[28] Burchiel SW, Edwards BS, Kuckuck FW, Lauer FT, Prossnit ER, Ransom JT, Sklar LA (2000) Analysis of free intracellular calcium by flow cytometry: multiparameter and pharmacologic applications. *Methods 21:* 221-230.

[29] Schönknecht G, Bauer CS, Simonis W (1998) Light-dependent signal transduction and transient changes in cytosolic Ca^{2+} in a unicellular green alga. *J Exp Bot 49:* 1-11 .

[30] Jepras RI, Carter J, Pearson SC, Paul FE, Wilkinson MJ (1995) Development of a robust flow cytometric assay for determining numbers of viable bacteria. *Appl Environ Microbiol 61:* 2696-2701.

INDEX

A

abiotic, 346, 452
abnormalities, 389
absorption, 248, 310, 332, 333, 335, 336, 445, 455
acceptor, 286, 290, 291, 292, 293, 294
accidental, 351, 352, 353, 354
accounting, viii, 35, 36, 130, 149, 231
accreditation, 388
accuracy, 271, 284, 316
acetate, viii, 7, 19, 36, 37, 38, 82, 83, 120, 125, 128, 129, 130, 136, 192, 228, 235, 274, 275, 276, 278, 361
acetic acid, 119, 124, 278, 318
acetone, 37, 38, 128, 276, 346, 469
acetonitrile, viii, x, 19, 20, 21, 22, 23, 24, 25, 27, 31, 36, 37, 38, 82, 83, 128, 130, 148, 269, 276, 278, 284, 446
acetylcholine, ix, 5, 6, 11, 233, 235, 442
acetylcholinesterase, ix, 4, 5, 9, 12, 233, 234, 252, 256, 257, 260, 261, 264, 265, 266, 267, 387, 450, 451
acidic, 277, 291, 436, 440, 442, 443, 460
acidity, 462
acoustic, 425
acrylate, xiii, 453, 454, 457
activation, 248, 251, 255, 256, 257, 260, 486
active site, 234, 235, 280, 324
active transport, 501
activity level, 5
acute, ix, xi, 8, 233, 234, 235, 236, 242, 257, 258, 263, 265, 350, 351, 352, 353, 354, 386, 389, 437
additives, xiv, 36, 433, 437, 475
adducts, 147
adjustment, 291, 294, 392
administration, 254, 265
adsorption, 280, 326, 327, 440, 485
adult, 4, 8, 13, 355, 358, 388, 389, 390, 395, 402, 406, 407
adulthood, 389, 392, 408
adults, 4, 6, 8, 9, 262, 389, 409
aerobic, 243
Africa, 353
age, xi, 345, 350, 381, 388, 389, 391, 406, 408, 423
ageing, 327
agent, vii, 352, 353, 393
agents, 234, 262, 276, 286, 289, 309, 353, 388, 389, 406, 433, 443, 485
aggregation, 318
aggression, 234
aging, ix, 233, 235, 327, 328, 437
aging process, ix, 233, 235
agonist, 396
agricultural, vii, xiv, 1, 3, 16, 36, 249, 257, 259, 308, 309, 331, 334, 340, 345, 346, 352, 388, 389, 390, 412, 413, 422, 430, 442, 455, 476, 477, 491, 492
agricultural crop, vii, 1, 352
agricultural sector, 388
agriculture, 16, 36, 270, 351, 354, 378, 388, 405, 413, 433
agrochemicals, 389, 450, 451
agroindustrial, 455
aid, 394
air, xiv, 7, 20, 255, 265, 281, 287, 294, 356, 451, 475
airborne particles, 342
air-dried, 7
albumin, 7
alcohols, 243, 356, 477
Aldrin, xi, 39, 59, 126, 131, 151, 195, 282, 309, 326, 330, 331, 337, 338, 349, 350, 351, 372, 374, 376, 377, 380, 382, 463
algae, xiv, 340, 491, 492, 497, 504
alkaline, 436, 440, 443, 460

alkaloids, 343
Allium cepa, 450
alpha, 137, 139, 155, 199, 205
alternative, vii, x, xiv, 1, 4, 235, 236, 242, 271, 272, 276, 278, 280, 292, 298, 299, 307, 310, 312, 313, 315, 327, 337, 356, 441, 444, 455, 457, 464, 476, 492, 493
alternatives, x, 4, 269, 276
alters, 408
aluminium, 121, 123, 353
ambient pressure, 18
amine, 276
amines, 498
amino, 482
ammonium, 121, 470, 493
amphetamine, 342
amphibia, 266
amphibians, 248
analog, 443, 444
analysis of variance, 397, 401
analysts, 37
analytical techniques, xiii, 37, 273, 287, 430
analytical tools, 450
androgen, 396
androgens, 390
anesthetics, 241
aniline, xiii, 454, 455
animal husbandry, 36
animal studies, 11, 353, 409
animal tissues, 37
animals, 7, 235, 351, 352, 353, 354, 355, 389, 391, 392, 393, 402, 406
anodes, 353
ANOVA, 397, 400, 401
antagonist, 396
anthropogenic, 345
anti-estrogenic, 390
antioxidant, ix, 233, 234, 235, 242, 243, 244, 245, 246, 247, 248, 249, 250, 252, 253, 256, 260, 261, 262, 263, 265, 267
anxiety, 11, 392, 395, 396, 407, 408, 409
apoptosis, ix, 234, 242, 244, 250, 252, 253, 256, 257
application, vii, x, xii, xiii, 2, 235, 241, 250, 253, 256, 260, 270, 289, 297, 309, 320, 321, 328, 331, 335, 382, 429, 430, 431, 433, 436, 444, 447, 448, 450, 464, 470, 482, 487, 493
aquaculture, 266
aquatic systems, 436, 493
aqueous solution, 313, 315, 316, 317, 346, 432, 436, 440, 449, 450, 451, 457, 493
aqueous solutions, 317, 436, 440
arachidonic acid, 251
Argentina, 233, 254, 258, 259, 266
argon, 493
aromatic compounds, 477
aromatic hydrocarbons, 312, 346
arthritis, 353
aryl hydrocarbon receptor, ix, 234, 248, 256, 262
ascorbic, 243
ascorbic acid, 243
ash, 424
Asia, 353
assessment, 8, 11, 371, 376, 386, 406, 471, 493, 495
assimilation, 4
assumptions, 11, 253
ataxia, 352, 354
atherosclerosis, 356
athletes, 356
Atlantic, 358
Atlantic Ocean, 358
Atlantis, 73, 148
atmosphere, 312, 390
atmospheric pressure, 324, 361
atoms, 3, 354
Australia, 340, 485
autofluorescence, 493, 498, 499
automation, x, 269, 280, 287, 289, 299, 362, 455, 456, 469
availability, 16, 286, 476, 484
avoidance, 242, 396
azo dye, 488

B

Bacillus subtilis, 480, 486
bacteria, vii, 477, 505
bacterial, 12, 486
barley, viii, 35, 126, 278
barriers, 497, 501
basidiomycetes, 483, 486
batteries, 402
battery, 11, 251
beetles, 351
behavior, xii, 130, 242, 245, 257, 265, 266, 345, 386, 389, 390, 391, 392, 394, 395, 396, 397, 399, 400, 401, 405, 407, 409, 436
behavioral change, 389
benefits, 339, 356
benzene, 272
benzoquinone, 482
B-esterases, 235, 255
beverages, 455
bias, xii, 386, 391, 394, 396
bilirubin, 451

binding, ix, 5, 234, 250, 251, 252, 257, 264, 345, 449, 481, 486, 501
Bioanalytical, 448, 470
bioassay, 12, 261, 443, 447, 497
biocatalysts, xiv, 475
biochemistry, 258
biodegradability, 313
biodegradable, 309, 430
biodegradation, xiv, 430, 475, 476, 482, 483, 485, 489
bioindicators, 12, 254
biological interactions, 256, 257
biological processes, 387
biological responses, 390, 393
biomarker, 6, 7, 12, 234, 235, 241, 242, 245, 249, 250, 253, 254, 256, 258, 259, 260, 261, 263, 264, 389
biomarkers, vii, ix, 2, 4, 7, 9, 13, 234, 235, 241, 242, 249, 250, 253, 254, 256, 259, 261, 262, 263, 264, 266, 387, 395, 402
biomass, 483, 493
biomonitoring, 253, 267
bioreactor, 482, 488
bioremediation, 486, 489
biosciences, 264
biosensors, 450, 451
biosynthesis, 390, 486
biota, 340
biotransformation, 234, 244, 248, 263, 488
Biotransformation, 260, 261, 487
birds, vii, 355
birth, 16, 352, 392
bisphenol, 486
bivalve, 425
black tea, 463, 464
bladder, 352, 353
bladder cancer, 353
bleaching, 477, 480
blends, 280
blood, 278, 340, 342, 354, 387, 472
blood dyscrasias, 354
boats, 359
body weight, 351, 352, 353, 354
boiling, 17, 23, 280, 324, 460, 463
bonding, 294
bottleneck, 270
bovine, 7
bradycardia, 236
brain, ix, 233, 235, 236, 237, 238, 239, 240, 241, 247, 250, 255, 256, 258, 259, 264, 265, 266, 387, 390, 392, 395
Brazil, 385, 405
Brazilian, 256, 473
breakdown, 309, 353, 354, 412, 432, 439
breast milk, 352, 388, 473
breathing, 265
breeding, 391
Bromide, 119, 128
Brussels, 300
bubbles, 281, 288, 294
Buenos Aires, 233, 258, 259
buffer, 494, 497
butyric, 476
by-products, 432

C

Ca^{2+}, 250, 264, 501, 505
cabbage, 21, 29, 30
cadmium, 264
calcium, xiv, 121, 125, 492, 497, 501, 502, 503, 505
calibration, 27, 58, 86, 298, 327, 394, 442
calorie, 356
CAM, 12, 264
cAMP, 250, 262, 264
Canada, 37, 86, 353, 489
cancer, 351, 352, 353
candidates, 400
capacity, 16, 37, 244, 245, 251, 264, 280, 310, 313, 314, 316, 322, 457, 483
capillary, x, xiii, 16, 33, 58, 269, 270, 280, 293, 340, 341, 343, 371, 453, 454, 456, 466, 467, 469
CAR, 280, 455, 457, 462
carbamic acid, 264
carbohydrate, 260
carbon, 3, 17, 37, 261, 278, 289, 313, 424, 441, 470
carbon dioxide, 17
carbon tetrachloride, 261
Carbonyl, 128
carboxylic, 235
carcinogen, 352
carcinogenesis, xi, 350
carcinogenic, 352, 354, 382
carcinogenicity, 352, 353, 354
carcinogens, 16
cardiovascular disease, 356
carotene, 249
carotenoids, 278
carrier, 16, 17, 20, 23, 58, 85, 291, 362
CAT, 243, 244, 245, 246, 247, 252
catalase, ix, 234, 243, 252, 261, 266
catalyst, 436
catalytic activity, 477
catfish, 260, 263
cDNA, 253, 483
cell death, 493

cell division, 495, 497, 505
cell fate, ix, 234
cell line, ix, 234
cell lines, ix, 234
cell membranes, 493, 494
cell signaling, ix, 234, 250, 252, 253, 255, 265
cellular homeostasis, 234
cellular signaling pathway, 251
central nervous system, 389
cereals, viii, 4, 35, 37, 340, 497
c-Fos, 250, 252
CFSE, 497, 498, 499
channels, 273, 414, 423
chelators, 501
chemical oxidation, 448
chemical properties, 86, 147
chemical reactions, 430
chemical structures, 3, 274, 433, 434, 436
chemicals, viii, xii, xiv, 6, 8, 11, 12, 35, 36, 38, 86, 118, 124, 126, 127, 128, 129, 130, 147, 148, 231, 257, 259, 270, 345, 350, 351, 355, 361, 382, 386, 387, 389, 391, 393, 396, 408, 429, 475, 476, 480, 486, 487, 491, 492, 493
children, 353, 356, 388, 389, 402, 406, 408, 409
China, 35, 37, 86, 340, 427, 430
Chinook salmon, 240, 249, 257, 258, 267
Chl, 42, 62, 123, 133, 150
chlordane, 41, 61, 142, 200, 352, 353, 476
chloride, 120, 121, 129, 138, 141, 188, 193, 226, 440, 482
chloride anion, 440
chlorine, 2, 440, 448
chlorobenzene, 352
chloroform, 7
chlorophyll, xiv, 277, 278, 491, 492, 493, 495, 496, 497
chloroplast, 501
chlorpromazine, 480
cholinergic, vii, ix, 1, 233, 235, 236, 242, 250, 252, 253, 260, 264, 408
cholinesterase, 5, 234, 235, 241, 242, 255, 258, 259, 260, 261, 263, 265, 266, 267, 309, 442, 449, 450, 451
cholinesterase inhibitors, 263, 309
chromatograms, 23, 31, 85, 86, 118, 274, 276, 281, 282, 285, 337, 362, 366, 442, 444, 446, 461, 462, 463, 464, 465, 466, 467, 468
chromatographic technique, xiii, 299, 339, 430, 441, 451
chromatography, ix, xiii, 16, 33, 269, 270, 319, 331, 334, 340, 341, 342, 343, 344, 347, 418, 431, 433, 440, 447, 449, 451, 452, 453, 454, 468, 471, 472, 473
cis, 40, 41, 42, 60, 61, 135, 139, 187, 198, 199, 202, 351, 371, 374, 380, 435, 450
citrus, 342, 497
CL, 263
clams, 416, 418
classes, xi, 249, 385, 386, 460, 469
classical, xiii, 276, 311, 341, 430, 431, 454, 464
clay, 327
cleaning, 469
cleanup, viii, 35, 37, 38, 82, 83, 122, 126, 128, 130, 149, 231, 297, 310
cleavage, 235, 244
clinical trial, 405
clinical trials, 405
clusters, 495
CMC, 308, 315
c-Myc, 252
CNS, xi, 385
Co, 341, 481, 484, 494
coastal areas, 340
coatings, 280, 281, 286, 456, 457, 472
cocaine, 409
coefficient of variation, 396
coffee, 462, 463, 473
cognition, 390
cognitive, 395, 406
cognitive dysfunction, 395
cognitive function, 406
cohort, 352
Colorado, 4
coma, 235
combined effect, 386
commodity, 383
communication, xi, 349, 424
communities, 389
community, 12, 418
compatibility, 27, 313, 314, 322
compensation, 298
competition, 336, 339
competitiveness, 339
complement, xiv, 402, 423, 475, 476, 480, 486, 488
complexity, vii, 2, 6, 10, 13, 253, 298, 310, 381, 387, 393, 495
compliance, 27, 371, 383
components, x, xi, 17, 276, 284, 308, 310, 334, 341, 349, 386, 391, 394, 395, 455, 471, 500
composition, xiii, 335, 352, 356, 453, 467
concentrates, 442
concentration ratios, 420
conditioning, 83, 297, 299, 327, 335
conduction, 440
conductivity, 326
confidence, 331, 393, 417

confidence interval, 331
configuration, 292, 293, 444
conformity, 36, 394
confounders, 391, 395
confrontation, 396
confusion, 352
conjugation, 244, 248, 254
connective tissue, 356
conservation, 261
construct validity, 391, 396
consumers, 359
consumption, 8, 9, 264, 273, 276, 279, 299, 311, 339, 356, 382, 392, 430, 431
contact time, 327
contaminant, 256, 262, 266, 355, 362
contaminants, 244, 253, 262, 271, 275, 298, 322, 345, 355, 357, 371, 378, 379, 454, 470, 472, 493
contamination, xi, 15, 31, 36, 241, 244, 261, 265, 270, 321, 340, 349, 350, 351, 352, 354, 355, 356, 357, 361, 376, 378, 379, 381, 382, 412, 483, 484
continental shelf, xii, 411, 412, 414, 421, 422, 426
control, vii, x, xii, 1, 4, 7, 236, 243, 248, 250, 251, 307, 308, 309, 351, 352, 353, 386, 388, 391, 393, 394, 396, 397, 398, 399, 401, 406, 412, 422, 426, 452, 494, 495, 496, 497, 498, 499, 500, 501, 502, 503
control condition, 498
control group, 7, 396
controlled trials, 408
conversion, 347, 439, 442, 443, 485
convulsion, 353
cooling, 17, 20, 315
coordination, 264, 406
copolymers, 280
copper, 477, 479, 480, 485, 486, 505
correlation, 149, 150, 173, 234, 242, 253, 255, 323, 328, 351, 352, 382, 405, 485
correlation coefficient, 149, 150, 173, 328
correlations, 395
corticosterone, 408
cortisol, 257
cosmetics, 313
costs, 271, 278
cotton, 4, 352
coupling, x, xiii, 269, 308, 331, 333, 334, 335, 341, 346, 430, 469, 480, 484, 486
covalent, 345, 481, 482, 484
covalent bond, 484
coverage, 421
covering, 282, 322, 358
crack, 37
critical micelle concentration, 315
critical period, xii, 385, 390, 391, 395
critical temperature, 315
criticism, xii, 386
crops, vii, 1, 350, 352
cross-sectional, 310
cryogen, 17
crystallisation, 485
CT, 361
culture, xiv, 7, 259, 492, 494, 495, 496, 497, 499, 500, 502, 503
cycles, 128, 292
cycling, 82, 147, 265
cyclohexane, viii, 19, 20, 21, 22, 23, 24, 25, 27, 31, 36, 38, 82, 83, 128, 130, 294
cypermethrin, 122, 125, 131, 163, 205, 388, 432, 452
cysteine, 262, 484
cytochrome, ix, 234, 243, 248, 254, 257, 258, 259, 265
cytokine, 242
cytokines, 251
cytokinesis, 264
cytometry, xiv, 491, 492, 493, 497, 503, 504, 505
cytoplasm, 500
cytoplasmic membrane, 501, 503
cytoprotective, 252
cytosol, 251
cytosolic, 505
Cytotoxic effects, xiv, 491, 492
cytotoxicity, 265, 503

D

DAD, xiii, 275, 429, 431, 444, 445, 463, 464, 466
dairy, 356
dairy products, 356
data analysis, 270
data collection, 390
database, viii, 11, 35, 38, 126, 362, 365, 393, 421
dating, 423
daughter cells, 498, 499
de novo, 239
death, 5, 235, 236, 242, 244, 256, 352
decay, 437
decision making, 11
decomposition, 13, 309, 327, 432, 436, 451
deep-sea, 260, 412, 414
defects, 16
defense, 243, 260, 261, 386, 396
defense mechanisms, 261, 386
defenses, 244
deficits, 406
deformation, 426

degradation, xii, xiii, xiv, 277, 278, 309, 318, 324, 327, 345, 346, 351, 352, 429, 430, 431, 432, 433, 436, 437, 439, 440, 441, 443, 447, 448, 449, 450, 451, 470, 475, 480, 481, 482, 483, 484, 486, 487, 488
degradation mechanism, 327, 432
degradation pathway, 447
degradation process, 327, 441, 447, 483
degradation rate, 309, 437
degrading, 327, 430
dehalogenation, 484
dehydrogenase, 12, 254
delivery, 21, 423
demand, 359
Denmark, 387
density, 18, 316, 318, 395, 495
Department of the Interior, 383
dependent variable, 324
depolarization, 501
deposition, 333, 354, 357, 424, 457
deposits, 423, 424
depressed, 259
derivatives, 3, 318, 382, 460, 467, 482, 484, 501
desertion, 324
desorption, 280, 281, 284, 298, 310, 331, 332, 335, 337, 345, 455, 456, 460, 463, 467, 468, 470
destruction, 430
detection, vii, ix, x, 15, 16, 36, 82, 86, 231, 270, 271, 274, 279, 281, 282, 283, 289, 292, 308, 319, 320, 327, 328, 329, 332, 334, 337, 340, 343, 347, 356, 418, 431, 442, 443, 447, 448, 450, 451, 452, 460, 468, 469, 472, 473
detergents, 241
detoxification, 235, 236, 244, 248, 256, 483, 484, 486, 488
detoxifying, 234, 235, 242, 244, 245, 248, 249, 250, 251, 253, 258, 484
detritus, 6
developed countries, x, 307, 308
developing brain, 389
developing countries, 234, 351
developmental delay, 391
deviation, 27, 372
diabetes, 356
diacylglycerol, 255
Diamond, 487
dibenzofurans, 344
diet, 356, 386, 391, 402, 454
dietary, 4, 6, 259, 356, 387, 408
dietary habits, 356
diets, 391
differentiation, 250, 256, 389
diffusion, 287, 290, 291, 333, 354, 355, 357
diffusion rates, 287
diffusivity, 326
digestibility, 356
digestion, 321, 322
dimer, 480
dioxin, 248, 257, 259, 261
dioxins, 251
directives, 270
discharges, xiv, 491, 492, 501
discomfort, 353
discrimination, 20, 23
diseases, 350, 356, 389
disorder, 353
dispersion, xii, xiii, 271, 273, 276, 394, 411, 453, 470
dissociation, 294
dissolved oxygen, xiv, 475
distilled water, 7, 284
distribution, 281, 298, 316, 340, 395, 418, 422, 423, 427, 459, 493, 495, 497, 501
diuron, 256, 476
diversity, 5, 16, 249
diving, 414
division, 495, 498, 505
dizziness, 352, 353
DNA, 244, 251, 254, 260, 267, 476, 500, 504
DNA damage, 244, 254
docosahexaenoic acid, 356
donor, 290, 291, 293, 294
dosage, 436, 440, 441
dose-response relationship, 406
dosimetry, 406
dosing, 391
drainage, 413
drinking, xiii, 358, 386, 407, 430, 431, 432, 450, 451, 454, 455, 467
drinking water, xiii, 358, 386, 407, 430, 431, 432, 450, 451, 454, 455, 467
drug exposure, 395
drugs, 36, 408, 470
drying, 276
duration, 250, 322, 352, 389
dust, 356
dyes, 480, 486, 488, 493

E

early warning, 242
earth, 275
earthworms, 6
eating, 355
ecological, 5, 8, 12, 242, 253, 257, 484, 493
ecosystem, 256, 378

ecosystems, xiv, 242, 476, 491, 492, 493
ecotoxicology, 6, 253, 266
Education, 15
effluent, 425
effluents, 147, 426
Egypt, 254
eicosapentaenoic acid, 356
elderly, 356
electrolyte, 294
electron, 418, 431, 464, 473, 479, 480, 484, 495, 501
electron microscopy, 495
electron paramagnetic resonance, 479
electrons, 440, 477, 480
electrophoresis, x, xiii, 269, 270, 343, 453, 454, 469
ELISA, 340, 344
embryo, 254, 261
embryogenesis, 266
embryos, 267
emission, 340, 444, 462, 464, 468, 472, 473, 493, 494, 499, 500, 501, 502, 503
emotional, 395, 396
emotionality, 407
employees, 359
employment, 351
encoding, 249, 483
endocrine, 257, 389, 392, 396, 407, 409, 480, 486
endocrine system, 389, 392
energy, xiv, 58, 73, 74, 75, 311, 312, 314, 342, 355, 358, 475, 497
enlargement, 353
entrapment, 430
environment, vii, x, xi, xii, xiv, 2, 6, 10, 11, 12, 36, 234, 254, 270, 299, 307, 309, 312, 321, 339, 349, 350, 351, 353, 382, 387, 390, 391, 396, 402, 405, 412, 426, 429, 430, 431, 449, 469, 475, 476, 484, 491, 492, 493, 501
environmental change, xiv, 491, 492
environmental conditions, xiv, 386, 447, 483, 491, 492, 503
environmental contaminants, xiv, 241, 259, 389, 390, 475
environmental control, 391
environmental degradation, 351
environmental effects, xiv, 491, 492
environmental factors, 276
environmental impact, 270
environmental protection, 477
Environmental Protection Agency, 270, 308, 309, 340, 345, 412, 425, 426, 427
environmental regulations, xiv, 475, 476
enzymatic, 258, 260, 351, 442, 476
enzymatic activity, 442
enzyme induction, 259
enzymes, ix, xiv, 233, 234, 236, 241, 242, 243, 244, 245, 248, 249, 250, 251, 253, 254, 258, 260, 261, 262, 263, 265, 443, 475, 476, 477, 480, 482, 484, 485, 486
EPA, 2, 3, 4, 8, 9, 10, 11, 13, 270, 308, 309, 330, 340, 413, 419, 426
epidemics, 350
EPR, 477, 479
equilibrium, x, 235, 244, 270, 280, 284, 287, 297, 298, 310, 311, 332, 355, 457, 472
equipment, x, xi, 21, 32, 82, 270, 286, 349, 391, 393, 394, 430
erbium, 343
erosion, 418, 422
erythrocytes, 255, 260
ester, 46, 51, 54, 65, 66, 75, 78, 121, 123, 126, 127, 140, 143, 166, 167, 176, 184, 208, 216, 222, 434, 435, 437, 438, 497, 505
esterase, 235, 239, 250, 253, 254, 266, 442
esterases, ix, 233, 234, 235, 241, 255, 498
esters, 3, 235, 242, 264, 432, 437, 439, 443
estrogen, 262, 391, 396
estrogens, 390
estuaries, 261, 492
estuarine, 241, 259, 265
ethane, 119, 123, 412, 464, 476, 483
ethanol, 7, 265, 463, 464, 473
Ethanol, 464
Ether, 308, 323, 361
ethyl acetate, viii, 19, 36, 37, 38, 82, 83, 128, 130, 274, 275, 276
ethylene, 315, 412
ethylene oxide, 315
etiology, 396
EU, xiii, 11, 37, 86, 270, 278, 430, 454, 469
Europe, 4, 304, 352, 353, 383
European Commission, 232
European Community, 376, 454
European Parliament, 383
European Union, 2, 16, 270, 300, 405, 443, 451, 454
evaporation, 18, 23, 83, 128, 271, 272, 312, 322, 354
evolution, 497, 499
excitation, 444, 445, 493
experimental condition, vii, 2, 339, 391, 394, 396, 432, 439, 442
experimental design, 327, 344
expert, iv
extraction process, x, xiii, 278, 293, 307, 308, 316, 319, 322, 324, 332, 335, 453, 454, 460, 469
extrapolation, 11

F

face validity, 396
facies, 423
factorial, 323, 324
failure, 414
false negative, viii, 15, 19, 31, 442
family, 327, 356, 405, 488
farming, 357, 376
fasciculation, 235
fat, xi, 128, 130, 275, 278, 350, 355, 356, 357, 361, 372, 374, 376, 377, 378, 379, 380, 381
fatigue, 352
fatty acid, 356, 455, 471, 500
fatty acids, 356, 455, 471, 500
FDA, 374, 376, 380
February, 417, 426
feeding, 4, 9, 10, 242
females, 392
fenarimol, 396, 405, 409
fertility, 389
ferulic acid, 481, 482
fetal, 392, 408
fetal brain, 392
fetus, 409
fiber, x, xiii, 270, 271, 279, 280, 281, 282, 286, 290, 291, 292, 293, 294, 297, 299, 310, 311, 331, 332, 333, 453, 454, 455, 456, 457, 459, 460, 461, 462, 463, 464, 465, 466, 467, 468, 469
fiber membranes, 292
fibers, 280, 281, 286, 292, 457, 460, 461, 466, 469, 470
FID, 20, 23, 27, 431
film, 18, 58, 279, 280, 287, 362
film thickness, 280, 362
filtration, 272, 289, 312, 322, 340, 482
financial support, 504
fish, vii, ix, xi, 233, 234, 235, 236, 237, 238, 239, 241, 242, 244, 245, 246, 247, 248, 249, 250, 251, 252, 253, 254, 255, 256, 258, 259, 260, 261, 262, 263, 264, 265, 266, 267, 349, 350, 355, 356, 357, 358, 362, 366, 376, 377, 381, 383, 463, 464, 473
fishing, 359
fitness, 390
FL, 504
flame, 431, 447, 461, 462
flame ionization detector, 447
flank, xii, 411, 415, 421, 422, 424, 425
flavor, 471
floating, 497
flow, xiii, xiv, 17, 18, 20, 21, 23, 24, 25, 27, 58, 73, 82, 147, 286, 361, 362, 414, 423, 426, 430, 439, 442, 446, 492, 493, 497, 503, 504, 505
flow rate, 58, 73, 82, 147, 439, 446
flow value, 24
fluid, xiii, 37, 311, 312, 341, 342, 344, 345, 453, 454, 470
fluid extract, xiii, 37, 311, 312, 341, 342, 344, 345, 453, 454, 470
fluorescence, xiv, 340, 477, 480, 491, 492, 493, 495, 496, 497, 498, 499, 500, 501, 502, 503
fluorogenic, 500
fluvial, 413
focusing, 388
food, vii, viii, ix, x, xi, xiii, 4, 6, 7, 8, 9, 12, 15, 16, 19, 21, 27, 31, 242, 258, 264, 269, 270, 271, 276, 278, 279, 281, 296, 298, 299, 300, 309, 341, 349, 350, 352, 354, 355, 356, 371, 388, 391, 392, 407, 430, 431, 448, 450, 451, 452, 453, 454, 455, 457, 458, 460, 461, 462, 463, 464, 466, 467, 468, 469, 470, 473, 477, 493
Food and Drug Administration, 383
food commodities, 454
food products, 16
food safety, xi, xiii, 349, 453, 454
foodstuffs, 36, 270, 431, 442, 443
Ford, 489
Forest Service, 12
fortification, 19, 122, 126, 148, 149, 231
FP, 504
fractionation, 271, 273
France, 382, 427
free radical, 243, 244, 432, 500
free radicals, 243, 244, 432
free-radical, 255
freezing, 361
fresh water, 266
freshwater, 235, 254, 260, 262, 263, 492, 497, 505
frontal cortex, 266
fruit juice, xiii, 37, 272, 274, 283, 442, 451, 453, 455, 457, 459, 460, 461, 462, 466, 471
fruit juices, xiii, 37, 272, 274, 283, 451, 453, 455, 457, 459, 460, 461, 462, 466, 471
fruits, 4, 37, 270, 272, 274, 276, 278, 340, 342, 356, 442, 450, 467, 470, 471, 472
FTD, 461, 462
fumarate, 434, 438
functional changes, 389, 390
fungal, 480, 483, 485, 486, 487, 488
fungi, xiv, 353, 443, 475, 477, 481, 483, 485, 486, 487, 489
fungicide, 258, 263, 396, 409, 456, 470, 482, 483, 484, 487
fungus, 480, 481, 482, 483, 484, 486, 488
furs, 11

G

G protein, 252
Gamma, 43, 63, 125
gas, vii, ix, xiii, 15, 16, 17, 18, 20, 23, 24, 33, 58, 73, 85, 128, 147, 269, 270, 278, 281, 309, 310, 314, 315, 331, 340, 341, 342, 343, 344, 347, 362, 414, 418, 425, 426, 431, 433, 436, 447, 448, 449, 451, 452, 453, 454, 470, 471, 472, 473
gas chromatograph, vii, ix, 15, 16, 17, 20, 24, 33, 269, 270, 310, 331, 340, 341, 342, 343, 344, 347, 418, 431, 433, 447, 449, 451, 452, 470, 471, 472, 473
gases, 3
gastrointestinal, 352
gauge, 18
Gaussian, 395
GCS, 244, 252
gel, 16, 37, 130, 457
gel permeation chromatography, 16, 130
gender, 406, 409
gender differences, 409
gene, ix, 234, 250, 251, 253, 257, 260, 262, 264, 402
gene expression, ix, 234, 250, 251, 253, 257, 260, 262
generation, xiv, 249, 254, 430, 491, 492, 493, 498, 505
genes, 249, 251
genetics, 392
Geneva, 427
genome, 483
genomics, 253, 254
genotoxic, 259, 341
Germany, 3, 19, 32, 37, 86, 282, 292, 353, 361, 480, 481
gestation, xi, 385, 405, 409
gill, 236, 244, 245, 247, 255, 266
GIS, 340
glass, 83, 282, 284, 297, 332, 361, 436, 452
glucocorticoids, 408
glutamate, 407
glutathione, ix, 4, 7, 9, 233, 234, 243, 244, 252, 254, 255, 258, 261, 262, 263, 265, 484
glutathione peroxidase, 243, 255
glycol, 344
GPC, viii, 16, 21, 35, 36, 37, 38, 82, 83, 130
GPx, 243, 244, 245, 246, 247, 252
grain, viii, 35, 36, 85, 128, 130, 149, 231, 422
grains, viii, 35, 37, 130, 149, 195, 209, 214, 231
grape juice, 274
graphite, 353
grasses, 493, 497
gravitational force, 287
gravity, 423
green tea, 286
groundwater, 309, 321, 407
grouping, 38, 39, 49, 375
groups, xii, xiv, 5, 38, 85, 118, 126, 147, 245, 277, 312, 313, 371, 376, 387, 391, 394, 395, 400, 401, 429, 439, 449, 477, 483, 491, 492, 498
growth, 244, 353, 426, 477, 483, 493, 497, 499, 504
growth rate, 493
GST, 4, 243, 244, 246, 247, 248, 251, 252
guava, 462, 473
guidance, 388
guidelines, 356, 371
Gulf of Mexico, 358

H

H_2, 20, 275
HA, 472
habitat, 378
habituation, 409
half-life, 239, 412, 437, 476, 484
halogenated, 473, 480
handling, 272, 279, 297, 314, 334, 433
harmful effects, 386
harvest, 389
hazardous wastes, xiv, 475, 476
hazards, 4, 11, 271, 389, 414
H-bonding, 327
HE, 260, 500
headache, 352, 354
health, xiii, 12, 234, 270, 271, 276, 350, 356, 386, 387, 389, 390, 392, 402, 405, 408, 431, 453, 454, 467, 476
health effects, 386, 387, 408
heat, 17, 23, 242, 251, 257, 311, 322, 324, 386
heat shock protein, 242, 251, 257
heating, 17, 82, 128, 280
helium, 362
hematological, 242
heme, 245
hepatocytes, 251, 257, 259, 263, 265
heptane, 275
herbal, 464, 473
herbicide, 248, 260, 340, 346, 482, 484, 487, 493, 497, 504
herbicides, xi, xiii, xiv, 241, 270, 274, 276, 278, 283, 292, 296, 346, 385, 449, 454, 455, 467, 470, 471, 472, 473, 480, 491, 492, 493, 497, 503, 505
Herbicides, 2, 275
herbs, 455
heterodimer, 251

heterogeneity, 5, 345
heterogeneous, xiv, 491, 492, 497
hexachlorobenzene, 278, 464
hexachlorobiphenyl, 264
hexane, viii, 36, 38, 83, 128, 274, 275, 287
high fat, 376
high pressure, 311, 436
high resolution, 37
high temperature, 288, 311, 312, 322, 333
high-level, viii, xi, 36, 349
high-performance liquid chromatography, 340, 344, 450, 471, 473
histogram, 494, 495, 496, 497, 498, 499, 500, 503
histological, 242, 261
histology, 256
Holocene, 424, 426, 427
homeostasis, 244, 253, 261, 505
homogenized, 16, 373
homogenous, 130
honey, 272, 281, 282, 284, 431, 448, 449, 469, 473
honey bees, 431
horizon, 418
hormones, 389, 485
House, 426
housing, 391, 393, 409
HPA, 392
HPLC, x, xiii, 16, 17, 37, 38, 232, 308, 310, 313, 314, 319, 322, 327, 329, 331, 334, 336, 337, 339, 343, 346, 347, 429, 430, 431, 433, 437, 440, 444, 445, 446, 447, 460, 463, 464, 466, 467, 468, 469, 470
HPLC-MS/MS, 466, 467, 468
HR, 248, 257, 262
HSP90, 251
human, xii, 11, 36, 256, 257, 262, 265, 270, 340, 341, 352, 353, 354, 356, 382, 385, 386, 387, 389, 390, 396, 431, 472, 476, 493, 503
human activity, 354
human behavior, 389, 396
human exposure, 11, 353
humane, 391
humans, vii, xii, 234, 309, 351, 352, 353, 354, 355, 382, 385, 387, 396, 429
humic acid, 272, 327, 484, 487
humic substances, 327
humidity, 7
Hungary, 15, 16, 19, 32
husbandry, 36, 393, 409
hydrate, 120, 125, 137, 425, 426
hydro, 291, 312, 313, 319, 322, 341, 342, 344, 345, 346, 455, 480, 482, 487
hydrocarbon, ix, 234, 248, 251, 256, 262, 313, 315
hydrocarbons, 341, 342, 344, 488
hydrodynamic, 341
hydrogen, xiv, 7, 243, 245, 250, 294, 317, 448, 475, 505
hydrogen bonds, 317
hydrogen peroxide, xiv, 243, 245, 250, 448, 475, 505
hydrolases, 242
hydrolysis, 242, 248, 264, 327, 346, 436, 440, 442, 443
hydrolyzed, 443
hydroperoxides, 245
hydrophilic, 291, 313, 482
hydrophobic, xiv, 289, 291, 313, 475
hydrophobic properties, 291
hydrophobicity, 291, 309
hydroxyl, 235, 264, 436
hygienic, 270
hyperactivity, 235
hypercholesterolemia, 356
hypertrichosis, 353
hypertriglyceridemia, 356
hypothalamic, 396, 407
hypothalamus, 396
hypothesis, xi, 236, 242, 379, 381, 385, 400, 401

I

IARC, 352, 353, 382
ice, 361
identification, xiii, 86, 127, 235, 241, 253, 254, 281, 282, 288, 362, 373, 386, 387, 389, 391, 392, 429, 433, 456, 468, 497
identity, 389
ILAR, 406, 407, 409
image analysis, 504
images, 417
immersion, xiii, 279, 281, 286, 332, 453, 455, 457, 459, 462, 464, 466, 467, 468
immobilization, 5, 292
immune system, 353
immunological, 260
implementation, xiv, 12, 402, 475, 476, 484
impurities, 16, 273, 354, 438
in situ, 476
in vitro, 255, 256, 259, 262, 390, 481, 484, 486
in vivo, xiv, 236, 251, 254, 255, 256, 264, 265, 266, 390, 405, 409, 486, 491, 492, 498
inactivation, 234, 235, 266
incidence, 356
incubation, 494
incubation period, 494
indication, 359, 423
indicators, 250, 264, 266, 405, 501

indices, 8, 402
indigenous, 256
indirect effect, 253
inducer, 484
induction, ix, 234, 242, 244, 245, 248, 249, 252, 253, 259, 265, 484, 486, 500
industrial, 351, 447, 448, 450, 476, 484, 488
industry, 353, 483, 485
inert, 280, 318, 433, 443, 487
infants, 353, 389
infection, 386
infrastructure, 387, 390
infusions, 473
ingestion, 352
inhibition, ix, 4, 5, 6, 233, 234, 235, 236, 237, 238, 239, 240, 241, 242, 245, 250, 252, 253, 255, 256, 257, 258, 259, 260, 261, 263, 264, 387, 390, 442, 443, 445, 447, 449, 450, 451
inhibitors, ix, 233, 263, 309, 442, 501
injection, vii, viii, xiii, 15, 16, 17, 18, 19, 20, 23, 24, 26, 27, 31, 58, 73, 82, 86, 276, 278, 280, 284, 292, 337, 371, 391, 430, 442, 451, 460
injections, 17, 18, 20, 25, 31, 83, 362, 371
injuries, 245
injury, iv
inorganic, 326, 386, 431, 492
inositol, 251
insecticides, vii, x, xi, xii, xiii, 1, 2, 3, 4, 5, 234, 235, 236, 239, 241, 244, 246, 248, 249, 250, 251, 252, 253, 255, 259, 261, 265, 267, 270, 277, 307, 312, 342, 354, 385, 388, 390, 429, 446, 447, 448, 449, 451, 462, 471, 472, 480
insects, vii, 1, 4, 351, 352, 353, 412, 477
insight, 432
inspection, 37
Inspection, 35, 232
inspections, xi, 349
instability, 286
instruments, x, 84, 269, 270, 271
insults, 389
integration, 147, 387, 392, 402
integrity, 243, 494, 497
intensity, 147, 250, 371, 493
interaction, 266, 322, 323, 327, 387, 391, 402
interactions, 256, 257, 262, 325, 330, 345, 386, 388
interface, xiii, 16, 58, 261, 280, 291, 310, 453, 455, 460, 463, 467, 469
interference, 147, 289, 294, 298, 365
International Agency for Research on Cancer, 352, 353, 382
International Labour Organisation, 382, 383
interpretation, xii, 354, 386, 387, 388, 390, 391, 393, 406
interval, 18, 328, 420, 421
intervention, 387, 391
intervention strategies, 387
intravascular, 254
invertebrates, vii, ix, 1, 3, 5, 6, 233, 258, 259
Investigations, 383, 409, 447
iodine, 442
ionic, xiv, 272, 281, 287, 291, 294, 310, 313, 315, 318, 319, 322, 332, 333, 343, 365, 371, 457, 475
ionization, 58, 124, 291, 341, 342, 371, 431, 447
ionization energy, 58
ions, 58, 59, 60, 69, 84, 124, 126, 147, 294, 366, 371, 373, 434, 477, 479, 480, 486, 488, 501
IR, 254
iron, 245
irradiation, 440, 441
island, 321
isoenzymes, 241, 249, 258
isoforms, 241, 251, 257
isolation, 271, 273, 276, 280, 310, 386, 406
isomers, 120, 135, 351, 354, 360, 374, 380, 381
isopods, 6, 7, 8, 9, 10, 11, 12
isotope, 147, 424
isotropic, 311, 315
Italy, xi, 256, 341, 349, 350, 357, 358, 359, 361, 379, 453

J

January, 406
Japan, 4, 36, 37, 86, 352, 481, 484
Japanese, 36, 359
JNK, 251
judge, 352, 353
juices, 275, 289, 296
Jun, 251
Jung, 431, 449
juveniles, 6, 8, 9, 259

K

K^+, 501
kidney, 244, 247, 255, 261, 265, 387
kidneys, 353
kinase, 251, 253, 255, 257, 264, 265
kinase activity, 265
kinases, 242, 250, 251, 255
kinetic model, 406
kinetic parameters, 265
kinetics, xiii, 429, 433, 439, 440, 442, 448, 451, 484, 485, 499
KL, 255

L

L1, 400
L2, 400
LA, 257, 258, 259, 413, 505
labor, ix, 269, 271
laboratory studies, 6
labour, 431
laccase, xiv, 475, 477, 480, 482, 483, 484, 485, 486, 487, 488, 489
lactate dehydrogenase, 12, 254
lactation, 352, 396, 409
lakes, xi, 350, 355, 357, 358, 376, 377, 379, 381, 467
lambda, 204
land, 413, 414, 422
larvae, 238
larval, 255, 257, 263, 266
laser, 444, 445, 446, 447, 493
Late Quaternary, 423
Latino, 258, 259
law, 11
laws, 12, 300
leach, 309
leaching, xiv, 491, 492, 493
lead, xii, 234, 235, 244, 245, 250, 252, 253, 257, 354, 385, 387, 424
leakage, 254, 293
legislation, vii, xiii, 15, 278, 300, 430, 443, 466, 469
legumes, 497
lens, xiii, 430, 442, 444, 447, 448, 449, 450, 451
Lepidoptera, 352
lesions, 244, 353
lettuce, 277, 278
leukocytes, 405
lifetime, 279
ligand, 248, 251, 257
ligands, 248
lignin, 477, 484, 487, 489
limbic system, 392
limitation, 405
limitations, xii, 271, 273, 330, 386, 387, 392, 393
lindane, 274, 476, 481, 482, 488
linear, 86, 148, 149, 150, 173, 280, 328, 372, 457
linkage, 6, 482
links, 387, 409
lipase, 255
lipid, 244, 245, 251, 255, 262, 340, 353, 354, 378, 493, 500, 505
lipid peroxidation, 244, 251, 255, 262, 493, 500, 505
lipid peroxides, 244
lipids, 276, 356, 501
lipophilic, 278, 309, 361
liposomes, 505
liquid chromatography, x, xiii, 270, 308, 310, 315, 319, 340, 341, 343, 344, 346, 431, 433, 444, 447, 448, 449, 450, 451, 453, 454, 470, 471, 472, 473
liquid nitrogen, 17
liquid phase, 16, 17, 271
lithologic, 422
liver, 235, 241, 244, 245, 247, 250, 251, 255, 257, 258, 259, 261, 262, 265, 275, 352, 353, 361, 378, 387
liver cancer, 352
LM, 256, 259
location, 413, 420
locomotion, 396, 397
locomotor activity, 406
LOD, 148, 149, 150, 151, 173, 308, 334, 373, 443, 445, 460, 461, 462, 463, 464, 466, 467, 468
long distance, 390
long period, 241, 327
long-term, vii, x, 2, 12, 307, 309, 340, 396, 414
Los Angeles, xii, 411, 412, 413, 421
losses, 23, 271, 278, 280, 310, 322
low molecular weight, 243
low temperatures, 17, 241
low-level, viii, 36
LPO, 245, 246, 247, 249
lumen, 290, 292, 294
lung, 352, 353
lung cancer, 352, 353
lungs, 353
LV, 471
lymph, 353
lymph node, 353
lymphocyte, 497
lymphocytes, 505
lysis, 500

M

magnesium, 19
magnet, 282
magnetic, iv, 282
magnets, 282
maintenance, viii, 15, 19, 31, 356, 394
maize, viii, 35, 126, 350, 352, 497
malaria, 308, 351
Malathion, 10, 19, 28, 29, 30, 39, 57, 59, 81, 140, 152, 193, 196, 228, 238, 309, 316, 319, 329, 330, 436, 437, 438
males, 386
mammalian cell, ix, 234, 250

mammalian cells, 250
mammals, vii, ix, 233, 248, 355, 392
management, xi, 349
manipulation, 284, 310, 390, 391, 392
manufacturing, 351, 352, 412, 430
MAPK, 251, 262
mapping, 425
marine environment, 375, 426
marine mammals, 344, 355
market, 13, 21, 357, 381, 382, 454
markets, 36
mask, 395
mass spectrometry, 270, 275, 299, 340, 341, 342, 343, 347, 431, 433, 447, 448, 450, 451, 452, 470, 471, 472, 473
mass transfer, 280, 281, 287, 333
maternal, xi, 385, 388, 392, 393, 395, 405, 407, 409
maternal care, 405
matrix, xiii, 15, 16, 31, 58, 84, 85, 86, 271, 273, 276, 279, 280, 281, 289, 298, 299, 310, 311, 312, 322, 324, 325, 327, 330, 331, 332, 334, 362, 365, 371, 448, 453, 454, 455, 457, 460, 461, 462, 463, 464, 466, 467, 468, 469, 470, 480, 485, 486
maturation, 395
MDA, 245, 246, 247
measurement, 258, 259, 298, 398, 422, 445, 457, 493, 497, 502
measures, 255, 298, 355, 358, 379, 382, 395, 396
meat, 356
media, x, 239, 307, 322, 386, 442, 450
mediators, 477, 484, 486, 487, 488
medical products, 354
medicinal plants, 346
medicine, 260
Mediterranean, xi, 263, 349, 358
membranes, 264, 495, 497, 500, 501
memory, 390
menstruation, 352
Merck, 7, 361
messages, 147
meta-analysis, 393, 409
metabolic, 242, 244, 260, 355, 356, 379, 386, 500, 504
metabolic disorder, 260, 356
metabolic pathways, 386
metabolism, 243, 244, 248, 249, 252, 254, 255, 258, 263, 265, 353, 501
metabolite, 482, 483
metabolites, x, 248, 265, 307, 309, 327, 341, 351, 354, 374, 380, 381, 382, 431, 442, 462, 467, 472, 480, 481, 482, 483, 484, 486, 487, 488
metabolizing, 260
metal ions, 488
metal oxide, 327
metal oxides, 327
metals, 241, 386, 407
methamphetamine, 342, 409
methanol, 7, 38, 271, 284, 332, 335, 337, 467, 468
methionine, 484
methyl group, 432
methyl groups, 432
$MgSO_4$, 37, 276
mice, 255, 265, 354, 386, 408, 409
micelle formation, 313
micelles, 313, 315, 317, 323
microalgae, xiv, 491, 492, 493, 497, 501, 503, 504
microarray, 253, 254, 260
microarray technology, 253, 260
microbes, vii
microbial, 327
microcosm, 6
microcosms, 13
microelectrodes, 501
microemulsions, 294
microflora, 340, 393
micrograms, 373, 379
microorganisms, 430, 476, 492, 501
Micropollutants, 341
microscopy, 495
microwave, x, 296, 307, 310, 311, 312, 314, 321, 322, 323, 324, 331, 332, 335, 339, 342, 343, 344, 345, 346, 347, 449, 473
microwave radiation, 322
microwaves, 312
migration, 242, 293
milk, 272, 278, 296, 352, 388, 462, 463, 464, 473
mineralization, 432, 433, 436, 440, 441, 450
minerals, 356
miniaturization, x, 269, 276, 286, 299
Ministry of Education, 485
miosis, 235
mitochondria, 502
mitogen, 251
mitosis, 498
mixing, 58, 83, 346, 421, 422, 424, 425
modalities, 395
modeling, 397, 406, 448
models, 250, 253, 260, 298, 387, 392, 395, 397
modulation, ix, 234, 390, 396, 408, 446
moisture, 37, 128, 495
molar ratio, 442
molar ratios, 442
molecular biology, 265
molecular mechanisms, ix, 233, 235, 250, 253, 254
molecular oxygen, 243, 440, 477, 480
molecular structure, 313

molecular weight, 18, 243, 482
molecules, xiv, 253, 280, 290, 294, 313, 436, 440, 460, 475
mollusks, vii
monomer, 312, 313
monomeric, 252
monomers, 480, 482, 487, 488
Montenegro, 32
morphology, 392, 406
mortality, 234, 235, 236, 238, 242, 253, 261, 351, 392
mother cell, 498, 499
mothers, 353, 397, 399, 400
mouth, 423
movement, 414, 423
muscarinic receptor, 242, 250, 255, 260
muscle, 235, 236, 241, 244, 245, 247, 259, 261, 359, 361, 378, 381
muscle tissue, 241, 261
mutant, 504

N

Na^+, 501
Na2SO4, 457, 458
N-acety, 263
NaCl, 37, 276, 316, 318, 319, 333, 457, 458
national, 37, 270
natural, xiv, 4, 36, 345, 346, 350, 355, 375, 391, 396, 447, 467, 472, 473, 477, 484, 485, 486, 488, 492, 497
natural environment, 350, 391
nausea, 353, 354
NC, 266, 481, 482
ND, 257
nebulizer, 73, 147
neglect, 354
nematodes, vii
neonatal, 262, 392, 406, 407, 408
neonicotinoids, vii, xiii, 1, 2, 4, 5, 6, 8, 429, 430, 431, 444, 445, 446, 447
neoplastic, 353
nerve, 3, 5, 432
nerve agents, 432
nerve gas, 3
nervous system, xii, 5, 385, 387, 389, 390
Netherlands, 362
neurobehavioral, xii, 386, 388, 389, 392, 394, 401
neuroblastoma, 256
neurochemistry, 408
neuroendocrine, 392, 405, 408
neurological disease, 257
neuronal circuits, 395
neurons, 264
neuropathological, 393
neuropathology, 406
neurophysiology, 260
neurotoxic, xi, 257, 261, 265, 385, 407, 409, 451
neurotoxic effect, 407, 409
neurotoxicity, 253, 255, 256, 260, 262, 264, 265, 389, 392, 393, 396, 406, 407, 408
neurotransmitter, ix, 233, 235, 250, 265
New South Wales, 340
New York, iii, iv, 12, 258, 259, 260, 261, 263, 267, 302, 340, 344, 470, 504, 505
New Zealand, 37, 481, 483, 489
NF-kB, 251
Ni, 302
Nicotiana tabacum, 483
nicotine, 3, 4, 5
Nicotine, 119, 129, 142, 180, 217
Nile, 245, 255, 261
NIST, 360, 362, 365
nitrate, 432, 440
nitrogen, 3, 17, 82, 83, 128, 432, 441, 462, 467
nitrogen gas, 128
NO, 9
NOAA, 413
noise, 147, 148, 149, 319, 336, 371, 373, 386
nonequilibrium, 472
nonlinear, 395
non-random, 396
nontoxic, 389
non-uniform, 399
normal, 241, 242, 272, 390, 392, 394, 395, 396, 399, 495
normal distribution, 395
North Atlantic, 358
North Carolina, 409
novel biomarkers, 253, 254, 263
novelty, 395, 396, 409
Nrf2, 251, 260, 261
NS, 238, 246, 247
nuclear, ix, 234, 251, 256, 257, 477, 479
nucleic acid, 494
nucleus, 251, 407
null hypothesis, 400, 401
nutrition, 392

O

obesity, 356
observations, xii, 396, 397, 398, 399, 401, 411, 417, 418, 422, 423
observed behavior, 396
occupational, 351, 353

OCs, 283
octane, 289
offshore, xii, 411, 413, 414, 424, 425, 426
oil, 272, 275, 341, 461, 462, 463, 472
olfactory, 242, 260
oligomeric, 480
oligomeric products, 480
oligomers, 484
olive, 272, 275, 341, 461, 462, 472
olive oil, 272, 275, 341, 461, 462, 472
olives, 278
onion, xiii, 450, 454, 467
online, 406
on-line, vii, x, 15, 269, 272, 275, 299, 340, 341
on-line, 447
on-line, 456
open-field, 396, 398, 405
operator, 272
Ops, 283
optical, 470, 501
optical fiber, 470
optimization, x, xi, 20, 24, 148, 267, 270, 294, 297, 308, 318, 319, 323, 324, 332, 346, 455, 457, 458, 460
oral, 10, 352, 353, 387, 391, 407, 408
orange juice, 465
Oregon, 409, 418, 426
organ, xi, 245, 353, 385, 389, 390, 392
organelles, 501
organic compounds, 310, 311, 312, 315, 319, 322, 339, 342, 343, 456
organic matter, 327, 345, 346, 355, 412, 414, 441, 484, 504
organic peroxides, 243
organic solvent, x, xiv, 271, 272, 278, 279, 284, 287, 288, 290, 291, 292, 293, 294, 296, 297, 299, 307, 309, 310, 311, 312, 315, 322, 331, 335, 337, 355, 412, 431, 475
organic solvents, x, xiv, 278, 279, 292, 307, 309, 310, 311, 312, 315, 412, 475
organism, xiv, 6, 248, 387, 388, 389, 390, 394, 475
organization, 242, 388, 390
organizations, 270
organochlorine compounds, 344, 464
organophosphates, 2, 3, 5, 260, 265, 327, 430
organophosphorous, 5, 315, 320, 329, 330, 344, 390, 408
Organophosphorus pesticides, 12, 263, 275, 283, 289, 296, 309
osmotic, 483, 486
osteoporosis, 353
ovary, 255
oxidants, 234, 243, 442
oxidation, 245, 248, 327, 339, 430, 432, 436, 437, 440, 442, 443, 444, 445, 446, 447, 448, 449, 477, 480, 481, 482, 484, 487, 489, 500
oxidation products, 432, 481
oxidative, ix, xiv, 233, 234, 242, 243, 244, 245, 248, 249, 250, 251, 252, 253, 256, 257, 260, 261, 262, 263, 264, 407, 430, 450, 475, 477, 482, 492, 495, 497, 500
oxidative damage, 242, 243, 244, 251, 260, 261
oxidative reaction, 248
oxidative stress, ix, xiv, 233, 234, 244, 245, 249, 250, 251, 252, 253, 256, 257, 262, 263, 264, 407, 492, 497, 500
oxide, 120, 126, 138, 294, 315
oxygen, xiv, 3, 243, 440, 475, 477, 480, 493, 500, 505
oyster, 254
ozonation, 430, 431, 433, 436, 439, 440, 447, 449
ozone, 432, 436, 439, 440, 441, 448, 449, 452

P

PA, xiii, 280, 283, 361, 453, 454, 457, 461, 462, 463, 464, 466, 467, 468, 469
Pacific, 358, 418
PAHs, 322, 480
pancreatic, 352
pancreatic cancer, 352
paper, xiv, 35, 263, 287, 432, 475, 478, 480, 485
Paper, 448, 486
paralysis, 235, 236
paramagnetic, 479
parameter, 18, 21, 24, 294, 319, 322, 324, 333, 355, 376, 381, 397, 457, 460, 493, 494, 495, 497
particles, xii, 290, 326, 327, 342, 411, 413, 497
partition, 37, 276, 279, 280, 287, 291, 310, 333, 355, 400, 448, 457, 472
pathogens, vii
pathways, xii, 250, 253, 262, 265, 327, 386, 390, 411, 423, 432, 447, 448, 451
PCBs, 322, 340, 344, 351, 371, 480
PCP, 482, 483, 489
PCR, 253
PD, 266, 505
PDMS, xiii, 280, 281, 282, 283, 284, 333, 453, 454, 456, 457, 459, 461, 462, 463, 464, 466, 467, 468, 469
PE, x, 266, 307, 334, 335, 347, 448
peanuts, 4
peptide, 244
performance, x, 27, 269, 272, 308, 310, 346, 371, 388, 389, 392, 394, 395, 401, 433, 449, 450, 466, 470, 473

perinatal, xii, 386, 388, 390, 396, 402, 405, 408
peripheral nervous system, 235
permeability, 505
permeation, 16, 128, 130
permit, 469, 495
Perna perna, 254
peroxide, 243, 245, 442, 500
peroxynitrite, 264
Persistent Organic Pollutants, xi, 349
personal, 486
pest control, 309
pests, vii, xiv, 1, 270, 309, 350, 351, 389, 491, 492
petroleum, 276
Petroleum, 361
pH, xiv, 264, 277, 288, 291, 294, 316, 318, 326, 332, 436, 440, 441, 442, 443, 448, 457, 458, 460, 475, 482, 484, 494, 497
pH values, 436, 443, 460, 484
pharmaceuticals, 278
pharmacokinetic, 386, 406
pharmacological, 313
phenolic, 251, 318, 343, 344, 477, 480, 485, 487
phenolic compounds, 343, 344
phosphate, 7, 46, 52, 55, 57, 66, 77, 80, 81, 146, 163, 167, 182, 190, 194, 205, 209, 219, 224, 227, 288, 432, 437, 440
phosphoinositides, 265
phospholipase C, 251
phosphonates, 3
phosphorus, 3, 264, 289, 450, 461, 462, 467
phosphorylation, 250, 251, 252, 264
photocatalysis, 430, 431, 432, 433, 436, 437, 440, 447, 448, 450, 451
photocatalysts, 431
Photocatalytic, 436, 448, 449, 450
photochemical, 431, 436
photodegradation, xiii, 341, 429, 433, 437, 439, 440, 446, 448, 452
photolysis, 431, 437, 448, 451
photosynthesis, xiv, 491, 492, 497, 504
photosynthetic, 493, 501
physical properties, 287, 412, 457
physicochemical, 270, 298
physicochemical properties, 298
physiological, xiv, 6, 235, 243, 264, 386, 387, 390, 396, 402, 492, 493, 497, 501
physiology, 242, 258, 389, 392
phytoplankton, 449, 493, 497
PI3K, 251
pigments, 276, 449, 493
pituitary, 396
PKC, 251
placental, 265
planar, 278
planning, 388, 390, 391
plants, xiv, 270, 272, 339, 342, 346, 430, 477, 481, 483, 491, 492
plasma, 241, 387, 501
plasma membrane, 501
plastic, 391
plastics, 272
platelet, 356
play, 244, 326, 333, 386, 396, 497, 501
PLC, 38, 251
pleasure, 464
Pleistocene, 426
Pleurotus ostreatus, 481, 482, 488
plurality, 37
plus-maze, 396
plus-maze test, 396
PN, 450
PO, 137, 138
poisoning, 235, 352, 353, 354
poisonous, vii
polarity, 24, 270, 274, 278, 281, 287, 457
pollutants, ix, x, xiv, 12, 245, 249, 254, 264, 269, 307, 309, 310, 311, 312, 326, 331, 341, 342, 343, 345, 346, 351, 354, 355, 356, 357, 361, 362, 373, 376, 379, 381, 402, 430, 431, 446, 448, 449, 455, 480, 483, 488, 491, 492, 494, 498
pollution, 12, 254, 255, 256, 265, 354, 355, 376, 378, 381, 430, 476
poly(dimethylsiloxane), xiii, 453, 454, 457
polyamine, 266
polyaromatic hydrocarbons, 488
polybrominated diphenyl ethers, 409
polychlorinated biphenyls (PCBs), 344, 480
polychlorinated dibenzofurans, 344
polycyclic aromatic hydrocarbon, 342, 344, 345, 455, 480, 487
polydimethylsiloxane, 280
polyelectrolytes, 485
polymer, 37, 280, 477, 480
polymers, 272, 315, 482
polypropylene, 292
Polysiloxane, 283
polyunsaturated fat, 356
polyunsaturated fatty acid, 356
polyunsaturated fatty acids, 356
pools, 373, 375
poor, xi, 122, 279, 299, 310, 350, 356, 388, 457
population, xiii, 6, 242, 351, 358, 387, 394, 395, 402, 453, 454, 493, 497, 499
pore, 292
pores, 290, 291, 292, 293
porosity, 353

porous, 292, 330, 443
Portugal, 259
postnatal exposure, 408
postsynaptic, 235
potassium, 7
potato, 4, 446, 450
potatoes, xiii, 4, 454, 455, 462, 463, 467, 473, 497
power, 311, 322, 323, 324, 325, 332, 444, 446
predators, 7, 355
predatory fishes, 355
prediction, 341, 387
predictive validity, 391
prefrontal cortex, 392, 406
pregnancy, 397, 398
pregnant, 389
pregnant women, 389
preschool, 388
preservatives, 2
pressure, 18, 20, 23, 24, 85, 128, 312, 313, 324, 361, 420, 436
prevention, 13, 351, 484
primary products, 432
printing, 480
probability, 391
probe, 444, 501, 505
process control, xiv, 475
producers, 17, 497
production, xii, 16, 245, 249, 251, 261, 270, 350, 353, 389, 411, 429, 442, 457, 486, 488, 493, 497
productivity, 271
proficiency testing, 371, 372
progeny, 497, 498, 499
program, 18, 20, 23, 24, 414
programming, 408
proliferation, ix, xiv, 234, 250, 252, 253, 492, 497, 498
promote, 245, 293, 389, 402, 436
pro-oxidant, 248, 249
property, iv, vii, 313
prostaglandins, 356
protection, vii, 2, 12, 244, 252, 262, 290, 300, 402, 454, 469, 477
protein, 242, 243, 245, 248, 249, 250, 251, 252, 253, 254, 255, 257, 264, 265, 356
protein kinase C, 250, 255, 257
protein kinase C (PKC), 250
protein kinases, 242, 251
proteins, 251, 272, 284, 356, 498
proteomics, 253, 254, 256
protic, 271
protocol, 277, 393, 402
protocols, 6, 21, 388
PSA, 19, 37, 276, 277, 278
pseudocholinesterase, 235
Pseudomonas, 340
psychostimulants, 408
PTFE, 286, 292
public, xii, 270, 308, 386, 390, 408, 430, 454
public health, xii, 270, 308, 386, 390, 408
pulse, 241
pure water, 335, 336
purification, 273, 275, 361
P-value, 401
pyrene, 261, 264
pyrethrin, 144
pyrethroids, 4, 275, 278, 281, 291

Q

quality control, xii, 386, 388, 396
quantitative technique, 469
quartz, 16
quaternary ammonium, 470, 493
quinones, 484

R

RA, 261, 264, 448
radiation, 244, 322, 436
radical, 245, 255, 261, 500, 505
radiolabeled, 498
rain, ix, 233
rainwater, 294
random, 394, 397, 399
range, xiv, 16, 19, 86, 148, 149, 270, 271, 274, 276, 278, 279, 289, 292, 299, 315, 316, 317, 331, 334, 336, 353, 372, 376, 391, 393, 396, 397, 422, 430, 442, 443, 444, 459, 475, 477, 493
rat, 251, 253, 255, 257, 258, 262, 266, 392, 396, 405, 408
rats, xii, 254, 262, 263, 265, 385, 387, 395, 396, 397, 398, 399, 401, 405, 406, 407, 408, 409
RB, 260, 265, 449
RC, 260, 263, 413, 415, 416
reaction medium, 443, 484
reaction rate, xiv, 475
reactive oxygen, ix, 234, 243, 252, 254, 261, 500
reactive oxygen species, ix, 234, 243, 252, 254, 261, 500
reactive oxygen species (ROS), 500
reactivity, 395, 396
reading, 254
reagent, 7, 294, 321, 346, 432, 442
reagents, 394
real time, 417

reality, 388, 401
reasoning, 355
receptors, ix, 11, 233, 234, 235, 242, 250, 252, 260, 266, 407
recognition, 454, 485
recombination, 440
recovery, viii, 27, 31, 36, 149, 235, 239, 240, 255, 258, 261, 264, 272, 278, 317, 318, 319, 321, 322, 323, 324, 326, 327, 334, 337, 352, 357, 372, 459, 460, 469
recycling, 450, 477
red meat, 356
redox, 234, 242, 243, 244, 245, 250, 251, 263, 264, 265, 477, 480, 482, 484, 488
Redox, 243, 255, 260, 486
reduction, x, 5, 9, 243, 244, 245, 248, 269, 271, 273, 276, 278, 299, 333, 339, 477, 480, 487, 488
regular, xi, 349, 356, 422
regulation, 11, 251, 254, 260, 261, 264, 265, 392, 407, 454, 495, 501
regulations, xi, 300, 349
regulatory bodies, 270
reinforcers, 391
rejection, 394
relationship, xi, 235, 236, 237, 263, 280, 316, 328, 350, 354, 376, 379, 388, 392, 420, 457
relationships, 392, 395, 406
relatives, 3
relevance, 235, 242, 250, 354, 356, 358, 376, 393, 442
reliability, 11, 31, 271, 393
remediation, xiv, 475, 483
repeatability, 287
replication, 250, 256
repressor, 251
reproduction, 390, 402, 408, 409
research, vii, ix, xii, xiii, xiv, 6, 234, 235, 309, 354, 355, 378, 382, 385, 390, 391, 393, 395, 402, 408, 429, 431, 433, 446, 447, 448, 467, 487, 491, 492
researchers, 312, 322, 457, 501, 503
reservoir, 83, 358
residential, 386
residuals, 431
residues, v, xii, xiii, 16, 18, 23, 35, 36, 37, 83, 149, 195, 209, 214, 231, 251, 270, 276, 281, 309, 339, 340, 341, 342, 347, 356, 387, 407, 411, 426, 427, 429, 441, 448, 450, 451, 453, 454, 455, 460, 463, 469, 470, 471, 473, 476
resin, 457
resistance, 353, 354
resolution, 310, 371, 486, 488
resources, 393, 469
retention, 16, 17, 27, 38, 58, 59, 83, 85, 86, 126, 147, 273, 311, 335, 354, 356, 362, 364, 366, 367, 370, 371, 438, 455, 465
reversed-phase high performance liquid chromatography, 319
rewards, 391
rhizosphere, 483
rice, viii, 4, 35, 126, 130, 350, 352
risk, vii, xi, xii, 2, 4, 8, 9, 10, 11, 12, 257, 266, 280, 349, 351, 352, 382, 385, 386, 387, 389, 392, 405, 406, 407, 454, 467
risk assessment, vii, xi, 2, 8, 10, 11, 12, 257, 266, 349, 386, 387, 392, 405, 406, 407
risk management, xi, 349
risks, 8, 12, 242, 262, 382, 386, 387, 390, 402
river systems, 265
rivers, 263, 355, 390, 413
robustness, 272
rodent, 396
rodents, 249, 354, 390, 396, 409
Rome, 349
room temperature, 7, 16, 284, 312, 458, 494
ROS, 243, 244, 245, 249, 250, 252, 500
roundworms, vii
Royal Society, 451, 470
rubber, 353
runoff, xiv, 491, 492, 493
rural, 388

S

SA, 12, 37, 86, 257, 258, 259, 260, 261, 263, 267, 450, 470
safety, xi, 11, 258, 313, 349, 350, 354, 376, 377, 378, 381, 402, 454
salmon, 240, 249, 257, 258, 260, 262, 264, 267
salt, 281, 287, 295, 316, 318, 319, 332, 333, 457, 501
salts, 255, 272, 281, 294, 315, 457
sampling, ix, x, xiii, 269, 270, 271, 279, 281, 282, 284, 286, 291, 297, 298, 357, 359, 379, 394, 417, 420, 422, 426, 453, 454, 471
sand, 7, 327, 418, 422, 423, 426
SAS, 470
saturated fat, 356
saturated fatty acids, 356
scatter, 419, 493, 494
scavenger, 244, 440
Schiff, 413, 419, 421, 422, 424, 426
Schmid, 451
scientific validity, 391
scientists, 350, 387, 391, 401
scores, 371

SD, viii, 11, 36, 255, 267, 286, 372, 396
SE, 342, 448
sea level, 358
seafood, 356
search, xiv, 3, 242, 333, 356, 359, 397, 475
searching, ix, 234
seawater, 281, 294, 344
seaweed, 335, 337, 338, 341, 347
second generation, 309
secretion, 235, 257
sediment, xii, 310, 322, 390, 411, 412, 413, 414, 417, 418, 421, 422, 423, 424, 425, 426
sedimentation, 424
sediments, x, 296, 307, 315, 340, 341, 342, 343, 344, 345, 414, 420, 421, 422, 423, 424, 425, 426, 427
seed, 4, 351
seeds, 353
selecting, 276, 322
selectivity, x, 5, 37, 270, 279, 280, 286, 290, 293, 308, 310, 311
semen, 352
semiconductor, 440
semiconductors, 450
senescence, 505
Senescence, 500, 505
sensitivity, vii, x, xiv, 15, 16, 37, 38, 84, 86, 119, 126, 147, 148, 234, 236, 237, 241, 243, 254, 280, 281, 284, 286, 287, 308, 310, 337, 356, 371, 373, 409, 457, 459, 460, 461, 463, 466, 491, 492, 504
sensitization, 407
separation, ix, 16, 148, 269, 270, 272, 279, 280, 310, 313, 316, 317, 318, 319, 339, 340, 341, 343, 361, 407, 431, 484
septum, 279
series, viii, 35, 36, 83, 86, 122, 315, 317, 338, 354, 371, 383
serine, 235, 242, 264
serum, 7, 235, 256, 260
serum albumin, 7
serum cholinesterase, 256
services, iv
sewage, 413, 425
sex, 379, 382, 391, 392, 393, 408
sexual behavior, 390
sexual identity, 389
shape, 23, 292, 293
shellfish, 340
shock, 242, 251, 257
short period, 327
short supply, 3
short-term, 12, 261, 267, 354, 409
shoulder, 479
SI, 1, 260
SIGMA, 361
signal transduction, 250, 501, 505
signaling, ix, 234, 244, 250, 251, 252, 253, 255, 260, 265, 402
signaling pathway, 234, 250
signaling pathways, 234, 250, 251, 252, 253
signalling, 262, 501
signals, x, 308, 392, 393, 493
signal-to-noise ratio, 147, 148, 149, 319
significance level, 400
silica, xiii, 37, 272, 275, 279, 280, 331, 453, 454, 470
similarity, 3, 330
Singapore, 487
SIS, 362, 365
sites, 5, 280, 324, 355, 358, 386, 414, 417, 418, 422, 424, 483, 485
skills, 406
skin, 352, 353, 354, 488
Slovenia, 1, 2, 7, 12, 429
sludge, 272
SMS, 416
social behavior, 406
social competence, 392
social isolation, 406
social stress, 395, 408
SOD, 243, 245, 246, 247, 251, 252
sodium, 83, 119, 120, 121, 122, 126, 127, 128, 129, 131, 136, 144, 145, 189, 226, 276, 278, 442, 443
software, 32, 362, 391, 493
soil, vii, xiv, 1, 3, 4, 5, 7, 12, 13, 255, 296, 309, 310, 311, 322, 326, 327, 328, 329, 330, 331, 332, 334, 340, 341, 342, 343, 344, 345, 346, 347, 350, 351, 353, 356, 413, 432, 449, 450, 451, 452, 472, 475, 476, 480, 483, 484, 486, 493
soil particles, 413
soil pollution, 255
soils, x, 272, 281, 307, 309, 311, 315, 326, 327, 334, 335, 340, 342, 344, 345, 346, 413, 483, 484
solar, 244, 450
solid matrix, 448, 464
solid phase, x, 269, 271, 279, 308, 331, 332, 334, 335, 336, 343, 346, 347, 361, 450, 470, 472, 473
solid waste, 278
solid-state, 488
solubility, 19, 38, 86, 281, 287, 291, 293, 294, 312, 322, 326, 333, 353, 354, 412, 466
solutions, 19, 25, 31, 38, 58, 86, 122, 126, 130, 147, 296, 298, 313, 315, 322, 323, 327, 343, 347, 361, 362, 408, 436, 437, 438, 448
solvent, vii, viii, x, 15, 16, 17, 18, 20, 21, 23, 26, 27, 31, 36, 37, 38, 39, 49, 82, 83, 128, 130, 269,

270, 271, 272, 273, 274, 276, 278, 280, 286, 287, 290, 291, 292, 293, 296, 297, 299, 310, 311, 312, 322, 324, 332, 339, 342, 371, 451, 454, 456, 460, 467, 468, 469
solvents, ix, x, 18, 19, 23, 24, 37, 128, 269, 271, 272, 274, 276, 278, 279, 286, 287, 292, 297, 307, 310, 311, 312, 339, 431, 455, 476
soman, 3, 262
soot, 265
sorbents, 272, 275, 341
sorption, 281, 326, 327, 330, 345
sorption process, 326, 327, 330
South America, 353
soy, 391
soybean, 130
Spain, 232, 269, 307, 321, 475, 491
spatial, 354, 409, 425
species, ix, xi, xiv, 5, 10, 11, 233, 234, 235, 236, 241, 243, 244, 245, 247, 248, 249, 252, 253, 254, 256, 259, 260, 261, 265, 266, 267, 275, 280, 291, 314, 350, 355, 357, 358, 376, 377, 378, 381, 382, 389, 391, 407, 431, 440, 491, 492, 493, 494, 497, 500
specificity, 12, 241, 262, 366, 371, 443, 477
spectrophotometric, 343
spectrophotometry, 431
spectroscopy, 451
spectrum, 84, 126, 352, 362, 364, 365, 366, 367, 368, 388, 477, 480, 493
speed, 18, 20, 21, 24, 27, 242, 281, 286, 288, 292, 294, 299, 324, 457
sperm, 408
spices, 455
spine, 406
stability, 12, 122, 126, 270, 287, 290, 292, 299, 431, 432, 440, 447, 457, 476, 484, 488
stabilize, 23
stages, 6, 265, 266, 327, 447, 455
standard deviation, viii, 36, 275, 319, 328, 329, 396, 399
standardization, 393
standards, 19, 37, 38, 84, 85, 86, 118, 298, 354, 361, 371, 438, 450
Standards, 371
statistical analysis, 388, 393, 406
statistics, 401
steel, 292
steroids, 390
sterols, 277, 462
stimulus, 401
stock, 19, 38
stomach, 352
storage, 241, 270, 366
strain, 391, 393, 409
strains, 340, 483
strategies, x, 307, 387, 405, 485
streams, 265
strength, xiv, 281, 287, 294, 315, 333, 475
stress, ix, xiv, 233, 234, 244, 245, 249, 250, 251, 252, 253, 255, 256, 257, 259, 260, 261, 262, 263, 264, 265, 267, 390, 392, 395, 407, 409, 483, 486, 492, 493, 497, 500, 501
stress level, 500
stressors, 252, 257, 386, 396, 408
structural characteristics, 390
structural defect, 389
structural defects, 389
styrene, 272, 353
subacute, 258
subgroups, 3
substances, vii, x, xi, xii, 2, 13, 235, 272, 274, 308, 334, 349, 350, 356, 362, 383, 386, 389, 433, 454, 469, 477
substitutes, 3
substrates, 477, 483
subtraction, 147
sugar, 352, 497
sugar cane, 497
sugars, 276, 356
suicidal, 352
suicide, 234, 354
sulfate, 38, 51, 76, 83, 123, 136, 137, 179, 201, 218
sulfur, 3, 294, 432
sulphate, 19, 276, 361, 432, 440
sulphur, 440
supercritical, xiii, 37, 311, 312, 341, 342, 344, 345, 453, 454
superoxide, ix, 234, 243, 245, 255, 257, 500, 505
Superoxide, 261, 505
superoxide dismutase, ix, 234, 243, 257, 505
supply, 394, 493
supported liquid membrane, x, 270, 271, 279, 291
surface area, 7, 282, 292, 327
surface water, 343, 344, 446, 467, 470
surface wave, 414
surfactant, 311, 312, 313, 314, 315, 316, 317, 318, 319, 322, 323, 324, 326, 327, 332, 335, 336, 339, 343
surfactants, 241, 310, 312, 313, 315, 317, 318, 319, 323, 339, 344, 345, 476
surgical, 391
surgical intervention, 391
surveillance, 409
survival, 4, 9, 36, 241, 242, 253, 263, 390, 392, 408
survivors, 351
susceptibility, 236, 250, 258, 263, 386, 389

suspensions, 449, 451, 500
sustainability, 6, 12
sustainable development, 258
Sweden, 339
swelling, 495
Switzerland, 361
symptomatic treatment, 354
symptoms, 235, 353, 354
synapses, vii, 1, 5, 6
synergistic, xi, 11, 385, 401, 440
synergistic effect, xi, 385, 440
synthesis, 239, 244, 442, 487

T

tabun, 3
tandem mass spectrometry, 340, 342, 470
Tanzania, 340
target organs, 387
targets, ix, 233, 234, 235, 236, 242, 250, 253, 256, 260, 266, 386, 396
taxonomic, 5, 449
tea, 296, 462, 463, 464, 473
technician, 278
technology, xiv, 253, 321, 475, 476
Teflon, 282, 436
telencephalon, 261
temperature, viii, xiv, 7, 15, 16, 17, 18, 19, 20, 21, 22, 23, 24, 58, 73, 128, 147, 241, 279, 281, 284, 288, 311, 312, 315, 316, 317, 322, 324, 327, 332, 333, 345, 346, 362, 457, 458, 459, 460, 475, 485
temporal, 425
temporal distribution, 425
tension, 395, 405
teratogens, 389
teratology, 405
termites, 4, 351, 352, 353
test efficiency, 130
test procedure, xii, 386, 394
testes, viii, 35
testis, 387
testosterone, 390
textile, 351, 477, 480
theory, 302, 343, 470
thermal lens, xiii, 430, 442, 444, 447, 448, 451
thermal stability, 270
thermodynamic equilibrium, 280
thermostability, 486
thin film, 18, 280
thioredoxin, 252
threat, xii, xiii, 8, 234, 429, 436, 453, 454
threatening, 396
three-dimensional, 480
threonine, 264
threshold, 483
thresholds, 264
thymus, 387
thyroid, 353, 387
thyroid gland, 353
time consuming, ix, x, 269, 271, 272, 307, 309, 431
time periods, 327
timing, 18, 297
TiO_2, 431, 433, 436, 437, 440, 441, 448, 449, 450, 451
tissue, xi, 235, 241, 245, 247, 261, 344, 356, 385, 406, 473
titania, 440
titanium, 448
Titanium, 449
titanium dioxide, 448
TJ, 505
TM, 449, 451, 473
tobacco, 343, 455, 471, 483, 488
TOC, 441
tolerance, 236, 244, 245, 254, 257, 263, 310, 374, 376, 378, 380
toluene, viii, 36, 38, 58, 83, 278, 287, 294
tomato, 278
topographic, 421
total product, 350
toxic, vii, x, xiv, 2, 3, 8, 11, 13, 16, 234, 235, 243, 245, 253, 261, 265, 267, 271, 279, 307, 309, 310, 311, 312, 314, 339, 351, 352, 354, 386, 389, 391, 392, 430, 433, 443, 447, 475, 476, 481, 483, 484, 491, 492, 493, 497
toxic effect, xiv, 245, 314, 351, 354, 386, 389, 391, 491, 492, 493
toxic substances, 13, 433
toxicities, 258
toxicity, vii, ix, x, xi, xii, xiv, xv, 1, 2, 3, 4, 5, 6, 7, 8, 10, 11, 12, 233, 234, 235, 236, 242, 243, 250, 254, 255, 256, 257, 262, 263, 264, 266, 270, 287, 307, 349, 352, 354, 385, 386, 387, 388, 389, 390, 392, 393, 407, 408, 409, 432, 437, 443, 447, 448, 449, 477, 484, 487, 488, 491, 492, 493, 503, 504
toxicological, 251, 313, 352, 354, 386, 388, 391
toxicology, 13, 253, 257, 258, 263, 354, 386, 387, 391, 392, 395, 405, 504
toxicology studies, 391
toxins, 244
tracers, 424, 426
tracking, 498
training, 272
trans, 40, 41, 51, 59, 60, 61, 75, 135, 139, 142, 146, 173, 187, 208, 215, 351, 374, 380, 450
transcription, ix, 234, 242, 250, 251, 252, 256, 264

transcription factor, ix, 234, 242, 250, 251, 252, 256
Transcription factor, 253
transcription factors, 234, 242, 250, 251, 252
transcriptional, 253, 265
transfer, 20, 83, 292, 294, 297, 457, 480
transformation, 340, 433, 436, 437, 438, 449, 452, 476, 482, 484
transformation product, 433, 437, 438, 484
transformations, 327, 447
transgenic, 483
Transgenic, 488
transition, 17, 255
transition metal, 255
transitions, 73, 74, 75
translocation, 251, 252, 501
transmembrane, 501
transmission, 495
transmission electron microscopy, 495
transport, 11, 354, 357, 412, 413, 414, 422, 423, 424, 426, 427, 501
traps, 423
treatment methods, 339
tremor, 352, 354
trend, 234, 250, 278, 310
trichloroethylene, 265
triggers, 500
tripeptide, 244
trout, 236, 239, 240, 248, 249, 250, 252, 253, 254, 255, 257, 258, 259, 260, 261, 263, 265, 267
tubular, 280
tumours, 352
Turkey, 353
typhus, 351
typology, 379
tyrosine, 251, 264

U

U.S. Geological Survey, 383
ubiquitous, 243, 418
UK, 19, 261, 383, 486, 505
ultrasound, 340
uncertainty, 11, 394, 402
underlying mechanisms, 235
uniform, 354, 399
United Nations, 382, 383
United Nations Environment Program, 382, 383
United States, 331, 360, 412, 414
urea, 315, 482, 484
urine, 391, 472
USDA, 12
USEPA, 128, 232, 331
UV, 283, 296, 331, 334, 343, 346, 347, 436, 440, 441, 444, 449, 466, 467, 468
UV irradiation, 440

V

validation, 27, 387, 388, 390, 393, 394
validity, 391, 396
values, 8, 10, 18, 27, 272, 278, 291, 294, 315, 319, 322, 324, 329, 331, 337, 353, 373, 376, 377, 395, 413, 419, 421, 436, 443, 460, 461, 463, 484
Van der Waals, 327
vapor, 279
variability, xii, 236, 248, 263, 265, 266, 326, 386, 387, 388, 391, 392, 393, 394, 396, 401, 409, 427
variable, ix, 233, 247, 250, 318, 323, 324, 455, 460
variables, xi, 6, 308, 323, 324, 332, 333, 388, 391, 394, 406, 455
variance, 397, 401
variation, 248, 272, 331, 393, 394, 395, 396, 422
vasodilator, 356
vector, vii, x, 307, 308, 351
vegetable oil, 450
vegetables, x, 4, 37, 270, 272, 274, 275, 276, 277, 278, 283, 296, 307, 340, 350, 356, 432, 446, 447, 450, 467, 470, 471
versatility, 334
vertebrates, ix, 3, 5, 233, 242, 250, 253, 258
vessels, 312, 324, 332
victims, 355
virus, vii
viscera, 247, 361
viscosity, 323
visible, 261, 444, 493, 501
vitamin E, 243, 249
vitamins, 356
volatility, xiii, 270, 278, 284, 293, 331, 453, 460
vomiting, 354
vulnerability, 389

W

Wales, 340
waste products, xi, 349, 354
waste treatment, 486
waste water, 321, 431
wastes, xiv, 475, 476
wastewater, 281, 294, 295, 426, 430, 432, 477, 482, 485, 487, 489
wastewater treatment, 426, 432
wastewaters, 412, 430
water quality, 13

water-soluble, 356
weakness, 353
weight changes, 390
weight gain, 4, 9, 392
well-being, 409
Western countries, 351
wetting, 433
wheat, viii, 35, 84, 85, 86, 126, 130, 350
wild type, 504
wildlife, 234, 259, 431
wine, xiii, 272, 283, 298, 454, 455, 461, 462, 463, 466, 467, 469, 471, 473
winter, 497
Wistar rats, 388, 407
women, 389
wood, 2, 351, 353
workers, 351, 352, 354, 388
workplace, 353
World Health Organization, 36, 353, 382, 383, 412, 427
World War, 3, 412
World War I, 3, 412
World War II, 3, 412

X

xanthophyll, 449
xenobiotic, ix, 234, 244, 248, 249, 250, 254, 345, 382, 389
Xenobiotic, 259
xenobiotics, 244, 248, 249, 250, 253, 266, 405
X-ray diffraction, 485
xylene, 45, 65, 125, 142, 206

Z

zebrafish, 248, 252, 261, 264, 267
zinc, 257